THE BIRTH OF STRING THEORY

String theory is currently the best candidate for a unified theory of all forces and all forms of matter in nature. As such, it has become a focal point for physical and philosophical discussions. This unique book explores the history of the theory's early stages of development, as told by its main protagonists.

The book journeys from the first version of the theory (the so-called Dual Resonance Model) in the late 1960s, as an attempt to describe the physics of strong interactions outside the framework of quantum field theory, to its reinterpretation around the mid-1970s as a quantum theory of gravity unified with the other forces, and its successive developments up to the superstring revolution in 1984. Providing important background information to current debates on the theory, this book is essential reading for students and researchers in physics, as well as for historians and philosophers of science.

ANDREA CAPPELLI is a Director of Research at the Istituto Nazionale di Fisica Nucleare, Florence. His research in theoretical physics deals with exact solutions of quantum field theory in low dimensions and their application to condensed matter and statistical physics.

ELENA CASTELLANI is an Associate Professor at the Department of Philosophy, University of Florence. Her research work has focussed on such issues as symmetry, physical objects, reductionism and emergence, structuralism and realism.

FILIPPO COLOMO is a Researcher at the Istituto Nazionale di Fisica Nucleare, Florence. His research interests lie in integrable models in statistical mechanics and quantum field theory.

PAOLO DI VECCHIA is a Professor of Theoretical Physics at Nordita, Stockholm, and at the Niels Bohr Institute, Copenhagen. He has worked on several aspects of theoretical particle physics, and has contributed to the development of string theory since its birth in 1968.

THE BIRTH OF STRING THEORY

Edited by

ANDREA CAPPELLI
INFN, Florence

ELENA CASTELLANI
Department of Philosophy, University of Florence

FILIPPO COLOMO
INFN, Florence

PAOLO DI VECCHIA
Nordita, Stockholm and Niels Bohr Institute, Copenhagen

CAMBRIDGE
UNIVERSITY PRESS

University Printing House, Cambridge CB2 8BS, United Kingdom

Published in the United States of America by Cambridge University Press, New York

Cambridge University Press is part of the University of Cambridge.

It furthers the University's mission by disseminating knowledge in the pursuit of education, learning and research at the highest international levels of excellence.

www.cambridge.org
Information on this title: www.cambridge.org/9780521197908

© Cambridge University Press 2012

This publication is in copyright. Subject to statutory exception and to the provisions of relevant collective licensing agreements, no reproduction of any part may take place without the written permission of Cambridge University Press.

First published 2012

A catalogue record for this publication is available from the British Library

Library of Congress Cataloguing in Publication data
The birth of string theory / edited by Andrea Cappelli, INFN, Florence; Elena Castellani, Department of Philosophy, University of Florence; Filippo Colomo, INFN, Florence; Paolo Di Vecchia, Niels Bohr Institute, Copenhagen and Nordita, Stockholm.
p. cm.
Includes bibliographical references and index.
ISBN 978-0-521-19790-8
1. String models. 2. Duality (Nuclear physics) I. Cappelli, Andrea, editor of compilation.
II. Castellani, Elena, 1959– editor of compilation. III. Colomo, F., editor of compilation.
IV. Di Vecchia, P. (Paolo), editor of compilation.
QC794.6.S85B57 2012
539.7'258 – dc23   2011052388

ISBN 978-0-521-19790-8 Hardback

Cambridge University Press has no responsibility for the persistence or accuracy of URLs for external or third-party internet websites referred to in this publication, and does not guarantee that any content on such websites is, or will remain, accurate or appropriate.

Contents

List of contributors	page x
Photographs of contributors	xiv
Preface	xxi
Abbreviations and acronyms	xxiv

Part I Overview   1

1 Introduction and synopsis   3

2 Rise and fall of the hadronic string   17
   GABRIELE VENEZIANO

3 Gravity, unification, and the superstring   37
   JOHN H. SCHWARZ

4 Early string theory as a challenging case study for philosophers   63
   ELENA CASTELLANI

EARLY STRING THEORY

Part II The prehistory: the analytic S-matrix   81

5 Introduction to Part II   83
   5.1 Introduction   83
   5.2 Perturbative quantum field theory   84
   5.3 The hadron spectrum   88
   5.4 S-matrix theory   91
   5.5 The Veneziano amplitude   97

6 Particle theory in the Sixties: from current algebra to the Veneziano
   amplitude   100
   MARCO ADEMOLLO

7 The path to the Veneziano model   116
   HECTOR R. RUBINSTEIN

8	Two-component duality and strings PETER G.O. FREUND	122
9	Note on the prehistory of string theory MURRAY GELL-MANN	129

Part III The Dual Resonance Model — **133**

10	Introduction to Part III	135
	10.1 Introduction	135
	10.2 N-point dual scattering amplitudes	137
	10.3 Conformal symmetry	145
	10.4 Operator formalism	147
	10.5 Physical states	150
	10.6 The tachyon	153
11	From the S-matrix to string theory PAOLO DI VECCHIA	156
12	Reminiscence on the birth of string theory JOEL A. SHAPIRO	179
13	Personal recollections DANIELE AMATI	191
14	Early string theory at Fermilab and Rutgers LOUIS CLAVELLI	193
15	Dual amplitudes in higher dimensions: a personal view CLAUD LOVELACE	198
16	Personal recollections on dual models RENATO MUSTO	202
17	Remembering the 'supergroup' collaboration FRANCESCO NICODEMI	208
18	The '3-Reggeon vertex' STEFANO SCIUTO	214

Part IV The string — **219**

19	Introduction to Part IV	221
	19.1 Introduction	221
	19.2 The vibrating string	223
	19.3 The rotating rod	226
	19.4 The relativistic point particle	228

	19.5 The string action	230
	19.6 The quantum theory of the string	231
20	From dual models to relativistic strings PETER GODDARD	236
21	The first string theory: personal recollections LEONARD SUSSKIND	262
22	The string picture of the Veneziano model HOLGER B. NIELSEN	266
23	From the S-matrix to string theory YOICHIRO NAMBU	275
24	The analogue model for string amplitudes DAVID B. FAIRLIE	283
25	Factorization in dual models and functional integration in string theory STANLEY MANDELSTAM	294
26	The hadronic origins of string theory RICHARD C. BROWER	312

TOWARDS MODERN STRING THEORY

Part V	Beyond the bosonic string	329
27	Introduction to Part V	331
	27.1 Introduction	331
	27.2 Chan–Paton factors	333
	27.3 The Lovelace–Shapiro amplitude	334
	27.4 The Ramond model	335
	27.5 The Neveu–Schwarz model	338
	27.6 The Ramond–Neveu–Schwarz model	339
	27.7 World-sheet supersymmetry	341
	27.8 Affine Lie algebras	344
28	From dual fermion to superstring DAVID I. OLIVE	346
29	Dual model with fermions: memoirs of an early string theorist PIERRE RAMOND	361
30	Personal recollections ANDRÉ NEVEU	373
31	Aspects of fermionic dual models EDWARD CORRIGAN	378

32	The dual quark models KORKUT BARDAKCI AND MARTIN B. HALPERN	393
33	Remembering the dawn of relativistic strings JEAN-LOUP GERVAIS	407
34	Early string theory in Cambridge: personal recollections CLAUS MONTONEN	414

Part VI The superstring — 419

35	Introduction to Part VI	421
	35.1 Introduction	421
	35.2 The field theory limit	423
	35.3 Unification of all interactions	427
	35.4 The QCD string	431
	35.5 A detour on spinors	433
	35.6 Spacetime supersymmetry	434
	35.7 The GSO projection	437
	35.8 The Kaluza–Klein reduction and supersymmetry breaking	439
	35.9 The local supersymmetric action for the superstring	442
	35.10 Supergravity	444
36	Supersymmetry in string theory FERDINANDO GLIOZZI	447
37	Gravity from strings: personal reminiscences of early developments TAMIAKI YONEYA	459
38	From the Nambu–Goto to the σ-model action LARS BRINK	474
39	Locally supersymmetric action for the superstring PAOLO DI VECCHIA	484
40	Personal recollections EUGÈNE CREMMER	490
41	The scientific contributions of Joël Scherk JOHN H. SCHWARZ	496

Part VII Preparing the string renaissance — 509

42	Introduction to Part VII	511
	42.1 Introduction	511
	42.2 Supergravity unification of all interactions	512
	42.3 A novel light-cone formalism	514
	42.4 Modern covariant quantization	518

42.5	Anomaly cancellation	521
42.6	A new era starts or, maybe better, continues	525

43 From strings to superstrings: a personal perspective 527
MICHAEL B. GREEN

44 Quarks, strings and beyond 544
ALEXANDER M. POLYAKOV

45 The rise of superstring theory 552
ANDREA CAPPELLI AND FILIPPO COLOMO

Appendix A	*Theoretical tools of the Sixties*	569
Appendix B	*The Veneziano amplitude*	579
Appendix C	*From the string action to the Dual Resonance Model*	586
Appendix D	*World-sheet and target-space supersymmetry*	604
Appendix E	*The field theory limit*	620
Index		626

Contributors

Marco Ademollo
Dipartimento di Fisica, Università di Firenze, and INFN, Sezione di Firenze,
Via G. Sansone 1, 50019 Sesto Fiorentino (FI), Italy

Daniele Amati
SISSA, Trieste, and INFN, Sezione di Trieste, via Bonomea 265, 34136 Trieste, Italy

Korkut Bardakci
Department of Physics, University of California, and Theoretical Physics Group,
Lawrence Berkeley National Laboratory, University of California, Berkeley, CA 94720,
USA

Lars Brink
Department of Fundamental Physics, Chalmers University of Technology,
S-412 96 Göteborg, Sweden

Richard C. Brower
Physics Department, Boston University, 590 Commonwealth Avenue, Boston, MA 02215,
USA

Andrea Cappelli
INFN, Sezione di Firenze, Via G. Sansone 1, 50019 Sesto Fiorentino (FI), Italy

Elena Castellani
Dipartimento di Filosofia, Università di Firenze, Via Bolognese 52, 50139 Firenze, Italy

Louis Clavelli
Department of Physics and Astronomy, University of Alabama, Tuscaloosa, AL
35487-0324, USA

Filippo Colomo
INFN, Sezione di Firenze, Via G. Sansone 1, 50019 Sesto Fiorentino (FI), Italy

Edward Corrigan
Department of Mathematical Sciences, Durham University, Durham, DH1 3LE, UK

Eugène Cremmer
Laboratoire de Physique Théorique, École Normale Supérieure, 24 rue Lhomond, 75231 Paris Cedex 05, France

Paolo Di Vecchia
Niels Bohr Institute, Blegdamsvej 17, 2100 Copenhagen, Denmark, and Nordita, Roslagstullsbacken 23, 10691 Stockholm, Sweden

David B. Fairlie
Department of Mathematical Sciences, Durham University, Durham, DH1 3LE, UK

Peter G. O. Freund
Enrico Fermi Institute and Department of Physics, University of Chicago, 5720 S. Ellis Avenue, Chicago, IL 60637, USA

Murray Gell-Mann
Santa Fe Institute, 1399 Hyde Park Road, Santa Fe, NM 87501, USA

Jean-Loup Gervais
Laboratoire de Physique Théorique, École Normale Supérieure, 24 rue Lhomond, 75231 Paris Cedex 05, France

Ferdinando Gliozzi
Dipartimento di Fisica Teorica, Università di Torino, and INFN, Sezione di Torino, Via P. Giuria 1, 10125 Torino, Italy

Peter Goddard
Institute for Advanced Study, Olden Lane, Princeton, NJ 08540, USA

Michael B. Green
DAMTP, Wilberforce Road, Cambridge, CB3 0WD, UK

Martin B. Halpern
Department of Physics, University of California, and Theoretical Physics Group, Lawrence Berkeley National Laboratory, University of California, Berkeley, CA 94720, USA

Claud Lovelace
Department of Physics and Astronomy, Rutgers University, 136 Frelinghuysen Road, Piscataway, NJ 08854-8019, USA

Stanley Mandelstam
Department of Physics, University of California, and Lawrence Berkeley Laboratory, University of California, Berkeley, CA 94720, USA

Claus Montonen
Department of Physics, University of Helsinki, P.O. Box 64, 00014 Helsinki, Finland

Renato Musto
Dipartimento di Scienze Fisiche, Università di Napoli Federico II, and INFN, Sezione di Napoli, 80126, Napoli, Italy

Yoichiro Nambu
Department of Physics, University of Chicago, 5720 S. Ellis Avenue, Chicago, IL 60637, USA

André Neveu
Laboratoire de Physique Théorique et Astroparticules, Case 070, CNRS, Université Montpellier II, 34095 Montpellier, France

Francesco Nicodemi
Dipartimento di Scienze Fisiche, Università di Napoli Federico II, and INFN, Sezione di Napoli, 80126, Napoli, Italy

Holger B. Nielsen
Niels Bohr Institute, Blegdamsvej 17, 2100, Copenhagen, Denmark

David I. Olive
Department of Physics, Swansea University, Singleton Park, Swansea, SA2 8PP, UK

Alexander M. Polyakov
Joseph Henry Laboratories, Princeton University, Princeton, NJ 08544, USA

Pierre Ramond
Institute for Fundamental Theory, Physics Department, University of Florida, Gainesville, FL 32611, USA

Hector R. Rubinstein
AlbaNova University Center, Royal Institute of Technology, Stockholm University, 106 91 Stockholm, Sweden

John H. Schwarz
Department of Physics, Mathematics and Astronomy, California Institute of Technology, 456 Lauritsen Laboratory Caltech 452-48, Pasadena, CA 91125, USA

Stefano Sciuto
Dipartimento di Fisica Teorica, Università di Torino, and INFN, Sezione di Torino, Via P. Giuria 1, 10125 Torino, Italy

Joel A. Shapiro
Department of Physics and Astronomy, Rutgers University, 136 Frelinghuysen Road, Piscataway, NJ 08854-8019, USA

Leonard Susskind
Department of Physics, Stanford University, Stanford, CA 94305-4060, USA

Gabriele Veneziano
Theory Division, CERN, CH-1211 Geneva 23, Switzerland, and Collège de France, 11 place M. Berthelot, 75005 Paris, France

Tamiaki Yoneya
Institute of Physics, University of Tokyo, Komaba, Meguro-ku, Tokyo 153-8902, Japan

Photographs of contributors

From left to right and from top to bottom: Marco Ademollo, Daniele Amati, Korkut Bardakci, Lars Brink, Richard C. Brower and Louis Clavelli.

From left to right and from top to bottom: Edward Corrigan, Eugène Cremmer, Paolo Di Vecchia, David B. Fairlie, Peter G. O. Freund, Sergio Fubini (1928–2005), Murray Gell-Mann, Jean-Loup Gervais and Ferdinando Gliozzi.

From left to right and from top to bottom: Peter Goddard, Michael B. Green, Martin B. Halpern, Claud Lovelace, Stanley Mandelstam, Claus Montonen, Renato Musto, Yoichiro Nambu and André Neveu [photograph of Goddard by Cliff Moore].

Photographs of contributors

From left to right and from top to bottom: Francesco Nicodemi, Holger B. Nielsen, David I. Olive, Alexander M. Polyakov, Pierre Ramond, Hector R. Rubinstein (1933–2009), Bunji Sakita (1930–2002), Joël Scherk (1946–1980) and John H. Schwarz.

From left to right and from top to bottom: Stefano Sciuto, Joel A. Shapiro, Leonard Susskind, Gabriele Veneziano and Tamiaki Yoneya.

Participants at the meeting 'The Birth of String Theory', 18–19 May 2007, GGI, Arcetri, Florence.

Preface

In May 2007 we organized a workshop on the origin and early developments of string theory at the Galileo Galilei Institute for Theoretical Physics in Arcetri (Florence). A fair number of researchers who had contributed to the birth of the theory participated and described, according to their personal recollections, the intriguing way in which the theory developed from hadron phenomenology into an independent field of research. It was the first occasion on which they had all been brought together since the 1975 conference in Durham, which represented the last meeting on string theory as applied to hadronic physics.

The workshop in Arcetri was a success: the atmosphere was enthusiastic and the participants showed genuine pleasure in discussing the lines of thought developed during the years from the late Sixties to the beginning of the Eighties, mutually checking their own reminiscences. This encouraged us to go on with the project we had been thinking of for some time, of an historical account of the early stages of string theory based on the recollections of its main exponents. We were fortunate enough to have on board practically all the physicists who developed the theory. While some of the contributions to this Volume originated from the talks presented at the meeting, most of them have been written expressly for this book.

In starting this project we were motivated by the observation that the history of the beginnings and early phases of string theory is not well accounted for: apart from the original papers, the available literature is rather limited and fragmentary. A book devoted specifically to the historical reconstruction of these developments – the formulation of a consistent and beautiful theory starting from hadron phenomenology, its failure as a theory of strong interactions, and, finally, its renaissance as a unified theory of all fundamental interactions – was not available. This Volume aims to fill the gap, by offering a collection of reminiscences and overviews, each one contributing from the Author's own perspective to the general historical account. The collection is complemented with an extended editorial apparatus (Introductions, Appendices and Editors' Chapters) according to criteria explained below.

In addition to the historical record, this book is of interest for several reasons. First, by showing the dynamics of the ideas, concepts and methods involved, it offers precious background information for a better understanding of the present status of string theory, which has recently been at the centre of a widespread debate. Second, it provides an illustration of the fruitfulness of the field, from both a physical and a mathematical perspective. A number of ideas that are central to contemporary theoretical physics of fundamental interactions, such as supersymmetry and extra spacetime dimensions, originated in this context. Furthermore, some theoretical methods, for example two-dimensional conformal symmetry, found important physical applications in various domains outside the original one. Finally, from a philosophical point of view, early string theory represents a particularly interesting case study for reflections on the construction and evaluation of physical theories in modern physics.

In the following, we illustrate the structure of the book and offer some guidelines to the reader. The Volume is organized into seven Parts: the first one provides an overview of the whole book; the others correspond to significant stages in the evolution of string theory from 1968 to 1984 and are accompanied by specific introductory Chapters.

In Part I, the Introduction summarizes the main developments and contains a temporal synopsis with a list of key results and publications. The following two Chapters, by Veneziano and by Schwarz, offer a rather broad overview on the early (1968–1973) and later (1974–1984) periods of the history of string theory, respectively. They introduce all the themes of the book that are then addressed in detail in the following Parts. The last Chapter of Part I, by Castellani, presents some elements for the philosophical discussion of the early evolution of the theory and the scientific methodology employed in it.

The Introductions to the other Parts and the Appendices are meant to fit the needs of undergraduate/early graduate students in theoretical physics, as well as of historians and philosophers, who have a background in quantum mechanics and quantum field theory, but lack the specific vocabulary to appreciate fully the Authors' contributions. The Introductions and Appendices, taken together with the final Chapter, can also be used as an entry-level course in string theory, presenting the main physical ideas with a minimum of technique.

For a broader audience, we suggest beginning with the first, nontechnical paragraph in each Introduction, and then approaching the less technical and more comprehensive Authors' Chapters which are located first in each Part. The rich material presented in the Chapters, together with the original literature, can be the starting point for in-depth historical study of the many events that took place in the development of string theory. The final Chapter of the book, by Cappelli and Colomo, provides a nontechnical overview of string theory from 1984 up to the present time, which complements the historical and scientific perspective.

We hope that the book can be read at different levels and, as such, will be useful for scientific, historical and philosophical approaches to this fascinating, but complex, subject.

The book has associated the webpage

$$\text{http://theory.fi.infn.it/colomo/string-book/}$$

which gives access to the original talks of the 2007 GGI workshop and to additional material already provided by some Authors or to be collected in the future.

We are very grateful to all those who have helped us in preparing this Volume. First and foremost, our thanks go to all the Authors who agreed to contribute their reminiscences. Many thanks go also to all those who gave us valuable comments and suggestions during the preparation of the Volume, in particular Leonardo Castellani, Camillo Imbimbo, Yuri Makeenko, Raffaele Marotta, Giulio Peruzzi, Igor Pesando, Franco Pezzella, Augusto Sagnotti, John H. Schwarz, Domenico Seminara, Gabriele Veneziano, Guillermo R. Zemba and Hans v. Zur-Mühlen. We are indebted to the Galileo Galilei Institute for hosting the 2007 workshop. We also wish to thank the staff of Cambridge University Press for assistance and Sara De Sanctis for helping with the bibliography. Finally, we are grateful to our collaborators and to our families for their patience and support.

Abbreviations and acronyms

AdS	Anti de Sitter (spacetime)
AdS/CFT	Anti de Sitter/conformal field theory (correspondence)
APS	American Physical Society
BRST	Becchi–Rouet–Stora–Tyutin (quantization)
Caltech	California Institute of Technology, Pasadena, CA
CERN	European Centre for Nuclear Research, Geneva
CFT	conformal field theory
CNRS	Centre National de la Recherche Scientifique, France
CP	Chan–Paton (factors)
CPT	charge conjugation, parity, time reversal (symmetries)
DAMTP	Department of Applied Mathematics and Theoretical Physics, Cambridge
DDF	Del Giudice–Di Vecchia–Fubini (states, operators)
DHS	Dolen–Horn–Schmid (duality)
Dp-brane	Dirichlet p-dimensional membrane
DRM	Dual Resonance Model
ENS	École Normale Supérieure, Paris
Fermilab	Fermi National Accelerator Laboratory (or FNAL), Illinois
FESR	finite energy sum rule
FNAL	Fermi National Accelerator Laboratory (or Fermilab), Illinois
GGI	Galileo Galilei Institute, Florence
GGRT	Goddard–Goldstone–Rebbi–Thorn (quantization)
GR	general relativity
GSO	Gliozzi–Scherk–Olive (projection)
GUT	Grand Unified Theories
IAS	Institute of Advanced Study, Princeton, NJ
ICTP	International Center for Theoretical Physics, Trieste
IHES	Institut des Hautes Études Scientifiques, Bures-sur-Yvette
IMF	infinite momentum frame
INFN	Istituto Nazionale di Fisica Nucleare, Italy
IR	infrared
ISR	Intersecting Storage Ring, CERN

ITP	(Kavli) Institute of Theoretical Physics, Santa Barbara, CA
KK	Kaluza–Klein (compactification)
KM	Kac–Moody (algebra)
KN	Koba–Nielsen (amplitudes)
KZ	Knizhnik–Zamolodchikov (equation)
LEP	Large Electron–Positron (collider), CERN
LHC	Large Hadron Collider, CERN
LPTENS	Laboratoire de Physique Théorique, École Normale Supérieure, Paris
LPTHE	Laboratoire de Physique Théorique et Hautes Energies, Orsay
MIT	Massachusetts Institute of Technology, Boston, MA
MSSM	Minimal Supersymmetric Standard Model
M-theory	matrix (or membrane) theory
NAL	National Accelerator Laboratory (FNAL after 1972), Illinois
NATO	North Atlantic Treaty Organization
Nordita	Nordic Institute for Theoretical Physics, Stockholm
NS	Neveu–Schwarz (model, sector)
NSF	National Science Foundation, USA
NYU	New York University
PCAC	partially conserved axial current
PS	Proton Synchrotron, CERN
QCD	quantum chromodynamics
QED	quantum electrodynamics
QFT	quantum field theory
R	Ramond (model, sector)
RIMS	Research Institute for Mathematical Sciences, Kyoto
RNS	Ramond–Neveu–Schwarz (model)
SISSA	Scuola Internazionale Superiore di Studi Avanzati, Trieste
SLAC	Stanford Linear Accelerator Center
S-matrix	scattering matrix
SM	Standard Model
SSC	Superconducting Super Collider
SSR	superconvergence sum rule
SUGRA	supergravity
SUSY	supersymmetry
SVM	Shapiro–Virasoro model
TOE	Theory of Everything
UV	ultraviolet
WZ	Wess–Zumino (model)
WZWN	Wess–Zumino–Witten–Novikov (model)
YM	Yang–Mills (gauge theory)

Part I
Overview

1
Introduction and synopsis

String theory describes one-dimensional systems, like thin rubber bands, that move in spacetime in accordance with special relativity. These objects supersede pointlike particles as the elementary entities supporting microscopic phenomena and fundamental forces at high energy.

This simple idea has originated a wealth of other concepts and techniques, concerning symmetries, geometry, spacetimes and matter, that still continue to astonish and puzzle the experts in the field. The question 'What is string theory?' is still open today: indeed, the developments in the last fifteen years have shown that the theory also describes higher-dimensional extended objects like membranes, and, in some limits, it is equivalent to quantum field theories of point particles.

Another question which is also much debated outside the circle of experts is: 'What is string theory good for?' In its original formulation, the theory could not completely describe strong nuclear interactions; later, it was reproposed as a unified theory of all fundamental interactions including gravity, but it still needs experimental confirmation.

This book will not address these kinds of questions directly: its aim is to document what the theory *was* in the beginning, about forty years ago, and follow the threads connecting its development from 1968 to 1984. Over this period of time, the theory grew from a set of phenomenological rules into a consistent quantum mechanical theory, while the concepts, physical pictures and goals evolved and changed considerably. These developments are described by the direct narration of thirty-five physicists who worked in the field at the time. From this choral ensemble, an interesting 'scientific saga' emerges, with its ups and downs, successes and frustrations, debates, striking ideas and preconceptions.

String theory started from the general properties of scattering amplitudes and some experimental inputs; it then grew as an independent theory, by progressive generalization and through the exploitation of symmetries and consistency conditions. It required plenty of imagination and hard work in abstract formalisms, and was very appealing to young researchers in the early Seventies. They collectively undertook the enterprise of

---

The Birth of String Theory, ed. Andrea Cappelli, Elena Castellani, Filippo Colomo and Paolo Di Vecchia.
Published by Cambridge University Press. © Cambridge University Press 2012.

understanding the Dual Resonance Model, as string theory was originally called, attracted by its novelty, beauty and deep intricacy. They were helped by some mentors, senior theorists who supported them, often against general opinion. Among them, we mention Amati (CERN), Fubini (MIT and CERN), Gell-Mann (Caltech), Mandelstam (Berkeley) and Nambu (Chicago).

The evolution of physical ideas in this field is fascinating. Let us just underline that in early string theory we can find the seeds of many new concepts and mathematical methods of contemporary theoretical physics, such as supersymmetry, conformal symmetry and extra spacetime dimensions. The mathematical methods helped to refine the tools and scope of quantum field theory and were also applied to condensed matter physics and statistical mechanics. The new concepts of supersymmetry and extra dimensions have been introduced in the theories of fundamental interactions beyond the Standard Model, which are awaiting experimental testing by the Large Hadron Collider now operating at CERN, Geneva.

A brief overview of early string history and the book

The book is divided into seven Parts that correspond to major steps in the development of the theory, arranged in logical/chronological order. The first Chapter in each Part is an Editors' Introduction to the main topics discussed, which helps the reader to understand the Authors' Chapters and follow the line of ideas.

Part I provides an introduction to the whole book: the present Chapter includes a synopsis of early string history and points to the essential references. Chapters 2 and 3, by Veneziano and Schwarz respectively, introduce the first (1968–1973) and second (1974–1984) periods into which the evolution of early string theory can be divided. They are followed by the Chapter by Castellani, which highlights some of the main aspects of philosophical interest in the developments narrated in the Volume.

Part II, 'The prehistory: the analytic S-matrix', discusses the panorama of theoretical physics in the Sixties from which the Veneziano amplitude, the very beginning of string theory, originated. The first steps of the theory were made in close connection with the phenomenology of strong interactions: experiments showed a wealth of particles, the hadrons, that could not all be considered elementary and had large couplings among themselves. The methods of perturbative quantum field theory, developed in earlier studies of the electromagnetic force, could not be applied since they relied on the existence of only a few, weakly interacting, elementary particles.

The dominant approach was the theory of the S-matrix (scattering matrix), which only involved first-principle quantum mechanics and empirical data, as originally advocated by Heisenberg. Approximated solutions to the scattering matrix were searched for starting from some phenomenological assumptions on particle exchanges and asymptotic behaviour, and then solving self-consistently the general requirements of relativistic quantum mechanics. A simplified form of these conditions, called Dolen–Horn–Schmid duality, allowed for the closed-form solution of the famous Veneziano four-meson scattering amplitude in 1968.

Veneziano's result had a huge impact because it provided a simple, yet rich and elegant solution after many earlier attempts. It was immediately clear that a new structure had been found, involving infinite towers of particles organized in linearly rising Regge trajectories.

Part III, 'The Dual Resonance Model', describes the intense activity taking place in the period 1969–1973: the Veneziano model was generalized to the scattering of any number of mesons and the structure of the underlying quantum theory was understood, separating the physical states from the unphysical states. The operator formalism was introduced and first loop corrections were computed in open and closed string theories, at the time called the Dual Resonance Model (DRM) and the Shapiro–Virasoro Model (SVM), respectively. Some theoretical methods were imported from the study of quantum electrodynamics, while others were completely new. It is surprising how far the theory was developed before a clear understanding of the underlying string dynamics, i.e. before the quantization of the string action.

The consistency conditions in the quantum theory of the DRM brought two striking results. First, the linear Regge trajectories were uniquely fixed, leading to the presence of tachyons (unphysical particles with negative mass squared) with spin zero, and of massless particles with spins one and two in the open and closed string theories, respectively. Second, unitarity of the theory required $d = 26$ spacetime dimensions, in particular for loop corrections, as observed by Lovelace in 1971. On the one hand, these results showed the beauty of the theory, stemming from its high degree of consistency and symmetry; on the other hand, they were in clear contradiction with hadron phenomenology, requiring $d = 4$ dimensions and no massless particle with spin.

Part IV, 'The string', illustrates how the DRM was eventually shown to correspond to the quantum theory of a relativistic string. The analogy between the DRM spectrum and the harmonics of a vibrating string was soon noticed in 1969, independently by Nambu, Nielsen and Susskind. The string action, proportional to the area of the string world-sheet, was also proposed by Nambu and then by Goto in analogy with the action of the relativistic point particle, proportional to the length of the trajectory.

Although the string action was introduced rather early, its quantization was not straightforward. Goddard, Goldstone, Rebbi and Thorn eventually worked it out in 1973, using the so-called light-cone gauge, involving the $(d - 2)$ transverse string coordinates. After quantization, they showed that Lorentz invariance was maintained only in $d = 26$ spacetime dimensions, where the DRM spectrum of physical states was recovered.

Part V, 'Beyond the bosonic string', collects the Authors' Chapters describing the addition of extra degrees of freedom to the DRM in the quest for a better agreement with hadron phenomenology. The addition of fermions, i.e. half-integer spin hadrons, was achieved by Ramond, while a new dual model for pions was developed by Neveu and Schwarz. These models were recognized as the two sectors of the Ramond–Neveu–Schwarz (RNS) fermionic string. This theory had a rich spectrum of states, including both bosons and fermions, and required $d = 10$ spacetime dimensions.

The RNS theory was the starting point for many modern developments. Gervais and Sakita observed a symmetry of the theory corresponding to transformations mapping fermionic and bosonic degrees of freedom among themselves: this was the beginning of supersymmetry. Moreover, the introduction of additional symmetries allowed for non-Abelian gauge symmetries in the massless spectrum and extended current-algebra invariances.

Part VI, 'The superstring', describes the transformation of string theory into its modern formulation. Around 1974, the application to hadron physics was abandoned in favour of the successful description provided by quantum chromodynamics (QCD), a non-Abelian gauge field theory. At the same time, it was understood by Scherk, Neveu, Schwarz and Yoneya that the presence of the massless spin one (two) states in the open (closed) string spectrum meant that the theory could reproduce gauge theories and Einstein gravity in the low energy limit, where all other states in the Regge trajectories become infinitely massive and decouple. Therefore, string theory could be considered as an extension of field theory rather than an alternative to it, as originally thought.

This result led Scherk and Schwarz to propose in 1974 the unification within string theory of all four fundamental interactions: the electromagnetic, weak and strong forces, described by gauge theories, together with gravity, described by Einstein's general relativity theory. This remarkable idea was much ahead of its time and could not be appreciated immediately: the theoretical physics community was mostly busy developing the gauge theories that form the Standard Model. Other ingredients of modern string theory, such as the Kaluza–Klein compactification of the extra dimensions and a mechanism for supersymmetry breaking, were also introduced by Scherk and Schwarz.

In the meanwhile, supersymmetry was formulated by Wess and Zumino in quantum field theory, independently of strings, as a spacetime symmetry relating particle spectra in four dimensions. Furthermore, the Ramond–Neveu–Schwarz string was proved to be spacetime supersymmetric by Gliozzi, Scherk and Olive in 1976, upon performing a projection of its spectrum that also eliminated the unwanted tachyon. To sum up, by 1976 open superstring theory was fully developed in its modern formulation of a unifying theory. However, it was left aside in favour of gauge theories, which were more economical and concrete.

Part VII, 'Preparing the string renaissance', describes the 'dark age' of string theory, between 1977 and 1983, when only a handful of people continued to work at it. They nevertheless obtained further results that were instrumental for its comeback in 1984. Towards the end of the Seventies, the main theoretical and experimental features of the Standard Model were being settled, and the issue of further unification became relevant in the theoretical physics community. Unification of electro-weak and strong interactions above the Standard Model energy scale, and unification with gravity, were addressed in the context of supersymmetric field theories and supergravities, respectively. Supergravity theories were the supersymmetric generalization of Einstein's general relativity, offering greater consistency and extra dimensions. Although they are low energy limits of superstring theories, they were mostly developed and analyzed within field theory.

The abrupt change of attitude that brought superstring theories back in focus is then described. The type I superstring was more appropriate and sound than the supergravity theories considered so far: it could describe the Standard Model spectrum of particles, requiring chiral fermions in four dimensions as well as the cancellation of the associated chiral anomalies, as shown remarkably by Green and Schwarz. Moreover, it provided a consistent quantum theory of gravity free of unwanted infinite quantities. On the other hand, supergravity theories, in particular the most fundamental theory in eleven dimensions, were still plagued with infinities.

These developments led to a new boom in string theory after 1984; since then the theory has been actively investigated till the present time. Recent findings show that string theory contains further degrees of freedom in addition to strings, i.e. membranes and D-branes, and that the five consistent superstring theories unify in a single theory called 'M-theory'. Furthermore, a novel relation between string and gauge theories has brought new insight into the hadronic string picture. A summary of these contemporary developments is presented in the last Chapter of Part VII.

Finally, the Volume contains five Appendices that provide more technical presentations of some key features of string theory: the S-matrix approach of the Sixties, the properties of the Veneziano amplitude, the full quantization of the bosonic string action, supersymmetry and the field theory limit.

Here we list the main books and review articles on early string theory. The Introductions to the Parts also provide general references on the topics discussed therein.

References

[Fra86] Frampton, P. H. ([1974] 1986). *Dual Resonance Models and Superstrings* (World Scientific, Singapore).

[GSW87] Green, M. B., Schwarz, J. H. and Witten, E. (1987). *Superstring Theory* (Cambridge University Press, Cambridge).

[Jac74] Jacob, M. ed. (1974). *Dual Theory*, Physics Reports Reprints Book Series, Vol. 1 (North Holland, Amsterdam).

[Sch75] Scherk, J. (1975). An introduction to the theory of dual models and strings, *Rev. Mod. Phys.* **47**, 123–163.

[Sch85] Schwarz, J. H. ed. (1985). *Superstrings: the First 15 Years of Superstring Theory* (World Scientific, Singapore).

Synopsis: 1968–1984

In the following we list the main developments in the early history of string theory, organized according to the Parts of the book in which they are described. Each topic is associated with some key references that are just a sample of the relevant literature. Complete lists of references can be found at the end of each Author's Chapter; a comprehensive guide to the bibliography on early string theory is given at the end of the textbook by Green, Schwarz and Witten, listed above.

Part II – The prehistory: the analytic S-matrix

Developments up to 1968

- The S-matrix approach to strong interactions, originally formulated in [Whe37] and [Hei43], is fully developed [Che61, ELOP66].
- Dolen, Horn and Schmid introduce an hypothesis on the structure of scattering amplitudes [DHS67], the so-called DHS duality, later called planar duality [Fre68, Har69, Ros69]; this is implemented in the superconvergence sum rules [ARVV68].
- Veneziano proposes a scattering amplitude obeying DHS duality: this is the beginning of the Dual Resonance Model [Ven68].

Other developments in theoretical physics

- The theory of weak nuclear interactions is developed.
- The spontaneous breaking of a symmetry is recognized as being a general phenomenon in many-body systems and quantum field theory.

References

[ARVV68] Ademollo, M., Rubinstein, H. R., Veneziano, G. and Virasoro, M. A. (1968). Bootstrap of meson trajectories from superconvergence, *Phys. Rev.* **176**, 1904–1925.

[Che61] Chew, G. F. (1961). *S-Matrix Theory of Strong Interactions* (W. A. Benjamin, New York).

[DHS67] Dolen, R., Horn, D. and Schmid, C. (1967). Prediction of Regge parameters of ρ poles from low-energy $\pi-N$ data, *Phys. Rev. Lett.* **19**, 402–407.

[ELOP66] Eden, R. J., Landshoff, P. V., Olive, D. I. and Polkinghorne, J. C. (1966). *The Analytic S Matrix* (Cambridge University Press, Cambridge).

[Fre68] Freund, P. G. O. (1968). Finite-energy sum rules and bootstraps, *Phys. Rev. Lett.* **20**, 235–237.

[Har69] Harari, H. (1969). Duality diagrams, *Phys. Rev. Lett.* **22**, 562–565.

[Hei43] Heisenberg, W. (1943). Die beobachtbaren grossössen in der theorie der elenemteilchen: I, *Z. Phys.* **120**, 513–538.

[Ros69] Rosner, J. L. (1969). Graphical form of duality, *Phys. Rev. Lett.* **22**, 689–692.

[Ven68] Veneziano, G. (1968). Construction of a crossing-symmetric, Reggeon behaved amplitude for linearly rising trajectories, *Nuovo Cimento* **A57**, 190–197.

[Whe37] Wheeler, J. A. (1937). On the mathematical description of light nuclei by the method of resonating group structure, *Phys. Rev.* **52**, 1107–1122.

Part III – The Dual Resonance Model

Developments during 1969–1973

- The Veneziano amplitude is generalized to the scattering of N particles [Cha69, CT69, GS69]; in particular, the string world-sheet first appears in Koba and Nielsen's work [KN69].

- Shapiro and Virasoro extend the Veneziano formula and obtain the first amplitudes of closed string theory [Vir69, Sha70].
- The residues of the poles of the N-point amplitude are shown to be given by a sum of factorized terms and their number is shown to increase exponentially with the mass [BM69, FV69].
- Fubini, Gordon and Veneziano introduce an operator formalism of harmonic oscillators that allows for the analysis of the theory spectrum [FGV69, FV70]; additional decoupling conditions are obtained if the intercept of the Regge trajectory is $\alpha_0 = 1$ [Vir70]; in this case the lowest state of the spectrum is a tachyon. Fubini and Veneziano obtain the algebra of the Virasoro operators and Weis finds its central extension [FV71].
- The equations characterizing the on-shell physical states are derived [DD70] and an infinite set of physical states, called DDF states after Del Giudice, Di Vecchia and Fubini, is found [DDF72]; the Dual Resonance Model has no ghosts if $d \leq 26$ [Bro72, GT72]; for $d = 26$ the DDF states span the whole physical subspace.
- One-loop diagrams are computed to restore perturbative unitarity [ABG69, BHS69, KSV69]; Lovelace shows that the nonplanar loop diagram complies with unitarity only for 26 spacetime dimensions [Lov71].
- The three-Reggeon vertex is constructed [CSV69, Sci69] and generalized to N external particles [Lov70a]; the N-Reggeon vertex is used to compute multiloop diagrams [KY70, Lov70b, Ale71, AA71].
- Vertex operators for excited states of the string are constructed [CFNS71, CR71].
- Brink and Olive obtain the physical state projection operator and clearly show that only $(d - 2)$ transverse oscillators contribute to one-loop diagrams [BO73].

Other developments in theoretical physics

- The non-Abelian gauge theory describing weak and electromagnetic interactions is formulated; this is the first step towards the Standard Model of particle physics.
- Experiments on deep inelastic scattering show the existence of pointlike constituents inside hadrons.

References

[Ale71] Alessandrini, V. (1971). A general approach to dual multiloop amplitudes, *Nuovo Cimento* **A2**, 321–352.

[AA71] Alessandrini, V. and Amati, D. (1971). Properties of dual multiloop amplitudes, *Nuovo Cimento* **A4**, 793–844.

[ABG69] Amati, D., Bouchiat, C. and Gervais, J. L. (1969). On the building of dual diagrams from unitarity, *Lett. Nuovo Cimento* **2S1**, 399–406.

[BHS69] Bardakci, K., Halpern, M. B. and Shapiro, J. A. (1969). Unitary closed loops in Reggeized Feynman theory, *Phys. Rev.* **185**, 1910–1917.

[BM69] Bardakci, K. and Mandelstam, S. (1969). Analytic solution of the linear-trajectory bootstrap, *Phys. Rev.* **184**, 1640–1644.

[BO73] Brink, L. and Olive, D. I. (1973). The physical state projection operator in dual resonance models for the critical dimension of the spacetime, *Nucl. Phys.* **B56**, 253–265.
[Bro72] Brower, R. C. (1972). Spectrum-generating algebra and no-ghost theorem in the dual model, *Phys. Rev.* **D6**, 1655–1662.
[CFNS71] Campagna, P., Fubini, S., Napolitano, E. and Sciuto, S. (1971). Amplitude for n nonspurious excited particles in dual resonance models, *Nuovo Cimento* **A2**, 911–928.
[CSV69] Caneschi, L., Schwimmer, A. and Veneziano, G. (1970). Twisted propagator in the operatorial duality formalism, *Phys. Lett.* **B30**, 351–356.
[Cha69] Chan, H. M. (1969). A generalized Veneziano model for the N-point function, *Phys. Lett.* **B22**, 425–428.
[CT69] Chan, H. M. and Tsou, S. T. (1969). Explicit construction of the N-point function in the generalized Veneziano model, *Phys. Lett.* **B28**, 485–488.
[CR71] Clavelli, L. and Ramond, P. (1971) Group-theoretical construction of dual amplitudes, *Phys. Rev.* **D3**, 988–990.
[DD70] Del Giudice, E. and Di Vecchia, P. (1970). Characterization of the physical states in dual-resonance models, *Nuovo Cimento* **A70**, 579–591.
[DDF72] Del Giudice, E., Di Vecchia, P. and Fubini, S. (1972). General properties of the dual resonance model, *Ann. Phys.* **70**, 378–398.
[FGV69] Fubini, S., Gordon, D. and Veneziano, G. (1969). A general treatment of factorization in dual resonance models, *Phys. Lett.* **B29**, 679–682.
[FV69] Fubini, S. and Veneziano, G. (1969). Level structure of dual-resonance models, *Nuovo Cimento* **A64**, 811–840.
[FV70] Fubini, S. and Veneziano, G. (1970). Duality in operator formalism, *Nuovo Cimento* **A67**, 29–47.
[FV71] Fubini, S. and Veneziano, G. (1971). Algebraic treatment of subsidiary conditions in dual resonance model, *Ann. Phys.* **63**, 12–27.
[GT72] Goddard, P. and Thorn, C. B. (1972). Compatibility of the pomeron with unitarity and the absence of ghosts in the dual resonance model, *Phys. Lett.* **B40**, 235–238.
[GS69] Goebel, C. J. and Sakita, B. (1969). Extension of the Veneziano form to N-particle amplitude, *Phys. Rev. Lett.* **22**, 257–260.
[KY70] Kaku, M. and Yu, L. (1970). The general multiloop Veneziano amplitude, *Phys. Lett.* **B33**, 166–170.
[KSV69] Kikkawa, K., Sakita, B. and Virasoro, M. (1969). Feynman-like diagrams compatible with duality. I. Planar diagrams, *Phys. Rev.* **184**, 1701–1713.
[KN69] Koba, Z. and Nielsen, H. B. (1969). Manifestly crossing invariant parameterization of n meson amplitude, *Nucl. Phys.* **B12**, 517–536.
[Lov70a] Lovelace, C. (1970). Simple n-reggeon vertex, *Phys. Lett.* **B32**, 490–494.
[Lov70b] Lovelace, C. (1970). M-loop generalized Veneziano formula, *Phys. Lett.* **B52**, 703–708.
[Lov71] Lovelace, C. (1971). Pomeron form factors and dual Regge cuts, *Phys. Lett.* **B34**, 500–506.
[Sci69] Sciuto, S. (1969). The general vertex function in generalized dual resonance models, *Lett. Nuovo Cimento* **2**, 411–418.
[Sha70] Shapiro, J. A. (1970). Electrostatic analogue for the Virasoro model, *Phys. Lett.* **B33**, 361–362.
[Vir69] Virasoro, M. (1969). Alternative constructions of crossing-symmetric amplitudes with Regge behavior, *Phys. Rev.* **177**, 2309–2311.

[Vir70] Virasoro, M. (1970). Subsidiary conditions and ghosts in dual-resonance models, *Phys. Rev.* **D1**, 2933–2936.

Part IV – The string

Developments during 1970–1973

- Nambu, Nielsen and Susskind suggest independently that the dynamics underlying the dual model is that of a relativistic string [Nie69, Sus69, Nam70a, Nam70b, Nie70, Sus70].
- Nambu and then Goto write the string action [Nam70b, Got71].
- The analogue model, proposed by Fairlie and Nielsen and related to the string picture, is used to compute dual amplitudes [FN70, FS70].
- Goddard, Goldstone, Rebbi and Thorn quantize the string action in the light-cone gauge; the spectrum is found to be in complete agreement with that of the Dual Resonance Model for $d = 26$ [GGRT73]; apart from the tachyon, string theory is now a consistent quantum-relativistic system.
- The computation by Brink and Nielsen [BN73] of the zero-point energy of the string gives a relation between the dimension of spacetime and the mass of the lowest string state.
- The interaction among strings is introduced within the light-cone path-integral formalism [Man73a] and within the operator approach by letting the string interact with external fields [ADDN74]; the coupling between three arbitrary physical string states is computed both in the path-integral [Man73a, CG74] and operator [ADDF74] formalisms, finding agreement.

References

[ADDN74] Ademollo, M., D'Adda, A., D' Auria, R., Napolitano, E., Sciuto, S., Di Vecchia, P., Gliozzi, F., Musto, R. and Nicodemi, F. (1974). Theory of an interacting string and dual-resonance model, *Nuovo Cimento* **A21**, 77–145.

[ADDF74] Ademollo, M., Del Giudice, E., Di Vecchia, P. and Fubini, S. (1974). Couplings of three excited particles in dual-resonance model, *Nuovo Cimento* **A19**, 181–203.

[BN73] Brink, L. and Nielsen, H. B. (1973). A simple physical interpretation of the critical dimension of the spacetime in dual models, *Phys. Lett.* **B45**, 332–336.

[CG74] Cremmer, E. and Gervais, J. L. (1974). Combining and splitting relativistic strings, *Nucl. Phys.* **B76**, 209–230.

[FN70] Fairlie, D. B. and Nielsen, H. B. (1970). An analog model for KSV theory, *Nucl. Phys.* **B20**, 637–651.

[FS70] Frye, G. and Susskind, L. (1970). Non-planar dual symmetric loop graphs and pomeron, *Phys. Lett.* **B31**, 589–591.

[GGRT73] Goddard, P., Goldstone, J., Rebbi, C. and Thorn, C. B. (1973). Quantum dynamics of a massless relativistic string, *Nucl. Phys.* **B56**, 109–135.

[Got71] Goto, T. (1971). Relativistic quantum mechanics of one dimensional mechanical continuum and subsidiary condition of the dual resonance model, *Prog. Theor. Phys.* **46**, 1560–1569.

[Man73a] Mandelstam, S. (1973). Interacting string picture of dual-resonance models, *Nucl. Phys.* **B64**, 205–235.

[Nam70a] Nambu, Y. (1970). Quark model and the factorization of the Veneziano amplitude, in *Proceedings of the International Conference on Symmetries and Quark Models, Wayne State University, June 18–20, 1969*, ed. Chand, R. (Gordon and Breach, New York), 269–277, reprinted in *Broken Symmetry, Selected Papers of Y. Nambu*, ed. Eguchi, T. and Nishijima, K. (World Scientific, Singapore, 1995), 258–277.

[Nam70b] Nambu, Y. (1970). Duality and hadrodynamics, lecture notes prepared for Copenhagen summer school, 1970, reproduced in *Broken Symmetry, Selected Papers of Y. Nambu*, ed. Eguchi, T. and Nishijima, K. (World Scientific, Singapore, 1995), 280.

[Nie69] Nielsen, H. B. (1969). A physical interpretation of the n-point Veneziano model (Nordita preprint, 1969).

[Nie70] Nielsen, H. B. (1970). An almost physical interpretation of the integrand of the n-point Veneziano model, paper presented at the *15th International Conference on High Energy Physics, Kiev, 26 August to 4 September, 1970* (Nordita preprint, 1969).

[Sus69] Susskind, L. (1969). Harmonic-oscillator analogy for the Veneziano model, *Phys. Rev. Lett.* **23**, 545–547.

[Sus70] Susskind, L. (1970). Structure of hadrons implied by duality, *Phys. Rev.* **D1**, 1182–1186.

Part V – Beyond the bosonic string

Developments during 1970–1974

- The Dual Resonance Model is generalized to spacetime fermions by Ramond [Ram71]; an extension of the Dual Resonance Model for pions is constructed by Neveu and Schwarz [NS71]; the two models are recognized as the two sectors of the Ramond–Neveu–Schwarz model [Tho71].
- The fermion emission vertex is constructed by Corrigan and Olive [CO72]; the scattering amplitude involving four fermions is computed within the light-cone path-integral [Man73b] and operator [CGOS73, SW73] formalisms.
- The one-loop [GW71] and multiloop [Mon74] amplitudes of the Ramond–Neveu–Schwarz model are computed.
- Gervais and Sakita find that the RNS model possesses a symmetry relating bosons to fermions, the world-sheet supersymmetry [GS71].
- Further extensions of the bosonic string involve the introduction of internal symmetry groups [CP69], current algebra symmetries [BH71], and extended supersymmetries [ABDD76].

Other developments in theoretical physics

- The gauge theory of quarks and gluons, quantum chromodynamics, is proposed for strong interactions; it is shown to be weakly interacting at high energy (asymptotic freedom).
- The proof of renormalization of non-Abelian gauge theories is completed.

- The renormalization group is understood as a general method to relate the physics at different energy scales in quantum field theory.

References

[ABDD76] Ademollo, M., Brink, L., D'Adda, A., D'Auria, R., Napolitano, E., Sciuto, S., Del Giudice, E., Di Vecchia, P., Ferrara, S., Gliozzi, F., Musto, R. and Pettorino, R. (1976). Supersymmetric strings and color confinement, *Phys. Lett.* **B62**, 105–110.

[BH71] Bardakci, K. and Halpern, M. B. (1971). New dual quark models, *Phys. Rev.* **D3**, 2493–2506.

[CP69] Chan, H. M. and Paton, J. E. (1969). Generalized Veneziano model with isospin, *Nucl. Phys.* **B10**, 516–520.

[CGOS73] Corrigan, E. F., Goddard, P., Olive, D. I. and Smith, R. A. (1973). Evaluation of the scattering amplitude for four dual fermions, *Nucl. Phys.* **B67**, 477–491.

[CO72] Corrigan, E. F. and Olive, D. I. (1972). Fermion-meson vertices in dual theories, *Nuovo Cimento* **A11**, 749.

[GS71] Gervais, J. L. and Sakita, B. (1971). Field theory interpretation of supergauges in dual models, *Nucl. Phys.* **B34**, 632–639.

[GW71] Goddard, P. and Waltz, R. E. (1971). One-loop amplitudes in the model of Neveu and Schwarz, *Nucl. Phys.* **B34**, 99–108.

[Man73b] Mandelstam, S. (1973). Manifestly dual formulation of the Ramond model, *Phys. Lett.* **B46**, 447–451.

[Mon74] Montonen, C. (1974). Multiloop amplitudes in additive dual resonance models, *Nuovo Cimento* **A19**, 69–89.

[NS71] Neveu, A. and Schwarz, J.H. (1971). Factorizable dual model of pions, *Nucl. Phys.* **B31**, 86–112.

[Ram71] Ramond, P. (1971). Dual theory for free fermions, *Phys. Rev.* **D3**, 2415–2418.

[SW73] Schwarz, J. H. and Wu, C. C. (1973). Evaluation of dual fermion amplitudes, *Phys. Lett.* **B47**, 453–456.

[Tho71] Thorn, C. (1971). Embryonic dual model for pions and fermions, *Phys. Rev.* **D4**, 1112–1116.

Part VI – The superstring

Developments during 1974–1977

- In the limit of infinite string tension, string theory reduces to quantum field theory [Sch71]: the open string leads to non-Abelian gauge theories [NS72, Yon74] and the closed string to gravity [Yon73, SS74]; therefore, string theory provides a framework for unifying all fundamental interactions [SS74, SS75].
- Wess and Zumino extend the world-sheet supersymmetry of the Ramond–Neveu–Schwarz model to four-dimensional field theory [WZ74]; supersymmetric extensions of all known quantum field theories are obtained.

- By performing a projection of states in the Ramond–Neveu–Schwarz model, Gliozzi, Scherk and Olive construct the first string theory that is supersymmetric in spacetime [GSO76]. This theory is free of tachyons and unifies gauge theories and gravity: modern superstring theory is born.
- To cope with experiments, the six extra dimensions can be compactified by using the Kaluza–Klein reduction [CS76], that also provides a mechanism for supersymmetry breaking [SS79].
- Supergravity, the supersymmetric extension of Einstein's field theory of gravitation, is formulated [DZ76a, FNF76].
- The supersymmetric action for the Ramond–Neveu–Schwarz string is obtained [BDH76, DZ76b].

Other developments in theoretical physics

- Quantum chromodynamics is widely recognized as the correct theory of strong interactions.
- The Standard Model of electro-weak and strong interactions is completed and receives experimental verification.
- Attempts are made to unify electro-weak and strong interactions beyond the Standard Model; the Grand Unified Theory is formulated.

References

[BDH76] Brink, L., Di Vecchia, P. and Howe, P. S. (1976). A locally supersymmetric and reparameterization invariant action for the spinning string, *Phys. Lett.* **B65**, 471–474.

[CS76] Cremmer, E. and Scherk, J. (1976). Dual models in four dimensions with internal symmetries, *Nucl. Phys.* **B103**, 399–425.

[DZ76a] Deser, S. and Zumino, B. (1976). Consistent supergravity, *Phys. Lett.* **B62**, 335–337.

[DZ76b] Deser, S. and Zumino, B. (1976). A complete action for the spinning string, *Phys. Lett.* **B65**, 369–373.

[FNF76] Freedman, D. Z., van Nieuwenhuizen, P. and Ferrara, S. (1976). Progress toward a theory of supergravity, *Phys. Rev.* **D13**, 3214–3218.

[GSO76] Gliozzi, F., Scherk, J. and Olive, D. (1976). Supergravity and the spinor dual model, *Phys. Lett.* **B65**, 282–286.

[NS72] Neveu, A. and Scherk, J. (1972). Connection between Yang–Mills fields and dual models, *Nucl. Phys.* **B36**, 155–161.

[Sch71] Scherk, J. (1971). Zero-slope limit of the dual resonance model, *Nucl. Phys.* **B31**, 222–234.

[SS74] Scherk, J. and Schwarz, J. H. (1974). Dual models for nonhadrons, *Nucl. Phys.* **B81**, 118–144.

[SS75] Scherk, J. and Schwarz, J. H. (1975). Dual field theory of quarks and gluons, *Phys. Lett.* **B57**, 463–466.

[SS79] Scherk, J. and Schwarz, J. H. (1979). Spontaneous breaking of supersymmetry through dimensional reduction, *Phys. Lett.* **B82**, 60–64.

[WZ74] Wess, J. and Zumino, B. (1974). Supergauge transformations in four dimensions, *Nucl. Phys.* **B70**, 39–50.

[Yon73] Yoneya, T. (1973). Quantum gravity and the zero slope limit of the generalized Virasoro model, *Lett. Nuovo Cimento* **8**, 951–955.

[Yon74] Yoneya, T. (1974). Connection of dual models to electrodynamics and gravitodynamics, *Prog. Theor. Phys.* **51**, 1907–1920.

Part VII – Preparing the string renaissance

Developments during 1978–1984

- Using techniques developed in non-Abelian gauge theories, Polyakov quantizes the string by covariant path-integral methods, opening the way to modern treatments of string theories [Pol81a, Pol81b]; the Polyakov approach is further developed [DOP82, Fri82, Fuj82, Alv83].
- The unique and most symmetric supergravity in eleven dimensions is constructed [CJS78].
- Green and Schwarz introduce a new light-cone formalism where the fermionic coordinate is an $SO(8)$ spinor [GS81, GS82a, GS82b]; they construct type IIA and IIB closed string theories [GS82c] and write the covariant spacetime supersymmetric action for the superstring [GS84a].
- The contribution of chiral fields to the gauge and gravitational anomalies is computed and shown to vanish in type IIB supergravity [AW84].
- Type I superstring and supergravity theories with gauge group $SO(32)$ are shown to be free from gauge and gravitational anomalies [GS84b, GS85].
- Two other anomaly-free superstring theories are constructed, the heterotic strings with $E_8 \times E_8$ and $SO(32)$ groups [GHMR85].
- Calabi–Yau compactifications of the $E_8 \times E_8$ heterotic string give supersymmetric four-dimensional gauge theories with realistic features for the description of the Standard Model and gravity [CHSW85].

Other developments in theoretical physics 1976–1984

- The Standard Model of electro-weak and strong interactions is fully confirmed by experiments.
- Attempts aiming at the unification of all interactions including gravity are based on supergravity theories, which are extensively studied.
- Phenomenological consequences of supersymmetry are investigated; the Minimal Supersymmetric Standard Model is formulated.
- This is the 'golden era' of modern quantum field theory, with several results in gauge theories: nonperturbative methods, numerical simulations, the study of anomalies and the interplay with mathematical physics.

References

[Alv83] Alvarez, O. (1983). Theory of strings with boundaries: fluctuations, topology and quantum geometry, *Nucl. Phys.* **B216**, 125–184.

[AW84] Alvarez-Gaumé, L. and Witten, E. (1984). Gravitational anomalies, *Nucl. Phys.* **B234**, 269–330.

[CHSW85] Candelas, P., Horowitz, G., Strominger, A. and Witten, E. (1985). Vacuum configurations for superstrings, *Nucl. Phys.* **B258**, 46–74.

[CJS78] Cremmer, E., Julia, B. and Scherk, J. (1978). Supergravity theory in 11 dimensions, *Phys. Lett.* **B76**, 409–412.

[DOP82] Durhuus, B., Olesen, P. and Petersen, J. L. (1982). Polyakov's quantized string with boundary terms, *Nucl. Phys.* **198**, 157–188.

[Fri82] Friedan, D. (1982). Introduction to Polyakov's string theory, in *Recent Advances in Field Theory and Statistical Mechanics, Proc. 1982 Les Houches Summer School*, ed. Zuber, J. B. and Stora, R. (Elsevier, Amsterdam).

[Fuj82] Fujikawa, K. (1982). Path integrals of relativistic strings, *Phys. Rev.* **D25**, 2584–2592.

[GS81] Green, M. B. and Schwarz, J. H. (1981). Supersymmetric dual string theory, *Nucl. Phys.* **B181**, 502–530.

[GS82a] Green, M. B. and Schwarz, J. H. (1982). Supersymmetric dual string theory. 2. Vertices and trees, *Nucl. Phys.* **B198**, 252–268.

[GS82b] Green, M. B. and Schwarz, J. H. (1982). Supersymmetric dual string theory. 3. Loops and renormalization, *Nucl. Phys.* **B198**, 441–460.

[GS82c] Green, M. B. and Schwarz, J. H. (1982). Supersymmetrical string theories, *Phys. Lett.* **B109**, 444–448.

[GS84a] Green, M. B. and Schwarz, J. H. (1984). Covariant description of superstrings, *Phys. Lett.* **B136**, 367–370.

[GS84b] Green, M. B. and Schwarz, J. H. (1984). Anomaly cancellation in supersymmetric $d = 10$ gauge theory and superstring theory, *Phys. Lett.* **B149**, 117–122.

[GS85] Green, M. B. and Schwarz, J. H. (1985). The hexagon gauge anomaly in type I superstring theory, *Nucl. Phys.* **B255**, 93–114.

[GHMR85] Gross, D. J., Harvey, J. A., Martinec, E. J. and Rohm, R. (1985). Heterotic string, *Phys. Rev. Lett.* **54**, 502–505.

[Pol81a] Polyakov, A. M. (1981). Quantum geometry of bosonic strings, *Phys. Lett.* **B103**, 207–210.

[Pol81b] Polyakov, A. M. (1981). Quantum geometry of fermionic strings, *Phys. Lett.* **B103**, 211–213.

2
Rise and fall of the hadronic string

GABRIELE VENEZIANO

Abstract

A personal account is given of the six to seven years (1967–1974) during which the hadronic string rose from down-to-earth phenomenology, through some amazing theoretical discoveries, to its apotheosis in terms of mathematical beauty and physical consistency, only to be doomed suddenly by the inexorable verdict of experiments. The a posteriori reasons for why the theorists of the time were led to the premature discovery of what has since become a candidate Theory of Everything, are also discussed.

2.1 Introduction and outline

In order to situate historically the developments I will be covering in this Chapter, let me start with a picture (see Figure 2.1) illustrating, with the help of Michelin-guide-style grading, the amazing developments that took place in our understanding of elementary particle physics from the mid-Sixties to the mid-Seventies. Having graduated from the University of Florence in 1965, I had the enormous luck to enter the field just at the beginning of that period which, a posteriori, can rightly be called the 'golden decade' of elementary particle physics.

The theoretical status of the four fundamental interactions was very uneven in the mid-Sixties: only the electromagnetic interaction could afford an (almost[1]) entirely satisfactory description (hence a 3-star status) according to quantum electrodynamics (QED), the quantum-relativistic extension of Maxwell's theory. Gravity too had a successful theoretical description, this time according to Einstein's general relativity, its 2-star rating being related to the failure of all attempts to construct a consistent quantum extension. In the middle part of Figure 2.1 I have put the other two interactions, the weak and the strong,

---

[1] The 'almost' refers to what is now known as the triviality problem, meaning that QED cannot be extended without changes to arbitrarily high energies.

The Birth of String Theory, ed. Andrea Cappelli, Elena Castellani, Filippo Colomo and Paolo Di Vecchia.
Published by Cambridge University Press. © Cambridge University Press 2012.

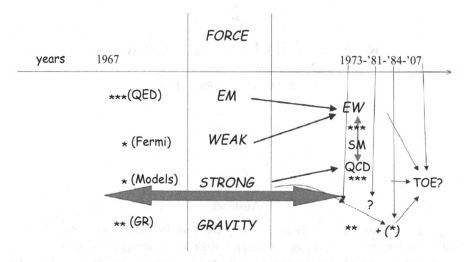

Figure 2.1 Evolution of our theories of fundamental forces during the 'golden decade'.

which, being still in search of something deserving the name of a theory, could only afford a single star.

If we look again at the situation some six or seven years later (say around 1973) we cannot fail to notice a striking improvement: the theory of weak interactions had been upgraded and unified with the electromagnetic theories in what became known as the electro-weak sector of the Standard Model, while the strong interaction had also found a beautiful description in the second sector of the Standard Model, quantum chromodynamics (QCD). Thus, by the mid-Seventies, all three nongravitational interactions had reached 3-star status, while the situation had basically remained unchanged for gravity.

This Chapter covers a parallel attempt to find a good theory for the strong interactions which, instead of succeeding in its goal, gave rise to a completely different theoretical framework, string theory. The string theory of hadrons met with phenomenological difficulties and could not compete eventually with QCD; it was abandoned as such for about a decade until it was brought back into the spotlight by the daring proposal that, upon a huge rescaling of the string characteristic size, it could become a serious candidate for a unified quantum theory of all interactions, including gravity, a Theory of Everything (TOE), as indicated at the far right of the figure.

The rest of the Chapter is organized as follows: in Section 2.2 I will recall the developments that, within a year of exciting work, culminated in the first example of a Dual Resonance Model, the beta-function ansatz, a mathematical expression later understood to describe quantum mechanically a reaction in which two open strings collide to give rise to two other open strings. In Sections 2.3 and 2.4 I will recall the striking developments which, in spite of many apparent theoretical obstacles, led to what appeared to be a fully consistent mathematical and physical framework for dealing with strong interactions. In Section 2.5 I will describe how those developments gave, at the same time, numerous hints for a string-like structure lying at the basis of Dual Resonance Models; largely missed for some time, these hints led eventually to the formulation of (super)string theory as it is basically known even today. Finally, in Sections 2.6 and 2.7, I will recall how, on the basis of several experimental tests, string theory had to concede defeat to its quantum-field-theory competitor, QCD. I will also outline the reasons why a crucial property of QCD, quark confinement, together with the idea of a $1/N$ expansion, does imply the existence of narrow string-like excitations in the hadronic world as well as the duality property of their mutual interactions, and thus explain the discovery of string theory many years before it could find an even more ambitious potential application in physics.

2.2 String prehistory

I will start my account from around 1967. At that time the status of the theory of strong (nuclear) interactions was not very brilliant. Data were abundant, but we could only confront them with a handful of models, each one capturing one or another aspect of the complicated hadronic world (hadron is a generic name for any particle feeling the strong force). The following features were noteworthy.

- Strong interactions are short ranged (implying the absence of massless hadrons).
- They are characterized by several (exact or approximate) symmetries and the corresponding conservation laws.
- Many hadrons had been identified, most of them metastable (resonances), and with large mass and angular momentum (spin): the 'hadronic zoo' seemed to be increasing in size every day.

The problem was to find a way to put some order and simplicity in this complex situation.

Today, with hindsight, we can easily assert that, in the late Sixties, we took the wrong way by rejecting, a priori, a description of these phenomena based on quantum field theory (QFT), the framework that had already been so successful for the electromagnetic interactions via quantum electrodynamics (QED). There were (at least) two very good excuses for having chosen this wrong way:

- unlike in QED, the theory of interacting photons and electrons, there were too many particles to deal with, actually, as I just said, an ever increasing number;
- QFTs of particles with high angular momentum were known to be very difficult, if not impossible, to deal with in a QFT framework.

Instead, a so-called S-matrix approach looked much more promising. Let's describe it briefly. The idea, originally due to J. H. Wheeler [Whe37] and W. Heisenberg [Hei43] (see Eden et al. [ELOP66] for a more modern account), is that a good theory should only deal with quantities that can be measured directly. When one considers a reaction in which a given initial state evolves into a final state according to the laws of quantum mechanics, the relevant object is the so-called probability amplitude, a complex number whose absolute

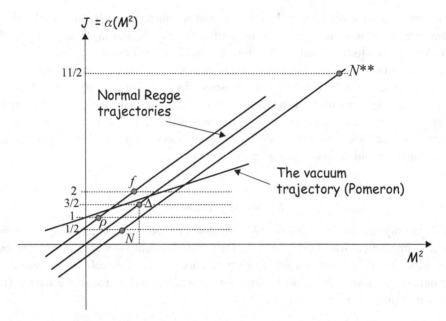

Figure 2.2 Regge trajectories at positive and negative values of M^2.

value squared gives the usual probability for that particular process to occur. Considering all possible initial and final states, the set of all these complex numbers defines a huge (actually infinite) matrix called the S-matrix (S for scattering) which, by conservation of probability, should be unitary.

The constraints of relativistic causality force the S-matrix elements to be analytic functions of the kinematical variables they depend upon, like the energy of the collision. Also, the symmetries of the strong interactions can easily be implemented at the level of the S-matrix. These symmetries could also be used to put some order in the hadronic zoo by grouping particles with the same spin into families.

Finally, the recently developed Regge theory (see Collins [Col77]) was also able to assemble together particles of different angular momentum. One amazing empirical observation at the time was that the masses M and angular momenta J of particles lying on the same 'Regge trajectory' satisfied a simple relation:

$$J = \alpha(M^2) = \alpha_0 + \alpha' M^2, \qquad (2.1)$$

with α_0 a parameter depending on the particular Regge family under consideration and α' a universal constant ($\alpha' \sim 0.9 \text{ GeV}^{-2}$ in natural units where $c = \hbar = 1$).

Regge theory had a second important facet, pointed out later by Chew and Mandelstam [Col77]: it could be used to describe the behaviour of the S-matrix at high energy. These two uses of Regge theory are illustrated in Figure 2.2, where we see the linear and parallel Regge trajectories (with one exception, the so-called vacuum or Pomeranchuk trajectory)

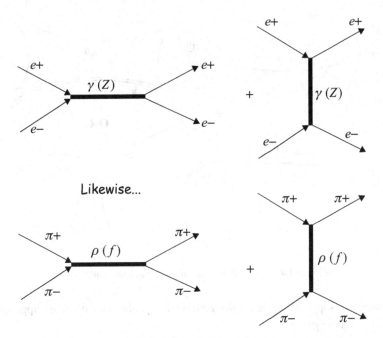

Figure 2.3 Feynman diagrams for QED and for strong interactions.

and the fact that the trajectory interpolates among different particles at positive J, M^2 while it determines high energy scattering at negative M^2.

Chew [Che66] had invoked these two appealing features of Regge's theory to formulate what I will call (for reasons that will become clear later) an 'expensive bootstrap'. Chew's idea was to add to the already mentioned constraints (unitarity, analyticity, symmetry) the assumption of 'nuclear democracy' according to which:

- *all* hadrons, whether stable or unstable, lie on Regge trajectories (at $M^2 \geq 0$) and are on the same footing;
- the high energy behaviour of the S-matrix is *entirely* given in terms of the same Regge trajectories (at $M^2 \leq 0$).

The hope was that an essentially unique S-matrix would come out after imposing this set of constraints. Actually, in Chew's programme Regge trajectories would appear twice in determining the structure of the S-matrix: once by giving the set of (unstable) intermediate states through which the process could proceed; and once by giving the set of (virtual) particles that could be exchanged. In this sense the situation would mimic that of QED. In Figure 2.3 we show the lowest order QED Feynman diagrams for electron–positron scattering (one-photon exchange) as well as the analogous diagrams for elastic $\pi^+\pi^-$ scattering through formation or exchange of a ρ^0-meson. The difference was that while electrons and

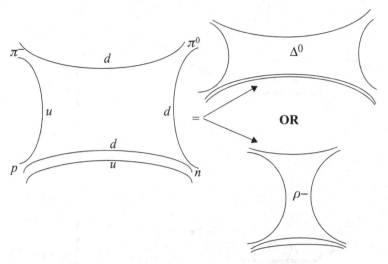

Figure 2.4 Duality diagrams for pion–nucleon scattering.

photons were not supposed to lie on Regge trajectories, the opposite was supposedly true for the pions and the ρ-meson.

However, an interesting surprise came out in 1967 through a fundamental observation made by Dolen, Horn and Schmid [DHS68, Sch68] who, after looking carefully at some pion–nucleon scattering data, concluded that contributions from resonance formation and those from particle exchange should *not* be added but were actually each one a complete representation of the process. This property became known as DHS duality.

In the spring of 1967 several attempts were made to couple DHS duality with another tool, developed by Sergio Fubini and collaborators, the so-called superconvergence equations [DFFR66, Fub66]. The combination of both ideas led Logunov, Soloviev and Tavkhelidze [LST67] and Igi and Matsuda [IM67] to write down the so-called 'finite energy sum rules' (FESR). In the summer of 1967, at a summer school in Erice, I was strongly influenced by a talk given by Murray Gell-Mann on DHS duality, stressing that such a framework could lead to what he defined as a 'cheap bootstrap' as opposed to Chew's expensive one. In order to get interesting constraints on the Regge trajectories themselves it was enough to require that the two dual descriptions of a process would produce the same answer.

I will open a short parenthesis here. Around the same time Harari [Har69] and Rosner [Ros69] gave an interesting graphical representation of DHS duality by drawing 'duality diagrams' (see Figure 2.4) where hadrons are represented by a set of quark lines (two for the mesons, three for the baryons) and the scattering process is described in terms of the flow of these quark lines through the diagram. By looking at the diagram in different directions (channels), the process is seen to proceed in different – but equivalent in the sense of DHS duality – ways. Duality diagrams are therefore very different from the Feynman diagrams of Figure 2.3. Notice that in those days quarks were just a mnemonic to keep track of quantum

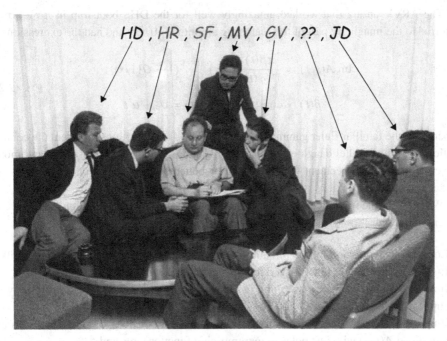

Figure 2.5 At the Weizmann Institute in 1967. From left to right: Hans Dahmen, Hector Rubinstein, Sergio Fubini, Miguel Virasoro, Gabriele Veneziano, unknown, Joe Dothan.

numbers and internal symmetries: they were not considered as having any real substance. Also, the duality diagrams were supposed to represent processes dominated by Regge trajectories other than the one carrying the vacuum quantum number: that exchange was supposed not to be dual to any resonances but rather to some non-resonating multiparticle background (see Harari [Har68] and Freund [Fre68]).

Coming back to the 'cheap bootstrap' the problem was that of finding a simple way to implement it. The original pion–nucleon process considered by Dolen, Horn and Schmid looked too complicated; it also represented a relation between mesons and baryons rather than a self-consistency condition among mesons only. Instead, in the fall of 1967, Ademollo, Rubinstein, Virasoro and myself (ARVV) [ARVV67, ARVV68] decided to apply the idea to a theoretically easier (even if experimentally unpractical) process: $\pi\pi \to \pi\omega$. This reaction has the property of being the same in all three channels and of allowing only very selective quantum numbers in each of them, basically those of the ρ-meson and its orbital excitations.

Between the fall of 1967 and the summer of 1968 ARVV (with the help of M. Bishari and A. Schwimmer and the advice and encouragement of S. Fubini, see Figure 2.5) made much progress in finding approximate solutions to this 'cheap bootstrap'. A rather simple ansatz was working remarkably well provided the ρ-Regge trajectory was taken to be straight and to be accompanied by lower parallel 'daughter' trajectories. I recall thinking (and telling people) that something even simpler was probably hiding behind all that . . .

The ARVV ansatz that worked amazingly well for the DHS bootstrap in $\pi\pi \to \pi\omega$ referred to the imaginary part of the scattering amplitude $A(s,t)$ and had the expression:

$$\text{Im } A(s,t) = \frac{\pi\beta(t)}{\Gamma(\alpha(t))} (\alpha's)^{\alpha(t)-1} (1 + O(1/s)),$$

$$\beta(t) \sim \text{constant}, \qquad \alpha(t) = \alpha_0 + \alpha't,$$
(2.2)

where Γ is the familiar Euler gamma function and the $1/s$ corrections represent the contribution from the parallel daughter trajectories. By tuning such corrections more and more we could make the agreement with DHS duality better and better and could extend it to a larger and larger range of t.

Which were the ingredients that led from that ansatz to an 'exact solution'? They were essentially three.

- Look for an expression for the full amplitude A rather than for its imaginary part; unlike Im A, A is an analytic function and thus easier to work with.
- Impose crossing symmetry, meaning in this case $A(s,t) = A(t,s)$; no such property is supposed to hold for Im A.
- Emphasize the resonance side of DHS duality rather than the Regge side, i.e. look for an amplitude $A(s,t)$ with just poles (a meromorphic function) in s and t.

At this point one notices that the analytic function whose imaginary part is given by Eq. (2.2) is simply:

$$A(s,t) = \beta(t)\Gamma(1-\alpha(t))(-\alpha's)^{\alpha(t)-1}(1 + O(1/s)).$$
(2.3)

However, this function does not obey crossing symmetry. It has poles in t but only a branch point in s. In order to introduce poles in s that follow exactly the pattern of those in t we can replace the factor $(-\alpha's)^{\alpha(t)-1}$ by $\Gamma(1-\alpha(s))$ and, in order not to change the high energy behaviour, we divide by a third Γ-function:

$$(-\alpha's)^{\alpha(t)-1}(1 + O(1/s)) \to \frac{\Gamma(1-\alpha(s))}{\Gamma(2-\alpha(s)-\alpha(t))},$$
(2.4)

and note, not without satisfaction, that imposing the right asymptotic behaviour has automatically provided a crossing-symmetric amplitude. Indeed the end result [Ven68] is the well-known Euler beta function:

$$A(s,t) = \beta \frac{\Gamma(1-\alpha(s))\Gamma(1-\alpha(t))}{\Gamma(2-\alpha(s)-\alpha(t))} \equiv \beta B(1-\alpha(s), 1-\alpha(t)).$$
(2.5)

The full scattering amplitude for $\pi\pi \to \pi\omega$ is actually obtained by adding to Eq. (2.5) the same object with $s \leftrightarrow u$ and the one with $t \leftrightarrow u$, so as to make it completely symmetric in all three Mandelstam variables.

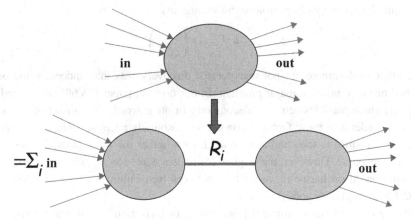

Figure 2.6 Counting states via factorization.

2.3 Dual Resonance Models

There was a big worry based on previous experience: possibly, in order to satisfy all the constraints, the beta function model had to contain 'ghosts', i.e. states produced with negative probability. If so the model would be inconsistent. Sergio Fubini was particularly insistant on this crucial test.

To answer that question one had to identify first all the states/resonances. The way to do this was to use a property of the S-matrix known as factorization. It is basically what unitarity reduces to in the single-particle-exchange approximation. The problem could be formulated as follows: how many terms were necessary and sufficient in order to write the residue of a pole in the S-matrix as a sum of products of a coupling to the 'initial' and 'final' states (see Figure 2.6)?

This question could not be answered by using just the beta function, but, fortunately, in the fall of 1968 several groups (Bardakci and Ruegg [BR69a, BR69b], Virasoro [Vir69a], Goebel and Sakita [GS69], Chan and Tsou [CT69], Koba and Nielsen [KN69], Chan and Paton [CP69]) had found a (pretty unique) generalization of the original ansatz to multiparticle initial and final states. Sergio Fubini and I at MIT (where I had just arrived for my first postdoc), as well as Bardakci and Mandelstam at Berkeley, started to look into this rather complex counting problem.

The result (Fubini and Veneziano [FV69] and Bardakci and Mandelstam [BM69]) turned out to be very surprising. Because of the parallel daughters, we were expecting a mild degeneracy (increasing, say, like a power of energy). Instead, the number of states grew much faster, like $\exp(bM)$, with b some known constant (with dimensions of 1/mass and of order $\sqrt{\alpha'}$).

Although unexpected, this was just the behaviour postulated by R. Hagedorn a few years earlier [Hag65] on more phenomenological grounds (for example, in order to get a Boltzmann-like factor in the particle spectra produced in high energy hadronic collisions). Taken at face value, such a density of states leads to a limiting (maximal, Hagedorn) temperature T_H given by (k_B is Boltzmann's constant):

$$k_B T_H = b^{-1} = O\left(\frac{1}{\sqrt{\alpha'}}\right). \tag{2.6}$$

The other (unfortunate but not unexpected) discovery was that, indeed, some of the states had negative norm, a major problem. In our original paper [FV69] we noticed that the lightest ghost particles actually 'decoupled'. In other words the full set of states was sufficient in order to achieve factorization but not absolutely necessary. Unfortunately, our original formalism was too cumbersome to check whether the more massive ghost states would also decouple. However, the original formalism was soon replaced by a much more handy one, based on harmonic oscillator operators (see Fubini, Gordon and Veneziano [FGV69] and Nambu [Nam69]).

In the new formalism a sufficient (vis-à-vis of factorization) set of states consisted of the energy eigenstates of an infinite set of decoupled harmonic oscillators with quantized frequencies, i.e.

$$|N_{n,\mu}\rangle \sim \prod_{n,\mu} (a_{n,\mu}^\dagger)^{N_{n,\mu}} |0\rangle, \quad (n = 1, 2, \ldots ; \mu = 0, 1, 2, 3),$$

$$\alpha' M^2 = \sum_{n,\mu} n\, a_{n,\mu}^\dagger a_n^\mu \equiv L_0 - \alpha' p^2, \tag{2.7}$$

where

$$[a_{n,\mu}, a_{m,\nu}^\dagger] = \delta_{n,m} \eta_{\mu\nu}, \quad \eta_{\mu\nu} = \text{diag}(-1, 1, 1, 1). \tag{2.8}$$

Because of the 'wrong' sign of the timelike commutation relation (2.8), states created by an odd number of timelike operators were ghosts. Was the DRM doomed? Well, almost. The only hope was that all those states were sufficient but perhaps only a (ghost-free) subset was necessary.

In my original paper with Fubini the following (so-called 'spurious') states were found to be unnecessary:

$$L_{-1}|X\rangle \equiv \left(p \cdot a_1^\dagger + \sum_n \sqrt{n(n+1)}\, a_{n+1}^\dagger \cdot a_n\right) |X\rangle, \tag{2.9}$$

with $|X\rangle$ any state.

This was probably sufficient to eliminate the ghosts created by the time component of $a_{1,\mu}$. But what about all the others? The situation looked almost desperate... until Virasoro [Vir70] made a crucial discovery. If $\alpha(0) = 1$ one could enlarge enormously the space of 'spurious' states to:

$$L_{-m}|X\rangle \equiv \left(p \cdot a_m^\dagger + \sum_n \sqrt{n(n+m)}\, a_{n+m}^\dagger \cdot a_n\right) |X\rangle, \tag{2.10}$$

with $m = 1, 2, \ldots$

Thus, for $\alpha(0) = 1$, there was a chance to eliminate all the ghosts! Unfortunately, $\alpha(0) = 1$ meant having a massless $J = 1$ state, something unwanted, but people kept hoping that such a problem could be solved, perhaps through loop corrections, and kept working enthusiastically.

2.4 Further developments

Between the summer of 1969 and the spring of 1970 several important developments took place in the operator formalism.

- Sciuto [Sci69] (see also Della Selva and Saito [DS70]) constructed a vertex describing the coupling of three arbitrary harmonic oscillator states, while Caneschi, Schwimmer and myself simplified Sciuto's work through the introduction of the twist operator [CSV69, CS70].
- Soon after, Gliozzi [Gli69] and, independently, Chiu, Matsuda and Rebbi [CMR69] and Thorn [Tho70] discovered that the operators L_0 and $L_{\pm 1}$ satisfy an $SU(1, 1)$ noncompact algebra.
- Fubini and myself [FV70], and independently Gervais [Ger70], constructed the field operator $Q(z)$ and the so-called 'vertex operators', $V(k)$, discussing their nontrivial correlators and transformations under the above mentioned $SU(1, 1)$ group. This opened the way to further important developments.
- Duality, factorization and the conditions characterizing spurious and physical states all came out in an elegant algebraic way.
- Then, Fubini and I [FV71] extended all this framework to the whole set of Virasoro's L_n operators [Vir70] and figured out the action of such operators on $Q(z)$ and $V(k)$. As a result we could (too) 'quickly' guess the algebra of the Virasoro operators ... missing its crucial 'central charge', a quantum effect soon discovered by the late Joe Weis (see Note added in proofs in [FV71]). This led to what has been known since then as the Virasoro algebra.

At this point the machinery was almost ready for a final assault at the ghost-killing programme. An essential step turned out to be the construction of the DDF (Di Vecchia, Del Giudice, Fubini) positive norm states [DDF72]. They were in one-to-one correspondence with $D - 2$ sets of harmonic oscillators (D being the dimensionality of spacetime naturally taken, at the time, to be 4). These states were physical and had positive norm, but did not look sufficient to span the whole Hilbert space. I remember well a talk given by Fubini and myself to the mathematicians at MIT where the mathematical problem at hand was formulated: no proof of the absence of ghosts came out of that attempt, though. Instead, mathematicians got quite interested in some of the mathematical aspects of the DRM, like the vertex operators and the Virasoro algebra.

The no-ghost theorem was proven instead soon after by Brower [Bro72] and by Goddard and Thorn [GT72]. It only worked for $\alpha(0) = 1$, of course, but, curiously enough, only for

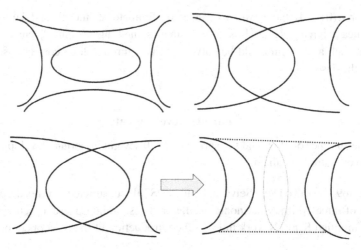

Figure 2.7 Planar and nonplanar loops; the latter give rise to new states for $D = 26$.

$D \leq 26$! At $D = 26$ the DDF states were both necessary *and* sufficient, while at $D < 26$ some other positive norm states were needed. At $D > 26$ ghosts were still present among the physical states. This basic result represented the happy conclusion of a long process and gave much confidence in the belief that the DRM was a theoretically sound and consistent starting point for a new theory of hadrons.

In what sense was it a first-order approximation? The DRM was the analogue of the tree-level approximation of a QFT. In order to implement unitarity fully (for example to give finite widths to the resonances), loop corrections had to be added. Having identified the physical states, this was (almost) a technical problem. One just had to be careful of not letting ghosts circulate in the loops and then factorization (unitarity) and duality would lead to basically unique answers (see Kikkawa, Sakita and Virasoro [KSV69], Bardakci, Halpern and Shapiro [BHS69], Amati, Bouchiat and Gervais [ABG69], Neveu and Scherk [NS70], Frye and Susskind [FS70] and Gross, Neveu, Scherk and Schwarz [GNSS70]).

Both planar and nonplanar loops (see Figure 2.7) were needed, the latter in order to describe the peculiarities of vacuum exchange processes. However, the nonplanar loop had a big surprise up its sleeve: Lovelace [Lov71] discovered that, for $D \neq 26$, this loop gave nonsensical singularities in the vacuum channel. By contrast, for $D = 26$, it gave new poles that could be interpreted as a new set of positive norm physical states with vacuum quantum numbers. Furthermore, those new states would interact just as the (already known) states of another DRM, the one invented by Virasoro [Vir69b] and Shapiro [Sha70] (and later reinterpreted as describing closed, rather than open, strings). Thus, the magic number 26 was again making its appearance[2] in a completely independent way! For a theory of hadrons these new states were very good candidates for hadrons lying on the Pomeron

---

[2] Actually this observation came before the one based on imposing the absence of ghosts.

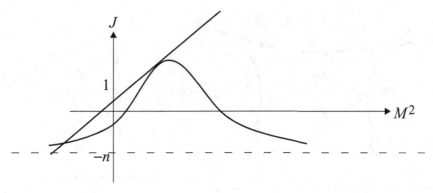

Figure 2.8 Regge trajectories in potential scattering (straight line) and in hadronic physics (curved line).

trajectory, except that the trajectory's intercept was 2, once more about a factor 2 larger than the experimental value!

2.5 Hints of a string

The (more or less vague) idea that the DRM had a physical interpretation in terms of some extended object came soon after they were invented. With hindsight we can find many (mostly missed) hints that such an underlying object had to be a string. I can mention at least five of them.

- From linear Regge trajectories. We have:

$$\alpha' = \frac{dJ}{dM^2} \sim 10^{-13} \text{ cm GeV}^{-1}. \qquad (2.11)$$

Its inverse, $T \sim 10^{13}$ GeV cm^{-1}, has dimensions of a string tension (where $c = 1$ but no $\hbar$ is needed)! One can add to this that linearly rising Regge trajectories are very different from those originally found by T. Regge in potential scattering. The latter would rise up to some point and then inevitably fall down at large M (see Figure 2.8).
- From duality and duality diagrams. The duality diagrams of Harari and Rosner can be further decorated with little springs connecting the quark lines and then they describe pictorially (see Figure 2.9) the joining and splitting of open strings. The intermediate states of the nonplanar loop would realize a DHS duality between two open strings in one channel and a single closed string in another.
- From the harmonic oscillators. This was certainly one decisive hint: indeed a string can be described as an infinite set of independent harmonic oscillators whose characteristic frequencies are a multiple of a fundamental (lowest) frequency.
- From $Q(z)$ and its correlators. These indicated the existence of one effective 'coordinate' z labelling points on a one-dimensional object; the correlators themselves, behaving logarithmically in the distance, were characteristic of a $(1 + 1)$-dimensional field theory.

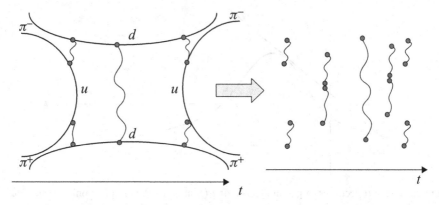

Figure 2.9 Duality diagrams interpreted in terms of the joining and splitting of strings.

- From DDF 'transverse' states. The fact that only $D-2$ sets of oscillators were enough corresponds to the statement that only vibrations of the string that are orthogonal to the string itself have physical meaning.

Indeed these last three hints were not missed, and the proposal that a string was lying at the basis of all those magic properties that had been found was finally made, particularly by Nambu [Nam70a], Nielsen [Nie70] and Susskind [Sus70]. The identification remained qualitative for some time until Nambu [Nam70b] and Goto [Got71] first formulated in a precise way the classical action of a relativistic string, and then the work of Goddard, Goldstone, Rebbi and Thorn [GGRT73] established the connection between the DRM spectrum and that of a quantized string. In doing so, GGRT established for the third time the necessity of the $\alpha(0) = 1$ and $D = 26$ constraints. But, paradoxically, now that the DRM had been raised to the level of a respectable theory, it became apparent that it was not the right theory for strong interactions.

2.6 Good and bad news

The good news was essentially theoretical (see also the Chapters by André Neveu, Pierre Ramond, Ferdinando Gliozzi and Michael Green, in this Volume):

- the Neveu–Schwarz [NS71] and Ramond [Ram71] fermionic extensions introducing fermions and lowering the critical dimension from $D = 26$ to $D = 10$;
- the discovery of supersymmetry in the West (found independently in Russia);
- the Gliozzi–Scherk–Olive [GSO76, GSO77] projection leading to the elimination of the tachyon and to fully consistent superstring theories;
- the Scherk–Schwarz proposal [SS74] that string theory should be reinterpreted as a theory of quantum gravity;

- the Green–Schwarz [GS84] anomaly cancellation, showing that consistent and realistic superstring theories unifying all interactions may exist.

The bad news (for the hadronic string) was instead phenomenological, i.e. related to experimental data:

- $D \neq 4$, i.e. the dimensionality of spacetime required by string theory is not the one we observe;
- massless states with $J = 0, 1/2, \ldots, 2$ were a big embarrassment for any theory of strong interactions;
- the softness of string theory did not allow for sizeable cross-sections for events with large momentum transfers, whereas
  - scaling in $R = \sigma(e^+ e^- \to \text{hadrons})/\sigma(e^+ e^- \to \mu^+ \mu^-)$,
  - Bjorken scaling in $e + p \to e + \text{hadrons}$,
  - large-p_t events at CERN's newest accelerator, the Intersecting Storage Ring (ISR),
  were all showing evidence for pointlike structure in the hadrons – alas, there was no such pointlike structure in the Nambu–Goto string!

2.7 QCD takes over

At about the same time a strong competitor to the DRM and strings came out: quantum chromodynamics (QCD). It was a consistent theory in $D = 4$. With its property of ultraviolet (asymptotic) freedom it could account for those hard events that string theory had difficulties dealing with. Also, the (at the time just conjectured) property of infrared slavery could explain why quarks (and also gluons) could not be seen directly in experiments, as well as the existence of a mass gap (no massless states).

Clearly, it was not the kind of QFT we had discarded early on. Less revolutionary than string theory, it had just the right amount of novelty to be right. Did we need more to be convinced and abandon (reluctantly) our strings? Personally, I still kept trying some phenomenology with string theory using its topological structure, apparently very unlike that of any QFT.

I gave up around 1974, when 't Hooft [tHo74] showed that even the topology of duality diagrams comes out of QCD, provided one considers a $1/N_c$ expansion, where N_c is the number of colours ($N_c = 3$ in real life). Indeed:

- in large-N_c QCD duality diagrams take up a precise meaning, they are planar Feynman diagrams bounded by quark propagators and filled with a 'fishnet' of gluon propagators and vertices (see Figure 2.10);
- they provide naturally a justification for the narrow-resonance approximation[3] that we had been using all the time;

---

[3] The fact that in the limit $N_c \to \infty$ mesons become stable and not interacting had been pointed out before the advent of QCD by Lipkin [Lip68].

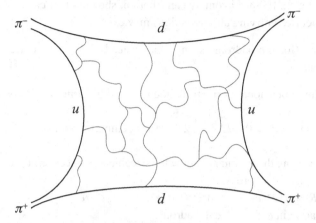

Figure 2.10 Large-N QCD interpretation of duality diagrams.

- at sub-leading order the nonplanar diagrams give new bound states, the glueballs, and presumably the Pomeron as the Regge trajectory these new states lie on;
- the Hagedorn temperature is reinterpreted as a deconfining temperature for quarks and gluons.

It all seems to fall beautifully into place ...

And finally the property of confinement (for which there is by now overwhelming numerical evidence and even a compelling theoretical picture based on an analogy with superconductivity) explains why string-like objects (such as chromoelectric flux tubes) and excitations do really exist in the hadronic world. This is no doubt the reason why, by trying to put some order in the complicated world of strong interactions, we did end up with a theory of strings, albeit with the wrong one. After 40 years we still do not know which is the correct string theory description of hadrons (a subject that has become fashionable once more in the modern string community), a kind of mirage employed by Nature to deceive many of us, but which led to the premature discovery of a framework that could one day answer even greater questions.

Acknowledgements

I wish to acknowledge the wonderful atmosphere provided by the Galileo Galilei Institute on the occasion of the May 2007 Conference on the early history of string theory. It gave a decisive push to the idea of preserving the testimony of one of the most extraordinary adventures in twentieth century physics.

Gabriele Veneziano was born in Florence in 1942. He received his 'Laurea' in physics at the University of Florence in 1965, and his PhD at the Weizmann Institute in 1967. He has been Research Associate, and then visiting scientist, at MIT (1968–1972), full Professor at the Weizmann Institute (1972–1977), and staff member in the Theory Division at CERN (1977–2007). He is currently Professor at the Collège de France, Paris. Honours received include the Pomeranchuk Prize (Moscow), Heineman Prize (APS), Einstein Medal (Bern), Fermi Medal (Italy), Oskar Klein Medal (Stockholm), James Joyce award (Dublin), and others.

References

[ARVV67] Ademollo, M., Rubinstein, H. R., Veneziano, G. and Virasoro, M. A. (1967). Bootstraplike conditions from superconvergence, *Phys. Rev. Lett.* **19**, 1402–1405.

[ARVV68] Ademollo, M., Rubinstein, H. R., Veneziano, G. and Virasoro, M. A. (1968). Bootstrap of meson trajectories from superconvergence, *Phys. Rev.* **176**, 1904–1925.

[ABG69] Amati, D., Bouchiat, C. and Gervais, J. L. (1969). On the building of dual diagrams from unitarity, *Lett. Nuovo Cimento* **2**, 399–406.

[BHS69] Bardakci, K., Halpern, M. B. and Shapiro, J. A. (1969). Unitary closed loops in reggeized Feynman theory, *Phys. Rev.* **185**, 1910–1917.

[BM69] Bardakci, K. and Mandelstam, S. (1969). Analytic solution of the linear-trajectory bootstrap, *Phys. Rev.* **184**, 1640–1644.

[BR69a] Bardakci, K. and Ruegg, H. (1969). Meson resonance couplings in a fivepoint Veneziano model, *Phys. Lett.* **B28**, 671–675.

[BR69b] Bardakci, K. and Ruegg, H. (1969). Reggeized resonance model for arbitrary production processes, *Phys. Rev.* **181**, 1884–1889.

[Bro72] Brower, R. C. (1972). Spectrum-generating algebra and no-ghost theorem in the dual model, *Phys. Rev.* **D6**, 1655–1662.

[CS70] Caneschi, L. and Schwimmer, A. (1970). Ward identities and vertices in the operatorial duality formalism, *Lett. Nuovo Cimento* **3**, 213–217.

[CSV69] Caneschi, L., Schwimmer, A. and Veneziano, G. (1969). Twisted propagator in the operatorial duality formalism, *Phys. Lett.* **B30**, 351–356.

[CP69] Chan, H. M. and Paton, J. E. (1969). Generalized Veneziano model with isospin, *Nucl. Phys.* **B10**, 516–520.

[CT69] Chan, H. M. and Tsou, S. T. (1969). Explicit construction of the N-point function in the generalized Veneziano model, *Phys. Lett.* **B28**, 485–488.

[Che66] Chew, G. F. (1966). *The Analytic S Matrix* (W. A. Benjamin, New York).

[CMR69] Chiu, C. B., Matsuda, S. and Rebbi, C. (1969). Factorization properties of the dual resonance model – a general treatment of linear dependences, *Phys. Rev. Lett.* **23**, 1526–1530.

[Col77] Collins, P. D. B. (1977). *An Introduction to Regge Theory and High-Energy Physics* (Cambridge University Press, Cambridge), and references therein.

[DFFR66] De Alfaro, V., Fubini, S., Furlan, G. and Rossetti, C. (1966). Sum rules for strong interactions, *Phys. Lett.* **21**, 576–579.

[DDF72] Del Giudice, E., Di Vecchia, P. and Fubini, S. (1972). General properties of the dual resonance model, *Ann. Phys.* **70**, 378–398.

[DS70] Della Selva, A. and Saito, S. (1970). A simple expression for the Sciuto three-Reggeon vertex-generating duality, *Lett. Nuovo Cimento* **4**, 689–692.

[DHS68] Dolen, R., Horn, D. and Schmid, C. (1968). Finite-energy sum rules and their application to pin charge exchange, *Phys. Rev.* **166**, 1768–1781.
[ELOP66] Eden, R. J., Landshoff, P. V., Olive, D. I. and Polkinghorne, J. C. (1966). *The Analytic S Matrix* (Cambridge University Press, Cambridge).
[Fre68] Freund, P. G. O. (1968). Finite-energy sum rules and bootstraps, *Phys. Rev. Lett.* **20**, 235–237.
[FS70] Frye, G. and Susskind, L. (1970). Removal of the divergence of a planar dual-symmetric loop, *Phys. Lett.* **B31**, 537–540.
[Fub66] Fubini, S. (1966). Equal-time commutators and dispersion relations, *Nuovo Cimento* **A43**, 475–482.
[FGV69] Fubini, S., Gordon, D. and Veneziano, G. (1969). A general treatment of factorization in dual resonance models, *Phys. Lett.* **B29**, 679–682.
[FV69] Fubini, S. and Veneziano, G. (1969). Level structure of dual resonance models, *Nuovo Cimento* **A64**, 811–840.
[FV70] Fubini, S. and Veneziano, G. (1970). Harmonic-oscillator analogy for the Veneziano model, *Nuovo Cimento* **A67**, 29–47.
[FV71] Fubini, S. and Veneziano, G. (1971). Algebraic treatment of subsidiary conditions in dual resonance models, *Ann. Phys.* **63**, 12–27.
[Ger70] Gervais, J. L. (1970). Operator expression for the Koba–Nielsen multi-Veneziano formula and gauge identities, *Nucl. Phys.* **B21**, 192–204.
[Gli69] Gliozzi, F. (1969). Ward-like identities and twisting operator in dual resonance models, *Lett. Nuovo Cimento* **2**, 846–850.
[GSO76] Gliozzi, F., Scherk, J. and Olive, D. (1976). Supergravity and the spinor dual model, *Phys. Lett.* **B65**, 282–286.
[GSO77] Gliozzi, F., Scherk, J. and Olive, D. (1977). Supersymmetry, supergravity theories and the dual spinor model, *Nucl. Phys.* **B122**, 253–290.
[GGRT73] Goddard, P., Goldstone, J., Rebbi, C. and Thorn, C. B. (1973). Quantum dynamics of a massless relativistic string, *Nucl. Phys.* **B56**, 109–135.
[GT72] Goddard, P. and Thorn, C. B. (1972). Compatibility of the Pomeron with unitarity and the absence of ghosts in the dual resonance model, *Phys. Lett.* **B40**, 235–238.
[GS69] Goebel, C. J. and Sakita, B. (1969). Extension of the Veneziano form to n-particle amplitudes, *Phys. Rev. Lett.* **22**, 257–260.
[Got71] Goto, T. (1971). Relativistic quantum mechanics of one-dimensional mechanical continuum and subsidiary condition of dual resonance model, *Prog. Theor. Phys.* **46**, 1560–1569.
[GS84] Green, M. B. and Schwarz, J. H. (1984). Anomaly cancellation in supersymmetric $d = 10$ gauge theory and superstring theory, *Phys. Lett.* **B149**, 117–122.
[GNSS70] Gross, D. J., Neveu, A., Scherk, J. and Schwarz, J. H. (1970). Renormalization and unitarity in the dual-resonance model, *Phys. Rev.* **D2**, 697–710.
[Hag65] Hagedorn, R. (1965). Statistical thermodynamics of strong interactions at high-energies, *Nuovo Cimento Suppl.* **3**, 147–186.
[Har68] Harari, H. (1968). Pomeranchuk trajectory and its relation to low-energy scattering amplitudes, *Phys. Rev. Lett.* **20**, 1395–1398.
[Har69] Harari, H. (1969). Duality diagrams, *Phys. Rev. Lett.* **22**, 562–565.
[Hei43] Heisenberg, W. (1943). Die beobachtbaren grossössen in der theorie der elenemteilchen: I, *Z. Phys.* **120**, 513–538.
[IM67] Igi, K. and Matsuda, S. (1967). New sum rules and singularities in the complex j plane, *Phys. Rev. Lett.* **18**, 625–627.

[KSV69] Kikkawa, K., Sakita, B. and Virasoro, M. A. (1969). Feynman-like diagrams compatible with duality. I: Planar diagrams, *Phys. Rev.* **184**, 1701–1713.

[KN69] Koba, Z. and Nielsen, H. B. (1969). Reaction amplitude for n-mesons a generalization of the Veneziano–Bardakci–Ruegg–Virasoro model, *Nucl. Phys.* **B10**, 633–655.

[Lip68] Lipkin, H. J. (1968). In *Physique Nucléaire, Proceedings, Ecole d'Eté de Physique Théorique, Les Houches, France*, ed. de Witt, C. and Gillet, V. (Gordon and Breach, New York), 585.

[LST67] Logunov, A. A., Soloviev, L. D. and Tavkhelidze, A. N. (1967). Dispersion sum rules and high energy scattering, *Phys. Lett.* **B24**, 181–182.

[Lov71] Lovelace, C. (1971). Pomeron form factors and dual Regge cuts, *Phys. Lett.* **B34**, 500–506.

[Nam69] Nambu, Y. (1969). University of Chicago Preprint EFI 69-64.

[Nam70a] Nambu, Y. (1970). *Proceedings of the International Conference on Symmetries and Quark Models, Wayne State University, June 18–20, 1969*, ed. Chand, R. (Gordon and Breach, New York), 269–277.

[Nam70b] Nambu, Y. (1970). Lectures at the Copenhagen Symposium, unpublished. See also Takabayasi, T. (1970). Internal structure of hadron underlying the Veneziano amplitude, *Prog. Theor. Phys.* **44**, 1117–1118.

[NS70] Neveu, A. and Scherk, J. (1970). Parameter-free regularization of one-loop unitary dual diagram, *Phys. Rev.* **D1**, 2355–2359.

[NS71] Neveu, A. and Schwarz, J. H. (1971). Factorizable dual model of pions, *Nucl. Phys.* **B31**, 86–112.

[Nie70] Nielsen, H. B. (1970). An almost physical interpretation of the integrand of the n-point Veneziano model, paper submitted to the *15th International Conference on High Energy Physics, Kiev, 26 August to 4 September, 1970* (Nordita preprint, 1969).

[Ram71] Ramond, P. (1971). Dual theory for free fermions, *Phys. Rev.* **D3**, 2415–2418.

[Ros69] Rosner, J. L. (1969). Graphical form of duality, *Phys. Rev. Lett.* **22**, 689–692.

[SS74] Scherk, J. and Schwarz, J. H. (1974). Dual models for nonhadrons, *Nucl. Phys.* **B81**, 118–144.

[Sch68] Schmid, C. (1968). Direct-channel resonances from Regge-pole exchange, *Phys. Rev. Lett.* **20**, 689–691.

[Sci69] Sciuto, S. (1969). The general vertex function in dual resonance models, *Lett. Nuovo Cimento* **2**, 411–418.

[Sha70] Shapiro, J. A. (1970). Electrostatic analogue for the Virasoro model, *Phys. Lett.* **B33**, 361–362.

[Sus70] Susskind, L. (1970). Dual-symmetric theory of hadrons 1, *Nuovo Cimento* **A69**, 457–496.

[tHo74] 't Hooft, G. (1974). A planar diagram theory for strong interactions, *Nucl. Phys.* **B72**, 461–473. See also Veneziano, G. (1976). Some aspects of a unified approach to gauge, dual and Gribov theories, *Nucl. Phys.* **B117**, 519–545.

[Tho70] Thorn, C. B. (1970). Linear dependences in the operator formalism of Fubini, Veneziano, and Gordon, *Phys. Rev.* **D1**, 1693–1696.

[Ven68] Veneziano, G. (1968). Construction of a crossing-symmetric, Reggeon-behaved amplitude for linearly rising trajectories, *Nuovo Cimento* **A57**, 190–197.

[Vir69a] Virasoro, M. (1969). Generalization of Veneziano's formula for the five-point function, *Phys. Rev. Lett.* **22**, 37–39.

[Vir69b] Virasoro, M. (1969). Alternative constructions of crossing symmetric amplitudes with Regge behaviour, *Phys. Rev.* **177**, 2309–2311.
[Vir70] Virasoro, M. (1970). Subsidiary conditions and ghosts in dual-resonance models, *Phys. Rev.* **D1**, 2933–2936.
[Whe37] Wheeler, J. A. (1937). On the mathematical description of light nuclei by the method of resonating group structure, *Phys. Rev.* **52**, 1107–1122.

3
Gravity, unification, and the superstring

JOHN H. SCHWARZ

Abstract

This Chapter surveys some of the highlights in the development of string theory through to the first superstring revolution in 1984. The emphasis is on topics in which the author was involved, especially the observation that critical string theories provide consistent quantum theories of gravity and the proposal to use string theory to construct a unified theory of all fundamental particles and forces.

3.1 Introduction

I am happy to have this opportunity to reminisce about the origins and development of string theory from 1962 (when I entered graduate school) through to the first superstring revolution in 1984. Some of the topics were discussed previously in three papers that were written for various special events in 2000 [Sch00a, Sch00b, Sch01]. Also, some of this material was reviewed in the 1985 reprint volumes [Sch85], as well as string theory textbooks (Green, Schwarz and Witten [GSW87] and Becker, Becker and Schwarz [BBS07]). In presenting my experiences and impressions of this period, it is inevitable that my own contributions are emphasized.

Some of the other early contributors to string theory present their recollections elsewhere in this Volume. Taken together, these contributions should convey a fairly accurate account of the origins of this remarkable subject. Since the history of science community has shown little interest in string theory, it is important to get this material on the record. There have been popular books about string theory and related topics, which serve a useful purpose, but there remains a need for a more scholarly study of the origins and history of string theory.

The remainder of this Chapter is divided into the following Sections:

- 1960–1968: the analytic S-matrix,
- 1968–1970: the Dual Resonance Model,

The Birth of String Theory, ed. Andrea Cappelli, Elena Castellani, Filippo Colomo and Paolo Di Vecchia.
Published by Cambridge University Press. © Cambridge University Press 2012.

- 1971–1973: the Ramond–Neveu–Schwarz model,
- 1974–1975: gravity and unification,
- 1975–1979: supersymmetry and supergravity,
- 1979–1984: superstrings and anomalies.

For each topic, I will give a brief account of the evolution of the research in the corresponding subject. For most of them, more detailed accounts will be provided by other contributors to this Volume. The main focus in this contribution will therefore be on one of the above topics (gravity and unification).

3.2 1960–1968: the analytic S-matrix

In the early Sixties there existed a successful quantum theory of the electromagnetic force (QED), which was completed in the late Forties, but the theories of the weak and strong nuclear forces were not yet known. In UC Berkeley, where I was a graduate student during the period 1962–1966, the emphasis was on developing a theory of the strong nuclear force.

I felt that UC Berkeley was the centre of the Universe for high energy theory at the time. Geoffrey Chew (my thesis advisor) and Stanley Mandelstam were highly influential leaders. Also, Steve Weinberg and Shelly Glashow were impressive younger faculty members. David Gross was a contemporaneous Chew student with whom I shared an office.

Geoffrey Chew's approach to understanding the strong interactions was based on several general principles [Che62, Che66]. He was very persuasive in advocating them, and I was strongly influenced by him. The first principle was that quantum field theory, which was so successful in describing QED, was inappropriate for describing a strongly interacting theory, where a weak-coupling perturbation expansion would not be useful. A compelling reason for holding this view was that none of the hadrons (particles that have strong interactions) seemed to be more fundamental than any of the others. Therefore a field theory that singled out some subset of the hadrons did not seem sensible. Also, it was clearly not possible to formulate a quantum field theory with a fundamental field for every hadron. One spoke of 'nuclear democracy' to describe this situation. The quark concept arose during this period, but the prevailing opinion was that quarks were just mathematical constructs. The SLAC deep inelastic scattering experiments in the late Sixties made it clear that quarks and gluons are physical (confined) particles. It was then natural to try to base a quantum field theory on them, and QCD was developed a few years later with the discovery of asymptotic freedom.

For these reasons, Chew argued that field theory was inappropriate for describing strong nuclear forces. Instead, he advocated focussing attention on physical quantities, especially the S-matrix, which describes on-mass-shell scattering amplitudes. The goal was therefore to develop a theory that would determine the S-matrix. Some of the ingredients that went into this were properties deduced from quantum field theory, such as unitarity and maximal analyticity of the S-matrix. These basically encode the requirements of causality and nonnegative probabilities.

Another important proposal, due to Chew and Frautschi, whose necessity was less obvious, was maximal analyticity in angular momentum [CF61, CF62]. The idea is that partial wave amplitudes $a_l(s)$, which are defined in the first instance for angular momenta $l = 0, 1, \ldots$, can be uniquely extended to an analytic function of l, $a(l, s)$, with isolated poles called Regge poles. The Mandelstam invariant s is the square of the invariant energy of the scattering reaction. The position of a Regge pole is given by a Regge trajectory $l = \alpha(s)$. The values of s for which l takes a physical value correspond to physical hadron states. The necessity of branch points in the l plane, with associated Regge cuts, was established by Mandelstam. Their role in phenomenology was less clear.

The theoretical work in this period was strongly influenced by experimental results. Many new hadrons were discovered in experiments at the Bevatron in Berkeley, the Alternating Gradient Synchrotron in Brookhaven, and the Proton Synchrotron at CERN. Plotting masses squared versus angular momentum (for fixed values of other quantum numbers), it was noticed that the Regge trajectories are approximately linear with a common slope:

$$\alpha(s) = \alpha(0) + \alpha's, \qquad \alpha' \sim 1.0 \, (\text{GeV})^{-2}. \tag{3.1}$$

Using the crossing-symmetry properties of analytically continued scattering amplitudes, one argued that exchange of Regge poles (in the t-channel) controlled the high energy, fixed momentum transfer, asymptotic behaviour of physical amplitudes:

$$A(s, t) \sim \beta(t)(s/s_0)^{\alpha(t)}, \qquad s \to \infty, \quad t < 0. \tag{3.2}$$

In this way one deduced from data that the intercept of the trajectory of the ρ-meson, for example, was $\alpha_\rho(0) \sim 0.5$. This is consistent with the measured mass $m_\rho = 0.76 \, \text{GeV}$ and the Regge slope $\alpha' \sim 1.0 \, (\text{GeV})^{-2}$.

The ingredients discussed above are not sufficient to determine the S-matrix, so one needed more. Therefore, Chew advocated another principle called the 'bootstrap'. The idea was that the exchange of hadrons in crossed channels provides forces that are responsible for causing hadrons to form bound states. Thus, one has a self-consistent structure in which the entire collection of hadrons provides the forces that makes their own existence possible. It was unclear for some time how to formulate this intriguing property in a mathematically precise way. As an outgrowth of studies of 'finite energy sum rules' in 1967 (Dolen, Horn and Schmid [DHS67, DHS68], Igi and Matsuda [IM67a, IM67b], Logunov, Soloviev and Tavkhelidze [LST67]) this was achieved in a certain limit in 1968 (Freund [Fre68], Harari [Har69] and Rosner [Ros69]). The limit, called the 'narrow resonance approximation' was one in which the inverse lifetimes of the resonances are negligible compared to their masses. The observed linearity of Regge trajectories suggested this approximation, since otherwise pole positions would have significant imaginary parts. In this approximation branch cuts in scattering amplitudes, whose branch points correspond to multiparticle thresholds, are approximated by a sequence of resonance poles.

The bootstrap idea had a precise formulation in the narrow resonance approximation, which was called 'duality'. This is the statement that a scattering amplitude can be expanded in an infinite series of s-channel poles, and this gives the same result as its expansion in an infinite series of t-channel poles. To include both sets of poles, as usual Feynman diagram techniques might suggest, would amount to double counting.

3.3 1968–1970: the Dual Resonance Model

I began my first postdoctoral position at Princeton University in 1966. For my first two and a half years there, I continued to do work along the lines described in the previous Section (Regge pole theory, duality, etc.). Then Veneziano dropped a bombshell – an exact analytic formula that exhibited duality with linear Regge trajectories [Ven68]. Veneziano's formula was designed to give a good phenomenological description of the reaction $\pi + \pi \to \pi + \omega$ or the decay $\omega \to \pi^+ + \pi^0 + \pi^-$. Its structure was the sum of three Euler beta functions

$$T = A(s,t) + A(s,u) + A(t,u), \tag{3.3}$$

$$A(s,t) = \frac{\Gamma(-\alpha(s))\Gamma(-\alpha(t))}{\Gamma(-\alpha(s) - \alpha(t))}, \tag{3.4}$$

where α is a linear Regge trajectory:

$$\alpha(s) = \alpha(0) + \alpha' s. \tag{3.5}$$

An analogous formula appropriate to the reaction $\pi + \pi \to \pi + \pi$ was quickly proposed by Lovelace [Lov68] and Shapiro [Sha69]. A rule for building in adjoint $SU(N)$ quantum numbers was formulated by Chan and Paton [PC69]. This symmetry was initially envisaged to be a global (flavour) symmetry, but it later turned out to be a local gauge symmetry.

The Veneziano formula gives an explicit realization of duality and Regge behaviour in the 'narrow resonance approximation'. The function $A(s,t)$ can be expanded in terms of the s-channel poles or the t-channel poles. The motivation for writing down this formula was mostly phenomenological, but it turned out that formulae of this type describe tree amplitudes in a perturbatively consistent quantum theory!

Very soon after the appearance of the Veneziano amplitude, Virasoro proposed an alternative formula [Vir69],

$$T = \frac{\Gamma(-\tfrac{1}{2}\alpha(s))\Gamma(-\tfrac{1}{2}\alpha(t))\Gamma(-\tfrac{1}{2}\alpha(u))}{\Gamma(-\tfrac{1}{2}\alpha(t) - \tfrac{1}{2}\alpha(u))\Gamma(-\tfrac{1}{2}\alpha(s) - \tfrac{1}{2}\alpha(u))\Gamma(-\tfrac{1}{2}\alpha(s) - \tfrac{1}{2}\alpha(t))}, \tag{3.6}$$

which has similar virtues. Since this formula has total stu symmetry, it is only applicable to particles that are singlets of the Chan–Paton group.

Over the course of the next year or so, string theory (or 'dual models', as the subject was then called) underwent a sudden surge of popularity, marked by several remarkable discoveries. One was the discovery of an N-particle generalization of the Veneziano formula (Bardakci and Ruegg [BR69], Goebel and Sakita [GS69], Chan and Tsou [CT69], Koba and Nielsen [KN69a, KN69b]):

$$A_N(k) = g_{open}^{N-2} \int d\mu_N(y) \prod_{i<j} (y_i - y_j)^{\alpha' k_i \cdot k_j}, \tag{3.7}$$

where $y_1, y_2, \ldots, y_N$ are real coordinates, any three of which are y_A, y_B, y_C, and

$$d\mu_N(y) = |(y_A - y_B)(y_B - y_C)(y_C - y_A)| \prod_{i=1}^{N-1} \theta(y_{i+1} - y_i)$$

$$\times \delta(y_A - y_A^0)\delta(y_B - y_B^0)\delta(y_C - y_C^0) \prod_{i=1}^{N} dy_i. \quad (3.8)$$

The formula is independent of y_A^0, y_B^0, y_C^0 because of its $SL(2, \mathbb{R})$ symmetry, which allows them to be mapped to arbitrary real values. This formula has cyclic symmetry in the N external lines.

Soon thereafter Shapiro formulated an N-particle generalization of the Virasoro formula [Sha70]:

$$A_N(k_1, k_2, \ldots, k_N) = g_{closed}^{N-2} \int d\mu_N(z) \prod_{i<j} |z_i - z_j|^{\alpha' k_i \cdot k_j}, \quad (3.9)$$

where $z_1, z_2, \ldots, z_N$ are complex coordinates, any three of which are z_A, z_B, z_C, and

$$d\mu_N(z) = |(z_A - z_B)(z_B - z_C)(z_C - z_A)|^2$$

$$\times \delta^2(z_A - z_A^0)\delta^2(z_B - z_B^0)\delta^2(z_C - z_C^0) \prod_{i=1}^{N} \delta^2 z_i. \quad (3.10)$$

The formula is independent of z_A^0, z_B^0, z_C^0 because of its $SL(2, \mathbb{C})$ symmetry, which allows them to be mapped to arbitrary complex values. This amplitude has total symmetry in the N external lines.

Both of these formulae were shown to have a consistent factorization on a spectrum of single-particle states described by an infinite number of harmonic oscillators (Fubini, Gordon and Veneziano [FGV69, FV69, FV70], Bardakci and Mandelstam [BM69], and Nambu [Nam70a]),

$$a_m^\mu, \quad \mu = 0, 1, \ldots, d-1, \quad m = 1, 2, \ldots, \quad (3.11)$$

with one set of such oscillators in the Veneziano case and two sets in the Virasoro case. These results were interpreted as describing the scattering of modes of a relativistic string (Nambu [Nam70a, Nam70b], Susskind [Sus70, FLS70], Fairlie and Nielsen [FN70, Nie70]): open strings in the first case and closed strings in the second case. Amazingly, the formulae preceded the interpretation. Although, we did not propose a string interpretation, Gross, Neveu, Scherk and I did realize that the relevant diagrams of the loop expansion were classified by the possible topologies of two-dimensional manifolds with boundaries [GNSS70a].

Having found the factorization, it became possible to compute radiative corrections by joining tree diagrams to form loop amplitudes. This was initiated by Kikkawa, Sakita and Virasoro [KSV69] and followed up by many others. Let me describe my role in this. I was at Princeton, where I collaborated with Gross, Neveu and Scherk in computing one-loop amplitudes. In particular, we discovered unanticipated singularities in the 'nonplanar' open string loop diagram [GNSS70b]. The world-sheet is a cylinder with two external particles attached to each boundary. Our computations showed that this diagram gives branch points that violate unitarity. This was a very disturbing conclusion, since it seemed to imply that the classical theory does not have a consistent quantum extension. This was also discovered by Frye and Susskind [FS70]. (The issue of quantum consistency turned out to be a recurring theme, which reappeared many years later, as discussed in Section 3.7.)

Soon thereafter Claud Lovelace pointed out [Lov71] that these branch points become poles provided that:

$$\alpha(0) = 1 \quad \text{and} \quad d = 26. \tag{3.12}$$

Until Lovelace's work, everyone assumed that the spacetime dimension was $d = 4$. As we were not yet talking about gravity, there was no reason to consider anything else. Later, these poles were interpreted as closed string modes in a one-loop open string amplitude. Nowadays this is referred to as open string–closed string duality. The word 'duality' appears in many different places in string theory. Here it applies to amplitudes that are well approximated by open string excitations at short distances and by closed string excitations at long distances. Lovelace's analysis also required there to be an infinite number of decoupling conditions. These turned out to be precisely the Virasoro constraints, which were discovered at about the same time (Virasoro [Vir70], Fubini and Veneziano [FV71]). A couple of years later Brink and Olive constructed a physical-state projection operator [BO73a], which they used to verify Lovelace's conjecture that the nonplanar loop amplitude actually contains closed string poles when the decoupling conditions in the critical dimension are imposed [BO73b].

Thus, quantum consistency was restored, but the price was high: a spectrum with a tachyon and 22 extra dimensions of space. In 1973, the origins of the critical dimension and the intercept condition were explained in terms of the light-cone gauge quantization of a fundamental string by Goddard, Goldstone, Rebbi and Thorn [GGRT73]. Prior to this paper the string interpretation of Dual Resonance Models was not relevant to most developments, serving mostly as a pictorial analogy. The GGRT approach was extended to interacting strings by Mandelstam [Man73].

3.4 1971–1973: the Ramond–Neveu–Schwarz model

In January 1971 Pierre Ramond constructed a Dual Resonance Model generalization of the Dirac equation [Ram71]. He reasoned as follows: just as the total momentum of a string, p^μ, is the zero mode of a momentum density $P^\mu(\sigma)$, so should the Dirac matrices γ^μ be the zero modes of densities $\Gamma^\mu(\sigma)$. Then he defined the modes of $\Gamma \cdot P$:

$$F_n = \int_0^{2\pi} e^{-in\sigma} \Gamma \cdot P d\sigma, \qquad n \in \mathbb{Z}. \tag{3.13}$$

In particular,

$$F_0 = \gamma \cdot p + \text{oscillator terms}. \tag{3.14}$$

He proposed the wave equation

$$(F_0 + m)|\psi\rangle = 0, \tag{3.15}$$

which is now known as the 'Dirac–Ramond equation'. Its solutions give the spectrum of a noninteracting fermionic string.

Ramond also observed that the Virasoro algebra generalizes to:

$$\{F_m, F_n\} = 2L_{m+n} + \frac{c}{3}m^2\delta_{m,-n},$$
$$[L_m, F_n] = \left(\frac{m}{2} - n\right)F_{m+n}, \tag{3.16}$$
$$[L_m, L_n] = (m-n)L_{m+n} + \frac{c}{12}m^3\delta_{m,-n}.$$

Note that central terms were added later. The free fermion spectrum should be restricted by the super-Virasoro constraints $F_n|\psi\rangle = L_n|\psi\rangle = 0$ for $n > 0$.

André Neveu and I proposed a new bosonic dual model, which we called the 'dual pion model', in March 1971 [NS71b]. It has a similar structure to Ramond's free fermion theory, with the periodic density $\Gamma^\mu(\sigma)$ replaced by an antiperiodic one $H^\mu(\sigma)$. Then the modes

$$G_r = \int_0^{2\pi} e^{-ir\sigma} H \cdot P d\sigma, \qquad r \in \mathbb{Z} + 1/2, \tag{3.17}$$

satisfy a similar super-Virasoro algebra. The free particle spectrum is given by the wave equation $(L_0 - 1/2)|\psi\rangle = 0$ supplemented by the constraints $G_r|\psi\rangle = 0$ for $r > 0$. (These formulae are appropriate in the $\mathcal{F}_2$ picture discussed below.) We also constructed N-particle amplitudes analogous to those of the Veneziano model.

The $\pi + \pi \to \pi + \pi$ amplitude computed in the dual pion model turned out to have exactly the form that had been proposed earlier by Lovelace and Shapiro. However, the intercepts of the π and ρ Regge trajectories were $\alpha_\pi(0) = 1/2$ and $\alpha_\rho(0) = 1$. These were half a unit higher than was desired in each case. This implied that the pion was tachyonic and the ρ-meson was massless.

Soon after our paper appeared, Neveu travelled to Berkeley, where there was considerable interest in our results. This led to Charles Thorn (a student of Stanley Mandelstam at the time) joining us in a follow-up project in which we proved that the super-Virasoro constraints were fully implemented [NST71]. This required recasting the original description of the string spectrum (called the $\mathcal{F}_1$ picture) in a new form, which we called the $\mathcal{F}_2$ picture. The three of us then assembled these bosons together with Ramond's fermions into a unified interacting theory of bosons and fermions [NS71a, Tho71], thereby obtaining an early version of what later came to be known as superstring theory.

The string world-sheet action that gives this spectrum of bosons and fermions is

$$S = \int d\sigma d\tau \left(\partial_\alpha X^\mu \partial^\alpha X_\mu - i \bar{\psi}^\mu \rho^\alpha \partial_\alpha \psi_\mu \right), \quad (3.18)$$

where ψ^μ are two-dimensional Majorana spinors and ρ^α are two-dimensional Dirac matrices. Later in 1971 Gervais and Sakita observed [GS71] that this action has 'two-dimensional global supersymmetry' described by the infinitesimal fermionic transformations:

$$\delta X^\mu = \bar{\varepsilon} \psi^\mu,$$
$$\delta \psi^\mu = -i \rho^\alpha \varepsilon \partial_\alpha X^\mu. \quad (3.19)$$

This is actually part of a much larger 'superconformal symmetry', which includes dilatation transformations. There are two possible choices of boundary conditions for the Fermi fields ψ^μ, one of which gives the boson spectrum (Neveu–Schwarz sector) and the other of which gives the fermion spectrum (Ramond sector). Five years later, a more fundamental world-sheet action with local supersymmetry was discovered (Brink, Di Vecchia and Howe [BDH76], Deser and Zumino [DZ76b]). It has the virtue of accounting for the super-Virasoro constraints as arising from covariant gauge fixing.

Bruno Zumino explored the RNS string's gauge conditions associated with the two-dimensional superconformal algebra [Zum74]. Following that, he and Julius Wess began to consider the possibility of constructing four-dimensional field theories with analogous features. This resulted in their famous work [WZ74] on globally supersymmetric field theories in four dimensions. As a consequence of their paper, supersymmetry quickly became an active research topic. (Note that the work of Golfand and Likhtman [GL71] in the USSR, which was the first to introduce the four-dimensional super-Poincaré group, was not known in the West at that time.)

The dual pion model has a manifest $\mathbb{Z}_2$ symmetry. Since the pion is odd and the ρ-meson is even, this symmetry was identified with G parity, a hadronic symmetry that is a consequence of charge conjugation invariance and isotopic spin symmetry. It was obvious that one could make a consistent truncation (at least at tree level) to the even G-parity sector and that then the model would be tachyon free. Because of the desired identification with physical hadrons, there was no motivation (at the time) to do that. Rather, considerable effort was expended in the following years attempting to modify the model so as to lower the intercepts by half a unit. As discussed in the next Section, none of these constructions was entirely satisfactory.

One of the important questions in this period was whether all the physical string excitations have a positive norm. States of negative norm (called 'ghosts') would represent a breakdown of unitarity and causality, so it was essential that they would not be present in the string spectrum. The first proof of the 'no-ghost theorem' for the original bosonic string theory was achieved by Brower [Bro72], building on earlier work by Del Giudice, Di Vecchia and Fubini [DDF72]. This work showed that a necessary condition for the absence of ghosts is $d \leq 26$, and that the dimension value $d = 26$ has especially attractive features, as we already suspected based on the earlier observations of Lovelace.

I generalized Brower's proof of the no-ghost theorem to the RNS string theory and showed that $d = 10$ is the critical dimension and that the ground state fermion should be massless [Sch72a]. This was also done by Brower and Friedman a bit later [BF73]. An alternative, somewhat simpler, proof of the no-ghost theorem for both of the string theories was given by Goddard and Thorn at about the same time [GT72]. Other related work included Gervais and Sakita [GS73], Olive and Scherk [OS73a], and Corrigan and Goddard [CG74].

Later in 1972, thanks to the fact that Murray Gell-Mann had become intrigued by my work with Neveu, I was offered a senior research appointment at Caltech. I think that the reason Gell-Mann became aware of our work was because he spent the academic year 1971–1972 on a sabbatical at CERN, where there was an active dual models group. I felt very fortunate to receive such an offer, especially in view of the fact that the job market for theoretical physicists was extremely bad at the time. Throughout the subsequent years at Caltech, when my work was far from the mainstream, and therefore not widely appreciated, Gell-Mann was always very supportive. For example, he put funds at my disposal to invite visitors. This facilitated various collaborations with Lars Brink, Joël Scherk, and Michael Green among others.

One of the first things I did at Caltech was to study the fermion–fermion scattering amplitude. Using the physical-state projection operator [BO73a], Olive and Scherk had derived a formula that involved the determinant of an infinite matrix [OS73b]. C. C. Wu and I [SW73] discovered that this determinant is a simple function. We derived the result analytically in a certain limit and then verified numerically that it is exact everywhere. (The result was subsequently verified analytically by Corrigan, Goddard, Olive and Smith [CGOS73].) To our surprise, the fermion–fermion scattering amplitude ended up looking very similar to the bosonic amplitudes. This might have been interpreted as a hint of spacetime supersymmetry, but this was before the Wess–Zumino paper, and that was not yet on my mind.

String theory is formulated as an on-shell S-matrix theory in keeping with its origins discussed earlier. However, the SLAC deep inelastic scattering experiments in the late Sixties made it clear that the hadronic component of the electromagnetic current is a physical off-shell quantity, and that its asymptotic properties imply that hadrons have hard pointlike constituents. With this motivation, I tried for the next year or so to construct off-shell amplitudes. Although some intriguing results were obtained [Sch73, Sch74, SW74], this was ultimately unsuccessful. Moreover, all indications were that strings were too soft to describe hadrons with their pointlike constituents.

At this point there were many good reasons to stop working on string theory: a successful and convincing theory of hadrons (QCD) was discovered, and string theory had many severe problems as a hadron theory. These included an unrealistic spacetime dimension, an unrealistic spectrum, and the absence of pointlike constituents. Also, convincing theoretical and experimental evidence for the Standard Model was rapidly falling into place. Understandably, given these successes and string theory's shortcomings, string theory rapidly fell out of favour. What had been a booming enterprise involving many theorists rapidly came to a grinding halt.

Given that the world-sheet descriptions of the two known string theories have conformal invariance and superconformal invariance, it was a natural question whether one could obtain new string theories described by world-sheet theories with extended superconformal symmetry. The $\mathcal{N} = 2$ case was worked out in Ademollo et al. [ABDD76]. The critical dimension is four, but the signature has to be (2, 2). For a long time it was believed that the critical dimension of the $\mathcal{N} = 4$ string is negative, but in 1992 Siegel argued that (due to the reducibility of the constraints) the $\mathcal{N} = 4$ string is the same as the $\mathcal{N} = 2$ string [Sie92].

3.5 1974–1975: gravity and unification

The string theories that were known in the Seventies (the bosonic string and the RNS string) had many shortcomings as a theory of hadrons. The most obvious of these was the necessity for an unrealistic spacetime dimension (26 or 10). Another is the occurrence of tachyons in the spectrum, which implies that the vacuum is unstable. However, the one that bothered us the most was the presence of massless particles in the spectrum, which do not occur among the hadrons.

In both string theories the spectrum of open strings contains massless spin one particles, and the spectrum of closed strings contains a massless spin two particle as well as other massless particles (a dilaton and an antisymmetric tensor in the case of oriented bosonic strings). These particles lie on the leading Regge trajectories in their respective sectors. Thus, the leading open string Regge trajectory has intercept $\alpha(0) = 1$, and the leading closed string Regge trajectory has intercept $\alpha(0) = 2$. It was tempting (in the RNS model) to identify the leading open string trajectory as the one for the ρ-meson. In fact, it had most of the properties expected for that case except that the empirical intercept is about $\alpha_\rho(0) = 1/2$. The leading closed string trajectory carries vacuum quantum numbers, as expected for the 'Pomeron' trajectory. Moreover, the factor of two between the open string and closed string Regge slopes is approximately what is required by the data. However, the Pomeron intercept was also double the desired value $\alpha(0) = 1$, which is the choice that could account for the near constancy (up to logarithmic corrections) of hadronic total cross-sections at high energy.

For these reasons we put considerable effort in the years 1972–1974 into modifying the RNS theory in such a way as to lower all open string Regge trajectories by half a unit and lower all closed string Regge trajectories by one unit. Some successes along these lines actually were achieved, accounting for some aspects of chiral symmetry and current algebra (Neveu and Thorn [NT71], [Sch72b]). However, none of the schemes was entirely consistent. The main problem was that changing the intercepts and the spacetime dimensions meant that the Virasoro constraints were not satisfied, and so the spectrum was not ghost free. Also, the successes for non-strange mesons did not extend to mesons made from heavy quarks, and the Ramond fermions did not really look like baryons.

The alternative to modifying string theory to get what we wanted was to understand better what the theory was giving without modification. String theories in the critical dimension clearly were beautiful theories, with a remarkably subtle and intricate structure, and they ought to be good for something. The fact that they were developed in an attempt to understand hadron physics did not guarantee that this was necessarily their appropriate physical application. Furthermore, the success of QCD made the effort to formulate a string theory of hadrons less pressing.

The first indication that such an agnostic attitude could prove worthwhile was a pioneering work by Neveu and Scherk [NS72], which studied the interactions of the massless spin one open string particles at low energies (or, equivalently, in the 'zero-slope limit' [Sch71]) and proved that their interactions agreed with those of Yang–Mills gauge particles in the adjoint representation of the Chan–Paton group. In other words, open string theory is Yang–Mills gauge theory modified by higher scale-dimension interactions. This implies that the Chan–Paton group is actually a Yang–Mills gauge group. Prior to this work by Neveu and Scherk, it was always assumed that the Chan–Paton symmetry is a global symmetry. I am struck by the fact that Yang and Mills in their original paper on $SU(2)$ gauge theory [YM54], tried to identify the gauge symmetry with isotopic spin symmetry and the gauge fields with ρ-mesons. In our failed efforts to describe hadrons, we had been making essentially the same mistake.

I arranged for Joël Scherk, with whom I had collaborated in Princeton, to visit Caltech in the winter and spring of 1974. Our interests and attitudes in physics were very similar, and so we were anxious to start a new collaboration. Each of us felt that string theory was too beautiful to be just a mathematical curiosity. It ought to have some physical relevance. We had frequently been struck by the fact that string theories exhibit unanticipated miraculous properties. What this means is that they have a very deep mathematical structure that is not fully understood. By digging deeper one could reasonably expect to find more surprises and then learn new lessons. Therefore, despite the fact that the rest of the theoretical high energy physics community was drawn to the important project of exploring the Standard Model, we wanted to explore string theory.

Since my training was as an elementary particle physicist, gravity was far from my mind in early 1974. Traditionally, elementary particle physicists had ignored the gravitational force, which is entirely negligible under ordinary circumstances. For these reasons, we were not predisposed to interpret string theory as a physical theory of gravity. General relativists, the people who did study gravity, formed a completely different community. They attended different meetings, read different journals, and had no need for serious communication with particle physicists, just as particle physicists felt they had no need for relativists who studied topics such as Black Holes or the early Universe.

Despite all this, we decided to do what could have been done two years earlier: we explored whether it is possible to interpret the massless spin two state in the closed string spectrum as a graviton. This required carrying out an analysis analogous to the earlier one of

Neveu and Scherk. This time one needed to decide whether the interactions of the massless spin two particle in string theory agree at low energy with those of the graviton in general relativity (GR). Success was inevitable, because GR is the only consistent possibility at low energies (i.e. neglecting corrections due to higher-dimension operators), and critical string theory certainly is consistent. At least, it contains the requisite gauge invariances to decouple all but the transverse polarizations. Therefore, the harder part of this work was forcing oneself to ask the right question. Finding the right answer was easy. In fact, by invoking certain general theorems, due to Weinberg [Wei65], we were able to argue that string theory agrees with general relativity at low energies [SS74]. Although we were not aware of it at the time, Tamiaki Yoneya had obtained the same result somewhat earlier [Yon73, Yon74].

In [SS74] Scherk and I proposed to interpret string theory as a quantum theory of gravity, unified with the other forces. This meant taking the whole theory seriously, not just viewing it as a framework for deriving GR and Yang–Mills theory as limits. Our paper was entitled 'Dual models for non-hadrons', which I think was a poor choice. It emphasized the fact that we were no longer trying to describe hadrons and their interactions, but it failed to emphasize what we were proposing to do instead. A better choice would have been 'String theory as a quantum theory of gravity unified with the other forces'.

This proposal had several advantages. First, gravity was required by both of the known critical string theories: a forceful way of expressing this is 'the assumption that the fundamental physical entities are strings predicts the existence of gravity'. In fact, even today, this is really the only direct experimental evidence that exists in support of string theory, though there are many other reasons to take string theory seriously.

Second, string theories are free from the UV divergences that typically appear in point-particle theories of gravity. The reason for this can be traced to the extended structure of strings. Specifically, string world-sheets are smooth, even when they describe interactions. So they do not have the short-distance singularities that are responsible for UV divergences. These divergences always occur when one attempts to interpret general relativity (with or without matter) as a quantum field theory, since ordinary Feynman diagrams do have short-distance singularities. This can also be expressed forcefully by saying 'the assumption that the fundamental physical entities are point particles predicts that gravity does not exist!' I found this line of reasoning very compelling. In light of subsequent developments in string theory, the distinction made here no longer seems quite so sharp. A single theory can have dual descriptions based on different fundamental entities. One description is weakly coupled when the other one is strongly coupled. For example, certain AdS/CFT duality relates string theory in a particular background geometry to a more conventional quantum field theory.

Third, extra dimensions could be a very good thing, rather than a problem, since in a gravity theory the geometry of spacetime is determined by the dynamics. (Prior to our proposal, their appearance as critical dimensions of string theories was viewed as a shortcoming of the theories rather than as a reason to study their possible implications.) In the gravitational setting one could imagine that the equations of motion would require (or at least allow) the extra dimensions to form a very small compact manifold. In 1974 there had been essentially no work on Kaluza–Klein theory for many years (and certainly none in the particle physics community), so the notion of extra dimensions seemed very bizarre to most particle theorists. It is discussed so much nowadays that it is easy to forget this fact. Since the only length scale in string theory is the string scale, determined by the string tension, that would be the natural first guess for the size of the compact space. Given this assumption, the value of Newton's constant in four dimensions can be deduced. The observed strength of gravity requires a Regge slope $\alpha' \sim 10^{-38}\,\text{GeV}^{-2}$ instead of $\alpha' \sim 1\,\text{GeV}^{-2}$, which is the hadronic value. Thus, the change in interpretation meant that the tension of the strings, which is proportional to the reciprocal of α', needed to be increased by 38 orders of magnitude. Equivalently, the size of the strings decreased by 19 orders of magnitude. This was a big conceptual leap, though the mathematics was unchanged.

Fourth, unification of gravity with other forces described by Yang–Mills theories was automatic when open strings are included. Of course, it was immediately clear that the construction of a realistic ground state (low energy theory) would be a great challenge. Indeed, that is where much of the effort these days is focussed. Other ways of incorporating gauge interactions in string theory were discovered many years later. These include heterotic string theory, where gauge fields appear as closed string modes in ten dimensions, coincident D-brane world-volume theories, and certain types of singularities in M-theory or F-theory (see, e.g., [BBS07]).

Scherk and I were very excited by the possibility that string theory could be the Holy Grail of unified field theory, overcoming the problems that had stymied other approaches. In addition to publishing our work in scholarly journals, we gave numerous lectures at conferences and physics departments all over the world. We even submitted a paper entitled 'Dual model approach to a renormalizable theory of gravitation' to the 1975 essay competition of the Gravity Research Foundation [SS75]. (It would have been better to say 'ultraviolet finite' instead of 'renormalizable'.) The first paragraph of that paper reads as follows: 'A serious shortcoming of Einstein's theory of gravitation is the nonrenormalizability of its quantum version when considered in interaction with other quantum fields. In our opinion this is a genuine problem requiring a modification of the theory. It is suggested in this essay that dual resonance models may provide a suitable framework for such a modification, while at the same time achieving a unification with other basic interactions.'

For the most part our work was received politely. Yet, for a decade, very few experts showed much interest. Part of the problem may have been that some key people were unaware of our proposal. Unfortunately, Scherk passed away midway through this 10-year period, though not before making some other important contributions that are discussed in the following Sections. After the subject took off in the autumn of 1984, our paper [SS74] became much better known.

3.6 1975–1979: supersymmetry and supergravity

Following the pioneering work of Wess and Zumino, discussed earlier, the study of supersymmetric quantum field theories became a major endeavour. One major step forward was the realization that supersymmetry can be realized as a local symmetry. This requires including a gauge field, called the 'gravitino field', which is vector-spinor. In four dimensions it describes a massless particle with spin 3/2, which is the supersymmetry partner of the graviton. Thus, local supersymmetry only appears in gravitational theories, which are called supergravity theories.

The first example of a supergravity theory was $\mathcal{N} = 1$, $d = 4$ supergravity. It was formulated in a second-order formalism by Freedman, van Nieuwenhuizen and Ferrara [FNF76] and subsequently in a first-order formalism by Deser and Zumino [DZ76a]. The first-order formalism simplifies the analysis of terms that are quartic in Fermi fields.

The two-dimensional locally supersymmetric and reparameterization-invariant formulation of the RNS world-sheet action was constructed very soon thereafter (Brink, Di Vecchia, Howe, Deser and Zumino [BDH76, DZ76b]). This generalized the one-dimensional result obtained a bit earlier for a spinning point particle (Brink, Deser, Zumino, Di Vecchia and Howe [BDZD76]). This construction was generalized to the $\mathcal{N} = 2$ string of [ABDD76] by Brink and me [BS77]. Reparameterization-invariant world-sheet actions of this type are frequently associated with the name Polyakov, because he used them very skilfully five years later in constructing the path-integral formulation of string theory [Pol81a, Pol81b]. Since neither Polyakov nor the authors of [BDH76, DZ76b] are happy with this usage, the new textbook (Becker, Becker and Schwarz [BBS07]) refers to this type of world-sheet action as a 'string σ-model action'.

The RNS closed string spectrum contains a massless gravitino (in ten dimensions) in addition to the graviton discussed in the previous Section. More precisely, as was understood later, there are one or two gravitinos depending on whether one is describing a type I or type II superstring. Since this is a gauge field, the only way the theory could be consistent is if the theory has local supersymmetry. This requires, in particular, that the spectrum should contain an equal number of bosonic and fermionic degrees of freedom at each mass level. However, as it stood, this was not the case. In particular, the bosonic sector contained a tachyon (the 'pion'), which had no fermionic partner.

In 1976 Gliozzi, Scherk and Olive [GSO76, GSO77] proposed a projection of the RNS spectrum – the 'GSO projection' – that removes roughly half of the states (including the tachyon). Specifically, in the bosonic (NS) sector they projected away the odd G-parity states, a possibility that was discussed earlier, and in the fermionic (R) sector they projected away half the states, keeping only certain definite chiralities. Then they counted the remaining physical degrees of freedom at each mass level. After the GSO projection the masses of open string states, for both bosons and fermions, are given by $\alpha' M^2 = n$, where $n = 0, 1, \ldots$ Denoting the open string degeneracies of states in the GSO-projected theory by $d_{NS}(n)$ and $d_R(n)$, they showed that these are encoded in the generating functions:

$$f_{NS}(w) = \sum_{n=0}^{\infty} d_{NS}(n) w^n$$

$$= \frac{1}{2\sqrt{w}} \left[\prod_{m=1}^{\infty} \left(\frac{1 + w^{m-1/2}}{1 - w^m} \right)^8 - \prod_{m=1}^{\infty} \left(\frac{1 - w^{m-1/2}}{1 - w^m} \right)^8 \right], \quad (3.20)$$

and

$$f_R(w) = \sum_{n=0}^{\infty} d_R(n)w^n = 8 \prod_{m=1}^{\infty} \left(\frac{1+w^m}{1-w^m}\right)^8. \tag{3.21}$$

In 1829, Jacobi proved the remarkable identity [Jac29]

$$f_{NS}(w) = f_R(w), \tag{3.22}$$

though he used a different notation. Thus, there is an equal number of bosons and fermions at every mass level, as required. This was compelling evidence (though not a proof) for ten-dimensional spacetime supersymmetry of the GSO-projected theory. Prior to this work, one knew that the RNS theory has world-sheet supersymmetry, but the realization that the theory should have spacetime supersymmetry was a major advance.

Since a Majorana–Weyl spinor in ten dimensions has 16 real components, the minimal number of supercharges is 16. In particular, the massless modes of open superstrings at low energies are approximated by an $\mathcal{N}=1$, $d=10$ super Yang–Mills theory with 16 supersymmetries. This theory was constructed in [GSO77] and in Brink, Schwarz and Scherk [BSS77]. When this work was done, Brink and I were at Caltech and Scherk was in Paris. Brink and I wrote to Scherk informing him of our results and inviting him to join our collaboration, which he gladly accepted. Brink and I were unaware of the GSO collaboration, which was underway at that time, until their work appeared. Both papers pointed out that maximally supersymmetric Yang–Mills theories in less than ten dimensions could be deduced by dimensional reduction, and both of them constructed the $\mathcal{N}=4, d=4$ super Yang–Mills theory explicitly.

Having found the maximally supersymmetric Yang–Mills theories, it was an obvious problem to construct the maximally supersymmetric supergravity theories. Nahm showed [Nah78] that the highest possible spacetime dimension for such a theory is $d=11$. Soon thereafter, in a very impressive work, the Lagrangian for $\mathcal{N}=1, d=11$ supergravity was constructed by Cremmer, Julia and Scherk [CJS78]. It was immediately clear that eleven-dimensional supergravity is very beautiful, and it aroused a lot of interest. However, it was puzzling for a long time how it fits into the greater scheme of things and whether it has any connection to string theory. Clearly, supergravity in eleven dimensions is not a consistent quantum theory by itself, since it is very singular in the ultraviolet. Moreover, since superstring theory only has ten dimensions, it did not seem possible that it could serve as a regulator. It took more than fifteen years to find the answer to this conundrum (Townsend [Tow95] and Witten [Wit95]): at strong coupling type IIA superstring theory develops a circular eleventh dimension whose radius grows with the string coupling constant. In the limit of infinite coupling one obtains 'M-theory', which is presumably a well-defined quantum theory that has eleven noncompact dimensions. Eleven-dimensional supergravity is the leading low energy approximation to M-theory. In other words, M-theory is the UV completion of eleven-dimensional supergravity.

In 1978–1979, I spent the academic year at the École Normale Supérieure in Paris supported by a Guggenheim Fellowship. I was eager to work with Joël Scherk on supergravity, supersymmetrical strings, and related matters. After various wide-ranging discussions we decided to focus on the problem of supersymmetry breaking. We wondered how, starting from a supersymmetric string theory in ten dimensions, one could end up with a nonsupersymmetric world in four dimensions. The specific supersymmetry breaking mechanism that we discovered can be explained classically and does not really require strings, so we explored it in a field theoretic setting [SS79a, SS79b]. The idea is that in a theory with extra dimensions and global symmetries that do not commute with supersymmetry (R symmetries and $(-1)^F$ are examples), one could arrange for a twisted compactification, and that this would break supersymmetry. For example, if one extra dimension forms a circle, the fields when continued around the circle could come back transformed by an R-symmetry group element. If the gravitino, in particular, is transformed then it acquires mass in a consistent manner.

An interesting example of our supersymmetry breaking mechanism was worked out in a paper we wrote together with Eugène Cremmer [CSS79]. We were able to find a consistent four-parameter deformation of $\mathcal{N} = 8$ supergravity.

Even though the work that Joël Scherk and I did on supersymmetry breaking was motivated by string theory, we only discussed field theory applications in our articles. The reason I never wrote about string theory applications was that in the string theory setting it did not seem possible to decouple the supersymmetry breaking mass parameters from the compactification scales. This was viewed as a serious problem, because the two scales are supposed to be hierarchically different. In recent times, people have been considering string theory brane-world scenarios in which much larger compactification scales are considered. In such a context our supersymmetry breaking mechanism might have a role to play. Indeed, quite a few authors have explored various such possibilities.

3.7 1979–1984: superstrings and anomalies

Following Paris, I spent a month (July 1979) at CERN. There, Michael Green and I unexpectedly crossed paths. We had become acquainted in Princeton around 1970, but we had not collaborated before. In any case, following some discussions in the CERN cafeteria, we began a long and exciting collaboration. Our first goal was to understand better why the GSO-projected RNS string theory has spacetime supersymmetry.

Green, who worked at Queen Mary College, London at the time, had several extended visits to Caltech in the period 1980–1985, and I had one to London in the fall of 1983. We also worked together several summers in Aspen. On several of these occasions we also collaborated with Lars Brink, who had visited Caltech and collaborated with me a few times previously.

After a year or so of unsuccessful efforts, Green and I discovered a new light-cone gauge formalism for the GSO-projected theory in which spacetime supersymmetry of the spectrum and interactions was easily proved. This was presented in three papers [GS81, GS82a, GS82b]. The first developed the formalism, while the next two used this light-cone gauge formalism to compute various tree and one-loop amplitudes and elucidate their properties. At this stage only open string amplitudes were under consideration.

Our next project was to identify more precisely the possibilities for superstring theories. The GSO work had identified the proper projection for open strings, but it left unclear what one should do with the closed strings. Green and I realized that there are three distinct types of supersymmetry possible in ten dimensions and that all three of them could be realized by superstring theories. In [GS82c] we formulated the type I, type IIA, and type IIB superstring theories. (We introduced these names a little later.) The type I theory is a theory of unoriented open and closed strings, whereas the type II theories are theories of oriented closed strings only.

Brink, Green and I formulated d-dimensional maximally supersymmetric Yang–Mills theories and supergravity theories as limits of superstring theory with $10 - d$ of the ten dimensions forming a torus. By computing one-loop string theory amplitudes for massless gauge particles in the type I theory and gravitons in the type II theory and taking the appropriate limits, we showed that both the Yang–Mills and supergravity theories are ultraviolet finite at one loop for $d < 8$ [GSB82]. The toroidally compactified string-loop formulae exhibited T-duality symmetry, though this was not pointed out explicitly in the article.

We also spent considerable effort formulating superstring field theory in the light-cone gauge [GS83a, GSB83, GS84b]. This work became relevant about 20 years later, when the construction was generalized to the case of type IIB superstrings in a plane-wave background spacetime geometry.

The fact that our spacetime supersymmetric formalism was only defined in the light-cone gauge was a source of frustration. Brink and I had found a covariant world-line action for a massless superparticle in ten dimensions [BS81],[1] so it was just a matter of finding the suitable superstring generalization. After a number of attempts, Green and I eventually found a covariant world-sheet action with manifest spacetime supersymmetry (and nonmanifest kappa symmetry) [GS84a, GS84c]. This covariant action reduces to our previous one in the light-cone gauge, of course. It was natural to try to use it to define covariant quantization. However, due to a subtle combination of first-class and second-class constraints, it was immediately apparent that this action is extremely difficult to quantize covariantly. Numerous unsuccessful attempts over the years bear testimony to the truth of this assertion. More recently, Berkovits seems to have found a successful scheme [Ber00].

Another problem of concern during this period was the formulation of ten-dimensional type IIB supergravity, which is the leading low energy approximation to type IIB superstring theory. Some partial results were obtained in separate collaborations with Green [GS83b] and with Peter West [SW83]. A challenging aspect of the problem is the presence of a self-dual five-form field strength, which obstructs a straightforward construction of a manifestly covariant action. Therefore, I decided to focus on the equations of motion, instead, which I

---

[1] Casalbuoni considered similar superparticle systems in four dimensions several years earlier [Cas76].

presented in [Sch83]. Equivalent results were obtained in a superfield formalism by Howe and West [HW84].

Let me now turn to the issue of anomalies, which are violations of classical symmetries, due to quantization in relativistic theories. Type I superstring theory is a well-defined ten-dimensional theory at tree level for any $SO(n)$ or $Sp(n)$ gauge group (see [Sch82] and Marcus and Sagnotti [MS82]). However, in every case it is chiral (i.e. parity violating) and the $d = 10$ super Yang–Mills sector is anomalous. Evaluation of a one-loop perturbative correction leads to a hexagon Feynman diagram that exhibits explicit nonconservation of gauge currents of the schematic form

$$\partial_\mu J^\mu \sim \varepsilon^{\mu_1 \cdots \mu_{10}} F_{\mu_1\mu_2} \cdots F_{\mu_9\mu_{10}}. \tag{3.23}$$

This is a fatal inconsistency of the theory.

Alvarez-Gaumé and Witten derived general formulae for gauge, gravitational, and mixed anomalies in an arbitrary spacetime dimension [AW84], and they discovered that the gravitational anomalies (nonconservation of the stress tensor) cancel in type IIB supergravity. This result was not really a surprise, since the one-loop type IIB superstring amplitudes are ultraviolet finite. It appeared likely that type I superstring theory is anomalous for any choice of the gauge group, but an explicit computation was required to decide for sure. In this case there are divergences that need to be regulated, so anomalies are definitely possible.

Green and I explored the anomaly problem for type I superstring theory off and on for almost two years until the crucial breakthroughs were made in August 1984 at the Aspen Center for Physics. That summer I was the organizer of a workshop entitled 'Physics in Higher Dimensions' at the Aspen Center for Physics. This attracted many participants, even though string theory was not yet fashionable, because by that time there was considerable interest in supergravity theories in higher dimensions and Kaluza–Klein compactification. We benefitted from the presence of many leading experts including Bruno Zumino, Bill Bardeen, Dan Friedan, Steve Shenker, and others.

Green and I had tried unsuccessfully to compute the one-loop hexagon diagram in type I superstring theory using our supersymmetric light-cone gauge formalism, but this led to an impenetrable morass. In discussions with Friedan and Shenker the idea arose to carry out the computation using the covariant RNS formalism instead. At that point, Friedan and Shenker left Aspen, so Green and I continued on our own.

It soon became clear that both the cylinder and Möbius-strip diagrams contributed to the anomaly. Before a workshop seminar by one of the other workshop participants (I don't remember which one), I remarked to Green that there might be a gauge group for which the two contributions cancel. At the end of the seminar Green said to me '$SO(32)$', which was the correct result. Since this computation only showed the cancellation of the pure gauge part of the anomaly, we decided to explore the low energy effective field theory to see whether the gravitational and mixed anomalies could also cancel. Before long, with the help of the results of Alvarez-Gaumé and Witten and useful comments by Bardeen and others, we were able to explain how this works. The effective field theory analysis was written up first [GS84d], and the string loop analysis was written up somewhat later [GS85a]. We also showed that the UV divergences of the cylinder and Möbius-strip diagrams cancel for $SO(32)$ [GS85b]. Nowadays such cancellations are usually understood in terms of tadpole cancellations in a dual closed string channel.

The effective field theory analysis showed that $E_8 \times E_8$ is a second gauge group for which the anomalies could cancel for a theory with $\mathcal{N} = 1$ supersymmetry in ten dimensions. In both cases, it is crucial for the result that the coupling to supergravity is included. The $SO(32)$ case could be accommodated by type I superstring theory, but we did not know of a superstring theory with gauge group $E_8 \times E_8$. We were aware of the article by Goddard and Olive that pointed out (among other things) that there are just two even self-dual Euclidean lattices in 16 dimensions, and these are associated with precisely these two gauge groups [GO85]. However, we did not figure out how to exploit this fact before the problem was solved by others.

Before the end of 1984 there were two other major developments. The first one was the construction of the 'heterotic string' by Gross, Harvey, Martinec and Rohm [GHMR85a, GHMR85b, GHMR86]. Their construction actually accommodated both of the gauge groups. The second one was the demonstration by Candelas, Horowitz, Strominger and Witten that 'Calabi–Yau compactifications' of the $E_8 \times E_8$ heterotic string give supersymmetric four-dimensional effective theories with many realistic features [CHSW85].

By the beginning of 1985, superstring theory – with the goal of unification – had become a mainstream activity. In fact, there was a very sudden transition from benign neglect to unbounded euphoria, both of which seemed to me to be unwarranted. After a while, most string theorists developed a more realistic assessment of the problems and challenges that remained.

3.8 Postscript

The construction of a dual string theory description of QCD is still an actively pursued goal. It now appears likely that every well-defined (finite or asymptotically free) four-dimensional gauge theory has a string theory dual in a curved background geometry with five noncompact dimensions. The extra dimension corresponds to the energy scale of the gauge theory. The cleanest and best understood example of such a duality is the correspondence between $\mathcal{N} = 4$ supersymmetric Yang–Mills theory with an $SU(N)$ gauge group and type IIB superstring theory in an $AdS_5 \times S^5$ spacetime with N units of five-form flux threading the sphere (see Maldacena [Mal98]). In particular, the (off-shell) energy-momentum tensor of the four-dimensional gauge theory corresponds to the (on-shell) graviton in five dimensions.

Such possibilities were not contemplated in the early years, so it is understandable that success was not achieved. Moreover, the dual description of QCD is likely to be considerably more complicated than the example described above. For one thing, for realistic numbers of colours and flavours, the five-dimensional geometry is expected to have string-scale curvature, so that a supergravity approximation will not be helpful. However, it might still be possible to treat the inverse of the number of colours as small, so that a semi-classical string theory approximation (corresponding to the planar approximation to the gauge theory) can be used. If one is willing to sacrifice quantitative precision, one can already give constructions that have the correct qualitative features of QCD. One of their typical unrealistic features is that the Kaluza–Klein scale is comparable to the QCD scale. I remain optimistic that a correct construction of a string theory configuration that is dual to QCD exists. However, finding it and analyzing it might take a long time.

Acknowledgements

I am grateful to Lars Brink for reading the manuscript and making several helpful suggestions. I also wish to acknowledge the hospitality of the Galileo Galilei Institute and the Aspen Center for Physics. This work was partially supported by the U.S. Dept. of Energy under Grant No. DE-FG03-92-ER40701.

John H. Schwarz was born in North Adams, Massachusetts in 1941. He did his undergraduate studies at Harvard, and received his PhD in physics at Berkeley in 1966. He is currently the Harold Brown Professor of Theoretical Physics at the California Institute of Technology. He has worked on superstring theory for almost his entire professional career. Schwarz has been awarded a Guggenheim Fellowship and a MacArthur Fellowship. He is a Fellow of the American Physical Society as well as a member of the National Academy of Sciences and the American Academy of Arts and Sciences. He received the Dirac Medal in 1989, and the Dannie Heineman Prize in Mathematical Physics in 2002.

References

[ABDD76] Ademollo, M., Brink, L., D'Adda, A., D'Auria, R., Napolitano, E., Sciuto, S., Del Giudice, E., Di Vecchia, P., Ferrara, S., Gliozzi, F., Musto, R., Pettorino, R. and Schwarz, J. (1976). Dual string with $U(1)$ colour symmetry, *Nucl. Phys.* **B111**, 77–110.
[AW84] Alvarez-Gaumé, L. and Witten, E. (1984). Gravitational anomalies, *Nucl. Phys.* **B234**, 269–330.
[BM69] Bardakci, K. and Mandelstam, S. (1969). Analytic solution of the linear-trajectory bootstrap, *Phys. Rev.* **184**, 1640–1644.
[BR69] Bardakci, K. and Ruegg, H. (1969). Reggeized resonance model for arbitrary production processes, *Phys. Rev.* **181**, 1884–1889.
[BBS07] Becker, K., Becker, M. and Schwarz, J. H. (2007). *String Theory and M-Theory: A Modern Introduction* (Cambridge University Press, Cambridge).
[Ber00] Berkovits, N. (2000). Super-Poincaré covariant quantization of the superstring, *JHEP* **4**, 018–034.
[BDZD76] Brink, L., Deser, S., Zumino, B., Di Vecchia, P. and Howe, P. S. (1976). Local supersymmetry for spinning particles, *Phys. Lett.* **B64**, 435–438.

[BDH76] Brink, L., Di Vecchia, P. and Howe, P. S. (1976). A locally supersymmetric and reparameterization invariant action for the spinning string, *Phys. Lett.* **B65**, 471–474.
[BO73a] Brink, L. and Olive, D. (1973). The physical state projection operator in dual resonance models for the critical dimension of spacetime, *Nucl. Phys.* **B56**, 253–265.
[BO73b] Brink, L. and Olive, D. (1973). Recalculation of the unitary single planar loop in the critical dimension of spacetime, *Nucl. Phys.* **B58**, 237–253.
[BS77] Brink, L. and Schwarz, J. H. (1977). Local complex supersymmetry in two-dimensions, *Nucl. Phys.* **B121**, 285–295.
[BS81] Brink, L. and Schwarz, J. H. (1981). Quantum superspace, *Phys. Lett.* **B100**, 310–312.
[BSS77] Brink, L., Schwarz, J. H. and Scherk, J. (1977). Supersymmetric Yang–Mills theories, *Nucl. Phys.* **B121**, 77–92.
[Bro72] Brower, R. C. (1972). Spectrum-generating algebra and no-ghost theorem in the dual model, *Phys. Rev.* **D6**, 1655–1662.
[BF73] Brower, R. C. and Friedman, K. A. (1973). Spectrum-generating algebra and no-ghost theorem for the Neveu–Schwarz model, *Phys. Rev.* **D7**, 535–539.
[CHSW85] Candelas, P., Horowitz, G. T., Strominger, A. and Witten, E. (1985). Vacuum configurations for superstrings, *Nucl. Phys.* **B258**, 46–74.
[Cas76] Casalbuoni, R. (1976). The classical mechanics for Bose–Fermi systems, *Nuovo Cimento* **A33**, 389–431.
[CT69] Chan, H. M. and Tsou, S. T. (1969). Explicit construction of the N-point function in the generalized Veneziano model, *Phys. Lett.* **B28**, 485–488.
[Che62] Chew, G. F. (1962). *S-Matrix Theory Of Strong Interactions* (W. A. Benjamin, New York).
[Che66] Chew, G. F. (1966). *The Analytic S-Matrix: A Basis for Nuclear Democracy* (W. A. Benjamin, New York).
[CF61] Chew, G. F. and Frautschi, S. C. (1961). Principle of equivalence for all strongly interacting particles within the S matrix framework, *Phys. Rev. Lett.* **7**, 394–397.
[CF62] Chew, G. F. and Frautschi, S. C. (1962). Regge trajectories and the principle of maximum strength for strong interactions, *Phys. Rev. Lett.* **8**, 41–44.
[CG74] Corrigan, E. and Goddard, P. (1974). The absence of ghosts in the dual fermion model, *Nucl. Phys.* **B68**, 189–202.
[CGOS73] Corrigan, E. F., Goddard, P., Olive, D. and Smith, R. A. (1973). Evaluation of the scattering amplitude for four dual fermions, *Nucl. Phys.* **B67**, 477–491.
[CJS78] Cremmer, E., Julia, B. and Scherk, J. (1978). Supergravity theory in 11 dimensions, *Phys. Lett.* **B76**, 409–412.
[CSS79] Cremmer, E., Scherk, J. and Schwarz, J. H. (1979). Spontaneously broken $N = 8$ supergravity, *Phys. Lett.* **84**, 83–86.
[DDF72] Del Giudice, E., Di Vecchia, P. and Fubini, S. (1972). General properties of the dual resonance model, *Ann. Phys.* **70**, 378–398.
[DZ76a] Deser, S. and Zumino, B. (1976). Consistent supergravity, *Phys. Lett.* **B62**, 335–337.
[DZ76b] Deser, S. and Zumino, B. (1976). A complete action for the spinning string, *Phys. Lett.* **B65**, 369–373.
[DHS67] Dolen, R., Horn, D. and Schmid, C. (1967). Prediction of Regge parameters of ρ poles from low-energy πN data, *Phys. Rev. Lett.* **19**, 402–407.
[DHS68] Dolen, R., Horn, D. and Schmid, C. (1968). Finite-energy sum rules and their application to pin charge exchange, *Phys. Rev.* **166**, 1768–1781.

[FN70] Fairlie, D. B. and Nielsen, H. B. (1970). An analog model for KSV theory, *Nucl. Phys.* **B20**, 637–651.

[FNF76] Freedman, D. Z., van Nieuwenhuizen, P. and Ferrara, S. (1976). Progress toward a theory of supergravity, *Phys. Rev.* **D13**, 3214–3218.

[Fre68] Freund, P. G. O. (1968). Finite-energy sum rules and bootstraps, *Phys. Rev. Lett.* **20**, 235–237.

[FLS70] Frye, G., Lee, C. W. and Susskind, L. (1970). Dual-symmetric theory of hadrons. II. Baryons, *Nuovo Cimento* **A69**, 497–507.

[FS70] Frye, G. and Susskind, L. (1970). Non-planar dual symmetric loop graphs and Pomeron, *Phys. Lett.* **B31**, 589–591.

[FGV69] Fubini, S., Gordon, D. and Veneziano, G. (1969). A general treatment of factorization in dual resonance models, *Phys. Lett.* **B29**, 679–682.

[FV69] Fubini, S. and Veneziano, G. (1969). Level structure of dual resonance models, *Nuovo Cimento* **A64**, 811–840.

[FV70] Fubini, S. and Veneziano, G. (1970). Duality in operator formalism, *Nuovo Cimento* **A67**, 29–47.

[FV71] Fubini, S. and Veneziano, G. (1971). Algebraic treatment of subsidiary conditions in dual resonance models, *Ann. Phys.* **63**, 12–27.

[GS71] Gervais, J. L. and Sakita, B. (1971). Field theory interpretation of supergauges in dual models, *Nucl. Phys.* **B34**, 632–639.

[GS73] Gervais, J. L. and Sakita, B. (1973). Ghost-free string picture of Veneziano model, *Phys. Rev. Lett.* **30**, 716–719.

[GSO76] Gliozzi, F., Scherk, J. and Olive, D. (1976). Supergravity and the spinor dual model, *Phys. Lett.* **B65**, 282–286.

[GSO77] Gliozzi, F., Scherk, J. and Olive, D. (1977). Supersymmetry, supergravity theories and the dual spinor model, *Nucl. Phys.* **B122**, 253–290.

[GGRT73] Goddard, P., Goldstone, J., Rebbi, C. and Thorn, C. B. (1973). Quantum dynamics of a massless relativistic string, *Nucl. Phys.* **B56**, 109–135.

[GO85] Goddard, P. and Olive, D. (1985). Algebras, lattices and strings, in *Vertex Operators in Mathematics and Physics*, ed. Lepowsky, J., Mandelstam, S. and Singer, I. M., Mathematical Research Institute Publications (Springer-Verlag, New York).

[GT72] Goddard, P. and Thorn, C. B. (1972). Compatibility of the Pomeron with unitarity and the absence of ghosts in the dual resonance model, *Phys. Lett.* **B40**, 235–238.

[GS69] Goebel, C. J. and Sakita, B. (1969). Extension of the Veneziano form to n-particle amplitudes, *Phys. Rev. Lett.* **22**, 257–260.

[GL71] Golfand, Yu. A. and Likhtman, E. P. (1971). Extension of the algebra of Poincaré group generators and violation of P invariance, *JETP Lett.* **13**, 323–326 [*Pisma Zh. Eksp. Teor. Fiz.* **13**, 452 (1971)].

[GS81] Green, M. B. and Schwarz, J. H. (1981). Supersymmetrical dual string theory, *Nucl. Phys.* **B181**, 502–530.

[GS82a] Green, M. B. and Schwarz, J. H. (1982). Supersymmetrical dual string theory. 2. Vertices and trees, *Nucl. Phys.* **B198**, 252–268.

[GS82b] Green, M. B. and Schwarz, J. H. (1982). Supersymmetrical dual string theory. 3. Loops and renormalization, *Nucl. Phys.* **B198**, 441–460.

[GS82c] Green, M. B. and Schwarz, J. H. (1982). Supersymmetrical string theories, *Phys. Lett.* **B109**, 444–448.

[GS83a] Green, M. B. and Schwarz, J. H. (1983). Superstring interactions, *Nucl. Phys.* **B218**, 43–88.

[GS83b] Green, M. B. and Schwarz, J. H. (1983). Extended supergravity in ten dimensions, *Phys. Lett.* **B122**, 143–147.
[GS84a] Green, M. B. and Schwarz, J. H. (1984). Properties of the covariant formulation of superstring theories, *Nucl. Phys.* **B243**, 285–306.
[GS84b] Green, M. B. and Schwarz, J. H. (1984). Superstring field theory, *Nucl. Phys.* **B243**, 475–536.
[GS84c] Green, M. B. and Schwarz, J. H. (1984). Covariant description of superstrings, *Phys. Lett.* **B136**, 367–370.
[GS84d] Green, M. B. and Schwarz, J. H. (1984). Anomaly cancellation in supersymmetric $d = 10$ gauge theory and superstring theory, *Phys. Lett.* **B149**, 117–122.
[GS85a] Green, M. B. and Schwarz, J. H. (1985). The hexagon gauge anomaly in type I superstring theory, *Nucl. Phys.* **B255**, 93–114.
[GS85b] Green, M. B. and Schwarz, J. H. (1985). Infinity cancellations in $so(32)$ superstring theory, *Phys. Lett.* **B151**, 21–25.
[GSB82] Green, M. B., Schwarz, J. H. and Brink, L. (1982). $N = 4$ Yang–Mills and $N = 8$ supergravity as limits of string theories, *Nucl. Phys.* **B198**, 474–492.
[GSB83] Green, M. B., Schwarz, J. H. and Brink, L. (1983). Superfield theory of type II superstrings, *Nucl. Phys.* **B219**, 437–478.
[GSW87] Green, M. B., Schwarz, J. H. and Witten, E. (1987). *Superstring Theory* (Cambridge University Press, Cambridge).
[GHMR85a] Gross, D. J., Harvey, J. A., Martinec, E. J. and Rohm, R. (1985). Heterotic string, *Phys. Rev. Lett.* **54**, 502–505.
[GHMR85b] Gross, D. J., Harvey, J. A., Martinec, E. J. and Rohm, R. (1985). Heterotic string theory. 1. The free heterotic string, *Nucl. Phys.* **B256**, 253–284.
[GHMR86] Gross, D. J., Harvey, J. A., Martinec, E. J. and Rohm, R. (1986). Heterotic string theory. 2. The interacting heterotic string, *Nucl. Phys.* **B267**, 75–124.
[GNSS70a] Gross, D. J., Neveu, A., Scherk, J. and Schwarz, J. H. (1970). The primitive graphs of dual resonance models, *Phys. Lett.* **B31**, 592–594.
[GNSS70b] Gross, D. J., Neveu, A., Scherk, J. and Schwarz, J. H. (1970). Renormalization and unitarity in the dual-resonance model, *Phys. Rev.* **D2**, 697–710.
[Har69] Harari, H. (1969). Duality diagrams, *Phys. Rev. Lett.* **22**, 562–565.
[HW84] Howe, P. S. and West, P. C. (1984). The complete $n = 2$, $d = 10$ supergravity, *Nucl. Phys.* **B238**, 181–220.
[IM67a] Igi, K. and Matsuda, S. (1967). New sum rules and singularities in the complex j plane, *Phys. Rev. Lett.* **18**, 625–627.
[IM67b] Igi, K. and Matsuda, S. (1967). Some consequences from superconvergence for πN scattering, *Phys. Rev.* **163**, 1622–1626.
[Jac29] Jacobi, C. G. J. (1829). *Fundamenta Nova*, Könisberg.
[KSV69] Kikkawa, K., Sakita, B. and Virasoro, M. A. (1969). Feynman-like diagrams compatible with duality. I: Planar diagrams, *Phys. Rev.* **184**, 1701–1713.
[KN69a] Koba, Z. and Nielsen, H. B. (1969). Reaction amplitude for n-mesons a generalization of the Veneziano–Bardakci–Ruegg–Virasoro model, *Nucl. Phys.* **B10**, 633–655.
[KN69b] Koba, Z. and Nielsen, H. B. (1969). Manifestly crossing-invariant parameterization of n-meson amplitude, *Nucl. Phys.* **B12**, 517–536.
[LST67] Logunov, A. A., Soloviev, L. D. and Tavkhelidze, A. N. (1967). Dispersion sum rules and high energy scattering, *Phys. Lett.* **B24**, 181–182.
[Lov68] Lovelace, C. (1968). A novel application of Regge trajectories, *Phys. Lett.* **B28**, 264–268.

[Lov71] Lovelace, C. (1971). Pomeron form factors and dual Regge cuts, *Phys. Lett.* **B34**, 500–506.
[Mal98] Maldacena, J. M. (1998). The large N limit of superconformal field theories and supergravity, *Adv. Theor. Math. Phys.* **2**, 231.
[Man73] Mandelstam, S. (1973). Interacting string picture of dual resonance models, *Nucl. Phys.* **B64**, 205–235.
[MS82] Marcus, N. and Sagnotti, A. (1982). Tree level constraints on gauge groups for type I superstrings, *Phys. Lett.* **B119**, 97–99.
[Nah78] Nahm, W. (1978). Supersymmetries and their representations, *Nucl. Phys.* **B135**, 149–166.
[Nam70a] Nambu, Y. (1970). Quark model and the factorization of the Veneziano amplitude, in *Proceedings of the International Conference on Symmetries and Quark Models, Wayne State University, June 18–20, 1969*, ed. Chand, R. (Gordon and Breach, New York), 269–277, reprinted in *Broken Symmetry, Selected Papers of Y. Nambu*, ed. Eguchi, T. and Nishijima, K. (World Scientific, Singapore, 1995), 258–277.
[Nam70b] Nambu, Y. (1970). Duality and hadrodynamics, Notes prepared for the Copenhagen High Energy Symposium, unpublished; published in *Broken Symmetry, Selected Papers of Y. Nambu*, ed. Eguchi, T. and Nishijima, K. (World Scientific, Singapore, 1995), 280–301.
[NS72] Neveu, A. and Scherk, J. (1972). Connection between Yang-Mills fields and dual models, *Nucl. Phys.* **B36**, 155–161.
[NS71a] Neveu, A. and Schwarz, J. H. (1971). Quark model of dual pions, *Phys. Rev.* **D4**, 1109–1111.
[NS71b] Neveu, A. and Schwarz, J. H. (1971). Factorizable dual model of pions, *Nucl. Phys.* **B31**, 86–112.
[NST71] Neveu, A., Schwarz, J. H. and Thorn, C. B. (1971). Reformulation of the dual pion model, *Phys. Lett.* **B35**, 529–533.
[NT71] Neveu, A. and Thorn, C. B. (1971). Chirality in dual resonance models, *Phys. Rev. Lett.* **27**, 1758–1761.
[Nie70] Nielsen, H. B. (1970). An almost physical interpretation of the integrand of the n-point Veneziano model, paper submitted to the *15th International Conference on High Energy Physics, Kiev, 26 August to 4 September, 1970* (Nordita preprint, 1969).
[OS73a] Olive, D. and Scherk, J. (1973). No-ghost theorem for the Pomeron sector of the dual model, *Phys. Lett.* **B44**, 296–300.
[OS73b] Olive, D. and Scherk, J. (1973). Towards satisfactory scattering amplitudes for dual fermions, *Nucl. Phys.* **B64**, 334–348.
[PC69] Paton, J. E. and Chan, H. M. (1969). Generalized Veneziano model with isospin, *Nucl. Phys.* **B10**, 516–520.
[Pol81a] Polyakov, A. M. (1981). Quantum geometry of bosonic strings, *Phys. Lett.* **B103**, 207–210.
[Pol81b] Polyakov, A. M. (1981). Quantum geometry of fermionic strings, *Phys. Lett.* **B103**, 211–213.
[Ram71] Ramond, P. (1971). Dual theory for free fermions, *Phys. Rev.* **D3**, 2415–2418.
[Ros69] Rosner, J. L. (1969). Graphical form of duality, *Phys. Rev. Lett.* **22**, 689–692.
[Sch71] Scherk, J. (1971). Zero-slope limit of the dual resonance model, *Nucl. Phys.* **B31**, 222–234.

[SS74] Scherk, J. and Schwarz, J. H. (1974). Dual models for nonhadrons, *Nucl. Phys.* **B81**, 118–144.
[SS75] Scherk, J. and Schwarz, J. H. (1975). Dual model approach to a renormalizable theory of gravitation, submitted to the 1975 Gravitation Essay Contest of the Gravity Research Foundation. Reprinted in *Superstrings*, ed. Schwarz, J. H., Vol. 1 (World Scientific, Singapore, 1985), 218–222.
[SS79a] Scherk, J. and Schwarz, J. H. (1979). Spontaneous breaking of supersymmetry through dimensional reduction, *Phys. Lett.* **B82**, 60–64.
[SS79b] Scherk, J. and Schwarz, J. H. (1979). How to get masses from extra dimensions, *Nucl. Phys.* **B153**, 61–88.
[Sch72a] Schwarz, J. H. (1972). Physical states and Pomeron poles in the dual pion model, *Nucl. Phys.* **B46**, 61–74.
[Sch72b] Schwarz, J. H. (1972). Dual-pion model satisfying current-algebra constraints, *Phys. Rev.* **D5**, 886–891.
[Sch73] Schwarz, J. H. (1973). Off mass shell dual amplitudes without ghosts, *Nucl. Phys.* **B65**, 131–140.
[Sch74] Schwarz, J. H. (1974). Off mass shell dual amplitudes. III, *Nucl. Phys.* **B76**, 93–108.
[Sch82] Schwarz, J. H. (1982). Gauge groups for type I superstrings, in *Proc. Johns Hopkins Workshop*, 233.
[Sch83] Schwarz, J. H. (1983). Covariant field equations of chiral $N = 2$, $D = 10$ supergravity, *Nucl. Phys.* **B226**, 269–288.
[Sch85] Schwarz, J. ed. (1985). *Superstrings, The First Fifteen Years of Superstring Theory* (World Scientific, Singapore).
[Sch00a] Schwarz, J. H. (2000). Reminiscences of collaborations with Joël Scherk, arXiv:hep-th/0007117.
[Sch00b] Schwarz, J. H. (2000). String theory: the early years, arXiv:hep-th/0007118.
[Sch01] Schwarz, J. H. (2001). String theory origins of supersymmetry, *Nucl. Phys. Proc. Suppl.* **101**, 54–61.
[SW83] Schwarz, J. H. and West, P. C. (1983). Symmetries and transformations of chiral $n = 2$, $d = 10$ supergravity, *Phys. Lett.* **B126**, 301–304.
[SW73] Schwarz, J. H. and Wu, C. C. (1973). Evaluation of dual fermion amplitudes, *Phys. Lett.* **B47**, 453–456.
[SW74] Schwarz, J. H. and Wu, C. C. (1974). Off mass shell dual amplitudes. II, *Nucl. Phys.* **B72**, 397–412.
[Sha69] Shapiro, J. A. (1969). Narrow resonance model with Regge behavior for $\pi\pi$ scattering, *Phys. Rev.* **179**, 1345–1353.
[Sha70] Shapiro, J. A. (1970). Electrostatic analogue for the Virasoro model, *Phys. Lett.* **B33**, 361–362.
[Sie92] Siegel, W. (1992). $N = 4$ string is the same as the $N = 2$ string, *Phys. Rev. Lett.* **69**, 1493–1495.
[Sus70] Susskind, L. (1970). Dual-symmetric theory of hadrons. I, *Nuovo Cimento* **A69**, 457–496.
[Tho71] Thorn, C. B. (1971). Embryonic dual model for pions and fermions, *Phys. Rev.* **D4**, 1112–1116.
[Tow95] Townsend, P. K. (1995). The eleven-dimensional supermembrane revisited, *Phys. Lett.* **B350**, 184–188.

[Ven68] Veneziano, G. (1968). Construction of a crossing-symmetric, Reggeon behaved amplitude for linearly rising trajectories, *Nuovo Cimento* **A57**, 190–197.
[Vir69] Virasoro, M. (1969). Alternative constructions of crossing-symmetric amplitudes with Regge behavior, *Phys. Rev.* **177**, 2309–2311.
[Vir70] Virasoro, M. (1970). Subsidiary conditions and ghosts in dual-resonance models, *Phys. Rev.* **D1**, 2933–2936.
[Wei65] Weinberg, S. (1965). Photons and gravitons in perturbation theory: derivation of Maxwell's and Einstein's equations, *Phys. Rev.* **138**, 988–1002.
[WZ74] Wess, J. and Zumino, B. (1974). Supergauge transformations in four dimensions, *Nucl. Phys.* **B70**, 39–50.
[Wit95] Witten, E. (1995). String theory dynamics in various dimensions, *Nucl. Phys.* **B443**, 85–126.
[YM54] Yang, C. N. and Mills, R. L. (1954). Conservation of isotopic spin and isotopic gauge invariance, *Phys. Rev.* **96**, 191–195.
[Yon73] Yoneya, T. (1973). Quantum gravity and the zero slope limit of the generalized Virasoro model, *Lett. Nuovo Cimento* **8**, 951–955.
[Yon74] Yoneya, T. (1974). Connection of dual models to electrodynamics and gravidynamics, *Prog. Theor. Phys.* **51**, 1907–1920.
[Zum74] Zumino, B. (1974). Relativistic strings and supergauges, in *Renormalization and Invariance in Quantum Field Theory,* ed. Caianiello, E. (Plenum Press, New York), 367.

4
Early string theory as a challenging case study for philosophers

ELENA CASTELLANI

4.1 Introduction

The history of the origins and first developments of string theory, from Veneziano's formulation of his famous scattering amplitude in 1968 to the so-called first string revolution in 1984, provides invaluable material for philosophical reflection. The reasons why this episode in the history of modern physics – one still largely unknown to the philosophy of science community despite its centrality to theoretical physics – represents a particularly interesting case study are several and various in nature. It is the aim of the present Chapter to illustrate some of them.

In general, the story of the construction of a new scientific theory is of evident interest in itself, as a concrete example of how a particular theory has been discovered and developed by a given community and over a certain period of time. On the other hand, case studies taken from the history of science are commonly used, by those philosophers of science who pay attention to actual scientific practice, to provide some evidence for or against given positions on traditional epistemological or methodological issues. In other words, with respect to philosophical 'theories' on given aspects of the scientific enterprise, historical case studies are attributed a role analogous to that of the data of experience in scientific theories. These aspects can be of a very general character, such as those regarding the methodology, aim and evaluation of scientific theories; or they can be of a more specific kind, such as the significance of a certain principle, argument or concept.

The case study presented by early string theory, as narrated by the contributors to this Volume, is fruitful from both a general and a specific perspective. It provides, first of all, an illustration of the first steps of a scientific theory which has dominated a significant part of theoretical physics research over recent decades, thus feeding the philosophical reflection with valuable data on how scientific theories are constructed and selected. It sheds light, at the same time, on the original meaning of ideas – such as duality, supersymmetry and extra spacetime dimensions – and mathematical techniques that are basic ingredients in today's theoretical physics, as is also discussed in the Introduction to Part I.

The Birth of String Theory, ed. Andrea Cappelli, Elena Castellani, Filippo Colomo and Paolo Di Vecchia.
Published by Cambridge University Press. © Cambridge University Press 2012.

The philosophical interest of early string theory is examined here on the basis of the historical outline emerging from the collection of reminiscences and surveys contained in this Volume. In this history, both 'internal' factors (such as the form, content and logic of the theory) and 'external' factors (such as the psychological or sociological aspects influencing the scientific work), to use a traditional distinction, are well represented. It is usually considered controversial whether an historical reconstruction based mainly on personal records, viewpoints and experiences can indeed be objective. In this case, however, we are confident that the multifaceted account resulting from the gathering of the independent recollections of almost all the scientists involved, checked on the grounds of the original papers, can provide an accurate and balanced historical picture.

4.2 The case study

The discovery by Veneziano of his 'dual' amplitude for the scattering of four mesons is widely acknowledged as the starting point for the developments leading to string theory. In fact, as illustrated in the first Parts of the Volume, it immediately gave rise to the very intense theoretical activity that is known, in general, as the 'dual theory of strong interactions': from the first two models proposed – the Dual Resonance Model and the Shapiro–Virasoro model, respectively – to all the subsequent endeavours to extend, complete and refine the theory, including its string interpretation and the addition of fermions.

This was the first phase of early string theory; it was motivated by the aim of finding a viable theory of hadrons in the framework of the S-matrix theory as developed in the early Sixties. Chronologically, it extends from summer 1968, when Veneziano presented his formula at the Vienna Conference on High Energy Physics with strong impact on the theoretical physics community, to the end of 1973, when interest in dual models for describing hadronic physics began to decline. This happened primarily for the following two reasons: firstly, the presence of unphysical features in the theory (such as extra spacetime dimensions and an unrealistic particle spectrum); secondly, the evidence for pointlike constituents at short distance inside hadrons (subsequently to be identified with the quarks) that was obtained in deep inelastic scattering experiments at the Stanford Linear Accelerator Center (SLAC).[1] This feature of strong interactions could not be explained in terms of the dual theory, while the competing gauge field theory, quantum chromodynamics (QCD), was able to account for it.

This first phase has some remarkable and unusual characteristics and can be considered as a case study in itself. It provides a remarkable example of a revolutionary theoretical project rising, flourishing and apparently subsiding in a rather short and well-delimited period of time and with a rare confluence of people, intentions and places. As testified by most of its exponents, the atmosphere in which the project was first developed was

---

[1] It is worth noting that, although the SLAC deep inelastic scattering experiments were contemporary with the discovery of the Veneziano amplitude, it took some time to formulate an appropriate theoretical description, which eventually provided essential support for QCD. In this time span the dual models could be developed.

particularly enthusiastic and cooperative. The young age of the majority of the physicists involved – many were either graduate students or postdoctoral fellows – no doubt had something to do with it; but what counted most was the shared conviction that something new and exciting was being created.

Another 'external' factor was conducive to the creation of such an atmosphere, namely, the concentration of research in the field in just a few institutions across the USA and Europe. One of these in particular, the CERN Theory Division in Geneva, played a central role, mostly thanks to the charismatic presence of Amati (CERN staff member at the time). He was able to gather a strong group of research fellows and visitors and almost all those who were working on dual theory spent some time there during this period. Under Amati's guidance, the 'dual group' formed a sort of theoretical laboratory, with regular seminars and several collaborations going on, in an atmosphere that was not only stimulating but also very friendly. The outcome was impressive: in a few years, the theory advanced remarkably fast and many decisive results were obtained, such as the proof of the no-ghost theorem and the quantization of the string action.

In the course of 1974, however, the atmosphere changed radically. The fascination with the theory remained the same but, on account of the problems it encountered as a description of strong interactions, the interest of the high energy community focussed on other developments, such as QCD and Standard Model physics. The dual string theory of strong interactions seemed to be a failed programme, with the consequence that many of the young theorists involved felt forced to leave the field to retain any hope of pursuing their academic careers.

Some people, however, resisted the general trend and went on addressing the unresolved problems of the theory, thus preparing its later renaissance in the form of modern string theory. This second phase, illustrated in the last two Parts of the Volume, extends from 1974 to when, towards the end of 1984, the interest in string theory went up again thanks to crucial results obtained in the context of supersymmetric string theory. Following this 'first superstring revolution', as it is now usually referred to, string theory rapidly became a mainstream activity and a new phase began, one that continues right up to the present.

From a sociological point of view, the second period, starting with the general change of attitude in 1974, is less homogeneous than the first. The motivation for pursuing string theory was a shared one: the theory was regarded as so beautiful and had such a compelling mathematical structure, obtained in agreement with consistency conditions and deep physical principles, that it was expected to be in some way related to the physical world. This was the attitude adopted, in particular, by Scherk and Schwarz and, independently, by Yoneya. They took the view that those very features that were considered drawbacks in describing hadronic physics, such as the presence of spin one and spin two massless particles and extra dimensions, could instead reveal the true nature of string theory. This view led them to investigate in depth the connection between dual models and those field theories which turned out to be the relevant ones: Yang–Mills (non-Abelian) gauge theories, that are employed in the Standard Model of particle physics, and general relativity, describing gravity. The resulting remarkable proposal, advanced by Scherk and Schwarz, was that string theory should be considered as a unified quantum theory of all the fundamental interactions. It was a 'big conceptual leap', as Schwarz describes it in his Chapter (Part I). In particular, it implied a huge rescaling (by 19 orders of magnitude) of the theory's characteristic scale, in order to relate it to the Planck scale, that is, the scale associated with quantum gravity.

However remarkable and promising, such a change of the domain and goal of the theory was not really appreciated, outside of a limited circle, for almost a decade. String theory remained a side issue for several years, notwithstanding the two major results obtained in 1976: the formulation of the supersymmetric action for string theory and the construction, by means of the so-called GSO projection, of the first totally consistent supersymmetric string theory (i.e. superstring) in 10 dimensions.

The case study considered in this Chapter is focussed on the development of the theory up to the 'exile' period before its renaissance in 1984. In this last stage of dual string theory, covered in the final Part of the Volume, only a few adepts remained, with the consequence that progress slowed down considerably. In contrast with the flourishing early phase, the work on string theory was no longer the central activity of a consistent and intensively collaborating group, but rather the independent research of small groups or isolated individuals who were also actively engaged in the more popular area of supersymmetric extension of field theory and general relativity.

As discussed in Part V, supersymmetry – the symmetry relating bosonic and fermionic particles – was discovered in the context of early string theory. It was implicitly present in Ramond's construction of the fermionic string in 1971, and was explicitly remarked as a property of the string action by Gervais and Sakita in the same year. Supersymmetry was then extended to quantum field theory by Wess and Zumino, in 1974, as a global symmetry acting on four-dimensional spacetime. Following their work, supersymmetric field theories in four and higher dimensions were quickly developed. Meanwhile, theories with local supersymmetry were constructed and analyzed independently of string theory. Since these theories generalized Einstein's theory of general relativity, they were referred to as 'supergravity theories'.

These developments were also motivated by the growing interest, toward the end of the Seventies, in the programme of unification of electro-weak and strong interactions above the Standard Model energy scale and in that of unification with gravity. For several reasons that are illustrated, in particular, in the Introductions to Parts VI and VII (see also Chapter 45, Section 45.2), supersymmetric field theories and supergravity appeared to be promising candidates for realizing this programme. Despite these shifts toward quantum field theory and supergravity as the primary interest of many theoretical physicists, research in string theory continued during these years and led to important developments. As described in Part VII, a new Lorentz covariant quantization of string theory, based on path-integral methods, was obtained by Polyakov in 1981; furthermore, superstring theories, offering a consistent unified quantum theory in ten spacetime dimensions, were classified and their properties were analyzed in the quest for a convincing unification framework.

In fact, research activities in superstring and supergravity were deeply intertwined, and many ideas and techniques first motivated by the string theoretical context, such as Kaluza–Klein compactification of extra dimensions, were developed and applied in the supergravity context. The string theory workshop on 'Physics in Higher Dimensions' at the Aspen Center for Physics in August 1984 was particularly emblematic of the above situation: as recalled by Schwarz in his Chapter, many of the participants were working on supergravity theories.

This workshop can be considered the turning point leading to the renaissance of string theory, as it was on that occasion that Green and Schwarz obtained a crucial result in their study of chiral anomalies in type I superstring theory. Chiral anomaly cancellation was a necessary condition for a unifying theory to be realistic, that is, to incorporate the Standard Model spectrum of particles and interactions. As explained in Part VII, the weak-interacting fermions are chiral, i.e. occur in specific combinations of spin and momentum. The associated chiral symmetry may be violated at the quantum level (developing an anomaly); the endeavour was to check the absence of such a fatal violation.

Superstring theory could also provide a consistent quantum theory of gravity, in contrast with the supergravity theories that had so far been considered. The result of Green and Schwarz thus opened a concrete path toward string theory unification of Standard Model physics with gravity, with the effect of producing a radical change of attitude in the theoretical physics community. This achievement marks the end of early string theory: a history initiated with the presentation of Veneziano's formula in the late summer of 1968 and covering a period of 16 years.

4.3 Theory progress: generalizations, analogies and conjectures

As stressed by all the contributors to the Volume, Veneziano's formula represented a turning point in the physics of strong interactions developed in the Sixties in the context of the so-called 'analytic S-matrix' or 'S-matrix theory'. This approach, described in Part II, belongs to the prehistory of the case study considered here. The S-matrix programme, pursued by Chew and his collaborators, has been thoroughly investigated from a historical and philosophical point of view in the 1990 book *Theory Construction and Selection in Modern Physics: The S-Matrix* by Cushing [Cus90]. For the purposes of this Chapter, it will be sufficient to recall the approach's general strategy, setting the agenda for the developments leading to the birth of string theory.

Motivated by the difficulties arising in a field theoretic description of strong interactions, and inspired by earlier work of Heisenberg, the aim of S-matrix theory was to determine the relevant observable physical quantities, namely, the scattering amplitudes (which formed the elements of the S-matrix) on the basis of general principles, such as unitarity, analyticity and crossing symmetry, and a minimal number of additional assumptions. In this context, the problem to which Veneziano's result provided a first, brilliant solution was the following: to find a consistent scattering amplitude that also obeyed the duality principle known as Dolen–Horn–Schmid duality (DHS duality) or dual bootstrap. As illustrated in detail in the Chapters by Veneziano, Schwarz and Ademollo, this was the assumption, suggested by the experimental data, that the contributions from resonance intermediate states and from particle exchange each formed a complete representation of the scattering process (so that they should not be added to one another in order to obtain the total amplitude). In terms of Mandelstam's variables and using the framework of the so-called Regge theory, the duality principle (as initially stated) established direct relations between a low energy and a high energy description of the hadronic scattering amplitude $A(s, t)$: namely, the low energy description in terms of direct channel (s-channel) resonance poles, and the high energy description in terms of the exchange of Regge poles in the crossed channel (t-channel), could each be obtained from the other by analytic continuation. In this sense, the duality principle represented an explicit and 'cheaper' formulation of the general bootstrap idea dominating the S-matrix programme:[2] that is, the idea of a self-consistent hadronic structure in which the entire ensemble of hadrons provided the forces (by hadron exchange) making their own existence (as intermediate states) possible.

The task of finding a formula for a scattering amplitude $A(s, t)$ that could be expanded as a series of poles in either the s-channel or the t-channel, and thus embody the duality principle, initially seemed very daunting. This explains the impact of Veneziano's result and the excitement it generated: his formula realized the DHS duality in a simple and clear way in the case of the four-meson scattering process $\pi\pi \to \pi\omega$ (a detailed description of the properties of Veneziano's amplitude is given in Appendix B). It is worth noting, in this regard, that the reconstruction offered by Veneziano and Ademollo, in their respective Chapters, of the steps leading to this achievement in around one year – from the collective works of Ademollo, Rubinstein, Veneziano and Virasoro on superconvergence sum rules to the discovery of the Veneziano formula – provides an illuminating example of the rationale of a scientific progress, one that is characterized by the close interplay of mathematically driven creativity and physical constraints (both theoretical and experimental). In fact, this modality of theory building is predominant in the history of early string theory, as following examples will also show.

Veneziano's four-particle amplitude represented a particular solution to the problem of constructing an S-matrix with the required properties for describing hadrons and their interactions. In this sense, it was a 'model' of the general S-matrix 'theory'. Indeed, the whole research activity to which the Veneziano formula gave rise, going under the label of 'dual models', can be seen as a theoretical process evolving from initially specific models (in the above sense) towards a general and consistent theory (of strong interactions first, of all four fundamental interactions later).[3]

---

[2] See Veneziano's Chapter, Section 2.2, for details on the distinction between Chew's 'expensive' bootstrap and the 'cheap' dual bootstrap.

[3] On the issue of the relation between models and theories, see later on in this Section, and Section 4.4.

The Veneziano amplitude, which started the above process, itself contained the germs for the successive evolution of the model. It was natural to try to extend it immediately, in order to overcome its limitations, for example the limited number of particles considered and their specific type. Moreover, the model violated unitarity because of the narrow-resonance assumption (the approximation corresponding to the fact that only single-particle stable intermediate states are allowed; see Ademollo's Chapter, Section 6.7). It was natural, as well, to search for other models that could include neglected but important physical features, and, more generally, to try to reach a better understanding of the physical theory underlying the models that were being constructed: in other words, to search for a satisfying physical interpretation of the mathematical structures obtained.

Methodologically, the theoretical work initiated with the Veneziano formula mainly advanced by the concurrent action of generalizations, analogies, and conjectures. The following part of this Section is devoted to highlighting illustrative significant threads in this process. As is stressed by the majority of those who participated in the building of the theory, this was essentially a bottom-up activity that had more the character of a patchwork than of a coherent construction. Results were obtained by following various alternative paths and (apparently) side issues, and their convergence often seemed almost miraculous. In fact, with hindsight, the story is much less surprising or serendipitous: the cohesiveness of the description obtained can be understood as a consequence of the strong constraints put on the theory by the underlying symmetry, the infinite-dimensional conformal symmetry, to be discussed later.

4.3.1 Generalizations

Following Veneziano's paper, generalizations in various directions immediately led to progress both with respect to obtaining a complete S-matrix and to constructing more realistic models. The most salient of these developments were the following.

- The generalization of the amplitude to the scattering of more than four particles (initially five and subsequently an arbitrary number N of particles), in order to implement the S-matrix consistency condition of 'factorization' (to which unitarity reduces in the narrow-resonance approximation) and, by means of it, analyze the full spectrum of the physical states. It was necessary to identify all the physical states to ensure the absence of negative-norm or ghost states (leading to unphysical negative probabilities) and hence the consistency of the theory. This is the line of research that led to the construction of the Dual Resonance Model (DRM), also called the generalized or multiparticle Veneziano model; it was later understood as the bosonic open string theory, as illustrated in particular in the contributions to the Volume by Veneziano, Schwarz, Di Vecchia (Chapter 11), Goddard, Mandelstam and Brower.
- The generalization toward more realistic models. These efforts extended from the independent attempts of Lovelace and Shapiro to obtain the amplitude for the scattering of four pions, recalled in their respective Chapters, to the construction of dual models for

the scattering of particles with internal symmetry and with spin. By these means, the first dual model including fermions was obtained by Ramond, as well as the immediately following model proposed by Neveu and Schwarz for extending the Lovelace–Shapiro amplitude to an arbitrary number of pions. The Ramond and Neveu–Schwarz models were soon recognized as the two sectors, fermionic and bosonic, of the same model, called the Ramond–Neveu–Schwarz (RNS) model. The steps leading to these results are reconstructed in detail in the Chapters by Ramond, Neveu and Schwarz.

- The generalization of the N-particle amplitudes, regarded as the lowest-order or 'tree' Feynman diagrams of a perturbative expansion, to include 'loops' and thus fulfil the S-matrix unitarity condition beyond the narrow-resonance approximation. This 'unitarization programme' was based on the analogy between the narrow-resonance approximation in dual theory and the Born approximation (involving tree diagrams only) in conventional quantum field theory. Once the general dual amplitudes and the couplings were known, the theoretical activity could focus on the construction of loop amplitudes. From the first attempt in 1969 by Kikkawa, Sakita and Virasoro, this programme was actively pursued in the flourishing period of early string theory.

- Another parallel development implying generalization started immediately after Veneziano's formula with the introduction by Virasoro of a different representation of the four-particle amplitude. The Virasoro amplitude was then generalized by Shapiro to the case of N particles using the idea of the electrostatic 'analogue model' of Fairlie and Nielsen, as illustrated in the Chapters by Shapiro and Fairlie (on the analogue model see also Section 4.3.2 below). This line of research resulted in the alternative dual model known as the Shapiro–Virasoro model that was later understood to describe the scattering of closed strings. In fact, the generalized Veneziano model (interpreted as a theory of open relativistic strings) and the Shapiro–Virasoro model (interpreted as a theory of closed strings) were themselves parts of the same theory (as found by considering loop corrections).[4]

The above general 'guidelines' provided the research framework in which the dual theory of hadrons was initially investigated and extended. At this point, one could be tempted to interpret the first phase of early string theory in the sense of what philosophers of science call 'normal science activity', after the terminology and scheme introduced by Thomas Kuhn in his 1962 book *The Structure of Scientific Revolutions*: that is, a problem solving activity under a well-established paradigm. The above framework, however, was far from being fixed: the first generalizing steps immediately gave rise, in their turn, to other generalizations and different issues as well as to the introduction of new ideas, methods and formalisms. The history of dual models offers plenty of examples in this sense. Here, we focus on a choice of illustrative cases and analyze some of the most significant analogies, conjectures and discoveries that played an important role in the construction of early string theory.

---

[4] As discussed, for example, by Di Vecchia in his review [DiV08], Section 8.

4.3.2 Analogies

As well as generalizations, analogies were also extensively used in the development of dual theory from its very beginning. They were generally inspired by the properties and progress of the mathematical formalism, such as the following (limiting the examples to those directly relevant to the cases examined below): the Koba–Nielsen integral representation of dual amplitudes, suggesting their relation with two-dimensional surfaces and their conformal invariance; the operator formalism (introduced as a way of exhibiting the factorization properties of multiparticle amplitudes and thus simplifying the study of the spectrum of states), suggesting the string analogy; and the so-called Virasoro conditions (involving the Virasoro generators of conformal coordinate transformations), suggesting the connection with gauge conditions. A significant part of analogical reasoning, on the other hand, was based also on pictorial components, as in the case of the (planar and nonplanar) duality diagrams (discussed, in particular, in the Chapter by Freund), as well as on conceptual similarities. In this last respect, let us emphasize the leading and continuous influence exercised by quantum field theory.

The role of analogical reasoning in extending scientific knowledge, providing explanations and generating new predictions, is a traditional issue in the philosophy of science. In particular, analogy is discussed in connection with scientific models and issues such as the relation between a theory and its models or between a model and the portion of the physical world it is intended to represent. With respect to models, analogy is first of all the means by which one constructs (what philosophers call) 'analogical models', i.e. representations based on similarity of properties, structures or functions. In the most interesting cases, analogies transform into real identifications: what was initially taken to be merely an analogy and used accordingly is later understood as indicating an underlying essential aspect of the theory. In other words, the analogy provides an 'interpretation' and thus plays a decisive role in the transformation process from initial incomplete descriptions (the 'models') into a full-fledged theory. This is exactly what happened in the case of the 'string analogy', that is, the analogy leading to the crucial conjecture that the underlying structure of the Dual Resonance Model was that of a quantum-relativistic string. In fact, early string theory offers relevant illustrations both of purely analogical models and of deeper analogies. Here, we focus on three especially representative examples of the role of analogy in the theory's construction: the (emblematically named) 'analogue model' proposed by Fairlie and Nielsen in 1970, and the mentioned string and gauge analogies.

The analogue model. This model, the idea and motivations of which are described in the Chapters by Fairlie and Nielsen, is a paradigmatic instance of the concurrence of formal, figurative and conceptual elements in analogical reasoning. The analogy on which it is based is with two-dimensional electrostatics. The dual amplitude, in its Koba–Nielsen integral form, is described by means of the picture of external currents (the analogues, in the model, of the momenta of the external particles) fed into a conducting disc. The amplitude integrand is then interpreted and calculated in terms of the heat generated by the currents inside the disc, with the important property that the result does not depend on the shape of the conducting surface, in agreement with conformal invariance. The use of this electrostatic representation was purely analogical but it led to important developments, especially by suggesting appropriate mathematical techniques for loop calculations. As underlined by Di Vecchia and Schwimmer in their historical review [DS08], it represented 'the first appearance of the two-dimensional world-sheet in a mathematical role rather than just as a picture in the duality diagram'.

Although this is a typical example of a purely analogical model (in the philosophical sense), its original motivations were deeper. For Nielsen, in particular, the electric analogue was connected with his 1969 view of hadrons as 'threads' or 'strings', the propagation of which was described by two-dimensional surfaces or 'fishnet diagrams' (a fishnet diagram being approximated, in the analogy, by a planar homogeneous conductor). This view resulted from his idea, motivated by the search for a physical interpretation of the N-particle Veneziano amplitude, that strong interactions should be treated by very high-order Feynman diagrams (given the strength of the coupling constants) in field theory and his consequent attempt to visualize such diagrams with many lines.

The string analogy. In the course of 1969, Nambu, Nielsen and Susskind all arrived at the conjecture that the dynamics of the Dual Resonance Model could be represented by that of an oscillating string, though each in an independent way as underlined in their contributions to the Volume. A feature that undoubtedly bore great responsibility for the influence of analogical reasoning in this interpretation of dual theory was the similarity that could be established between the DRM spectrum and that of a vibrating string. The analogy, based on the harmonic oscillator, was clear: on the one hand, the DRM states had been described in terms of an infinite number of creation and annihilation operators of the harmonic oscillator; on the other hand, a vibrating string could also be described by harmonic oscillators whose frequencies, i.e. harmonics, are multiples of a fundamental frequency.

The analogy with the string was, indeed, very deep and came to reveal the very nature of the theory. In this respect, it is worth emphasizing that the string conjecture originated in an attempt to arrive at a deeper understanding of the physics described by dual amplitudes. In other words, although the dominant framework was that of the S-matrix theory based on the observable scattering amplitudes, a year after the formulation of the Veneziano amplitude some people were already searching for a physical interpretation of the dual amplitudes in terms of an underlying dynamics and an appropriate Lagrangian. This is just one of the many examples in the history of early string theory that speak against the appropriateness of interpreting a theory's construction according to rigid schemes.

Although the breakthrough came in 1969, the process leading to the full acceptance of the string interpretation took some time. The correct string action – formulated in terms of the area of the surface swept out by a one-dimensional extended object moving in spacetime (the string world-sheet), in analogy with the formulation of the action of a point particle in terms of the length of its trajectory – was proposed by Nambu in 1970 and then by Goto. But this interpretation was only applied effectively after the quantization of the string action obtained by Goddard, Goldstone, Rebbi and Thorn (GGRT) in 1973. With this result it became possible to derive, in a clear and unified way, all that had previously been discovered regarding the DRM spectrum by proceeding along various paths and according to a bottom-up approach. It is worth stressing the following methodological point: the above process shows clearly that neither the framework of the S-matrix (implemented with DHS duality), nor the string Lagrangian framework, if taken separately, could work as an appropriate dominating paradigm for an accurate reconstruction of the evolution of early string theory in its first, flourishing phase.

The gauge analogies. The analogies with gauge field theory were initially investigated and used in the attempt to overcome a problematic feature in the dual theory's spectrum of states, namely, the presence of unphysical negative-norm states, that were called 'ghosts' at the time. The ghost elimination programme was implemented on the basis of an analogy suggested by the quantization procedure of electrodynamics, where the unphysical negative-norm states were removed by means of a condition (the 'Fermi condition') following from gauge invariance of the theory, as described in full detail by Di Vecchia in Chapter 11.

This route to ghost elimination in turn generated new developments and also new problems. The analogue, in the DRM, of the Fermi condition for quantum electrodynamics was found to be given by the so-called Virasoro conditions, providing the necessary infinite number of gauge conditions. These were associated with the infinite-dimensional symmetry corresponding to the conformal transformations of the two-dimensional world-sheet, that is, the complex plane of the Koba–Nielsen variables. However, for the Virasoro conditions to hold, the intercept of the leading Regge trajectory had to be taken equal to unity ($\alpha_0 = 1$). This implied, in particular, that the lowest state of the DRM spectrum had negative mass squared – that is, it was an unphysical particle called 'tachyon'. In addition to this problem (only solved by Gliozzi, Scherk and Olive in 1976), there was another puzzling aspect in the spectrum at this intercept value:[5] the presence of massless particles, not observed to occur amongst the hadrons (a massless spin one particle in open string theory; a massless spin two particle in closed string theory).

However, what initially appeared to be a drawback turned out to be a decisive factor in the theory's progress, as already mentioned in Section 4.2. The presence of massless particles with spin in the spectrum suggested that these particles could be identified with gauge bosons and gravitons, respectively. This is underlined by Yoneya in particular, who devotes special attention to the issue of the relation between dual models and quantum field theory in his Chapter.

---

[5] The other condition that had to be assumed for the elimination of ghosts in the Dual Resonance Model, namely, that the number of spacetime dimensions d had to be equal to 26, is discussed later on in Section 4.3.3.

In fact, the correspondence with gauge field theories – a correspondence in which the massless particles played the role of gauge bosons and the gauge symmetry proper of string theory was given by reparameterization invariance of the string world-sheet (implying the conformal invariance of the amplitudes) – revealed itself as much more than a simple analogy. That the analogy had a deeper meaning was indeed suggested by the connection that could be established between dual string theory considered in the low energy limit and quantum field theory. In the pioneering work by Scherk in 1971, the Dual Resonance Model was studied in the limit of vanishing slope α' of the Regge trajectories, which is equivalent to the low energy limit (in string terminology this corresponds to the limit of infinite string tension; see the Introduction to Part VI). Then Scherk and Neveu could show, in 1972, how the massless spin one states interacted in agreement with gauge theory in that limit. Thus, dual models, which were 'originally very close to the S-matrix approach', had gone 'closer and closer towards field theory', to quote from the review article on dual models and strings written (in the academic year 1973–1974) by Scherk [Sch75] (his crucial scientific contributions to early string theory are illustrated in detail by Schwarz in Chapter 41). It is also worth recalling how, at the same time, in his introduction to the 1974 *Physics Reports* reprints volume on *Dual Theory* [Jac74], Fubini pointed out that the very striking analogy with conventional field theory led one to think about a 'strong photon' and a 'strong graviton' in the framework of 'strong gauge theories'.

The above remarks by Scherk and Fubini were made just before the change of perspective according to which string theory was viewed no longer as a description of hadronic physics but 'as a quantum theory of gravity unified with the other forces' (see Schwarz, Chapter 3). In fact, as recalled by Schwarz and Yoneya in their Chapters, the decisive step toward this 'revolution' in the theory's interpretation was precisely the idea of extending the earlier analysis of Scherk and Neveu for the case of open strings to the case of closed strings: this led to exploration of whether the interactions of the massless spin two particle in string theory, considered in the low energy limit, agreed with those of the graviton in general relativity. The result was quite remarkable: both dual string theories (open and closed) could be viewed as short-distance modifications of their field theory analogues (Yang–Mills theory and general relativity, respectively), thus opening the possibility of interpreting string theory as a unified theory of all four fundamental interactions.

4.3.3 Discoveries, alternative ways and convergent results

As the above developments already illustrate, the history of early string theory provides many examples of how decisive conjectures or discoveries originated and of how important results were obtained and corroborated. From a philosophical point of view, it thus offers novel data for discussing traditional issues, such as, in particular:

- the nature of scientific discovery and the controversial distinction between 'discovery' and 'justification' as two separate moments of scientific activity, and

- the role and characteristics of the evidential support (empirical as well as extra-empirical) in the construction process of a scientific theory.

As regards the issue of discovery, the detailed reconstructions available of the rational steps leading to many of the impressive ideas and conjectures characterizing the development of dual string theory undoubtedly speak in favour of some 'rationality in scientific discovery'. However bold and full of unusual physical intuition as some of the principal ideas may have been, they did not come out of nowhere or emerge in a purely irrational way. A clear example is offered in the Chapters by Susskind, Nielsen and Nambu, where they retrace their own paths to the conjecture that the underlying dynamics of dual theory was that of a string. Moreover, the very fact that they arrived at the same conjecture by proceeding in independent ways can be used as a further argument for the rationality of their discovery.

The same can be said for the case of the 'big conceptual leap' leading to a new interpretation of string theory as a theory of all fundamental interactions. This was a true change of paradigm, in that it compelled physicists to view the scope and the domain of the theory in a radically different way. It represented a discontinuity also from a sociological point of view, since high energy physicists and general relativists formed two separate communities at the time. Nevertheless, this paradigm change was the result of theorists working through independent theoretical processes. These steps are described precisely in the Chapters by Schwarz and Yoneya.

As well as being fertile ground for discussions of issues of discovery, this case also provides an illustrative example of a form of evidential support similar to that provided by (what philosophers call) an 'inference to the best explanation'. Here, the inference is more or less the following: interpreting string theory as a unified theory including gravity, the assumption that the fundamental physical entities are strings implies, and thus explains the existence of gravity, which is an empirical fact (see Schwarz, Section 3.5). It is worth noting how this kind of inference, not rare in the history of physics, is very similar to that used by Dirac in support of his theory of magnetic monopoles. In that case, the experimental fact was that of the quantization of electric charge, to which the theory of monopoles provided the unique theoretical explanation (at the time), and the hypothesis in need of support was that of the existence of unobserved magnetic charges [Dir48].

Another significant 'discovery case', illustrative of both the rationale leading to apparently bold guesses and the kind of evidential support motivating a theory's progress, is that of the so-called critical dimension: that is, the discovery that consistency conditions of the Dual Resonance Model required the value $d = 26$ for the spacetime dimension (reducing to $d = 10$ in the case of the Ramond–Neveu–Schwarz model).

The critical value $d = 26$ was originally obtained in two independent ways. The first one was by Lovelace who, in a work published in 1971, addressed a problematic singularity arising in the construction of the nonplanar one-loop amplitude in the framework of the unitarization programme discussed in Section 4.3.2. As Lovelace describes in his Chapter, the problem posed by the singularity was solved by turning it into a pole, that is, by interpreting it as being due to the propagation of a new intermediate particle state. This he conjectured to be the Pomeron – the particle that was later understood as the graviton.[6] This solution was arrived at through adjusting the theoretical description by considering the possibility that the spacetime dimension d might be different from 4 and treating it as a free parameter (together with leaving arbitrary the number d' of degrees of freedom circulating in the loop): the result was that the singularity became a pole only for $d = 26$ (and $d' = 24$).

The same result for the critical dimension issued through another route: namely, from the examination of the DRM physical spectrum of states in the context of the ghost elimination programme (discussed above in this Section). As described in detail in the Chapters by Goddard and Brower, using the infinite set of positive-norm states found by Del Giudice, Di Vecchia and Fubini (the so-called DDF states) it was possible to prove that the DRM had no ghosts if the spacetime dimension d was less than or equal to 26. This result, known as the 'no-ghost theorem', was obtained by Brower, and by Goddard and Thorn, in 1972. For $d = 26$, it was shown that the DDF states could span the entire Hilbert space of physical states, as illustrated in detail in Chapter 11 by Di Vecchia.

In the critical dimension, consistency was thus satisfied – but at the high price of 22 extra space dimensions. This was a rather unrealistic feature, especially for a theory that was intended to describe hadronic physics. Nonetheless, the extra dimensions became gradually accepted, owing to the fact that the critical dimension result received further evidential support from successive theoretical developments. In particular, the 1973 work of Goddard, Goldstone, Rebbi and Thorn provided additionial decisive evidence by obtaining the $d = 26$ condition from the canonical quantization of the string in the light-cone gauge: together with the condition $\alpha_0 = 1$ for the intercept value, it resulted from the requirement of Lorentz invariance in the quantum theory, as described thoroughly by Goddard in his Chapter, Section 20.7 (see also the Introduction to Part IV, Section 19.6.1).

In fact, with hindsight, the critical dimension is a consequence of the conformal symmetry of string theory. More precisely, an 'anomaly', i.e. a violation of symmetry, arises in the implementation of this symmetry in the quantum theory unless $d = 26$. Prior to a definitive understanding of the conformal anomaly, reached by Polykov in his 1981 work, the critical dimension condition could be found only on the basis of 'side effects': that is, as a condition required by unitarity of the theory or by Lorentz invariance in the quantization of the string action.[7] Thus, what had appeared to be a surprising convergence of different calculational procedures to one and the same result, could be seen as a natural consequence of the theory in its fully fledged form. It is worth stressing that the convergence of results obtained in alternative ways and from different starting points provided important evidential support, that encouraged perseverance with the theory notwithstanding the presence of unrealistic features such as extra space dimensions.

---

[6] See the Introduction to Part III, Section 10.2.3 Detailed recollections of how Lovelace arrived at his result can be found in the Chapters by Lovelace and Olive.
[7] For more details on this point, see for example [DS08], Section 5.

4.4 Conclusion

To sum up, many fruitful lessons may be drawn from the developments of dual string theory for philosophical reflections on both scientific practice and theoretical progress. The history of early string theory is, of course, far richer than has been possible to highlight in this Chapter. Nevertheless, the hope is that some points of philosophical interest might come out even from such a partial presentation; let us list them.

- Pluralistic scientific methodology. Early string theory shows, once again, how traditional methodological schemes for describing scientific progress (inductivism, falsificationism, normal science activity intertwined with revolutionary changes of paradigms, competing research programmes, and so on) are both too rigid and too limited to account appropriately for the actual dynamics of a theory building process, at least if taken separately.
- Models and theories. The development of string theory from the original dual models offers a valuable example for the current philosophical discussion on how to characterize a scientific model, a scientific theory, and the relation between models and theories – in particular, for the so-called semantic conception of scientific theories, according to which a theory is defined in terms of the collection of its models.
- The role and nature of evidential support in scientific progress. The phenomenological origin of dual models, initially developed in the context of the S-matrix ideology and on the basis of the experimental hadronic data available, as well as the falsification of dual string theory as a description of hadronic physics, both show the relevance of experimental evidence in the evolution of early string theory. On the other hand, cases such as those discussed in Section 4.3.3 clearly illustrate the decisive role of extra-empirical evidential support in the theory building process.
- The influence of analogical reasoning in theoretical developments and, in particular, in the discovery process, as described in Section 4.3.2.
- The nature of new ideas, such as that of the string, in the light of their evolution. In the case of duality and its role in string theory, for example, it is important to clear out the different meanings that are attributed to this notion. The historical reconstruction of early string theory allows one to understand how the original DHS duality or 'dual bootstrap', at the core of the dual models, is in fact a consequence of the conformal symmetry of the theory.

Acknowledgements

I am very grateful to Andrea Cappelli, Filippo Colomo and Paolo Di Vecchia for their constant support. I would also like to thank Roberto Casalbuoni, Leonardo Castellani, Kerry McKenzie, Antigone Nounou, Giulio Peruzzi and Christian Wüthrich for their very helpful comments and suggestions.

Note on the references

As regards the philosophical issues mentioned in the Chapter, the reader will find updated introductions and relevant references in the related articles of the Stanford Encyclopedia of Philosophy (at the webpage http://plato.stanford.edu).

References

[Cus90] Cushing, J. T. (1990). *Theory Construction and Selection in Modern Physics: The S-Matrix* (Cambridge University Press, Cambridge).

[Dir48] Dirac, P. M. A. (1948). The theory of magnetic poles, *Phys. Rev.* **47**, 817–830.

[DiV08] Di Vecchia, P. (2008). The birth of string theory, in *String Theory and Fundamental Interactions*, ed. Gasperini, M. and Maharana, J. (Springer, Berlin), Lecture Notes in Physics, Vol. 737, 59–118.

[DS08] Di Vecchia, P. and Schwimmer, A. (2008). The beginning of string theory: a historical sketch, in *String Theory and Fundamental Interactions*, ed. Gasperini, M. and Maharana, J. (Springer, Berlin), Lecture Notes in Physics, Vol. 737, 119–136.

[Jac74] Jacob, M. ed. (1974). *Dual Theory*, Physics Reports Reprints Book Series, Vol. 1 (North Holland, Amsterdam).

[Sch75] Scherk, J. (1975). An introduction to the theory of dual models and strings, *Rev. Mod. Phys.* **47**, 123–163.

EARLY STRING THEORY

Part II
The prehistory: the analytic S-matrix

5
Introduction to Part II

5.1 Introduction

Part II deals with theoretical particle physics in the Sixties and the developments that led to the Veneziano formula for the scattering amplitude of four mesons – the very beginning of string theory. In this Introduction, we provide some background for these developments.

We begin by recalling some aspects of quantum field theory, the basic theory for describing elementary particles and fundamental interactions. Quantum field theory was fully developed during the Fifties and successfully applied to the description of the electromagnetic force, leading to quantum electrodynamics (QED), the theory of photons, electrons and positrons. Remarkably, QED predicted new quantum-relativistic effects, such as the anomalous magnetic moment of electrons and the Lamb shift in atomic spectra, which were confirmed experimentally to very high precision.

A crucial aspect of the theory was the presence of a small parameter, the fine-structure coupling constant $\alpha \sim 1/137$: all quantities could be expanded in power series of α, the so-called perturbative expansion, and the first few terms were precise enough to be compared with the experimental data. This approximation allowed the complexity of field interactions to be disentangled.

At the beginning of the Sixties, quantum field theory methods were also applied to the weak nuclear force (responsible for radioactive decays) and the strong force (gluing protons and neutrons inside nuclei), the names clearly indicating their strength. Weak interactions could be described within perturbative quantum field theory, although with the limitations due to the incompleteness of the Fermi theory (see Appendix A for more details).

On the contrary, the strong interactions could not be described perturbatively: for example, the pion–nucleon coupling constant was of order one. Furthermore, many new strongly interacting particles, the hadrons, were found as short-lived states (resonances) created in scattering experiments. Their spectra could be organized in families of increasing mass M and spin J, obeying $M^2 \propto J$, up to values of $J = 6$.

The hadrons were too many to be considered as elementary, independent quanta appearing in the Lagrangian of strong interactions. Some of their properties were well explained by assuming that they were bound states of more elementary constituents, called quarks, but this description was still unclear. Even considering hadrons as elementary particles, their field theory could not be analyzed owing to the large coupling constants – nonperturbative methods had not been developed yet. Because of these problems, quantum field theory of the Sixties could not cope with hadron physics.

This led many people to search for alternative methods to describe strong interactions. They considered quantities directly related to experiments, that did not necessarily require the field theory description, in particular the scattering matrix (S-matrix). This is the quantum mechanical amplitude whose absolute square value gives the probability of a certain scattering process, and thus the cross-section.

The S-matrix should obey some general conditions coming from the probabilistic interpretation (unitarity), the Lorentz invariance, and the symmetries under charge conjugation, parity and time-reversal transformations (crossing symmetry). Furthermore, relativistic causality implies that the S-matrix is an analytic function of the kinematic variables (analyticity). Besides these rather general requirements, other ideas were introduced to describe the features of hadrons. One main input was Regge scattering theory, describing the behaviour of the S-matrix at high energy in terms of the exchange of the so-called Regge poles.

The Veneziano formula for the S-matrix of four-meson scattering was the first complete solution satisfying the above conditions (except unitarity) and the additional assumption of Dolen–Horn–Schmid duality. The formula represented the culmination of the efforts to build an S-matrix without field-theoretic input: remarkably, it contained all the properties envisaged thus far and agreed reasonably with experimental data. This result led to great excitement, and to the desire to understand the complete structure behind it. This was the beginning of the Dual Resonance Model, discussed in Part III of this Volume, that was later understood as the quantum theory of the relativistic string (see Part IV).

In this Introduction, we provide some basic vocabulary of quantum field theory and the S-matrix approach. A more detailed discussion of these topics is given in Appendix A.

5.2 Perturbative quantum field theory

In quantum field theory, the interactions are described by local, pointwise couplings between the fields that represent incoming and outgoing particles and other quanta (mediators) that are created and annihilated within the scattering process. The different patterns are depicted by means of Feynman diagrams.

Let us consider the case of QED. The theory is specified by the following Lagrangian, involving the photon field A_μ and the fermionic field ψ describing the electron and positron:

$$\mathcal{L} = \mathcal{L}_0 + \mathcal{L}_{int},$$
$$\mathcal{L}_0 = \bar{\psi}\left(i\gamma^\mu \partial_\mu - m\right)\psi - \frac{1}{4}\left(\partial_\mu A_\nu - \partial_\nu A_\mu\right)^2, \quad (5.1)$$
$$\mathcal{L}_{int} = e A_\mu J^\mu = e A_\mu \bar{\psi}\gamma^\mu \psi.$$

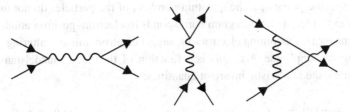

Figure 5.1 Leading Feynman diagrams for electron scattering in QED.

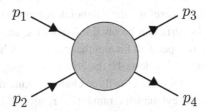

Figure 5.2 The scattering process.

The kinetic terms in $\mathcal{L}_0$ describe the free propagation of elementary particles with spin $J = 1/2$ (the electron and positron) and $J = 1$ (the photon). They are represented by straight and curly segments in Feynman diagrams, respectively (examples are given in Figure 5.1). The cubic term in $\mathcal{L}_{int}$ describes the interaction of the fermion current $J^\mu = \bar\psi \gamma^\mu \psi$ with the photon field: one can think of it as a process of destroying one electron (ψ) at one point and creating another electron ($\bar\psi$) and a photon (A_μ) at the same point. Such a process is represented by a Feynman diagram in which three lines (two straight and one curly) join at one point, the vertex. One power of the electric charge e is associated with each vertex, and its square is related to the fine-structure constant $\alpha = e^2/(4\pi\hbar c)$. Note that the photon is a spin one particle represented by the vector field A_μ, while the spin one-half Dirac fermions are described by spinors with four components, $\psi = \{\psi_i\}$, $\bar\psi = \{\bar\psi_i\}$, $i = 1, \ldots, 4$, corresponding to the degrees of freedom of the electron (particle) and positron (antiparticle) with spin up and spin down, respectively. The Dirac matrices $\gamma^\mu = \{(\gamma^\mu)_{ij}\}$ in the J^μ current (5.1) act in spinor space and make the connection with spacetime vectors.

Entering into more detail, let us consider the scattering of two particles with incoming momenta, $p_1^\mu = (E_1, \mathbf{p}_1)$, $p_2^\mu = (E_2, \mathbf{p}_2)$, and outgoing momenta $p_3^\mu = (E_3, \mathbf{p}_3)$ and $p_4^\mu = (E_4, \mathbf{p}_4)$ (see Figure 5.2). The scattering is said to be elastic if the outgoing particles are the same as the incoming ones. In general, this is only one of the possible results of interaction; others may involve more than two outgoing particles, due to the relativistic creation of matter, but they will not be considered here for simplicity.

In QED, the scattering to lowest order in the fine-structure constant α is given by the Feynman diagrams drawn in Figure 5.1: they correspond to the three possible ways of combining vertices with the four external fermion lines. Some of these diagrams may be absent for a specific process, if the quantum numbers of the particles do not match at the vertices. For example, the first diagram corresponds to electron–positron annihilation and does not occur for two incoming electrons or, say, in electron–muon scattering.

The scattering amplitude $A(s, t, u)$ is a function of the three Mandelstam variables, s, t, u, which are the relativistic invariant quantities:

$$s = -(p_1 + p_2)^2, \qquad t = -(p_2 - p_3)^2, \qquad u = -(p_2 - p_4)^2. \qquad (5.2)$$

These are related by $s + t + u = \sum_{i=1}^{4} m_i^2$, owing to momentum conservation, $p_1 + p_2 = p_3 + p_4$, and the energy-momentum relations, $p_i^2 = -m_i^2$ (in the notation $p^2 = p^\mu \eta_{\mu\nu} p^\nu$, with $\eta_{\mu\nu} = \text{diag}(-1, 1, 1, 1)$). Therefore, the u dependence can be omitted in the scattering amplitude $A(s, t, u)$. The S-matrix can be obtained from the amplitude by multiplying it by known kinematic and spin dependent factors that are omitted here for simplicity.

The QED processes in Figure 5.1 describe the creation, propagation and annihilation of a photon that mediates the interaction in the three channels named after the corresponding Mandelstam variable. The first Feynman diagram is the creation of a photon in the s-channel: its momentum is $P = p_1 + p_2$, due to conservation of the four-momentum at the vertex, and obeys $-P^2 = s > 0$. Note that the external particles obey the energy-momentum relation $p_i^2 = -m_i^2$ (mass-shell condition), for $i = 1, \ldots, 4$: they are real, or on-shell, particles. On the other hand, the internal photon does not obey the corresponding relation $P^2 = 0$ for a massless particle: it is a virtual, or off-shell, particle. Indeed, quantum mechanics requires summing over all possible intermediate processes, or particle trajectories, of arbitrary energy.

We now discuss the crossing symmetry of scattering amplitudes. A general property of quantum field theory is CPT symmetry: the Lagrangian, for example (5.1), is left invariant by the combined change of each particle with its antiparticle (charge conjugation C) and the space (P) and time (T) coordinate reflections. One can say that antiparticles are like particles going backwards in time (and space). CPT symmetry implies that the scattering amplitudes are related to one another by permuting pairs of particles, one incoming and one outgoing, and by changing the corresponding arrows, i.e. the signs of their momenta. The amplitude in Figure 5.2 describes both the original process, $(p_1, p_2) \to (p_3, p_4)$, with positive energies, $E_i > 0, i = 1, \ldots, 4$, as well as the crossed ones $(p_1, -p_3) \to (-p_2, p_4)$, $(p_1, -p_4) \to (p_3, -p_2)$, where the negative energies correspond to antiparticles. In the original process, the conditions $E_i > 0$ imply the range, $s > 4m^2, t < 0, u < 0$ for the Mandelstam variables (where we restrict our discussion to the case of equal masses, $m_i = m$, for simplicity). In the crossed amplitudes, these variables are permuted among themselves and their range changes accordingly: therefore, the crossed processes are described by the same scattering amplitude in different domains of the (s, t, u) variables.

In conclusion, crossing symmetry can be used to connect amplitudes of different processes and to constrain the form of those amplitudes that are mapped into themselves. For example, the amplitude for electron–positron scattering in QED is $s \leftrightarrow t$ symmetric because it is described by the sum of the first two Feynman diagrams in Figure 5.1, which

Figure 5.3 Perturbative correction of photon self-energy in QED.

map one into the other under the interchange of s and t. In the following, we shall mostly discuss $s \leftrightarrow t$ crossing symmetry.

Each of the three graphs in Figure 5.1 yields a term in the amplitude with a characteristic pole singularity when the intermediate virtual particle, the photon, becomes real; for example, the first graph has a pole at $s = 0$. In the more general case of the propagation in the s-channel of a particle of spin J and mass M, we get the following behaviour of the scattering amplitude for $s \sim M^2$ and for large values of $-t$:

$$A(s, t) \sim -\frac{g^2 (-t)^J}{s - M^2} \qquad (s\text{-channel}), \qquad (5.3)$$

where g^2 is the square of the coupling constant ($g^2 \sim \alpha$ and $M = 0$ in QED). In general, the complete amplitude is obtained by summing terms of the form (5.3) for each channel and for each exchanged particle, leading to a sum of pole singularities in the s and t variables.

We now consider the next terms in the perturbative expansion of QED: they are of order α^2 and are represented by Feynman diagrams involving four vertices. For example, the loop correction to the photon propagation is given in Figure 5.3: it describes the virtual process of creation of an electron–positron pair by the photon, and its successive annihilation. Since the momenta of the pair are unconstrained by the external momenta, they are summed over up to arbitrary large energies leading to ill-defined infinite expressions. During the Fifties, the theory of renormalization was developed for regularizing these quantities and obtaining physical (finite) results for the photon interaction and propagation. Renormalization is a well-understood aspect of quantum field theory with several physical implications, but it is not directly related to the present discussion of early hadronic physics and will not be addressed here.

The loop amplitude presents the following features.

(i) It possesses a characteristic singularity associated with the possibility of the virtual pair becoming real, i.e. being produced. This occurs for $s > 4m^2$, that is for energies above the rest mass of the pair. It leads to a singular line in the complex s plane, called a cut singularity, that starts at $s = 4m^2$ and continues on the real positive axis of s up to infinity.
(ii) The amplitude is better understood by extending its variables to complex values. One can prove under general assumptions, such as relativistic causality, that $A(s, t)$ is an analytic function of s and t, apart from poles and cut singularities. It follows that properties of complex analysis can be used, such as the analytic continuation of

functions from one region of parameters to another and the reconstruction of functions from knowledge of their singularities. In general, the simple poles are associated with single-particle exchanges (5.3) and cut singularities correspond to pair productions as in the loop corrections (Figure 5.3). Direct and crossed channels of scattering pertain to different ranges of Mandelstam variables and can be obtained by analytic continuation of $A(s, t)$ from one region to another of the s and t complex planes.

5.3 The hadron spectrum

High energy physics of the Sixties was characterized by a wealth of experimental results: several accelerators were built in large-scale facilities in both the United States and Soviet Union, and in Western Europe at the CERN laboratory in Geneva. Since nuclear physics had been so important for military applications, subnuclear fundamental studies were highly valued by funding agencies and governments. The field was dubbed 'Big Science' to express the fact that its economic size and organization went beyond anything conceived in physics so far. These developments are well accounted in the volume *The Rise of the Standard Model* [HBRD97].

Between the end of the Sixties and the beginning of the Seventies, more than 20 meson (integer spin) and 50 baryon (half-integer spin) particles were found in accelerator experiments. They formed a rich phenomenology that took some time to be organized: as outlined above, field theory was used as the basic language, but its perturbative expansions did not fit the strong interaction.

Let us introduce some elements of hadron classification. Hadrons were characterized by their spacetime features, such as the spin J, and even-odd properties under parity P and charge conjugation C, in the notation J^{PC}. For example, the pions $\pi^0, \pi^\pm$, with charges $0, \pm 1$, are pseudoscalar mesons with $J^{PC} = 0^{-+}$. The particles were also characterized by 'internal' quantum numbers that were conserved, or approximately conserved, in the scattering and production processes. Besides the baryon number B, vanishing for mesons, there was the isospin I, introduced earlier in nuclear physics in analogy with the spin and described by the $SU(2)$ unitary group; isospin was combined with the strangeness S into the larger $SU(3)$ group by Gell-Mann and Ne'eman in the mid-Sixties. Other quantum numbers, for example 'charm', are known today and are collectively named hadron 'flavours', $I, S, C, \ldots$, corresponding to flavour symmetry groups.

Examples of the classification by the $SU(3)$ symmetry are given in Figures 5.4 and 5.5: these diagrams describe multiplets (representations) of the $SU(3)$ group with dimensions **8** (octet) and **10** (decuplet). While particles with $SU(2)$ spin J should form multiplets of $2J + 1$ components, corresponding to the possible values of the spin projection J_z, $SU(3)$ multiplets fit two-dimensional diagrams in the isospin I_z and hypercharge $Y = B + S$ coordinates. One sees that the diagrams are parts of a regular triangular lattice, i.e. are made by joining triangular tiles; the basic three-dimensional representation **3**, and its charge conjugate **3̄**, correspond to the basic tiles, pointing upward and downward, respectively. In

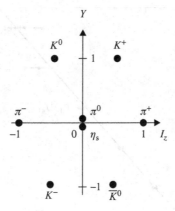

Figure 5.4 Octet of pseudoscalar mesons $J^{PC} = 0^{-+}$: the axes show the isospin I_z and hypercharge Y values, the latter being proportional to the strangeness quantum number S.

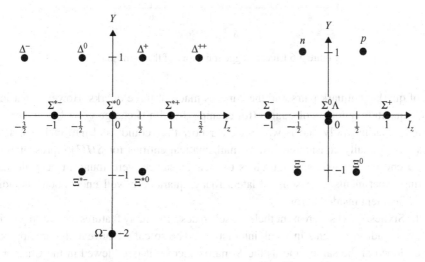

Figure 5.5 Decuplet and octet of baryons with $J^P = \frac{3}{2}^+$ and $\frac{1}{2}^+$, respectively.

mathematical terms, the octet and decuplet are obtained respectively by the product of two and three basic representations, as follows:

$$3 \times \bar{3} = 1 + 8,$$
$$3 \times 3 \times 3 = 1 + 8 + 8 + 10. \quad (5.4)$$

The whole hadron spectrum could be described by putting mesons in octets and baryons in octets and decuplets: no other representation was needed. This remarkable fact led to the introduction of particles associated with the **3** and **3̄** representations, called 'quarks' by Gell-Mann (and 'aces' by Zweig), which could be the hadron constituents: the mesons

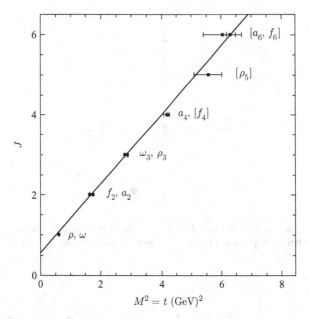

Figure 5.6 Linear Regge trajectory of the ρ-meson.

made of quark–antiquark pairs and the baryons made of three quarks. However, particles with the quantum numbers of quarks (for example fractional charge $Q = \pm 1/3$) were not observed experimentally and it was not clear how they could bind into hadrons. Thus, quarks were initially considered just as mathematical entities for $SU(3)$ representations, until the end of the Sixties, when clues to their existence were found in deep inelastic scattering experiments, as discussed later. Today, quarks are well understood, including their confinement inside hadrons.

In the Sixties, quarks were nonetheless used to describe many features of hadron physics, such as the hadron currents in weak interactions. The so-called current algebra approach was developed at the same time as the S-matrix theory: it is reviewed in the Chapter by Ademollo and further described in Appendix A.

Another relevant property of hadrons was the regularity of the mass spectrum of the resonances, the short-lived particles that were observed as peaks in the scattering amplitudes, as discussed in Section 5.4.1. Resonances with the same quantum numbers could be arranged in families, involving values of spin J and mass M obeying the linear law:

$$J = \alpha(s) = \alpha_0 + \alpha' s, \qquad s = M^2. \tag{5.5}$$

Such families of particles were called 'linear Regge trajectories'. An example is shown in Figure 5.6 for the resonances with the quantum numbers of the ρ-meson: even parity, isospin $I = 1$ and strangeness $S = 0$; the spin values are $J = 1, 2, \ldots, 6$.

An interesting feature was that the slope α' of Regge trajectories was the same for all meson and baryon families, $\alpha' \sim 0.86$ $(GeV)^{-2}$ – only the intercept α_0 varied ($\alpha_0 \sim 0.55$ for the ρ trajectory). That was a first hint of string dynamics, namely that quarks are bound together by an elastic string with universal tension $T \sim 1/\alpha'$, as described in Part IV.

5.4 S-matrix theory

The S-matrix approach grew out of the problems of perturbative quantum field theory, namely its inability to describe hadrons either as fundamental or as composite particles, and its inability to cope with their large coupling constants. The rather radical proposal was to forget field theories and Lagrangians and to consider observable quantities only – the scattering amplitudes. However, no precise method was given for constructing these amplitudes apart from the bootstrap principle, which was rather vague originally. Additional conditions came from Regge theory. In Appendix A, further details on the S-matrix theory are provided; for more complete discussions, see [Chew66], [ELOP66] and [DFFR73].

The bootstrap principle was introduced as a self-consistency condition. In the S-matrix approach, trial amplitudes were built by assuming certain particle exchanges, consistent with the selection rules of the process, leading to pole singularities in given channels (cf. Eq. (5.3)): the resulting expressions were then checked for unitarity and crossing symmetry. The bootstrap hypothesis required that the same set of hadrons should occur in external and intermediate states. Starting from a small set of hadrons as external particles, the amplitude was written so as to include intermediate particles of the same set. Next, the consistency with unitarity and crossing symmetry required the inclusion of other intermediate particles. One then iterated the process until a closed self-consistent set of hadrons was found.

This approach obeyed the principle of 'nuclear democracy', introduced by Chew, that no hadron was more fundamental than the others. This was sometimes formulated by saying that each hadron consisted of all the other hadrons and was not a bound state of some more fundamental objects, or that all hadrons lie on Regge trajectories (see Section 5.4.2). The idea that hadrons were composed of quarks was not considered useful for describing the dynamics.

As stressed by Gell-Mann in his contribution to this Volume, the S-matrix approach was not in contradiction with the field theory methods, as they both yield S-matrices satisfying the same conditions. However, quantum field theory provided clear rules for computing the S-matrix through Feynman diagrams, while the S-matrix approach only provided a set of self-consistency conditions and some basic inputs to build a solution. Indeed, the bootstrap principle was too general and a practical implementation only came with the introduction of the Dolen–Horn–Schmid duality to be discussed below.

The inputs for scattering amplitudes of hadronic processes were obtained from the abundant experimental data of the time. These amplitudes presented two standard behaviours: at low energy, they were reproduced by a sum of resonances described by Breit–Wigner amplitudes, while at high energy they were analyzed in terms of Regge poles. Let us now describe these two behaviours in some detail.

5.4.1 Resonances

We already mentioned that a pole in the scattering amplitude $A(s, t)$ manifests itself as a bump in the cross-section, $\sigma \sim |A(s, t)|^2$, when the energy reaches the value for producing a resonance with mass M_R, say in the s-channel for $s \sim M_R^2$. This behaviour is described by the following Breit–Wigner amplitude in the neighbourhood of $s = M_R^2$ (consider here $J = 0$ for simplicity):

$$A(s, t) \sim \frac{M_R A_{if}}{M_R^2 - s - i M_R \Gamma} \qquad \left(s \sim M_R^2\right). \tag{5.6}$$

In this equation, Γ is the width of the bump ($\Gamma \ll M_R$), inversely proportional to the resonance lifetime, $\tau \sim 1/\Gamma$, and A_{if} is the 'residue' at the resonance pole, whose indices i and f stand for the initial particles that produce the resonance and the final states in which the resonance decays. As discussed in Appendix A, unitarity relations imply that the residue is proportional to the probability of creating such an intermediate particle and should thus be positive. The Breit–Wigner behaviour (5.6) follows from the general properties of unitarity and analyticity of the scattering amplitude, and it is also reproduced by lowest-order perturbative field theory (cf. Eq. (5.3)), apart from the width Γ that is introduced by the loop corrections.

The hadronic scattering amplitudes measured at low energy (small s) were well described by a sum of Breit–Wigner amplitudes, each one corresponding to one intermediate hadron with proper quantum numbers to be created in the s-channel. As the energy increased, more and more hadrons were created.

Let us now discuss the high energy behaviour of $A(s, t)$, corresponding to large values of $s \gg 1$ GeV^2 (the hadronic scale). The t variable is related to the scattering angle θ in the centre-of-mass reference frame by:

$$t = -s \sin^2 \frac{\theta}{2}. \tag{5.7}$$

We can consider two regimes as $s \to \infty$:

(i) fixed and negative t, of the order of 1 GeV^2; this corresponds to forward scattering, $\theta \to 0$, where most of the particles are produced;
(ii) fixed θ, i.e. $|t| \sim s \to \infty$; in this regime, the cross-section is very small.

The second regime was considered much later, starting with the SLAC deep inelastic scattering experiment in 1968: it was found that strong interactions become weaker and can be described by perturbative quantum field theory. This fact opened the way to the theory of quantum chromodynamics, fully developed in the mid-Seventies [HBRD97] (see Part VI). Before the SLAC result, experiments were done in the first asymptotic regime of fixed $|t|$, where the behaviour of the amplitude could not be accounted for by field theory.

On the other hand, Regge theory did provide a framework to explain the scattering data, as well as the patterns of resonances with linearly increasing mass squared (5.5).

5.4.2 Regge poles

In his approach, Regge considered the general expansion of the scattering amplitudes with respect to the angular momentum ℓ of the output matter waves: a given ℓ value underpins a specific θ dependence of the amplitude, the two variables being related by a generalized Fourier transformation. The scattering amplitude $a(\ell, t)$ was obtained by transformation of the original $A(s, t)$, seen as a function of (θ, t) according to (5.7).

The analytic extension of the functions $A(s, t)$ and $a(\ell, t)$ to complex values of their arguments was rather natural: in particular, $a(\ell, t)$ may have pole and cut singularities in the complex ℓ plane. The contribution of a simple pole (Regge pole) at a point $\ell = \alpha(t)$ yields the following nice asymptotic behaviour of the original amplitude for $s \gg |t|$ [FR97]:

$$A(s, t) \sim \beta(t) \left(1 \pm e^{-i\pi\alpha(t)}\right) \Gamma(1 - \alpha(t)) \, s^{\alpha(t)}, \tag{5.8}$$

where $\alpha(t)$ is so far an arbitrary function of t called the 'Regge trajectory'. This singularity is assumed to yield the leading asymptotic contribution to $A(s, t)$ (other trajectories can also give subleading contributions as explained later). The function $\beta(t)$ is called the residue and $\Gamma(z)$ is the Euler gamma function (see Appendix B for its properties). The signature factor corresponding to the $\pm$ appearing in Eq. (5.8) applies to the case of even or odd spins lying on a Regge trajectory. Let us now illustrate the main features of this formula and show that it possesses a double match with experimental data.

In the t-channel, i.e. for $t > 0$, the amplitude develops simple poles as a function of t for $\alpha(t) = J = 1, 2, \ldots$, due to the behaviour $\Gamma(z) \sim 1/(z + n)$ for $n = 0, 1, \ldots$. As said, the poles correspond to exchanges of particles with spin J in the t-channel (cf. Eq. (5.3)). These (infinitely many) poles lie on the Regge trajectory $\alpha(t)$ still to be determined; several independent trajectories can also contribute. The empirical mass formula (5.5) (see Figure 5.6) indicated that the Regge trajectories were linear functions of t; they also had a unique value of the slope (except for the so-called Pomeron trajectory which will be discussed later).

Therefore, the Regge formula (5.8) was applied with linear $\alpha(t)$. We remark that the Breit–Wigner expression describes a single resonance, while a Regge pole corresponds to all the resonances that lie on its Regge trajectory.

Once the function $\alpha(t)$ has been fixed from the behaviour for positive t, it can be extrapolated for $t < 0$, i.e. in the physical region of the s-channel, where it yields the following large-s asymptotic behaviour for the scattering amplitude and for the differential cross-section:

$$A(s, t) \sim s^{\alpha_0 - \alpha'|t|}, \qquad \frac{d\sigma}{dt} \sim s^{2\alpha_0 - 2\alpha'|t| - 2}, \qquad s \gg |t|. \tag{5.9}$$

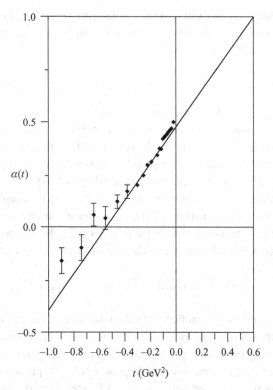

Figure 5.7 Fit of the data of the $\pi^- + p \to \pi^0 + n$ cross-section with the linear Regge trajectory of the ρ-meson extrapolated at $t < 0$.

These results were in very good agreement with the experimental data: for example, the ρ trajectory has the appropriate quantum numbers for being exchanged in the t-channel in the scattering $\pi^- + p \to \pi^0 + n$. Figure 5.7 shows the matching of large-s experimental data of the cross-section with the Regge behaviour (5.9), where the straight line is the extrapolation to $t < 0$ values of the ρ trajectory in Figure 5.6. In summary, the Regge pole behaviour interpolates between the high energy behaviour in the s-channel (small negative values of t) and the low energy behaviour in the t-channel (small positive values of t).

We now discuss the total cross-section, where one sums over all possible final states. Using unitarity relations, this is obtained from the imaginary part of the elastic scattering amplitude in Eq. (5.9), leading to the following behaviour:

$$\sigma_{tot} \sim s^{\alpha_0 - 1}. \tag{5.10}$$

It turned out that all Regge trajectories had intercepts $\alpha_0 < 1$, implying vanishing cross-sections at large energies; in the cases where the quantum numbers of the scatterers allowed the exchanges of more than one Regge trajectory, the one with the highest intercept dominated at large s (the leading Regge trajectory).

Experimental data for $\pi^- - p$ scattering were fitted by the power-law behaviour [FR97]:

$$\sigma_{tot} \sim As^{0.08} + Bs^{-0.45}, \tag{5.11}$$

where A, B are numerical constants. The subleading power was in agreement with the ρ trajectory contribution, but the leading trajectory with $\alpha_0 = \alpha_P = 1.08$ was missing, i.e. no hadron was known to be exchanged in the t-channel for $t > 0$. This leading trajectory was observed in several processes where particles with quantum numbers of the vacuum could be exchanged in the t-channel. Therefore, an additional Regge pole was introduced, called the Pomeron after the physicist I. Y. Pomeranchuk, carrying the quantum numbers of the vacuum and with intercept $\alpha_0 \sim 1$. Its slope was much smaller than that of the other Regge poles, and the particles on its Regge trajectory were yet to be identified.

In conclusion, the Regge behaviour was able to reproduce the asymptotic behaviour of cross-sections with the exchange of hadrons lying on linearly rising Regge trajectories, as observed experimentally. This required particles with arbitrary large J values in each trajectory.

This phenomenological achievement should be contrasted with the description by perturbative quantum field theory, where particle exchanges must be introduced one by one. Moreover, by Eq. (5.3), such individual exchange of particles with spin $J > 1$ in the t-channel would lead to asymptotically rising amplitudes, $A(s,t) \sim s^J$, in contradiction with experiments and general theorems based on unitarity. Therefore the coherent contribution of an infinite number of particles lying on a Regge trajectory was needed to obtain the observed constant or decaying behaviour.

5.4.3 Finite energy sum rules

We have seen that the low energy behaviour of the scattering amplitude is well described by a sum of Breit–Wigner terms (5.6) for exchanges of resonances in the s-channel, while the high energy behaviour is given by the exchanges of Regge poles in the t-channel. The important question was, at that time, whether the two contributions, resonances and Regge poles, had to be summed or not. Comparison with the experimental data of the Sixties suggested a negative answer, namely that the sum over the resonances would automatically generate the Regge behaviour and vice versa. This property was called Dolen–Horn–Schmid (DHS) duality.

Dolen, Horn and Schmid considered the so-called finite energy sum rules (FESR) and found the relation:

$$\int_0^L v^n \, \text{Im} A_{RES}(v,t) dv = \int_0^L v^n \, \text{Im} A_{REGGE}(v,t) dv. \tag{5.12}$$

In this equation, the integration is over the variable $v = (s-u)/4$ and goes up to a maximal energy scale L. The left-hand side of this equation contains the imaginary part of the amplitude with the contribution of the resonances, while the right-hand side contains the

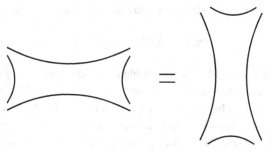

Figure 5.8 DHS duality represented pictorially by double-line duality diagrams.

contribution of the Regge poles (each pole yields a finite contribution to the integral of the imaginary part, see Appendix A).

The equality between the two sides of Eq. (5.12) expresses the fact that the sum of the contributions of the resonances in the s-channel is equal to the sum of the contributions of the Regge poles in the t-channel. A comparison of this relation with experimental data is given in Figure 6.4 of the Chapter by Ademollo. These sum rules imposed conditions between the parameters of the resonances and those of the Regge trajectories for all strong interaction processes, thus providing an explicit realization of the bootstrap idea.

The duality between resonances and Regge poles had important phenomenological consequences, as discussed in the Chapter by Freund. Consider, for instance, the elastic scattering $\pi^+\pi^+ \to \pi^+\pi^+$: it has no resonances in the s-channel because there is no particle with charge 2 that can be exchanged. This means that the left-hand side of Eq. (5.12) vanishes: DHS duality thus implies that the right-hand side should vanish too. This is possible if in the t-channel there are two Regge poles with the same trajectory and with residues having opposite signatures in (5.8): in this case, the two contributions to the imaginary part of the scattering amplitude cancel among themselves and Eq. (5.12) can be verified. This was called the exchange degeneracy for Regge poles. In the previous example, the Regge trajectories are those of the ρ, with spin one, and of the f_0, with spin two, which indeed have opposite signatures. Exchange degeneracy was a successful prediction of the Dolen–Horn–Schmidt duality.

Furthermore, Harari and Rosner introduced the 'duality diagrams', providing a pictorial description of the duality between resonances and Regge poles in which the mesons were represented by double lines describing the quark and the antiquark. The duality diagram for the scattering of four mesons is drawn in Figure 5.8. It can be deformed in two different ways to show either the exchange of meson resonances in the s-channel or the exchange of Regge poles in the t-channel. This illustrates the fact that the two corresponding sums are equal, as expressed by the FESR equation (5.12).

We have seen that the Pomeron Regge trajectory, needed to explain the behaviour of the total cross-section, differs from the others in having no particles lying on it as well as a much smaller slope α'. Freund and Harari conjectured that the Pomeron contribution to

Eq. (5.12) was DHS dual to the background, rather than to the resonances as in the case of the other Regge poles.

Freund and Harari described the Pomeron by using a nonplanar duality diagram that had Pomeron exchange in the t-channel, but no resonance exchange in the s-channel (see Figure 2.7 in the Chapter by Veneziano, and the Introduction to Part III). This explained the difference between the cross-sections involving the resonance ρ, bound state of up and down quarks, and the resonance ϕ, bound state of strange quarks.

In today's language, duality diagrams are nothing other than open string diagrams: the string extends between the quarks and creates a two-dimensional surface in spacetime, the string world-sheet. However, as we shall see, it took some years to understand this point! The nonplanar diagram corresponds to the one-loop nonplanar diagram of open string theory that contains a closed string propagating in-between (see Figure 10.8 in the Introduction to Part III). In conclusion, the mesons and their Regge poles correspond to open string states, while the Pomeron corresponds to closed string states.

5.5 The Veneziano amplitude

The fundamental result by Veneziano was to find a complete, analytic expression for the scattering amplitude of four mesons, $\pi\pi \to \pi\omega$, that explicitly realizes the ideas discussed above, namely:

(i) it implements the DHS duality;
(ii) it has an infinite set of poles in the s- and t-channels, lying on linearly rising Regge trajectories;
(iii) it reproduces the asymptotic Regge behaviour (5.8) for large s and small negative t;
(iv) it is completely crossing symmetric, i.e. invariant under permutations of (s, t, u), as required by the quantum numbers of the chosen meson scattering $\pi\pi \to \pi\omega$.

A first hint of the Veneziano result comes from the duality diagrams in Figure 5.8: they suggest that a single amplitude should contain the resonances in both the s-channel and the t-channel. This can be realized only if the number of resonances exchanged in both channels is infinite, leading in fact to an expression that is symmetric in their exchange. Further insights that inspired the derivation of the amplitude are discussed in the Chapters by Ademollo, Rubinstein and Veneziano.

The celebrated Veneziano formula reads:

$$A(s, t, u) = B\left[1 - \alpha(s), 1 - \alpha(t)\right] + B\left[1 - \alpha(s), 1 - \alpha(u)\right] \\ + B\left[1 - \alpha(t), 1 - \alpha(u)\right], \tag{5.13}$$

$$B\left[1 - \alpha(s), 1 - \alpha(t)\right] = \frac{\Gamma\left[1 - \alpha(s)\right]\Gamma\left[1 - \alpha(t)\right]}{\Gamma\left[2 - \alpha(s) - \alpha(t)\right]}, \tag{5.14}$$

where $\alpha(s) = \alpha_0 + \alpha' s$ is the linear Regge trajectory described earlier and $B[z, w]$ is the Euler beta function, a ratio of gamma functions (see Appendix B for a more detailed analysis of the amplitude).

Let us focus on the first of the three terms, involving the (s, t) variables. The infinite set of poles in the t-channel follows from the term $\Gamma(1 - \alpha_0 - \alpha' t)$, owing to the poles of $\Gamma(z)$ for $z = 0, -1, -2, \ldots$. Moreover, the asymptotic Regge behaviour (5.8) at large s is reproduced thanks to the properties $\Gamma(z)\Gamma(1 - z) = \pi / \sin(\pi z)$ and $\Gamma(z + a)/\Gamma(z) \sim z^a$ for Re $z \gg a > 0$ (see Appendix B). Therefore, the (s, t) term of the amplitude (5.13) realizes the DHS duality by having infinite poles in one channel and Regge behaviour in the other, in agreement with the diagrams of Figure 5.8.

Actually, DHS duality is achieved here by an expression that is completely invariant under the exchange of the s- and t-channels. An amplitude with this property is said to satisfy 'planar duality'. Full crossing symmetry is then obtained by summing the expressions for the three pairs of channels, as in Eq. (5.14). We stress again the difference with field theory, where an (s, t) crossing-symmetric amplitude is obtained by summing the first two Feynman diagrams in Figure 5.1, while here a single term $A(s, t)$ already contains both the poles in the s-channel and those in the t-channel. Another difference with field theory is the presence of an infinite set of poles in both channels.

Although the ingredients were all present in earlier S-matrix studies, the Veneziano formula was remarkable for its simplicity and clarity and it caused great excitement in the research community. The amplitude had properties that were different from those of any perturbative field theory. Therefore, it was very interesting to investigate the underlying structure and to generalize it in several directions, as described in Part III. The Veneziano amplitude, extended to multiparticle scattering, opened up a new research field, that of the Dual Resonance Models (DRM). After intense research activity, its intriguing properties were understood and eventually a new theory was discovered, that of a relativistic string.

Was this theory really an alternative to field theory? As said, it looked very different from the perturbative field theory given by Feynman diagrams. It was later understood (see Part VI) that string theory actually reproduces some specific field theories in the limit of infinite string tension; therefore, it is a generalization of quantum field theory, rather than a different structure. Nowadays, another relation between string theory and field theory has been found: the AdS/CFT correspondence, formulated by Maldacena in the mid-Nineties, establishes a nonperturbative equivalence between a particular string theory in ten dimensions and a gauge field theory in four dimensions [BBS07].

Let us conclude with a final remark about the unitarity of the scattering process described by the Veneziano amplitude. We have seen that the singularities of the amplitude are simple poles occurring for real values of the Mandelstam variables: their vanishing imaginary parts imply infinite lifetimes of the exchanged particles. Comparison with the general Breit–Wigner expression (5.6) shows that this result is in contradiction with unitarity. Moreover, there are no cut singularities, corresponding to the production of pairs of intermediate particles or of multiparticle states, which should be possible in general.

For these reasons the Veneziano amplitude is not unitary. This was not considered a serious problem, because it also occurs in quantum field theory in the contribution of tree Feynman diagrams: loop diagrams should be added to satisfy unitarity relations. Analogously, the DRM amplitudes were considered as the 'tree' approximation (called 'narrow-width approximation') to the full theory, and loop diagrams were also considered that generated the required nonzero widths for resonances and multiparticle cuts. However, a new feature occurred that is not present in field theory: the nonplanar loop diagrams were found to violate unitarity unless the spacetime dimension is $d = 26$. This new aspect of the DRM will be discussed in the Introduction to Part III.

References

[BBS07] Becker, K., Becker, M. and Schwarz, J. H. (2007), *String Theory and M-Theory: A Modern Introduction* (Cambridge University Press, Cambridge).

[Chew66] Chew, G. F. (1966). *The Analytic S Matrix: a Basis for Nuclear Democracy* (W. A. Benjamin, New York).

[Col77] Collins, P. D. B. (1977). *An Introduction to Regge Theory and High Energy Physics* (Cambridge University Press, Cambridge).

[DFFR73] De Alfaro, V., Fubini, S., Furlan, G. and Rossetti, C. (1973). *Currents in Hadron Physics* (North Holland, Amsterdam).

[ELOP66] Eden, R. J., Landshoff, P. V., Olive, D. I. and Polkinghorne, J. C. (1966). *The Analytic S Matrix* (Cambridge University Press, Cambridge).

[FR97] Forshaw, J. R. and Ross, D. A. (1997). *Quantum Chromodynamics and the Pomeron* (Cambridge University Press, Cambridge).

[HBRD97] Hoddeson, L., Brown, L., Riordan, M. and Dresden, M. (1997). *The Rise of the Standard Model: Particle Physics in the 1960s and 1970s* (Cambridge University Press, Cambridge).

6
Particle theory in the Sixties: from current algebra to the Veneziano amplitude

MARCO ADEMOLLO

6.1 Introduction

During the Fifties it became clear that quantum electrodynamics (QED), constructed by applying the quantization rules to classical electrodynamics, was in very good agreement with experimental results. In particular, QED accounted for the deviations observed in experiments with respect to the Dirac theory of electrons and positrons. This was considered a major success of field theory.

In the following years, many efforts were made in the attempt to build a suitable theory for the description of the weak and strong interactions of elementary particles. These efforts culminated in the formulation of the electro-weak theory of Weinberg and Salam in 1967, and quantum chromodynamics in 1973. These are the two fundamental blocks of the Standard Model.

In the Sixties, when the theoretical community was still seeking a satisfying theory of strong interactions, several alternative approaches were explored, to overcome the encountered difficulties. In particular, one of these, the S-matrix, complemented with the bootstrap hypothesis and Dolen–Horn–Schmid duality, was to give rise to dual models and string theory.

The aim of the present Chapter is to recall some of the relevant results which originated in this context, in the years between 1961 and 1968, and which led to the Veneziano formula. In the next Section I will sketch the $SU(3)$ symmetry and the quark model. In Section 6.3 I will recall current algebra, with a glance at its exploitation by means of sum rules. After a brief summary of strong interactions in Section 6.4, in Section 6.5 I will summarize the method of the superconvergence sum rules for strong interactions, a method that was fruitfully extended to the so-called 'finite energy sum rules'. In Section 6.6 I will review my personal work in collaboration with H. R. Rubinstein, G. Veneziano and M. A. Virasoro, which contributed to the concept of duality, explained briefly in Section 6.7. Finally in Section 6.8 the famous Veneziano formula will be presented.

The Birth of String Theory, ed. Andrea Cappelli, Elena Castellani, Filippo Colomo and Paolo Di Vecchia.
Published by Cambridge University Press. © Cambridge University Press 2012.

6.2 SU(3) symmetry and the quark model

Let me start in 1961, which was very important for at least three reasons:

(i) the first attempt by Glashow [Gla61] to unify the weak and electromagnetic interactions, with the introduction of the weak isospin;
(ii) the introduction by Ne'eman [Nee61] and Gell-Mann [Gel61] of $SU(3)$ as an approximate symmetry of strong interactions; this model gave a simple classification scheme of the known hadrons in terms of octets (or nonets) and decuplets;
(iii) the proposal by Gell-Mann [Gel62] of the charge algebra.

In this important paper Gell-Mann proposed that the hadronic electromagnetic current and the vector currents of the weak interactions belong to the same $SU(3)$ octet V_i^μ ($i = 1, \ldots, 8$):

$$\text{electromagnetic current} \quad V_{em}^\mu = V_3^\mu + \frac{1}{\sqrt{3}} V_8^\mu,$$

$$\text{weak current } \Delta S = 0, \ \Delta Q = +1 \quad V_{\Delta S=0}^\mu = V_1^\mu + i V_2^\mu, \quad (6.1)$$

$$\text{weak current } \Delta S = \Delta Q = +1 \quad V_{\Delta S=1}^\mu = V_4^\mu + i V_5^\mu.$$

Similarly, the axial vector currents A_i^μ form another octet.

The weak interactions, responsible for example for the baryonic β-decay, were given by the Feynman–Gell-Mann $V - A$ theory of 1958 [FG58], in the current by current form $\mathcal{L}_w = G J^\mu l_\mu$, where J^μ and l_μ are the hadronic and the leptonic currents, respectively. In the framework of $SU(3)$, the hadronic current was given by Cabibbo [Cab63] in 1963 by his famous formula:

$$J^\mu = \cos\theta [(V_1^\mu + i V_2^\mu) - (A_1^\mu + i A_2^\mu)] + \sin\theta [(V_4^\mu + i V_5^\mu) - (A_4^\mu + i A_5^\mu)]. \quad (6.2)$$

The absence of candidates for the fundamental **3** representation of $SU(3)$ suggested to Gell-Mann [Gel64a] and Zweig [Zwe64] in 1964 the hypothesis of the 'quarks'. According to this model, all the known particles are bound states of three quarks, named u (for 'up'), d (for 'down') and s (for 'strange') – they are the 'light' quarks of today. They have fractional charges, namely electric charges $Q = +\frac{2}{3}, -\frac{1}{3}, -\frac{1}{3}$, respectively, in units of the proton charge, and baryonic charge $B = +\frac{1}{3}$. In terms of quarks, we have:

$$\text{mesons } q\bar{q} \quad (\text{e.g. } \pi^+ = u\bar{d}, \quad K^- = s\bar{d}),$$
$$\text{baryons } qqq \quad (\text{e.g. } p = uud, \quad \Lambda = [ud]s). \quad (6.3)$$

The quark model also supplies a field theoretic expression for the currents:

$$V_i^\mu(x) = \bar{q}(x) \gamma^\mu \lambda_i q(x), \quad A_i^\mu(x) = \bar{q}(x) \gamma^\mu \gamma_5 \lambda_i q(x), \quad (6.4)$$

where $q(x)$ is the quark field, with spinor and $SU(3)$ triplet indices omitted, γ^μ are the Dirac matrices and λ_i are the eight 3×3 Gell-Mann matrices of $SU(3)$ generators.

6.3 Current algebra and sum rules

In his paper [Gel62], Gell-Mann proposed that the $SU(3)$ vector and axial vector charges, defined by the space integral of the first components of the currents,

$$F_i(t) = \int V_i^0(x)\,d^3x, \qquad F_i^5(t) = \int A_i^0(x)\,d^3x, \qquad (6.5)$$

obey the equal time commutation relations,

$$[F_i(t), F_j(t)] = if_{ijk}F_k(t),$$
$$[F_i(t), F_j^5(t)] = if_{ijk}F_k^5(t), \qquad (6.6)$$
$$[F_i^5(t), F_j^5(t)] = if_{ijk}F_k(t).$$

These relations represent the $SU(3)\otimes SU(3)$ Lie algebra, where f_{ijk} are the $SU(3)$ structure constants. The first equation corresponds to the $SU(3)$ algebra, of which the vector charges $F_i(t)$ are assumed to be the generators. (The charges F_i are time independent only if the vector currents $V_i^\mu(x)$ are exactly conserved, i.e. $\partial_\mu V_i^\mu(x) = 0$, as follows from Eq. (6.5). In this case the $SU(3)$ symmetry is exact.) The algebra (6.6) can be easily derived from a field theory model like the quark model. However, at the time Gell-Mann used the so-called 'symmetric Sakata model', where the role of quarks was played by the proton, neutron and Λ.

Gell-Mann pointed out that this algebra could give important predictions of the hadronic form factors (i.e. the matrix elements of charges between hadronic states). In fact, unlike other linear relations such as the dispersion relations, the above commutators are nonlinear in the charges and can fix the magnitude of the charges themselves. This was in fact the result obtained by Adler and Weisberger, with the sum rule mentioned below.

In 1964, after the quark hypothesis, Gell-Mann [Gel64b] extended the charge algebra (6.6) to the 'current algebra', i.e. the local equal time commutators of the time components of the currents, given by:

$$[V_i^0(\mathbf{x}, t), V_j^0(\mathbf{y}, t)] = if_{ijk}\,V_k^0(\mathbf{x}, t)\,\delta(\mathbf{x} - \mathbf{y}),$$
$$[V_i^0(\mathbf{x}, t), A_j^0(\mathbf{y}, t)] = if_{ijk}\,A_k^0(\mathbf{x}, t)\,\delta(\mathbf{x} - \mathbf{y}), \qquad (6.7)$$
$$[A_i^0(\mathbf{x}, t), A_j^0(\mathbf{y}, t)] = if_{ijk}\,V_k^0(\mathbf{x}, t)\,\delta(\mathbf{x} - \mathbf{y}).$$

These relations are easily derived from the expressions (6.4) of the vector and axial vector currents of the quark model.

Commutators including the space components of the currents can also be considered, but they could contain the so-called Schwinger terms, involving the derivatives of the delta function, and for this reason they were considered less reliable.

The exploitation of the charge algebra required almost three years. The first idea was to take the commutators between one-particle states and to insert a complete set of intermediate states. Let us take for example the commutator,

$$[F_+^5, F_-^5] = 2F_3, \qquad (6.8)$$

where $F_\pm^5 = F_1^5 \pm i F_2^5$ are the weak axial charges and $F_3 = I_3$ is the isotopic spin operator. Taking the expectation value in a single-particle state $|a\rangle$, we get the sum rule:

$$\sum_n \left[|\langle a|A_+^0(0)|n\rangle|^2 - |\langle a|A_-^0(0)|n\rangle|^2 \right] (2\pi)^3 \delta(\mathbf{p}_n - \mathbf{p}_a) = 4E_a I_{3a}. \qquad (6.9)$$

However, this formula (for finite p_a) has two basic drawbacks: (i) the squared momentum transferred to the axial charge, $q^2 = (p_a - p_n)^2 = (E_a - E_n)^2$, depends on the intermediate state energy E_n, making it difficult to use the sum rule; (ii) q^2 increases without limit with the mass of the intermediate state n, and this makes it difficult to guess the convergence of the series.

A good idea, devised by Fubini and Furlan in 1964 [FF65], was to take the limit $p_a \to \infty$. Calling $\mathbf{p}_a = \mathbf{p}_n = \mathbf{p}$, for $|\mathbf{p}| \to \infty$ we have:

$$q^2 = (E_a - E_n)^2 = \left[(\mathbf{p}^2 + m_a^2)^{1/2} - (\mathbf{p}^2 + m_n^2)^{1/2} \right]^2 \to 0, \qquad (6.10)$$

for any intermediate state. This changes the sum rule (6.9), where q^2 depends on the intermediate state, into a fixed q^2 sum rule, namely at $q^2 = 0$. This method was then called 'the infinite momentum limit' and opened the way to a number of important applications of current algebra.

An important ingredient of the sum rules is the hypothesis of the 'partial conservation of the axial current' (PCAC), formulated by Gell-Mann and Lévy [GL60] in 1960. This relates the divergence of the isotopic spin components of the axial vector current to the pion field by the formula:

$$\partial_\mu A_i^\mu(x) = f_\pi m_\pi^2 \phi_i(x), \qquad i = 1, 2, 3, \qquad (6.11)$$

where f_π is a constant related to the pion lifetime and $\phi_i(x)$ is the pion field. In this way the matrix elements, $\langle a|A_\pm^0(0)|n\rangle$, can be written in the form:

$$\langle a|A_\pm^0(0)|n\rangle = -i(E_a - E_n)^{-1} \langle a|\partial_\mu A_\pm^\mu(0)|n\rangle$$
$$= -i f_\pi m_\pi^2 (E_a - E_n)^{-1} \langle a|\phi_\pm(0)|n\rangle, \qquad (6.12)$$

and the result is proportional to the a–π–n vertex.

We want just to mention some important results obtained from current algebra in the years 1964–1966. The first and most important result was obtained independently by Adler [Adl65a, Adl65b] and Weisberger [Wei65, Wei66] in 1964. Starting from the axial charge commutator (6.8) and using the PCAC relation and the infinite momentum limit, they were able to calculate the form factor of the axial vector coupling of the nuclear β decay, in terms of the total cross-section of pion–proton scattering. The calculated value was in good agreement with experiment (within 5%).

Other important relations are:

- the Callan–Treiman relation [CT66], connecting the leptonic decays of the K-meson $K \to \pi l \nu$ and $K \to \pi \pi l \nu$;
- the Cabibbo–Radicati sum rule [CR66], giving a combination of the electromagnetic form factors of the nucleon in terms of the photon–nucleon total cross-section; this was the first sum rule tested at $q^2 \neq 0$;
- the Weinberg [Wei66a] calculation of the $K \to \pi \pi l \nu$ form factors, from which the decay rate of $K^+ \to \pi^+ \pi^- e^+ \nu$ results in excellent agreement with experiment;
- the Weinberg theory of multiple pion production [Wei66b] and the calculation of the pion scattering lengths [Wei66c].

6.4 Strong interactions

During the Sixties, i.e. before the formulation of QCD, a fundamental theory of strong interactions did not exist. The general theories known at the time were quantum field theory and S-matrix theory.

Quantum field theory. This approach was very successful in describing electromagnetic and, after 1967, weak interactions. Its application to strong interactions, however, encountered some basic difficulties.

(i) There were too many hadrons and too many forces to construct a sensible interaction Lagrangian. Models of infinite component field theories were also considered, but the arbitrariness was too high to build a useful model. On the other hand, the idea of taking a few elementary particles to explain the spectrum of the hadrons and their interactions was at the time far from feasible.

(ii) Quantum field theory is based on a perturbative expansion for small coupling constant, leading to Feynman diagrams; however, the strong interaction couplings were large, therefore the perturbative expansion was meaningless.

S-matrix theory. The ambitious programme proposed by Chew [Che66] in the years 1960–1965 can be condensed in the principle of the 'bootstrap of strong interactions'. It affirms that the experimental data are self-consistently determined through the study of the S-matrix, by enforcing its properties of analyticity and symmetry, as outlined in the following. These conditions would give an infinite set of coupled nonlinear equations, whose solution is of course hopeless. Therefore, even if the principles are correct, in practice some drastic approximations are necessary in order to obtain useful results.

In this context the Dual Resonance Model, which is based on the narrow-resonance approximation, was welcome as a brave theory of strong interactions. In order to outline the way toward duality, it may be useful to summarize the basic concepts of the scattering theory.

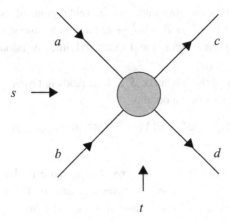

Figure 6.1 Graphic representation of $a + b \to c + d$ scattering and the Mandelstam variables.

6.4.1 Basics of scattering theory

Consider the two-body scattering (see Figure 6.1):

$$a + b \longrightarrow c + d. \tag{6.13}$$

The scattering process is described by a 'scattering amplitude' defined as follows.

(i) We know from quantum mechanics that the scattering matrix S is a unitary operator transforming the initial state $|a, b\rangle$ (at $t \to -\infty$) into the state evolved at $t \to +\infty$, which is a superposition of all possible final states. The amplitude for the process in Figure 6.1 is:

$$S_{ab}^{cd} = \langle c, d| \, S \, |a, b\rangle. \tag{6.14}$$

(ii) For a process described by a connected diagram we define the transition matrix T by $S = I + iT$, where the identity I accounts for the case of no interaction.

(iii) For a scattering process $|i\rangle \to |f\rangle$ we define the scattering amplitude M_{fi} by the relation

$$\langle f|T|i\rangle = (2\pi)^4 \delta(p_f - p_i) M_{fi}, \tag{6.15}$$

where the Dirac δ-function takes into account energy-momentum conservation.

(iv) The amplitude M_{fi} is then expanded as:

$$M_{fi} = \sum_r K_{fi}^r A_r(s, t, u), \tag{6.16}$$

where K_{fi}^r are covariant factors depending on spin, isotopic spin or $SU(3)$ indices and momentum components of the external particles and $A_r(s, t, u)$ are invariant amplitudes, depending only on the Mandelstam variables s, t and u (see Figure 6.1).

We assume that invariant amplitudes $A_r(s, t, u)$ obey the following fundamental properties.

Crossing symmetry. The same amplitude (in different regions of its parameters) describes all the reactions $ab \to cd$, $b\bar{d} \to \bar{a}c$, $a\bar{d} \to c\bar{b}$ and their inverses, obtained by the CPT transformation, i.e. charge conjugation and space and time inversion.

Unitarity. The unitarity of the S-matrix, $S^\dagger S = I$, reads in terms of the transition matrix: $T - T^\dagger = iT^\dagger T$. Using (6.15), we obtain

$$M_{fi} - M_{if}^* = i(2\pi)^4 \sum_n M_{nf}^* M_{ni} \delta(p_n - p_i). \tag{6.17}$$

Analyticity. The amplitude $A(s,t,u)$ at fixed t is analytic in the complex plane of the variable $v = (s-u)/4$, with singularities (poles and cuts) on the real axis. Since analytic functions can be expressed as Cauchy integrals, one obtains the so-called 'dispersion relations', of the form:

$$A(v,t) = \text{poles} + \frac{1}{\pi} \int_{-\infty}^{+\infty} \frac{\text{Im } A(v',t)}{v' - v} dv'. \tag{6.18}$$

In this equation, the pole terms come from single-particle intermediate states in the s- and u-channels and Im A, called the absorptive part of the amplitude, is different from zero only above a certain threshold corresponding to intermediate states of two or more particles.

Asymptotic behaviour. In the Regge pole model (see Frautschi [Fra63]), for $s \to \infty$ and fixed t we have:

$$A(s,t) \to \beta(t)\xi(t)s^{\alpha(t)}, \tag{6.19}$$

where $\alpha(t)$ is the Regge trajectory, $\beta(t)$ is called the residue and is an entire function of t and $\xi(t) = (1 \pm e^{-i\pi\alpha(t)})\Gamma(-\alpha(t))$, where Γ is the Euler gamma function. In the following we shall always consider linear Regge trajectories, i.e. of the form $\alpha(t) = \alpha_0 + \alpha' t$, with a constant slope α'.

The goal of the S-matrix approach is to obtain relations between the invariant amplitudes coming from symmetries, kinematic and unitarity relations and the natural requirement of analytic extension of the amplitudes for complex values of the s, t and u variables. These relations should provide phenomenological constraints on the observed scattering processes. No reference is made to an underlying field theory description, except for the existence of conserved currents associated with the symmetries.

6.5 Superconvergence and finite energy sum rules

6.5.1 Superconvergence

A special subject developed in the years 1966–1968 was superconvergence. This property was discovered by De Alfaro, Fubini, Furlan and Rossetti [DFFR66] in the study of the sum rules from the local current algebra commutators. A similar result was obtained at the same time by Drell and Hearn [DH66]. Although the authors started by considering current algebra sum rules, superconvergence has nothing to do with weak interaction currents, but is only concerned with strong interactions. In particular, it exploits the asymptotic behaviour of the invariant amplitudes in special cases.

Consider the scattering of particles with spin, for example the $\pi\rho$ forward ($t = 0$) scattering. There are three kinds of invariant amplitudes, with angular momentum transfer in the t-channel of 0, 1 and 2, which we shall call A_0, A_1 and A_2. Their large-s behaviours are s^α, $s^{\alpha-1}$, $s^{\alpha-2}$ respectively, where $\alpha = \alpha(0)$ is the intercept of the leading Regge trajectory. Consider in particular the amplitudes with isospin 1 in the t-channel, which are dominated by the ρ-trajectory, with $\alpha(0) \approx 0.5$. The amplitudes $A_1(\nu)$ and $A_2(\nu)$ are convergent for $\nu \to \infty$, are odd functions of ν by crossing symmetry, hence with an even absorptive part Im $A_2(\nu)$, and obey the dispersion relation (6.18) at $t = 0$. Now for large ν ($\approx \frac{1}{2}s$) we have $\nu A_2(\nu) \to 0$. Then from Eq. (6.18), multiplying by ν and letting $\nu \to \infty$ we obtain the important relation [DFFR66]:

$$\text{pole residues} + \int_0^\infty \text{Im } A_2(\nu, t=0)\, d\nu = 0, \tag{6.20}$$

which is called a superconvergence relation. When Im A_2 is expressed by unitarity as a sum of contributions of the intermediate states in the s- and u-channels we get a relation which is called a 'superconvergence sum rule' (SSR).

6.5.2 Major applications

The first important application was the sum rule obtained by Drell and Hearn [DH66] in 1966. They considered the spin-flip amplitude of the photon–proton Compton scattering and obtained a relation giving the anomalous magnetic moment of the proton (from the single proton intermediate state) in terms of the γp total cross-section.

De Alfaro, Fubini, Furlan and Rossetti, in their original SSR paper [DFFR66], studied the $\rho\pi$ forward scattering and considered two SSR, one for the amplitude A_2 seen before and the other for A_1 and isospin 2 in the t-channel. Saturating the SSR with the low-lying resonances ω and ϕ only, they obtained a relation between the coupling constants $g_{\omega\rho\pi}$ and $g_{\phi\rho\pi}$, in good agreement with experiment.

Igi and Matsuda [IM67] and independently Logunov, Soloviev and Tavkhelidze [LST67] and Raoul Gatto [Gat67] in 1967 analyzed the πN scattering with charge exchange ($\pi^- p \to \pi^0 n$). This is not an SSR, because the integral does not converge, due to asymptotic ρ-trajectory contribution. Separating this contribution, they obtained a relation between the πp total cross-sections and the parameters of the ρ-trajectory.

6.5.3 Finite energy sum rules

An important step forward was made by Dolen, Horn and Schmid [DHS67, DHS68] in 1967. Inspired by the preceding authors, they extended the superconvergence relation (6.20) in the following way. The integral over ν is split into two parts, one up to a finite value $\nu = N$ and the other for $\nu > N$ is evaluated in terms of the asymptotic Regge behaviour. The authors obtained the following relation, called a 'finite energy sum rule' (FESR):

$$\int_0^N \text{Im}\, A(\nu, t)\, d\nu = \sum_i \frac{\beta_i(t) N^{\alpha_i+1}}{\Gamma(\alpha_i + 2)}. \qquad (6.21)$$

Here N is a large but finite value of ν and the right-hand side is the sum of all the Regge terms, where $\alpha_i = \alpha_i(t)$ are the Regge trajectories and $\beta_i(t)$ are the residues. Notice that the SSR is reobtained in the limit $N \to \infty$, if $\alpha_i < -1$. This relation, however, holds for any value of α_i.

The authors applied the FESR to the scattering $\pi^- p \to \pi^0 n$. From the low energy πN data they predicted the parameters of the ρ-meson Regge trajectory as functions of t.

The importance of the FESR lies in the property that it relates the low energy data in the s- and u-channels to the Regge trajectories in the t-channel. The authors argued that this kind of relation represents a sort of bootstrap for the scattering amplitude, in the sense that it relates the low energy to the high energy behaviour.

6.6 The ARVV collaboration

Let me now come to my personal recollections. In the spring of 1967, I went for two weeks to Israel, to the Weizmann Institute, to collaborate with Gabriele Veneziano on the new idea of superconvergence. There I met Hector Rubinstein and Miguel Virasoro, and all together we started a collaboration that went on for more than one year.

6.6.1 Superconvergence sum rules for meson–meson scattering

In a first paper [ARVV67a] we discussed the saturation of the SSR at $t \neq 0$. Separating the resonance from the Regge contribution, we obtained an FESR that can be continued analytically to the region where the integral of the SSR would diverge. Studying in particular the $\rho\pi$ scattering, we recognized the necessity of the Regge contribution to satisfy the SSR at all t. In successive papers [ARVV67b, ARVV68a, ARVV68b] we analyzed the saturation of SSR for several processes of meson–meson scattering, of the type $PP \to PV$, $PP \to PT$, $PV \to PV$, where P, V, T stand for pseudoscalar, vector and tensor (spin two) mesons.

6.6.2 The scattering $\pi\pi \to \pi\omega$

The most interesting case was the scattering $\pi\pi \to \pi\omega$ [ARVV68b], having the nice property that the s-, t- and u-channels are identical (see Figure 6.2). Considering the spin, parity and isotopic spin of the particles there is only one invariant amplitude $A(s, t, u)$,

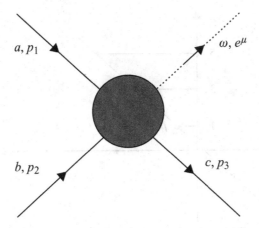

Figure 6.2 Scattering $\pi\pi \to \pi\omega$. The solid lines represent the pions, with momenta p_1, p_2, p_3 and isospin components a, b, c, and the dashed line represents the ω-meson, with polarization vector e^μ.

according to the formula

$$\langle e; c, p_3 | M | a, p_1; b, p_2 \rangle = \epsilon_{abc}\, \epsilon_{\mu\nu\rho\sigma}\, e^\mu p_1^\nu p_2^\rho p_3^\sigma\, A(s, t, u), \qquad (6.22)$$

where a, b, c are the pion isospin components, p_i are the pion momenta and e^μ is the ω polarization vector. The amplitude $A(s, t, u)$ is fully symmetric in the Mandelstam variables s, t and u and is dominated in each channel by the ρ-meson.

We considered the amplitude $A(\nu, t)$ as a function of the independent variables $\nu = (s - u)/4$ and t. The behaviour for large ν, according to the Regge model, is given by:

$$A(\nu, t) = \beta(t)\xi[\alpha(t)](\nu/\nu_1)^{\alpha(t)-1}, \qquad (6.23)$$

where

$$\xi(\alpha) = (1 - e^{-i\pi\alpha})/\sin(\pi\alpha),$$
$$\beta(t) = \bar{\beta}/\Gamma[\alpha(t)], \qquad (6.24)$$

and $\bar{\beta}$ is a constant.

We considered the generalized sum rule:

$$\int_0^{\bar{\nu}} \nu^n\, \mathrm{Im}\, A(\nu, t)\, d\nu = \frac{\beta(t)}{\alpha(t) + n} \left(\frac{\bar{\nu}}{\nu_1}\right)^{\alpha(t)-1} \bar{\nu}^{n+1}, \qquad (6.25)$$

where the moment n is a nonnegative integer, $\bar{\nu}$ is the limit of the resonance region and ν_1 is an arbitrary scale parameter. Since $A(\nu, t)$ is an even function of ν, due to s–u crossing symmetry, and then $\mathrm{Im}\, A(\nu, t)$ is odd, the first nontrivial moment in the sum rule is $n = 1$. For the resonance part on the left-hand side we considered the ρ-meson and the higher resonances ρ_3 and ρ_5 with spins $j = 3$ and $j = 5$ respectively, lying on the leading ρ-trajectory, and for the high energy on the right-hand side the ρ-trajectory itself (see Figure 6.3).

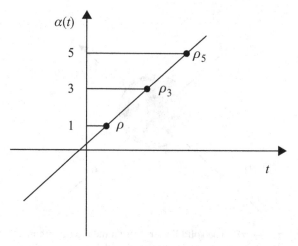

Figure 6.3 Meson states on the leading Regge trajectory of the ρ-meson. Here ρ represents the usual ρ vector meson, ρ_3 and ρ_5 are higher resonances with spins 3 and 5 respectively.

The numerical results were good even with the insertion of the ρ-resonance alone and became much better with the contributions of the higher resonances, as is shown in Figure 6.4, reporting figures from [ARVV68b]. However, it was realized that a better agreement between the resonance and the asymptotic parts of the sum rule would require the contributions of the lower-lying Regge trajectories.

In conclusion, since the resonant part and the high energy part of the sum rule (6.25) involve only the ρ-trajectory, Eq. (6.25) can be seen as a bootstrap of the ρ-trajectory itself, that is a self-consistency condition for its parameters.

6.7 The concept of duality

The idea of duality was born from the finite energy sum rules of Eq. (6.21), relating the low energy to the high energy behaviour of the scattering amplitude. Although the FESR involves only the absorptive part of the amplitude, the idea was that the whole amplitude should exhibit only resonance poles in the physical region of the s-, u- and t-channels, and at the same time the right asymptotic behaviour. More precisely, and still considering the $2 \to 2$ particle scattering, we can define a dual model as a theoretical model with the following two properties: narrow resonance approximation and duality.

- 'Narrow resonance approximation' means that in each channel only single particles are exchanged, or better that in the physical region of each channel the scattering amplitude has only poles, corresponding to narrow resonances, and no continuum in its absorptive part.
- 'Duality' means that the resonances in a given channel also determine the correct asymptotic behaviour in the same channel and the resonances in the crossed channel.

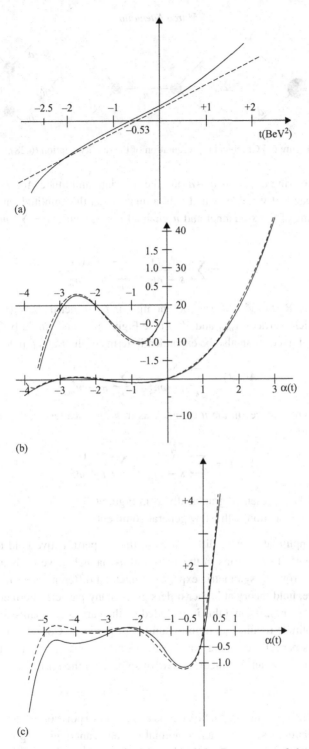

Figure 6.4 Comparison between the resonance part and the Regge part of the sum rule (6.25), represented by solid and dashed lines. For the saturation of the resonance part, the ρ resonance alone is used in the upper graph, the ρ_3 resonance is added in the middle graph and the ρ_5 resonance is also included in the lower graph. The graphs are taken from [ARVV68b].

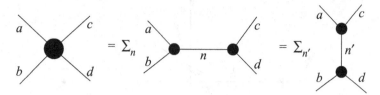

Figure 6.5 Graphical representation of the duality relation (6.28).

Consider the scattering $a + b \to c + d$ and the invariant amplitude $A(s, t)$, as a function of the two independent variables s and t. In a dual model the amplitude at fixed t can be expanded in terms of the s-channel and u-channel poles, where $u = \sum_i m_i^2 - s - t$ (see Figure 6.1):

$$A(s, t) = \sum_n \frac{R_n(t)}{s - m_n^2} + \sum_m \frac{\tilde{R}_m(t)}{u - m_m^2}, \qquad (6.26)$$

where the residues $R_n(t)$ and $\tilde{R}_m(t)$ are polynomials in t. Furthermore, $R_n(t)$ is the product of the three-particle vertices V_{abn} and V_{ncd} (see Figure 6.5) and similarly for $\tilde{R}_m(t)$. The same amplitude at fixed s can also be expanded in terms of the t and u poles:

$$A(s, t) = \sum_k \frac{S_k(s)}{t - m_k^2} + \sum_m \frac{\tilde{S}_m(s)}{u - m_m^2}. \qquad (6.27)$$

In the case of no resonances in the u channel, as in $\pi^+\pi^-$ scattering, the duality relation has the simple form

$$A(s, t) = \sum_n \frac{R_n(t)}{s - m_n^2} = \sum_{n'} \frac{S_{n'}(s)}{t - m_{n'}^2}. \qquad (6.28)$$

This relation can be represented graphically as in Figure 6.5.

We conclude this Section with a few general comments.

(i) The dual amplitude is quite different from that of perturbative field theory: at first order, the pole (Born) terms in the s-, t- and u-channels have to be added together leading to a crossing-symmetric expression made of different terms in each channel. Furthermore, field theory at higher orders gives many-particle intermediate states, as is required by unitarity, and this is excluded by the narrow resonance approximation.

(ii) A dual amplitude requires an infinite number of particles. Consider the scattering of two spinless particles of equal mass in the centre-of-mass system of the s-channel. The contribution of an intermediate state of spin l is of the form:

$$A_l(s) P_l(\cos\theta), \qquad \cos\theta = 1 + 2t/(s - 4m^2), \qquad (6.29)$$

where θ is the scattering angle. Since P_l is a (Legendre) polynomial, a finite number of these contributions still gives a polynomial in t and cannot give a pole. Furthermore, the asymptotic behaviour in s would be of the form s^{-1}, and not $s^{\alpha(t)}$.

(iii) The dual model is not unitary. This can be seen in at least two ways. First, the unitarity relation gives Im A as a sum of contributions of all the physical intermediate states, including many-particle states that would give rise to a continuum, excluded by the narrow resonance approximation. Secondly, the coupling of the intermediate particles to the external ones implies that they can decay (when allowed) with a certain rate. This gives the resonance a finite width and places a pole in the complex s plane out of the real axis.

6.8 The Veneziano formula

Soon after our collaboration on superconvergence and the bootstrap of the ρ-trajectory, Gabriele Veneziano had the brilliant idea of his famous formula [Ven68], which was the starting point for the Dual Resonance Models. He considered the simple case of the scattering $\pi\pi \to \pi\omega$ discussed in Section 5.2 and was looking for an amplitude with the following analytic properties:

(i) an infinite set of poles in the s- and u-channels at fixed t;
(ii) asymptotic behaviour for large s and fixed t like $s^{\alpha(t)}$;
(iii) complete crossing symmetry in the three channels.

He started considering the large-s behaviour of the amplitude given by Eqs. (6.23) and (6.24), which can be written in the form:

$$A(s, t, u) \to \overline{\beta}\,\Gamma[1 - \alpha(t)][\alpha(s)]^{\alpha(t)-1}. \tag{6.30}$$

The idea was to hold the first factor, giving the poles for $\alpha(t) = n \geq 1$, and to consider the second factor as valid only asymptotically. With a suitable choice of the residue function, he then replaced this factor as follows:

$$[\alpha(s)]^{\alpha(t)-1} \to \Gamma[1 - \alpha(s)]/\Gamma[2 - \alpha(s) - \alpha(t)]. \tag{6.31}$$

This gives for the s–t part of the amplitude the expression

$$A(s, t) = \overline{\beta}\,\frac{\Gamma[1 - \alpha(s)]\Gamma[1 - \alpha(t)]}{\Gamma[2 - \alpha(s) - \alpha(t)]}. \tag{6.32}$$

This expression has the following nice properties:

(i) it is symmetric in s and t;
(ii) it has poles for $\alpha(s) = n \geq 1$ and also for $\alpha(t) = n \geq 1$;
(iii) it avoids simultaneous poles in s and t;
(iv) it has the right asymptotic behaviour.

Finally, the complete crossing-symmetric Veneziano amplitude [Ven68] is given by:

$$A(s, t, u) = \overline{\beta}\,\{B[1 - \alpha(s), 1 - \alpha(t)] + B[1 - \alpha(t), 1 - \alpha(u)]$$
$$+ B[1 - \alpha(s), 1 - \alpha(u)]\}, \tag{6.33}$$

where $B(x, y)$ is the Euler beta function,

$$B(x, y) = \frac{\Gamma(x)\Gamma(y)}{\Gamma(x+y)} = \int_0^1 t^{x-1}(1-t)^{y-1}\,dt. \qquad (6.34)$$

The Veneziano amplitude (6.33) has all the required properties for a dual amplitude. In particular, it has infinite simple poles in each channel and the correct asymptotic behaviour.

It turns out that the poles are located not only on the leading Regge trajectory, but also on the lower trajectories. The generalization of the Veneziano four-particle amplitude to five- and N-particle dual amplitudes and the problem of the spectrum of the intermediate states were the next problems to be investigated. This was the beginning of the Dual Resonance Model and its later realization as the quantum theory of the relativistic string.

Marco Ademollo was born in Florence in 1936. He graduated in physics from Florence University in 1958, and received his PhD in 1965. He was Lecturer at Florence University in 1959, and has been Professor of Theoretical Physics since 1976. His research activity has concerned particle phenomenology, symmetries in particle physics, current algebra, superconvergence sum rules, dual models, string and superstring theory.

References

[ARVV67a] Ademollo, M., Rubinstein, H. R., Veneziano, G. and Virasoro, M. A. (1967). Saturation of superconvergent sum rules at nonzero momentum transfer, *Nuovo Cimento* **A51**, 227–231.

[ARVV67b] Ademollo, M., Rubinstein, H. R., Veneziano, G. and Virasoro, M. A. (1967). Bootstraplike conditions from superconvergence, *Phys. Rev. Lett.* **19**, 1402–1405.

[ARVV68a] Ademollo, M., Rubinstein, H. R., Veneziano, G. and Virasoro, M. A. (1968). Reciprocal bootstrap of the vector and tensor trajectories from superconvergence, *Phys. Lett.* **B27**, 99–102.

[ARVV68b] Ademollo, M., Rubinstein, H. R., Veneziano, G. and Virasoro, M. A. (1968). Bootstrap of meson trajectories from superconvergence, *Phys. Rev.* **176**, 1904–1925.

[Adl65a] Adler, S. L. (1965). Calculation of the axial-vector coupling constant renormalization in β decay, *Phys. Rev. Lett.* **14**, 1051–1055.

[Adl65b] Adler, S. L. (1965). Sum rules for the axial-vector coupling-constant renormalization in β decay, *Phys. Rev.* **140**, B736–B747.

[Cab63] Cabibbo, N. (1963). Unitary symmetry and leptonic decays, *Phys. Rev. Lett.* **10**, 531–533.

[CR66] Cabibbo, N. and Radicati, L. A. (1966). Sum rule for the isovector magnetic moment of the nucleon, *Phys. Lett.* **19**, 697–699.

[CT66] Callan, C. G. and Treiman, S. B. (1966). Equal-time commutators and K-meson decays, *Phys. Rev. Lett.* **16**, 153–157.

[Che66] Chew, G. F. (1966). *The Analytic S Matrix* (W. A. Benjamin, New York).

[DFFR66] De Alfaro, V., Fubini, S., Furlan, G. and Rossetti, C. (1966). Sum rules for strong interactions, *Phys. Lett.* **21**, 576–579.

[DHS67] Dolen, R., Horn, D. and Schmid, C. (1967). Prediction of Regge parameters of ρ poles from low-energy πN data, *Phys. Rev. Lett.* **19**, 402–407.

[DHS68] Dolen, R., Horn, D. and Schmid, C. (1968). Finite energy sum rules and their application to π–N charge exchange, *Phys. Rev.* **166**, 1768–1781.

[DH66] Drell, S. D. and Hearn, A. C. (1966). Exact sum rule for nucleon magnetic moments, *Phys. Rev. Lett.* **16**, 908–911.
[FG58] Feynman, R. P. and Gell-Mann, M. (1958). Theory of the Fermi interaction, *Phys. Rev.* **109**, 193–198.
[Fra63] Frautschi, S. C. (1963). *Regge Poles and S-Matrix Theory* (W. A. Benjamin, New York).
[FF65] Fubini, S. and Furlan, G. (1965). Renormalization effects for partially conserved currents, *Physics* **1**, 229–247.
[Gat67] Gatto, R. (1967). New sum rules for superconvergence, *Phys. Rev. Lett.* **18**, 803–806.
[Gel61] Gell-Mann, M. (1961). Report CTSL 20, unpublished.
[Gel62] Gell-Mann, M. (1962). Symmetries of baryons and mesons, *Phys. Rev.* **125**, 1067–1084.
[Gel64a] Gell-Mann, M. (1964). A schematic model of baryons and mesons, *Phys. Lett.* **8**, 214–215.
[Gel64b] Gell-Mann, M. (1964). The symmetry group of vector and axial vector currents, *Physics* **1**, 63–75.
[GL60] Gell-Mann, M. and Lévy, M. (1960). The axial vector current in beta decay, *Nuovo Cimento* **16**, 705–726.
[Gla61] Glashow, S. L. (1961). Partial-symmetries of weak interactions, *Nucl. Phys.* **22**, 579–588.
[IM67] Igi, K. and Matsuda, S. (1967). New sum rules and singularities in the complex J plane, *Phys. Rev. Lett.* **18**, 625–627.
[LST67] Logunov, A. A., Soloviev, L. D. and Tavkhelidze, A. N. (1967). Dispersion sum rules and high energy scattering, *Phys. Lett.* **B24**, 181–182.
[Nee61] Ne'eman, Y. (1961). Derivation of strong interactions from a gauge invariance, *Nucl. Phys.* **26**, 222–229.
[Ven68] Veneziano, G. (1968). Construction of a crossing-symmetric, Reggeon-behaved amplitude for linearly rising trajectories, *Nuovo Cimento* **A57**, 190–197.
[Wei66a] Weinberg, S. (1966). Current-commutator calculation of the Kl4 form factors, *Phys. Rev. Lett.* **17**, 336–340.
[Wei66b] Weinberg, S. (1966). Current-commutator theory of multiple pion production, *Phys. Rev. Lett.* **16**, 879–883.
[Wei66c] Weinberg, S. (1966). Pion scattering lengths, *Phys. Rev. Lett.* **17**, 616–621.
[Wei67] Weinberg, S. (1967). A model of leptons, *Phys. Rev. Lett.* **19**, 1264–1266.
[Wei65] Weisberger, W. I. (1965). Renormalization of the weak axial-vector coupling constant, *Phys. Rev. Lett.* **14**, 1047–1051.
[Wei66] Weisberger, W. I. (1966). Unsubtracted dispersion relations and the renormalization of the weak axial-vector coupling constants, *Phys. Rev.* **143**, 1302–1309.
[Zwe64] Zweig, G. (1964). CERN Report no. 8182, unpublished.

7
The path to the Veneziano model

HECTOR R. RUBINSTEIN

Abstract

Marco Ademollo, in his contribution to this Volume, has given a review of early work that led to string theory. I will complement this view of that period and its origin. My main purpose is to go further back in time and put the appearance of string theory work in the context of the evolution of the theory of elementary particles.

7.1 Introduction

Just after the end of the second world war, the resounding confirmation of relativistic field theory (more precisely, quantum electrodynamics) took place with Lamb's measurement of the level shifts in hydrogen and Schwinger's calculation of the electron magnetic moment, which differed from the value predicted from Dirac's equation. Renormalization theory with American, European, and Japanese contributions gave remarkable results with accuracies beyond one part in one hundred million.

All of these results were obtained using perturbation theory and though the agreement was remarkable, questions of principle on whether the series are convergent and divergences exist, requiring a modification of the theory, loomed in the discussions.

Soon after, with the discovery of mesons, and later other baryons, besides the proton and neutron, attention concentrated on the strong interactions. Pauli, Wentzel, Tamm, Dancoff, Dyson and others worked on the problem (for a short review, see Salpeter [Sal08]).

The natural thing to do was to write a Lagrangian for mesons and nucleons, and try all possible couplings in perturbation theory. It soon became clear that these calculations did not describe the experimental data.

Given that at the time nonperturbative methods had not been developed, and enough information from experiment was not available, the task was to find results that did not depend on detailed dynamical Lagrangians but on fundamental principles such as unitarity and causality.

The Birth of String Theory, ed. Andrea Cappelli, Elena Castellani, Filippo Colomo and Paolo Di Vecchia.
Published by Cambridge University Press. © Cambridge University Press 2012.

7.2 The new lines of attack: analyticity and symmetry

Two schools started to develop: the symmetries school and the analyticity school. The former became a very important subject: discrete symmetries like C, P and T, and combinations, internal symmetries like isospin and with the proliferation of particles larger groups like $SU(3)$ and later others became sources of important results. The analyticity approach avoided specific dynamical models. Using the results of causality, expressed on analyticity conditions and using dispersion relations, people looked in detail at the analytic properties of scattering amplitudes. This was in the spirit of S-matrix theory that was to dominate research for a decade. Using this approach important results emerged. One of the most important results along these lines is the remarkable work of Tullio Regge [Reg59], which established the analytic properties in the angular momentum plane. In this way, families of particles were identified.

Experimentally, much information accumulated concerning the strong interactions. Families of particles, Regge trajectories, classified all known states and a new concept, the bootstrap, was developed by Chew and Frautschi [CF61, CF62] and many others. The idea was to think that all particles are equivalent, there are no fundamental states. Their properties can be established by starting with any set of particles and deducing the properties of all the others using analyticity and consistency conditions. Chew called this approach 'nuclear democracy'. The calculations of Chew and Mandelstam on pion–pion scattering [CM60] looked very promising and a large effort was invested in determining the analyticity framework required to write these amplitudes. Establishing the analytic properties of the spin carrying process was a nontrivial task and many papers were written on the subject. This took much work in the early Sixties.

In a more phenomenological vein, a method introduced by Zachariasen [Zac61] was used to determine the position and couplings of resonances by exchanging the nearest singularity. This looked promising, but we know today that low lying states are determined, at least partly, by nonperturbative contributions. The calculational method was the N/D method in which left- and right-hand singularities were separated.

Enormous enthusiasm arose when Dashen [Das64] using the N/D method showed that the neutron was heavier than the proton. In fact the calculation had the wrong sign! The mass difference between the proton and neutron was a puzzle that preoccupied the best theorists. It had a simple and unexpected answer: the d quark is heavier than the u quark, and the proton, to the surprise of many, is a composite state. This was one of the early signs that the dynamics of strong interactions had to be understood at a deeper level than that of the hadrons.

Another line of thought that also proved to be very promising was current algebra. The hope that it would be the basis of a dynamical theory did not materialize though some interesting results came out using the method. Low energy theorems determining the pion scattering lengths (Weinberg [Wei66]), weak interaction processes (Callan and Treiman [CT66]) and the understanding of G_A/G_V coupling constants in weak interaction currents (Adler [Adl65] and Weisberger [Wei65]) gave hope of much more to come. This did not happen. None of these approaches had a large impact on string theory. The analytic thinking was useful for finding bounds on the growth of cross-sections with energy, the Froissart bound [Fro61], and on angular distributions, the Martin bound [Mar63]. The Pomeranchuk theorem proving equality of particle–antiparticle and particle–particle cross-sections was also a very appealing result.

The group theoretic people were studying classification of states and relations between cross-sections involving particles in given multiplets. It is at this stage, after the discovery of the Ω-particle, that $SU(3)$ was fully established. The history of the classification of particles is also complicated since several wrong experiments confused the issue for some time. $O(10)$ and $G(2)$ were competitors at the beginning and even $SU(3)$ was first used logically for a baryon triplet, the Sakata model, but the octet model proved more rewarding. Some of the evidence was not foolproof and the puzzle of the fundamental representation not appearing in nature was bothersome.

Soon after, Gell-Mann [Gel61], Ne'eman [Nee61] and Zweig [Zwe64] asked why the fundamental representation was absent and predicted quarks. The search for these fractional states proved to be a failure.

In Western Europe, Soviet Union of the time, and Israel quarks were taken seriously even if they were not observed. Low energy proton–antiproton annihilation (Levin and Frankfurt [LF65]), the scattering length of baryons, ratios of cross-sections of mesons and baryons (Rubinstein and Stern [RS66]), decay properties of meson multiplets (Elitzur *et al.* [ELRS66]) and spectroscopy of baryon states (Amati *et al.* [AJRV68]) gave very strong evidence of their presence inside the nucleons. America joined the believers later, after the SLAC experiments proposed by Bjorken showed that pointlike objects were inside the nucleons. Form factors going like $1/p^4$ indicated three quarks inside the nucleon [AJRV68].

These developments proved fatal to Chew's democracy. Strong indications that in its simplest form the idea was in trouble existed even before. Models with nearby forces failed to explain why only octets of mesons existed and the 27 representation did not bind. Quark–antiquark and three-quark bound states led automatically to octets and not 27 representation states. Experiments at SLAC in deep inelastic scattering removed all doubts of the reality of quarks and the theoretical struggle was to understand how these light states are not produced.

In view of this evidence the phenomenological work continued. The tools that became useful were the superconvergence and the finite energy sum rules introduced mainly by Fubini, and discussed in the Chapter by Ademollo. Amazingly enough this led to string theory and all the work described elsewhere in this Volume. The phenomenology of duality first pointed out by Dolen, Horn and Schmid [DHS68] opened new vistas on the nature of the strong interactions. Soon it was discovered that the dual model predicted parallel Regge trajectories (Rubinstein, Schwimmer, Veneziano and Virasoro [RSVV68]) and much phenomenology. It all culminated with the Veneziano model [Ven68]. I will describe here some early phenomenological results that seemed to be the proper explanation of the behaviour of strong interactions. As we all know, this is one of the most frustrating chapters of the theory of elementary particles. Experiments confirmed the almost linearity of meson and baryon trajectories as predicted by the finite energy sum rules, and couplings also fell in line. Soon after, good and bad results started to accumulate. At the time of the international Vienna conference in 1968 more than two hundred papers had appeared on the subject.

Theoretically, serious inconsistencies were shown to exist unless the number of dimensions was different than four. Even then, not all diseases could be cured.

Phenomenologically, things started looking promising. Using the n-point function for $n = 5$ and 6 legs in its crudest form, the multiparticle amplitudes showed a pattern of presence and absence of resonances that fit experiment qualitatively without effort. In a few words, not every peak in the distributions was a particle, many were dual reflections.

The showcase however was the study of the annihilation of an antiproton at rest with a neutron into three charged pions first proposed by Lovelace [Lov68] (see also [AR69]). This reaction showed some features that are quite spectacular. First, analyzed naively, it shows a low energy exotic resonance (doubly charged state); second, the Dalitz plot shows a total depletion of events at its centre. The natural interpretation is a resounding success of the Veneziano formula. First, the resonance is just a dual reflection of the cross channel and not a new state; second, the zero reflects a simple property of the scattering amplitude

$$A(s, t) = \frac{\Gamma(\alpha(s))\Gamma(\alpha(t))}{\Gamma(\alpha(s) + \alpha(t) - 1)}. \tag{7.1}$$

This simple amplitude embodies all the necessary conditions for the scattering amplitude. It is exotic and therefore has no resonances in the direct channel, and has the appropriate asymptotic behaviour. The antiproton–neutron initial state at rest is equivalent to a heavy pion and therefore the amplitude has spinless external legs. This configuration avoids the complications that the theory has to face in the presence of fermions. The amplitude vanishes for $\alpha(s) = \alpha(t) = 1/2$. In the relevant units it is the centre of the Dalitz plot! This is what is measured experimentally (Anninos et al. [AGHK68]). This result may be accidental but gives a clue of the possible relevance of the Veneziano amplitude for hadronic physics.

This is very encouraging but we know that quantum chromodynamics (QCD) gives a natural explanation of strong processes even if no rigorous proof of confinement exists yet. The question is still whether QCD will be embedded in some form of string theory. Recent work using the Maldacena conjecture is interesting but not definitive.

The early Seventies led to a new branching: the return to field theory and the search for a field theory of the weak interactions. In a remarkable twist the quark confinement had led also to a field theory of the strong interactions: QCD. This theory and the electro-weak model have been accurately tested in Geneva by LEP and have restored the preeminence of field theory.

The early significant results of string theory, though interesting, were eclipsed by these developments. Nevertheless, Green and Schwarz amongst others continued working on it, and their discovery of anomaly cancellations in 1984 gave new impetus to the theory.

7.3 Conclusions

After the outstanding success of quantum electrodynamics, the period 1950–1960 was dominated by the discovery of a rich variety of particles, both strongly and weakly interacting. Some were predicted for theoretical reasons, like the heavy bosons a bit later. Some, like the triplication of generations, remain a mystery today. On the theoretical side, phenomenology played a role. Though historically, understanding the high energy limit was the difficult part, the new paradigm, asymptotic freedom, made the low energies difficult to calculate. Regge classification of the hadrons is very successful but incorporating it in QCD has not been possible until now.

Of the two competing approaches to particle physics, field theory and S-matrix theory, both had some rewards but not complete success. Group theory led to quarks, but grand unification is still not established and certainly its simplest form is ruled out. Analyticity and S-matrix theory led to string theory and a possible unification of all interactions. However, the hopes seem far from being realized.

It is field theory that has revived, but its form beyond the Standard Model is unclear. The new element in the game is cosmology, which has introduced new complications. Can string theory accommodate naturally inflation and other aspects of cosmology? This is certainly not impossible, but no obvious model has been proposed. The situation makes even more challenging the search for a Theory of Everything.

At the present time the most promising aspect of string theory is the understanding of gravity. The progress is slow but steady, and hopes are still high.

Hector Rubinstein (1933–2009) was born in Buenos Aires. He studied in Argentina and received his PhD at Columbia, New York. He worked on quarks and early string theory at the Weizmann Institute, in collaboration with A. Schwimmer, G. Veneziano, M.Virasoro and other students. Later, his main interest was QCD and in recent years cosmology, in particular the nagging question of magnetic fields in the Universe.

References

[Adl65] Adler, S. L. (1965). Calculation of the axial vector coupling constant renormalization in beta decay, *Phys. Rev. Lett.* **14**, 1051–1055.

[AR69] Altarelli, G. and Rubinstein, H. R. (1969). Dalitz plots including duality, *Phys. Rev.* **183**, 1469–1471.

[AJRV68] Amati, D., Jengo, R., Rubinstein, H. R., Veneziano, G. and Virasoro, M. (1968). Compositeness as a clue for the understanding of the asymptotic behaviour of form factors, *Phys. Lett.* **B27**, 38–41.

[AGHK68] Anninos, P., Gray, L., Hagerty, P., Kalogeropoulos, T., Zenone, S., Bizzarri, R., Ciapetti, G., Gaspero, M., Laakso, I., Lichtman, S. and Moneti, G. C. (1968). Production of three charged pions in $\bar{p} + n$ annihilation at rest, *Phys. Rev. Lett.* **20**, 402–406.

[CT66] Callan, C. G. and Treiman, S. B. (1966). Equal time commutators and K meson decays, *Phys. Rev. Lett.* **16**, 153–157.

[CF61] Chew, G. F. and Frautschi, S. C. (1961). Principle of equivalence for all strongly interacting particles within the S matrix framework, *Phys. Rev. Lett.* **7**, 394–397.
[CF62] Chew, G. F. and Frautschi, S. C. (1962). Regge trajectories and the principle of maximum strength for strong interactions, *Phys. Rev. Lett.* **8**, 41–44.
[CM60] Chew, G. F. and Mandelstam, S. (1960). Theory of low-energy pion–pion interactions, *Phys. Rev.* **119**, 467–477.
[Das64] Dashen, R. F. (1964). Calculation of the proton–neutron mass difference by s-matrix methods, *Phys. Rev.* **B135**, 1196–1202.
[DHS68] Dolen, R., Horn, D. and Schmid, C. (1968). Finite energy sum rules and their application to $\pi-N$ charge exchange, *Phys. Rev.* **166**, 1768–1781.
[ELRS66] Elitzur, M., Lipkin, H. J., Rubinstein, H. R. and Stern, H. (1966). Spin 2+ meson decays in the quark model, *Phys. Rev. Lett.* **17**, 420–422.
[Fro61] Froissart, M. (1961). Asymptotic behavior and subtractions in the Mandelstam representation, *Phys. Rev.* **123**, 1053–1057.
[Gel61] Gell-Mann, M. (1961). The eightfold way, preprint CTSL-20, Caltech, Pasadena, CA.
[LF65] Levin, A. M. and Frankfurt, L. L. (1965). The quark hypothesis and relations between cross-sections at high-energies, *JETP Lett.* **2**, 65–70.
[Lov68] Lovelace, C. (1968). A novel application of Regge trajectories, *Phys. Lett.* **B28**, 264–268.
[Mar63] Martin, A. (1963). Unitarity and high-energy behavior of scattering amplitudes, *Phys. Rev.* **129**, 1432–1436.
[Nee61] Ne'eman, Y. (1961). Derivation of strong interactions from a gauge invariance, *Nucl. Phys.* **26**, 222–229.
[Reg59] Regge, T. (1959). Introduction to complex angular momentum, *Nuovo Cimento* **14**, 951–976.
[RSVV68] Rubinstein, H. R., Schwimmer, A., Veneziano, G. and Virasoro, M. (1968). Generation of parallel daughters from superconvergence, *Phys. Rev. Lett.* **21**, 491–495.
[RS66] Rubinstein, H. R. and Stern, H. (1966). Nucleon–antinucleon annihilation in the quark model, *Phys. Lett.* **21**, 447–449.
[Sal08] Salpeter, E. E. (2008). Bethe–Salpeter equation – the origins, preprint arXiv:0811.1050.
[Ven68] Veneziano, G. (1968). Construction of a crossing-symmetric, Regge behaved amplitude for linearly rising trajectories, *Nuovo Cimento* **A57**, 190–197.
[Wei66] Weinberg, S. (1966). Pion scattering length, *Phys. Rev. Lett.* **17**, 616–621.
[Wei65] Weisberger, W. I. (1965). Renormalization of the weak axial vector coupling constant, *Phys. Rev. Lett.* **14**, 1047–1051.
[Zac61] Zachariasen, F. (1961). Self-consistent calculation of the mass and width of the $J=1, T=1, \pi-\pi$ resonance, *Phys. Rev. Lett.* **7**, 112–113.
[Zwe64] Zweig, G. (1964). CERN Report no. 8182, unpublished.

8
Two-component duality and strings

PETER G.O. FREUND

The Veneziano model [Ven68], the starting point of string theory, addressed the at that point much studied and phenomenologically successful idea of 'two-component duality'. Here I would like to recall this idea and give its meaning in modern terms. At the risk of letting the cat out of the bag at too early a stage, let me say right away that the two components in question will turn out to be the open and the closed hadronic strings.

The argument for two-component duality is the following. Unlike quarks, hadrons (mesons and baryons) are obviously not elementary objects. Yet the particles appearing in the initial, final and intermediate states of the hadronic S-matrix, describing all quantum processes involving hadrons, are precisely these composite objects and not their constituent quarks. With elementary particles it is obvious how to calculate the S-matrix, just use the Feynman rules. For instance, when studying lowest order (tree-level) Bhabha scattering (i.e. electron–positron scattering) in QED, we are instructed to *add* the so-called s- and t-channel photon-pole Feynman diagrams, as in Figure 8.1. But there we have a Lagrangian and the photon is an 'elementary' particle whose field appears in this Lagrangian.

When dealing with composite states, these are represented by an infinite sum of Feynman diagrams in the quantum field theory of the elementary fields out of which the composite particles are built. In the simplest case we can think of these diagrams as Bethe–Salpeter ladder-diagrams.[1] If only one kind of line (field) is involved in these ladder-diagrams, as for instance in the theory of a single scalar field $\Phi(x)$ with a $g\Phi^3$ Lagrangian, then adding the diagrams in which the composite particle pole appears in the s- and in the t-channels would lead to the double counting of the one-box diagram as in Figure 8.2.

If, instead of these Bethe–Salpeter ladder-diagrams, we were summing over all planar 'fishnet' diagrams of say a scalar QFT with cubic and quartic self-interaction (an example of such a planar fishnet diagram is given in Figure 8.3), then not only the one-box diagram, but each and every diagram would be counted twice. This partial or total double counting is

---

[1] Editors' note: i.e., the ladder diagrams that occur in the Bethe–Salpeter self-consistent equation for relativistic bound states.

The Birth of String Theory, ed. Andrea Cappelli, Elena Castellani, Filippo Colomo and Paolo Di Vecchia.
Published by Cambridge University Press. © Cambridge University Press 2012.

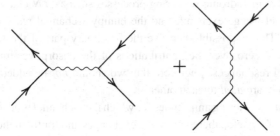

Figure 8.1 Tree-level Bhabha scattering diagrams.

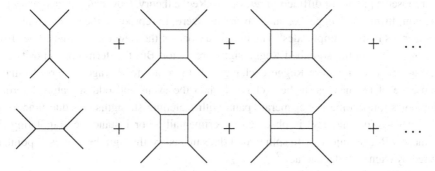

Figure 8.2 Partial double counting when adding s- and t-channel Bethe–Salpeter diagrams.

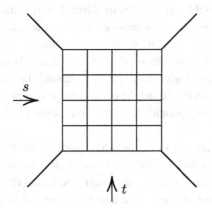

Figure 8.3 Full double counting when adding s- and t-channel planar fishnet diagrams.

the essential difference between a theory of elementary particles and a theory of composite particles.

In their seminal paper, Dolen, Horn and Schmid [DHS67] showed that for the composite hadrons studied in the laboratory such a full double counting would be involved. I distinctly remember the excitement with which we learned of their paper from Murray Gell-Mann, when he visited us in 1967 before the circulation of the preprint. Dolen, Horn and Schmid carefully showed that in hadronic scattering processes such as πN charge-exchange, both the smooth t-channel Regge exchange and the bumpy s-channel resonances account for the *full* amplitude. This is possible because the imaginary part of the t-channel (Regge) exchange averages to zero over the contributions of the absorptive (imaginary) parts of the direct s-channel resonances. Therefore the two should *not* be added, to avoid double counting. Rather they are *dual* to each other.

But even for elastic scattering amplitudes for which the s-channel is devoid of resonances, such as $\pi^+\pi^+$, pp, K^+p, etc., there are Regge exchanges and for them there is also diffraction. Diffraction is described by exchanging a Regge pole (or maybe a more complicated Regge singularity) called the Pomeron after the Russian physicist I. Ya. Pomeranchuk, who first proposed a picture of diffraction relevant to Regge theory. How can the imaginary part of the non-diffractive Regge exchange average to zero, to reproduce the observed absence of resonances in these amplitudes? The imaginary part of the Regge exchange comes from the term $e^{-i\pi\alpha(t)}$ in the so-called Regge signature factor. But this term has opposite sign for even- and odd-signature Regge exchanges. If these are to average to zero, to match the absence of resonances in the s-channel, then the even- and odd-signature hadronic Regge poles must come in degenerate pairs with matching strengths (residue functions). This 'exchange degeneracy' is observed experimentally. For instance, extrapolating the rectilinear ρ Regge trajectory to spin two, it does indeed go through the f-meson point, as required by exchange degeneracy.[2]

But all this still does not take the Pomeron into account. As a consequence of unitarity, diffraction should correspond not to tree-level but to higher order processes and it was conjectured by Freund [Fre68] and by Harari [Har68] that, unlike the other Regge poles, the Pomeron is dual not to s-channel resonances, but to s-channel non-resonant background. With this FH conjecture a two-component picture has thus emerged in which, besides the mesonic and baryonic Regge trajectories dual to resonances, there is a second component, the Pomeron, dictated by unitarity as a largely t-channel flavour singlet trajectory, dual to non-resonant s-channel background. This two-component picture accounted for a vast body of data. The remaining question was how to account for the crucial features of the Pomeron this way.

Mesonic Regge poles and their dualities were modelled by Rosner [Ros69] and Harari [Har69] with what were called 'duality diagrams' – such as the one in Figure 8.4 – and what are in retrospect clearly open string diagrams, Chan–Paton [PC69] rules and all that. For the Pomeron, loop diagrams are dictated by unitarity. The simplest planar diagram of Figure 8.5 is clearly not of the right type, for neither does it select the flavour singlet in its t-channel, as its dominant part, nor does it correspond to nonresonant background in the s-channel. In fact it is nothing more than a loop correction to t-channel Regge pole exchange, or equivalently to direct s-channel resonances. Freund, Rivers and Jones [FR69, FJR71] found the diagram which selects the flavour singlet to be the one in Figure 8.6.

---

[2] Editors' note: see for instance Figure 2.2 in the Chapter by Gabriele Veneziano.

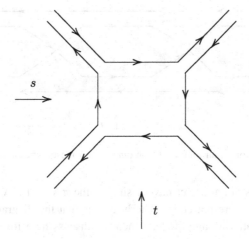

Figure 8.4 A duality diagram.

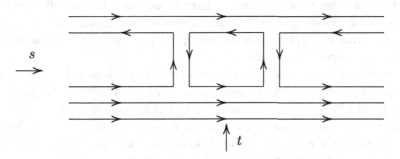

Figure 8.5 Planar loop meson–baryon scattering duality diagram.

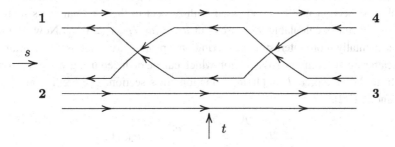

Figure 8.6 Nonplanar loop Pomeron diagram for meson–baryon scattering.

It has no three-quark intermediate state in the s-channel, in other words no resonances and, as such, only non-resonant background as required by the FH conjecture. Moreover, this diagram obviously selects the t-channel singlet: the quarks of the hadrons labelled 1 and 4 in Figure 8.6 are always the same, as are those for hadrons 2 and 3.

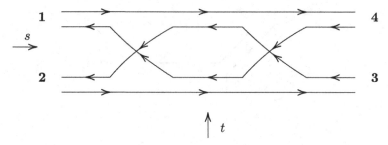

Figure 8.7 Nonplanar loop Pomeron diagram for meson–meson scattering.

But the Pomeron is not an exact flavour singlet: the πN and KN total cross-sections observed in experiment are not equal. The beauty is that the diagram of Figure 8.7 for the meson–meson scattering amplitude, which we discuss here for simplicity, accounts quantitatively for the different values of say the $\rho\pi$ and $\phi\pi$ total cross-sections. Essentially, as follows from the work of Carlitz, Green and Zee [CGZ71], this amounts to noticing that, viewed from the t-channel, the diagram of Figure 8.7 corresponds to the following picture. The two mesonic open strings 2 and 3 merge into a unique open string, which then closes up into a closed string, which propagates and then opens up into an open string and ultimately breaks up into the open strings 1 and 4.

We see that the Pomeron is different because it corresponds to a closed string which couples differently to ρ-mesons and to ϕ-mesons. This is so because, as we just saw, the Pomeron couples by first opening up into a string that has non-strange quarks at its ends in the case of the ρ-meson and strange quarks at its ends in the case of the ϕ-meson. The energy denominators associated with these two cases are different and it is clear that the larger denominator will appear in the ϕ case since strange quarks are heavier than non-strange quarks and thus the difference between the Pomeron and vector meson trajectories' intercepts is larger in the case of the ϕ-meson. We immediately understand that the $\phi\pi$ total cross-section will be suppressed with respect to the $\rho\pi$ total cross-section; the suppression factor is calculable and turns out to be $(m_\rho/m_\phi)^2 \simeq 0.57$. Now returning to the experimentally more interesting scattering on a proton target, let us concentrate on the $J/\psi p$ scattering differential cross-section which can be deduced using vector meson dominance from the measured J/ψ-photoproduction cross-section. The prediction of Carlson and Freund [CF72],

$$\frac{d\sigma_{J/\psi p}}{dt} \bigg/ \frac{d\sigma_{\rho p}}{dt} = \frac{8}{9} \frac{m_\rho^4}{m_{J/\psi}^4} \simeq 0.0035,$$

was confirmed experimentally (see the review of Gaillard, Lee and Rosner [GLR75]).

In the meantime, Veneziano's four-point amplitude was generalized to N-point amplitudes, and people started calculating loop amplitudes. The diagram of Figure 8.7 could now be calculated, and this was done by Frye and Susskind [FS70] and by Gross, Neveu, Scherk and Schwarz [GNSS70], and its careful analysis by Lovelace [Lov71] revealed an amplitude with singularities which violate the well-known analyticity properties of scattering amplitudes. Lovelace noticed that this serious problem could be avoided for bosonic strings if spacetime had precisely 26 instead of 4 dimensions. Critical string theory was thus started. Schwarz [Sch72] and Goddard and Thorn [GT72] then found that for superstrings the critical dimension was reduced to 10.

The advent of QCD put strings on the back burner, until Schwarz and Scherk [SS74] and Yoneya [Yon74] made the bold proposal that string theory is much more than a theory of hadrons; it contains the graviton and could be the ultimate physical theory. Then in 1984, things started moving, but that is no longer early string theory history.

Peter G. O. Freund was born in 1936 in Timisoara, Romania. He obtained his PhD at the University of Vienna. Since 1965 he has been on the faculty of the University of Chicago, currently as Professor Emeritus. He has worked on compactification of extra dimensions, strings, magnetic monopoles and phenomenology. He is author of the general interest book 'A Passion for Discovery'.

References

[CGZ71] Carlitz, R., Green, M. and Zee, A. (1971). A model for pomeranchukon couplings, *Phys. Rev. Lett.* **26**, 1515–1518.
[CF72] Carlson, C. and Freund, P. G. O. (1972). The case for a quartet model of hadrons, *Phys. Lett.* **B39**, 349–352.
[DHS67] Dolen, R., Horn, D. and Schmid, C. (1967). Prediction of Regge parameters of ρ poles from low-energy πN data, *Phys. Rev. Lett.* **19**, 402–407.
[Fre68] Freund, P. G. O. (1968). Finite-energy sum rules and bootstraps, *Phys. Rev. Lett.* **20**, 235–237.
[FJR71] Freund, P. G. O., Jones, H. F. and Rivers, R. J. (1971). Dynamics versus selection rules in diffraction dissociation, *Phys. Lett.* **B36**, 89–92.
[FR69] Freund, P. G. O. and Rivers, R. J. (1969). Duality, unitarity and the Pomeranchuk singularity, *Phys. Lett.* **B29**, 510–513.
[FS70] Frye, G. and Susskind, L. (1970). Non-planar dual symmetric loop graphs and pomeron, *Phys. Lett.* **B31**, 589–591.
[GLR75] Gaillard, M. K., Lee, B. W. and Rosner, J. L. (1975). Search for charm, *Rev. Mod. Phys.* **47**, 277–310.
[GT72] Goddard, P. and Thorn, C. B. (1972). Compatibility of the pomeron with unitarity and the absence of ghosts in the dual resonance model, *Phys. Lett.* **B40**, 235–238.
[GNSS70] Gross, D. J., Neveu, A., Scherk, J. and Schwarz, J. H. (1970). Renormalization and unitarity in the dual-resonance model, *Phys. Rev.* **D2**, 697–710.
[Har68] Harari, H. (1968). Pomeranchuk trajectory and its relation to low-energy scattering amplitudes, *Phys. Rev. Lett.* **20**, 1395–1398.
[Har69] Harari, H. (1969). Duality diagrams, *Phys. Rev. Lett.* **22**, 562–565.
[Lov71] Lovelace, C. (1971). Pomeron form factors and dual Regge cuts, *Phys. Lett.* **B34**, 500–506.
[PC69] Paton, J. F. and Chan, H. M. (1969). Generalised Veneziano model with isospin, *Nucl. Phys.* **B10**, 516–520.
[Ros69] Rosner, J. L. (1969). Graphical form of duality, *Phys. Rev. Lett.* **22**, 689–692.

[SS74] Scherk, J. and Schwarz, J. H. (1974). Dual models for nonhadrons, *Nucl. Phys.* **B81**, 118–144.
[Sch72] Schwarz, J. H. (1972). Physical states and pomeron poles in the dual pion model, *Nucl. Phys.* **B46**, 61–74.
[Ven68] Veneziano, G. (1968). Construction of a crossing-symmetric, Reggeon-behaved amplitude for linearly rising trajectories, *Nuovo Cimento* **A57**, 190–197.
[Yon74] Yoneya, T. (1974). Connection of dual models to electrodynamics and gravidynamics, *Prog. Theor. Phys.* **51**, 1907–1920.

9
Note on the prehistory of string theory

MURRAY GELL-MANN

I thought it might be interesting to mention here some personal recollections of the prehistory and early history of string theory. These reminiscences are presented in an informal manner, as if they were a contribution to oral history, without the usual footnotes and references of a scientific article.

I have always been a strong supporter of string theory, although (especially early on) I did not know exactly, any more than others did, how it would be useful.

During the Seventies and Eighties, in accordance with my role as an ardent conservationist, I set up at Caltech a nature reserve for endangered superstring theorists. I brought John Schwarz and Pierre Ramond to Caltech and encouraged André Neveu to visit.

Over the next few years we hosted Joël Scherk and Michael Green and a number of other brilliant long-term visitors. Some of our graduate students became distinguished superstring theorists. Between 1972 and 1984, a significant fraction of the work on superstrings was done at Caltech, but I myself did not carry out original research on superstrings. Earlier, however, I did have a connection with the prehistory of string theory.

During the Sixties I regarded somewhat favourably the bootstrap approach to the theory of hadrons and the strong interaction, as put forward by Chew and Frautschi. It was connected with the mass-shell formulation of quantum field theory. I had proposed that formulation in the mid-Fifties and described it at the Rochester Conference on High Energy Physics in 1956. In my talk, I referred casually to Heisenberg's notions about the importance of the S-matrix.

During the next few years I tried to convince Geoffrey Chew of the value of the mass-shell formulation, but with only limited success until, in 1961, at the La Jolla meeting that Keith Brueckner and I organized, Chew proposed it himself, labelling it 'S-matrix theory' and claiming that it was somehow different from field theory. He thought that field theory was not applicable to hadrons, but S-matrix theory was. I have never understood that claim. For traditional field theories, especially gauge theories, the mass-shell approach is just a reformulation, using analyticity, crossing symmetry, and extended unitarity, of the

---

The Birth of String Theory, ed. Andrea Cappelli, Elena Castellani, Filippo Colomo and Paolo Di Vecchia.
Published by Cambridge University Press. © Cambridge University Press 2012.

scattering amplitudes evaluated on the mass shell (i.e. all external four-momenta satisfying the mass-shell condition $p_i^2 = m_i^2$, $i = 1, \ldots, n$).

For superstring theory, the situation may be different. It is always presented using the mass-shell approach and has never been reformulated as a field theory in a completely satisfactory way. Perhaps the long-sought unified theory of all the particles and forces does require the mass-shell formulation. Of course there is always the tricky point that in a theory that includes QCD the colour nonsinglet objects, such as quarks and gluons, are confined and do not really have a mass shell. The S-matrix actually deals only with the colour singlets.

Geoffrey Chew laid great stress on the behaviour of elementary particles in the complex angular momentum plane. They were supposed to be fixed poles, in contrast to the Regge poles that characterized composite objects. In the bootstrap scheme of Chew and Frautschi, all the hadrons were to be composites of one another and thus represented by Regge poles. But then 'Murph' Goldberger and I, together with our collaborators, presented evidence that, in gauge theories, the elementary particles turned from fixed poles into Regge poles when radiative corrections were included. The bootstrap proposal using mass-shell theory gave similar behaviour to that of conventional field theory (at least for gauge theories).

As mentioned before, I liked the bootstrap idea. But I was not happy about the problems that were studied as examples; for instance the pion–pion scattering amplitude using just the ρ-meson as an intermediate state in the s-, t-, and u-channels. I thought that was far too restrictive. Even if baryons were excluded for the moment, I would have preferred to see an infinite number of meson states as incoming and outgoing particles and as intermediate states as well, in all the channels. Those mesons would all be treated initially as stable, zero-width resonances and then later, in a higher approximation, as decaying states with finite widths. That seemed to me to be the direction in which to go rather than beating to death the simple π and ρ system. My postdocs Dolen, Horn and Schmid ran with this idea in their ingenious paper on what was called 'duality'.

Then we heard about the remarkable formula of Veneziano, the result of his search for a concrete realization of the dual bootstrap. It achieved exactly what we wanted, namely consistent behaviour in all three channels with the same infinite set of mesons. It was not long before the Veneziano model was shown to be a string theory. But that string theory had some important defects. The need for 26 dimensions instead of four could perhaps be tolerated if the extra 22 spatial dimensions were curled up into a tiny ball, as was suggested later. More serious was the inclusion, as one of the mesons, of an unphysical tachyon state. And of course the suppression of baryons was something that would have to be altered.

Then came the work of Neveu and Schwarz and of Ramond, resulting in the superstring theory in ten dimensions, with fermions as well as bosons, related by supersymmetry, which would have to be violated to agree with observation. Superstring theory appeared at first to have unphysical states like those of the purely bosonic string theory, but it turned out later that those states could be consistently omitted.

Still, as a theory of hadrons, the superstring picture was seriously flawed. For one thing, the state of zero mass and spin two was an embarrassment. Then suddenly, by changing the squared masses in the theory by a factor of ten to the thirty-eighth or so, that problem was solved. The mysterious spin two particle was the graviton and the theory was now a candidate for a unified theory of all the particles and all the forces rather than a theory of the hadrons.

In a suitable approximation, the superstring theory yields Einstein's general-relativistic theory of gravitation. Whereas quantized Einsteinian gravitation gave rise to unrenormalizable infinite radiative corrections, superstring theory, by what seemed like a miracle, had the infinities cancel out, at least for the lowest radiative corrections. That remarkable result persuaded me that superstring theory may well have something to do with the ultimate unified theory.

Murray Gell-Mann was born in New York in 1929. He is currently Distinguished Fellow at the Santa Fe Institute (which he helped to found) as well as the Robert Andrews Millikan Professor Emeritus at the California Institute of Technology, where he joined the faculty in 1955. In 1969 he was awarded the Nobel Prize in physics for his work on the theory of elementary particles. Gell-Mann's contributions to physics include work on the renormalization group, the theory of the weak interaction, quantum field theory on the mass shell, and the theory of quarks and gluons, which are the fundamental building blocks of strongly interacting particles.

Part III
The Dual Resonance Model

10

Introduction to Part III

10.1 Introduction

Part III deals with the developments of string theory in the period 1969–1973, during which the Veneziano amplitude was generalized to the scattering of an arbitrary number of particles and the spectrum of intermediate states was obtained. This S-matrix theory, implementing the Dolen–Horn–Schmid (DHS) planar duality, was called the Dual Resonance Model (DRM) and was later recognized to correspond to open bosonic string theory.

At the same time, the four-point amplitude of another model with nonplanar duality was constructed and then generalized to the scattering of N particles. This model was called the Shapiro–Virasoro model and its amplitudes described the scattering of closed strings. Remarkably, most of the properties of bosonic string theory were established in that context with no reference to the string dynamics and Lagrangian; the connection was firmly established only in 1973 with the work of Goddard, Goldstone, Rebbi and Thorn (see Part IV).

Through a number of brilliant guesses, the N-particle amplitude was uniquely determined and its Möbius and conformal symmetries were identified as characteristic of an associated two-dimensional space, later recognized as the world-sheet of the string. An operator description of the N-point amplitude was introduced, involving an infinite set of harmonic oscillators a_n^μ and $a_n^{\dagger\mu}$, with spacetime index μ and integer mode n, and the associated Fubini–Veneziano field operator $Q_\mu(z)$, later identified with the string oscillation modes and string coordinate, respectively. The amplitudes were written as correlators of two-dimensional quantum fields called vertex operators. Through the analysis of the factorization properties of the residues of the simple poles of N-point amplitudes, and the use of the operator formalism, the structure of the Hilbert space spanned by the intermediate states was determined: this involved the elimination of unwanted states with negative norm, called ghosts, using theoretical methods that were new at the time.

The main results concerning the physical spectrum were that: (i) it was consistent only for the values $\alpha_0 = 1$ of the intercept of the Regge trajectory, and $d = 26$ of the spacetime

---

The Birth of String Theory, ed. Andrea Cappelli, Elena Castellani, Filippo Colomo and Paolo Di Vecchia.
Published by Cambridge University Press. © Cambridge University Press 2012.

dimension; (ii) it contained a massless spin one field (later identified with a gauge field) in the Dual Resonance Model (open string) and a massless spin two field (later, the graviton) in the Shapiro–Virasoro model (closed string); (iii) in both cases, it contained a tachyon, a particle with negative mass squared and thus moving faster than light.

As explained in the Introduction to Part II, unitarity was violated by the DRM amplitudes, owing to their zero-width resonances; this problem was cured by adding loop diagrams. The corresponding duality diagrams involved two-dimensional surfaces of growing topology with boundaries, such as the annulus at one-loop in the case of the DRM, and without boundaries, such as the torus at one-loop in the Shapiro–Virasoro model. Higher loops involved higher-genus Riemann surfaces with and without boundaries, that were later understood as corresponding to open and closed string world-sheets. In summary, the perturbative expansion of interacting string theory was completely set up during this period.

Concerning these developments, it is worth making some remarks.

(i) The research in this period focussed on the study of the underlying structure of the Dual Resonance Model. One motivation for investigating the theory behind the Veneziano amplitude was that this could suggest ways of avoiding tachyons and finding better contact with hadron phenomenology. Another motivation was the deep and beautiful structure that the theory seemed to possess. This second motivation prevailed as the results of the theory were gradually found to be incompatible with hadron physics, owing to the presence of massless particles with spins one and two in the spectrum, the need for 26 dimensions, and the prediction of too soft scattering at large t values. This change of perspective also had an impact on the careers of those who continued to work on the DRM.

(ii) The interpretation of DRM amplitudes in terms of propagating relativistic strings was proposed in 1969, soon after the Veneziano formula, by Nambu, Nielsen and Susskind, independently. The correct form of the relativistic Lagrangian was found in 1970 by Nambu and later by Goto (see Part IV); however, it was not really used until 1973. One reason for this delay was that it was not clear how to quantize the Lagrangian: its nonlinear form and the local reparameterization symmetry, similar to the gauge symmetry of quantum electrodynamics, required new theoretical tools that took some time to be developed.

(iii) Notwithstanding the brilliant intuitions and discoveries, it is rather astonishing that the amplitude involving many particles could be uniquely found without any further assumptions and that the full spectrum of the theory could be determined. For the theoretical physicists involved, the uniqueness of the dual model was a source of beauty and motivated their efforts. Today we clearly understand that this success of the S-matrix approach was due to the strong constraints put on the quantum theory by the infinite-dimensional conformal symmetry.

In this Chapter we provide some background to the developments of the Dual Resonance Model described in Part III. The discovery of the string dynamics underlying the dual models will be described in Part IV; a detailed account of the quantization of the

Nambu–Goto action is given in Appendix C. Further extensions of the Dual Resonance Model that took place during the same years 1969–1973, such as the generalization to fermionic degrees of freedom and the inclusion of internal symmetries, are treated in Part V.

10.2 N-point dual scattering amplitudes

Let us start from the four-point Veneziano amplitude and discuss the features that allowed for its generalization. The amplitude contains three terms,

$$A(s, t, u) = A(s, t) + A(s, u) + A(t, u), \tag{10.1}$$

involving the beta function,

$$A(s, t) = \frac{\Gamma(-\alpha(s))\Gamma(-\alpha(t))}{\Gamma(-\alpha(s) - \alpha(t))}$$

$$= \int_0^1 dx \, x^{-\alpha(s)-1}(1-x)^{-\alpha(t)-1}, \tag{10.2}$$

where the three Mandelstam variables s, t, u are given by:

$$s = -(p_1 + p_2)^2, \qquad t = -(p_2 + p_3)^2, \qquad u = -(p_2 + p_4)^2. \tag{10.3}$$

Here $\alpha(s) = \alpha_0 + \alpha' s$ is a linearly rising Regge trajectory and the total momentum is conserved: $p_1 + p_2 + p_3 + p_4 = 0$ (we use the convention that all particles are incoming). The shift by one in the argument of the gamma functions with respect to the amplitude discussed in the Introduction to Part II is due to the fact that here we consider four spinless particles (no helicity flip), while the earlier amplitude described three spinless particles, i.e. pions, and a spin one particle, the ω (helicity flip by one unit).

The integral representation of the beta function in (10.2) is originally defined for the values $\alpha(s), \alpha(t) < 0$: it becomes singular at $\alpha(s) = 0$ and $\alpha(t) = 0$ from the integration regions $x \to 0^+$ and $x \to 1^-$, respectively. The other singularities of gamma functions for positive integer values of $\alpha(s)$ and $\alpha(t)$ can be obtained by extending (analytically continuing) the integral representation to $\alpha(s), \alpha(t) > 0$ through integration by parts.

Each term of Eq. (10.1) is represented by a planar duality diagram, as shown in Figure 10.1, where the double lines correspond to the world-lines of the quark–antiquark pairs forming the mesons. The amplitude contains pole singularities that can be inferred as follows. The diagram in Figure 10.1 can be deformed in two possible ways while keeping it on the plane. The deformation on the left-hand side shows the propagation of intermediate mesons in the s-channel, leading to a series of poles in the s variable; the deformation on the right-hand side indicates the t-channel poles. This means that the scattering amplitude must contain the poles both in the s-channel and in the t-channel. They are, in fact, obtained by 'pinching' the amplitude, that is by expanding its integral representation (10.2) around $x = 0$ and $x = 1$, respectively. The two series of poles are not

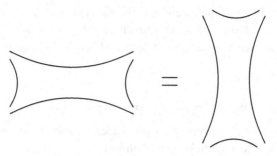

Figure 10.1 Four-point duality diagram: deformations showing (s, t) duality.

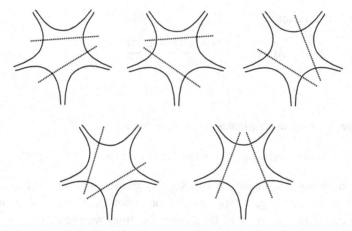

Figure 10.2 Dashed lines denote the pinchings of the five-point duality diagram on two simultaneous channels.

present simultaneously, because the series expansions of the integrand are done around two distinct points of the integration region.

Furthermore, a single duality diagram, as the one in Figure 10.1, is not fully crossing symmetric, because the corresponding amplitude is not invariant under all s, t, u exchanges. This symmetry is obtained by summing over the three noncyclic permutations of the four external particles as in Eq. (10.1). The corresponding duality diagrams are all independent, as they cannot be deformed into each other while keeping them on the plane.

The previous analysis can be generalized to the scattering of more than four particles: the N-particle amplitude is obtained by considering the corresponding planar duality diagrams and by generalizing the integral representation of the beta function in (10.2); this was determined by requiring the correct singularity pattern as described by the allowed deformations of the diagrams.

In the following we consider, as an example, the scattering of five external particles. Upon drawing the duality diagram for five particles, we find that it can be pinched simultaneously in two channels; furthermore, this can be done in five different ways (see Figure 10.2).

Thus planar duality implies a five-particle amplitude with two simultaneous series of pole singularities. This behaviour can be realized by generalizing the integral representation (10.2) to two variables x_1, x_2 and finding an integrand that contains all the poles of the corresponding duality diagram; there are five distinct singular integration regions, each one realizing pairs of pole series. In the Chapter by Di Vecchia in Part III, the amplitude associated with an N-particle diagram is similarly found in terms of an $(N-3)$-dimensional integral representation. The complete crossing-symmetric scattering amplitude is finally obtained by summing over terms with noncyclic permutations of the external particles, each one associated with a noncoplanar duality diagram.

10.2.1 Koba–Nielsen form

We now discuss the Koba–Nielsen representation of the amplitude, that suggested the relation between duality diagrams and two-dimensional surfaces (the open string worldsheet, see Part IV). We present the main steps that led to the association of DRM amplitudes with two-dimensional geometry and unveiled the Möbius and conformal symmetries that played a very important role in all the subsequent developments of the theory.

In the Koba–Nielsen representation of the amplitude, a real coordinate z_i is associated with each meson (e.g. $i = 1, \ldots, 4$ for the four-point amplitude), and the scattering amplitude is written as an integral over the Koba–Nielsen variables z_i. A potential problem could be that, if we integrate over the four independent variables z_i, we may generate simultaneous poles that are not allowed by the corresponding duality diagram (Figure 10.1). The way out is to show that the integrand of the scattering amplitude is invariant under the coordinate transformations of the three-parameter Möbius group, such that only one integration is independent as in Eq. (10.2). Let us enter into more details.

The Möbius transformations,

$$z \to z' = \frac{\alpha z + \beta}{\gamma z + \delta}, \qquad \alpha\delta - \beta\gamma = 1, \qquad (10.4)$$

map the complex plane into itself and are invertible, i.e. they do not introduce any singularity. They are also known as regular conformal transformations. The coefficients $\alpha, \beta, \gamma, \delta$ form a two-dimensional matrix which gives a representation of the group $SL(2, \mathbb{R})$. There are three basic transformations, translations, dilatations and inversions, respectively given by:

$$z \to z + \beta, \qquad z \to \alpha z, \qquad z + \delta \to -\frac{1}{z + \delta}. \qquad (10.5)$$

The Möbius transformations act on the variables z_i of the integral representation of scattering amplitude: if the integrand is Möbius invariant, it can be properly associated with the duality diagram.

To this effect, we introduce the anharmonic ratio $\eta_{ab,cd}$ and integration measure dV_{abc},

$$\eta_{ab,cd} = \frac{z_{ab} z_{cd}}{z_{ac} z_{bd}}, \qquad dV_{abc} = \frac{dz_a\, dz_b\, dz_c}{z_{ab} z_{ac} z_{bc}} \qquad (z_{ab} = z_a - z_b), \qquad (10.6)$$

which are invariant under the transformations (10.4), as can be easily checked; z_a, z_b, z_c, z_d are the Koba–Nielsen variables associated with four particles. Using these quantities, we can rewrite the Veneziano amplitude (10.2) in a manifest Möbius invariant form, which is the Koba–Nielsen representation.

This reads:

$$A(s,t) = \frac{1}{V} \int \prod_{i=1}^{4} dz_i \prod_{1 \leq i < j \leq 4} (z_{ij})^{2\alpha' p_i \cdot p_j}, \tag{10.7}$$

where the variables z_i associated with each meson are integrated on the real axis with the ordering $z_1 > z_2 > z_3 > z_4$, and V is a normalization constant.

By changing variables from the z_i to the Möbius invariant ones, see Eq. (10.6), and using momentum conservation $p_1 + p_2 + p_3 + p_4 = 0$, we can rewrite Eq. (10.7) in the following form:

$$A(s,t) = \frac{1}{V} \int dV_{124} \int_0^1 d\eta_{12,34} \, (\eta_{12,34})^{2\alpha' p_1 \cdot p_2} (1 - \eta_{12,34})^{2\alpha' p_1 \cdot p_4}. \tag{10.8}$$

We choose $\alpha_0 = 1$ for simplicity: more general formulae for $\alpha_0 \neq 1$ can be found in the Chapter by Di Vecchia in Part III, although, as we shall see later, the unit value is required by unitarity and other consistency conditions. For this value the mesons have unphysical negative mass squared, $-p_i^2 = M^2 = -1/\alpha'$, i.e. they are tachyons (this was a problem for the theory, but we shall deal with it later).

The expression (10.8) uses variables $\eta_{12,34}$ and dV_{124} that are manifestly Möbius invariant. In particular, the integral in dV_{124} can be traded for an integral over the $SL(2, \mathbb{R})$ group, with constant Jacobian $|dV_{124}/d(\alpha, \beta, \gamma)|$, because the dV measure is Möbius invariant. Thus, the two integrals in Eq. (10.8) decouple and the first provides a constant (infinite) factor V (the volume of the $SL(2, \mathbb{R})$ group), which is cancelled by the normalization factor $1/V$.

Finally, the original expression of the Veneziano amplitude (10.2) is recovered from (10.8) by using the property of Möbius transformations to set three arbitrary points to fixed values. Let us choose:

$$z_1 = \infty, \quad z_2 = 1, \quad z_4 = 0, \tag{10.9}$$

such that $\eta_{12,34} = z_3$ becomes the variable in the integral representation of the beta function.

The Möbius invariant form of the amplitude (10.8) can be used to write it for different geometries; for example, the transformation,

$$z' = \frac{z-i}{z+i}, \tag{10.10}$$

maps the upper half-plane, Im $z > 0$, to the interior of the unit disc, $|z'| < 1$. On this geometry, the Koba–Nielsen variables are integrated on the boundary circle, as shown in Figure 10.3. Therefore, the Veneziano amplitude can eventually be associated with the two-dimensional geometry of the disc.

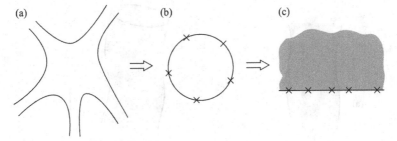

Figure 10.3 (a) Duality diagram; Koba–Nielsen amplitude (b) on the disc and (c) on the upper half-plane.

As described in the Chapter by Di Vecchia, the Koba–Nielsen representation (10.8) was generalized to N particles, making apparent the associated two-dimensional space, later understood as the string world-sheet, and the Möbius symmetry. Actually, the integrand of the amplitudes and the underlying theory possess the larger, infinite-dimensional symmetry under arbitrary analytic coordinate transformations of the complex plane; these are the conformal transformations to be described in Section 10.3. In the next Section, we briefly discuss other results concerning closed strings and loop corrections.

10.2.2 Shapiro–Virasoro model and closed strings

Besides the Dual Resonance Model, another S-matrix theory was developed at the same time. The four-point amplitude was originally written by Virasoro and its generalization to N points was derived by Shapiro by extending the 'analogue model' described in the Chapter by Fairlie (Part IV). These developments are recalled in the Chapter by Shapiro. As understood later, the Shapiro–Virasoro model described the scattering of closed strings.

In the following, the Virasoro four-point amplitude and its duality diagram will be introduced with hindsight, leaving aside technical aspects. We note that the time evolution of an open string is described by a strip with quark and antiquark trajectories at the boundaries (Figure 10.4). On the other hand, the world-sheet of a closed string is a tube: thus, the duality diagram of the corresponding four-point amplitude is a closed (spherical) surface with four holes (Figure 10.5).

The Virasoro amplitude for the scattering of four spinless particles has the following form:

$$A(s,t,u) = \frac{\Gamma\left(-\frac{\alpha(s)}{2}\right)\Gamma\left(-\frac{\alpha(t)}{2}\right)\Gamma\left(-\frac{\alpha(u)}{2}\right)}{\Gamma\left(-\frac{\alpha(s)+\alpha(t)}{2}\right)\Gamma\left(-\frac{\alpha(s)+\alpha(u)}{2}\right)\Gamma\left(-\frac{\alpha(t)+\alpha(u)}{2}\right)}$$
$$\propto \int d^2z\, |z|^{-4-\frac{\alpha'}{2}s} |1-z|^{-4-\frac{\alpha'}{2}t}. \tag{10.11}$$

This expression shows that the closed-string Regge trajectory is $\alpha(s) = 2 + \frac{\alpha'}{2}s$, i.e. it has half slope and twice intercept, i.e. $\alpha_0 = 2$, with respect to that of the Dual Resonance Model

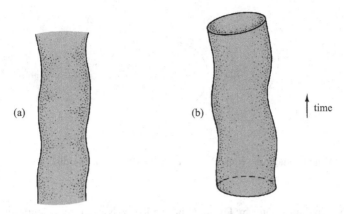

Figure 10.4 World-sheets of the open (a) and closed (b) string.

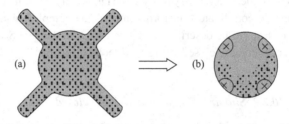

Figure 10.5 (a) Duality diagram for the closed-string amplitude; (b) associated integration region of the sphere.

(the value of α_0 is again determined by consistency conditions, as explained later). The amplitude (10.11) was later understood to correspond to the scattering of four closed string tachyons with mass $M^2 = -4/\alpha'$.

Originally, the DRM and the Shapiro–Virasoro model were considered two independent models. Later on, it was recognized that the DRM contained extra states, belonging to the spectrum of the Shapiro–Virasoro model that occurred in nonplanar loop diagrams at the critical dimension $d = 26$. These states were interpreted as those of the Pomeron, the trajectory with vacuum quantum numbers mentioned in the Introduction to Part II. In fact, as seen from Figure 10.5, the closed string has no end-points to attach quarks and thus the intermediate states carry no flavour quantum numbers.

The integral representation in (10.11) involves one coordinate z and its complex conjugate $\bar{z}$ and the domain of integration is the whole complex plane, equivalent to the sphere (Figure 10.5). Three of the four complex points can again be fixed to $(0, 1, \infty)$. An interesting property of the Virasoro amplitude is that a single expression obeys the DHS duality over all the three scattering channels (complete symmetry under exchanges of s, t, u). This follows from the fact that, because the sphere is a closed surface, the duality diagram for

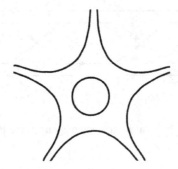

Figure 10.6 Example of a planar one-loop diagram.

the sphere amplitude allows for the three deformations corresponding to the three channels. This property was one of the original motivations for Virasoro's ansatz.

10.2.3 Loop diagrams

The dual amplitudes described so far contain zero-width resonances that are not consistent with unitarity, as discussed in the Introduction to Part II. In analogy with field theory, it was suggested that this problem could be solved by adding loop diagrams. The dual diagrams were considered as the tree diagrams in field theory, and from their factorization the states of the theory were found; next, the loop diagrams were constructed by allowing these states to circulate in the loops (see for example Figure 10.6). More complete presentations of the Dual Resonance Model and loop calculations can be found in the reviews [DiV08] and [DS08], that also include extensive references to the original papers. Aspects of loop calculations are discussed in the Chapters by Amati, Lovelace, Susskind, Neveu and Montonen.

A number of interesting facts arose in the loop calculations.

(i) As in gauge field theories, there are states that do not have positive norm. In order for the quantum theory to be consistent, they should decouple from the physical spectrum and, in particular, they should not circulate in the loops.
(ii) Closed strings appeared as intermediate states in loop diagrams of open strings.
(iii) Unitarity conditions for loop diagrams were only obeyed for a particular value of the spacetime dimension, $d = 26$, the so-called critical dimension. This was the first time that such a requirement appeared in the theory.

The physical and unphysical states can be studied most directly by using the operator formulation in terms of an infinite set of harmonic oscillators, which we shall discuss later. This led to the determination of the values $\alpha_0 = 1$ and 2 for the intercepts of the Dual Resonance and Shapiro–Virasoro models, respectively.

The two other aspects (ii) and (iii) occurred in the study of the nonplanar loop amplitudes. For example, let us discuss the nonplanar loop diagram shown in Figure 10.7, also described

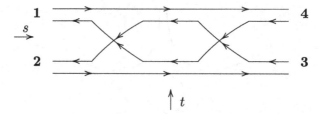

Figure 10.7 Nonplanar one-loop diagram of the Pomeron.

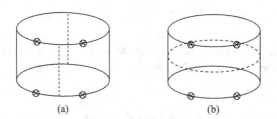

Figure 10.8 Pomeron diagram: (a) open string propagating in one channel; (b) closed string propagating in the crossed channel.

in the Chapter by Freund: it can be continuously deformed in the cylinders of Figure 10.8. The surface of the diagram in Figure 10.8 extends between the two edge circles. Using modern language, it can be described in two ways: (a) in one channel, called the open string channel, it can be seen as an open string loop correction; (b) in the other, called the closed string channel, it corresponds to a tree diagram of the closed string (with vacuum quantum numbers). This twofold interpretation of duality diagrams, not present in perturbative quantum field theory, implies that the open and closed strings should be considered as being part of the same theory. In short, open string loops introduce additional states that correspond to excitations of the closed string.

As discussed in the Chapters by Lovelace and Olive, the calculation of the nonplanar loop diagram in Figure 10.8 led to the discovery of the critical dimension $d = 26$. Due to their phenomenological origin, dual model amplitudes had only been considered in four spacetime dimensions, where the amplitude had a cut singularity in the closed string channel, incompatible with unitarity. Nevertheless, Lovelace extended this calculation to arbitrary spacetime dimension d and number d' of degrees of freedom circulating in the loops. He found that for $d = 26$ and $d' = 24$ the cut turned into poles, which could be interpreted as intermediate states of the closed string propagating in the loop.

A spacetime of 26 dimensions was a big leap of imagination in those times; it was very hard to accept in a model of hadrons. Nonetheless Lovelace's result was confirmed by subsequent studies and therefore gradually accepted. As anticipated, this and other results in disagreement with the phenomenology of strong interactions led some researchers to abandon the subject, while others continued to investigate it for its intriguing and beautiful

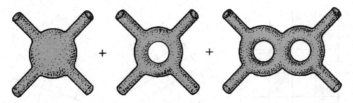

Figure 10.9 Loop expansion in closed string theory involving Riemann surfaces of higher genus.

theoretical structure. The extra dimensions became more plausible when string theory was applied to the unification of all fundamental forces including gravity, see Part VI.

Multiloop diagrams were also computed in 1970 using the so-called N-Reggeon vertex, discussed in the Chapters by Lovelace and Sciuto. Closed string loop amplitudes are depicted in Figure 10.9. At that time, the Abelian differentials, the period matrix and other elements of the geometry of two-dimensional Riemann surfaces were obtained without yet knowing that the underlying dynamics was that of a string. With these results, the dual models had developed into a consistent theory, apart from the problem of tachyons that was solved a few years later (Part VI), and of the integration measure in higher loop diagrams that was determined completely after the first 'string revolution' in 1984 [Ma88].

As we know today, the fundamental reason for the critical dimension $d = 26$ in string theory is the presence of an 'anomaly': this is an obstruction to the extension of a classical symmetry to the quantum level in a system with many degrees of freedom, as in the case of a relativistic theory. As fully understood by Polyakov in 1981, the conformal symmetry of the classical string Lagrangian is anomalous in the quantum theory unless $d = 26$ (see the Introduction to Part VII). Before this work, the anomaly manifested itself through inconsistencies with unitarity or in the implementation of Lorentz symmetry (see Part IV).

10.3 Conformal symmetry

The Möbius transformations can be generalized to a larger set of coordinate reparameterizations of the plane. Consider an infinitesimal, analytic change of the complex coordinate z:

$$z \to z' = z + \epsilon(z), \qquad \epsilon(z) = \sum_{n=-\infty}^{\infty} \epsilon_n z^{n+1}. \tag{10.12}$$

The small increment $\epsilon(z)$ is a generic analytic function, i.e. it obeys $\partial \epsilon(z)/\partial \bar{z} = 0$; it can be expanded in Laurent series around one point, say $z = 0$, with coefficients ϵ_n, $n \in \mathbb{Z}$. The Möbius transformations (10.5) correspond, in infinitesimal form, to the subset of the terms $\epsilon_0, \epsilon_1, \epsilon_{-1}$ in (10.12).

We now consider the transformation of two infinitesimal vectors $dv(z, \bar{z}), dw(z, \bar{z})$: they may be tangent to a family of curves in the plane at the point $(z, \bar{z})$, as in Figure 10.10(c). The relative angle θ between the vectors can be computed from their scalar product, which

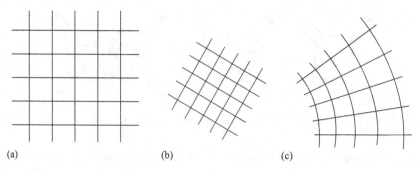

Figure 10.10 (a) Coordinates in the plane; (b) global rotation and dilatation; (c) local rotation and dilatation, i.e. conformal transformation.

in complex notation reads

$$\cos\theta = \frac{\mathrm{Re}\,(\overline{dw}\,dv)}{|dw|\,|dv|}. \tag{10.13}$$

Using the transformation law of vectors under coordinate changes we find:

$$dw(z) = dw'(z')\frac{\partial z'}{\partial z} + \overline{dw'(z')}\frac{\partial \bar{z}'}{\partial z} = dw'(z')\left(1 + \frac{d\epsilon(z)}{dz}\right). \tag{10.14}$$

Under the analytic reparameterizations (10.12) we see that the second, antianalytic term in (10.14) vanishes and thus the vector only acquires the scale factor $(1 + \epsilon'(z))$, which cancels out in the expression for $\cos\theta$ in (10.13). This shows that the relative angle remains invariant. Therefore, the analytic transformations (10.12) have the geometrical property of keeping invariant the relative angle between vectors and are thus called 'conformal'.

In the neighbourhood of each point, the conformal transformations are equivalent to an infinitesimal translation, rotation and dilatation; on the other hand, Möbius transformations correspond to global translations, rotations and dilatations of the whole plane. Their respective actions on a coordinate grid are shown in Figure 10.10. Note that local transformations require an infinite number of parameters as in (10.12); moreover, they cannot be invertible in the whole plane, because the function $\epsilon(z)$, being analytic, must have some singular point somewhere (or be constant). In contrast, Möbius transformations are always invertible. In Section 10.2.1, we have shown that the Möbius transformations leave invariant the integrand of the DRM amplitudes, cf. (10.8); in the next Section, we shall see that the conformal coordinate reparameterizations transform this integrand 'covariantly', i.e. modify it in a definite way.

The algebra of conformal generators is readily obtained. The infinitesimal transformation of a generic function $f(z)$ under (10.12) is found by Taylor expansion:

$$f(z) \to f(z + \epsilon(z)) = f(z) + \epsilon(z)\frac{df}{dz}. \tag{10.15}$$

Upon substituting the Laurent series in (10.12), we find:

$$f(z) \to \left[1 + \sum_n \epsilon_n z^{n+1} \frac{d}{dz}\right] f(z). \tag{10.16}$$

This formula lets us identify the operators multiplying the coefficients ϵ_n as the generators of conformal transformations: $L_n = -z^{n+1} d/dz$. These differential operators obey the Virasoro algebra, which reads:

$$[L_n, L_m] = (n - m) L_{n+m} + \frac{c}{12} n (n^2 - 1) \delta_{n+m,0}. \tag{10.17}$$

The first term on the right-hand side can be verified by commuting the derivatives. The second term cannot be found using this simple argument: it is not an operator but a number, called the 'central extension' of the Virasoro algebra, and is parameterized by the constant c, the 'central charge'.

The presence of the central extension of the algebra is a consequence of the already mentioned conformal anomaly. The central charge c takes a specific value for each quantum theory considered: for the bosonic string $c = d - 26$, where d is the number of spacetime dimensions in which the string propagates, while the number 26 comes from the contribution of Faddeev–Popov ghosts (not described here). As anticipated, the critical dimension corresponds to vanishing anomaly, as required for the proper quantization of the theory (see Part IV). However, this way of understanding the central extension only became clear after Polyakov's work in 1981 (see Part VII).

10.4 Operator formalism

The spectrum of states of the Dual Resonance Model was found by factorizing the N-point amplitude in a given channel and by studying the residues of the poles: these involved a finite number of terms that yielded the degeneracy of states (see Appendix B for an example). As discussed in detail in the Chapter by Di Vecchia, the spectrum was studied by introducing the operator formalism. This amounted to rewriting the N-point amplitudes as quantum-mechanical expectation values of creation–annihilation operators of the harmonic oscillator.

Let us first recall some general features of these operators. For a single degree of freedom, one can introduce the pair $(a, a^\dagger)$ which obeys the commutation relations:

$$\left[a, a^\dagger\right] = 1. \tag{10.18}$$

The physical meaning of these operators is that $a^\dagger$ adds (creates) one quantum, i.e. an excitation, to any state $|\psi\rangle$, while a removes (annihilates) it; the vacuum (ground state) has no excitations, i.e. $a|0\rangle = 0$. The operator $N = a^\dagger a$ counts the number of excitations in any state, as follows from the commutators $[N, a] = -a$ and $\left[N, a^\dagger\right] = a^\dagger$. In fact, the last one implies that:

$$N|j\rangle = j|j\rangle, \qquad |j\rangle \propto \left(a^\dagger\right)^j |0\rangle. \tag{10.19}$$

The harmonic oscillator is described by the Hamiltonian $H = \omega a^\dagger a + E_0$, where ω is the energy of one quantum. The ground state is the vacuum of quanta and has energy $E_0 = \omega/2$, while the excited states are given in (10.19) and their energy is $E_j = \omega j + E_0$, $j = 1, 2, \ldots$.

Harmonic oscillators also appear in quantum field theory where the field operator describes the creation and annihilation of freely propagating quanta; for example, for a scalar particle,

$$\varphi(\vec{x}, t) \sim \sum_{\vec{k}} \left[e^{-i\vec{k}\vec{x} - iEt} a_{\vec{k}} + e^{i\vec{k}\vec{x} + iEt} b_{\vec{k}}^\dagger \right]. \tag{10.20}$$

The field φ is expanded in terms of $a_{\vec{k}}$ operators, each one multiplied by the particle wave function of momentum $k^\mu = (E, \vec{k})$; the $b_{\vec{k}}$ are analogous operators for antiparticles. When applied to a quantum state, the field $\varphi(\vec{x}, t)$ annihilates a particle (with any momentum) and creates an antiparticle.

In the DRM, the scattering amplitude could be written in terms of an infinite set of creation–annihilation operators α_n^μ, with Lorentz index $\mu = 0, \ldots, d-1$ and 'moding' $n \in \mathbb{Z}$. Their commutation relations are:

$$[\alpha_n^\mu, \alpha_m^\nu] = n \delta_{n+m,0} \, \eta^{\mu\nu}, \qquad n, m \in \mathbb{Z} \neq 0, \quad \mu, \nu = 0, \ldots, d-1, \tag{10.21}$$

where $\eta^{\mu\nu} = \mathrm{diag}(-1, 1, \ldots, 1)$ is the Minkowski metric in d dimensions.

The relation with the standard $(a, a^\dagger)$ operators defined earlier is:

$$\alpha_n^\nu = \sqrt{n}\, a_n^\mu, \qquad \alpha_{-n}^\nu = \sqrt{n}\, a_n^{\mu\dagger}, \qquad n > 0. \tag{10.22}$$

In the DRM, the spectrum of intermediate states at the level N consisted of particles with mass M_N^2, expressed in terms of the number operators $\alpha_{-n}^\mu \alpha_n^\mu$ of each oscillator pair (μ, n) as follows:

$$\alpha(M_N^2) \equiv \alpha_0 + \alpha' M_N^2 = \sum_{\mu=0}^{d-1} \sum_{n=1}^{\infty} \alpha_{-n}^\mu \alpha_{n\mu} = N. \tag{10.23}$$

Because of Eq. (10.21), each creation oscillator α_n^μ of mode n contributes with n units to the mass squared. Thus, the number of states at level N, i.e. their degeneracy, is obtained by counting the number of partitions of integers ℓ_n^μ satisfying the equation:

$$\sum_{\mu=0}^{d-1} \sum_{n=1}^{\infty} n\, \ell_n^\mu = N. \tag{10.24}$$

A known mathematical result is that the number of partitions grows exponentially for large N. Note also that in Eq. (10.23) the intercept of the Regge trajectory α_0 was left arbitrary, but it is fixed to $\alpha_0 = 1$ by consistency conditions.

The following operator was introduced by Fubini and Veneziano:

$$Q_\mu(z) = \hat{q} - 2i\alpha' \hat{p} \log(z) + i\sqrt{2\alpha'} \sum_{n=1}^{\infty} \frac{1}{n} \left(\alpha_n z^{-n} - \alpha_{-n} z^n \right). \tag{10.25}$$

This is a spacetime vector that was later identified as the string coordinate; it actually describes the spacetime evolution of one of the end-points of the open string, as shown in Figure 10.4(a). Its expression is very similar to that of the field operator in Eq. (10.20), but there is a conceptual difference: the quantity in (10.25) describes the coordinate of a single string moving in spacetime, while the quantity in Eq. (10.20) represents a quantum field in Minkowski spacetime. However, the coordinate of the string can also be considered as a field defined on the two-dimensional world-sheet of the string (as discussed further in the Introduction to Part IV).

Using the oscillator algebra (10.21), it can be shown that the two-point correlator of $Q_\mu(z)$ has the standard logarithmic behaviour,

$$\langle 0|Q_\mu(z)Q_\nu(z')|0\rangle \sim \alpha'\eta_{\mu\nu}\log(z-z'), \tag{10.26}$$

like the Green function of a massless two-dimensional scalar field. This implies that the Veneziano amplitude in Eqs. (10.2) and (10.8) can be rewritten as the correlator of exponential fields, known as the Fubini–Veneziano 'vertex operators',

$$V(z;p) =: \exp\left(ip^\mu Q_\mu(z)\right):, \tag{10.27}$$

as follows:

$$A(s,t) = \frac{1}{V}\int \prod_{i=1}^{4} dz_i \, \langle 0| \prod_{i=1}^{4} V(z_i;p_i)|0\rangle. \tag{10.28}$$

Evaluation of the expectation value of the four vertex operators yields the amplitude in the form (10.7) (see Appendix C). The representation (10.28) can be clearly generalized to N-particle amplitudes by considering the expectation value of N vertex operators. Note that the definition of the vertex operator in (10.27) involves the normal ordering procedure, denoted by two colons; this amounts to putting all operators α_n^μ on the right (left) if $n > 0$ ($n < 0$).

The transformation of vertex operators under conformal transformations is particularly simple. It reads:

$$V(z;p)\,dz = V'(z';p)\,dz', \qquad z' = z + \epsilon(z). \tag{10.29}$$

Namely, vertex operators transform like covariant vectors under analytic coordinate reparameterizations. Equation (10.29) implies that the integrand of the open string amplitudes, such as (10.28), maintains the same functional form under conformal transformations. This is the conformal symmetry of string theory: as in general relativity, covariant quantities and their local relations have the same form in any coordinate system.

In the modern terminology of conformal field theory, fields transforming as in (10.29) are called 'primary fields' of scaling dimension $\Delta = 1$ [DMS97].

The conformal transformations can be realized within the operator formalism; the Virasoro generators L_n can be written as bilinear expressions of creation–annihilation operators, as follows:

$$L_n = \frac{1}{2} \sum_{k \in \mathbb{Z}} : \alpha^\mu_{-k} \alpha_{\mu,k+n} : \qquad (10.30)$$

(see the Chapter by Di Vecchia for details). The Virasoro algebra (10.17) is then recovered by using the oscillator commutators (10.21); the central extension of the algebra is also found to be $c = d$.

Furthermore, the infinitesimal transformation of vertex operators in (10.29) is expressed in operator notation by the commutation relations:

$$[L_n, V(z; p)] = \frac{d}{dz}\left(z^{n+1} V(z; p)\right). \qquad (10.31)$$

The previous considerations can be extended to the Shapiro–Virasoro model corresponding to the scattering of closed strings. In this case, one introduces a 'field' operator $Q_\mu(z, \bar{z})$ analogous to (10.25) that corresponds to the string coordinate: for any point on the two-dimensional surface of the string (the world-sheet) parameterized by $(z, \bar{z})$, the vector $Q_\mu(z, \bar{z})$ assigns its position in spacetime (cf. Figure 10.4(b)).

10.5 Physical states

As described in the previous Section, cf. (10.19), the space of oscillator states can be obtained by applying powers of creation operators to the ground state $|0\rangle$. However, in the DRM, states with vanishing and negative norm are also obtained. They are not acceptable in quantum mechanics, owing to the probabilistic interpretation of the norm of states. These states were called 'ghosts' in the DRM, but nowadays they should rather be called 'unphysical states', since the term ghost is reserved for the unphysical fields arising in the Faddeev–Popov, or Becchi–Rouet–Stora–Tyutin (BRST), quantization procedure based on the path integral [BBS08].

In a consistent quantum theory, the unphysical states should decouple from the physical ones: the physical spectrum should span a positive definite subspace of the entire space of states, and the unphysical states should belong to the orthogonal subspace. These results were indeed found for the DRM in the early Seventies, by using the operator formalism described above, and they almost completely unveiled the dynamics of string theory. They were confirmed in 1973 by the alternative light-cone quantization of the string Lagrangian by Goddard, Goldstone, Rebbi and Thorn.

The occurrence of negative-norm states can be exemplified as follows. Since the operators carry a spacetime index, the commutation relations (10.21) of timelike components have opposite sign to those of spacelike ones. Using the Hermiticity rules in (10.22) and the vacuum property, $\alpha^\mu_n |0\rangle = 0$ for $n > 0$, it follows that

$$\langle 0| \left[\alpha^0_n, \alpha^0_{-n}\right] |0\rangle = \langle 0| \alpha^0_n \alpha^0_{-n} |0\rangle = -n\langle 0|0\rangle, \qquad n > 0. \qquad (10.32)$$

This equation shows that the state $\alpha^0_{-n}|0\rangle$ ($n > 0$) has negative norm, since the vacuum has positive norm by definition. The states $\alpha^i_{-n}|0\rangle$, $i = 1, \ldots, d$, have positive norm instead. In general, all states containing an odd number of oscillators with time component have negative norm.

We have just seen that states with negative norm are required by the relativistic invariance of the theory. The next question is: under which conditions do they decouple from the physical states and do not contribute as intermediate states? A set of auxiliary conditions was shown to identify the physical subspace of positive-norm states.

The analysis was done in analogy with quantum electrodynamics (QED), where a similar problem had already been encountered. Indeed, the canonical quantization of the gauge field A_μ also involves creation and annihilation operators a^μ_k with a Lorentz index μ (cf. (10.20)).

The following results were known in QED, being specific of its gauge-invariant dynamics.

(i) In the classical theory, not all the components of the A_μ field give rise to propagating degrees of freedom; out of the four components (in $d = 4$), only two correspond to photons with physical polarization. The scalar potential A_0 is completely determined by the sources and does not propagate, while the longitudinal component, $A_\mu(k) \propto k_\mu$, can be changed at will by (local) gauge transformations, which leave the Lagrangian invariant. These read:

$$A_\mu(x) \to A_\mu(x) + \partial_\mu \omega(x), \tag{10.33}$$

where $\omega(x)$ is an arbitrary function of the coordinate x^μ. In other words, a gauge theory has a redundant set of (field) variables. They are all needed if one wants to have a manifest relativistic invariant description of the theory, but the physical degrees of freedom consistent with quantum mechanics are a subset of them.

(ii) In order to quantize the theory, one should break the gauge symmetry, by imposing an auxiliary condition ('fixing a gauge choice'). There are basically two options: (a) first completely fix the gauge and then quantize only the physical degrees of freedom, or (b) quantize all four degrees of freedom and then impose the gauge condition in the quantum theory.

(iii) In the first case, the quantum states corresponding to physical variables all have positive norm; however, the formalism treats the four A_μ components in different ways, so it is not manifestly Lorentz invariant. Thus, the covariance of physical quantities should be checked after quantization.

(iv) In the second case ('covariant quantization'), one deals with quantities that are manifestly Lorentz invariant; however, the fact of having more degrees of freedom than those physically needed manifests itself in the presence of negative norm states. It is nevertheless possible to isolate and eliminate these from the physical spectrum, which thus only contains positive-norm states.

The second case, relevant for the Dual Resonance Model, is the so-called Gupta–Bleuler quantization of QED. The gauge condition is $\partial_\mu A^\mu(x) = 0$, respecting Lorentz invariance; at the quantum level, it becomes an operator equation to be carefully analyzed. Since the quantity A_μ, as well as its derivative, obey Heisenberg's uncertainty conditions with the corresponding conjugate variables, the gauge condition cannot be imposed at the operator level, or the conjugate quantity would be infinite (and indefinite). A weaker, classically equivalent condition can be imposed on expectation values only. For any pair of physical states, $|\phi\rangle, |\psi\rangle$, we require,

$$\langle \phi | \partial_\mu A^\mu | \psi \rangle = 0, \qquad \text{equivalent to} \qquad \partial_\mu A^{(+)\mu} | \psi \rangle = 0. \tag{10.34}$$

In this equation, $A^{(+)}$ denotes the positive-energy part in the mode expansion (cf. (10.20)): since the field is real, one has $A_\mu^{(-)} = (A_\mu^{(+)})^\dagger$, and the vanishing condition is verified on either side of the expectation value in Eq. (10.34).

It can be shown that the solutions to Eq. (10.34), called the 'Fermi condition', correspond to the physical states with two transverse photon polarizations. The scalar and longitudinal polarizations do appear as intermediate states at the pole of the photon and in loop amplitudes, but their contributions cancel among themselves, leaving only physical propagating states [IZ80].

Going back to the Dual Resonance Model, it was found by Del Giudice and Di Vecchia that the analogue of the physical-state condition (10.34) could be written in terms of the L_n operators of Section 10.2.1, leading to the so-called Virasoro conditions (for $\alpha_0 = 1$):

$$(L_0 - 1)|\psi\rangle = 0, \qquad L_n |\psi\rangle = 0, \qquad \forall \, n > 0. \tag{10.35}$$

As described in the Chapter by Di Vecchia in this Part, the analysis of the low-lying states obeying these conditions showed that they had positive norm for $d \leq 26$. An infinite set of states was then found, the so-called Del Giudice–Di Vecchia–Fubini (DDF) states, that also had positive norm for $d \leq 26$. After the proof of the no-ghost theorem, described in the Chapters by Goddard and Brower in Part IV, it was understood that the DDF states provide a complete basis of the physical space for $d = 26$. Furthermore, additional positive-norm states exist for $d < 26$, which become zero-norm and decouple from the DDF states at $d = 26$, the value of critical dimension also required by unitarity of loop amplitudes. This structure of the Hilbert space, involving positive states and decoupled null states, was understood in the Eighties to be a general feature of theories with gauge symmetry.

Summing up, the operator approach gave a satisfactory understanding of the physical states at $d = 26$ and was then applied to the construction of loop amplitudes (see the Chapters by Musto and Nicodemi).

The other quantization approach involving only physical degrees of freedom was later developed by Goddard, Goldstone, Rebbi and Thorn in 1973; since it uses the string Lagrangian, it is described in the Introduction to Part IV and Appendix C. It was found that, in the so-called light-cone gauge, the theory has $(d-2)$ physical degrees of freedom corresponding to the space components $Q_i(z), i = 2, \ldots, d-2$, and the states constructed by the corresponding oscillators all have positive norm. Lorentz symmetry is broken by this gauge choice and has to be checked after quantization: remarkably, it was found to be realized only for $d = 26$ and $\alpha_0 = 1$. Therefore, an agreement between the noncovariant and covariant approaches was eventually found. After these results, the bosonic string theory was recognized to be a consistent quantum system (apart from a tachyon); moreover, the necessity of a spacetime dimension $d \neq 4$ was accepted, at least within the string community.

Let us conclude this Section with a couple of remarks on the results of covariant quantization. The physical-state conditions (10.35) have an interpretation in terms of conformal symmetry. Consider the states obtained by applying the vertex operator (10.27) to the vacuum, i.e. $|\psi\rangle = \lim_{z \to 0} V_\psi(z)|0\rangle$; having scaling dimension $\Delta = 1$, they do satisfy the physical state conditions in (10.35), as is shown by using the commutator (10.31) and the vacuum conditions,

$$L_n|0\rangle = 0, \qquad n = -1, 0, 1, 2, \ldots, \tag{10.36}$$

where the cases $n = 0, \pm 1$ express Möbius invariance. This characterization extends to other physical states besides the tachyon, each corresponding to a conformal covariant ($\Delta = 1$) vertex operator (examples are found in the Chapters by Clavelli, Di Vecchia and Ramond). In modern language, the physical states are called highest-weight states for their role in representation theory of the Virasoro algebra [DMS97]. It is interesting to notice that this state/vertex-operator correspondence relevant for modern applications of conformal field theory was basically understood in the early Seventies.

The correspondence between QED and DRM Hilbert spaces of states found in covariant quantization was more than an analogy: for $\alpha_0 = 1$, the open string spectrum contains a massless spin one state that can be identified with a gauge boson (as indeed done later, when string theory was proposed as a model unifying all interactions, see Part VI). The conformal invariance of the DRM also follows from a gauge symmetry of string theory: in Part IV, we shall see that the string Lagrangian is invariant under two-dimensional local reparameterizations of the world-sheet coordinates $(z, \bar{z})$. Conformal analytic reparameterizations are then found to be the subset of transformations that respect the gauge condition called 'orthonormal gauge'.

10.6 The tachyon

Once established that $\alpha_0 = 1$ and $d = 26$ were the values required to give both a positive definite physical Hilbert space and unitarity, one problem remained: the lowest spin zero state in the Regge trajectory had negative mass squared $M^2 = -1/\alpha'$ (respectively, $\alpha_0 = 2$ and $M^2 = -4/\alpha'$ in the closed string). The presence of a tachyon causes a violation of causality, the basic fact that effects should take place after their cause, which has important consequences in relativistic theories. However, this problem was not considered as serious as the presence of negative-norm states, which would have made the model completely inconsistent. Let us briefly explain this point.

According to special relativity, we can associate any point x^μ of spacetime with its 'lightcone' surface: this cone has its tip at the point x^μ and contains the points y^μ that are at null

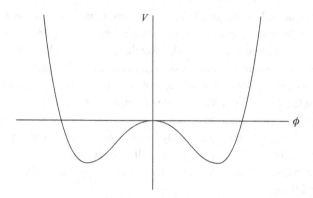

Figure 10.11 Scalar field potential V with unstable ground state at $\phi = 0$ and stable ground states at $\phi = \pm v$.

square distance, $d^2 = -(x-y)^2 = 0$. Since a particle passing by x^μ cannot travel faster than light, its future and past world-line are contained inside the cone ($d^2 > 0$). Moreover, any light signal interacting with the point x^μ is also contained ($d^2 = 0$): it follows that the region outside the cone (at spacelike distances $d^2 < 0$) cannot be causally related with the particle. In quantum theory, two fields commute at spacelike distances, lacking causal connection, and many other properties of S-matrix amplitudes can be deduced.

Tachyonic particles behave differently: for imaginary mass, $M = iq$, the mass-energy formula is modified:

$$E = \frac{Mc^2}{\sqrt{1 - \frac{v^2}{c^2}}} \quad \longrightarrow \quad E = \frac{qc^2}{\sqrt{\frac{v^2}{c^2} - 1}}. \tag{10.37}$$

Therefore, the tachyon travels faster than light and its world-line stays outside the lightcone: the tachyon could transport information between spacelike distant points and leads to violations of causality.

Nevertheless, quantum field theory can accommodate tachyons to some extent, owing to its ability to describe off-shell states: the tachyon should not be considered as a real particle but as an unstable state, a signal of the instability of the vacuum, as will be illustrated.

Let us consider a scalar field with Lagrangian:

$$\mathcal{L} = -\frac{1}{2}\partial_\mu \phi \partial^\mu \phi - V(\phi), \quad V(\phi) = \frac{1}{2}M^2\phi^2 + \lambda\phi^4, \tag{10.38}$$

corresponding to a self-interacting spinless particle. The classical potential $V(\phi)$ describes the static properties of the theory: the stationary points identify the ground states (vacua) and the small fluctuations around them are the low energy excitations (particles) of the theory. If $M^2 > 0$, then the potential has a minimum at $\phi = 0$, and a particle with positive mass squared M^2 is found upon expanding around this point.

For negative mass squared, $M^2 < 0$, the expansion around $\phi = 0$ yields a tachyonic particle instead. In this case, $\phi = 0$ is a maximum of the potential (Figure 10.11) and the presence of a tachyon just indicates that one has expanded around an unstable ground state. The two stable ground states of the theory are located at the two minima, $\phi = \pm v$: the expansion around them leads to a positive quadratic term. In conclusion, the tachyon at $\phi = 0$ is a signal of instability of the point $\phi = 0$, corresponding to a false vacuum of the theory that should decay into one of the two true minima $\phi = \pm v$ [IZ80].

This field theory argument was known at the time of the DRM and suggested that the tachyon was an instability of the theory; there were hopes that a better vacuum of string theory could be found.

In the mid-Seventies, the supersymmetric extension of string theory was shown to be free of tachyons (see Part VI); this was one indication that supersymmetry is an important ingredient of the theory, as it helps to stabilize it. However, this solution of the tachyon problem is not completely satisfactory, as it amounts to removing the problem from the start, rather than understanding it. In the field theory example, supersymmetry simply forbids a scalar potential $V(\phi)$ of the form in Figure 10.11. Since supersymmetry should eventually be broken at low energies, the problem of the tachyon may reappear.

Moreover, string theory, being a first quantization theory, is defined for on-shell states and, therefore, can determine the tachyon potential $V(\phi)$ only in the neighbourhood of $\phi = 0$. As discussed in recent literature, the approach of string field theory could determine the complete form of $V(\phi)$ and describe the decay of the false vacuum through the process of 'tachyon condensation' [BBS08].

References

[BBS08] Becker, K., Becker, M. and Schwarz, J. H. (2007). *String Theory and M-Theory: a Modern Introduction* (Cambridge University Press, Cambridge).

[DMS97] Di Francesco, P., Mathieu, P. and Sénéchal, D. (1997). *Conformal Field Theory* (Springer Verlag).

[DiV08] Di Vecchia, P. (2008). The birth of string theory, in *String Theory and Fundamental Interactions*, ed. Gasperini, M. and Maharana, J. (Springer, Heidelberg), Lecture Notes in Physics, Vol. 737, 59–118.

[DS08] Di Vecchia, P. and Schwimmer, A. (2008). The beginning of string theory: a historical sketch, in *String Theory and Fundamental Interactions*, ed. Gasperini, M. and Maharana, J. (Springer, Heidelberg), Lecture Notes in Physics, Vol. 737, 119–136.

[GSW87] Green, M. B., Schwarz, J. H. and Witten, E. (1987). *Superstring Theory* (Cambridge University Press, Cambridge).

[IZ80] Itzykson, C. and Zuber, J. B. (1985). *Quantum Field Theory* (McGraw Hill, New York).

[Ma88] Mandelstam, S. (1986). The interacting-string picture and functional integration, in *Unified String Theories*, ed. Green, M. and Gross, D. (World Scientific, Singapore).

11
From the S-matrix to string theory

PAOLO DI VECCHIA

Abstract

In this Chapter I describe the ideas and developments that led to the recognition that the structure underlying the N-point scattering amplitude of the Dual Resonance Model was that of a quantum-relativistic string.

11.1 Introduction

The history of the origin of string theory is very peculiar and difficult to understand if one looks at it with today's eyes. In particular, the fact that one could understand so many string properties without making reference to any Lagrangian, and without even knowing that one was studying a string, seems almost miraculous.

In fact, the starting point for describing the property of a relativistic string is today the string Lagrangian, which is invariant under reparameterizations of the world-sheet coordinates. From it, using techniques developed for quantizing theories with local gauge invariance, such as the Faddeev–Popov and the BRST quantization, one derives the spectrum of physical states and their scattering amplitudes. However, at the end of the Sixties and at the beginning of the Seventies, these techniques were not yet known and, even though the Nambu–Goto action had already been written, it was not clear how to use it to deduce its physical consequences. Therefore, the historical path that led us to understand the properties of a relativistic string was quite different from the path that one follows today, for instance, when teaching string theory in a university course. But at that time, where did string theory come from?

In order to answer this question one should first discuss the theoretical ideas that were dominant and the physical problems that one wanted to solve at that time.

The Sixties were a period in which a large number of hadronic states were found in the newly constructed accelerators and one wanted to reach an understanding of their spectrum

---

The Birth of String Theory, ed. Andrea Cappelli, Elena Castellani, Filippo Colomo and Paolo Di Vecchia.
Published by Cambridge University Press. © Cambridge University Press 2012.

and interaction. But, because of the large number of such particles, and also because of the strength of their coupling,[1] it was believed that a Lagrangian field theory, so successful in the case of quantum electrodynamics (QED), was unable to describe the experiments involving strongly interacting particles.

The dominant idea (expressed for instance in the books by Chew [Che66] and by Eden, Landshoff, Olive and Polkinghorne [ELOP66]) was to forget objects as Lagrangians and actions that are not observed directly in the experiments and, instead, construct an observable quantity as the S-matrix, by using its properties and the bootstrap principle (which, however, had no precise formulation). As discussed in the Chapters by Ademollo and Veneziano, this approach based on the S-matrix theory, together with the Harari–Rosner duality diagrams [Har69, Ros69], Dolen–Horn–Schmid duality [DHS68, Sch68], and an intensive study of the finite energy sum rules [ARVV68], were the basic ingredients that led, in the summer of 1968, to the formulation of the Veneziano model [Ven68] for the scattering of four scalar particles and, during 1969, to its extension to the scattering of N scalar particles [BR69a, BR69b, CT69, GS69, KN69a, KN69b, Vir69]. This model for the scattering of N scalar particles was called at that time the Dual Resonance Model (DRM) and, for the value of the Regge intercept $\alpha_0 = 1$, is what we call today 'the N-tachyon scattering amplitude of the bosonic open string'. In other words, guided by these theoretical ideas, it was possible to write down directly first the four-point amplitude and then the N-point amplitude, as explained in the Chapters by Ademollo and Veneziano.

At this point, the DRM was constructed and the research moved in the direction of investigating its properties, such as its spectrum of physical states and their interaction. In this Chapter I will discuss the developments that led to the recognition that the theory underlying the N-point scattering amplitude of the DRM was that of a quantum-relativistic string. As we will see, this process took a bit more than three years, from the formulation of the DRM S-matrix to the complete proof that it corresponded to the quantization of a relativistic string. In particular, I will explain the basic concepts, giving also the corresponding formulae which, however, will not be derived in detail; for detailed derivations, see [DiV08, DS07]. Other developments of the same period, such as the construction of multiloop diagrams for implementing the unitarity of the S-matrix and the attempts to construct more realistic models, such as the Ramond–Neveu–Schwarz model, will not be covered, as they are discussed by other authors in this Volume.

The Chapter is organized as follows. In Section 11.2, the N-point amplitude of the DRM is presented and, using the operator formalism, its factorization properties are discussed. In particular, it is shown how the spectrum of states of the DRM was constructed starting from the basic property of S-matrix theory that a state (particle) corresponds to a simple pole in the scattering amplitude with factorized residue. However, it turned out that, as in any relativistic theory, many of these states had negative norm. These states are called 'ghosts' and are not allowed in a quantum theory, because of the probabilistic interpretation of the norm of a state.

---

[1] For example, the value of the pion–nucleon coupling constant is $g_{\pi NN}^2/4\pi \sim 13.5$.

In Section 11.3 the problem of ghosts in an arbitrary quantum-relativistic theory is discussed and it is shown how it is solved in QED. This is achieved using gauge invariance, which implies some relations for the scattering amplitudes with photon emission. From these, one can deduce the equation, called the 'Fermi condition', that characterizes the on-shell physical photon states. In four dimensions these are the two states corresponding to the two possible values of helicity of the photon.

In Section 11.4 we introduce the Virasoro conditions that generalize those resulting from gauge invariance in QED. They are conditions on the scattering amplitudes involving an arbitrary excited particle of the spectrum. In other words, what is done in QED for the photon is generalized in the DRM to an arbitrary excited state.

Section 11.5 is devoted to the equations that characterize the physical states and that are a generalization of the 'Fermi condition' in QED. In Section 11.6 the vertex operators corresponding to the physical states are introduced. Although the equations characterizing the physical states and their interaction had been written down, it was not yet clear that the spectrum of physical states was ghost free.

A first step in this direction is made in Section 11.7, where we describe how, by using the vertex operator of the massless gauge field, an infinite set of physical states with positive norm was constructed. It turned out that this set spans the entire space of physical states if the spacetime dimension is $d = 26$. This implies that the DRM is ghost free if $d = 26$.

Section 11.8 describes how it was first guessed and then proven that the structure underlying the DRM was that of a quantum-relativistic string.

Finally, my recollections on how I started to work in the DRM at the beginning of my research career are presented in Section 11.9 followed by some conclusions in Section 11.10.

11.2 The N-point amplitude and its factorization properties

The starting point of our discussion is the explicit form of the N-point amplitude and its factorization properties. As mentioned in the introduction, this amplitude was constructed following the principle of planar duality and the axioms of S-matrix theory. In the Koba–Nielsen form the amplitude reads [KN69b]:

$$B_N(p_1, p_2, \ldots, p_N) = \int_{-\infty}^{\infty} \frac{\prod_1^N dz_i \theta(z_i - z_{i+1})}{dV_{abc}}$$

$$\times \prod_{i=1}^{N} \left[|z_i - z_{i+1}|^{\alpha_0 - 1}\right] \prod_{i<j} (z_i - z_j)^{2\alpha' p_i \cdot p_j}, \quad (11.1)$$

where $z_{N+1} \equiv z_1$, and

$$dV_{abc} = \frac{dz_a dz_b dz_c}{(z_a - z_b)(z_a - z_c)(z_b - z_c)} \quad (11.2)$$

is the volume of the Möbius group, as we are going to explain.

Equation (11.1) represents the scattering amplitude involving N scalar particles and treats all of them in a symmetric way by associating a Koba–Nielsen variable z_i to each of them. On the other hand, as follows from the duality diagrams, the amplitude cannot have more than $N-3$ simultaneous poles and, since each integral in general generates a pole, one expects that only $N-3$ Koba–Nielsen variables should be integrated over, by fixing three of them in some way. It turns out also that the integrand in (11.1) is invariant under an arbitrary projective transformation acting simultaneously on all Koba–Nielsen variables,

$$z_i \to \frac{\alpha z_i + \beta}{\gamma z_i + \delta}, \qquad i = 1, \ldots, N, \qquad \alpha\delta - \beta\gamma = 1, \qquad (11.3)$$

and therefore, if we integrate over all Koba–Nielsen variables, the integral will be infinite. Both previous arguments tell us that we must integrate only over $N-3$ Koba–Nielsen variables, but we cannot just fix three Koba–Nielsen variables and integrate over the others. The three integration variables should be removed in a uniform way, by factoring out the infinite volume of the Möbius group from the N-dimensional z_i integration space. This can be done as follows: we express the infinitesimal volume element in terms of the three independent parameters of the Möbius transformation, say (α, β, γ), and then change variables to three arbitrary integration coordinates, say z_a, z_b, z_c. The Möbius transformation is the one that maps them respectively into $\infty, 1, 0$:

$$z_a \to \infty, \qquad z_b \to 1, \qquad z_c \to 0, \qquad (11.4)$$

and the parameters of this transformation are given by

$$\alpha = f(z_a, z_b, z_c)(z_a - z_b), \qquad \beta = -f(z_a, z_b, z_c)(z_a - z_b)z_c,$$
$$\gamma = -f(z_a, z_b, z_c)(z_b - z_c), \qquad \delta = f(z_a, z_b, z_c)(z_b - z_c)z_a, \qquad (11.5)$$

where

$$f(z_a, z_b, z_c) \equiv \frac{1}{\sqrt{(z_a - z_b)(z_a - z_c)(z_b - z_c)}}. \qquad (11.6)$$

Using the previous expressions it is easy to express the infinitesimal volume of the Möbius group in terms of the three Koba–Nielsen variables z_a, z_b, z_c. We get:

$$dV_{\text{Möbius}} \equiv 2d\alpha\, d\beta\, d\gamma\, d\delta\, \delta(\alpha\delta - \beta\gamma - 1) = \frac{2d\alpha\, d\beta\, d\gamma}{\alpha}$$
$$= \det\left(\frac{\partial(\gamma, \alpha, \beta)}{\partial(z_a, z_b, z_c)}\right) \frac{dz_a dz_b dz_c}{\alpha} = \frac{dz_a dz_b dz_c}{(z_a - z_b)(z_a - z_c)(z_b - z_c)}$$
$$= dV_{abc}, \qquad (11.7)$$

which is equal to the quantity in (11.2), after computing the determinant

$$\det\left(\frac{\partial(\gamma, \alpha, \beta)}{\partial(z_a, z_b, z_c)}\right) = -f^2(z_a - z_b)\left[f + (z_a - z_c)f_{z_a} + (z_b - z_c)f_{z_b}\right], \qquad (11.8)$$

where $f_x = \partial f/\partial x$. In conclusion, by dividing the integrand in (11.1) with the infinitesimal volume of the Möbius group, we have made the N-point amplitude finite, being expressed only in terms of $N - 3$ integrals. The three Koba–Nielsen variables z_a, z_b, z_c can of course be fixed in an arbitrary way because the integrand in (11.1) does not depend on them. The calculation of the amplitude is, however, greatly simplified by the following convenient choice (cf. (11.4)):

$$z_a = z_1 = \infty, \qquad z_b = z_2 = 1, \qquad z_c = z_N = 0. \qquad (11.9)$$

The N-point amplitude was constructed by requiring that its only singularities were simple poles occurring at nonnegative integer values m of the linearly rising Regge trajectory:

$$\alpha(s) = \alpha_0 + \alpha' s = m, \qquad (11.10)$$

where s is the Mandelstam variable corresponding to an arbitrary planar channel.

The N-point amplitude (11.1) is invariant only under cyclic permutations of the external legs. Since the scattering amplitude of N identical particles should be symmetric under any exchange of the external particles, it can be constructed by summing the amplitudes (11.1) over noncyclic permutations of the external legs. Furthermore, one can make the external particle transform under an internal symmetry group, for instance $U(M)$, by introducing the Chan–Paton factors [CP69]. With the Chan–Paton factors the amplitude (11.1) becomes:

$$B_N^{a_1\ldots a_N}(p_1,\ldots,p_N) = \mathrm{Tr}(\lambda^{a_1}\cdots\lambda^{a_N})\frac{1}{g^2(2\alpha')^{d/2}}\left[2g\left(2\alpha'\right)^{(d-2)/4}\right]^N$$

$$\times \int \frac{\prod_{i=1}^N dz_i \theta(z_i - z_{i+1})}{dV_{abc}} \prod_{i=1}^M |z_i - z_{i+1}|^{\alpha_0 - 1} \prod_{i<j}(z_i - z_j)^{2\alpha' p_i \cdot p_j}, \qquad (11.11)$$

where λ^a are matrices of $U(M)$ in the fundamental representation, d is the dimension of spacetime and g is a quantity related to the string coupling constant g_s. We have also introduced the overall factors that make the amplitude correctly normalized to satisfy the factorization properties. The Chan–Paton factors were originally introduced to describe the flavour degrees of freedom of the mesons, for instance the isospin. In today's string interpretation, they correspond instead to the colour degrees of freedom.

Finally, the complete amplitude, totally symmetric under the exchange of external particles, is given by the following sum over the $(N - 1)!$ noncyclic permutations P of the external legs:

$$B_N^{symm} = \sum_P B_N^{a_1\ldots a_N}(p_{a_1},\ldots,p_{a_N}). \qquad (11.12)$$

The N-point amplitude was constructed within one and a half years after the Veneziano model and then the question was: what is the meaning of this amplitude? where does it come from? or, more modestly, what is the spectrum of particles predicted by it?

Since, following S-matrix theory, a particle corresponds to a simple pole in the scattering amplitude with a factorized residue, the 'obvious' thing to do was to study the factorization

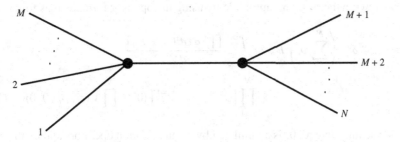

Figure 11.1 Factorization of the amplitude at the pole.

properties of the N-point amplitude at an arbitrary pole in an arbitrary planar channel. By factorized residue we mean that, as shown in Figure 11.1, the residue of the simple pole must be the product of two terms corresponding, respectively, to the amplitude for producing the excited particle from a bunch of initial particles and to the decay amplitude of the excited particle into a bunch of final particles. It was found that the residue at an arbitrary pole is the sum of a finite number of factorized terms. From this sum the spectrum of particles [BM69, FV69] was extracted by matching each term of the sum with a different excited state.

The most convenient way of studying the factorization properties of the N-point amplitude was to rewrite it by means of an operator formalism. One introduces an infinite set of harmonic oscillators $(a_{n\mu}, a_{n\mu}^\dagger)$ [FGV69, Nam69, Sus69, Sus70], having both a positive integer index n and a spacetime coordinate $\mu = 1, \ldots, d$, and the 'centre-of-mass' coordinates $(\hat{p}_\mu, \hat{q}_\mu)$ satisfying the following commutation relations:

$$[a_{n\mu}, a_{m\nu}^\dagger] = \eta_{\mu\nu}\delta_{nm}, \qquad [\hat{q}_\mu, \hat{p}_\nu] = i\eta_{\mu\nu}, \qquad \eta_{\mu\nu} = (-1, 1, \ldots, 1). \qquad (11.13)$$

The oscillators and the centre-of-mass coordinates commute with each other. In terms of them, one constructs the Fubini–Veneziano operator [FV70, FV71],[2]

$$Q_\mu(z) = Q_\mu^{(+)}(z) + Q_\mu^{(0)}(z) + Q_\mu^{(-)}(z), \qquad (11.14)$$

where

$$Q^{(+)} = i\sqrt{2\alpha'}\sum_{n=1}^{\infty}\frac{a_n}{\sqrt{n}}z^{-n}, \qquad Q^{(-)} = -i\sqrt{2\alpha'}\sum_{n=1}^{\infty}\frac{a_n^\dagger}{\sqrt{n}}z^n,$$
$$Q^{(0)} = \hat{q} - 2i\alpha'\hat{p}\log z, \qquad (11.15)$$

which is actually the coordinate of one of the two end-points of an open string (but this was only recognized later). One then introduces the vertex operator,

$$V(z; p) =: e^{ip\cdot Q(z)} := e^{ip\cdot Q^{(-)}(z)}e^{ip\cdot \hat{q}}e^{+2\alpha'\hat{p}\cdot p\log z}e^{ip\cdot Q^{(+)}(z)}, \qquad (11.16)$$

where $:\cdots:$ indicates the normal ordering, defined in the last term of the previous equation.

---

[2] This operator was also introduced by Gervais in [Ger70].

We can rewrite the N-point amplitude by using the operator formalism as follows:

$$A_N \equiv (2\pi)^d \delta^{(d)}\left(\sum_{i=1}^N p_i\right) B_N = \int_{-\infty}^\infty \frac{\prod_1^N dz_i \theta(z_i - z_{i+1})}{dV_{abc}}$$

$$\times \prod_{i=1}^N [(z_i - z_{i+1})^{\alpha_0 - 1}] \langle 0, 0 | \prod_{i=1}^N V(z_i, p_i) | 0, 0 \rangle, \quad (11.17)$$

where the vacuum state $|0, 0\rangle$ is annihilated by all annihilation operators and by the centre-of-mass momentum

$$a_{n\mu}|0, 0\rangle = \hat{p}_\mu |0, 0\rangle = 0. \quad (11.18)$$

The proof of Eq. (11.17) can be obtained by bringing all the annihilation operators on the right and the creation operators on the left, and all the operators containing $\hat{q}$ on the left of all operators containing $\hat{p}$ by using the commutation relations (11.13) and the Baker–Campbell–Hausdorff formula,

$$e^A e^B = e^B e^A e^{[A,B]}, \quad (11.19)$$

which is valid if the commutator $[A, B]$ is a c-number (which is indeed the case here). Performing this operation on the two vertex operators, one gets:

$$V(z; p)V(w; k) =: V(z; p)V(w; k) : (z - w)^{2\alpha' p \cdot k}. \quad (11.20)$$

By iterating this procedure for all vertex operators one gets Eq. (11.17).

In the following, we will restrict ourselves to the case $\alpha_0 = 1$, which is the most important case as we will see shortly. By introducing the propagator:

$$D = \int_0^1 dx \, x^{L_0 - 1} = \frac{1}{L_0 - 1} = \frac{1}{\alpha' \hat{p}^2 + R - 1}, \quad (11.21)$$

where

$$L_0 = \alpha' \hat{p}^2 + \sum_{n=1}^\infty n a_n^\dagger \cdot a_n \equiv \alpha' \hat{p}^2 + R, \quad (11.22)$$

and using the property $z^{L_0} V(1, p) z^{-L_0 - 1} = V(z, p)$, we can further rewrite the N-point amplitude as:

$$A_N \equiv \langle 0, p_1 | V(1, p_2) \ldots V(1, p_M) D \, V(1, p_{M+1}) \ldots V(1, p_{N-1}) | 0, p_N \rangle. \quad (11.23)$$

Equivalently, we can also rewrite

$$A_N(p_1, p_2, \ldots, p_N) = \langle P_{(1,M)} | D | P_{(M+1,N)} \rangle, \quad (11.24)$$

where

$$\langle P_{(1,M)} | = \langle 0, p_1 | V(1, p_2) D V(1, p_3) \ldots V(1, p_M), \quad (11.25)$$

and

$$|p_{(M+1,N)}\rangle = V(1, p_{M+1})D \ldots V(1, p_{N-1})|p_N, 0\rangle. \tag{11.26}$$

This last form of the amplitude is the most convenient for deriving the spectrum of physical states. This is achieved by studying the factorization properties of the amplitude at each pole, as shown in Figure 11.1. By inserting in (11.24) two complete sets of states, one gets:

$$A_N = \sum_{\lambda,\mu} \langle p_{(1,M)}|\lambda, P\rangle\langle\lambda, P|\frac{1}{R - \alpha(s)}|\mu, P\rangle\langle\mu, P|p_{(M+1,N)}\rangle, \tag{11.27}$$

where $R = \sum_{n=1}^{\infty} n a_n^\dagger \cdot a_n$.

It is easy to see that the propagator in (11.27) develops a simple pole when

$$\alpha(s) \equiv 1 - \alpha' P^2 \equiv 1 - \alpha'(p_1 + \cdots + p_M)^2 = m \tag{11.28}$$

is a nonnegative integer ($m \geq 0$).

The residue at the pole $\alpha(s) = m$ factorizes in a finite sum of terms corresponding to the states $|\mu, P\rangle$ satisfying the condition:

$$R|\mu, P\rangle \equiv \sum_{n=1}^{\infty} n a_n^\dagger \cdot a_n |\mu, P\rangle = m|\mu, P\rangle. \tag{11.29}$$

The lowest state, corresponding to $m = 0$, is the vacuum of oscillators: $|0, P\rangle$ with $1 - \alpha' P^2 = 0$. This is a tachyon with mass M given by $\alpha' M^2 = -1$.

At the next level, where $m = 1$, we have the state $a_{1\mu}^\dagger|0, P\rangle$ corresponding to a massless vector.

At the level $m = 2$, we have the following states ($1 - \alpha' P^2 = 2$):

$$a_{1\mu}^\dagger a_{1\nu}^\dagger |0, P\rangle, \qquad a_{2\mu}^\dagger |0, P\rangle. \tag{11.30}$$

At the level $m = 3$, we have the following states ($1 - \alpha' P^2 = 3$):

$$a_{1\mu}^\dagger a_{1\nu}^\dagger a_{1\rho}^\dagger |0, P\rangle, \qquad a_{2\mu}^\dagger a_{1\nu}^\dagger |0, P\rangle, \qquad a_{3\mu}^\dagger |0, P\rangle, \tag{11.31}$$

and so on.

11.3 The problem of ghosts and its solution in QED

Once the factorization properties of the N-point amplitudes were understood, a problem immediately appeared. The Lorentz invariance of the N-point amplitude forces one to factorize it by introducing the space spanned by the harmonic oscillators (11.13); but this space is not positive definite because the Lorentz metric has a negative sign for the time component. Thus the states with an odd number of time components have negative norm.

This is, however, in contradiction with the fact that in a quantum theory the Hilbert space must be positive definite, due to the probabilistic interpretation of the norm of a state. We face here a general problem of quantum-relativistic theories: how to put quantum theory and special relativity together in a consistent way?

It was immediately clear that, if one wanted to shed some light on this problem, one should go back to a consistent quantum-relativistic theory such as QED, that was fully developed at the time, and analyze how the problem was solved there. Since the DRM consisted of an infinite sum of simple poles, it would be sufficient to study the behaviour of a scattering amplitude in QED near the pole of the photon and see whether the residue at the photon pole factorizes in a sum of positive definite terms. It was well known that a scattering amplitude in QED near a photon pole ($P^2 \sim 0$) could be written:

$$A^\mu(p_1,\ldots,p_M, P) \frac{\eta_{\mu\nu}}{P^2} B^\nu(P, p_{M+1},\ldots, p_N), \tag{11.32}$$

where the quantity $\eta_{\mu\nu}/P^2$ is the photon propagator, with $\eta_{\mu\nu} = (-1, 1, 1, 1)$. At first sight it seems that the residue at the photon pole consists of four terms, one for each component of the metric tensor $\eta_{\mu\nu}$, and that one of them has a negative sign, corresponding to the exchange of a negative norm state given by the timelike component of the photon field. However, the theory possesses a local symmetry, called gauge invariance: this implies that not all the components of the photon field are physical. In particular, gauge invariance implies local identities among the amplitudes, that enforce a reduction of the degrees of freedom. In the case of the factorized amplitude in (11.32), these conditions imply:

$$P_\mu A^\mu(p_1,\ldots,p_M, P) = P_\mu B^\mu(P, p_{M+1},\ldots, p_N) = 0. \tag{11.33}$$

In a reference frame suitable for a massless particle ($P^2 = 0$), we can take $P_\mu = E(1, 0, 0, 1)$ and thus obtain:

$$A_3 - A_0 = B_3 - B_0 = 0. \tag{11.34}$$

These two equations imply that the residue at the photon pole actually contains only two terms, because the other two cancel out in (11.32):

$$\sum_{i,j=1}^{2} A^i(p_1,\ldots,p_M, P) \frac{\delta_{ij}}{P^2} B^j(P, p_{M+1},\ldots, p_N), \quad i, j = 1, 2. \tag{11.35}$$

The two remaining contributions correspond to the two helicities ± 1 of the photon. This analysis shows that the negative norm state is actually decoupled.

We see that in QED gauge invariance eliminates the potential conflict between special relativity and quantum theory: the amplitude at the photon pole can be written in a Lorentz covariant way in a space containing negative norm states in agreement with special relativity, but then gauge invariance eliminates the unwanted states and the spectrum of physical states is positive definite in agreement with the quantum theory. In other words, special relativity requires the introduction of a space with an indefinite norm, but then the space of the physical states is a subspace of it with positive definite norm in agreement with the quantum theory.

In conclusion, in QED the invariance under gauge transformations prevents a potential conflict between special relativity and the quantum theory and makes the theory consistent.

In QED the physical states are characterized by the 'Fermi condition':

$$\partial^\mu A_\mu^{(+)}|\text{Phys.}\rangle = 0, \tag{11.36}$$

where (+) stands for the positive frequency modes corresponding to the annihilation operators. This condition is also called the Lorentz condition (see, for instance, Section II.10.3 of [Hei36]).

It can be seen that the two components of the photon transverse to its three-momentum $P_\mu = E(1, 0, 0, 1)$ satisfy the Fermi condition and they span a positive definite Hilbert space. In fact, the Fermi condition in momentum space acting on the one-photon states reads:

$$P \cdot a|\text{Phys.}\rangle \equiv P^\nu a_\nu \alpha^\mu a_\mu^\dagger |0\rangle = 0, \tag{11.37}$$

where α^μ are arbitrary parameters. The most general solution of the previous equation is given by:

$$|\text{Phys.}\rangle = \left[\alpha^i a_i^\dagger + \alpha(a_3^\dagger - a_4^\dagger)\right]|0\rangle, \tag{11.38}$$

where $i = 1, 2$. The first two states correspond to the two physical helicities ± 1 of a photon, while the last state is a zero-norm state which, however, does not couple to the external states in Eq. (11.32) – because of Eq. (11.33), which is due to gauge invariance – and therefore does not belong to the physical spectrum. In other words, it satisfies the Fermi condition, but it must be eliminated by hand from the physical spectrum because it is not coupled to the external particles.

11.4 The Virasoro conditions

Going back to the Dual Resonance Model, we proceed in analogy with QED and look for relations of the type of (11.33) in order to be able to eliminate the negative-norm states. Do we have such relations in the DRM? This is what we are going to discuss in the following.

One relation was found immediately:

$$W_1|p_{(1,M)}\rangle = 0, \qquad W_1 = L_1 - L_0, \tag{11.39}$$

in terms of the operators L_0 and L_1 [CMR69, Gli69, Tho70], which can be written in terms of the harmonic oscillators. This relation was used to show that there was no negative-norm state at the first excited level and, more in general, on the leading Regge trajectory [FV71]. It was clear, however, that Eq. (11.39) was not enough to eliminate all the nonpositive-norm states.

Then Virasoro [Vir70] realized in 1969 that, if $\alpha_0 = 1$, one can find an infinite number of such conditions:

$$W_n|p_{1...M}\rangle = 0, \qquad n = 1, \ldots, \infty, \qquad W_n = L_n - L_0 - (n-1). \tag{11.40}$$

The existence of the above relations, generalizing the conditions (11.33) obtained from gauge invariance in QED, gave some hope that all the nonpositive-norm states could be eliminated from the spectrum of physical states. But it took a few more years to show that this was indeed the case. We want to stress here that, while the relations in (11.33) were a consequence of the gauge invariance of QED, it was not at all clear where the relations in (11.40) came from. But their existence maintained the hope that the DRM was indeed free of ghosts.

Soon after the paper by Virasoro it became clear that the 'Virasoro operators' L_n satisfy the algebra of the conformal group in two dimensions:

$$[L_n, L_m] = (n - m)L_{n+m} + \frac{d}{12}n(n^2 - 1)\delta_{n+m,0}. \tag{11.41}$$

This was first constructed by Fubini and Veneziano [FV71] without the central charge, whose presence was then recognized by Weis (see note added in proof in [FV71]). It is important to stress that this was the original derivation of the Virasoro algebra.

As discussed in [FV71], the geometrical meaning of the transformations generated by the L_n can be understood as follows. Let us consider infinitesimal coordinate transformations of the complex (Koba–Nielsen) variable z,

$$z \to z + \epsilon(z), \qquad \epsilon(z) = \sum_{n \in \mathbb{Z}} \epsilon_n z^{n+1}, \tag{11.42}$$

where $\epsilon(z)$ is an analytic function of z that can be expanded locally in a Laurent series. Under these transformations, the two-dimensional flat metric is modified by an overall scale factor

$$ds^2 = dz\, d\bar{z} \quad \to \quad ds^2 = \left|1 + \frac{d\epsilon}{dz}\right|^2 dz\, d\bar{z}. \tag{11.43}$$

This means that these are the conformal mappings preserving the angles between vectors. The generators of transformations on classical functions $f(z)$ under analytic reparameterization are obtained as follows:

$$f(z) \to f(z + \epsilon(z)) = [1 + \epsilon_n L_n] f(z), \qquad L_n = -z^{n+1}\frac{d}{dz}. \tag{11.44}$$

It can be checked readily that these differential operators satisfy the Virasoro algebra (11.41) without the second term on the right-hand side. This was precisely the derivation in [FV71].

The second term on the right-hand side of the algebra (11.41) is the so-called central extension, which later on was recognized to correspond to the conformal anomaly, a relativistic quantum effect. Here the central charge c is equal to d, the number of spacetime dimensions.

The central charge was found by Weis by using the representation of the Virasoro operators L_n in terms of the infinite set of harmonic oscillators introduced in Section 11.2

and, for $n \neq 0$, given by:

$$L_n = \sqrt{2\alpha' n}\, \hat{p} \cdot a_n + \sum_{m=1}^{\infty} \sqrt{m(n+m)}\, a_{n+m} \cdot a_m^{\dagger}$$

$$+ \frac{1}{2} \sum_{m=1}^{n-1} \sqrt{m(n-m)}\, a_{n-m} \cdot a_m, \qquad n \geq 0, \qquad (11.45)$$

and obeying $L_n = L_n^{\dagger}$. L_0 was given in (11.22).

11.5 Characterization of physical states

Virasoro found the relations in Eq. (11.40) that are the analogues of the conditions imposed by gauge invariance in Eq. (11.33). On the other hand, the physical states in QED are characterized by the fact that they satisfy the Fermi condition in Eq. (11.36). Do we have in the DRM conditions generalizing the Fermi condition of QED? These conditions were found by proceeding in analogy with QED and are [DD70]:

$$(L_0 - 1)|\alpha, P\rangle = 0, \quad L_n |\alpha, P\rangle = 0, \qquad n > 0, \qquad (11.46)$$

where $|\alpha, P\rangle$ denotes an arbitrary physical state with momentum P. In [DD70] the first three levels were analyzed in detail and it was shown that one was left with a spectrum of physical states spanning a positive definite Hilbert space. In particular, it was shown that, at the level $m = 1$, the analysis reduces to that in QED for the photon because, at this level, only the condition $L_1 |\alpha, P\rangle = 0$ has to be imposed and it reduces to the 'Fermi condition' in (11.36).

At the level $m = 2$, in the frame where $P_\mu = (M, 0, 0, 0)$, the physical states are a spin two state,

$$|\alpha_1, P\rangle = \left[a_{1,i}^{\dagger} a_{1,j}^{\dagger} - \frac{1}{(d-1)} \delta_{ij} \sum_{k=1}^{d-1} a_{1,k}^{\dagger} a_{1,k}^{\dagger} \right] |0, P\rangle, \qquad (11.47)$$

with positive norm (i, j are space indices), and a spin zero state,

$$|\alpha_2, P\rangle = \left[\sum_{i=1}^{d-1} a_{1,i}^{\dagger} a_{1,i}^{\dagger} + \frac{d-1}{5} (a_{1,0}^{\dagger 2} - 2 a_{2,0}^{\dagger}) \right] |0, P\rangle, \qquad (11.48)$$

with norm equal to

$$2(d-1)(26-d). \qquad (11.49)$$

This analysis shows that the state in Eq. (11.48) is a ghost if $d > 26$, has positive norm if $d < 26$, and has zero norm[3] and decouples from the spectrum of physical states (like the

---

[3] The analysis of the zero-norm states is too technical for the present discussion. For further details, see [DD70, DF72].

longitudinal zero-norm state in QED, discussed in Section 11.3) if $d = 26$. Unfortunately, in [DD70] this analysis was done while keeping the value of $d = 4$, because at that time it was not obvious that the dimension of spacetime d could be left arbitrary. Indeed, the DRM was a theory for mesons living in four dimensions.

In conclusion, in 1970 it was clear that the first three levels of the DRM were free of negative norm states, but at that time it was difficult to proceed further. We had to wait a couple more years to have a general proof of the absence of ghosts.

As in the case of the relations in (11.40), the origin of the conditions in (11.46) was also obscure. By analogy with Eq. (11.36), one could imagine that they were the quantum version of classical conditions that required $L_n = 0$ for any integer n, but it was unclear where this constraint came from. In particular, it was unclear that a gauge invariance implied the vanishing of L_n.

11.6 Scattering amplitudes for physical states

In the previous Section we have written the equations (11.46) that characterize the physical states of the DRM. In order to compute the scattering amplitudes involving an arbitrary number of physical states and not only the ground state as we have done in the previous Sections, we have to introduce for each physical state $|\alpha, P\rangle$ the corresponding vertex operator $V_\alpha(z, P)$.

In [CFNS71, CR71] it was found that the vertex operators, corresponding to the physical states, are conformal or 'primary' fields with conformal dimension $\Delta = 1$. This implies that these operators satisfy the following commutation relations:

$$[L_n, V_\alpha(z, P)] = \frac{d}{dz}\left(z^{n+1} V_\alpha(z, P)\right). \tag{11.50}$$

Furthermore, they are related to the corresponding physical states by the relations [CFNS71]

$$\lim_{z \to 0} V_\alpha(z, P)|0, 0\rangle \equiv |\alpha; P\rangle, \qquad \langle 0; 0| \lim_{z \to \infty} z^2 V_\alpha(z, P) = \langle \alpha, P|, \tag{11.51}$$

and satisfy the Hermiticity relation [CFNS71]:

$$V_\alpha^\dagger(z, P) = V_\alpha\left(\frac{1}{z}, -P\right)(-1)^m, \qquad 1 - \alpha' P^2 = m. \tag{11.52}$$

In terms of these vertices one can write the most general amplitude involving physical states [CFNS71]:

$$(2\pi)^4 \delta\left(\sum_{i=1}^N p_i\right) B_N^{ex} = \int_{-\infty}^\infty \frac{\prod_1^N dz_i \theta(z_i - z_{i+1})}{dV_{abc}} \langle 0, 0| \prod_{i=1}^N V_{\alpha_i}(z_i, p_i)|0, 0\rangle. \tag{11.53}$$

We see that many basic concepts of conformal field theories that are very much used today, had their origin already in 1970!

Before proceeding further, it is important to note that, although the ground state scalar particle seemed originally to play a privileged role, we have obtained equations characterizing the physical states and their vertex operators that treat them all on the same footing. This means that there is complete democracy among the physical states as advocated by the followers of S-matrix theory [Che66].

We will see, however, that there is a vertex operator associated with the massless vector state that is somewhat special and will play a very important role in the proof of the no-ghost theorem. It is given by:

$$V_\epsilon(z, k) \equiv \epsilon \cdot \frac{dQ(z)}{dz} e^{ik \cdot Q(z)}, \qquad k \cdot \epsilon = k^2 = 0. \tag{11.54}$$

11.7 DDF states and no ghosts

By using the vertex operator corresponding to the massless gauge field one can define the Del Giudice–Di Vecchia–Fubini (DDF) operator [DDF72]:

$$A_{i,n} = \frac{i}{\sqrt{2\alpha'}} \oint_0 dz \epsilon_i^\mu P_\mu(z) e^{ik \cdot Q(z)}, \qquad 2\alpha' p \cdot k = n, \tag{11.55}$$

where p_μ is the four-momentum of the state on which it acts and

$$P(z) \equiv \frac{dQ(z)}{dz} = -i\sqrt{2\alpha'} \sum_{n=-\infty}^{\infty} \alpha_n z^{-n-1}. \tag{11.56}$$

They are physical operators,

$$[L_m, A_{n,i}] = 0, \tag{11.57}$$

and satisfy the algebra of the harmonic oscillators:

$$[A_{n,i}, A_{m,j}] = n\delta_{ij}\delta_{n+m;0}, \qquad i, j = 1, \ldots, d-2. \tag{11.58}$$

In terms of this infinite set of transverse oscillators we can construct an orthonormal set of states:

$$|i_1, N_1; i_2, N_2; \ldots; i_m, N_m\rangle = \prod_h \frac{1}{\sqrt{\lambda_h!}} \prod_{k=1}^m \frac{A_{i_k, -N_k}}{\sqrt{N_k}} |0, p\rangle. \tag{11.59}$$

After the construction of this infinite set of physical states spanning a positive definite Hilbert space, it was checked whether they were complete, i.e. whether they spanned the entire space of physical states. Unfortunately the check was again done for $d = 4$ spacetime dimensions, where it turned out that they were not complete. We know today that for $d < 26$ there are additional physical states, called Brower states. At that point of the investigation it seemed very difficult to construct the entire physical spectrum and we had to wait another year before it was realized that the DDF states are indeed complete but only if $d = 26$! This observation made it immediately clear that the subspace of the physical states was positive definite if $d = 26$, as follows directly from Eq. (11.58), where on the right-hand side we have a positive definite metric. This also opened the way to proving the absence of ghosts for $d < 26$ [Bro72, GT72].

It must be said that $d = 26$ had already appeared in a paper by Lovelace [Lov71] in 1970, where it was shown that, if $d = 26$, there is no violation of unitarity in the nonplanar loop. But one must also say that almost nobody took it seriously because it was very difficult, at a psychological level, to think of a theory for strong interactions in $d \neq 4$! However, after the proof of the no-ghost theorem, everybody started to accept it.

In 1972, after four years of hard work since the appearance of the Veneziano model, the basic properties of the DRM, such as the spectrum of physical states and their scattering amplitudes at the tree level, were well understood. Moreover, loop diagrams necessary to implement unitarity were constructed starting from the N-Reggeon vertex [Lov70a] and using the sewing procedure. In particular, already in 1970 it was found that, as a result of the sewing procedure, functions which were well defined on Riemann surfaces were generated [Lov70b, AA71, Ale71].

In 1972, however, it was still unclear what the underlying structure was.

11.8 From the DRM to string theory

The approach presented in the previous Sections is a real bottom-up approach. The hadronic experimental data were the driving force in the construction of the Veneziano model for the scattering of four ground state particles and of its generalization to the N-point scattering amplitude. The rest of the work described in this Chapter consisted in extracting its properties. The result was, except for a tachyon, a fully consistent quantum-relativistic model that was a source of fascination for those who worked in the field. Although the model grew out of S-matrix theory, according to which the scattering amplitude is the only observable object and the action or Lagrangian are not relevant, some people [Nam69, Nie69, FN70, Nam70, Sus70] nevertheless started to investigate, already in 1969, the underlying microscopic structure that gave rise to such a consistent and beautiful model (see also [Tak70, Har71, CM72, MN72, Min72]).

They guessed correctly that this underlying structure is that of a quantum-relativistic string. However, the process of connecting the Dual Resonance Model to string theory took several years from the original idea in 1969 to a complete and convincing proof of the conjecture in 1973. In this Section I want to discuss the main steps of this process.

From what I have been able to recollect, the idea that the underlying structure of the DRM was that of a relativistic string was formulated already in 1969 [Nam69, Nie69, Nam70, Sus70] and was suggested by the fact that the DRM contained an infinite number of harmonic oscillators.

A Lagrangian was also written that was a generalization to two dimensions of the Lagrangian for a pointlike particle in the proper time gauge (τ is the proper time of the particle):

$$L \sim \frac{1}{2} \frac{dX}{d\tau} \cdot \frac{dX}{d\tau} \implies L \sim \frac{1}{2} \left[\frac{dX}{d\tau} \cdot \frac{dX}{d\tau} - \frac{dX}{d\sigma} \cdot \frac{dX}{d\sigma} \right]. \tag{11.60}$$

Since the Lagrangian was conformal invariant, the generators of the conformal group were also constructed and they were identified with the operators L_n appearing in (11.40) and (11.41). This identification was very appealing, but, at that time, not completely convincing because the conformal symmetry of the Lagrangian (11.60) was not a gauge symmetry and, therefore, it did not imply the vanishing of the classical generator,

$$L_n = 0, \tag{11.61}$$

as Eq. (11.46) seemed to imply. I remember that at that time I was fascinated by the analogies between the properties of this early string theory and those of the DRM, but I also got frustrated when I tried to go beyond this and prove that a relativistic string was the structure underlying the DRM.

A nonlinear string Lagrangian was also proposed [Nam70, Got71] that was invariant under arbitrary reparameterizations of the world-sheet coordinates σ and τ (see Part IV for further details):

$$S = -cT \int_{\tau_i}^{\tau_f} d\tau \int_0^\pi d\sigma \sqrt{(\dot{X} \cdot X')^2 - \dot{X}^2 X'^2}. \tag{11.62}$$

But it took three more years to show [GGRT73] that the spectrum of the DRM and the critical dimension ($d = 26$) followed from it. The expression (11.62) was introduced as a generalization of the nonlinear action for a free point particle, which is proportional to the length of the path described by the particle in its motion. It can be seen that the action in Eq. (11.62) is proportional to the area spanned by the motion of the string and, being a geometrical quantity, is invariant under an arbitrary change of the world-sheet coordinates σ and τ:

$$(\sigma, \tau) \to (\sigma', \tau'), \quad \sigma' = \sigma'(\sigma, \tau), \quad \tau' = \tau'(\sigma, \tau). \tag{11.63}$$

This is the reparameterization invariance of the world-sheet coordinates.

When one learned how to treat the nonlinear Lagrangian it became clear that the role that in QED is played by gauge invariance is, in string theory, taken by the invariance under arbitrary reparameterization of the string world-sheet coordinates σ and τ. By choosing the orthonormal gauge, which nowadays is called the conformal gauge, Chang and Mansouri [CM72] showed that the invariance under reparameterizations implies Eq. (11.61), opening the way to the solution of the problems mentioned above. They realized that, by fixing the conformal gauge, one does not fix the gauge completely. There are still reparameterizations, given by conformal transformations, that can be performed without leaving the conformal gauge. Being gauge transformations, their generators must be zero. This finally explained the origin of the constraints $L_n = 0$ in Eqs. (11.61) and (11.46).

However, the presence of a dimension $d = 26$ was not yet clear in their formulation. The realization that one could fix the gauge completely under reparameterizations by going into the light-cone gauge finally helped to find the critical dimension and connect the DRM in a satisfactory way to string theory. This is what we are going to explain in the following.

In string theory one starts with the string coordinate that has d components as required by special relativity, but then, because of reparameterization invariance, two of them – the time and one of the space components – are redundant and can be gauged away. This can be seen most clearly by choosing the light-cone gauge where one eliminates, already at the classical level, all the redundant unphysical degrees of freedom by imposing the condition [GGRT73]

$$X^+ \equiv \frac{X^0 - X^d}{\sqrt{2}} = 2\alpha' p^+ \tau, \tag{11.64}$$

that identifies the world-sheet coordinate τ with the light-cone coordinate X^+. This gauge choice in string theory corresponds to the Coulomb gauge in QED.

However, in this formulation one loses the manifest Lorentz invariance, and thus after quantization one must check that the Lorentz algebra is still satisfied. It turns out that this holds only at $d = 26$. This is the way in which the critical dimension was derived for the first time from the string Lagrangian [GGRT73].

Immediately after, the scattering amplitudes of the DRM were also derived from string theory [Man73, ADDN74]. In particular, it was shown [ADDN74] that the Fubini–Veneziano operator is the open string coordinate,

$$Q^\mu(z) \to X^\mu(e^{i\tau}, \sigma = 0), \qquad z = e^{i\tau}, \tag{11.65}$$

thus proving a guess which had already appeared in [Sus70].

11.9 How I started to work on string theory

I studied physics at the University of Rome and during my fourth year of University in 1965, as a preparation for my thesis, I decided to follow a course on quantum electrodynamics, held by Bruno Touschek at the Scuola di Perfezionamento. In this course Bruno was discussing the method of Block–Nordsieck for the emission of soft photons and, later on, he started to discuss the quantization of the electromagnetic field. When he arrived at the Gupta–Bleuler formalism he decided to stop the course for a couple of weeks, because he was not satisfied with the fact that he had to work in a space with indefinite norm. He used these two weeks to try to develop an alternative method to the Gupta–Bleuler formalism, where he could use a positive definite Hilbert space. After two weeks of hard work he was not able to find this alternative method and, without his usual enthusiasm, he described to us the Gupta–Bleuler formalism. Then he went on to describe the quantization of the electromagnetic field and its interaction with electrons and positrons.

I was so fascinated by his way of explaining that, after having followed his course, I went to him and asked him to be my advisor for the thesis. He accepted and told me to use the Block–Nordsieck method, valid for soft photons, for computing the total cross-section for the process of double bremsstrahlung (electron–positron scattering with the emission of two photons), because he was thinking of using it as a monitor for the storage ring ADONE that was under construction at the 'Laboratori Nazionali di Frascati'. It did not take much time to do this calculation and, when I went back to him, he gave me a paper by two Russian physicists from Novosibirsk, Bayer and Galitsky [BG65], in which they had computed the total cross-section in QED for the double bremsstrahlung with photons of arbitrary energy without giving many details, and he asked me to check their result. I worked very hard for more than one year and only in the spring 1966, in collaboration with Mario Greco, were we able to check that the Russian calculation was correct [DG67]. In the meantime, in February 1966 I had become 'Dottore in Fisica' at the University of Rome. Although the period of my thesis was not easy because the problem that I had to solve was very complicated, I enjoyed enormously working together with Bruno. I liked him as a physicist for his deep and clear thinking, as a teacher for his patient, logical and pedagogical explanations, and as a person for his warm and informal way of dealing with everybody (both professors and students).

After finishing my work on the double bremsstrahlung, I moved in the direction of Regge poles, finite energy sum rules and S-matrix theory and, when Gabriele Veneziano wrote his model, I immediately realized that his was an important contribution and I wanted to work on it. But in Rome or in Frascati, where in the meantime I had got a permanent position, there was nobody working in the field. Therefore, I decided to apply for a NATO fellowship to go to the United States. I had visited Caltech for three months in the fall of 1968 and before going back to Rome I had visited Berkeley and I liked the atmosphere there. Therefore, I decided to use my fellowship to go there, if I were able to get it. In June 1969 I went to a small meeting in Naples, where Sergio Fubini was giving a talk on the factorization of the N-point amplitude of the DRM by counting the states and showing that the leading Regge trajectory was ghost free. I was so fascinated by his talk that, after it, I introduced myself to him and I asked him if I could use my NATO fellowship, which in the meantime was awarded to me, at MIT in Cambridge where he was working. His answer was immediately positive and in December 1969 I left for Cambridge.

There I met Gabriele Veneziano, who was working with Fubini, and Emilio Del Giudice, visiting for one year from Naples. I started to work with Emilio on the problem of DRM physical states and, in this connection, it was useful to know the Gupta–Bleuler formalism of QED that I had learned in the course given by Touschek. Actually, we looked at it with different eyes than in the course of Touschek. It was no longer just an 'ugly trick', but a necessary ingredient to have a manifestly Lorentz invariant formulation of QED. I remember having read these considerations in an article by Heisenberg,[4] that I found in the MIT library, where he pointed out that any manifestly Lorentz invariant formulation of a quantum-relativistic theory necessarily requires an indefinite norm space. But then, the absence of nonpositive norm states from the physical spectrum, necessary for a consistent

---

[4] I looked for it again, while writing the present contribution, but I could not retrieve it.

quantum theory, required the existence of decoupling conditions. We understood that these decoupling conditions were provided in QED by gauge invariance through the so-called Fermi condition. Having understood the mechanism in QED, it was then easy to generalize it to the DRM and arrive at the conditions [DD70]:

$$(L_0 - 1)|\psi, P\rangle = 0, L_n|\psi, P\rangle = 0, \qquad n > 0, \qquad (11.66)$$

characterizing the on-shell physical states of the DRM. I consider the paper in which these conditions are derived as one of my best papers, but, for some reasons that are not clear to me, it has been largely ignored in the literature.

When Sergio came back from Geneva and Turin after the summer 1970 he showed us a paper [CFNS71] where he had extended our previous physical state conditions from the states to their vertex operators and we started to work together on trying to construct explicitly the physical states of the DRM. Being exposed to his thoughts day after day, we really learned a lot of physics from him. We got a unified view of the status of theoretical particle physics. He was using arguments from both field theory and S-matrix theory in a very concrete and direct way that was very fascinating for us. He had a very vivid imagination, which he used to connect things that at a first sight had nothing to do with each other. With Touschek, he is the person who taught me most on how to do research in physics. Like Bruno, he had a very warm, pleasant and informal personality. Sometimes it was not easy to contribute to our common papers, because he did not proceed in a deductive way, but used to jump around and therefore it was not easy to follow his train of thought. He was enthusiastic about the Veneziano model which, he thought, was the solution to many problems he had in current algebra. But he was a bit disappointed when it was realized that the whole thing just reduced to quantizing a relativistic string.

Through Emilio and Sergio I got to know Marco Ademollo from Florence, Francesco Nicodemi, Roberto Pettorino and Renato Musto from Naples and Alessandro D'Adda, Riccardo D'Auria, Nando Gliozzi, Ernesto Napolitano and Stefano Sciuto from Turin, with whom I had the great pleasure of collaborating in the years to come. This collaboration is described in other Chapters of this Volume. Here I just want to add a few more comments. Since we were all working along the same direction, and the problems that we were trying to solve were rather complicated, it seemed very natural to us that it would be better and that the problems would be solved much faster if, instead of competing, we would join forces and share all our calculations, publishing the papers together. An important ingredient of our collaboration was that it should be kept open to others who were willing to join even if their ideas were not the same as ours, and this happened without any problem. It was also clear that this was an experiment and that it would last a limited amount of time, because the world was, and still is, not mature enough for such experiments. For me, this was a very important experience, from which I learned quite a lot not only about physics but, most importantly, also about human relations and the difficulties that one encounters in making a better and more efficient society.

11.10 Conclusions

In this Chapter, I have described the developments that led from the construction of the N-point amplitude of the DRM to the recognition that a quantum-relativistic string is its underlying structure. In particular, in the years 1968–1973 a complete understanding of the perturbative properties of what is today called the bosonic string was achieved, including the spectrum of physical states and their scattering amplitudes at the tree and multiloop levels. Only the integration measure in multiloop diagrams was determined later. At one-loop level, it was determined in 1973 using the Brink–Olive projection operator [BO73a, BO73b].

It must also be said that, although the idea of an underlying string theory had appeared already in 1969, it played a very minor role in the period 1969–1973 described in this Chapter.

Apart from the integration measure in multiloop diagrams, in 1973 we were faced with two main unsolved problems. One was the presence of a tachyon and the other was the spacetime dimension $d = 26$ that was difficult to accept in a theory of hadrons. The first problem was solved only in 1976, with the GSO projection [GSO76, GSO77] in the Ramond–Neveu–Schwarz model, while the second was solved in 1974 by thinking of string theory as a theory unifying gravity with the other interactions [SS74, Yon74]. In this new framework the presence of extra dimensions can be considered a prediction of string theory.

Paolo Di Vecchia was born in Terracina (Italy) in 1942. He obtained his 'Laurea' in physics at the University of Rome in 1966, under the supervision of Bruno Touschek. He then worked in Frascati, MIT, CERN, Nordita and Berlin. He became Professor at Wuppertal University in 1980 and at Nordita in Copenhagen in 1986. Since the move of Nordita to Stockholm, he has divided his time between the Niels Bohr Institute in Copenhagen and Nordita in Stockholm. He has worked on several aspects of theoretical particle physics using both perturbative and nonperturbative methods in field and string theory.

References

[ADDN74] Ademollo, M., D'Adda, A., D'Auria, R., Di Vecchia, P., Gliozzi, F., Musto, R., Napolitano, E., Nicodemi, F. and Sciuto, S. (1974). Theory of an interacting string and dual resonance model, *Nuovo Cimento* **A21**, 77–145.

[ARVV68] Ademollo, M., Rubinstein, H. R., Veneziano, G. and Virasoro, M. A. (1968). Bootstrap of meson trajectories from superconvergence, *Phys. Rev.* **176**, 1904–1925.

[Ale71] Alessandrini, V. (1971). A general approach to dual multiloop diagrams, *Nuovo Cimento* **A2**, 321–352.

[AA71] Amati, D. and Alessandrini, V. (1971). Properties of dual multiloop amplitudes, *Nuovo Cimento* **A4**, 793–844.

[BM69] Bardakci, K. and Mandelstam, S. (1969). Analytic solution of the linear-trajectory bootstrap, *Phys. Rev.* **184**, 1640–1644.

[BR69a] Bardakci, K. and Ruegg, H. (1969). Reggeized resonance model for arbitrary production processes, *Phys. Rev.* **181**, 1884–1889.

[BR69b] Bardakci, K. and Ruegg, H. (1969). Meson resonance couplings in a five-point Veneziano model, *Phys. Lett.* **B28**, 671–675.

[BG65] Bayer, V. N. and Galitsky, V. M. (165). Double bremsstrahlung in electron collisions, *JETP Lett.* **2**, 165–167.

[BO73a] Brink, L. and Olive, D. I. (1973). The physical state projection operator in dual resonance models for the critical dimension of spacetime, *Nucl. Phys.* **B56**, 253–265.

[BO73b] Brink, L. and Olive, D. I. (1973). Recalculation of the unitary single planar dual loop in the critical dimension of space time, *Nucl. Phys.* **58**, 237–253.

[Bro72] Brower, R. C. (1972) Spectrum generating algebra and no ghost theorem for the dual model, *Phys. Rev.* **D6**, 1655–1662.

[CFNS71] Campagna, P., Fubini, S., Napolitano, E. and Sciuto, S. (1971). Amplitude for n nonspurious excited particles in dual resonance models, *Nuovo Cimento* **A2**, 911–928.

[CP69] Chan, H. M. and Paton, J. E. (1969). Generalized Veneziano model with isospin, *Nucl. Phys.* **B10**, 516–520.

[CT69] Chan, H.-M. and Tsou, S. T. (1969). Explicit construction of the n-point function in the generalized Veneziano model, *Phys. Lett.* **B28**, 485–488.

[CM72] Chang, L. N. and Mansouri, F. (1972). Dynamics underlying duality and gauge invariance in the dual resonance models, *Phys. Rev.* **D5**, 2535–2542.

[Che66] Chew, G. F. (1966). *The Analytic S-Matrix* (W. A. Benjamin, New York).

[CMR69] Chiu, C. B., Matsuda, S. and Rebbi, C. (1969). Factorization properties of the dual resonance model – a general treatment of linear dependences, *Phys. Rev. Lett.* **23**, 1526–1530.

[CR71] Clavelli, L. and Ramond, P. (1971). Group theoretical construction of dual amplitudes, *Phys. Rev.* **D3**, 988.

[DD70] Del Giudice, E. and Di Vecchia, P. (1970). Characterization of the physical states in dual-resonance models, *Nuovo Cimento* **A70**, 579–591.

[DDF72] Del Giudice, E., Di Vecchia, P. and Fubini, S. (1972). General properties of the dual resonance model. *Ann. Phys.* **70**, 378–398.

[DiV08] Di Vecchia, P. (2008). In *String Theory and Fundamental Interactions*, ed. Gasperini, M. and Maharana, J. (Springer, Berlin), Lecture Notes in Physics, Vol. 737, 59–118.

[DF72] Di Vecchia, P. and Fubini, S. (1972). Recent progress in dual models, in *Erice 1972, Proceedings, Highlights in Particle Physics*, Vol. 10, 351–373.

[DG67] Di Vecchia, P. and Greco, M. (1967). Double photon emission in e^+e^- collisions, *Nuovo Cimento* **A50**, 319–332.

[DS07] Di Vecchia, P. and Schwimmer, A. (2008). In *String Theory and Fundamental Interactions*, ed. Gasperini, M. and Maharana, J. (Springer, Berlin), Lecture Notes in Physics, Vol. 737, 119–136.

[DHS68] Dolen, R., Horn, D. and Schmid, C. (1968). Finite energy sum rules and their application to $\pi-N$ charge exchange, *Phys. Rev.* **166**, 1768–1781.

[ELOP66] Eden, R. J., Landshoff, P. V., Olive, D. I. and Polkinghorne, J. C. (1966). *The Analytic S-Matrix* (Cambridge University Press, Cambridge).

[FN70] Fairlie, D. and Nielsen, H. B. (1970). An analog model for KSV theory, *Nucl. Phys.* **B20**, 637–651.

[FGV69] Fubini, S., Gordon, D. and Veneziano, G. (1969). A general treatment of factorization in dual resonance models, *Phys. Lett.* **B29**, 679–682.

[FV69] Fubini, S. and Veneziano, G. (1969). Level structure of dual-resonance models, *Nuovo Cimento* **A64**, 811–840.

[FV70] Fubini, S. and Veneziano, G. (1970). Duality in operator formalism, *Nuovo Cimento* **67**, 29–47.

[FV71] Fubini, S. and Veneziano, G. (1971). Algebraic treatment of subsidiary conditions in dual resonance models, *Ann. Phys.* **63**, 12–27.

[Ger70] Gervais, J. L. (1970). Operator expression for the Koba–Nielsen multi-Veneziano formula and gauge identities, *Nucl. Phys.* **B21**, 192–204.

[Gli69] Gliozzi, F. (1969). Ward-like identities and twisting operator in dual resonance model, *Lett. Nuovo Cimento* **2**, 846.

[GSO76] Gliozzi, F., Scherk, J. and Olive, D. I. (1976). Supergravity and the spinor dual model, *Phys. Lett.* **B65**, 282.

[GSO77] Gliozzi, F., Scherk, J. and Olive, D. I. (1977). Supersymmetry, supergravity theories and the dual spinor model, *Nucl. Phys.* **B122**, 253–290.

[GGRT73] Goddard, P., Goldstone, J., Rebbi, C. and Thorn, C. B. (1973). Quantum dynamics of a massless relativistic string, *Nucl. Phys.* **B56**, 109–135.

[GT72] Goddard, P. and Thorn, C. B. (1972). Compatibility of the dual Pomeron with unitarity and the absence of ghosts in the dual resonance model, *Phys. Lett.* **B40**, 235–238.

[GS69] Goebel, C. G. and Sakita, B. (1969). Extension of the Veneziano form to N-particle amplitudes, *Phys. Rev. Lett.* **22**, 257–260.

[Got71] Goto, T. (1971). Relativistic quantum mechanics of one-dimensional mechanical continuum and subsidiary condition of dual resonance model, *Prog. Theor. Phys.* **46**, 1560–1569.

[Har71] Hara, O. (1971). On origin and physical meaning of ward-like identity in dual-resonance model, *Prog. Theor. Phys.* **46**, 1549–1559.

[Har69] Harari, H (1969). Duality diagrams, *Phys. Rev. Lett.* **22**, 562–565.

[Hei36] Heitler, W. (1960). *The Quantum Theory of Radiation* (Clarendon Press, Oxford).

[KN69a] Koba, Z. and Nielsen, H. B. (1969). Reaction amplitude for n mesons: a generalization of the Veneziano–Bardakci–Ruegg–Virasoro model, *Nucl. Phys.* **B10**, 633–655.

[KN69b] Koba, Z. and Nielsen, H. B. (1969). Manifestly crossing invariant parameterization of n meson amplitude, *Nucl. Phys.* **B12**, 517–536.

[Lov70a] Lovelace, C. (1970). Simple n-reggeon vertex, *Phys. Lett.* **B32**, 490–494.

[Lov70b] Lovelace, C. (1970). M-loop generalized Veneziano formula, *Phys. Lett.* **B32**, 703–708.

[Lov71] Lovelace, C. (1971). Pomeron form-factors and dual Regge cuts, *Phys. Lett.* **B34**, 500–506.

[Man73] Mandelstam, S. (1973). Interacting string picture of dual resonance models, *Nucl. Phys.* **B64**, 205–235.

[MN72] Mansouri, J. and Nambu, Y. (1972). Gauge conditions in dual resonance models, *Phys. Lett.* **B39**, 357.

[Min72] Minami, M. (1972). Plateau's problem and the Virasoro conditions in the theory of duality. *Prog. Theor. Phys.* **48**, 1308–1323.

[Nam69] Nambu, Y. (1969). Quark model and factorization of the Veneziano amplitude, in *Proc. Int. Conf. on Symmetries and Quark Models, Wayne State University 1969* (Gordon and Breach, 1970), 269.

[Nam70] Nambu, Y. (1970). Lectures at the Copenhagen Symposium, 1970, unpublished.

[Nie69] Nielsen, H. B. (1969). Paper submitted to the *15th Int. Conf. on High Energy Physics, Kiev, 1970* (Nordita preprint, 1969).
[Ros69] Rosner, J. L. (1969). Graphical form of duality, *Phys. Rev. Lett.* **22**, 689–692.
[Sch68] Schmid, C. (1968). Direct-channel resonances from Regge-pole exchange, *Phys. Rev. Lett.* **20**, 689–691.
[SS74] Schwarz, J. H. and Scherk, J. (1974). Dual models for nonhadrons, *Nucl. Phys.* **B81**, 118–144.
[Sus69] Susskind, L. (1969). Harmonic-oscillator analogy for the Veneziano model, *Phys. Rev. Lett.* **23**, 545–547.
[Sus70] Susskind, L. (1970). Dual-symmetric theory of hadrons. 1, *Nuovo Cimento* **A69**, 457–496.
[Tak70] Takabayasi, T. (1970). Internal structure of hadron underlying the Veneziano amplitude, *Prog. Theor. Phys.* **44**, 1117–1118.
[Tho70] Thorn, C. B. (1970). Linear dependences in the operator formalism of Fubini, Veneziano, and Gordon, *Phys. Rev.* **D1**, 1693–1696.
[Ven68] Veneziano, G. (1968). Construction of a crossing-symmetric, Regge behaved amplitude for linearly rising trajectories, *Nuovo Cimento* **A57**, 190–197.
[Vir69] Virasoro, M. A. (1969). Generalization of Veneziano's formula for the five-point function, *Phys. Rev. Lett.* **22**, 37–39.
[Vir70] Virasoro, M. A. (1970). Subsidiary conditions and ghosts in dual resonance models, *Phys. Rev.* **D1**, 2933–2936.
[Yon74] Yoneya, T. (1974). Connection of dual models to electrodynamics and gravidynamics, *Prog. Theor. Phys.* **51**, 1907–1920.

12
Reminiscence on the birth of string theory

JOEL A. SHAPIRO

Abstract

These are my personal impressions of the environment in which string theory was born, and what the important developments affecting my work were during the hadronic string era, 1968–1974. I discuss my motivations and concerns at the time, particularly in my work on loop amplitudes and on closed strings.

12.1 Introduction

It is not unusual in theoretical physics for conceptual frameworks to ride roller-coasters, but few have had as extreme highs and lows as string theory from its beginnings in 1968 to the present. In fact, string theory was so dead in the mid to late Seventies that it is a common assumption of many articles in the popular press, and of many younger string theorists, that the field originated in the Eighties, completely ignoring the period we are discussing here, which is primarily 1968–1974. So it was pleasantly surprising to be invited to reminisce about the early days of string theory. Research results from that era have been extensively presented and reviewed, so I will try to give my impression of the atmosphere at the time, and what questions we were trying to settle, rather than review the actual results.

12.2 The placenta

In the mid-Sixties, the framework for understanding fundamental physics was very different from what it is now. We still talk about the four fundamental interactions, but we know that the weak and electromagnetic interactions are part of a unified gauge field theory, that strong interactions are also described by a gauge field theory which might quite possibly unify with the others at higher energy, and that even general relativity is a form of gauge field theory. In the Sixties things were very different. Not only were the four interactions considered to be of completely different natures, but for the most part the physicists who worked on them were divided into groups by the interactions on which they worked.

The Birth of String Theory, ed. Andrea Cappelli, Elena Castellani, Filippo Colomo and Paolo Di Vecchia.
Published by Cambridge University Press. © Cambridge University Press 2012.

Of course, every budding particle theorist learned quantum field theory (QFT) and how wonderfully successful it was in treating quantum electrodynamics (QED). But one also learned how these perturbative methods could not be used for strong interactions because the coupling constant was too large, and that for the weak interactions one could only work at the Born approximation, because all existing field theories for the weak interactions were nonrenormalizable. So particle theorists were divided into separate groups: one working on strong interactions, one on weak interaction phenomenology, and one doing high order, esoteric QED calculations. Each group had very different techniques and styles.

Even more removed from the world of a strong interaction physicist was the fourth interaction, gravity, which was studied, if at all, by general relativists. (How separated general relativity and particle physics were in the Sixties is discussed by David Kaiser [Kai07] who argues that funding cuts in particle theory in the late Sixties and Seventies played a large role in the subsequent bringing together of particle theorists and general relativists.) When, as a very naive graduate student who knew nothing of the fields of physics research (I had just received my ScB in applied mathematics), I was asked by my future advisor what I might be interested in, I replied 'unified field theory'. Nonetheless, it was never suggested that I take a course in general relativity!

So the context into which string theory was born was not so much theoretical fundamental physics or even particle theory, but rather strong interaction theory/phenomenology. The principal recent successes in that field had been in searching for patterns and fitting simple models[1] to scattering data. Scattering cross-sections were dominated by resonance peaks and the high energy asymptotic behaviour described by Regge trajectories. A huge number of particles and resonances had been found and were listed in the particle data tables (Rosenfeld *et al.* [RBPP67]). The organization of these particles into (flavour) $SU(3)$ multiplets was the most impressive thing understood about the strong interactions.

The quasi-stable hadrons and the resonances fell beautifully into patterns which could be understood by treating baryons *as if* they consisted of three quarks and mesons as if they consisted of a quark and an antiquark. Even though this described very successfully the dominant experimental observations, theorists were very reluctant to think of the quarks as real constituents of hadrons.

Fits to the data were done by treating the scattering amplitude as a sum of resonance production and decay, together with an additional contribution due to the exchange of the same particles in the form of Regge poles to describe the high energy behaviour. This sum was called the interference model. But the experimentalists kept finding more and more resonances, and they were joined by phase-shift analysts. It soon appeared that the sequence of resonances continued indefinitely to higher masses and spins, in what clearly looked

---

[1] I am using 'model' and 'theory' with a distinction that is perhaps not generally accepted. To me, a theory is a comprehensive approach to explaining part of physics in a way which will at least have features which are fundamentally correct, while a model tries, with less ambition, to fit aspects of the data, but cannot be taken as the fundamental truth, even as an approximation of the truth. Thus QED, QCD, and general relativity are theories, even though the last clearly needs modification to include quantum mechanics, while the interference model, DHS (Dolen–Horn–Schmid) duality [DHS68], and my thesis are models. The Dual Resonance Model might be taken to have evolved into a theory when we started calculating unitary corrections in the form of loop graphs.

like linearly rising Regge trajectories. In fact, my PhD thesis [Sha68] was a very naive nonrelativistic model using PCAC, which rather successfully explained the experimental [RBPP67] π-nucleon decay widths of a tower of five Δ resonances with spins ranging from 3/2 to 19/2. Unfortunately the top two of these resonances have subsequently dissolved [Yao06]. This infinite sequence of resonances suggested the idea of duality [DHS68], that the amplitude could be described *either* in terms of a sum of resonances *or* in terms of a series of Regge poles. The possibility that a scattering amplitude $A(s, t)$ could be given as a sum of resonant poles in s or alternatively as a sum of Regge poles in t caused great excitement, but also scepticism that such a function could exist.

12.3 Conception and the embryonic period

Thus it seemed miraculous when Veneziano [Ven68] discovered that Euler had given us just such a function in 1772, to describe the $\pi\pi \to \pi\omega$ scattering amplitude. This paper arrived at the Lawrence Radiation Lab in Berkeley in the summer of 1968 while I was away on a short vacation, and I returned to find the place in a whirlwind of interest. Everyone had stopped what they were doing, and were asking if this idea could be extended to a more accessible interaction, such as $\pi\pi \to \pi\pi$. I suggested the very minor modification necessary to remove the tachyon,

$$\frac{\Gamma(-\alpha(s))\Gamma(-\alpha(t))}{\Gamma(-\alpha(s)-\alpha(t))} \longrightarrow \frac{\Gamma(1-\alpha(s))\Gamma(1-\alpha(t))}{\Gamma(1-\alpha(s)-\alpha(t))},$$

and Joel Yellin and I investigated whether this could be taken as a realistic description [SY68] for $\pi\pi$ scattering. It had a lot of good qualitative features, including resonance dominance, Regge behaviour, and full duality. We were forced to have exchange degeneracy between the $I=0$ and $I=1$ trajectories, which was well fit by the data. We noticed the problem that such an amplitude can wind up with ghosts, with a negative decay width for the ϵ' meson, the 0^+ partner of the f, but also that this problem disappeared if the ρ trajectory intercept exceeded 0.496, very close to the value of 0.48 which we got from fitting the low energy phase shifts. A much more serious problem was that we predicted a ρ' degenerate with the f, which seemed to be ruled out by experimental data. That the simplest function did not produce a totally acceptable model was discouraging, especially to Yellin, although we realized that there was no compelling reason not to add subsidiary terms to the simple ratio of gamma functions, except that to do so removed all predictive power! This convinced Yellin that he did not want to coauthor the full version of our paper [Sha69]. But Lovelace [Lov68], who discovered the same amplitude independently, managed to do a favourable comparison to experiment.

There were a number of papers attempting to do phenomenology with dual models, mostly describing two-body scattering processes. In general the results had, as did our paper, nice qualitative features but unsatisfactory fitting of the data. At the same time, the formal model was becoming much more serious, as great progress was made in extending the narrow resonance approximation amplitude, first to the five-point function, as done by

Bardakci and Ruegg [BR68], and then the N-particle amplitudes, as done by Bardakci and Ruegg [BR69], by Chan [Cha69], by Chan and Tsou [CT69] and by Goebel and Sakita [GS69]. A very elegant formulation of these amplitudes was given by Koba and Nielsen [KN69a, KN69b, Nie70], in which the external particles correspond to charges given by their momentum, entering on the boundary of a unit circle, and the amplitude is given by an integral, over relative positions of the particles, of the two-dimensional electrostatic energy. Here the conformal invariance was seen to play a crucial role, and in particular the Möbius invariance explained the cyclic symmetry. From the N-point amplitude for ground state particles, one could factor in multiparticle channels to extract the scattering amplitudes for all the particles which occurred in intermediate states, thus determining all amplitudes in what could be considered the equivalent of the tree approximation in a Lagrangian field theory. We took the attitude that the particles of the theory should be all and only those which arose from N-point scattering amplitudes of the ground state particles, as intermediate states in the n-point tree function. The amplitude for an arbitrary particle X connected to p ground states could be found by factoring the $p + q$ ground state amplitude, as discussed by Bardakci and Mandelstam [BM69], and amplitudes involving more arbitrary states could come from factoring that. Thus one could determine, in the tree approximation, the arbitrary N-particle amplitude. In a sense, this was a form of bootstrap, as the set of particles generated as intermediate states were added to form a consistent set, with the same particles as intermediate and external states.

12.4 Birth of string theory

Of course a set of tree amplitudes is not a unitary theory. In perturbative field theory, the Feynman rules are guaranteed to implement unitarity by specifying loop graphs whose discontinuities give the required sum over intermediate states, because these all come from a Hermitian Lagrangian. The possibility of advancing dual models to a unitary theory became feasible once we had the tree amplitudes for arbitrary single particle states, as one could sew together the loop graphs to give a perturbative (in the number of loops) theory satisfying unitarity. In perturbative quantum field theory, loop graphs give the appropriate contributions to the optical theorem, satisfying the unitarity of the S-matrix. Bardakci, Halpern and I [BHS69] defined the one-loop graph by the requirement of two-particle unitarity. An earlier attempt by Kikkawa, Sakita and Virasoro [KSV69] defined the planar loop graph by extending duality to the internal legs, which gave most of the factors in the loop integrand. But to get the full expression, the one-loop amplitude for N ground state particles σ should be required to have the correct two-particle discontinuity, a sum over all possible two-particle intermediate states. Starting with the tree amplitude for N σ plus $X(p)$ plus $X(-p)$, and summing over all possible states X and momentum p, as shown by the stitch marks in Figure 12.1, one is summing not only over X but also over all particles in the left arm, because these are all included in the tree graph.

Of course we called this process 'sewing', which led to an amusing battle with Sy Pasternack, the editor of *Physical Review*, on a subsequent paper by Halpern, Klein and

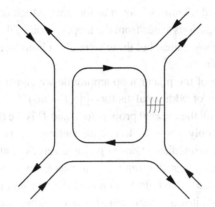

Figure 12.1 Sewing a tree to make a loop [HKS69].

Shapiro [HKS69]. Pasternack thought he needed to uphold a certain formality, and was responsible for 'Pomeronchukon' rather than 'Pomeron'. He wrote us a very witty letter arguing that 'sewing' would lead inevitably to 'weaving', 'braiding', 'darning', 'knitting' and 'sew-on' [Pas69]. We objected, however, that the actual thread lines were shown in the figures. Redrawing figures in those days was a major undertaking. That won the argument.

I should point out that at the time our description of the intermediate states and the amplitudes was quite clumsy, using the rather messy techniques of the Bardakci–Mandelstam factorization [BM69]. While we were working on deriving the loop graph, Nambu [Nam70], and Fubini, Gordon and Veneziano [FGV69] were developing the elegant operator formalism, in which the states of the system are described by harmonic oscillator excitation operators $a_n^{\mu\dagger}$, $n = 1, \ldots, \infty$, acting on a ground state $|0\rangle$. Here n corresponds to the number of half-wavelengths along the string, while μ is a Lorentz index indicating the direction of the excitation. The amplitudes can then be written as a matrix element with vertex functions for each external particle, and propagators integrated $\int_0^1 du$ over an internal variable $u \sim e^{-\tau}$, where τ acts like a time describing how long an intermediate state propagates. Thus resonance poles in a tree, or two-particle intermediate states in a loop, come from the $\tau \to \infty$, $u \to 0$ limit for the corresponding propagator. This formalism made the calculations much easier. It enabled the authors of [KSV69] to discover independently from us the correct extra factors that get two-particle unitarity, except for spurious states. The new formalism was so superior (see Amati, Bouchiat and Gervais [ABG69]) that few people were encouraged to read our paper.

The operator formalism made clearer two problems that had already been vaguely seen. In this formalism, the amplitudes appear to lose the Möbius invariance, but they do not, due to the existence of Ward identities. That is, there are spurious states, combinations of excitations which decouple from all N ground state amplitudes, and therefore by our philosophy should not be included among the states. Secondly, the timelike creation operators create ghosts, particles with negative widths, which clearly should not be there. The set of these ghosts produced by the lowest node operator, $a_1^{0\dagger}$, were precisely those that could be exorcised by those Ward identities known at the time. Much of our effort in [BHS69] involved excluding these spurious states from the loop. Of course the higher timelike modes $a_n^{0\dagger}$, $n > 1$ also produce ghost states, but these were also eliminated by the Ward identities found later by Virasoro [Vir70].

One unpleasant feature of the planar loop amplitude we constructed from two-particle unitarity was the presence of additional factors $\prod_r (1 - w^r)^{-D}$, where $w = \prod u_j$ is the product of the u factors of all the internal propagators, and D is the dimension of spacetime, which at that point we simply wrote as 4. The discontinuities we were building into the loops come from several intermediate states u_i going to zero, so only the $w \to 0$ end-point contributes, but the natural integration range for w was from 0 to 1. Of course at $w = 1$ this factor has extremely bad behaviour. Eliminating the one set of spurious states known about at the time eliminated just one power of $(1 - w)^{-1}$, which did not help much. Later, Virasoro [Vir70] discovered that if the Reggeon intercept $\alpha(0) = 1$, there was an infinite set of generators of spurious states, and eliminating those gets rid of all the ghosts (for $D \leq 26$), and one full set of $\prod_r (1 - w^r)^{-1}$. Still, there is a very serious divergence as $w \to 1$, which will turn out to be connected to the Pomeron/closed string. Before that was realized, there was speculation about whether this divergence was removable, and whether the two-particle discontinuity had the expected two-Reggeon cut asymptotic form (Shapiro and Thorn [ST72]) as $s \to \infty$.

It should be mentioned that this effort to raise dual models to the same level of legitimacy as perturbative QFT was a departure which made many uncomfortable. Strong interaction theorists had been divided into field theorists and S-matrix folk, and dual models were generally considered the domain of S-matrix people, but here they were adopting the moral values of field theorists, even if the context was different. The phenomenologically inclined thought it would be better simply to assign imaginary parts to the Regge trajectories in the dual amplitudes to go beyond a narrow resonance fit. As there had been no real data-fitting successes, many had great scepticism about the value of dual model research. One such sceptic asked me why I would work on something so unlikely to be the real physical truth. I recall saying that even though the probability that dual models would be part of the real answer was small, perhaps 10%, at least there was a chance of working towards the truth, while fitting elastic scattering data to Regge poles, to me, seemed not to have any chance of leading to fundamental physical understanding. I mention this because in 1987, in Aspen where string theory was the superhot 'Theory of Everything', I asked some of the younger researchers what their estimate was of the probability that string theory would be part of the true theory of physics, and was rather astounded to hear answers upwards of 50%.

Anyway, let's get back to the construction of loop graphs for a complete, unitary dual resonance theory. This was a very active field. Neveu and Scherk [NS70] used their superior French mathematical education to express the planar loop in terms of elegant Jacobi theta-functions, enabling them to extract the divergent behaviour. The operator formalism by Fubini and Veneziano [FV69], Fubini, Gordon and Veneziano [FGV69] and Susskind [Sus70] made tractable the calculation of nonplanar loops (Kikkawa, Klein, Sakita and Virasoro [KKSV70], Gross, Neveu, Scherk and Schwarz [GNSS70] and Kaku and Thorn

[KT70]) and multiloop amplitudes (Kaku and Yu [KY70]). Abelian integrals were used by Lovelace [Lov70, Lov71] and Alessandrini [Ale71]. The former suggested that experimentalists deprived of funding for higher energy machines could 'still construct duality diagrams in tinfoil and measure induced charges' as their contribution to understanding particle physics.

12.5 Closed strings

My second postdoc appointment was at the University of Maryland, which had a pleasant and active high energy theory group, but no one doing dual models. I felt quite isolated, and while I was able to write some technical papers [Sha70a, Sha71], I missed the stimulating environment I had had in Berkeley. In particular, the first of these papers was a rather misguided attempt to get rid of the spurious states given by the Ward identity with L_1, before I became aware that Virasoro had found, for the 'unrealistic choice of $\alpha(0) = 1$', that there was an infinite set of such Ward identities, enough to get rid of all the ghosts produced by $a_n^{0\dagger}$. Fortunately I was free to visit Berkeley and Aspen during the summers. During my Berkeley stay in 1970 I spent a day at SLAC where, in a discussion with Nussinov and Schwimmer, they asked me a very interesting question. At the time, the N-point Veneziano formula was best described by the Koba–Nielsen picture of external charges (or currents) on the circumference of a disc. The integrand could be interpreted as the electrostatic energy of the charges or as the heat generated by the currents, inside the disc. There was also much interest in this being an approximation of very complex Feynman diagrams called fishnets within the disc. The question Nussinov asked is what would happen if the external particles, instead of residing on the circumference of a disc, were on the surface of a sphere. Nussinov answered his own question with 'I bet one would get the Virasoro formula'. This is because, with the particles integrated over the surface, there is no cyclic order constraint, and any collection of particles are free to approach each other and produce a singularity in P^2, where P^μ is the sum of their momenta. This is what happens in the Virasoro formula [Vir69].

Should the fishnets, or electric fields, or currents, fill the ball, or should they be confined to the surface with the external particles? In my view, the new Virasoro identities were associated with the local conformal invariance of analytic functions, a much richer group than conformal transformations in Euclidean spaces of higher dimension. Thus filling the three-dimensional ball was unlikely to work, but putting fishnets on the surface might be very interesting. So the three of us began to work out electrostatics on the surface of the sphere.

In the Fairlie and Nielsen approach [FN70, Nie70] one needs the electrostatic energy of a configuration of point charges at arbitrary locations, and then one integrates over the charges' positions.[2] We can solve Poisson's equation for each configuration of charges on a 2-sphere, but we cannot define the electrostatic energy as the integral of $(\vec{\nabla}\phi)^2$,

---

[2] Editors' note: see also the Chapters by David Fairlie and by Holger Nielsen in this Volume.

because that includes the infinite self-energy of each charge. Instead we might define $E = \frac{1}{2}\sum_{i\neq j} q_i\phi_j(\vec{r}_i)$, but to do so one needs to be able to find the electrostatic potential of a single charge. We cannot have a source of electric flux without a sink, and we seemed to hit an impasse and let the matter drop. Several weeks later, after we had all gone our separate ways, while I was (I think) in Aspen, I decided to look at this again, and I realized that putting an arbitrary sink for all the fields would do no harm. After all, for two-dimensional electrostatics one takes $\phi \propto \ln|\vec{r} - \vec{r}_0|$ without worrying about the flux which goes off to infinity, and in two dimensions conformal invariance makes infinity no different from any other point. As the sphere is conformally equivalent to the complex plane, the potentials are just logarithms of $|(z_i, z_j, a, b)|$, the absolute value of the cross ratio, including the arbitrary sink point b and a point a at which we can set the potential to zero. The Möbius invariance, now extended to three *complex* parameters, again permits the positions of three charges to be fixed arbitrarily, and the others integrated over the sphere, or complex plane. We have consistency only for $\alpha(0) = 2$, but there we have a consistent n-point function for closed string scattering [Sha70b]. I speculated that this was equivalent to the Pomerons which appeared as a problem in the loop graphs of open strings, and later, with Clavelli [CS73], I showed that this is indeed the case.

There was at the time not much interest in closed strings, which have no ends. All the semi-successes of dual model phenomenology were based on Harari–Rosner diagrams [Har69, Ros69] being incorporated by Chan–Paton factors [CP69], which required string ends on which to attach quarks. Even I postponed looking at factorization and loop graphs in this model in favour of a paper [Sha71] showing that nonorientable graphs do not enter theories with $SU(n)$ flavours incorporated à la Chan–Paton. But the following summer, in Aspen again, with the factorization having been done by others (Yoshimura [Yos71] and Del Giudice and Di Vecchia [DD71]), I addressed the one-loop diagram [Sha72] in what I called the dual-tube model. The propagators now have an integral over the length of the tube and the angle of twist to get to the next particle, and the complex variable w has $|w| = \exp -T$, where T is the combined times of propagation, but w also has a phase given by the angle of twist in sewing the initial end of the tube to the final end. The amplitude involves

$$\int_{|w|\leq 1} d^2w \prod_{r=1}^{\infty} |1 - w^r|^{-2(D-E)},$$

where D is the dimension of spacetime and E is the number of factors assumed to disappear mysteriously if one removes the spurious states. E had already been shown to be 1 in arbitrary D dimensions, but we were hoping, as was found to be true later, that $E = 2$ in the right D. Still, $D - E$ is a positive number, and $1 - w^r$ vanishes for w any integral power of $e^{2\pi i/r}$, so we have a terrible divergence at every point of the unit circle for which the angle is 2π times a rational number.

Fortunately, by the time I was looking at this, I had read the elegant reformulation [NS70] of the open string loop in terms of Jacobi theta-functions. This encouraged us to look at $\tau = (\ln w)/(2\pi i)$. Of course the integrand is invariant under $\tau \to \tau + 1$, because w is unchanged, but it is also invariant under the Jacobi imaginary transformation, $\tau \to -1/\tau$, provided we have the magic dimensions $D = 26$, $E = 2$. The world-sheet of a loop of closed strings is a doughnut, conformally equivalent to a parallelogram with sides 1 and $i\tau$, with opposite sides identified. While a 'hula hoop' and a doughnut with a small hole may not look the same, multiplying the parallelogram by $-i/\tau$ maps one into the other. These transformations generate the modular group, so invariance shows that in integrating w over the unit disc we are including an infinite number of copies, while we wanted, for unitarity, only one copy of the region around $w \approx 0$. Thus the right thing to do is restrict our integration to the fundamental domain, which is $|\tau| \geq 1$, $-\frac{1}{2} < \operatorname{Re} \tau \leq \frac{1}{2}$. In terms of w, this is a subset of $|w| < 0.0044$, so we stay far away from the horrible divergences.

If closed strings aroused little interest at the time, loops of them really aroused none. Interest in dual models as models of the strong interactions was fading fast. Firstly, the evidence for partons, pointlike constituents of hadrons, found in deep inelastic scattering starting in 1969, was inconsistent with the soft, extended object picture of strings. Secondly, non-Abelian gauge theories were proven renormalizable ('t Hooft [tHo71]), explained neutral currents in a unified electro-weak theory, and gave quantum chromodynamics as a theory of the strong interactions. This greatly improved the appeal of conventional field theory at the expense of string theory. And within string theory, the inclusion of fermions by Ramond [Ram71] and Neveu–Schwarz [NS71] was more exciting than loops of Pomerons.

In the fall of 1971 I started an Assistant Professorship at Rutgers in a new high energy theory group headed by Lovelace, and including Clavelli as a postdoc. Lovelace had a very ambitious programme for describing arbitrary multiloop diagrams, and Clavelli and I looked at how the closed string intermediate state in the nonplanar loop interacts with the ordinary (open string) states. Then I struggled with understanding renormalization [Sha75], an effort which would have been totally trivial if I had only realized that $\sum_{n=1}^{\infty} n = -1/12$ (by analytic continuation of the Riemann ζ-function), but the many math classes I had taken did not encourage that kind of thinking.

The decision on my tenure was coming up, and dual models did not seem the best way to prove my worth, so I reluctantly got into several other endeavours, and was rather slow to get back into strings when they arose like a phoenix, or perhaps like a fire storm, in 1984. But that is after the period we are considering here.

12.6 A comment on impact

I want to say a few words about how this field was perceived within the physics community. In recent years there have been numerous attacks from some in the high energy theory community, and from experimentalists, that strings are, like the Pied Piper, leading the bright young theorists astray. String theory was not quite so dominant in the 1969–1974 era, though it did absorb the attention of a very large fraction of the young theorists. It did not gain similar acceptance by most of the more established people, though I think Europe was more receptive than America. In particular *Physical Review Letters* (of the American

Physical Society) published very few articles in this field, but *Physics Letters* (of Elsevier, The Netherlands) had many of the important papers.

The field did attract quite a bit of attention. In fact, in the early Seventies I was interviewed by a sociologist who wanted to do a study of what attracted so many people to working on dual models. Unfortunately she followed a narrow set of preprepared questions focussing on which experimental data encouraged me to continue working in the field, while it was really theoretical considerations that did so. I don't think anything came of that study, nor have I seen any such studies about the even more intense string theory activity of the mid-Eighties. There have been many questions raised by other physicists about the appropriateness of this focus of effort, so understanding the sociology would be useful, were it possible.

Is there any real physics in string theory, and should so many people be working on it? Undoubtedly there will be a shift towards more applied high energy theory as the LHC starts giving us more data to work with. It will be very interesting to see where the field goes.

Joel Shapiro received his PhD at Cornell in 1967. He started to work in string theory during his first postdoc at Berkeley. After a second postdoc at Maryland, he has been permanently at Rutgers, except for several sabbaticals, at MIT and IAS, Princeton. He also benefitted from summer visits to Aspen, Berkeley, Argonne, CERN, l'Ecole Normale, Orsay, and other places. His research interests shifted in the Nineties from string theory and supergravity to computer aids to elementary physics instruction.

References

[Ale71] Alessandrini, V. (1971). A general approach to dual multiloop diagrams, *Nuovo Cimento* **A2**, 321–352.

[ABG69] Amati, D., Bouchiat, C. and Gervais, J. L. (1969). On the building of dual diagrams from unitarity, *Lett. Nuovo Cimento* **2**, 399–406.

[BHS69] Bardakci, K., Halpern, M. B. and Shapiro, J. A. (1969). Unitary closed loops in reggeized Feynman theory, *Phys. Rev.* **185**, 1910–1917.

[BM69] Bardakci, K. and Mandelstam, S. (1969). Analytic solution of the linear trajectory bootstrap, *Phys. Rev.* **184**, 1640–1644.

[BR68] Bardakci, K. and Ruegg, H. (1968). Reggeized resonance model for production amplitude, *Phys. Lett.* **B28**, 342–347.

[BR69] Bardakci, K. and Ruegg, H. (1969). Reggeized resonance model for arbitrary production processes, *Phys. Rev.* **181**, 1884–1889.

[Cha69] Chan, H.-M. (1969). A generalised Veneziano model for the *N*-point function, *Phys. Lett.* **B28**, 425–428.

[CP69] Chan, H. M. and Paton, J. E. (1969). Generalized Veneziano model with isospin, *Nucl. Phys.* **B10**, 516–520.

[CT69] Chan, H.-M. and Tsou, T.-S. (1969). Explicit construction of the *N*-point function in the generalized Veneziano model, *Phys. Lett.* **B28**, 485–488.

[CS73] Clavelli, L. and Shapiro, J. A. (1973). Pomeron factorization in general dual models, *Nucl. Phys.* **B57**, 490–535.

[DD71] Del Giudice, E. and Di Vecchia, P. (1971). Factorization and operator formalism in the generalized Virasoro model, *Nuovo Cimento* **A5**, 90–102.

[DHS68] Dolen, R., Horn, D. and Schmid, C. (1968). Finite-energy sum rules and their application to pin charge exchange, *Phys. Rev.* **166**, 1768–1781.

[FN70] Fairlie, D. B. and Nielsen, H. B. (1970). An analog model for KSV theory, *Nucl. Phys.* **B20**, 637–651.

[FGV69] Fubini, S., Gordon, D. and Veneziano, G. (1969). A general treatment of factorization in dual resonance models, *Phys. Lett.* **B29**, 679–682.

[FV69] Fubini, S. and Veneziano, G. (1969). Level structure of dual resonance models, *Nuovo Cimento* **A64**, 811–840.

[GS69] Goebel, C. J. and Sakita, B. (1969). Extension of the Veneziano form to n-particle amplitudes, *Phys. Rev. Lett.* **22**, 257–260.

[GNSS70] Gross, D. J., Neveu, A., Scherk, J. and Schwarz, J. H. (1970). Renormalization and unitarity in the dual-resonance model, *Phys. Rev.* **D2**, 697–710.

[HKS69] Halpern, M. B., Klein, S. A. and Shapiro, J. A. (1969). Spin and internal symmetry in dual Feynman theory, *Phys. Rev.* **188**, 2378–2384.

[Har69] Harari, H. (1969). Duality diagrams, *Phys. Rev. Lett.* **22**, 562–565.

[Kai07] Kaiser, D. (2007). When fields collide, *Sci. Am.* **296**(6), 40–47.

[KT70] Kaku, M. and Thorn, C. B. (1970). Unitary nonplanar closed loops, *Phys. Rev.* **D1**, 2860–2869.

[KY70] Kaku, M. and Yu, L. (1970). The general multi-loop Veneziano amplitude, *Phys. Lett.* **B33**, 166–170.

[KKSV70] Kikkawa, K., Klein, S. A., Sakita, B. and Virasoro, M. A. (1970). Feynman-like diagrams compatible with duality. II. General discussion including nonplanar diagrams, *Phys. Rev.* **D1**, 3258–3366.

[KSV69] Kikkawa, K., Sakita, B. and Virasoro, M. A. (1969). Feynman-like diagrams compatible with duality. I. Planar diagrams, *Phys. Rev.* **184**, 1701–1713.

[KN69a] Koba, Z. and Nielsen, H. B. (1969). Reaction amplitude for n-mesons a generalization of the Veneziano–Bardakci–Ruegg–Virasoro model, *Nucl. Phys.* **B10**, 633–655.

[KN69b] Koba, Z. and Nielsen, H. B. (1969). Manifestly crossing-invariant parameterization of n-meson amplitude, *Nucl. Phys.* **B12**, 517–536.

[Lov68] Lovelace, C. (1968). A novel application of Regge trajectories, *Phys. Lett.* **B28**, 264–268.

[Lov70] Lovelace, C. (1970). M-loop generalized Veneziano formula, *Phys. Lett.* **B32**, 703–708.

[Lov71] Lovelace, C. (1971). Pomeron form factors and dual Regge cuts, *Phys. Lett.* **B34**, 500–506.

[Nam70] Nambu, Y. (1970). Quark model and the factorization of the Veneziano amplitude, in *Proceedings of the International Conference on Symmetries and Quark Models, Wayne State University, June 18–20, 1969*, ed. Chand, R. (Gordon and Breach, New York), 269–277, reprinted in *Broken Symmetry, Selected Papers of Y. Nambu*, ed. Eguchi, T. and Nishijima, K. (World Scientific, Singapore, 1995), 258–277.

[NS70] Neveu, A. and Scherk, J. (1970). Parameter-free regularization of one-loop unitary dual diagram, *Phys. Rev.* **D1**, 2355–2359.

[NS71] Neveu, A. and Schwarz, J. H. (1971). Factorizable dual model of pions, *Nucl. Phys.* **B31**, 86–112.

[Nie70] Nielsen, H. B. (1970). An almost physical interpretation of the integrand of the n-point Veneziano model, paper submitted to the *15th International Conference on High Energy Physics, Kiev, 26 August to 4 September, 1970* (Nordita preprint, 1969).

[Pas69] Pasternack, S. (1969). Private communication, letter of September 3, 1969; retrievable at http://www.physics.rutgers.edu/~shapiro/pastnk.pdf.
[Ram71] Ramond, P. (1971). Dual theory of free fermions, *Phys. Rev.* **D3**, 2415–2418.
[RBPP67] Rosenfeld, A. H., Barbaro-Galtieri, A., Podolsky, W. J., Price, L. R., Soding, P., Wolh, C. G., Ross, M. and Willis, W. J. (1967). Data on particles and resonant states, *Rev. Mod. Phys.* **39**, 1–51.
[Ros69] Rosner, J. L. (1969). Graphical form of duality, *Phys. Rev. Lett.* **22**, 689–692.
[Sha68] Shapiro, J. A. (1968). A quarks-on-springs model of the baryons, *Ann. Phys.* **47**, 439–467. Notice I was just one letter off from a really exciting and prescient idea.
[Sha69] Shapiro, J. A. (1969). Narrow resonance model with Regge behavior for $\pi\pi$ scattering, *Phys. Rev.* **179**, 1345–1353.
[Sha70a] Shapiro, J. A. (1970). General factorization of the n-point beta function into two parts in a Möbius invariant manner, *Phys. Rev.* **D1**, 3371–3376.
[Sha70b] Shapiro, J. A. (1970). Electrostatic analogue for the Virasoro model, *Phys. Lett.* **B33**, 361–362.
[Sha71] Shapiro, J. A. (1971). Nonorientable dual loop graphs and isospin, *Phys. Rev.* **D4**, 1249–1251.
[Sha72] Shapiro, J. A. (1972). Loop graph in the dual tube model, *Phys. Rev.* **D5**, 1945–1948.
[Sha75] Shapiro, J. A. (1975). Renormalization of dual models, *Phys. Rev.* **D11**, 2937–2942.
[ST72] Shapiro, J. A. and Thorn, C. B. (1972). Two particle discontinuities of one loop graphs in the dual resonance model, *Phys. Rev.* **D5**, 927–932.
[SY68] Shapiro, J. A. and Yellin, J. (1968). A model for $\pi\pi$ scattering, UCRL preprint 18500 (September 1968); *Yad. Fiz.* **11**, 443–447 (1970); *Sov. J. Nucl. Phys.* **11**, 247–249 (1970).
[Sus70] Susskind, L. (1970). Structure of hadrons implied by duality, *Phys. Rev.* **D1**, 1182–1186.
[tHo71] 't Hooft, G. (1971). Renormalizable Lagrangian for massive Yang–Mills fields, *Nucl. Phys.* **B35**, 167–188.
[Ven68] Veneziano, G. (1968). Construction of a crossing-symmetric, Reggeon-behaved amplitude for linearly rising trajectories, *Nuovo Cimento* **A57**, 190–197.
[Vir69] Virasoro, M. (1969). Alternative constructions of crossing-symmetric amplitudes with Regge behavior, *Phys. Rev.* **177**, 2309–2311.
[Vir70] Virasoro, M. (1970). Subsidiary conditions and ghosts in dual-resonance models, *Phys. Rev.* **D1**, 2933–2936.
[Yao06] Yao, W.-M. *et al.* (2006). Review of particle physics, *J. Phys.* **G33**, 1–1232.
[Yos71] Yoshimura, M. (1971). Operatorial factorization and symmetry of the Shapiro–Virasoro model, *Phys. Lett.* **B34**, 79–82.

13
Personal recollections

DANIELE AMATI

String theory was born under the guise of a dual scattering amplitude in the strong interaction community. I still remember the very young Gabriele Veneziano dropping into my office at CERN in 1968 to show me his beta function as an expression for an s–t dual scattering amplitude with only poles (bound states) in both channels and only Regge pole asymptotic behaviours. It looked to me an amusing mathematical realization of an idea that several groups were pursuing. But, I confess, I did not follow the general enthusiasm met by the function as well as its extentions to more particle amplitudes.

The revelation on the road to Damascus, and my subsequent conversion, came to me through the Fubini–Gordon–Veneziano paper in which dual amplitudes were factorized in terms of a well-defined – albeit infinite – set of operators. This implied an operatorial expression for tree diagrams thus ready to be extended so to produce the loop expansion as usually required by unitarity: a process that converted an expression for scattering amplitudes into a perturbative expansion of a full fledged field theory.

At that time, I was spending a sabbatical at Orsay and it took us little time to compute, with Claude Bouchiat and Jean-Loup Gervais, the first loop dual amplitude – and also to infect with enthusiasm for the new theory two bright students who were at the edge of getting their degrees and sailing to Princeton: André Neveu and Joël Scherk.

Back at CERN I had the chance to participate in a lively group in which the theory was discussed and developed. I remember the exciting atmosphere and the network of long lasting friendships and collaborations that grew there. The group was often considered a sect by both field theorists and phenomenologists, but we succeeded in maintaining a healthy exchange of ideas with them. I remember Bruno Zumino's interest in the two-dimensional world-sheet boson–fermion symmetry of the Ramond–Neveu–Schwarz model and the success of its generalization to four dimensions with which he and Julius Wess launched supersymmetry. And, on the other side, Lovelace's first proposal of extra spacetime dimensions (26 for the case) to produce the Pomeron as a pole in a dual loop amplitude.

Going back to the operatorial construction of loop diagrams, the duality of the theory produced a novel topological structure. With Alessandrini we succeeded in obtaining the general expression for arbitrary loop amplitudes in terms of Neumann functions – evaluated at all possible insertions of the external particles – defined over all conformally inequivalent Riemann surfaces. All this was before the string interpretation of the theory: we were thus obtaining from unitarity a bottom-up construction of the world-sheet formulation that only afterwards could have been the starting point of Polyakov's top-down formulation, in which the theory was defined by the string action, and the path integral by a sum over all Riemann surfaces describing the string world-sheet.

Another recollection regards the death of the old string theory and the long exile of the very few adepts left, until two of them, Mike Green and John Schwarz, resuscitated it in a new form. In the old approach the fundamental string scale was hadronic, implying a soft structure of hadrons (protons in particular) involving no elementary constituents; this view was finally killed by deep inelastic experiments. The early Eighties revolution in the field took place because string theory was considered as a theory for fundamental constituents with its basic scale related to the Planck scale. As suggested also by the fact, remarked by Scherk and Schwarz, and by Yoneya, that strings may contain gravity.

This change of paradigm led to a burst of unconditional optimism in claiming the advent of a 'Theory of Everything' and a correlated production of antibodies towards a pretension that would be hard to verify or confute in any foreseeable future experiment. The theory gave rise, nevertheless, to many novel developments that, despite remaining rather abstract, opened new ways of thinking that percolated beyond the string community, such as the idea of nontrivial extra spacetime dimensions, and the emergence of more complex extended objects that may have cosmological relevance.

> Daniele Amati received his degree in physics at Buenos Aires in 1952. He has worked in Rome, CERN and Trieste. He was Director of SISSA (1986–2001), where he is presently Honorary Professor. During his research career he worked on many aspects of particle physics, S-matrix, field and string theories and, more recently, on quantum aspects of gravitational collapse. He recently widened his interests to some aspects of cognitive neuroscience.

14
Early string theory at Fermilab and Rutgers

LOUIS CLAVELLI

Today there is a very noticeable tension between string theory and phenomenology in particle physics. It is interesting to muse over the roots of this situation in the very early days of string theory. In this personal account of a small part of the ancient history I review some developments in the string theory groups at Fermilab and Rutgers University in the early Seventies. The present conflict between string theory and phenomenology of particle physics is probably more easily understandable than that of the early days. At its base today is, if sometimes subterranean, a debate over the relative federal funding that the two areas should properly receive. Among phenomenologists there is the perception that progress in string theory has slowed noticeably while still attracting a disproportionate fraction of students and postdoc positions. On the other side there is the feeling that the killing of the Superconducting Super Collider (SSC) project has led to a pronounced drought in the qualitatively new experimental discoveries that could feed phenomenological work. An exception is the confirmation of neutrino masses with well-measured mass splittings and mixing angles. In the beginning, however, string theory (or dual model theory) was totally devoted to the building of a theory of strong interactions that could deal directly with the large amount of data on hadronic phenomena being accumulated at Brookhaven, CERN, and Serpukhov.

One might, therefore, have expected a long honeymoon between dual model theorists and phenomenologists. This however never materialized.

In the rustic environment of Fermilab (then called the National Accelerator Laboratory, NAL) there were, from fall of 1969 to summer of 1971, three theorists excited about dual models, Pierre Ramond, David Gordon, and myself. Pierre has given a valuable account of those days (see his contribution to this Volume), and I have presented my recollections elsewhere [Cla01]. Around the world in physics theory circles there was, in the years leading up to the dual models, a feeling of pessimism about the prospects of dealing with the rapidly increasing number of known hadronic resonances in terms of the simple Lagrangian models of the type that had been studied. Lagrangian field theory was at its nadir and had been widely discarded. Hadronic physics was dominated by Geoff Chew's ideas of democracy

---

The Birth of String Theory, ed. Andrea Cappelli, Elena Castellani, Filippo Colomo and Paolo Di Vecchia.
Published by Cambridge University Press. © Cambridge University Press 2012.

among resonances in an 'S-matrix bootstrap'. Veneziano's remarkable model which burst upon the field in 1968, and its N-point generalization that rapidly followed, seemed to have magical properties. However, what we considered fascinating and somewhat magical, others considered strange and other-worldly. We heard that Sam Treiman had said that no one working on string theory would ever get tenure at Princeton.

I composed a list of twenty phenomenological reasons to pursue the dual models as a theory of hadrons. These are still valid but progress in this direction has come to a standstill. We are no closer today to a predictive theory of hadrons in agreement with the observed trajectories and couplings of Regge poles than we were in 1970. Today's younger string theory experts may not even be too sure of the meaning of some of these ideas. Collectively, they represent a distillation of hundreds of forgotten experimental and theoretical papers of the Sixties. Without explaining each of them in detail let me list them here:

   (i) resonances build Regge behaviour;
  (ii) linear trajectories;
 (iii) suppression of exotics via exchange degeneracy;
  (iv) explicit crossing symmetry and analyticity;
   (v) finite energy sum rules;
  (vi) ghost-free saturation of superconvergence relations;
 (vii) quantization of masses and intercepts;
(viii) smooth connection from Regge to resonance region;
  (ix) Adler zeros;
   (x) N-point generalization, S-matrix bootstrap;
  (xi) wide angle behaviour, P_T cutoff;
 (xii) factorization, particle interpretation;
(xiii) exponential degeneracy;
 (xiv) ghost-free, relativistic harmonic oscillator spectrum;
  (xv) wholesale treatment of resonances;
 (xvi) generation of the Pomeron, half the slope and twice the intercept;
(xvii) f dominance and modified Wu–Yang behaviour of the Pomeron;
(xviii) quantization of the dimensionality of spacetime;
 (xix) abnormal duality in Pomeron–particle scattering;
  (xx) finite mass sum rules.

The experimental staff at Fermilab was well aware of the properties of hadronic scattering although they were too absorbed in building the machine to worry about whether the dual models were going to be the ultimate answer. The laboratory management on the other hand seemed to think we should have been working on more short term solutions. Ned Goldwasser, the deputy director, had told us of Director Bob Wilson's preference that the theory group be actively involved in analysis directly related to the experimental programme which at that point was a year or two away. He did, of course, temper his remarks with the concession that, if we made some formal advances of great significance, that would be OK also. Ned Goldwasser had previously assured us that we five theorists were on the same appointments as the experimental staff and that any decisions on our future at the lab were at least five years away.

Pierre Ramond and I had been working on generalizing the Veneziano model. Having dissected the group theoretical basis of the model [CR70, CR71] we were both investigating possibilities involving anticommuting oscillators though from different points of view [Cla71, Ram71]. The patience of the lab, however, came abruptly to an end. In the fall of 1970 we were summarily dismissed, with the explanation that interactions between us and the experimental physicists had not developed as fully as had been expected.

Even Pierre Ramond's remarkable construction of a supersymmetry on the string worldsheet was not considered sufficiently relevant phenomenologically nor sufficiently significant as a formal advance to save him. Nevertheless, Pierre's supersymmetry spread rapidly at more enlightened institutes around the world. Bunji Sakita, one of the active string theorists of that day, brought Pierre's work to the attention of Bruno Zumino and Julius Wess at CERN. The supersymmetric Wess–Zumino model that resulted brought supersymmetry to life as a field theory independent of string theory.

We left Fermilab in August of 1971, replaced by theorists of a more phenomenological bent. Pierre Ramond took up a position at Yale and I moved down the coast from him to Rutgers.

One year earlier, on the banks of the Raritan, the state of New Jersey had initiated a very strong effort in string theory. Our leader was Claud Lovelace who had surprised (and initially amused) string theorists around the world with the observation that the dual Pomeron would be a factorizable trajectory if spacetime had 26 dimensions. Claud was an admired guru in string theory. Also at Rutgers, fresh from very productive postdocs at Berkeley and Maryland, was Joel Shapiro. Joel was already well known for the Lovelace–Shapiro model [Lov68, Sha69] and the Shapiro–Virasoro model [Vir69, Sha70]. The Rutgers physics department was located in a relatively new but already cramped facility in Piscataway north of the river from the main campus. I naturally began working with Claud Lovelace and Joel Shapiro on strings.

In the fall of 1971, a war of wall posters broke out between Claud Lovelace and one of the Rutgers experimentalists. The experimentalist posted a challenge to the theorists to stop working on strings and tell him what they would see at Fermilab. Claud responded with a pictorial suggesting that he would see a jagged cross-section while the rest of the world would see a smooth Regge behaviour. This experimentalist had become well known for reporting that the A_2 resonance had a pronounced dip in the centre. This result, which was initially confirmed by another experiment and which had triggered a barrage of theory papers, later evaporated.

Joel Shapiro, Claud Lovelace, and I began work on the double twisted loop that contained the Pomeron behaviour. Claud gave us voluminous notes on a dual Reggeon calculus but for the most part we never saw him except at seminars. Claud worked mostly at night sometimes slipping notes under our doors about his results. Meanwhile Joel and I, following the harmonic oscillator formalism, began diverging from Claud's path. We attempted to keep Claud abreast of our progress and we always assumed we were going to write a joint paper. However, when the manuscript, ultimately 80 pages long, began to take form in the winter of 1973, the gap between Claud's methods and ours was too great to bridge. Claud claimed that no equation in our paper followed from any other, which we, of course, denied. Joel and I had laboured well over a year by then, and if we were to try to combine Claud's methods with ours, it would have certainly required at least another year.

The *Nuclear Physics* paper that came out with only Joel Shapiro and me as authors was called 'Pomeron factorization in general dual models' [CS73], but the results were much more general than might be suggested by that title. The paper survived the revolution under which string theory came to be considered a theory of the fundamental interactions rather than a theory of hadronic resonances and has continued to be referenced up to the present time.

The main result was a technique by which a string theory trace could be replaced by a simple expectation value of some rotated oscillators. This allowed us to demonstrate the factorization of arbitrary dual models including the Neveu–Schwarz model and the dual quark models of Bardakci and Halpern. John Schwarz expressed to me his previous belief that those models would not have a factorizable Pomeron. The factorization of the Pomeron in the general model implied the existence of an operator, Υ, that changed a closed string into an open string. This allowed one to replace the Pomeron exchange graph with a tree graph. I later foolishly spent a year using the still complicated method to calculate inclusive reactions in the double Pomeron region [Cla79]. Inclusive reactions were the final attempt to use dual models for hadronic physics. Veneziano devoted quite some time to this effort [Ven71]. The result was that the tree level dual amplitudes gave a fairly good description of the exponential cutoff in transverse momentum of the produced particle at intermediate transverse momenta, but could not reproduce the parton model power law cutoff observed at still higher transverse momenta. I finally arrived at a result from the double Pomeron contribution to the six-point function but still found no parton model behaviour.

The paper with Joel Shapiro on Pomeron factorization also noted how the extra dimensions of the dual models could be traded for anticommuting degrees of freedom. The dimensionality of the Veneziano model could be reduced in a ghost-free way from 26 to 4 by adding 22 dual quarks of the Bardakci–Halpern type. This later became crucial in the heterotic string, making the critical dimensionality of the Veneziano and Ramond–Neveu–Schwarz sides equal.

The twist operator for the general model was also presented. Ironically, both Joel and I forgot this result years later and, when the generalization of the Neveu–Schwarz model to dimensions lower than ten was under consideration in the next decade, it initially appeared that there would be no possibility of writing such a model in four or eight dimensions (Clavelli *et al.* [CCEH89]). Zvi Bern and David Dunbar eventually showed how to perform this reduction using the correct twist operator from Joel's and my paper (which they had independently rediscovered) [BD89].

The first incarnation of string theory ended about 1974. After this time there was little work on hadronic strings or on strings as fundamental interactions. The triumph of field theory inspired by the success of the Standard Model was complete. For the most part,

Pierre Ramond and I and the vast majority of other string theorists began working on what Pierre referred to as four-dimensional physics. It was the end of an era. Only a few hardy souls, notably Mike Green and John Schwarz, kept the theory alive until the renaissance of 1984.

> Louis Clavelli received his bachelor degree from Georgetown University and his doctoral degree from the University of Chicago. Since then he has been employed in teaching and research at a number of universities and national laboratories including Yale, Rutgers, Bordeaux, Bonn, and Fermilab. He joined the faculty of the University of Alabama in 1985.

References

[BD89] Bern, Z. and Dunbar, D. (1989). Construction of four dimensional open superstrings, in *Superstrings and Particle Theory*, ed. Clavelli, L. and Harms, B. (World Scientific, Singapore).

[Cla71] Clavelli, L. (1971). New dual N point functions, *Phys. Rev.* **D3**, 3166–3172.

[Cla79] Clavelli, L. (1979). Particle production in the double dual pomeron region, *Nucl. Phys.* **B154**, 47–61.

[Cla01] Clavelli, L. (2001). Physics at Weston, http://www.bama.ua.edu/~lclavell/Weston.

[CCEH89] Clavelli, L., Cox, P. H., Elmfors, P. and Harms, B. (1989). Supersymmetric open strings in fewer than ten dimensions, *Phys. Rev.* **D40**, 4078–4084.

[CR70] Clavelli, L. and Ramond, P. (1970). $SU(1, 1)$ analysis of dual resonance models, *Phys. Rev.* **D2**, 973–979.

[CR71] Clavelli, L. and Ramond, P. (1971). Group theoretical construction of dual amplitudes, *Phys. Rev.* **D3**, 988–990.

[CS73] Clavelli, L. and Shapiro, J. A. (1973). Pomeron factorization in general dual models, *Nucl. Phys.* **B57**, 490–535.

[Lov68] Lovelace, C. (1968). A novel application of Regge trajectories, *Phys. Lett.* **B28**, 264–268.

[Ram71] Ramond, P. (1971). Dual theory for free fermions, *Phys. Rev.* **D3**, 2415–2418.

[Sha69] Shapiro, J. A. (1969). Narrow resonance model with Regge behavior for $\pi\pi$ scattering, *Phys. Rev.* **179**, 1345–1353.

[Sha70] Shapiro, J. A. (1970). Electrostatic analogue for the Virasoro model, *Phys. Lett.* **B33**, 361–362.

[Ven71] Veneziano, G. (1971). Sum rules for inclusive reactions from discontinuity formulae, *Phys. Lett.* **B36**, 397–399.

[Vir69] Virasoro, M. (1969). Alternative constructions of crossing-symmetric amplitudes with Regge behavior, *Phys. Rev.* **177**, 2309–2311.

15
Dual amplitudes in higher dimensions: a personal view

CLAUD LOVELACE

This is a very personal view of the origin of string theory, and it also makes an interesting prediction.

When I first encountered the Veneziano model in 1968, I was at CERN working on the phenomenology of strong interactions. I saw an application to pion–pion scattering (Lovelace [Lov68] and Shapiro [Sha69]), which occurred independently to Shapiro and Yellin [SY70] at Berkeley. It attracted considerable attention, but now seems misguided. Later I combined dual amplitudes with Gribov's Reggeon calculus [Lov71b]. This application has been revived recently by people exploiting AdS/CFT duality, and should eventually succeed.

All through 1969 people were adding legs to the Veneziano amplitude, or chopping it in half. I will leave this story to other contributors. Amati and coworkers gave a series of seminars at CERN on their $SL(2, \mathbb{R})$ group theoretic formulation which I thought very elegant [AALO71]. This inspired me to drop phenomenology and join them. In modern language, the N-Reggeon vertex [Lov70a] is the open string disc amplitude with arbitrary excited states on all external lines. Several people had written down a complicated formula. I noticed it could be factorized by a single set of oscillators at the centre of the disc, and realized that an easy construction of the complete perturbation expansion to all orders in string loops would follow. Alessandrini and Amati [Ale71, AA71], and I [Lov70b] worked this out.

I am culexic – my favourite way of developing an idea is to cover a table with piles of possibly relevant books and papers, and rapidly turn pages till I catch a scent like the one in my mind. Then I read more carefully. Nielsen had emphasized that dual formulae can be visualized as momentum flowing across the world-sheet. It did not take Alessandrini [Ale71, AA71] and me [Lov70b] long to discover that hydrodynamics on multiply connected Riemann surfaces is solved by Abelian integrals, which can be constructed by Poincaré theta series (Burnside 1894), which perfectly matched the result of sewing together the external lines of the N-Reggeon vertex. Kaku and Yu [KY70] were also working on the multiloop problem under Mandelstam at Berkeley, but they did not find the relevant

---

The Birth of String Theory, ed. Andrea Cappelli, Elena Castellani, Filippo Colomo and Paolo Di Vecchia.
Published by Cambridge University Press. © Cambridge University Press 2012.

mathematics. I am still very proud of that 1970 paper, though few people seem to have read it.

That year I was fired by CERN for discovering too many nucleon resonances (all of which were subsequently confirmed), so I looked for a job in America. I settled on Rutgers. In 1970–1971, I was living in the Princeton Holiday Inn, commuting to Rutgers during the week, and working at IAS on weekends. My passport shows that I was in America from 13 November 1970 to 27 May 1971, when I returned to Geneva.

Gross *et al.* [GNSS70, GS70] at Princeton, and Frye and Susskind [FS70] at Yeshiva had both found a strange singularity in the one-loop amplitude. Like everyone else I thought that open strings were Reggeons, so this must be the Pomeron, which would be very interesting to a phenomenologist. Unfortunately it was tachyonic with a continuous spectrum. My notebooks show that I started redoing their calculations on 1 October 1970 at CERN. I needed a realistic model for phenomenology, so the Pomeron intercept had to be 1. By next day I had concluded that the intercept was $D/2$ in D spacelike dimensions (i.e. those with oscillators). However, the ensuing calculations turned the cut into a pole by arbitrarily deleting log R factors. They go on for 88 pages until a note written in Princeton in early January says 'I think we need 24 spacelike and 2 *timelike* dimensions to get complete cut cancellation.' I suspect the correct solution came to me suddenly at night, since this note is in different ink. Thus in 26 spacetime dimensions, and assuming that two sets of oscillators decoupled, the Pomeron spectrum became discrete [Lov71a]. There was still one tachyon, but the next particle had zero mass and spin two. This matched the Shapiro–Virasoro formula.

I gave a seminar at IAS circa 1 February 1971, which was attended by some powerful people as well as the dual model group. Treating the result as a joke, I said I had bootstrapped the dimension of spacetime but the result was slightly too big. Everyone laughed. A few years later, I was the only professor not being promoted despite the many citations of my papers. However, the jeers of the physics establishment did have one good consequence. When my discovery of the critical dimension turned out to be correct and significant, they remembered that I had said it first. One has to be very brave to suggest that spacetime has 26 dimensions.

In my 1971 paper I relegated it to a footnote. The paper thanks everyone with whom I had discussed it, in particular Alessandrini and Amati at CERN, Schwarz at Princeton, Shapiro at Maryland. It was received by *Physics Letters* on 15 February 1971 and published 29 March. In summer 1971 I discussed my $D = 26$ discovery with many people at CERN.

I taught myself general relativity and quantum mechanics at the age of sixteen (1950), when I was at school in Switzerland, and got interested in unified field theories soon after, so I knew about Kaluza–Klein. (However, I had no useful guidance till I met Salam in 1958.) In hindsight I was inexcusably stupid not to see in 1971 that my Pomeron was the graviton. In 1972 Thorn told me that his group had verified my wild conjecture that a second set of oscillators would decouple in the critical dimension [GT72]. Shapiro also found that one-loop modular invariance of what we now call closed strings gave the same 26 dimensions [Sha72]. Even after Scherk and Schwarz [SS74] in 1974 made the graviton connection, I did not take it seriously, and wasted the Seventies investigating what asymptotic freedom would do to Reggeons. I now wish I had stayed in contact with Salam, who had better taste. Between two feasible problems, always choose the more fundamental.

However, I did encourage Clavelli and Shapiro to complete the factorization in the Pomeron channel [CS73]. As we now know, open string loops are dual to closed string trees. Many years later (1987) I took this up again when working with Callan's group [CLNY87], and discovered the boundary state formalism after noticing one sentence in a 1975 paper by Ademollo et al. [ADDG75]. It took me a long time to realize that even much smarter people don't have this ability to scan print for hidden gems. Most physicists just read a few famous authors and follow the latest fashion. John Schwarz toiled for 10 years on quantum supergravity, working at Caltech without tenure, but few people noticed until Witten discovered him. That is the way to make a big contribution if one has the guts.

This Volume is about the past, but I would like to end with another prophecy. The fundamental problem in present-day string theory is not how to compactify the theories, but how to unify them. In particular, the AdS/CFT duality must be generalized. On the CFT side $\mathcal{N} = 4$, super-Yang–Mills is a compactification of open superstring theory in (9, 1) dimensions (the notation is (number of spaces, number of times)). This should be a boundary of something that compactifies to $AdS_5 \times S_5$. The only obvious candidate is FLAT (10, 2) spacetime. So the unifying theory is F not M, and the second time (T_2) is not a mathematical device, nor is it compactified. We don't notice a second clock ticking because our entire (9, 1) universe is in an asymptotic stationary state of the second Hamiltonian, just like the incoming beam in a scattering experiment! This seems to me the obvious way to exploit F-theory, but I could not find any previous author willing to take the mental leap. Holography is then just causality in T_2 (perhaps Black Holes are not quite stationary).

After scanning two hundred papers, I have a quite detailed vision of this ultimate theory, though some crucial mathematics is still missing. Our familiar (1, 1) world-sheets will become (2, 2) membranes. Our external particle states will become lightlike (1, 1) and timelike (2, 1) asymptotic limits of these membranes, targeting incoming (9, 1) and (10, 1) universes respectively. The free RNS fields on the old world-sheet will become self-dual Yang–Mills coupled to Majorana–Weyl fermions on the new membrane. Our central charge will become the loop anomaly that couples these to (2, 2) self-dual gravity. This anomaly must be cancelled by ghosts in the critical dimension, where the target is the bulk (10, 2) spacetime in which different one-time universes interact. There will probably exist a double light-cone gauge in which all longitudinal (2, 2) modes are constructed nonlinearly from transverse (8, 0) modes. Note that classical self-dual Yang–Mills theory is integrable. The authors to study are Siegel (very carefully), and Rudychev. To conclude with an apocalyptic pun, 'Who is worthy to open the book and to loose the seals thereof?'

Claud Lovelace was born in England in 1934. He started to study physics on his own at the age of 15. He received his BSc from Cape Town in 1954, but then switched to architecture. He returned to physics in 1958 at Imperial College under Salam. He moved to CERN in 1965, then to Rutgers in 1971.

References

[ADDG75] Ademollo, M., D'Adda, A., D'Auria, R., Gliozzi, F., Napolitano, E., Sciuto, S. and Di Vecchia, P. (1975). Soft dilatons and scale renormalization in dual theories, *Nucl. Phys.* **B94**, 221–259.

[Ale71] Alessandrini, V. (1971). A general approach to dual multiloop diagrams, *Nuovo Cimento* **A2**, 321–352.

[AA71] Alessandrini, V. and Amati, D. (1971). Properties of dual multiloop amplitudes, *Nuovo Cimento* **A4**, 793–844.

[AALO71] Alessandrini, V., Amati, D., Le Bellac, M. and Olive, D. (1971). The operator approach to dual multiparticle theory, *Phys. Rep.* **C1**, 269–346.

[CLNY87] Callan, C. G., Lovelace, C., Nappi, C. R. and Yost, S. A. (1987). Adding holes and crosscaps to the superstring, *Nucl. Phys.* **B293**, 83–113.

[CS73] Clavelli, L. and Shapiro, J. A. (1973). Pomeron factorization in general dual models, *Nucl. Phys.* **B57**, 490–535.

[FS70] Frye, G. and Susskind, L. (1970). Non-planar dual symmetric loop graphs and pomeron, *Phys. Lett.* **B31**, 589–591.

[GT72] Goddard, P. and Thorn, C. B. (1972). Compatibility of the pomeron with unitarity and the absence of ghosts in the dual resonance model, *Phys. Lett.* **B40**, 235–238.

[GNSS70] Gross, D. J., Neveu, A., Scherk, J. and Schwarz, J. H. (1970). Renormalization and unitarity in the dual-resonance model, *Phys. Rev.* **D2**, 697–710.

[GS70] Gross, D. J. and Schwarz, J. (1970). Basic operators of dual resonance model, *Nucl. Phys.* **B23**, 333–357.

[KY70] Kaku, M. and Yu, L. (1970). The general multi-loop Veneziano amplitude, *Phys. Lett.* **B33**, 166–170.

[Lov68] Lovelace, C. (1968). A novel application of Regge trajectories, *Phys. Lett.* **B28**, 264–268.

[Lov70a] Lovelace, C. (1970). Simple n-reggeon vertex, *Phys. Lett.* **B32**, 490–494.

[Lov70b] Lovelace, C. (1970). M-loop generalized Veneziano formula, *Phys. Lett.* **B32**, 703–708.

[Lov71a] Lovelace, C. (1971). Pomeron form factors and dual Regge cuts, *Phys. Lett.* **B34**, 500–506.

[Lov71b] Lovelace, C. (1971). Regge cut theories, *Amsterdam Conference 1971* (North Holland, Amsterdam), 141–151.

[SS74] Scherk, J. and Schwarz, J. H. (1974). Dual models for nonhadrons, *Nucl. Phys.* **B81**, 118–144.

[Sha69] Shapiro, J. A. (1969). Narrow resonance model with Regge behavior for $\pi\pi$ scattering, *Phys. Rev.* **179**, 1345–1353.

[Sha72] Shapiro, J. A. (1972). Loop graph in the dual tube model, *Phys. Rev.* **D5**, 1945–1948.

[SY70] Shapiro, J. A. and Yellin, J. (1970). University of California Lawrence Radiation Laboratory Report No. UCRL-18500 (unpublished), *Yad. Fiz.* **11**, 443–447 (1970).

16
Personal recollections on dual models

RENATO MUSTO

1968. We were young. Revolution was in the air. Gabriele Veneziano's paper [Ven68] was an event in that eventful year. It was immediately clear that it was an important result, but very few people would have thought of it as the beginning of an entirely new theoretical approach that was going to grow in the shell of the old one. It seems appropriate to me to try to combine in this Chapter my personal recollections of the days of the birth of string theory with a few observations on a change in perspective that occurred in theoretical physics. It is a change that we expected in society and that we have witnessed in the course of the years in physics.

It is enough to look at the high energy physics titles of the W. A. Benjamin series *Frontiers in Physics*, that started in 1961, to get the feeling of the early Sixties: in the first year, Geoffrey F. Chew's *S-Matrix Theory of Strong Interactions*; in 1963, *Regge Poles and S-Matrix* by S. C. Frautschi, *Mandelstam Theory and Regge Poles* by R. Omnès and M. Froissart, together with *Complex Angular Momenta and Particle Physics* by E. J. Squires; in 1964, *Strong-Interaction Physics* by M. Jacob and G. F. Chew. Chew's *Analytic S-Matrix*, with its openly programmatic subtitle *A Basis for Nuclear Democracy*, was published in 1966, again by Benjamin. In order to complete the list one has to add to these S-matrix inspired titles only a few books, among which is *The Eightfold Way* by M. Gell-Mann and Y. Ne'eman.

There were different and connected reasons for the dominant role of S-matrix theory and for the increasing suspicion of field theory methods. The mathematical foundation of field theory appeared unsound. For many physicists field theory meant only Feynman graphs. The renormalization programme, even if successful in curing the problem of infinities present in perturbative expressions, appeared a mathematical trick, as its physical meaning was not yet understood. This is very clearly expressed in the introduction of a classic textbook of the time (Berestetskii, Lifshitz and Pitaevskii [BLP71]). Even worse, the presence of the Landau pole, found in the behaviour of what today is called the running of the coupling constant in quantum electrodynamics, appeared to be a fatal blow to the full consistency of the theory.

The Birth of String Theory, ed. Andrea Cappelli, Elena Castellani, Filippo Colomo and Paolo Di Vecchia.
Published by Cambridge University Press. © Cambridge University Press 2012.

To quote from Landau and Lifshitz: 'Thus we reach the important conclusion that quantum electrodynamics as a theory with weak interaction is essentially incomplete. Yet the whole formalism of the existing theory depends on the possibility of regarding the electromagnetic interaction as a small perturbation' (Lifshitz and Pitaevskii [LP74]). To understand fully what that meant, one must realize that the results of quantum electrodynamics were thought of at that time as having universal validity in quantum field theory (an interesting discussion on these points may be found in [Gro05]).

But it was in the realm of strong interactions, where the most intense experimental activity was performed, that the use of quantum field theory appeared more problematic. At first, the scheme of quantum electrodynamics appeared successful with Yukawa description of nuclear forces in terms of nucleon–pion interaction. But, then, the increasing number of resonances that were found made such a scheme useless. Furthermore, the strength of the coupling was a fundamental obstacle, as nonperturbative methods were not available. To cope with these problems, a new perspective was developed. Instead of starting from a fundamental set of fields, the attempt was made to construct a self-consistent 'bootstrap' scheme 'with all strongly interacting particles (hadrons) enjoying equivalent status in a *nuclear democracy*' (Chew [Che66]). The correct framework for this programme was not quantum field theory, which requires the choice of a fundamental set of fields and the construction of a Hamiltonian spacetime evolution, but the S-matrix approach of Wheeler and Heisenberg: 'The simple framework of S-matrix theory and the restricted set of questions that it presumes to answer constitutes a major advantage over quantum field theory. The latter is burdened with a superstructure, inherited from classical electromagnetic theory, that seems designed to answer a host of experimentally unanswerable questions' [Che66].[1]

By the time of the revival of his S-matrix programme Heisenberg had long changed his point of view, looking for a fundamental *nonlinear* field theory: 'It is perhaps not exaggerated to say that the study of the S-matrix is a very useful method for deriving relevant results for collision processes by going around fundamental problems. But these problems should be solved... The S-matrix is an important but very complicated mathematical quantity that should be derived from the fundamental field equations...' [Hei57]. According to Chew, this change in perspective was due to the weakness of the original formulation of the S-matrix theory that lacked dynamical content: 'Heisenberg lost interest probably because in the forties he lacked the full analytic continuation that is required to give the S-matrix dynamical content' [Che61].

In the second half of the Fifties, analytic properties of scattering amplitudes were extensively studied in the framework of field theory, and, at the beginning of the Sixties, Regge poles and analytic continuation in angular momentum were taken from potential scattering into relativistic high energy physics. These were the two ingredients that, according to Chew, were needed to make S-matrix a full dynamical theory. It is worth noticing that in

---

[1] The term 'S-matrix' was introduced by Wheeler in 1937 [Whe37], the S-matrix programme was developed by Heisenberg starting in 1943 [Hei43, Hei44]. For a thorough discussion of the development of the S-matrix philosophy see Cushing [Cus90].

1961, at the Twelfth Solvay Conference, answering to a comment by Chew, Heisenberg recalled his reason for giving up the construction of a pure S-matrix theory: '... when one constructs a unitary S-matrix from simple assumptions (like a Hermitian η matrix by assuming $S = e^{i\eta}$), these S-matrices always become non analytical at places where they ought be analytical. But I found it very difficult to construct analytic S-matrices' [Hei85]. However, it was exactly the difficulty encountered in constructing an S-matrix satisfying the requirements of unitarity and analyticity and exhibiting the expected Regge behaviour that was the reason for hoping for an extraordinarily predictive power of such a programme, once accomplished.

In the spirit of Heisenberg's ideas of the Forties, the bootstrap programme was to give the possibility of a spacetime description of microscopic phenomena in favour of an even more ambitious goal, namely to obtain complete knowledge of masses and coupling constants, perhaps in terms of a unique constant with the dimension of a length: 'The belief is growing that a successful theory of strong interactions will not tolerate arbitrary parameters; perhaps no arbitrariness of any kind is possible...' [Che66]. This extremely attractive, though certainly disputable, ideal of total predictability, so much in tune with the spirit of the time, is probably the strongest heritage that string theory has received from the S-matrix programme.

It was impossible for a graduate student to escape the fascination of these ideas in a period when it was claimed that 'The times they are a-changin'' and that 'the old road is rapidly agin''. Youth appeared as a value by itself and quantum field theory appeared old, at least as it was taught to me. (Even worse, general relativity was taught as a branch of mathematics.) I was fascinated by the idea of giving up a spacetime description that seemed to be a fake theatrical perspective, deprived of any dynamical content. I was attracted by the possibility of being in touch with the latest developments in high energy physics, but I felt quite uneasy with the lack of a strong theoretical structure and the continuous guesswork that the S-matrix approach required. I was much more confident in the algebraic methods that, with the success of the Eightfold Way, became a promising substitute for a lacking sound dynamical principle. Symmetry and nonsymmetry groups had proven very effective in potential theory, as in the case of the hydrogen atom (Englefield [Eng72]); the algebra of currents, that Murray Gell-Mann had abstracted from field theory [Gel64], was opening a new path for understanding strong interactions. Under the guidance of Lochlainn O'Raifeartaigh I derived algebraic relations in the bootstrap approach, mainly by using finite energy sum rules.

Then everything changed. Dolen, Horn and Schmid duality [DHS68, Sch68] suggested that resonance and Regge pole contributions should not be added and Gabriele Veneziano implemented the idea explicitly in the case of a four-point scalar amplitude [Ven68]. He showed that the right starting point for the S-matrix programme was *not* a unitary S-matrix but rather a crossing-symmetric amplitude made up by an infinite sum of zero-width resonances lying on linearly rising Regge trajectories that build up Regge behaviour at high energy. While the attempt to saturate finite energy sum rules to obtain numerical or algebraic relations involved only a finite number of states, the essential feature of the

Veneziano amplitude was the presence of an infinite number of states. The theory of the Dual Resonance Model (DRM) had started.

By June 1969, when Sergio Fubini visited Naples, much progress had already been made in the DRM (see for instance [DiV08, DS08]). And Fubini shared with us not only his knowledge of the latest developments but also his enthusiasm and his vision of the new theoretical structure that was being unravelled. On that occasion I met some of the brilliant young physicists with whom I became coworker and friend. It was then that I started my activity in the DRM.

In 1971 I joined Emilio Del Giudice, Paolo Di Vecchia and Sergio Fubini at MIT, where they had completed their important work on transverse physical state operators [DDF72]. The structure of the DRM had been largely clarified but, in the meantime, the physical picture of strong interactions had been drastically changed by the deep inelastic scattering experiments, probing the short distance structure of nucleons. The question of the day was to explain the almost free behaviour of elementary constituents inside the nucleon. Could the DRM help in understanding this new physics? In order to clarify this point, we classified multiparton states by means of a subgroup of the conformal group, obtaining a picture of hadrons closely resembling the structure of physical states in the DRM. The hope was strong, but the resulting structure functions were not reasonable for *wee* partons [DDFM72].

The distance between the DRM, with its fundamental length, and scale invariant deep inelastic physics was not bridged. And the discrepancies between the DRM and the actual phenomenology of strong interactions were still present. As a phenomenological model, the DRM was too rigid. But this tight, beautiful theoretical structure was so attractive and the emerging string picture was so promising. So, back in Italy, I joined forces with a group of young physicists, all connected in one way or another with Fubini, all stubbornly convinced that the DRM had a rich future. It was an unusually large research group, with great enthusiasm and friendship, very much in the egalitarian spirit of the time. I only want to mention the first paper of the collaboration (Ademollo, D'Adda, D'Auria, Napolitano *et al.* [ADDN74]), where we extended the string picture of the DRM from the spectrum to the interaction, for which only few results had been obtained up to that point. The question was clearly stated: 'If the string is not only an analogue model of the DRM, one should be able to understand the interaction among the hadrons by starting from the string picture.' We gave a positive answer to the question, by showing that the possible interactions were determined by the reparameterization invariance of the string world-sheet.

I like to look at this result as a step away – at least for me, and I think also for all those involved in that collaboration – from an S-matrix approach toward a geometrical spacetime description, very much in the spirit of Einsteinian physics. The decisive step in this direction was, obviously, the discovery that the Virasoro–Shapiro dual model contains Einstein's theory of gravity (Yoneya [Yon74]) and the proposal by Scherk and Schwarz of a DRM approach to gravity [SS74a, SS74b], opening the way to the development of string theory as a unified theory of all interactions.

Thus one can look at the birth of string theory as the growth, out of the shell of the S-matrix approach, of a new geometrical programme that is the closest realization of the Einsteinian dream. And the correspondence of the results that may be obtained by starting from two such different points of view may be taken as the sign of the appearance of a new fundamental theory, as when matrix mechanics and wave mechanics were recognized as two formulations of quantum mechanics.

Acknowledgements

I wish to thank Dr Vincenzo De Luise, librarian of my Department, for his constant help.

Renato Musto graduated in Naples and received his PhD at Syracuse University. He is Professor of Theoretical Physics at Naples University 'Federico II'. He has visited CERN, DIAS, MIT and Nordita. His main interests are in quantum field theory and string theory. He has worked on the quantum Hall effect, quantum groups and complementarity. He has also published articles on the rise of electromagnetism in the romantic period, and on different subjects in the history of ideas. He is author of a book on industrial workers of the world and, together with E. Napolitano, of one on Mozart's Magic Flute.

References

[ADDN74] Ademollo, M., D'Adda, A., D'Auria, R., Napolitano, E., Sciuto, S., Di Vecchia, P., Gliozzi, F., Musto R. and Nicodemi, F. (1974). Theory of an interacting string and dual-resonance model, *Nuovo Cimento* **A21**, 77–145.

[BLP71] Berestetskii, V. B., Lifshitz, E. M. and Pitaevskii, L. P. (1971). *Relativistic Quantum Theory*, Vol. 4 of Landau and Lifshitz's *Course in Theoretical Physics*, Part 1, p. 4 (Pergamon Press, Oxford) (*Relyativistskaya kvantovaya teoriya*, chast' 1, 1968).

[Che61] Chew, G. F. (1961). *S-Matrix Theory of Strong Interactions* (W. A. Benjamin, New York), 1.

[Che66] Chew, G. F. (1966). *The Analytic S-Matrix* (W. A. Benjamin, New York), 2–3.

[Cus90] Cushing, J. T. (1990). *Theory Construction and Selection in Modern Physics. The S-Matrix* (Cambridge University Press, Cambridge).

[DDF72] Del Giudice, E., Di Vecchia, P. and Fubini, S. (1972). General properties of the dual resonance model, *Ann. Phys.* **70**, 378–398.

[DDFM72] Del Giudice, E., Di Vecchia, P., Fubini, S. and Musto, R. (1972). Light-cone physics and duality, *Nuovo Cimento* **A12**, 813–862; Light-cone physics and duality, in *Scale and Conformal Symmetry in Hadron Physics*, ed. Gatto, R. (Wiley-Interscience, New York, 1973), 153–166.

[DiV08] Di Vecchia, P. (2008). The birth of string theory, in *String Theory and Fundamental Interactions*, ed. Gasperini, M. and Maharana, J. (Springer, Berlin).

[DS08] Di Vecchia, P. and Schwimmer, A. (2008). The beginning of string theory: a historical sketch, in *String Theory and Fundamental Interactions*, ed. Gasperini, M. and Maharana, J. (Springer, Berlin).

[DHS68] Dolen, R., Horn, D. and Schmid, C. (1968). Finite energy sum rules and their application to πN charge exchange, *Phys. Rev.* **166**, 1768–1781.

[Eng72] Englefield, M. J. (1972). *Group Theory and the Coulomb Problem* (Wiley-Interscience, New York).

[Gel64] Gell-Mann, M. (1964). The symmetry group of vector and axial vectors currents, *Physics* **1**, 63–75.

[Gro05] Gross, D. J. (2005). Nobel Lecture: the discovery of asymptotic freedom and the emergency of QCD, *Rev. Mod. Phys.* **77**, 837–849.

[Hei43] Heisenberg, W. (1943). Die beobachtbaren grossössen in der theorie der elenemteilchen: I, *Z. Phys.* **120**, 513–538; Die beobachtbaren grossössen in der theorie der elenemteilchen: II, *Z. Phys.* **120**, 673–702.

[Hei44] Heisenberg, W. (1944). Die beobachtbaren grossössen in der theorie der elenemteilchen: III, *Z. Phys.* **123**, 93–112.

[Hei57] Heisenberg, W. (1957). Quantum theory of fields and elementary particles, *Rev. Mod. Phys.* **29**, 269–278.

[Hei85] Heisenberg, W. (1985–1993) *Gesammelte Werke*, Abt. A: *Wisswemschaftilische Original arbeiten*, ed. Blum, W., Durr, H. P. and Rechenberg, H. (Springer-Verlag, Berlin), Part III, 568.

[LP74] Lifshitz, E. M. and Pitaevskii, L. P. (1974). *Relativistic Quantum Theory*, Vol. 4 of Landau and Lifshitz's *Course in Theoretical Physics*, Part 2, p. 509 (Pergamon Press, Oxford) (*Relyativistskaya kvantovaya teoriya*, chast' 2, 1971).

[SS74a] Scherk, J. and Schwarz, J. H. (1974). Dual models and the geometry of spacetime, *Phys. Lett.* **B52**, 347–350; Dual model approach to a renormalizable theory of gravitation, submitted to the 1975 Gravitation Essay Contest of the Gravity Research Foundation, reprinted in *Superstrings*, Vol. 1, ed. Schwarz, J. H. (World Scientific, Singapore, 1985), 218–222.

[SS74b] Scherk, J. and Schwarz, J. H. (1974). Dual models for nonhadrons, *Nucl. Phys.* **B81**, 118–144.

[Sch68] Schmid, C. (1968). Direct-channel resonances from Regge-pole exchange, *Phys. Rev. Lett.* **20**, 689–691.

[Ven68] Veneziano, G. (1968). Construction of a crossing-symmetric, Reggeon behaved amplitude for linearly rising trajectories, *Nuovo Cimento* **A57**, 190–197.

[Whe37] Wheeler, J. A. (1937). On the mathematical description of light nuclei by the method of resonating group structure, *Phys. Rev.* **52**, 1107–1122.

[Yon74] Yoneya, T. (1974). Connection of dual models to electrodynamics and gravidynamics, *Prog. Theor. Phys.* **51**, 1907–1920.

17
Remembering the 'supergroup' collaboration

FRANCESCO NICODEMI

The birth of string theory took place in the years from 1968 (Veneziano formula) to 1973–1974. At the time I was a young physicist interested in the dynamics of strongly interacting particles. This problem was clearly far from being solved but, due to the everyday increase of 'elementary particles' and their 'strong' interaction, one thing seemed sure to me: the answer was not to be found in quantum field theory!

From the Goldberger–Treiman relation, $m_N g_A = f_\pi G_{\pi NN}$, one obtained $G_{\pi NN} \sim 10$ for the pion–nucleon coupling constant. So a perturbative expansion for the strong interaction was clearly out of the question. But for me it was more than that. With all its 'infinities' popping up everywhere, quantum field theory (QFT) appeared to me not only inadequate but just conceptually 'wrong' on the whole, because of the large number of 'unobservable' quantities it introduced.

In the spirit of Heisenberg the description of the physical world was to be based only on 'observables'! Since the only really 'measurable' quantities were the cross-sections in scattering processes, Chew's 'bootstrap' programme seemed the right conceptual framework and his book *S-Matrix Theory of Strong Interactions* [Che62] was (not only for me, I believe) the reference text.

Of course the calculation of the full S-matrix appeared too ambitious and the immediate objective was to obtain relations among a set of measurable parameters. In this context, a relevant method to obtain physical information was based on the idea, emphasized especially by Sergio Fubini and others, to use dispersion relations together with current algebra to obtain sum rules involving cross-sections. The latter were observable only in principle, because they involved integrals over energies up to infinity. Further progress occurred when it was realized that in some processes one could approximate the integral with a much more tractable sum over the dominant poles.

At the end of 1967 Dolen, Horn and Schmid [DHS68], analyzing some experimental data, suggested that the t-channel contribution was 'dual' to the s-channel one, in that each of them described the whole amplitude and the two were not to be added as in usual field theory calculations. In this way the idea of 'duality' was born.

The Birth of String Theory, ed. Andrea Cappelli, Elena Castellani, Filippo Colomo and Paolo Di Vecchia.
Published by Cambridge University Press. © Cambridge University Press 2012.

In such a context the construction by Gabriele Veneziano of his famous formula for the scattering of four mesons [Ven68], which fulfilled all the requirements of S-matrix theory (with the exception of unitarity, since in the tree approximation the resonances had zero width) appeared as the triumph of the S-matrix approach. As is well known, the Veneziano amplitude for a given ordering of the external particles can be written either as a sum over poles in the 'direct' channel or as a sum over poles in the crossed channel, thus fulfilling DHS duality. Moreover, the (infinite) sum over resonant states in one channel, each with a given angular momentum, yields a Regge pole behaviour in the crossed variable, which was really astounding at the time.

So, in collaboration with a large number of friends (most of them from Naples, Florence and Turin), I started working on the 'Veneziano model', which later was called the Dual Resonance Model (DRM). This was seen as a set of rules for constructing scattering amplitudes in accordance with the requirements of crossing symmetry, good asymptotic behaviour and Lorentz covariance without explicit reference to any Lagrangian. Nevertheless, the factorization and unitarity conditions were checked as if there was an underlying field theory.

The atmosphere of the collaboration was very pleasant and stimulating, and being before the Internet era, we had to travel a lot, meeting alternately in the various towns. The number of people involved was so large (around a dozen) that we used to refer to ourselves with the (of course prophetic!) name of 'supergroup'.

The attempt to build models reproducing at least some of the strong interaction phenomenology led to a large number of papers. However, at a more fundamental level, the field was hampered by the fact that in general the spectrum included 'ghost' states (of negative norm), even if the allowed 'physical' states were selected by the action of an infinite set, L_n, of gauge operators.

It was shown pretty soon by Virasoro [Vir70] that one could avoid having ghosts among the 'physical' states if the intercept of the Regge trajectory was taken to be $\alpha_0 = 1$, but this had two unpleasant consequences:

- in the spectrum there was a tachyon, i.e. an imaginary mass state;
- the lowest mass 'physical' particle ($m = 0$) had spin one, unlike the pion, which is a scalar particle.

Furthermore, there were indications that the DRM did not properly account for the strong interaction phenomenology. For instance, for large angle scattering it led to a behaviour which was too soft in comparison to observations. In particular the electro-production experiments carried out at Stanford pointed out that in the deep inelastic region the hadrons behave as if they have an internal structure and the constituents appear to be pointlike charges (although, even in 1973, Gell-Mann himself was not sure of the 'reality' of quarks).

So the hope of describing the observed world faded away somewhat, but there was continuously very fast and fascinating progress in understanding the properties of the DRM, which started to have its own life, independent of the phenomenology.

In 1969 the N-point amplitude was constructed and its factorization led to the study of the whole spectrum of intermediate states and to the discovery that the degeneracy of states (along the Regge trajectory) grows exponentially with the mass. Fubini and Veneziano realized that such a degeneracy could be accounted for by using an infinite number of harmonic oscillators which could be grouped together into the Fubini–Veneziano 'field' [FV70].

Then it was shown by Shapiro [Sha70] and Virasoro [Vir69] that one could also build another kind of amplitude which still satisfies the S-matrix paradigm but has nonplanar duality.

For the Shapiro–Virasoro model (SVM) it turned out that the intercept of the Regge trajectory had to be twice that of the Veneziano model. Taking the latter to be $\alpha_0 = 1$ (to avoid ghost states) one then finds $\alpha_0^{SV} = 2$. This corresponds to a fundamental meson of spin two, which was not observed. Only much later, when it was definitely clear that the DRM did not describe the phenomenology of strong interactions, Yoneya [Yon73] and, independently, Scherk and Schwarz [SS74], realized that this particle was to be identified with the graviton.

In 1970, Ramond [Ram71], and Neveu and Schwarz [NS71] found that one could add spin into the model introducing mysterious 'anticommuting' variables and the states in the Ramond model were recognized as 'bona fide' spacetime fermions.

Another intriguing feature of the DRM was that, for the first time in the history of physics, its internal logical coherence led to a limitation on the number of spacetime dimensions. The result was $d = 26$ for the Veneziano model but it reduced to $d = 10$ for the Neveu–Schwarz model. There was also an attempt, by part of the 'supergroup' collaboration, to reduce the 'critical' dimension further to $d = 4$ by increasing the number of supersymmetries, but this led to $d = 2$!

Pretty soon (1970) suggestions appeared (Nambu [Nam69], Nielsen [Nie70] and Susskind [Sus70]) that the DRM reflected the underlying structure of a more fundamental object, a relativistic string. In this way the 'gauge' operators L_n had a very simple physical interpretation: they corresponded to the generators of the group of invariance of the action under any change of coordinates on the surface spanned by the string.

Personally, at the time I was not enthusiastic about this idea, because it appeared to me again to introduce unobservable quantities into the game, starting once more with a Lagrangian and all the machinery of QFT.

Of course, from today's point of view it is clear that the 'string' idea really involved a great amount of intuition, but at the time and for several years after, it was not even certain that the two approaches were completely equivalent.

So, like many others, I preferred to continue working with the 'physical' states and with the operators which describe them in the so-called 'operator' approach. Only in 1973, through the work of Goddard, Goldstone, Rebbi and Thorn [GGRT73], was it definitely established that, for $d = 26$, the string approach reproduces all the features of the spectrum of the DRM.

These developments concerned free strings, giving no information on the interactions among them. On the other hand the DRM gave detailed information not only about the spectrum of 'hadrons' but also about the couplings among the various particles in the spectrum. If the string was not just an analogue model of the DRM, one should be able to understand the interaction among 'hadrons' starting from the string picture. This goal was achieved by Mandelstam [Man73] in 1973 and shortly afterward by the 'supergroup' collaboration (Ademollo et al. [ADDN74]), as I will now briefly describe.

In order to lead to the amplitudes of the DRM the interaction among strings had to satisfy a basic 'bootstrap' requirement: the intermediate set of states has to coincide with the spectrum of the free string. This fact suggested the by now standard pictorial model of interaction among strings: in, say, a four-point amplitude for open strings, two strings travel freely from time $t = -\infty$ to a certain time T when two end-points of the two strings join. The two strings 'coalesce' and propagate as a single free string up to a time T', when the 'composed' string splits into two new ones. In the spirit of quantum mechanics, the total amplitude is obtained by summing over all inequivalent ways in which the strings can join and split. A similar picture holds for closed strings.

The approach of the 'supergroup' collaboration was based on the idea that the interaction should not alter too drastically the free motion of the string. This was realized by allowing an open string to interact only at its end-points. The effect of the other strings was simulated by means of an external field.

For example, to study the interaction of an open string 'charged' at its end-points with a massless spin one field, $A_\mu(x)$, the following action was introduced:

$$S = \int_{\tau_1}^{\tau_2} d\tau \int_0^\pi d\sigma \left\{ \mathcal{L}_0 + A_\mu(x) j^\mu \right\}, \tag{17.1}$$

where $\mathcal{L}_0$ is the free string Lagrangian and

$$j_\mu = \dot{x}_\mu [q_0 \delta(\sigma) + q_\pi \delta(\sigma - \pi)], \tag{17.2}$$

q_0 and q_π being the 'charges' at the ends $\sigma = 0$ and $\sigma = \pi$ of the string.

It then turns out that, as one might expect, the string equations of motion are the same as in the free case whereas the boundary conditions are changed. In the particular case of an external monochromatic field $A_\mu(x) = \epsilon_\mu e^{ikx}$, all the equations can be solved exactly provided three important conditions are met. The external photon must be (1) on its mass shell, $k^2 = 0$, (2) physical, $k \cdot \epsilon = 0$, and (3) circularly polarized.

In the closed string case, since one cannot attach a charge anywhere on the surface swept by the string without violating reparameterization invariance, the interaction with the 'strong graviton' was put in a geometrical way (à la 'general relativity') by using the Lagrangian:

$$\mathcal{L} = \partial^a x^\mu \partial_a x^\nu g_{\mu\nu}(x), \quad \text{where } a = \sigma, \tau.$$

Taking the external 'gravitational' field to be monochromatic, i.e.

$$g_{\mu\nu}(x) = \eta_{\mu\nu} + \varepsilon_{\mu\nu}\exp\{ik_\rho x^\rho\},$$

it turned out that the interaction could be introduced consistently only if the graviton were physical, i.e. if it satisfied the free field conditions:

$$\varepsilon^\mu_\mu = k_\mu \varepsilon^{\mu\nu} = k_\mu k^\mu = 0.$$

This example led to the interesting result that an 'external' field can interact consistently with a string only if it satisfies the equations of motion dictated by the string dynamics. In other words internal coherence required strings to interact only with string states.

In addition, it was found that open strings also interact necessarily with closed strings (while the latter can also live on their own) and a new 'mixed' dual amplitude was constructed in which 'Veneziano' particles interact with 'Virasoro' particles. A few years later, once it was understood that the spin two particle of closed string theory 'is' the graviton, it was clear that this reflects just the universality of gravity.

I would say that the first period of string theory ends in 1976 when Gliozzi, Scherk and Olive [GSO76] showed that from the Ramond and the Neveu–Schwarz models one could obtain a theory free of tachyons and fully consistent, even if it did not have the initial phenomenological application to strong interactions. Of course, the study of string theory still goes on today, based only on its internal logical consistency, with extremely exciting discoveries and alternating periods of fortune. With today's insight this appears quite reasonable to me, since with string theory we are making 'gedanken' experiments probing the structure of spacetime many orders of magnitude smaller than what we can observe experimentally.

And, far from my initial and naive point of view that only 'observable' quantities are needed in the description of physical reality, I see no reason why Nature should be so kind as to allow us to describe (not to say: understand!) what happens at those extreme distances using concepts that we have formulated on the basis of the observation of the world at our scale. As Dirac once wrote, referring to quantum mechanics: 'It is becoming more and more apparent that in our description of the physical world we are forced to introduce non observable quantities.' No wonder that completely new and unexpected ideas have to be introduced in such a context. Rather one should be surprised at the results achieved in string theory on merely conceptual grounds.

Indeed I think that, as Sergio Fubini once told me very many years ago, 'String theory is a piece of twenty-first century physics fallen by chance into the twentieth century.'

Francesco Nicodemi received his degree in physics in 1962 in Naples. He obtained his PhD in 1967, discussing a thesis on current algebra and finite energy sum rules under the supervision of A. P. Balachandran. Back in Naples as Associate Professor he continued working on this subject and was captured by the 'avalanche' started by the Veneziano paper. He is currently Professor of Quantum Mechanics at the University of Naples.

References

[ADDN74] Ademollo, M., D'Adda, A., D'Auria, R., Napolitano, E., Sciuto, S., Di Vecchia, P., Gliozzi, F., Musto, R. and Nicodemi, F. (1974). Theory of an interacting string and dual-resonance model, *Nuovo Cimento* **A21**, 77–145.

[Che62] Chew, G. F. (1962). *S-Matrix Theory of Strong Interactions* (W. A. Benjamin, New York).

[DHS68] Dolen, R., Horn, D. and Schmid, C. (1968). Finite-energy sum rules and their application to $\pi - N$ charge exchange, *Phys. Rev.* **166**, 1768–1781.

[FV70] Fubini, S. and Veneziano, G. (1970). Duality in operator formalism, *Nuovo Cimento* **A67**, 29–47.

[GSO76] Gliozzi, F., Scherk, J. and Olive, D. (1976). Supergravity and the spinor dual model, *Phys. Lett.* **B65**, 282–286.

[GGRT73] Goddard, P., Goldstone, J., Rebbi, C. and Thorn, C. B. (1973). Quantum dynamics of a massless relativistic string, *Nucl. Phys.* **B56**, 109–135.

[Man73] Mandelstam, S. (1973). Interacting string picture of dual resonance models, *Nucl. Phys.* **B64**, 205–235.

[Nam69] Nambu, Y. (1970). Quark model and the factorization of the Veneziano amplitude, in *Proceedings of the International Conference on Symmetries and Quark Models, Wayne State University, June 18–20, 1969*, ed. Chand, R. (Gordon and Breach, New York), 269–277, reprinted in *Broken Symmetry, Selected Papers of Y. Nambu*, ed. Eguchi, T. and Nishijima, K. (World Scientific, Singapore, 1995), 258–277.

[NS71] Neveu, A. and Schwarz, J. H. (1971). Factorizable dual model of pions, *Nucl. Phys.* **B31**, 86–112.

[Nie70] Nielsen, H. B. (1970). An almost physical interpretation of the integrand of the n-point Veneziano model, paper submitted to the *15th International Conference on High Energy Physics, Kiev, 26 August to 4 September, 1970* (Nordita preprint, 1969).

[Ram71] Ramond, P. (1971). Dual theory of free fermions, *Phys. Rev.* **D3**, 2415–2418.

[SS74] Scherk, J. and Schwarz, J. H. (1974). Dual models and the geometry of spacetime, *Phys. Lett.* **B52**, 347–350.

[Sha70] Shapiro, J. A. (1970). Electrostatic analogue for the Virasoro model, *Phys. Lett.* **B33**, 361–362.

[Sus70] Susskind, L. (1970). Dual-symmetric theory of hadrons. 1, *Nuovo Cimento* **A69**, 457–496.

[Ven68] Veneziano, G. (1968). Construction of a crossing-symmetric, reggeon behaved amplitude for linearly rising trajectories, *Nuovo Cimento* **A57**, 190–197.

[Vir69] Virasoro, M. (1969). Alternative constructions of crossing-symmetric amplitudes with Regge behavior, *Phys. Rev.* **177**, 2309–2311.

[Vir70] Virasoro, M. (1970). Subsidiary conditions and ghosts in dual-resonance models, *Phys. Rev.* **D1**, 2933–2936.

[Yon73] Yoneya, T. (1973). Quantum gravity and the zero slope limit of the generalized Virasoro model, *Lett. Nuovo Cimento* **8**, 951–955.

18
The '3-Reggeon vertex'

STEFANO SCIUTO

The year 1968 was quite remarkable also in science: it was the year of the Veneziano formula, a real breakthrough towards new horizons in theoretical physics.

In July 1968 in Turin we heard a seminar by Gabriele Veneziano who came to submit his beta function amplitude to the attention of Sergio Fubini, an opinion leader in the theory of strong interactions after his works on current algebra (the $p \to \infty$ method, with G. Furlan) and on superconvergence (with V. De Alfaro, G. Furlan and C. Rossetti). Fubini encouraged Veneziano to publish his paper, which soon appeared [Ven68] in *Nuovo Cimento* (at that time the most important journal in high energy physics in Europe) and invited the young physicists in Turin to look at this promising model.

Actually my interest increased when the generalization of the Veneziano amplitude to N points was proposed by K. Bardakci, H. Ruegg, M. A. Virasoro, H. M. Chan, S. T. Tsou, C. J. Goebel, B. Sakita, Z. Koba and H. B. Nielsen. I was fascinated by the elegance of the 'duality rules': the poles in the various channels were exhibited by simple changes of integration variables, dictated by (elementary) geometry.

My enthusiasm burst in the spring of 1969, when Miguel Virasoro gave a seminar in Turin on his work with Keiji Kikkawa and Bunji Sakita [KSV69]: up to some (crucial) factors, the structure of the multiloop amplitudes was simply dictated by consistency with the duality rules.

As usual, also in 1969 Fubini came back to Turin at the end of May, bringing the latest news from MIT. But that year things were not as usual; in previous years, while in Turin, Fubini used to sit down and work with his collaborators in his office and, typically, after some days someone used to come out with a new current algebra sum rule, that graduate students like me had to saturate by adding the contribution of the resonance $N_{3,3}$.

In May 1969 it was much more stimulating; happily, Fubini's coworkers were busy writing '*il Libro*' (their book on current algebra) and Fubini was very excited about the new developments in the Dual Resonance Model (DRM). So he came to Turin with Veneziano and they explained to us directly the big achievement of the factorization by means of an

---

The Birth of String Theory, ed. Andrea Cappelli, Elena Castellani, Filippo Colomo and Paolo Di Vecchia.
Published by Cambridge University Press. © Cambridge University Press 2012.

infinite set of harmonic oscillators [FGV69] (at that time, no one among us realized that they were associated with the vibration modes of a relativistic string).

I remember that Fubini ended his enthusiastic speech by saying: 'Now we know the spectrum, we have the propagator and we have the vertex for the emission of the lowest lying states, we only miss the vertex for the emission of a generic state; if we were able to get it, we would have a **theory**, not only a model!'

I was rather shy, but I replied immediately: 'It seems to me that one just needs to make a duality transf...'; Fubini stopped me: 'Don't say more, rather sit down, make your calculations and come back only when you are done!' I followed Fubini's advice and some days later I showed him and Veneziano the 'General 3-Reggeon vertex'. Fubini helped me a lot in writing the paper [Sci69], but he generously let me sign it alone.

Actually the hope that the vertex was the only missing stone needed to build the theory was exceedingly optimistic. It was already clear to Fubini and Veneziano, as well as to all the people in the field, that the spectrum was affected by spurious states, due to the temporal and longitudinal components of the creation operators. Thus, sewing vertices and propagators was not enough to build multiloop corrections (necessary to recover unitarity), but one needed to project out spurious states. The complete answer in the operator formalism came only when BRST invariance was fully implemented in string theory in the Eighties.

The very first steps in understanding the symmetries of the DRM were already made in 1969: in Copenhagen by Ziro Koba and Holger Bech Nielsen [KN69], who discovered the projective invariance of the N-point amplitude; in Turin by Nando Gliozzi [Gli69], who realized it in the operator formalism, by writing the generators L_0, L_+ and L_- of the $SL(2, \mathbb{R})$ algebra, which was the first part of the Virasoro algebra, to be discovered later.

Intermediate steps towards building multiloop amplitudes were the four-point vertex [CS70], found by Gabriella Carbone and myself, and the N-point vertex, first written by Mike Kosterlitz and Dennis Wray, and then put in its most elegant form by Claud Lovelace [Lov70].

These quantities were obtained by sewing three-point vertices and propagators, with the essential help of a paper by Chiu, Matsuda and Rebbi [CMR69], which provided the suitable gauge transformations. This paper was only the first of the important contributions to string theory of my friend, and classmate in Turin, Claudio Rebbi, which culminated in his seminal paper with Goddard, Goldstone and Thorn on the quantum dynamics of a massless relativistic string [GGRT73].

The elegance of the Lovelace vertex derives also from its manifest cyclic symmetry, which was lacking in the original form of the three-point vertex I wrote in 1969 and which was implemented quite soon by Luca Caneschi, Adam Schwimmer and Veneziano [CSV69] and proved rigorously by Caneschi and Schwimmer [CS70], by using the gauge transformations of [CMR69].

The asymmetric vertex, however, had its own virtues, especially in the form [DS70] written in Naples by Angelo Della Selva and Satoru Saito in terms of the (quasi) conformal field introduced by Fubini and Veneziano [FV70] and Jean-Loup Gervais [Ger70] (it is worthwhile remarking that these two papers opened the route to conformal field theories,

14 years before the fundamental paper by Belavin, Polyakov and Zamolodchikov [BPZ84]). In fact such a vertex is the only tool that allows one to send any highest weight state in the corresponding primary field in an explicit and direct way [CFNS71].

My transition to 'conscious' string theory began with the works by Ademollo *et al.* In the first one [ADDN74] we extended to the interacting case the proof, given in [GGRT73], that the Dual Resonance Models are described by the dynamics of relativistic strings. But this is another story. Let me conclude by quoting the names of the people who started this huge collaboration, who are usually omitted in bibliographies: Marco Ademollo (Florence), Alessandro D'Adda, Riccardo D'Auria, Ernesto Napolitano, Stefano Sciuto (Turin), Paolo Di Vecchia, Ferdinando Gliozzi (CERN), Renato Musto, Francesco Nicodemi (Naples)[1] and John Schwarz [ABDD76b]. We had all collaborated with Sergio Fubini in different combinations and learned a lot from him; at that time we decided that we were grown up enough and that it was better to join forces, cooperating rather than competing with each other; this unconventional choice was also a fruit of the spirit of 1968.

Acknowledgements

I dedicate this contribution to Sergio Fubini (1928–2005).

Stefano Sciuto was born in Mondovì (Italy) in 1943. He has been Professor of Theoretical Physics since 1987, first in Naples, and currently in Turin. His main contributions to string theory have been the dual vertex, the interacting string in the operatorial formalism, the $\mathcal{N} = 2$ and $\mathcal{N} = 4$ superconformal algebras, the multiloop amplitudes of the bosonic string in the Schottky parameterization and the classical p-branes from boundary states.

References

[ABDD76a] Ademollo, M., Brink, L., D'Adda, A., D'Auria, R., Napolitano, E., Sciuto, S., Del Giudice, E., Di Vecchia, P., Ferrara, S., Gliozzi, F., Musto, R. and Pettorino, R. (1976). Supersymmetric strings and color confinement, *Phys. Lett.* **B62**, 105–110.

[ABDD76b] Ademollo, M., Brink, L., D'Adda, A., D'Auria, R., Napolitano, E., Sciuto, S., Del Giudice, E., Di Vecchia, P., Ferrara, S., Gliozzi, F., Musto, R., Pettorino, R. and Schwarz, J. (1976). Dual string with $U(1)$ colour symmetry, *Nucl. Phys.* **B111**, 77–110.

[ABDD76c] Ademollo, M., Brink, L., D'Adda, A., D'Auria, R., Napolitano, E., Sciuto, S., Del Giudice, E., Di Vecchia, P., Ferrara, S., Gliozzi, F., Musto, R. and Pettorino, R. (1976). Dual string models with nonabelian color and flavor symmetries, *Nucl. Phys.* **B114**, 297–316.

[ADDN74] Ademollo, M., D'Adda, A., D'Auria, R., Napolitano, E., Sciuto, S., Di Vecchia, P., Gliozzi, F., Musto, R. and Nicodemi, F. (1974). Theory of an interacting string and dual-resonance model, *Nuovo Cimento* **A21**, 77–145.

---

[1] Later 'Ademollo *et al.*' grew further, including Emilio Del Giudice and Roberto Pettorino (Naples); moreover extended supersymmetry was introduced by our group in collaboration with Lars Brink and Sergio Ferrara [ABDD76a, ABDD76b, ABDD76c] and John Schwarz.

[BPZ84] Belavin, A. A., Polyakov, A. M. and Zamolodchikov, A. B. (1984). Infinite conformal symmetry of critical fluctuations in two dimensions, *J. Stat. Phys.* **34**, 763–774.

[CFNS71] Campagna, P., Fubini, S., Napolitano, E. and Sciuto, S. (1971). Amplitude for n nonspurious excited particles in dual resonance models, *Nuovo Cimento* **A2**, 911–928.

[CS70] Caneschi, L. and Schwimmer, A. (1970). Ward identities and vertices in the operatorial duality formalism, *Lett. Nuovo Cimento* **3S1**, 213–217.

[CSV69] Caneschi, L., Schwimmer, A. and Veneziano, G. (1969). Twisted propagator in the operatorial duality formalism, *Phys. Lett.* **B30**, 351–356.

[CS70] Carbone, G. and Sciuto, S. (1970). On amplitudes involving excited particles in dual resonance models, *Lett. Nuovo Cimento* **3S1**, 246–252.

[CMR69] Chiu, C. B., Matsuda, S. and Rebbi, C. (1969). Factorization properties of the dual resonance model – a general treatment of linear dependences, *Phys. Rev. Lett.* **23**, 1526–1530.

[DS70] Della Selva, A. and Saito, S. (1970). A simple expression for the Sciuto three-reggeon vertex-generating duality, *Lett. Nuovo Cimento* **4S1**, 689–692.

[FGV69] Fubini, S., Gordon, D. and Veneziano, G. (1969). A general treatment of factorization in dual resonance models, *Phys. Lett.* **B29**, 679–682.

[FV70] Fubini, S. and Veneziano, G. (1970). Duality in operator formalism, *Nuovo Cimento* **67**, 29–47.

[Ger70] Gervais, J. L. (1970). Operator expression for the Koba–Nielsen multi-Veneziano formula and gauge identities, *Nucl. Phys.* **B21**, 192–204.

[Gli69] Gliozzi, F. (1969). Ward-like identities and twisting operator in dual resonance models, *Lett. Nuovo Cimento* **2**, 846–850.

[GGRT73] Goddard, P., Goldstone, J., Rebbi, C. and Thorn, C. B. (1973). Quantum dynamics of a massless relativistic string, *Nucl. Phys.* **B56**, 109–135.

[KSV69] Kikkawa, K., Sakita, B. and Virasoro, M. A. (1969). Feynman-like diagrams compatible with duality. I: Planar diagrams, *Phys. Rev.* **184**, 1701–1713.

[KN69] Koba, Z. and Nielsen, H. B. (1969). Manifestly crossing-invariant parameterization of n-meson amplitude, *Nucl. Phys.* **B12**, 517–536.

[KW70] Kosterlitz, J. M. and Wray, D. A. (1970). The general N-point vertex in a dual model, *Lett. Nuovo Cimento* **3S1**, 491–497.

[Lov70] Lovelace, C. (1970). Simple n-Reggeon vertex, *Phys. Lett.* **B32**, 490–494.

[Sci69] Sciuto, S. (1969). The general vertex function in dual resonance models, *Lett. Nuovo Cimento* **2**, 411–418.

[Ven68] Veneziano, G. (1968). Construction of a crossing-symmetric, Reggeon behaved amplitude for linearly rising trajectories, *Nuovo Cimento* **A57**, 190–197.

Part IV
The string

19
Introduction to Part IV

19.1 Introduction

The connection between the Dual Resonance Model (DRM) and the relativistic string, a one-dimensional extended system, was observed soon after Veneziano's fundamental paper. Indeed, the presence of an infinite set of harmonic oscillators, with frequencies that were multiples of a fundamental tune, was clearly suggestive of a vibrating string.

Part IV contains the contributions by the authors who proposed the string interpretation, found the action and studied the quantization. In the years 1969 and 1970, Nambu, Nielsen and Susskind, each from his own perspective, suggested independently that a string model was at the basis of the DRM. In 1970, Nambu, and later Goto, wrote the correct relativistic and reparameterization-invariant form of the string Lagrangian. The different perspectives can be summarized as follows.

(i) Susskind's starting point was the comparison of DRM scattering amplitudes with those for the relativistic harmonic oscillator. From the existence of many frequencies of oscillation, i.e. harmonics, Susskind had the idea of a 'rubber band' or a 'violin string'.

(ii) Nielsen's intuition came from an analogy of dual diagrams with high-order Feynman diagrams, called 'fishnet diagrams'; in this approximation particles interact approximately with nearest neighbours and effectively form a one-dimensional chain, i.e. a string.

(iii) Nambu first introduced the linear Lagrangian of the string, reproducing the infinite harmonic oscillators; then he proposed the correct nonlinear action, by analogy with the relativistic point-particle action (closely related results were also obtained soon after by Goto and Hara). The particle action has a simple geometrical meaning, being proportional to the length of the particle trajectory. The corresponding geometrical quantity for the string is the area of the world-sheet; accordingly, Nambu introduced an action proportional to the area.

A related picture was proposed by Fairlie and Nielsen, using the Koba–Nielsen world-sheet description of DRM amplitudes they developed, in analogy with classical

The Birth of String Theory, ed. Andrea Cappelli, Elena Castellani, Filippo Colomo and Paolo Di Vecchia.
Published by Cambridge University Press. © Cambridge University Press 2012.

electrostatics in two dimensions, the 'analogue model', relying on the fact that electric fields and strings obey the same equations of motion. As described by Fairlie in his Chapter, the analogue model was able to reproduce DRM amplitudes including loops. It was later used by Shapiro in his study of the closed string for determining the many-point extension of the Virasoro amplitude (see Part III).

Apart from the analogue model, the early results on string dynamics did not play a role in the developments of the DRM until 1973. The quantization of the Nambu–Goto Lagrangian was too difficult, owing to its nonlinearities and gauge symmetries, and the properties of the DRM, such as scattering amplitudes and the physical spectrum, could not be easily recovered. Thus, the research activity in the early Seventies focussed on the study of the properties of the DRM leaving aside the Nambu–Goto action.

As described in Goddard's Chapter, the quantization of the string was obtained by Goddard, Goldstone, Rebbi and Thorn in their 1973 paper: the results were in complete agreement with those obtained in the DRM by using the Fubini–Veneziano oscillator approach, in particular as regards the decoupling of the nonpositive norm states from the physical spectrum.

The canonical quantization of the string action led to two main developments. First, the invariance of the action under arbitrary reparameterizations of the world-sheet coordinates was understood as being the gauge symmetry underlying the structure of DRM physical states, as described in the Introduction to Part III. Second, the quantization of the string in the light-cone gauge, based on the $(d-2)$ transverse string coordinates, provided an independent derivation of the values of critical dimension $d=26$ and intercept $\alpha_0 = 1$, following from the requirement of relativistic invariance at the quantum level.

Summing up, all known results of the DRM fitted into place, showing that the bosonic string was a consistent relativistic quantum mechanical system – only tachyons were still to be understood. Nonetheless, the model had many shortcomings in describing hadron dynamics, leading to its abandonment in the mid-Seventies. How string theory was later reconsidered for unifying gravity with the other fundamental forces will be described in the following Parts of the book. Note, however, the Chapter by Brower in this Part, which discusses the modern resurgence of the hadronic string in the context of the Maldacena AdS/CFT correspondence.

The derivation of the DRM from an action principle was a fundamental result, instrumental to many further developments of string theory. Some were found soon after, such as the supersymmetry relating fermions to bosons (see Part V). Others were discovered later, in the Nineties, such as the existence of extended objects called D-branes.

In general, the study of the action allows one to discuss geometrical principles and symmetries, and to obtain the physical, semi-classical, intuition. In the case of the string, the action plays a particular role because dynamics is deeply related to geometry. For example, string interactions are completely determined by the geometry of surfaces and there is a unique coupling constant. On the contrary, there are many possible interactions

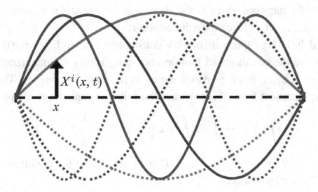

Figure 19.1 The vibrating string.

between point particles, since several terms can be added to the field theory Lagrangian, corresponding to different vertices in Feynman diagrams.

In the following, we shall introduce some elements of the string dynamics, construct the Nambu–Goto action and describe the main steps leading to its quantization and the appearance of the critical dimension and physical intercept. A detailed description of the quantization is given in Appendix C; the string action and its supersymmetric generalization are described further in Part VI.

An application of the string Lagrangian is discussed in Mandelstam's Chapter. He discusses the path-integral quantization of string theory in the light-cone gauge, including string interactions that lead to DRM amplitudes. Another approach to studying string interactions was that of adding the coupling with external fields to the string Lagrangian. It turned out that the unique external states allowed by the symmetry of the Lagrangian (conformal invariance in the covariant gauge) are those corresponding to the on-shell physical states of the string spectrum, in agreement with the uniqueness of interaction mentioned above. This approach is discussed in the Chapters by Nicodemi in Part III and by Gliozzi in Part VI. Further results were the couplings between three arbitrary string states found by Mandelstam and by Cremmer and Gervais, which matched those of the operator approach. This topic is discussed in the Chapter by Gervais in Part V.

19.2 The vibrating string

In this Section we discuss the classical and quantum properties of a nonrelativistic vibrating string and we show that it reproduces several features of the DRM. Let us consider a guitar string of length a, fixed at two end-points $x = 0, a$, along the first of D space dimensions. The string oscillations are described by the vector of transverse displacement $X^i(x, t)$, $i = 2, \ldots, D$, as shown in Figure 19.1. The oscillations are sinusoidal standing waves with a basic frequency (fundamental sound) ν, as well as its integer multiples (the harmonics), $\nu_n = n\nu$; these waves have nodes at intermediate points, $x = k/(na)$, $k = 1, \ldots, n-1$, for the nth harmonic.

In order to find the energy spectrum of the vibrating string, let us first write the Lagrangian of the system, solve the equations of motion and then discuss the quantization. This is a rather well-known subject, but it provides the basis for later discussions. The Lagrangian for the string displacement field $X^i(x, t)$ takes the following quadratic form:

$$S = \int dt\, L = \int dt \int_0^a dx \left(\frac{\mu}{2} \dot{X}^i \dot{X}^i - \frac{T}{2} X'^i X'^i \right), \tag{19.1}$$

where summation over the index $i = 2, \ldots, D$ is assumed, and the notations $\dot{X}^i = \partial X^i/\partial t$ and $X'^i = \partial X^i/\partial x$ are introduced.

The terms in the Lagrangian can be understood as follows. Let us imagine the string divided in bits of size Δx, each representing one degree of freedom, for example $X^i(x_o, t)$ at point $x = x_o$. Owing to discretization, the x-integration in (19.1) is replaced by a sum over Lagrangians pertaining to each bit. The first term thus provides the standard kinetic energy for the bits, with μ being the mass per unit length. The second term is a quadratic potential, $V\left(X^i(x_o)\right) \sim T\left(X^i(x_o + \Delta x) - X^i(x_o)\right)^2$, corresponding to the linear force that keeps attached consecutive bits along the string, T being the string tension.

The classical dynamics is obtained by finding the stationary points of the action. The variation of the action under the infinitesimal change $X^i \to X^i + \delta X^i$ reads:

$$\delta S \propto \int dt \int_0^a dx \left(\dot{X}^i \delta \dot{X}^i - c^2 X'^i \delta X'^i \right)$$

$$\propto \int dt \int_0^a dx \left[\left(-\ddot{X}^i + c^2 X''^i \right) \delta X^i - c^2 \frac{\partial}{\partial x} \left(X'^i \delta X^i \right) \right], \tag{19.2}$$

where in the second line we performed an integration by parts. From the vanishing of δS we obtain the equations of motion

$$\left[\left(\frac{\partial}{\partial x} \right)^2 - \frac{1}{c^2} \left(\frac{\partial}{\partial t} \right)^2 \right] X^i(x, t) = 0, \tag{19.3}$$

as well as the boundary conditions

$$\left(X'^i \delta X^i \right)\big|_{x=0} = \left(X'^i \delta X^i \right)\big|_{x=a} \quad \text{(no sum over } i\text{)}. \tag{19.4}$$

The general solution of the equations of motion describes waves propagating in two directions, as follows:

$$X^i = f^i(x + ct) + g^i(ct - x), \tag{19.5}$$

where the velocity c is related to the string parameters and to the fundamental frequency ν by:

$$c = \sqrt{\frac{T}{\mu}} = 2a\nu. \tag{19.6}$$

Let us now impose the boundary conditions (19.4). In the case of the guitar string, we should take fixed boundaries, $X^i|_{x=0,a} = 0$ and $\delta X^i|_{x=0,a} = 0$, called Dirichlet boundary conditions. These conditions imply:

$$f^i(ct) + g^i(ct) = 0, \qquad f^i(a+ct) + g^i(ct-a) = 0. \tag{19.7}$$

The first equation requires that the two functions f and g are equal, while the second implies that they are periodic functions with period equal to $2a$. In conclusion, the general solution of the equations of motion and boundary conditions for the guitar string can be written as follows:

$$X^i(t,x) = i\sqrt{\frac{c\hbar}{\pi T}} \sum_{n=1}^{\infty} \frac{1}{\sqrt{n}} \sin\left(\frac{\pi n x}{a}\right) \left[a_n^i e^{-2i\pi n v t} - a_n^{i*} e^{2i\pi n v t}\right]. \tag{19.8}$$

The arbitrary coefficients a_n^i and a_n^{i*} describe the amplitude and phase of the oscillation of frequency nv in the ith transverse direction. The normalization factor (including the factor $\sqrt{\hbar}$) has been conveniently chosen to ensure that a_n^i and a_n^{i*} satisfy the harmonic-oscillator algebra in the quantum theory (see Eq. (19.9)).

Summing up, we found stationary waves in the $(d-2)$ transverse dimensions of spacetime, $d = D + 1$, with frequencies $v_n = nv$, multiples of the fundamental frequency, with v given in Eq. (19.6). Let us now turn to the quantum theory and obtain the energy spectrum.

The canonical quantization of the system is straightforward: the quantity X^i becomes an operator and the quantization amounts to imposing equal-time commutation relations between X^i and its canonical momentum $P^i = \mu \dot{X}^i$. Upon substitution of the expansion (19.8) in these commutators, the coefficients a_n^i become creation–annihilation operators of independent harmonic oscillators:

$$[a_n^i, a_m^{j\dagger}] = \delta^{ij}\delta_{nm}. \tag{19.9}$$

The result (19.9) can also be inferred heuristically as follows. The displacement field X^i (19.8) can be viewed as a collection of independent waves with coordinates $q_{(n)}^i(t)$, $n = 1, \ldots, \infty$, propagating in the transverse directions: $X^i = \sum_{n=1}^{\infty} \sin\left(\frac{\pi n x}{a}\right) q_{(n)}^i(t)$. The nth wave corresponds to the oscillation with frequency nv, and obeys the harmonic-oscillator equation of motion, $d^2 q_{(n)}^i / dt^2 + (2\pi n v)^2 q_{(n)}^i = 0$.

We now recall (see the Introduction to Part III, Section 10.4) that the Hamiltonian of a quantum harmonic oscillator is expressed in terms of the creation–annihilation operators $a, a^\dagger$ as $H = hv(a^\dagger a + 1/2)$, and that the energy spectrum is given by integer-spaced energy levels $E_\ell = hv(\ell + 1/2)$ (h is Planck's constant). Using these formulae, we find that the nonrelativistic string quantum Hamiltonian and spectrum are given by:

$$H = hv \sum_{i=1}^{d-2} \sum_{n=1}^{\infty} n\, a_n^{i\dagger} a_n^i + E_0, \tag{19.10}$$

$$E = hv N + E_0, \qquad N = \sum_{i=1}^{d-2} (\ell_1^i + 2\ell_2^i + \cdots). \tag{19.11}$$

The spectrum (19.11) realizes the basic quantum mechanical relation between waves and quanta, namely that oscillations with frequency ν acquire energy $E = h\nu$. Note the strong similarity with the DRM spectrum discussed in Part III, and described by the operator formalism (recall $\alpha_n = \sqrt{n} a_n$, $n > 0$). Both systems contain an infinite set of harmonic oscillators, with the important difference that, in the case of a guitar string, we find only $D - 1$ vibrating spacelike directions, while in the DRM we have harmonic oscillators with a Lorentz index μ running over all directions, spacelike and timelike. Furthermore, the spectrum extracted from the DRM also contains the four-momentum operator p that generates spacetime translations, while in the case of the vibrating string translational invariance is broken by the Dirichlet boundary conditions. In spite of these differences, the presence of an infinite set of harmonic oscillators is suggestive that a string theory is underlying the DRM.

Before closing this Section, we note that the boundary conditions in Eq. (19.4) can be satisfied by imposing periodicity conditions, $X^i(0, t) = X^i(a, t)$, corresponding to a closed string. In this case, the periodic functions f^i and g^i are independent and their expansions involve two independent sets of harmonic oscillators. Finally, the boundary conditions in Eq. (19.4) can also be satisfied by imposing the so-called Neumann boundary conditions, $X'^i(0, t) = X'^i(a, t) = 0$. In the next Section we will present simple string motions satisfying these conditions.

19.3 The rotating rod

Although the guitar string contains an infinite set of harmonic oscillators like the DRM, its spectrum is neither relativistic nor translational invariant. The Lagrangian in Eq. (19.1) can be made relativistic invariant by adding two further components to the displacement field, X^0 and X^1 for the temporal and longitudinal directions, respectively. But, in so doing, we must introduce two new dimensionless variables σ and τ that parameterize the Lorentz covariant displacement field $X^\mu(\tau, \sigma)$, where $\mu = 0, 1, \ldots, D$, instead of the t and x variables used for the vibrating string in the previous Section.

With these modifications the Lagrangian (19.1) becomes:

$$S = \frac{T}{2c} \int d\tau \int_0^\pi d\sigma \left(\dot{X}^\mu \dot{X}_\mu - X'^\mu X'_\mu \right), \tag{19.12}$$

where for convenience the two end-points of the string are parameterized by the values $\sigma = 0$ and $\sigma = \pi$, respectively. In this equation, T is the string tension, c the speed of light and we use the notation $\dot{X}^\mu \equiv \partial X^\mu / \partial \tau$ and $X'^\mu \equiv \partial X^\mu / \partial \sigma$.

The previous Lagrangian is manifestly Lorentz invariant together with its equations of motion:

$$\left[\frac{\partial^2}{\partial \tau^2} - \frac{\partial^2}{\partial \sigma^2} \right] X^\mu(\sigma, \tau) = 0. \tag{19.13}$$

Translation invariant, actually Poincaré invariant, solutions can be found by imposing the Neumann boundary conditions, namely $X'^\mu(\sigma, \tau) = 0$ at $\sigma = 0, \pi$. In the following, we

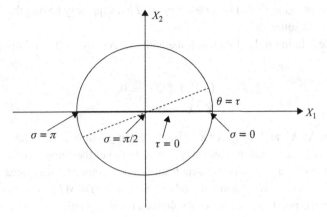

Figure 19.2 The rotating rod.

consider a particular solution, corresponding to a straight rod of length $2a$ rotating around its centre in the (X^1, X^2) plane (see Figure 19.2). It is given by:

$$X^0 = ct = a\tau,$$
$$X^1 = a\cos(\tau)\cos(\sigma), \quad X^2 = a\sin(\tau)\cos(\sigma), \quad (19.14)$$
$$X^i = 0, \quad i = 3, 4, \ldots, d-1.$$

It is easy to check that the end-points of the straight rod move at the speed of light. This string motion is different from that of the previous Section because it does not involve oscillations.

The energy spectrum is given by the relation between mass M and angular momentum J. From the expression for the string momentum density,

$$P_\mu = \frac{\partial L}{\partial \dot{X}^\mu} = \frac{T}{c} \dot{X}_\mu, \quad (19.15)$$

we obtain the total mass,

$$M = \frac{1}{c} \int_0^\pi d\sigma\, P^0 = \frac{T}{c^2} \int_0^\pi d\sigma\, \dot{X}^0 = \frac{Ta\pi}{c^2}, \quad (19.16)$$

and the angular momentum (pointing in the direction orthogonal to the plane of rotation),

$$J = \int_0^\pi d\sigma\, (X_1 P_2 - X_2 P_1) = \frac{\pi T a^2}{2c}. \quad (19.17)$$

Upon comparing the expressions (19.16) and (19.17), we find the following equation governing the classical string motion:

$$J = \frac{M^2 c^3}{2\pi T} = \alpha' \hbar (Mc)^2, \qquad \alpha' = \frac{c}{2\pi T \hbar}. \quad (19.18)$$

At the quantum level, the angular momentum is quantized, $J = \hbar l$, with l integer, and becomes the spin of resonances. We have thus obtained a linearly rising Regge trajectory with zero intercept, $\alpha_0 = 0$, and slope $\alpha' = c/(2\pi T\hbar)$, a quantity having the dimension of inverse momentum squared.

Note that the rotating rod solution automatically obeys two further relativistic invariant conditions:

$$(\dot{X})^2 + (X')^2 = 0, \tag{19.19}$$

$$\dot{X} \cdot X' = 0, \tag{19.20}$$

where $\dot{X} \cdot X' = \dot{X}^\mu X'_\mu$ and $\dot{X}^2 = \dot{X}^\mu \dot{X}_\mu$. Their meaning will become clear below.

Up to now, we have considered only a special solution of the equation of motion (19.13) with Neumann boundary conditions, which turns out to describe the states lying on the leading Regge trajectory. What about the other states of the DRM? It is reasonable to think that they will correspond to more general solutions involving string oscillations. However, starting from the Lagrangian in Eq. (19.12) and quantizing it, as done in the previous Section, we immediately face the problem that the space of the states is not positive definite owing to the time component carried by the harmonic oscillators. In other words, the Lagrangian (19.12) is relativistic invariant, but leads to an inconsistent quantum theory.

The solution to this puzzle is that the quadratic Lagrangian (19.12) does not provide a complete description of the relativistic quantum string. We anticipate that the conditions (19.19) and (19.20) should be added and they actually correspond to the DRM physical-state conditions (see Section 19.6). We shall find that the correct relativistic string Lagrangian possesses a gauge symmetry and these conditions arise as a gauge choice, the so-called 'orthonormal gauge'. Specifically, the rotating rod solution obeys these conditions and describes physical states (the leading trajectory); other oscillation solutions do not automatically obey these conditions and yield both physical and unphysical DRM states.

In the next Section, we discuss the Lagrangian describing a relativistic point particle as a preparation for writing the correct string Lagrangian.

19.4 The relativistic point particle

The action of a free relativistic particle is written in terms of the spatial vector $x^i(t)$, $i = 1, \ldots, D = d - 1$, describing the particle trajectory, or 'world-line', as follows:

$$S = -mc^2 \int dt \sqrt{1 - \frac{1}{c^2}\left(\frac{dx^i}{dt}\right)^2}. \tag{19.21}$$

As is well known, the familiar expression of the particle kinetic energy $E = m\vec{v}^2/2$ is recovered by expanding the square root for $c \to \infty$ (the leading constant term being the energy at rest).

This action has a geometrical meaning: it is proportional to the proper length of the trajectory, a relativistic, reparameterization invariant quantity. Let us explain this point. The expression (19.21) is invariant under Lorentz transformations, but not in an obvious, manifest way, because the time and $(d-1)$ space coordinates occur differently in it. In order

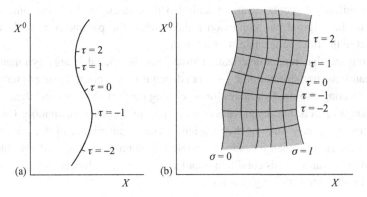

Figure 19.3 Parameterizations of (a) the particle world-line and (b) string world-sheet.

to treat all spacetime components on equal footing, we introduce the coordinate $x^\mu(\tau)$, with $\mu = 0, 1, \ldots, d-1$, where $x^0(\tau) = ct$ is proportional to time, and the evolution parameter τ is an arbitrary monotonic function of time, such that the change of variables from t to τ is invertible (see Figure 19.3). The dynamics should be independent of the arbitrary choice of parameterization $t = t(\tau)$; this is possible because the action is proportional to the length of the particle trajectory, as we now illustrate.

The infinitesimal length ds of the path is obtained from the metric,

$$ds^2 = -dx^\mu dx^\nu \, \eta_{\mu\nu}, \tag{19.22}$$

by expressing it in terms of the trajectory, $x^\mu = x^\mu(\tau)$, as follows:

$$ds^2 = -\frac{dx^\mu}{d\tau}\frac{dx_\mu}{d\tau} d\tau^2. \tag{19.23}$$

Therefore, the length of the trajectory spanned by the particle is given by:

$$\text{Length} = \int ds = \int d\tau \sqrt{-\dot{x}^\mu \dot{x}_\mu}, \tag{19.24}$$

where $\dot{x}^\mu = dx^\mu/d\tau$. Let us check that the length is indeed independent of the choice of the parameter τ. In fact, upon changing to another world-sheet parameter $\tau'(\tau)$, the particle coordinate, being a world-line scalar quantity, changes to $x'^\mu(\tau') = x^\mu(\tau)$. As a consequence, the integration measure is multiplied by the Jacobian $|d\tau/d\tau'|$, while the integrand acquires the inverse factor $|d\tau'/d\tau|$, and thus the action remains invariant:

$$d\tau \sqrt{-\left(\frac{dx(\tau)}{d\tau}\right)^2} = d\tau' \sqrt{-\left(\frac{dx'^\mu(\tau')}{d\tau'}\right)^2}. \tag{19.25}$$

We thus write the particle action in the form,

$$S = -mc \int ds = -mc \int d\tau \sqrt{-\dot{x}^2}, \tag{19.26}$$

which is manifestly Lorentz invariant and also reparameterization invariant; moreover it reduces to the original particle action (19.21) when the parameter τ is chosen to be proportional to the time coordinate, $\tau \sim x^0 = ct$.

Let us remark that reparameterization invariance is a local, gauge symmetry of the action, because the transformation $\tau \to \tau'(\tau)$ depends on the point. In simple terms, this is the result of having chosen d variables for describing the particle, one more than necessary.

A well-known fact is that any global symmetry implies a conserved quantity. On the other hand, a gauge symmetry, such as the reparameterization invariance of the particle action, implies – in the Hamiltonian formalism – a relation among the canonical variables that is independent of dynamics: this condition is called a 'constraint'. Indeed, reparameterization invariance implies the following constraint:

$$p^\mu(\tau) \, p_\mu(\tau) = -(mc)^2, \tag{19.27}$$

where the canonical momentum is obtained from the Lagrangian, $p^\mu \equiv \partial L/\partial \dot{x} = mc\dot{x}^\mu/\sqrt{-\dot{x}^2}$. The presence of this constraint on the particle four-momentum is a consequence of having introduced a redundant coordinate.

19.5 The string action

The analysis of the relativistic point particle in the previous Section can be generalized to the case of the string. Nambu and then Goto introduced an action proportional to the analogous geometrical quantity, the area swept by the string in spacetime,

$$S_{NG} = -\frac{T}{c} \int dA. \tag{19.28}$$

Following the same steps as in the case of the particle action, we now express the area element as a function of the string coordinate $X^\mu(\tau, \sigma)$.

A two-dimensional surface, like the string world-sheet, immersed in d-dimensional Minkowski flat space can be curved: for example, we can draw a sphere in three-dimensional Euclidean space. The (intrinsic) curvature is described by the metric of the surface $g_{ab}(\sigma, \tau)$. This is clearly a function of the embedding $X^\mu(\sigma, \tau)$, and it is obtained by expressing the Minkowski metric of the ambient space in terms of the surface parameterization, as follows (cf. the case of the trajectory in (19.23)):

$$ds^2 = -\frac{\partial X^\mu}{\partial \xi^a} \frac{\partial X_\mu}{\partial \xi^b} \, d\xi^a \, d\xi^b = -g_{ab}(\xi) \, d\xi^a \, d\xi^b. \tag{19.29}$$

In this equation, $g_{ab}(\xi)$ is the 'induced metric' on the world-sheet, and the parameters ξ^a, $a = 0, 1$, are a short-hand notation for $\xi^0 = \tau, \xi^1 = \sigma$.

From the metric we can compute the area element dA as follows. Consider the infinitesimal parallelogram $d\Sigma^{\mu\nu}(\xi)$ made by the two vectors tangent to the surface at the point ξ, i.e. $\dot{X}^\mu(\xi)$ and $X'^\nu(\xi)$, see Figure 19.3(b). The area of the parallelogram is given by the absolute value of the wedge product of the vectors:

$$dA = |\dot{X}||X'|\sin\theta|d\sigma d\tau. \tag{19.30}$$

A covariant expression can be obtained for its square: by substituting the sine square for the cosine square, and expressing the latter with the scalar product of tangent vectors, we obtain

$$dA^2 = \left[(\dot{X}\cdot X')^2 - \dot{X}^2 X'^2\right]\left(d^2\xi\right)^2. \tag{19.31}$$

We recognize that this expression is the determinant of the induced metric $g_{ab}(\xi)$ in (19.29):

$$dA^2 = -\det(g_{ab})\left(d^2\xi\right)^2. \tag{19.32}$$

(Note the minus sign needed to obtain a positive quantity in Minkowski space.)

Using Eq. (19.31) for the area element, we finally obtain the Nambu–Goto action of the relativistic string:

$$S_{NG} = -\frac{T}{c}\int d^2\xi\,\sqrt{(\dot{X}\cdot X')^2 - \dot{X}^2 X'^2}. \tag{19.33}$$

Let us now check the reparameterization invariance of this expression: under a change of world-sheet parameters,

$$\xi^a \to \xi'^a, \quad \xi^a = \xi^a(\xi'), \tag{19.34}$$

the integration measure $d^2\xi$ in (19.33) acquires the Jacobian factor $|\det(M_{ab})|$, where $M_{ab} = \partial\xi^a/\partial\xi'^b$ is the matrix for infinitesimal coordinate change. As regards the integrand $\sqrt{-\det(g_{ab}(\xi))}$, we need the transformation of the determinant. Owing to (19.29), the metric transforms by two inverse matrices, $g \to M^{-1}\cdot g \cdot M^{-1}$, and the determinant acquires the factor $1/(\det M)^2$. Therefore, the transformation of the integrand cancels that of the Jacobian, and the action takes the same form in the new parameterization. In conclusion, we have introduced the Nambu–Goto string action, which is proportional to a geometric quantity, and thus independent of the choice of world-sheet coordinates.

19.6 The quantum theory of the string

In the previous Sections we have discussed the properties of the guitar string and of the rotating rod and their similarity with those of the DRM that inspired, in particular, Nambu and Susskind in proposing the relativistic string model for the DRM. Then, in analogy with the case of the relativistic point particle, we have constructed the reparameterization invariant Nambu–Goto string action (19.33). The next step is to derive the DRM spectrum and the physical state conditions from the canonical quantization of the Nambu–Goto action. This programme was rather challenging, as the square-root nonlinearity of the action seemed to prevent any standard approach. On the other hand, the quadratic action (19.1) was easy to quantize, but did not yield a satisfactory description, because the physical-state conditions of the Dual Resonance Model could not be derived within the theory and had to be imposed by hand.

The key issue was to understand the consequences of reparameterization invariance (19.34) of the Nambu–Goto action. In the case of the point particle, we have seen that the local symmetry of the action, involving one arbitrary function, implies the constraint (19.27). From the Dirac procedure for quantizing systems with constraints, it was known that the constraints become physical-state conditions in the quantum theory. In particular, the constraint (19.27) becomes the Klein–Gordon equation for the relativistic point particle. In the case of the relativistic string, reparameterization invariance involves two arbitrary functions and, therefore, there are two constraints on the canonical variables X^μ and P_μ: these can be easily derived from the Nambu–Goto Lagrangian and read,

$$P^2 + \frac{T^2}{c^2} X'^2 = 0, \qquad P \cdot X' = 0, \qquad (19.35)$$

where $P_\mu \equiv \partial L / \partial \dot{X}^\mu$ is the string canonical momentum.

As for the point particle, the constraints (19.35) should become conditions on the physical states in the quantum theory. Their identification with the physical-state conditions of the DRM became possible after the choice of a particular gauge condition for reparameterization invariance, where the momentum is fixed to be proportional to $\dot{X}$, namely $P^\mu = (T/c)\dot{X}^\mu$. With this choice the constraints in Eq. (19.35) become exactly equal to the conditions (19.19) and (19.20) satisfied by a rotating rod.

These are called orthonormal gauge conditions owing to their geometrical meaning of putting the world-sheet metric (19.29) in diagonal form, as follows:

$$\begin{aligned} g_{ab}(\xi) &= \rho(\xi)\, \eta_{ab}, \\ g_{01} &= \dot{X} \cdot X' = 0, \qquad g_{00} + g_{11} = \dot{X}^2 + X'^2 = 0. \end{aligned} \qquad (19.36)$$

In general, two of the three independent components of the metric can always be fixed by a suitable choice of coordinates; in the present case, the remaining part is the scalar quantity, $\rho(\xi) = (-\dot{X}^2 + X'^2)/2$, called the conformal factor. Since the conditions (19.36) respect Lorentz invariance in the target space, they are also called '(old) covariant gauge'.

The consequences of these gauge conditions are the following.

(i) Using them, the Nambu–Goto action (19.33) reduces to the quadratic action (19.12), whose quantization is straightforward and leads to the Fubini–Veneziano oscillator states. Thus, the string dynamics derived from the Nambu–Goto action is equivalent to that of the quadratic action plus the orthonormal conditions $(\dot{X} \pm X')^2 = 0$.

(ii) As shown in Appendix C, after quantization the orthonormal conditions become operator expressions that are proportional to the Virasoro generators L_n. Therefore, they are indeed found to be the same as the conditions $L_n |\psi\rangle = 0$, for $n > 0$, that identify the physical states in the DRM. This result finally proved that the quantum theory of the string is completely equivalent to the DRM.

(iii) There is a residual gauge symmetry, i.e. there exists a set of coordinate changes that keep the world-sheet metric diagonal. Upon introducing Euclidean coordinates on the world-sheet by analytic continuation of time, $\xi^0 = -i\xi^2$, we can form the complex coordinate $\zeta = \xi^1 + i\xi^2$ and rewrite the diagonal metric (19.36) as follows:

$$ds^2 = \rho(\xi^a)\left(-(d\xi^0)^2 + (d\xi^1)^2\right) = \rho(\zeta,\bar{\zeta})d\zeta d\bar{\zeta}. \tag{19.37}$$

Under analytic reparameterizations, $\zeta = \zeta(\zeta')$, with $d\zeta/d\bar{\zeta}' = 0$, we see that the metric remains diagonal: thus, these transformations express the residual gauge symmetry of the theory in the covariant gauge. Since a change of the scalar factor ρ does not affect the measure of angles, the analytic reparameterizations are precisely the conformal transformations discussed in the Introduction to Part III. Conformal transformations are the residual gauge symmetry of string theory in the covariant gauge and characterize the multipoint amplitudes of the Dual Resonance Model.

19.6.1 Light-cone quantization

In the Introduction to Part III, we described two ways of quantizing theories with gauge symmetry, corresponding to the two options for removing the unphysical (gauge) degrees of freedom: either we maintain (part of) them in the quantum theory and then impose conditions that characterize the physical states, or we eliminate all unphysical degrees of freedom already at the classical level and then quantize the physical ones only.

The first possibility was extensively discussed above: the presence of unnecessary degrees of freedom in the DRM led to states with negative and zero norm that had to be removed, thus identifying the physical subspace. The quantization of the string in the covariant gauge showed that the DRM physical-state conditions corresponded to the gauge conditions eliminating two string coordinates.

The second possibility of solving the gauge conditions at the classical level leads to the quantization of only the physical $(d-2)$ transverse string coordinates. This approach was considered by Goddard, Goldstone, Rebbi and Thorn. Its analogue in QED is the quantization in the Coulomb gauge: a rather economic approach, since it deals only with physical variables and states. The price to pay is a formalism that is not manifestly Lorentz invariant. The quantization in noncovariant variables may spoil Lorentz symmetry, which should be checked afterwards. In the case of QED, the symmetry is verified, while in string theory it holds under some conditions, as we will now discuss.

The light-cone quantization of string theory proceeds as follows: the time coordinate and one of the space coordinates of the string are combined in light-cone variables,

$$X^\pm = \frac{X^0 \pm X^{d-1}}{\sqrt{2}}. \tag{19.38}$$

Next, the X^+ component is fixed to be proportional to the parameter τ, $X^+ = 2\alpha' P^+ \tau$, where P^+ is the momentum in the light-cone (+) direction. Then, from the orthonormal gauge conditions (19.19) and (19.20) one can determine X^- in terms of the transverse physical string coordinates, as follows:

$$\dot{X}^- = \frac{1}{4\alpha' P^+} \sum_{i=1}^{d-2} \left(\dot{X}_i^2 + X_i'^2 \right), \qquad X'^- = \frac{1}{2\alpha' P^+} \sum_{i=1}^{d-2} \dot{X}_i X_i'. \qquad (19.39)$$

Thus, in this procedure we are left with $(d-2)$ independent degrees of freedom at the classical level, which can be quantized. This results in an infinite set of $(d-2)$ spacelike oscillators that yield positive norm states only. In particular, the structure of the Del Giudice, Di Vecchia, Fubini physical states found earlier in the DRM is recovered in this approach (DDF states are discussed in the Introduction to Part III and in Di Vecchia's Chapter).

Next, Goddard, Goldstone, Rebbi and Thorn proceeded to check Lorentz invariance in this quantum theory. The best way to verify a quantum symmetry is to compute the algebra of symmetry generators, written as operator expressions in terms of the canonically quantized variables. In the present case, the generators of the d-dimensional Lorentz group (indicated as $O(d-1, 1)$) were written in terms of the $(d-2)$ canonical pairs X^i and P_i for transverse directions. While the algebra of rotations in the transverse, spacelike directions was obviously correct, the algebra involving Lorentz boosts, i.e. the other two directions, was nontrivial, owing to the nonlinear relations (19.39) implied by the solution of the constraints. The remarkable result was that the Lorentz algebra was verified only for the values $\alpha_0 = 1$ and $d = 26$ of the intercept and spacetime dimension.

In conclusion, these quantum conditions were found in two different ways: (i) by requiring positive-norm states and unitarity of loop amplitudes in a manifestly Lorentz invariant approach, as discussed in the Introduction to Part III; (ii) by quantizing the string in the light-cone gauge and requiring Lorentz symmetry, as we just described. (This is a simplified presentation: a complete agreement between covariant and light-cone quantizations was only achieved in the early Eighties within the modern covariant quantization, see Part VII.)

The fact that different quantization approaches reach the same conclusion for different reasons should not be puzzling: as understood more clearly in the Eighties, this is a characteristic feature of anomalous theories. In the Introduction to Part III, we said that an anomaly is an obstruction to implementing a classical symmetry at the quantum level; more precisely, two classical symmetries cannot be realized simultaneously. In string theory, for $d \neq 26$ or $\alpha_0 \neq 1$ the anomaly concerns either conformal symmetry (causing unitarity violations) or reparameterization invariance (implying Lorentz violations). Each quantization approach assumed one symmetry and found violation of the other.

The results presented in this Introduction show that the spectrum of the Dual Resonance Model (see Part III) could be derived straightforwardly from the canonical quantization of the Nambu–Goto action. A clear physical picture for the DRM as the theory of a relativistic quantum string was established. This was the starting point for many further developments.

References

[GSW87] Green, M. B., Schwarz, J. H. and Witten, E. (1987). *Superstring Theory* (Cambridge University Press, Cambridge).

[Zwi04] Zwiebach, B. (2004). *A First Course in String Theory* (Cambridge University Press, Cambridge).

20
From dual models to relativistic strings

PETER GODDARD

Abstract

A personal view is given of the development of string theory out of dual models, including the analysis of the structure of the physical states and the proof of the no-ghost theorem, the quantization of the relativistic string, and the calculation of fermion–fermion scattering.

20.1 A snapshot of the dual model at two years old

When particle physicists convened in Kiev at the end of August 1970, for the 15th International Conference on High Energy Physics, the study of dual models was barely two years old. Gabriele Veneziano, reporting on the precocious subject that he had in a sense created [Ven68], characterized it as 'a very young theory still looking for a shape of its own and for the best direction in which to develop. On the other hand duality has grown up considerably since its original formulation. Today it is often seen as something accessible only to the *initiated*. Actually theoretical ideas in duality have evolved and changed very fast. At the same time, they have very little in common with other approaches to particle physics' [Ven70]. Forty years later, his comments might still be thought by some to apply, at least in part.

In describing the theoretical developments, Veneziano stressed the progress that had been made in understanding the spectrum of the theory, obtained by factorizing the multiparticle generalizations of his four-point function in terms of the states generated from a vacuum state, $|0\rangle$, by an infinite collection of harmonic oscillators, a_n^μ, labelled by both an integral mode number, n, and a Lorentz index, μ (Fubini, Gordon and Veneziano [FGV69] and Nambu [Nam70a]),

$$[a_m^\mu, a_n^\nu] = mg^{\mu\nu}\delta_{m,-n}, \qquad a_n^{\mu\dagger} = a_{-n}^\mu, \qquad a_n^\mu|0\rangle = 0, \quad \text{for } n > 0. \qquad (20.1)$$

Here $g^{\mu\nu}$ is the (flat) spacetime metric, taken to have signature $(-, +, +, +)$ and, as Veneziano commented, the timelike modes, a_n^0, meant that theory potentially contained

---

The Birth of String Theory, ed. Andrea Cappelli, Elena Castellani, Filippo Colomo and Paolo Di Vecchia.
Published by Cambridge University Press. © Cambridge University Press 2012.

negative norm or 'ghost' states, which would lead to unphysical negative probabilities. (I shall use here the term 'ghost' to refer to a negative-norm state rather than the fields associated with gauge invariance in functional approaches to quantization.)

A similar potentiality exists in the covariant approach to quantum electrodynamics, but the ghost states there are removed by a condition that follows from the electromagnetic gauge invariance of the theory. A condition that removed some of the ghosts had been found (Fubini and Veneziano [FV69] and Bardakci and Mandelstam [BM69]), for the dual model, and this was associated with an $SO(2,1)$ invariance (Gliozzi [Gli68]) or, equivalently, Möbius invariance in terms of the complex variables introduced by Koba and Nielsen [KN69] to describe the n-point function in a symmetrical way. But, this condition is not sufficient to remove all the ghosts, essentially because an infinity of such conditions is needed corresponding to the infinite number of time components, a_n^0.

All was not lost, however. Virasoro had found [Vir70] that, when a parameter of the theory, α_0 (the intercept of the leading Regge trajectory), is such that $\alpha_0 = 1$, the Möbius invariance enlarges to an infinite-dimensional symmetry corresponding to the conformal group in two dimensions, the complex plane of the Koba–Nielsen variables. For $\alpha_0 = 1$, the physical states of the theory satisfy a corresponding infinity of conditions, which stood a chance of eliminating all of the unwanted ghost states, but a proof of this was lacking. Further, the choice $\alpha_0 = 1$ brought with it a problem: the masses, M, of the states in theory satisfy

$$\tfrac{1}{2}M^2 = \sum_{n=1}^{\infty} \sum_{\mu=0}^{3} N_n^\mu n - \alpha_0, \qquad (20.2)$$

for a state with occupation number N_n^μ in the nth mode corresponding to the oscillator a_n^μ, and so the lowest state, with all occupation numbers zero, has $M^2 = -2\alpha_0$. If $\alpha_0 = 1$, which we shall in general take to be the case in what follows, so that Virasoro's conditions obtain, the lowest state has negative mass squared, i.e. it is a tachyon: the cost of having a chance of eliminating ghosts was the presence of a tachyon!

In his report [Ven70], Veneziano also surveyed the attempts that had been made to interpret what had been learnt about the dual model and its spectrum in terms of a field theoretic, or some more directly physical, picture. Holger Nielsen, in a paper [Nie69] submitted to the conference and circulated the previous year, argued, first, that the integrand for the dual amplitude could in practice be calculated by using an analogue picture, formulated in terms of the heat generated in a metallic surface into which currents, corresponding to the momenta of the external particles, are fed and, second, that the dual amplitudes might be viewed as the approximate description of the contribution of a class of very complicated Feynman diagrams in a field theory, which he argued should be two dimensional in some sense. Hadrons he described as 'threads', the history of whose propagation was described by the two-dimensional surfaces.

'String' interpretations had also been developed by Yoichiro Nambu [Nam70a] and Leonard Susskind [Sus69, Sus70a, Sus70b]. They each deduced that the dual model could be pictured in some sense as an oscillating string, because the spectrum (20.2) of the dual model can be viewed as being formed from energies associated with a basic oscillator and all its higher harmonics (or more precisely with one such oscillator for each dimension of spacetime). At this stage, an intrinsically defined geometric action principle to govern the dynamics of the string had not been proposed. As time evolves, the string describes a surface (or 'world-sheet' [Sus70b]), $x^\mu(\sigma, \tau)$, where x^μ satisfies the wave equation as a function of σ and τ, seen as following from an action of a familiar quadratic form,

$$\mathcal{A}_O \propto \int \mathcal{L}_O \, d\sigma d\tau, \qquad \mathcal{L}_O = \frac{1}{2}\left(\frac{\partial x}{\partial \sigma}\right)^2 - \frac{1}{2}\left(\frac{\partial x}{\partial \tau}\right)^2. \qquad (20.3)$$

As Di Vecchia and Schwimmer put it, 'using plausible arguments [Nambu, Nielsen and Susskind] obtained expressions similar to the N-point (tree) amplitudes' [DS08] from this formalism, but, just before the Kiev conference, elsewhere Nambu was considering whether it might not be more satisfactory to replace the quadratic action (20.3) with one that was intrinsically geometric, and he proposed the (Lorentzian) area,

$$\mathcal{A}_{NG} = -\frac{T_0}{c}\int \mathcal{L}_{NG} \, d\sigma d\tau,$$

$$\mathcal{L}_{NG} = \sqrt{\left(\frac{\partial x}{\partial \sigma} \cdot \frac{\partial x}{\partial \tau}\right)^2 - \left(\frac{\partial x}{\partial \sigma}\right)^2 \left(\frac{\partial x}{\partial \tau}\right)^2}. \qquad (20.4)$$

Being intrinsically geometric, this action is of course invariant under arbitrary changes of the parameters σ and τ. Nambu's initial discussion of this action was in notes [Nam70b] prepared for a symposium in Copenhagen in August 1970 that, in the event, he was unable to attend. (Although knowledge of this aspect of the content of the notes spread by word of mouth over time, the notes were not generally available until the publication of Nambu's *Selected Papers* in 1995.) Nambu's proposal of (20.4) was made rather *en passant* and he did not discuss the consequences of the action in his notes beyond remarking that it obviously leads to nonlinear equations of motion. The action was discussed somewhat further by Goto [Got71] some months later, leading to $\mathcal{A}_{NG}$ being designated the Nambu–Goto action. We shall discuss the further development of the understanding of $\mathcal{A}_{NG}$ in Section 20.7.

Returning to Nielsen's analogue approach to calculating dual amplitudes, he had developed it further with David Fairlie in a published paper [FN70], which showed how one-loop amplitudes could be obtained, as well as the tree amplitudes we have been discussing up to now. The analogue approach was also used by Lovelace [Lov70] and by Alessandrini [Ale71] to motivate the mathematical constructions needed to calculate multiloop amplitudes. However, as the contributions of the previous Chapters have emphasized (see also Di Vecchia and Schwimmer [DS08]), the initial development of the theory was based within the framework of S-matrix theory, rather than the quantization of the dynamics of a 'string'.

20.2 Analyticity, asymptotics and the S-matrix

While much of the high energy physics community was gathered in Kiev, I was moving from Cambridge, where I had just completed my PhD, to Geneva, to begin a postdoctoral fellowship. I already knew CERN from the marvellous summer I had spent there three years earlier as a vacation student, assigned to the $g-2$ experiment, on the anomalous magnetic moment of the muon. It had been an extraordinary privilege and education to participate in this beautiful experiment as a very junior temporary colleague of Emilio Picasso, Leon Lederman and others.

Returning to Cambridge in October 1967 after my experimental vacation at CERN, I had started research in theoretical high energy physics under the supervision of John Polkinghorne. The prevailing intellectual ethos is well reflected in the book, *The Analytic S-Matrix* [ELOP66], which Eden, Landshoff, Olive and Polkinghorne had produced the previous year. I, and nearly all my fellow research students, worked on strong interaction physics. (One of us was trying to work out the correct Feynman rules for gauge field theories, but this tended to be regarded as a rather recondite or eccentric enterprise.)

The general strategy was to seek to determine and elucidate the analyticity properties of the S-matrix, i.e. the singularity structure implied and determined by unitarity, in the expectation or hope (depending on the strength of one's faith) that the particle spectrum and the scattering amplitudes might be determined, with some minimal number of further assumptions, by some sort of 'bootstrap' process. It seemed that ideas related to analyticity would need to be supplemented by assumptions about high energy behaviour, described in terms of Regge theory.

The guide for analyzing both analyticity structure and high energy behaviour was perturbative quantum field theory, usually a simple ϕ^3 scalar field theory, to avoid 'the inessential complications of spin'. It was not that anyone in Cambridge, in those days before asymptotic freedom, thought that perturbative quantum field theory would be directly relevant to strong interaction physics (as far as I know), rather that it provided a *formal* solution to the unitarity conditions and, in a certain sense, exhibited Regge high energy behaviour. Thus it provided a guide as to what to assume or what to try to prove in the more abstract context of S-matrix theory; it was a sort of theoretical laboratory.

The analyticity structure of the S-matrix in the physical region was already well understood: the singularities occur on curves specified by the Landau equations with discontinuities specified by Cutkosky rules. My own first steps were in the direction of trying to elucidate aspects of the singularity structure outside the physical region. Feeling somewhat isolated, after about eighteen months, I moved to work on Regge theory, collaborating with my contemporary Alan White, following the group theoretical approach pioneered by Marco Toller, using harmonic analysis on noncompact groups.

Toller was a staff member in the Theory Division at CERN at the time and, encouraged by David Olive, who was spending a couple of terms on leave there, he arranged a two-year postdoctoral position for me. David was then a Lecturer in DAMTP, Cambridge, and, when he returned from leave in May 1970, he gave a course of lectures on dual models, based in part on the work he had been doing with Alessandrini, Amati and Le Bellac at CERN. Presented with his characteristic simplicity, elegance and depth, the subject strongly interested me, because of the sense he conveyed of an emerging mathematical structure. Further, it offered a new theoretical laboratory, seemingly *somewhat* different from that of the Feynman diagrams of quantum field theory, with the Regge asymptotic behaviour present from the tree diagrams onwards (rather than contained in certain selectively chosen infinite sums of diagrams as in quantum field theory), in which one could study the analyticity and high energy properties of the S-matrix. It was this theoretical potential, rather than any phenomenological relevance, that engaged my interest.

20.3 Cuts and poles

Arriving at CERN in September 1970, I found an air of excitement surrounding these ideas, and a number of physicists, mainly younger ones, working on them, inspired in large part by the leadership of CERN staff member Daniele Amati. Over the coming two years nearly everyone making fundamental contributions to dual models, or dual theory as it was sometimes more portentously called, passed through or came to stay. David Fairlie had come on leave from Durham, a little later Ian Drummond arrived from Cambridge and, in July 1971, David Olive returned to take a post as a staff member.

I began my own efforts by trying to understand the details of how the Veneziano model, together with the higher loop amplitudes that were being constructed, would provide a formal perturbative solution to the unitarity equations [FGW71]. There was, however, what looked like a fly in the ointment. In analyzing the divergences that occurred in one-loop dual amplitudes, which one needed to try to regularize by a renormalization process, a singularity in momentum had been found, coming from the same part of the integration region as the divergences. This singularity was a cut, which it seemed would imply a violation of unitarity.

The one-loop amplitudes in the Veneziano models were associated with diagrams involving either an annulus (orientable case) or a Möbius band (nonorientable case), to the edges of which were attached lines associated with the initial and final state particles and, so, incoming and outgoing momenta. In the string picture which, as we have said, was used at the time as an analogue or figurative representation rather than a rigorous basis for calculation, the annulus or band corresponded to the world-sheet describing the history of the string or strings formed by the joining or splitting of the incoming string states. For the orientable case of the annulus, there were two subcases to consider: a planar case where all the momenta entered on the outside edge of the annulus; and a nonplanar case where the momenta entered on both edges. It was this last case that contained the unitarity-violating cut.

The divergence in the integral corresponding to the annulus came as the radius of the circle constituting its inner edge tended to zero. In the nonplanar case, where momenta followed into both edges of the annulus, the integral corresponding to the loop amplitude could be made finite by taking the momenta entering on the inner edge to sufficiently unphysical (spacelike) values and, from this region, the amplitude could be defined more generally by analytic continuation. Further, it seemed that it should be possible to obtain a finite result for the planar case, from the nonplanar case, by analytically continuing all the momenta on the inner edge to zero, leaving only those on the outer edge.

Using this observation, I sought to develop this analytic approach to renormalization of dual loops in a paper written in February 1971 [Godd71], but there was a problem:

as the momenta on the inner edge were taken to zero, through real but spacelike values, the onset of the divergence corresponded to a cut, the unitarity-violating cut to which we have referred. We could continue analytically past the cut to obtain a finite answer for zero momenta but, because of the cut, the value would inevitably be complex rather than real as desired, and the answer would depend on whether we continued above or below the singularity in the complex plane.

On the other hand, if the theory could be modified so that the cut became a pole, unitarity would no longer necessarily be violated, provided that the theory contained new particles corresponding to this pole, and the regularization technique I was pursuing might well work. Moreover, as noted in the paper [Godd71], Daniele Amati told me that Claud Lovelace had been considering such a possible modification of the theory. As we now know, in the 'correct' version of the theory the cut does indeed become a pole, corresponding to the propagation of a closed string, and the analytic renormalization procedure I was seeking to develop leads to an interpretation of the divergence of the planar loop in terms of the emission of closed string states into the vacuum, and its removal by a renormalization of the slope of the Regge trajectories (Shapiro [Sha75] and Ademollo et al.[ADDG75]).

Lovelace's considerations were in fact probably the most prophetic – perhaps most Delphic would be more appropriate – in the development of string theory. He had recently left CERN, because his post as a staff member had not been made permanent, and he was now at Rutgers working on extracting a calculus of Reggeons and Pomerons from dual models to describe high energy scattering. The Reggeons corresponded to the external particles of the original Veneziano model and their excited states, whilst the Pomerons corresponded to the singularity (initially a unitarity-violating cut) that we have been discussing. Lovelace needed this singularity to be a pole for consistency with unitarity, of course, and, at the beginning of November 1970 (see his contribution to this Volume), he started modifying the theory in an ad hoc fashion to achieve this. He considered the possibilities that the dimension, d, of spacetime might be different from 4 and that E infinite sets of oscillators might be cancelled by 'Virasoro-like' conditions, rather than just the one infinite set that one might naively expect. He found that he could remove the cut and obtain the desired pole by setting $d = 26$ and $E = 2$ [Lov71].

20.4 Searching for ghosts

At the same time as Lovelace, others were also considering what could be learnt from varying the spacetime dimension, d. In the summer of 1970, Richard Brower and Charles Thorn had begun a study of the physical states of the theory. As we have recounted, Virasoro had found that, when $\alpha_0 = 1$, the physical states, $|\psi\rangle$, satisfy an infinity of conditions,

$$L_0|\psi\rangle = |\psi\rangle, \qquad L_n|\psi\rangle = 0, \quad n > 0, \tag{20.5}$$

where L_n is given by the normal ordered expression

$$L_n = \frac{1}{2}\sum_n :a_m \cdot a_{n-m}:. \tag{20.6}$$

(Note that $L_n^\dagger = L_{-n}$.) Fubini and Veneziano [FV71] and Del Giudice and Di Vecchia [DD70] began the analysis of the Virasoro conditions (20.5) using the operator formalism (20.1), but Brower and Thorn [BT71] went further and set out a general algebraic approach to determining whether the conditions were sufficient to ensure the absence of the unwanted ghost states, characterizing the problem entirely in terms of the Virasoro algebra,

$$[L_m, L_n] = (m-n)L_{m+n} + \frac{c}{12}m(m^2-1)\delta_{m,-n}, \qquad (20.7)$$

where $c = 4$ for the Veneziano model as originally formulated in four-dimensional spacetime.

Virasoro [Vir69] had not actually written down the algebra that bears his name and the final c term was initially omitted in discussions until Joe Weis (who tragically died in a mountaineering accident in 1978) pointed out that it should be there [BT71, FV71]. Brower and Thorn set about analyzing the norms of physical states by considering the signature of the orthogonal space, spanned by states of the form $L_{-n}|\phi\rangle$, called spurious states. Their profound paper [BT71], now unfortunately not well known, introduced ad hoc in their essence the concepts of a Verma module and the contravariant form (see e.g. Wakimoto [Wak01]).

They noted that the condition for the absence of ghosts could be expressed in terms of the eigenvalues of the (contravariant form) matrices, $\mathcal{M}_N(c, h)$, formed from the scalar products of the states

$$L_{-1}^{n_1} L_{-2}^{n_2} \ldots L_{-m}^{n_m} |h\rangle, \qquad (20.8)$$

where the n_j are nonnegative integers such that

$$\sum_{j=1}^{m} n_j = N, \qquad (20.9)$$

and $|h\rangle$ is a normalized state satisfying $L_0|h\rangle = |h\rangle$ and $L_n|h\rangle = 0$ for $n > 0$. The condition was that, for $N > 1$, $h + N \leq 0$, and h integer, the signature of $\mathcal{M}_N(c, h)$ should be the same as that of the matrices obtained by replacing L_{-n} by a_{-n}^0 in the defining equations for $\mathcal{M}_N(c, h)$.

In all this, c equals the dimension of spacetime, d, so that, naturally, Brower and Thorn took $c = 4$ in their analysis. They showed that their condition held up to the ninth level of states in the model. By about November, as Thorn told me, they had started allowing d to vary in their analysis, with the objective of establishing the result for $d = 4$ by showing that no eigenvalues passed through zero as one continued from a region in which the result was true. Although this line of argument could not be completed at the time, it did lead to the discovery of a ghost amongst the physical states if the dimension of spacetime, $d > 26$, as they noted in a footnote added to their paper [BT71]. Some years later, Charles Thorn [Tho85] gave a proof of the absence of ghosts completing this original line of argument, using the famous determinant formula Kac had found for $\det \mathcal{M}_N(c, h)$ [Kac79] (Feigen and Fuchs [FF82]).

Thus, early in 1971, the suggestion that 26-dimensional spacetime had a particular significance in the context of dual models had come almost simultaneously from studies of two apparently unconnected aspects of the theory. This seemed deeply mysterious but hardly likely to be a complete coincidence. One thing was clear: the dimension of spacetime had to play a role in any proof of the absence of ghosts.

I learnt about the results on ghosts directly from Rich Brower, who had arrived at CERN as a postdoctoral fellow about the same time as I had. We had began to collaborate, initially looking for new dual models. Soon after Veneziano had produced the first dual model, Virasoro had produced a model with somewhat similar properties [Vir69]. In fact, it turned out to be closely related to the original model. Shapiro [Sha70] showed that the N-point Virasoro model could be obtained by a similar analogue approach but this time integrating the locations of the external momentum lines (Koba–Nielsen variables) over the whole Riemann sphere rather than just the circular boundary of a disc. Further, the model also had an infinite number of physical state conditions, which might well eliminate all the ghost states in suitable circumstances.

Brower and I sought to generalize the Virasoro model by replacing the two-dimensional integral by an n-dimensional integral, with the group responsible for duality generalizing from the $SO(2, 1)$ that Gliozzi had found for the Veneziano model ($n = 1$) and $SO(3, 1)$, which fulfilled the role for the Virasoro model ($n = 2$), to $SO(n + 1, 1)$. But the problem was that, whereas the algebras of these groups have suitable infinite-dimensional extensions for $n = 1, 2$ (being isomorphic to $SL(2, \mathbb{R})$ and $SL(2, \mathbb{C})$, respectively), this is not the case for general n. So our new models [BG71] lacked the infinite-dimensional algebra ((20.7) for $n = 1$ and two copies of it for $n = 2$) needed to have a chance of eliminating ghosts.

Still we thought it worth doing some simple ghost searches, in part motivated by the scary rumour that a student had found a ghost coupling in the four-point Veneziano model at about the 13th level. We spent a weekend doing a computer-aided search and found nothing to worry about down to the 30th level (for $1 \leq n \leq 45$), which scotched the rumour and restored our confidence about the conjectured absence of ghosts in the original Veneziano model.

Meanwhile, very much more interesting models were emerging elsewhere. The model of Veneziano was proposed as a model for mesons and had only bosonic states. In January, Pierre Ramond [Ram71] produced an equation describing a free dual fermion, generalizing in a sense the free Dirac equation. Soon after, André Neveu and John Schwarz [NS71] used fermion oscillators to construct a new dual model for mesons, which could interact with Ramond's fermions, and which possessed an infinite-dimensional superalgebra of physical state conditions [NST71] that offered the prospect of ghost elimination.

It was an interesting, if quick and straightforward, task to calculate one-loop amplitudes for the Neveu–Schwarz model, so Ron Waltz and I did this exercise in June 1971, and it was indeed fun to see the predictable elliptic functions emerging for the first time [GW71]. For one thing, it enabled one to perform immediately the analysis that Lovelace had performed for the Veneziano model and determine the value of the spacetime dimension, d, for which the unitarity-violating cut in the nonplanar loop became an acceptable pole, the answer being $d = 10$ instead of $d = 26$. I do not remember at what point someone did a calculation, analogous to that of Brower and Thorn, to show that the model possesses ghosts for $d > 10$, so that it became clear that $d = 10$ plays exactly the role for the Ramond–Neveu–Schwarz (RNS) model that $d = 26$ does for the Veneziano model.

20.5 Spectrum generating algebra

Thus the challenge was to prove the absence of ghosts in the Veneziano model for $d \leq 26$ and in the RNS model for $d \leq 10$, with the limiting or 'critical' values of the spacetime dimension being required for consistency with unitarity at the one-loop level. In June 1971, Del Giudice, Di Vecchia and Fubini (DDF) [DDF72] made a major advance in understanding the physical states by seeking to construct a basis of physical states, and hoping to prove their positivity, rather than just attempting to prove the absence of ghosts. Their approach was to consider the operator describing the coupling of the massless spin one particle, which exists in the theory under the assumption that $\alpha_0 = 1$, and might more naturally be identified, at least figuratively, with the photon than with a vector meson (belying the interpretation of the dual model as a theory of strong interactions).

To describe the DDF operators, we need to select a lightlike vector, k. DDF worked in four-dimensional spacetime but, with a view to subsequent developments, we consider general d-dimensional spacetime. The DDF operators are

$$A_n^i = \epsilon_\mu^i A_n^\mu, \qquad A_n^\mu = \frac{1}{2\pi i} \oint P^\mu(z) V(nk, z) dz, \qquad (20.10)$$

where the ϵ^i, $1 \leq i \leq d-2$, are orthonormal polarization vectors for a photon of momentum k, i.e. $\epsilon^i \cdot \epsilon^j = \delta^{ij}$, $k \cdot \epsilon^i = 0$, and n is an integer. In (20.10), V is the basic vertex operator,

$$V(\alpha, z) =: \exp\{i\alpha \cdot Q(z)\} :, \qquad (20.11)$$

where

$$Q^\mu(z) = q^\mu - ip^\mu \log z + i \sum_{n \neq 0} \frac{1}{n} a_n^\mu z^{-n}, \qquad P^\mu(z) = i \frac{dQ^\mu}{dz}, \qquad (20.12)$$

with $a_0^\mu = p^\mu$, the momentum operator, and $[q^\mu, p^\nu] = ig^{\mu\nu}$. The DDF operators, A_n^i, are well defined on states of momentum p, such that $k \cdot p = 1$ and they can be used to generate physical states because they commute with the Virasoro algebra,

$$[L_m, A_n^i] = 0. \qquad (20.13)$$

DDF showed that the A_n^i create an orthonormal set of positive norm physical states corresponding to $d - 2$ dimensions of oscillators. But, to get all the physical states (20.5), we need $d - 1$ dimensions of oscillators.

I started studying with Rich Brower the progress DDF had made. We focussed more directly on the algebraic properties of the DDF operators, noting that they really did behave like annihilation and creation operators, satisfying the appropriate commutation relations [BG72, BG73],

$$[A_m^i, A_n^j] = m\delta^{ij}\delta_{m,-n}, \qquad 1 \le i, j \le d-2. \qquad (20.14)$$

We set out to find the operators that would create the remaining physical states, with the hope that we could then prove the absence of ghosts using the algebraic properties of these additional operators.

Because the operators A_n^μ are associated with the coupling of a 'photon', one would not expect the components other than those in the directions of the polarization vectors ϵ^i to be physical. There are two other independent components, which we can take to be in the direction k of the 'photon' momentum and in a longitudinal direction defined by a lightlike vector $\tilde{k}$ satisfying $k \cdot \tilde{k} = -1$. The k components essentially vanish, $k \cdot A_n = k \cdot p\delta_{n,0}$, and the longitudinal components, $\tilde{A}_n = \tilde{k}_\mu A_n^\mu$, fail to commute with the Virasoro algebra,

$$[L_m, \tilde{A}_n] = -\frac{1}{2}m(m+1)n\Phi_m^n, \qquad \Phi_m^n = \frac{1}{2\pi}\oint e^{ink \cdot Q(z)} z^m \frac{dz}{z}, \qquad (20.15)$$

though they satisfy a neat Virasoro algebra themselves [BG72, BG73],

$$[\tilde{A}_m, \tilde{A}_n] = (m-n)\tilde{A}_{m+n} + m^3 \delta_{m,-n}. \qquad (20.16)$$

This suggested that we should seek a modification of the $\tilde{A}_n$ by some function F_n of the $K_m = k \cdot a_m$,

$$A_n^L = \tilde{A}_n + F_n, \qquad (20.17)$$

which would adjust $\tilde{A}_n$ so that the resulting longitudinal operator A_n^L commutes with L_m. Brower and I began to look for F_n.

Daniele Amati and Sergio Fubini arranged for me to talk about this work in progress at a summer school I was going to in Varenna in August [BG73]. An impressive list of lecturers gathered in that beautiful setting on Lake Como, including Callan, Dashen, Coleman, Goldberger, Maiani, Regge and Salam, as well as Amati and Fubini.

It was a memorable meeting and one particular vignette has stuck in my mind as an illustration of the prevailing attitude towards the use of modern mathematics in theoretical high energy physics. A senior and warmly admired physicist gave some lectures on the Regge theory of high energy processes. With great technical mastery, he was covering the board with special functions, doing manipulations that I knew from earlier work could be handled efficiently and elegantly using harmonic analysis on noncompact groups. Just as I was wondering whether it might be too impertinent to make a remark to this effect, the lecturer turned to the audience and said, 'They tell me that you can do this all more easily if you use group theory, but I tell you that, if you are strong, you do not need group theory'.

Brower and I continued our search for the modification F_n together in Varenna and on a subsequent visit to Cambridge, and at a distance after he returned to the USA to resume a postdoctoral fellowship at MIT, and I returned to CERN for my second year. But before

I went back to Geneva, I went to Durham to be interviewed for a faculty position in the Department of Mathematics. I had enjoyed discussions with David Fairlie, who had been on leave at CERN from Durham, however I did not really think I stood much chance of getting the job, because I was only one year postdoctoral.

Just a couple of years earlier this would not have seemed an obstacle, but now things were getting much more difficult. When I was called for interview, I was convinced that they would appoint the other, more senior, candidate. Perhaps because of this I was not nervous, and perhaps consequently got the job. Durham agreed that I could go back to CERN to complete the second of my two years there. Having a reasonably secure future, I could concentrate on my research. By September 1971, Brower was back in the USA and we continued our collaboration by post. I worked on some other ideas as well but my mind kept grappling with trying to understand the physical states and prove the absence of ghosts.

At the beginning of November, Brower and I submitted our paper [BG72] on the spectrum generating algebra for the dual model to *Nuclear Physics*, still lacking a form for F_n to make the correct physical states creating longitudinal operators, A_n^L.

20.6 No-ghost theorem

In January 1972, Charles Thorn arrived at CERN from Berkeley to spend a year and, after a while, we began to discuss our individual efforts to prove the absence of ghosts. Charles had been studying further the physical states as a function of the spacetime dimension, d. Brower and Thorn [BT71] had noted that a ghost state appears amongst the physical states when $d > 26$. The critical dimension, $d = 26$, marks a transition point at which the norm of a physical state moves from positive values through zero to negative values.

In each dimension d, states of the form $L_{-1}|\phi\rangle$ are null physical states, i.e. physical states orthogonal to all other physical states, provided that $|\phi\rangle$ satisfies the physical state conditions $L_n|\phi\rangle = 0, n > 0$, with the adjusted mass condition $L_0|\phi\rangle = 0$. Exactly for $d = 26$, we have more states like this: for example,

$$\left(L_{-2} + \tfrac{3}{2}L_{-1}^2\right)|\phi\rangle, \qquad (20.18)$$

are null physical states for any $|\phi\rangle$ satisfying $L_n|\phi\rangle = 0, n > 0$, and $L_0|\phi\rangle = -|\phi\rangle$. They correspond to a zero eigenvalue of $\mathcal{M}_2(d, -1)$ at $d = 26$, and this makes a transition to a negative eigenvalue, corresponding to a ghost, for $d > 26$. Simple counting shows that the null physical states (20.18) cannot all be independent of those of the form $L_{-1}|\phi\rangle$, because, if they were, the null physical states taken together with the DDF states would exceed the known number of physical states. So there must be relations between them, corresponding to further zero eigenvalues of the matrices $\mathcal{M}_N(d, h)$ for $d = 26, N > 2$ and $h \leq 1 - N$. From his detailed calculations, Charles began to uncover an intricate structure of such relations at $d = 26$.

Suddenly the realization came that perhaps all the physical states, other than the DDF states, which we knew to be positive definite, became null at $d = 26$. As I remember, Charles and I were walking into the CERN cafeteria to have lunch when lightning struck. This would mean that, at $d = 26$, only the transverse DDF modes would be needed to construct the physical states; the additional transverse modes A_n^L that Brower and I had been seeking would not correspond to physically distinct modes of motion in the critical dimension.

Moreover, this would make the prophecy of Lovelace come to pass, for he had written [Lov71] that not only should we have $d = 26$ but also $E = 2$, that is that two dimensions of oscillators should be effectively cancelled by the Virasoro conditions (20.5), because the $d - 2 = 24$ dimensions of oscillators described by the DDF operators create all the physical states apart from the null states that we would expect to drop out of the theory. If they do indeed decouple from loops, the unwanted cuts should become poles consistent with unitarity.

Thorn set about trying to find recursively all the null states necessary to establish the accuracy of this picture and so prove the absence of ghosts. Then we thought it might be easier to seek to characterize the DDF states algebraically and show that any physical state could be written as the sum of a DDF state and a null physical state. The space generated by the DDF states can be specified by imposing the conditions

$$K_n|\psi\rangle = 0, \qquad n > 0, \tag{20.19}$$

in addition to the physical state conditions (20.5). The K_n together with the L_n form an extension of the Virasoro algebra,

$$[L_m, L_n] = (m - n)L_{m+n} + \frac{d}{12}m(m^2 - 1)\delta_{m,-n},$$
$$[L_m, K_n] = -nK_{m+n}, \tag{20.20}$$
$$[K_m, K_n] = 0.$$

We also define the space of *null physical states* to be the states satisfying (20.5) that are linear combinations of states of the form $L_{-n}|\phi\rangle$, $n > 0$. Such states are obviously orthogonal to all physical states. Having formulated the problem like this, we were able to prove the no-ghost theorem [GT72]. For $d = 26$, if $|\psi\rangle$ is a physical state, i.e. a solution of (20.5), it can be written in the form

$$|\psi\rangle = |f\rangle + |n\rangle, \tag{20.21}$$

where $|f\rangle$ is a DDF state and $|n\rangle$ is a null physical state. The no-ghost theorem follows for $d < 26$ by embedding in the $d = 26$ case. We were also able to prove the no-ghost theorem for the Neveu–Schwarz dual model and the Veneziano model at the same time by essentially the same argument, confirming the critical dimension as $d = 10$ for the RNS model.

As we were working towards our proof, we heard indirectly that Brower had completed a proof of the absence of ghosts. Earlier, in January, soon after Thorn had arrived in CERN, I had heard from Brower that he was planning to write a paper about ideas he was working on for proving the no-ghost theorem, progressing further along the route we had been following, but that he then still lacked a construction for F_n. Charles and I thought we should publish our proof, because it was conceptually different and because it had been arrived at independently, but we did not want to do anything which might have the effect of scooping Rich, so we delayed circulating our paper for three weeks until the beginning of May and we adopted a low key, perhaps slightly obscure, title, 'The compatibility of the dual Pomeron with unitarity and the absence of ghosts in the dual resonance model'.

To frame his proof, Rich had realized that if the addition of F_n to $\tilde{A}_n$ to form A_n^L were to double the c number in (20.16), so that

$$[A_m^L, A_n^L] = (m-n)A_{m+n}^L + 2m^3 \delta_{m,-n}, \qquad (20.22)$$

then, for $d = 26$, there would be in essence an isomorphism between, on the one hand, the algebra of the A_n^i and A_n^L and, on the other, the algebra of Euclidean oscillators, a_n^i, $1 \leq i \leq 24$ and the Virasoro generators formed from them, L_n^a. Assuming that the A_n^i and A_n^L did create all the physical states, the isomorphism of the algebras could be used to calculate their norms using the operators a_n^i and L_n^a instead, and, since these act within a space with no negative norm states, there could be no ghosts.

Rich found the F_n we had been seeking [Bro72] (see also the Chapter by Brower),

$$F_n = \frac{n}{4\pi i} \oint \frac{k \cdot P'(z)}{k \cdot P(z)} V(nk, z) dz, \qquad (20.23)$$

for which it is indeed the case that $A_n^L = \tilde{A}_n + F_n$ satisfies (20.22) and

$$[L_m, A_n^L] = 0, \qquad [A_m^L, A_n^i] = -n A_{m+n}^i. \qquad (20.24)$$

Then, for $d = 26$, the algebra of the A_n^i, A_n^L is essentially isomorphic to that of the A_n^i, L_n^A, where the L_n^A are the Virasoro algebra formed from the A_n^i. It follows that the operators $A_n^L - L_n^A$ commute with the A_n^i and it is easy to show that they create null states, leading to the same conclusion [Bro72] about the structure of the physical states as in (20.21).

This approach to proving the no-ghost theorem was extended by John Schwarz [Sch72] to the Neveu–Schwarz model and this was also done by Richard Brower and Kenneth Friedman [BF73].

The validity of the no-ghost theorem had a profound effect on me. It seemed clear that this result was quite a deep mathematical statement, and that the proofs involved elegant argument and revealed a beautiful structure, but also that no pure mathematician would have written it down. It had been conjectured by theoretical physicists because it was a necessary condition for a mathematical model of particle physics not to be inconsistent with physical principles. Of course, it is reverse logic and perhaps dubious philosophy, but I could not help thinking that, in some sense, there would be no reason for this striking result to exist unless the dual model had something to do with physics, though not necessarily in the physical context in which it had been born.

The deep mathematical significance of the no-ghost theorem was demonstrated two decades later when it played an important role in Borcherds' proof [Bor92] of Conway and Norton's 'moonshine conjectures' about the Monster Group. I felt enormously flattered, but faced the most formidable challenge, when I was invited to deliver the laudation [Godd98] for Borcherds at the International Congress of Mathematicians in Berlin in 1998, when he was awarded a Fields Medal.

20.7 Quantizing the relativistic string

Claudio Rebbi, also at CERN on a postdoctoral fellowship, had been studying the physical states of the dual model by considering the limit in which the momentum of the states became infinite in a particular Lorentz frame. Charles Thorn and I began investigating with Claudio Rebbi what happened to the DDF states when a rotation is made about the momentum of the state. If E_{Li} is the generator of the little group of the momentum that generates a rotation in the ith longitudinal plane, we found

$$[E_{Li}, A_m^j] = \delta^{ij} A_m^L + \sum_{n=1}^{\infty} \frac{1}{n} \left(A_{-n}^i A_{m+n}^j - A_{m-n}^j A_n^i \right), \qquad (20.25)$$

with A_n^L as defined by Brower [Bro72]. Actually, (20.25) can be taken as a definition of A_n^L and it can be shown (Goddard, Rebbi and Thorn [GRT72]) then that the properties of A_n^L, namely (20.22) and (20.24), follow from this definition, avoiding the need to construct F_n, so that this provides a construction of all the physical states and a basis for Brower's proof of the absence of ghosts. A similar argument can be used in the Neveu–Schwarz model.

Now, because of the no-ghost theorem, we know that, in the critical dimension $d = 26$, the DDF operators generate all the physical states, provided that we disregard the null states, which do not contribute to tree amplitudes. Thus we must be able to define the action of all the generators of the little group of the momentum, including E_{Li}, on the transverse states generated by the DDF operators. We found the corresponding expression for E_{Li},

$$E_{Li} = \sum_{n=1}^{\infty} \frac{1}{n} \left(L_{-n}^A A_n^i - A_{-n}^i L_n^A \right), \qquad (20.26)$$

and noted that, so defined, E_{Li} closes with E_{ij}, the generators of rotations in the (i, j)-plane, to form the $SO(d - 1)$ algebra of the little group, if and only if $d = 26$.

In the limit $k \to 0$ (and $\tilde{k}$ becomes infinite), the DDF operators A_n^i become the basic oscillators a_n^i, the basic oscillators of the theory (suitably identifying axes), and the L_n^A become L_n^a, the Virasoro generators made from a_n^i. This suggests that, in a suitable limit, the description of the dynamics of the system might simplify. This took us back to reconsidering the attempts to understand the dual model in terms of strings.

As outlined in Section 20.1, very soon after the oscillator description of the dual model had been obtained [FGV69, Nam70a], it was realized that this implied that the model could be regarded as describing some sort of physical system specified in terms of a one-dimensional object or string [Nie69, Nam70a, Sus70b]. Initially, the main impact of this picture was through the analogue approach introduced by Nielsen [Nie69], because it suggested the appropriate mathematical techniques for loop calculations [Lov70, Ale71]. It was unclear whether this should be regarded merely as an analogy and calculational guide or as a deeper physical description.

In 1970, Nambu [Nam70b], and then Goto [Got71], had written down an action (20.4) that would prove suitable for describing the dynamics of the string, but they had not then developed the analysis of this action to derive its properties. Nambu's paper was not widely circulated and the Nambu–Goto action, as it later came to be called, was little discussed. I remember Jeffrey Goldstone, who had come on leave from Cambridge to CERN for the first part of 1971, writing down this action and asking me whether it was what people were talking about when the string picture of the dual model was referred to. I replied that I wasn't sure.

I had known Jeffrey since I had been a freshman at Trinity College, Cambridge, where he was then a Fellow. As an undergraduate, I had attended courses of lectures by him on theoretical physics. These did not always draw large audiences but I found them deeper than some other offerings. Certainly, there was the sense of the subject being developed in real time rather than something prepared earlier being warmed up. When I was a graduate student in DAMTP, Jeffrey Goldstone was not much in evidence in the department. A few times I went to seek him out for advice in his rooms in Trinity College. Jeffrey combined a lugubrious demeanour with a penetrating wit. I remember him counselling that the nature of research was not so much the answering of predetermined questions as 'What is the precise question to which this is the vague answer?' (At CERN, there was a visitor, presumably unofficial, dressed in white, wearing gold trainers, who came to seminars. It seemed he originated from the World Questions Institute, apparently a private house boat on the Hudson River. He was looking for questions rather than answers. He sat at the back of seminars, feeling the vibrations, which were most stimulating during talks on the ghost problem in dual models it seemed.)

Early in 1972, building on earlier work of Chang and Mansouri [CM72], Mansouri and Nambu [MN72] noted that the geometric character of the action (20.4) implies that it is independent of the choice of coordinates (σ, τ), and then showed that the Virasoro conditions (20.5) were related to a choice of coordinates that ensures the equality of the Lagrangians $\mathcal{L}_O$ of (20.3) and $\mathcal{L}_{NG}$ of (20.4).

As Thorn, Rebbi and I sought to understand how the covariance properties of the states of the dual model, which we had obtained [GRT72], related to the quantization of a string described by the Nambu–Goto action, we heard through Fairlie of work that Goldstone was doing back in Cambridge on the same topic. When we wrote to him at the end of July to exchange information about our work, his reply began characteristically: 'My questions are the same as your questions and I have no answers', before giving a succinct summary of the progress he had made, that overlapped considerably with what we had been doing. He explained that he intended to write up his results but he had 'been chasing a different hare, surfaces instead of strings, these past two weeks'.

Following Chang, Mansouri and Nambu [CM72, MN72], we made a choice of coordinates (σ, τ) on the world-sheet in which

$$\left(\frac{\partial x}{\partial \sigma}\right)^2 + \left(\frac{\partial x}{\partial \tau}\right)^2 = 0, \qquad \frac{\partial x}{\partial \sigma} \cdot \frac{\partial x}{\partial \tau} = 0. \qquad (20.27)$$

Such an *orthonormal* choice of coordinates not only ensures that $\mathcal{L}_{NG} = \mathcal{L}_O$ (for the particular solution), but it also linearizes the equations of motion that follow from variation of $\mathcal{A}_{NG}$, removing the nonlinearity that had initially deterred Nambu [Nam70b]. In an orthonormal coordinate system these nonlinear equations are equivalent to the wave equation for the world-sheet variable $x^\mu(\sigma, \tau)$.

This enables the introduction (at the classical level) of the harmonic oscillator variables a_n^μ to describe the normal modes of the solutions of the wave equation,

$$\frac{1}{\ell} x^\mu(\sigma, \tau) = q^\mu + p^\mu \tau + i \sum_{n \neq 0} \frac{a_n^\mu}{n} e^{-in\tau} \cos n\sigma. \qquad (20.28)$$

(Here the dimensional parameter ℓ is the characteristic string length, which will be specified in (20.29).) In terms of these variables, the conditions (20.27) are equivalent classically to the vanishing of the Virasoro generators, L_n [CM72, MN72].

It is possible to specialize further our choice of orthonormal world-sheet coordinate system by choosing τ to be the time coordinate in some Lorentz frame, $\tau = n \cdot x$, where n is a timelike vector. Then $n \cdot a_n = 0, n \neq 0$, so that the time coordinates $x^0(\sigma, \tau)$ do not describe independent oscillations. This removed an obscurity in the original spacetime description of the string, about what the apparent oscillations of the string in the time direction meant.

In addition, the classical equations $L_n = 0$, corresponding to (20.27), allow a further component of a_n^μ to be eliminated in principle, leaving a set of independent degrees of freedom corresponding to $d - 2$ dimensions of oscillators. Of course, this is what one should expect from the geometric properties of the Nambu–Goto action, $\mathcal{A}_{NG}$. It only depends on the area of the world-sheet, so that changes of $x^\mu(\sigma, \tau)$ tangential to the world-sheet will not change the action and are not dynamically significant; only the $d - 2$ transverse dimensions of oscillations are significant.

If the string picture of dual models, with dynamics specified by the action, $\mathcal{L}_{NG}$, suggested by Nambu [Nam70b], had been understood earlier, the path to understanding the structure of the physical states and the proof of the no-ghost theorem would have been clearer. For the degrees of freedom in the quantum theory to correspond to those in the classical theory (i.e. only transverse oscillations are physical), the Virasoro conditions must effectively eliminate two dimensions of oscillators. If this basic aspect of string dynamics had been understood, Lovelace's prophecy that E, the number of dimensions of oscillators removed, should be 2, as well as $d = 26$, would have seemed somewhat less obscure; and it would have been evident that the DDF construction [DDF72] should provide all the physical states, as it does in the critical dimension $d = 26$, pointing the way to the proof of the no-ghost theorem. In fact, this was all only evident with hindsight.

A difficulty with solving the classical constraint equations $L_m = 0$, after substituting $\mathrm{n} \cdot a_m = 0, m \neq 0$, is that the equations are quadratic. This difficulty is removed if n is taken to be lightlike rather than timelike, $\mathrm{n} = k$, because then $k \cdot a_m = 0, m \neq 0$ and the constraint equations $L_m = 0$ express $\bar{k} \cdot a_m$ in terms of the transverse components a_m^i of the oscillators: $\tilde{k} \cdot a_m = L_m^a, m \neq 0$, the Virasoro generators made from these transverse components. (Here $\tilde{k}^2 = 0, \tilde{k} \cdot k = -1, k \cdot p = 1$, as in Section 20.5.)

Using this light-cone approach, we canonically quantized the string, obtaining a spectrum built from only $d - 2$ dimensions of transverse oscillators (Goddard, Goldstone, Rebbi and Thorn [GGRT73]). The disadvantage with this approach is that the solution of the constraint conditions (20.27) has been achieved at the cost of the choice of k, which breaks manifest Lorentz invariance. To restore Lorentz invariance to the resulting theory, one must specify how to perform Lorentz transformations, particularly those generated by the E_{Li}. The straightforward expression for these is given by (20.26) with A_n^L replaced by $\tilde{k} \cdot a_n = L_n^a$ and A_n^i replaced by a_n^i. But, we knew from our previous work [GRT72] that these expressions only close to form the appropriate Lorentz algebra if $d = 26$; otherwise there is an anomaly term. It is only for $d = 26$ that the canonical quantization procedure is Lorentz invariant. Thus, the critical dimension condition was obtained from the process of canonically quantizing the relativistic string using light-cone coordinates.

An alternative approach to quantization, which preserves covariance at the sacrifice of being canonical, is to relax the constraints and treat all the oscillators a_n^μ as independent, imposing the constraints as weak conditions on the physical states in the quantum theory, i.e. the expectation values of (20.27) in physical states should vanish as $\hbar \to 0$. This approach is analogous to the Gupta–Bleuler method of quantizing electrodynamics, but electrodynamics is relatively straightforward because the Lorentz gauge constraint is linear whereas (20.27) is quadratic. The theory of quantizing constrained systems to handle this was developed in the Fifties by Paul Dirac [Dir58].

In this covariant approach, the conditions on the physical states take the form (20.5) and we knew from the no-ghost theorem [Bro72, GT72] that these only define physical states free from ghosts if $d \leq 26$. Further, unless $d = 26$, the states correspond broadly speaking to $d - 1$ dimensions of oscillators rather than just the $d - 2$ transverse dimensions that would correspond directly to the classical theory. For $d < 26$, although there are no negative norm states, there are 'additional' states containing longitudinal oscillations of the string present in the quantum theory that do not have classical analogues.

So, unless $d = 26$, in the canonical approach there is a breakdown of Lorentz invariance and, in the covariant approach, anomalous longitudinal modes of oscillation are present in the quantum theory. Lovelace's calculation [Lov71] showed that these extra modes would result in unitarity-violating cuts when interactions were introduced and loop contributions calculated (unless some way of modifying the theory could be found). For $d = 26$, the noncovariant canonical approach and the non-canonical covariant approach to quantizing agree, and give a consistent way of quantizing the relativistic string, yielding precisely the elegant spectrum found in dual models. Subsequently, Stanley Mandelstam [Man73] showed that interactions could be introduced in a simple way, by just allowing strings to split and to join at their ends. In this way, the original amplitude of Veneziano [Ven68] was obtained rigorously from a simple string picture.

The cost of all this was the seemingly unphysical requirement that the spacetime dimension $d = 26$ (and the tachyon that results from $\alpha_0 = 1$). But the critical dimension in the RNS model was $d = 10$ and the general view at the time was that one should search for a 'correct' model, with the sort of physical state structure that had been elucidated but with $d = 4$ and no tachyons.

In order for the quantum commutators of the oscillators a_n^μ to be normalized as in (20.1), the characteristic length ℓ must be given by

$$\ell = \sqrt{\frac{\hbar}{c} T_0 \pi}. \qquad (20.29)$$

With some ironic amusement, Goldstone pointed out that, for the quantum states of the string to have masses comparable with those of hadrons, the tension of the string, T_0, should have the somewhat macroscopic value of about 13 tons. The states on the leading Regge trajectory correspond to motions in which the string rotates like a rigid rod with the ends moving at the speed of light. Not everyone found our description of even the classical motion of the relativistic string plausible. One author characterized it as '*résultat évidemment faux*' (Souriau [Sou73]). Jeffrey's response was that it might conceivably be *faux* but it was surely not *évidemment faux*.

At the beginning of September 1972, I left CERN to take up my lectureship in Durham. On 28 August, just before I left, I gave a seminar on our work on the quantum mechanics of a relativistic string. The theorists working on dual models had been meeting at 2 pm most Thursdays throughout the academic year in a small discussion room, sitting on chairs round the periphery. The seminars were informal and there was a very free exchange of ideas. Amongst the regular participants at various times were Victor Alessandrini, Daniele Amati, Lars Brink, Edward Corrigan, Paolo Di Vecchia, Paul Frampton, David Olive, Claudio Rebbi, Joel Scherk and Charles Thorn. Evidently news of our work in progress must have spread – or perhaps it was just because Murray Gell-Mann, who was visiting CERN that year, had expressed interest in it – because a much larger audience than usual turned up and we had to relocate to the main theory conference room.

We drafted our paper on the 'Quantum Dynamics of a massless relativistic string' [GGRT73], with me already in Durham and Charles and Claudio still in CERN. The paper was finally produced in October. By then I was immersed in teaching my first lecture course, first-year applied mathematics.

The two years I had spent in CERN had built up to a crescendo of intellectual excitement and, though I have found much of my subsequent research work gripping and often extremely satisfying (when teaching duties and the largely self-inflicted wounds of administration have permitted), nothing has quite matched this period. In particular, I had the privilege of working closely for seven or eight months with Charles Thorn, whose combination of deep perception and formidable calculational power had provided the basis of what we managed to do. And, the exhilarating combination of the open and cooperative atmosphere that prevailed amongst (almost all) those working on dual models in CERN, the relative youth of most of those involved, the sense of elucidating a theory that was radically different, even the frisson of excitement that came from doing something that was regarded by some of those in power as wicked, because it might have nothing directly to do with the real world – this cocktail would never be offered to me again.

20.8 Fermions and fast forward

I arrived at Durham at the same time as Ed Corrigan, who had just completed his doctorate in Cambridge under the supervision of David Olive. He had been working with David on developing the understanding of amplitudes in the RNS model involving fermions. The model was then not yet complete because only tree amplitudes involving no more than two fermions (one fermion line) were known.

As a step towards constructing amplitudes for four fermions (i.e. fermion–fermion scattering) and more, the vertex describing the emission of a fermion, so converting a boson line into a fermion line, had been constructed [Tho71] (Corrigan and Olive [CO72]). We worked on understanding the gauge properties of this vertex [CG73] with a view to ensuring that the fermion–fermion scattering amplitude would only involve the exchange of physical boson states – otherwise the fermion–antifermion scattering might produce ghost boson states.

The calculation of the tree amplitude for fermion–fermion scattering in the operator formalism was comparable in complication with the calculation of loop amplitudes in the bosonic theory of Veneziano. In loop amplitudes, care had to be taken in the operator formalism that had been developed to ensure that ghost states do not circulate round the loop. As I mentioned in Section 20.6, the no-ghost theorem shows that the spectrum of physical states in the critical dimension agrees with the counting, corresponding to $d-2$ dimensions of oscillators, that Lovelace had deduced was necessary for loop amplitudes to be consistent with unitarity.

But, that did not in itself constitute a construction of the bosonic loop amplitude with only physical states circulating and agreeing with Lovelace's conjectured formula [Lov71]; this was achieved by Lars Brink and David Olive who constructed a projection operator onto the space of physical states [BO73], providing incidentally another proof of the no-ghost theorem. Using the projection operator of Brink and Olive, one could also ensure that only physical bosonic states were exchanged in fermion–fermion scattering. The net effect of doing this was to include a factor given by the determinant (Olive and Scherk [OS73])

$$\Delta(x) = \det\left(1 - M(x)^2\right), \tag{20.30}$$

where the matrix $M(x)$ has elements

$$M_{mn}(x) = (-x)^{\frac{1}{2}(m+n+1)} \frac{n + \frac{1}{2}}{m + n + 1} \binom{-\frac{1}{2}}{m} \binom{-\frac{1}{2}}{n}. \tag{20.31}$$

John Schwarz was particularly interested in the behaviour of $\Delta(x)$ near $x = 1$, because this governed the spectrum in the dual channel and he had suggested [Sch71] that the Neveu–Schwarz model should describe gluon states and the Ramond sector quark states; then $q\bar{q}$ states dual to the gluon states should describe the meson spectrum, which might be interestingly different from the original Neveu–Schwarz gluon spectrum. By numerical computer calculation, Schwarz and Wu [SW73] found that the leading behaviour of $\Delta(x)$, which determined the lowest mass in the dual channel, was $(1 - x)^{1/4}$. Indeed, they found that, within the limits of their numerical calculation,

$$\Delta(x) = (1 - x)^{1/4}, \qquad (20.32)$$

exactly.

This is how things stood when I reviewed dual models [Godd73] at the Aix conference in September 1973. Just after the meeting, Ed Corrigan and I produced a proof [CG74] of the absence of ghost state coupling in the fermion (Ramond) sector of the RNS model, provided that the lowest fermion mass in the model $m = 0$, a condition previously found to be necessary for consistency by Thorn [Tho73] and Schwarz [Sch73]. A little later, Ed Corrigan, Russell Smith, David Olive and I [CGOS73] were able to give an analytic proof of (20.32) and show that the spectra of states in the dual bosonic channels were the same apart from reversals of parity. These reversals of parity, implying a doubling of states, were unwelcome and, in fact, prior notice that the massless fermions should be chiral, indeed Majorana, as eventually realized by Gliozzi, Scherk and Olive [GSO76].

At the same time, Stanley Mandelstam had extended the interacting string picture to the Ramond–Neveu–Schwarz model [Man74] and obtained the same results on fermion scattering, including the reversals of parity, without having to face the calculation of $M(x)$. Then, and for some years to come, until the work of Polyakov [Pol81], Mandelstam was virtually alone in using the string picture and functional methods as a basis for calculating amplitudes (but see Giles and Thorn [GT77]).

Then it seemed that the challenges were to find a model for which the critical dimension would be 4, to understand better amplitudes involving fermions, and to calculate higher order loop contributions. The RNS model, as initially formulated, had a tachyon and a critical dimension $d = 10$. It was known that the tachyon could be removed by projecting onto a sector of the model but in a way that made the lowest fermion states, which were massless, chiral. The spectrum of the model necessarily had a massless spin one particle and, in the closed string sector, a massless spin two particle. Thus, although constructed with a view to describing the strong interactions, it had the characteristic massless particles, spin one, spin two and chiral spin one-half, of the other interactions: electromagnetism, gravity and the weak interactions, as Ed Corrigan emphasized to me. Ed, interpreting the Ramond fermion as a neutrino in an attempt to sort out the behaviour of the parity of the bosonic states under crossing, even made a calculation of the Weinberg angle, which he talked about at Aspen in August 1974, but did not publish. (See David Olive's review at the London conference [Oli74] for some comments on the interpretation of dual models in July 1974.)

By the end of 1973, as the fascination of dual models or string theory remained undimmed, though with ever increasing technical demands, the interest of many was shifting elsewhere. On 21 December, David Olive wrote to me, 'Very few people are now interested in dual theories here in CERN. Amati and Fubini independently made statements to the effect that dual theory is now the most exciting theory that they have seen but that it is too difficult for them to work with. The main excitement [is] the renormalization group and asymptotic freedom, which are indeed interesting.'

In the summer of 1974 I left Durham. At the beginning of 1973, Roger Dashen had invited me to spend one or two years at the Institute for Advanced Study, where André Neveu was then a long-term Member. I had obtained leave from Durham for the academic year 1974–1975 and then, in April 1974, I was offered and accepted a University Assistant Lectureship in Cambridge, and I was not allowed to delay taking it up beyond January 1975.

So, in July 1974, Helen and I, with our nine-month old daughter, Linda, set out for Aspen, where John Schwarz had assembled a good proportion of the world's dual model/string theorists. We were *en route* for Princeton via Berkeley, where I had secured an invitation from Stanley Mandelstam to spend six weeks before the Institute term began. In Berkeley, I wrote a largely cathartic paper [Godd75] on supersymmetry, which probably helped no one's understanding, except marginally my own. It had one memorable effect: namely, that when I reached Princeton I was invited to give a general seminar on supersymmetry, which most people did not know much about then. When I said I would rather talk about string theory, my offer was politely declined on the grounds that no one in Princeton was interested, a situation that has changed in the intervening years. Somewhat put out by this response, I did not give a seminar at all.

It was not quite true that no one was interested in strings. Tullio Regge, then a Professor at the Institute, was trying with Andy Hanson and Giorgio Ponzano (whom I never met – he was back in Torino) to quantize the relativistic string in a timelike rather than a lightlike gauge. We succeeded [GHP75] in doing this by introducing the DDF operators at the classical level and using the formalism of Dirac brackets. Working with Regge was exhilarating, even if he managed to avoid getting his name on the paper we wrote.

According to the report I submitted when I left the Institute, much of the remainder of my all too brief stay there was spent trying to quantize the motion of strings on spaces that were not flat, specifically group manifolds, as a way of compactifying the extra dimensions to produce non-Abelian symmetries. At the time, it seemed that compactification of the unwanted extra dimensions to form a torus (in the absence of a model with $d=4$) would only lead to Abelian symmetries. It was almost nine years later before David Olive and I began to see how compactifying on a torus could lead to non-Abelian symmetries in string theory, through the affine Kac–Moody algebras formed by vertex operators, and the preferred status of $E_8 \times E_8$ and $SO(32)$ (Goddard and Olive [GO85]).

Back in Cambridge in January 1975, I worked with my research student, Roger Horsley, to try to develop more powerful techniques for handling calculations in dual theory. However, not long after my next student, Nick Manton, began research, I started to realize that following my interests in strings or dual models might be a fine indulgence for me, but it was not going to help my students get jobs. (One of the great attractions of Cambridge at the time was that chances for promotion were so slim – Jeffrey Goldstone was still a Lecturer – that one did not need to be distracted by the prospects for advancement: they seemed negligible.)

I visited CERN in the summer of 1976 and there began learning from Jean Nuyts and David Olive and others about magnetic monopoles in gauge theory (I had attended Gerhard 't Hooft's seminal talk at the London conference in 1974 but I had not understood any of it at the time). The inverse relation between electric and magnetic charge, first found for electromagnetism by Dirac, generalized to the non-Abelian case, but we found that, in general, the magnetic charges were associated with a group somewhat different to the original 'electric' gauge group. This led to the introduction of a dual group for the magnetic charges [GNO77]. When I met Michael Atiyah for the first time at a conference on mathematical education at Nottingham in the spring of 1977, and told him about results on monopoles, he identified this dual group with the dual group that Robert Langlands had introduced for seemingly other reasons some years previously. Only recently are the connections between these occurrences of the dual group being fully understood (Kapustin and Witten [KW06]).

Even though I had felt I should move on from string theory to other topics, for my students' ultimate survival rather than any lack of my own interest, it turned out, perhaps ironically, perhaps inevitably, that the ideas David Olive and I worked on in the following ten years ended up being connected with string theory, even when this was not evident to begin with. Thus, I never escaped, even though I have been distanced by administration for much of the time in the last couple of decades. Many developments over the past forty years have definitively established the credentials of string theory as a deep and significant mathematical structure, while its ultimate status as a physical theory remains to be determined. It is disconcerting to think that this dénouement might be reached when one is no longer able to appreciate it.

Acknowledgements

I am grateful to Edward Corrigan, Louise Dolan, David Olive, Siobhan Roberts and Charles Thorn for helpful comments and corrections.

Peter Goddard was born in London in 1945. After studying mathematics and theoretical physics at the University of Cambridge, and a postdoctoral fellowship at CERN, he became a Lecturer in applied mathematics at the University of Durham in 1972. In 1975, he joined the faculty in Cambridge, becoming Professor of Theoretical Physics in 1991. He was founding deputy director of the Isaac Newton Institute for Mathematical Sciences and, until 2004, Master of St John's College, Cambridge. In 1997, together with David Olive, he was awarded the Dirac Medal of the International Centre for Theoretical Physics in Trieste. From 2002 to 2003, he was President of the London Mathematical Society. Since 2004 he has been Director of the Institute for Advanced Study in Princeton.

References

[ADDG75] Ademollo, M., D'Adda, A., D'Auria, R., Gliozzi, F., Napolitano, E., Sciuto, S. and Di Vecchia, P. (1975). Soft dilatons and scale renormalization in dual models, *Nucl. Phys.* **B94**, 221–259.

[Ale71] Alessandrini, V. (1971). A general approach to dual multiloop diagrams, *Nuovo Cimento* **A2**, 321–352.

[BM69] Bardakci, K. and Mandelstam, S. (1969). Analytic solution of the linear-trajectory bootstrap, *Phys. Rev.* **184**, 1640–1644.

[Bor92] Borcherds, R. E. (1992). Monstrous moonshine and monstrous Lie superalgebras, *Invent. Math.* **109**, 405–444.

[BO73] Brink, L. and Olive, D. (1973). Recalculation of the unitary single planar loop in the critical dimension of spacetime, *Nucl. Phys.* **B58**, 237–253.

[Bro72] Brower, R. C. (1972). Spectrum-generating algebra and no-ghost theorem in the dual model, *Phys. Rev.* **D6**, 1655–1662.

[BF73] Brower, R. C. and Friedman, K. A. (1973). Spectrum-generating algebra and no-ghost theorem for the Neveu–Schwarz model, *Phys. Rev.* **D7**, 535–539.

[BG71] Brower, R. C. and Goddard, P. (1971). Generalised Virasoro models, *Lett. Nuovo Cimento* **1**, 1075–1081.

[BG72] Brower, R. C. and Goddard, P. (1972). Collinear algebra for the dual model, *Nucl. Phys.* **B40**, 437–444.

[BG73] Brower, R. C. and Goddard, P. (1973). Physical states in the dual resonance model, in *Proceedings of the International School of Physics 'Enrico Fermi' Course LIV* (Academic Press, New York), 98–110.

[BT71] Brower, R. and Thorn, C. B. (1971). Eliminating spurious states from the dual resonance model, *Nucl. Phys.* **B31**, 163–182.

[CM72] Chang, L. N. and Mansouri, F. (1972). Dynamics underlying duality and gauge invariance in the dual-resonance models, *Phys. Rev.* **D5**, 2535–2542.

[CG73] Corrigan, E. F. and Goddard, P. (1973). Gauge conditions in the dual fermion model, *Nuovo Cimento* **A18**, 339–359.

[CG74] Corrigan, E. F. and Goddard, P. (1974). Absence of ghosts in the dual fermion model, *Nucl. Phys.* **B68**, 189–202.

[CGOS73] Corrigan, E. F., Goddard, P., Olive, D. and Smith, R. A. (1973). Evaluation of the scattering amplitude for four dual fermions, *Nucl. Phys.* **B67**, 477–491.

[CO72] Corrigan, E. and Olive, D. (1972). Fermion-meson vertices in dual theories, *Nuovo Cimento* **A11**, 749–773.

[DD70] Del Giudice, E. and Di Vecchia, P. (1970). Characterization of physical states in dual-resonance models, *Nuovo Cimento* **A70**, 579–591.

[DDF72] Del Giudice, E., Di Vecchia, P. and Fubini, S. (1972). General properties of the dual resonance model, *Ann. Phys.* **70**, 378–398.

[Dir58] Dirac, P. A. M. (1958). Generalized Hamiltonian dynamics, *Proc. R. Soc. London* **A246**, 326–332.

[DS08] Di Vecchia, P. and Schwimmer, A. (2008). *The Beginning of String Theory: a Historical Sketch*, arXiv:0708.3940.

[ELOP66] Eden, R. J., Landshoff, P. V., Olive, D. and Polkinghorne, J. C. (1966). *The Analytic S-Matrix* (Cambridge University Press, Cambridge).

[FN70] Fairlie, D. B. and Nielsen, H. B. (1970). An analog model for KSV theory, *Nucl. Phys.* **B20**, 637–651.

[FF82] Feigen, B. L. and Fuchs, D. B. (1982). Skew-symmetric differential operators on the line and Verma modules over the Virasoro algebra, *Funct. Anal. Appl.* **17**, 114.
[FGW71] Frampton, P. H., Goddard, P. and Wray, D. A. (1971). Perturbative unitarity of dual loops, *Nuovo Cimento* **A3**, 755–762.
[FGV69] Fubini, S., Gordon, D. and Veneziano, G. (1969). A general treatment of factorization in dual resonance models, *Phys. Lett.* **B29**, 679–682.
[FV69] Fubini, S. and Veneziano, G. (1969). Level structure of dual resonance models, *Nuovo Cimento* **A64**, 811–840.
[FV71] Fubini, S. and Veneziano, G. (1971). Algebraic treatment of subsidiary conditions in dual resonance models, *Ann. Phys.* **63**, 12–27.
[GT77] Giles, R. and Thorn, C. B. (1977). Lattice approach to string theory, *Phys. Rev.* **D16**, 366–386.
[Gli68] Gliozzi, F. (1968). Ward-like identities and twisting operator in dual resonance models, *Lett. Nuovo Cimento* **2**, 846–850.
[GSO76] Gliozzi, F., Scherk, J. and Olive, D. (1976). Supergravity and the spinor dual model, *Phys. Lett.* **B65**, 282–286.
[Godd71] Goddard, P. (1971). Analytic renormalisation of dual one loop amplitudes, *Nuovo Cimento* **A4**, 349–362.
[Godd73] Goddard, P. (1973). Dual resonance models, *Suppl. J. Phys. (Paris)*, **34**, 160–166.
[Godd75] Goddard, P. (1975). The connection between supersymmetry and ordinary Lie symmetry groups, *Nucl. Phys.* **B88**, 429–441.
[Godd98] Goddard, P. (1998). The work of R. E. Borcherds, in *Proceedings of the International Congress of Mathematicians, Berlin 1998* (Deutscher Mathematiker-Vereinigung), 99108, math/9808136.
[GGRT73] Goddard, P., Goldstone, J., Rebbi, C. and Thorn, C. B. (1973). Quantum dynamics of a massless relativistic string, *Nucl. Phys.* **B56**, 109–135.
[GHP75] Goddard, P., Hanson, A. J. and Ponzano, G. (1975). The quantization of a massless relativistic string in a time-like gauge, *Nucl. Phys.* **B89**, 76–92.
[GNO77] Goddard, P., Nuyts, J. and Olive, D. I. (1977). Gauge theories and magnetic charge, *Nucl. Phys.* **B125**, 1–28.
[GO85] Goddard, P. and Olive, D. (1985). Algebras, lattices and strings, in *Vertex Operators in Mathematics and Physics*, ed. Lepowsky, J., Mandelstam, S. and Singer, I. M. Mathematical Research Institute Publications (Springer-Verlag, New York).
[GRT72] Goddard, P., Rebbi, C. and Thorn, C. B. (1972). Lorentz covariance and the physical states in the dual resonance model, *Nuovo Cimento* **A12**, 425–441.
[GT72] Goddard, P. and Thorn, C. B. (1972). Compatibility of the pomeron with unitarity and the absence of ghosts in the dual resonance model, *Phys. Lett.* **B40**, 235–238.
[GW71] Goddard, P. and Waltz, R. E. (1971). One loop amplitudes in the model of Neveu and Schwarz, *Nucl. Phys.* **B34**, 99–108.
[Got71] Goto, T. (1971). Relativistic quantum mechanics of a one-dimensional mechanical continuum and subsidiary condition of dual resonance model, *Prog. Theor. Phys.* **46**, 1560–1569.
[Kac79] Kac, V. G. (1979). Contravariant form for the infinite-dimensional Lie algebras and superalgebras, *Lecture Notes in Physics* **94**, 441–445.
[KW06] Kapustin, A. and Witten, E. (2006). Electric-magnetic duality and the geometric Langlands program, arXiv: hep-th/0604151.
[KN69] Koba, Z. and Nielsen, H. B. (1969). Reaction amplitude for n-mesons a generalization of the Veneziano–Bardakci–Ruegg–Virasoro model, *Nucl. Phys.* **B10**, 633–655.

[Lov70] Lovelace, C. (1970). M-loop generalized Veneziano formula, *Phys. Lett.* **B32**, 703–708.

[Lov71] Lovelace, C. (1971). Pomeron form factors and dual Regge cuts, *Phys. Lett.* **B34**, 500–506.

[Man73] Mandelstam, S. (1973). Interacting string picture of dual resonance models, *Nucl. Phys.* **B64**, 205–235.

[Man74] Mandelstam, S. (1974). Interacting-string picture of Neveu–Schwarz–Ramond model, *Nucl. Phys.* **B69**, 77–106.

[MN72] Mansouri, F. and Nambu, Y. (1972). Gauge conditions in dual resonance models, *Phys. Lett.* **B39**, 375–378.

[Nam70a] Nambu, Y. (1970). Quark model and the factorization of the Veneziano amplitude, in *Proceedings of the International Conference on Symmetries and Quark Models, Wayne State University, June 18–20, 1969*, ed. Chand, R. (Gordon and Breach, New York), 269–277, reprinted in *Broken Symmetry, Selected Papers of Y. Nambu*, ed. Eguchi, T. and Nishijima, K. (World Scientific, Singapore, 1995), 258–277.

[Nam70b] Nambu, Y. (1970). Duality and hadrodynamics, unpublished notes prepared for the Copenhagen High Energy Symposium, published in *Broken Symmetry, Selected Papers of Y. Nambu*, ed. Eguchi, T. and Nishijima, K. (World Scientific, Singapore, 1995), 280–301.

[NS71] Neveu, A. and Schwarz, J. H. (1971). Factorizable dual model of pions, *Nucl. Phys.* **B31**, 86–112.

[NST71] Neveu, A., Schwarz, J. H. and Thorn, C. B. (1971). Reformulation of the dual pion model, *Phys. Lett.* **B35**, 529–533.

[Nie69] Nielsen, H. B. (1969). An almost physical interpretation of the integrand of the n-point Veneziano model, paper submitted to the *15th International Conference on High Energy Physics, Kiev, 26 August to 4 September, 1970* (Nordita preprint, 1969).

[Oli74] Olive, D. (1974). Plenary report on dual models, in *Proceedings of the XVII International Conference on High Energy Physics, July 1974, London* (Didcot), I, 269–280.

[OS73] Olive, D. and Scherk, J. (1973). Towards satisfactory scattering amplitudes for dual fermions, *Nucl. Phys.* **B64**, 334–348.

[Pol81] Polyakov, A. M. (1981). Quantum geometry of bosonic strings, *Phys. Lett.* **B103**, 207–210.

[Ram71] Ramond, P. (1971). Dual theory for free fermions, *Phys. Rev.* **D3**, 2415–2418.

[Sch71] Schwarz, J. H. (1971). Dual quark-gluon model of hadrons, *Phys. Lett.* **B37**, 315–319.

[Sch72] Schwarz, J. H. (1972). Physical states and pomeron poles in the dual pion model, *Nucl. Phys.* **B46**, 61–74.

[Sch73] Schwarz, J. H. (1973). Dual resonance theory, *Phys. Rep.* **8**, 269–335.

[SW73] Schwarz, J. H. and Wu, C. C. (1973). Evaluation of dual fermion amplitudes, *Phys. Lett.* **B47**, 453–456.

[Sha70] Shapiro, J. A. (1970). Electrostatic analogue for the Virasoro model, *Phys. Lett.* **B33**, 361–362.

[Sha75] Shapiro, J. A. (1975). Renormalization of dual models, *Phys. Rev.* **D11**, 2937–2942.

[Sou73] Souriau, J. M. (1973). Du bon usage des élastiques, *J. Relat.*, *Clermont-Ferrand*.

[Sus69] Susskind, L. (1969). Harmonic-oscillator analogy for the Veneziano model, *Phys. Rev. Lett.* **23**, 545–547.

[Sus70a] Susskind, L. (1970). Structure of hadrons implied by duality, *Phys. Rev.* **D1**, 1182–1186.
[Sus70b] Susskind, L. (1970). Dual-symmetric theory of hadrons: 1, *Nuovo Cimento* **A69**, 457–496.
[Tho71] Thorn, C. B. (1971). Embryonic dual model for pions and fermions, *Phys. Rev.* **D4**, 1112–1116.
[Tho73] Thorn, C. B. (1973). Private communication (April, 1973).
[Tho85] Thorn, C. B. (1985). A proof of the no-ghost theorem using the Kac determinant, in *Vertex Operators in Mathematics and Physics*, ed. Lepowsky, J., Mandelstam, S. and Singer, I. M. (Springer-Verlag, New York), 411–417.
[Ven68] Veneziano, G. (1968). Construction of a crossing-symmetric, Reggeon behaved amplitude for linearly rising trajectories, *Nuovo Cimento* **A57**, 190–197.
[Ven70] Veneziano, G. (1970). Duality and dual models, in *Proceedings of the 15th International Conference on High Energy Physics, Kiev, 26 August to 4 September, 1970*, 437–453.
[Vir69] Virasoro, M. (1969). Alternative constructions of crossing-symmetric amplitudes with Regge behavior, *Phys. Rev.* **177**, 2309–2311.
[Vir70] Virasoro, M. (1970). Subsidiary conditions and ghosts in dual-resonance models, *Phys. Rev.* **D1**, 2933–2936.
[Wak01] Wakimoto, M. (2001). *Infinite-Dimensional Lie Algebras*, (American Mathematical Society, Providence, RI) Translations of Mathematical Monographs, Vol. 195.

21
The first string theory: personal recollections

LEONARD SUSSKIND

Here is how I remember it. Supposedly I was a particle physicist: indeed I had spent three years as a graduate student in Cornell (1962–1965) and a year as a postdoc in Berkeley (1965–1966), learning about S-matrix theory, and hating every minute of it. The general opinion among leaders of the field was that hadronic length and time scales were so small that *in principle* it made no sense to probe into the guts of a hadronic process – the particles and the reactions were unopenable black boxes. Quantum field theory was out; Unitarity and Analyticity were in. Personally, I so disliked that idea that when I got my first academic job I spent most of my time with my close friend, Yakir Aharonov, on the foundations of quantum mechanics and relativity.

By 1967 I was convinced that the S-matrix black-box view was wrong-headed. I had spent a lot of time formulating relativistic quantum theory in what was then called the infinite-momentum frame (IMF), now called the light-cone frame. The idea that appealed to me was that by boosting a system to very large momentum you could slow down its internal motions. By stationing a sequence of detectors along the direction of the boost, one could imagine probing an interacting system of hadrons as it evolved. I wrote a lot of papers showing among other things that physics in the IMF had a Galilean symmetry so that the ordinary rules of nonrelativistic quantum mechanics must apply. In other words I was convinced that the IMF could be used to open the black box and find out what was inside.

Some time in 1968 Aharonov and myself had a visitor – I am pretty sure it was Hector Rubinstein – who was all excited about some fantastic new development in S-matrix theory: a young Italian, named Gabriele Veneziano, had solved the constraints of crossing symmetry and analyticity and come up with a formula [Ven68]. Neither Aharonov nor I understood where the formula came from, but it was so short and simple that I memorized it.

The thing that struck me most was that the poles were evenly spaced, not in the mass, but in mass squared, m^2. That resonated: in the Galilean formulation of the IMF, m^2 plays the role of energy. Equally spaced energy levels meant one thing – the system that the Veneziano formula described was some kind of harmonic oscillator. I wrote a paper called:

The Birth of String Theory, ed. Andrea Cappelli, Elena Castellani, Filippo Colomo and Paolo Di Vecchia.
Published by Cambridge University Press. © Cambridge University Press 2012.

'Gedanken experiments involving hadrons' [Sus68], explaining that if the Veneziano theory were correct, then you could watch an excited hadron oscillate in the IMF.

Sometime in late 1968 I began to experiment with IMF oscillator-based models that I hoped would give the Veneziano formula. The first one was simple: just imagine a charge attached to an ideal Hooke's law spring. A photon can scatter off the charge and I computed the amplitude. It looked vaguely similar to Veneziano's formula. Slightly simplified, Veneziano's formula had the form:

$$A(s,t) = \int_0^1 x^{-s}(1-x)^{-t} dx. \qquad (21.1)$$

My oscillator amplitude had the form:

$$A(s,t) = \int_0^1 x^{-s} (\exp(-x))^{-t} dx. \qquad (21.2)$$

Somehow the formulae were related by replacing $(1-x)$ by $\exp(-x)$.

I really did not know a great deal about particle physics or quantum field theory at that time, but what I did know was the harmonic oscillator. It was almost as easy to compute amplitudes involving several incoming and outgoing photons. The answers were multiple integrals with several factors similar to x^{-s} for the direct channels and factors like $(\exp(-x))^{-t}$ for the crossed channels. It was one of the most exciting moments in my physics career when I compared my amplitudes with the generalized Veneziano formulae that had been worked out by others. The results were identical except that wherever Veneziano had $(1-x)$ I had $\exp(-x)$. I wrote it up and sent it to the *Physical Review Letters* with the title 'Harmonic oscillator analogy for the Veneziano amplitude' [Sus69].

I was very puzzled by one feature of the Veneziano amplitude which the oscillator analogue did not share. In the simple oscillator case, a high energy collision would probe the shape of the oscillator ground state. In fact, at very high energy the amplitude is just $\exp(-t)$, which is the Fourier transform of the oscillator probability distribution. But when I tried to study the size and shape of a hadron by using the Veneziano amplitude in the same way, something strange happened. Instead of getting a definite answer, the size of the target grew indefinitely with the energy of the collision. I realized that this meant that when a hadron is probed at very short time scales it looks bigger than when it is probed slowly. This of course was nothing but the Regge behaviour manifesting itself, but I thought of it as a very rapid fluctuation of the size of the system that would be averaged out at low energy. It took a little while to realize what was going on – namely that Veneziano's hadron was not a simple spring, but that it had oscillations of all possible integer frequencies. As the energy increases, more and more harmonics reveal themselves in a growing structure. In other words, only the zero-point oscillations of an infinite number of oscillators could replace my $\exp(-x)$ by Veneziano's $(1-x)$. Then came the 'aha' moment when I realized that the model hadron was a continuous harmonic string.

I immediately wrote a paper for *Physical Review Letters*, including the notion of the world-sheet and the fact that the so-called duality was a consequence of conformal symmetry. The paper was rejected on the basis of not having any new experimental prediction. But I did get versions of it into the *Physical Review* [Sus70a] and the Italian journal, *Nuovo Cimento* [Sus70b]. It was also about that time that I became aware of Nambu's work which was so similar to my own [Nam70a, Nam70b].

When I did the mathematics I found a couple of unfortunate things. The first was that the ground state hadron was a tachyon. I'm not sure whether that was the first discovery of the open string tachyon or not. I also noticed that the first excitation was a massless vector particle. Today, of course, that is the great wonder of the emergence of gauge theory from open strings, but in 1969 the tachyon and the massless vector were a thorn in the side of string theory.

My colleague Graham Frye and I tried to compute closed loop amplitudes and realized that there had to be closed strings in the theory [FS69, FS70a]. We related them to Pomerons [FS70b]. Analogous results were obtained independently by Neveu and Scherk [NS70] and by Gross, Neveu, Scherk and Schwarz [GNSS70], respectively. But the closed strings were sick. They were branch points instead of poles and we did not know what to make of it. A little later Claud Lovelace noticed that the cuts became poles if one replaced four-dimensional spacetime by the 'absurdity' of 26 dimensions [Lov71]. Frankly I regarded that as too speculative to think about.

One of the most exciting things that happened was a correspondence that started when Holger Bech Nielsen sent me a handwritten letter explaining his 'conducting-disc' analogy (cf. Nielsen [Nie70], Fairlie and Nielsen [FN70]). Holger understood that the conducting disc was just the world-sheet and that the relation between his work and mine was simply momentum-position duality. Nielsen came to visit me in New York and we excitedly explored the possibility that the world-sheet is a dense planar Feynman diagram which we connected with Feynman's parton ideas. I believe the paper that we wrote with Aage Kraemmer was the first to contain the quadratic world-sheet action (Kraemmer, Nielsen and Susskind [KNS71]). Holger speculated that the density of the dense planar Feynman diagram should be integrated over and that it would create an extra dimension of space. For some reason we dropped that idea, but of course Polyakov picked it up again about ten years later [Pol81a, Pol81b].

What I most certainly did not have at that time was any inkling that string theory would ultimately provide much more than a theory of hadrons. The idea that the massless spin one and spin two particles would be connected with fundamental gauge and gravity forces was not part of my thinking. Or that more dimensions would prove to be an important element in giving string theory the flexibility to describe the spectrum of elementary particles.

Leonard Susskind was born in New York City in 1940. He attended public schools, received his BSc in physics at the City College of New York in 1962, and did his graduate work at Cornell University. He was a Professor at Yeshiva University from 1967 to 1977 and then moved to Stanford University where he is currently Felix Bloch Professor of Physics.

References

[FN70] Fairlie, D. B. and Nielsen, H. B. (1970). An analogue model for KSV theory, *Nucl. Phys.* **B20**, 637–651.

[FS69] Frye, G. and Susskind, L. (1969). Duality and Feynman graphs, preprint Yeshiva U. PRINT-69-1907.

[FS70a] Frye, G. and Susskind, L. (1970). Removal of the divergence of a planar dual-symmetric loop, *Phys. Lett.* **B31**, 537–540.

[FS70b] Frye, G. and Susskind, L. (1970). Non-planar dual symmetric loop graphs and pomeron, *Phys. Lett.* **B31**, 589–591.

[GNSS70] Gross, D. J., Neveu, A., Scherk, J. and Schwarz, J. H. (1970). Renormalization and unitarity in the dual-resonance model, *Phys. Rev.* **D2**, 697–710.

[KNS71] Kraemmer, A. B., Nielsen, H. B. and Susskind, L. (1971). A parton theory based on the dual resonance model, *Nucl. Phys.* **B28**, 34–50.

[Lov71] Lovelace, C. (1971). Pomeron form factors and dual regge cuts, *Phys. Lett.* **B34**, 500–506.

[Nam70a] Nambu, Y. (1970). Quark model and the factorization of the Veneziano amplitude, in *Proceedings of the International Conference on Symmetries and Quark Models, Wayne State University, June 18–20, 1969*, ed. Chand, R. (Gordon and Breach, New York), 269–277, reprinted in *Broken Symmetry, Selected Papers of Y. Nambu*, ed. Eguchi, T. and Nishijima, K. (World Scientific, Singapore, 1995), 258–277.

[Nam70b] Nambu, Y. (1970). Duality and hadrodynamics, lecture notes prepared for the Copenhagen summer school, 1970, reproduced in *Broken Symmetry, Selected Papers of Y. Nambu*, ed. Eguchi, T. and Nishijima, K. (World Scientific, Singapore, 1995), 280–301.

[NS70] Neveu, A. and Scherk, J. (1970). Parameter-free regularization of one-loop unitary dual diagram, *Phys. Rev.* **D1**, 2355–2359.

[Nie70] Nielsen, H. B. (1970). An almost physical interpretation of the integrand of the n-point Veneziano model, paper submitted to the *15th International Conference on High Energy Physics, Kiev, 26 August to 4 September, 1970* (Nordita preprint, 1969).

[Pol81a] Polyakov, A. M. (1981). Quantum geometry of bosonic strings, *Phys. Lett.* **B103**, 207–210.

[Pol81b] Polyakov, A. M. (1981). Quantum geometry of fermionic strings, *Phys. Lett.* **B103**, 211–213.

[Sus68] Susskind, L. (1968). Gedanken experiments involving hadrons, *Phys. Rev. Lett.* **21**, 944–945.

[Sus69] Susskind, L. (1969). Harmonic-oscillator analogy for the Veneziano model, *Phys. Rev. Lett.* **23**, 545–547.

[Sus70a] Susskind, L. (1970). Structure of hadrons implied by duality, *Phys. Rev.* **D1**, 1182–1186.

[Sus70b] Susskind, L. (1970). Dual-symmetric theory of hadrons. 1. *Nuovo Cimento* **A69**, 457–496.

[Ven68] Veneziano, G. (1968). Construction of a crossing-symmetric, Reggeon-behaved amplitude for linearly rising trajectories, *Nuovo Cimento* **A57**, 190–197.

22
The string picture of the Veneziano model

HOLGER B. NIELSEN

Abstract

This Chapter is about my memories of the discovery that the Veneziano model in fact describes interacting strings. Susskind and Nambu also found the string picture by following other, independent approaches. A characteristic feature of my approach was that I used very high-order 'fishnet' or planar Feynman diagrams to describe the development of the relativistic string. The arrangement of particles on a chain indeed leads to the dominance of planar diagrams, if only nearest neighbour interactions are relevant. I also mention the work of Ziro Koba and myself on extending the Veneziano model, first to five external particles – as also done by Bardakci and Ruegg, Chan and Tsou, and Goebel and Sakita – and subsequently to an arbitrary number of external mesons.

22.1 Introduction

In the unpublished preprint [Nie70], 'An almost physical interpretation of the integrand of the n-point Veneziano model', I proposed – independently of Nambu [Nam70] and Susskind [Sus70a, Sus70b] – that the dual model or Veneziano model [Ven68] was describing the scattering of relativistic strings.

In the present Chapter I present my recollections on how I came to understand the scattering of strings in the Veneziano model. The background that led to the string idea can be found in the works with Ziro Koba that extended dual models to an arbitrary number of external particles [KN69a, KN69b, KN69c]. This work is described in Section 22.4. It would, however, be natural in such a reminiscence to describe first what for me was crucial, namely the seminar by Hector Rubinstein given at the Niels Bohr Institute. Hector was very excited about the great discovery by Veneziano. These earliest memories will be presented in Sections 22.2 and 22.3. The line of thinking that ended with the insight that the Veneziano model describes interacting strings is recalled in Section 22.6. The hint came from my pet idea of using high-order Feynman diagrams for studying strong interactions: this is presented in Section 22.5. In Section 22.7, I would like to report the

---

The Birth of String Theory, ed. Andrea Cappelli, Elena Castellani, Filippo Colomo and Paolo Di Vecchia.
Published by Cambridge University Press. © Cambridge University Press 2012.

discussions with people at various conferences, such as the Lund conference, and my collaboration with David Fairlie. Finally I recall briefly two other later works, namely, in Section 22.8 the attempt (with P. Olesen) to make strings in a Higgs model, leading to the so-called Abrikosov–Nielsen–Olesen vortex lines, and in Section 22.9 the derivation, with Lars Brink, of critical dimensions in string theory from zero-point energy.

22.2 On the Veneziano model

Historically, Veneziano [Ven68] started by formulating a scattering amplitude for the four mesons π^+, π^0, π^- and ω. Owing to the analyticity properties of the Mandelstam representation, the amplitude is useful for describing the following physical scattering processes in the various channels:

$$\pi^+ + \pi^0 \to \pi^+ + \omega, \tag{22.1}$$
$$\pi^- + \pi^0 \to \pi^- + \omega, \tag{22.2}$$
$$\pi^+ + \pi^- \to \pi^0 + \omega, \tag{22.3}$$

and also the decay process of the unstable ω into three pions,

$$\omega \to \pi^+ + \pi^- + \pi^0. \tag{22.4}$$

The above list of processes can be extended by taking into account time reversal symmetry, which for example implies that the process $\pi^+ + \omega \to \pi^+ + \pi^0$ is related to $\pi^+ + \pi^0 \to \pi^+ + \omega$.

22.3 Introduction to the mathematical properties of the Veneziano amplitude

The monumental problem that Gabriele Veneziano succeeded in solving was that of finding a function of the Mandelstam variables s, t and u, with the property that there were no other singularities than the poles corresponding to the resonances connected with the Regge poles. Furthermore, in his construction the asymptotic behaviour was in agreement with the theory of Regge poles.

Now it should be pointed out that the Regge poles were not at first known experimentally, although one already knew about the ρ-meson Regge trajectory. Prior to Veneziano's breakthrough, other physicists had used the idea of introducing further trajectories in order to obtain a self-consistent amplitude with the combined Regge exchange and Regge resonance properties. These works must have been an extremely important source of inspiration for the Veneziano model. Veneziano then went on and proposed infinitely many Regge trajectories.

Amplitudes for scattering had been found before Veneziano, by imposing consistency of the exchange in various channels. However, Veneziano's choice of scattering process was very clever because the isospin properties implied that the amplitude must be totally antisymmetric under permutations of the four-momenta of the three pions. (The pions are isovectors that must couple to the ω-meson isosinglet.)

Antisymmetry was implemented by contracting the (totally antisymmetric) Levi-Civita symbol $\epsilon_{\mu\nu\tau\sigma}$ with the three linearly independent external four-momenta, $p_{\pi^+}^\nu$, $p_{\pi^0}^\tau$, $p_{\pi^-}^\sigma$, and the polarization vector ϵ^μ for the ω-particle, which is a spin one meson. This means that the scattering amplitude must contain this factor multiplied by a scalar function $A(s, t, u)$ of the Mandelstam variables s, t, and u:

$$\left(\epsilon_{\mu\nu\tau\sigma}\, \epsilon^\mu\, p_{\pi+}^\nu\, p_{\pi^0}^\tau\, p_{\pi-}^\sigma\right) \cdot A(s, t, u). \tag{22.5}$$

It is on the scalar function that analyticity properties are imposed in such a way as to ensure that only poles connected with the Regge trajectories are present. It was the discovery of this function, which is essentially the surprisingly simple Euler beta function, that was the great progress made by the Veneziano model.

The idea of considering a process involving just one scalar factor in (22.5) may be regarded as just a detail for simplifying the problem. One could alternatively have considered scalar particles without isospin complications. This case was analyzed by Koba and myself and also by other groups, including Chan and Tsou [Cha69, CT69], Bardakci and Ruegg [BR68] and Goebel and Sakita [GS69], and we found generalizations of the Veneziano model for five and more external particles. By choosing scalar particles, one departs from the experimentally relevant hadron scatterings: the lightest mesons are indeed pseudoscalars and vector mesons that mostly carry nontrivial isospins.

The scalar factor in the amplitude (22.5) should be symmetric under the permutation of the three external π-mesons. This means that Veneziano had to take a sum of three terms, related among themselves by exchanging the π-mesons. The really interesting question was how these three terms looked individually. In fact each of the three terms has only Regge poles in two out of the three channels, s, t and u. For example, it could be that one of the terms had resonances only in the s- and the t-channels, while there were no resonances corresponding to particles in the u-channel. One should notice that arrangements with resonances only in the t-channel but not in the u-channel would be impossible, if in the s-channel there were resonances with odd spins only, as indicated phenomenologically by the ρ-meson Regge trajectory. That would suggest that there would only be a sign difference when exchanging the t- and u-channels.

However, there is often the phenomenon of 'exchange degeneracy', meaning that there exists another trajectory with resonances of even spin together with the one with odd spins. By interference of the two trajectories, it becomes possible to arrange the asymmetry between the s- and the u-channels. So for the Veneziano model it was rather important to have this exchange degeneracy.

The reason why I have mentioned these aspects of exchange degeneracy and amplitude summation for permutation symmetry is that I believe that these little technical details were relevant in obtaining a relatively simple generalization of the Veneziano amplitude to an arbitrary number of external particles.

This gives me then an excuse for recollecting my 'Cand Scient' (PhD) thesis [Nie68], in which I studied the self-consistency of three-meson vertices described by the quark model. This may have inspired me to consider processes in which the mesons are composed of pairs of quarks and antiquarks, so arranged as to yield resonances of nonexotic mesons in some channels only: for example in the s- and t-channels but not in the u-channel, as in the case just described.

22.4 Developing dual models with arbitrary number of external particles with Ziro Koba

At the time that Hector Rubinstein gave his sensational seminar on the Veneziano model, I was rather urged to finish my thesis [Nie68]. Nevertheless, I managed to include a figure about the Veneziano model in it. After the thesis, I become so relaxed that I could begin to concentrate on the Veneziano model.

I did my first work together with Koba on the generalization to a five-point function. An integral representation of the amplitude was needed, with two independent variables that should then obey some algebraic relations so as to ensure the expected pole and Regge behaviour at the desired places in the amplitude as a function of the Regge pole variables $\alpha(s_{ij})$. These variables were put in the exponents of the dependent integration variables u_{ij}. Here I would like to report a little accident that may in fact have been of great significance for me in getting into the project in a fruitful way. At that time, I taught as an instructor – as I had done during my 'eternal' studies – a third year course in geometry, especially projective geometry, at the mathematics department. In that course, we had a problem about a pentagon in the projective plane, in which the students and the instructor had to prove a series of algebraic relations between the anharmonic ratios for the pentagon. Interestingly enough, these relations were exceedingly well suited for getting the relations between the dependent variables that would give the right properties for the dual scattering amplitude of five scalar particles.

During that period, the most important moments of my physics studies and research were the regular meetings with Ziro Koba and some younger people. At these meetings we presented ideas as they came to us: Koba was indeed very good at inspiring the students to be active. I remember that he once 'uninvited' me to participate, in order that an even younger group of students should feel more free to express themselves without being embarrassed by the presence of slightly more advanced people.

In generalizing from five to N particles, I first attempted the extension of the pentagon formulation. There is indeed such a possibility, but it is more complicated than the simplified version which Koba discovered: it was simpler to have the points on a line in the projective plane, the so-called Koba–Nielsen variables.

When we presented this work on the N-point Veneziano model at CERN, Koba gave the main talk and I spoke about the application of the Haar measure technology for making sense of the amplitude expressions, which would otherwise be divergent.

22.5 Ideas that arose in developing string theory

The attempts that led to strings from the Veneziano model were for me based on the idea of treating strong interactions by very high-order Feynman diagrams. It is of course a most natural thought that, if the coupling constants are strong, the higher order diagrams are dominant: they are expected to be of huge order. So I set about visualizing such diagrams.

If the Veneziano amplitude described hadronic scatterings, could it be reproduced by high-order Feynman diagrams?

Then I began to investigate whether I could get the limits of the integration variables in the Veneziano amplitude that give the asymptotic behaviour and poles from the high-order Feynman diagrams: I described the latter by their gross shape structures. It turned out that the approximation of diagrams by a speculative central limit theorem rendered everything Gaussian: when we convolute sufficiently many propagators, we can equivalently take them to be Gaussian functions of the momenta. It did not quite work out as expected, unless an extra assumption was made, that the diagrams should be planar. This means that they can be drawn on the plane without any over-crossing of lines.

This result was easily argued to be equivalent to assuming that the particles tend to sit in long chains. They really formed strings. Or, rather, I had to assume that they did sit in such string chains in order that the summation over the diagrams could lead to the Veneziano model; the summation over the different diagrams was identified with the integrations in the integral representation of the Veneziano model.

One question that I had to treat was how the external momenta would be distributed through the diagram in the asymptotic high energy limit, namely from which range in the loop integration space the dominant contribution came. This is analogous to the conduction of current through a network of resistances. Concerning this question, I corresponded with David Fairlie and our discussions led to interesting advancements.[1]

My belief was that if we have a very large – in numbers of loops and vertices – Feynman diagram, then there is an integration over a huge number of loop momenta and, therefore, the external momenta which are led through the diagram (formally in some arbitrarily specified way) will in some sense only give a small correction relative to the internal loop momenta. This led to the idea that there is some dominant region in the space of all the loop momenta from which one obtains the main contribution to the diagram. In thinking about the dominant or central point in the integration region, one can ask: how are the external momenta led through the diagram in a physically or mathematically sensible way? If one knows the dominant integration region, one can then ask: how are the contributions to the propagator momenta in this centre changed compared to the case of vanishing external momenta? This change can show the way the external momenta are led through the diagram.

If one plays around with the external momenta, the changes are expected to be very small compared to the loop momenta, and a Taylor expansion in the external momenta is reasonable. In the leading term of the Feynman diagram, it is natural to expect that the external four-momenta are led through in a distributed way, because there are so many ways the external momenta could find their way through the diagram. Thus, only a tiny fraction goes through each of the internal propagators. Even though it is not always correct to approximate the propagators by Gaussian functions, as suggested in my string paper

---

[1] Editors' note: for further details, see also the Chapter by David Fairlie.

[Nie70], it should work in very high-order complicated diagrams, because in this limit the Taylor expansion in the small external momenta is meaningful.

In this reasoning one can see some similarity to the arguments that in later years have been central to my pet theory of random dynamics: often one can use Taylor expansion and get essentially the same result regardless of the true underlying theory. So perhaps the true theory does not matter at all in the end.

22.6 Review of my string paper

My preprint(s) [Nie70] were poorly distributed. There was a first draft version with the title 'A physical model for the Dual model'. In the second slightly better distributed version the word 'almost' appeared in the title, 'An almost physical model for the Dual model', as a result of discussions with Knud Hansen. The concept of a string was put in by the assumption that the dominant, high-order Feynman diagrams were planar. In reality I expressed this assumption by calling them 'fishnet diagrams', and simply drew diagrams with lots of squares as if it were a $\lambda\phi^4$ theory. Of course you can easily argue that if the constituents of an object – in this case a hadron or rather a meson – sit sequentially along a chain forming a string, the only interaction will be between neighbours on this chain. The Feynman diagrams that are important would be obtained by first drawing the series of propagators representing the propagation of the constituents (particles) sitting side by side on the chain, and subsequently connecting them by exchanges that predominantly are between neighbours.

Next, I proposed to approximate the propagators $\Delta(p^2)$ by Gaussian expressions,

$$\Delta(p^2) \approx C \cdot \exp(\alpha \cdot p^2). \tag{22.6}$$

If one imagines after a transformation to have made it effectively into a Euclidean theory and one uses the $(+ - --)$ metric, there needs to be a minus sign in front of the 'α' in the Gaussian (here, C and α are just constants).

Gaussian loop integrals can be easily computed: it is not difficult to get an estimate of the dependence on the external momenta, which is the important thing. I corresponded with Fairlie, who provided important mathematical insights for this calculation. Actually, it is analogous to studying electrical networks with resistors. Then, the approximation of the fishnet diagram by a planar homogeneous conductor becomes rather obvious. One of the important points – also inspired by Fairlie – was that in a two-dimensional network the flow through of the external momenta is calculable even without knowing the resistance. Owing to conformal invariance, one can deform the conducting surface conformally without changing the result of the diagram. The external momentum-dependent factor for the diagram, being an exponential of a quadratic expression, thus has a coefficient proportional to α (identified with the resistivity).

There are divergences in the leading terms in the external momenta related to the connection points to the conducting disc. But the really important thing is that the variation of the positions on the edge of the disc – or the conformally deformed disc – does not matter due to conformal invariance. The result is the integrand in the multiple meson scattering amplitude determined by Koba and myself.

The summation over various large planar diagrams should then hopefully have led to the integration measure to be used in the formula by Koba and myself, but that part of the derivation – the correct measure – did not come (convincingly) out of my technology.

Later on, Fairlie and I wrote an article [FN70] in which we showed that the loop correction to dual models could also be understood in terms of string theory.

22.7 Discussions with many people

During the time that I developed the string interpretation of the Veneziano model, and especially when it was nearly finished, I met and discussed with many people. At the Lund conference – where my talk was about the multiparticle Veneziano model by Koba and myself – I talked with several people, and of course also with colleagues at the Niels Bohr Institute.[2]

At one meeting in Copenhagen, I should have talked about string theory but I began to tell the story somewhat privately – especially to Sakita – while the audience was more or less present. So I gave the talk in this half private way and did not really get to give the formal talk. Sakita must have learned my account very well because at the next big conference in Kiev he gave a special talk [Nie70] and subsequently produced the first published work about string theory using my approach with fishnet diagrams. Neither Susskind's nor Nambu's string theory versions were presented in Kiev and I believe Sakita had not heard about them.

I remember that I first became aware that Susskind was up to something similar to my work from a note at the end of one of his articles alluding to his string interpretation. Then I immediately asked my mother to send a copy of my preprint to him. I urged her so much, that she made remarks about the fact that some addressees of preprints enjoyed preferential treatment.

Nambu should have talked about the string at a Sinbi conference at the Niels Bohr Institute, but got stuck in a desert and could not come. But he did deliver his manuscript.

I also remember that I was honoured by being asked to give a talk about strings on the occasion of the visit of Heisenberg to the Niels Bohr Institute. There were two talks that day: mine and Heisenberg's. I do not think though that I managed to make Heisenberg extremely enthusiastic about strings.

22.8 The vortex line string

The vortex line, I suppose, should really not be counted as part of the 'beginning or birth' of string theory since we came to think of vortex lines – Poul Olesen and I – only after

---

[2] In the acknowledgement of my paper, I thanked Z. Koba, K. Hansen, J. Bjoernebo, J. Detlefsen, J. Hamilton, Fisbachin, J. Hoerup, D. B. Fairlie, M. Chaichan, for useful discussions, and M. Virasoro, H. Rubinstein and M. Jacob, for encouragement.

having already worked on and known about the string [NO73]. So we should rather say that we attempted to construct a physical model behind the string. The idea of vortex lines was at that time already known in the nonrelativistic case of superconductivity, but we were not familiar with this work. I believe that I learned about this work when presenting our work at CERN.

22.9 Memories from CERN on understanding the 26 and 10 dimensions

After our formulation of string theory described above, the period during which I was most occupied with string theory was during a nine-month stay at the Theory Division of CERN. Here there was a very good group working on dual models. I had strong contacts with Paul Frampton [BFN74]: at that time he was working on his book about dual models. With Lars Brink I published, among other things, a paper which I think provided the nicest way of seeing what was so special about the critical dimension 26 for the simple string with only the geometrical degrees of freedom, and 10 dimensions for the Ramond–Neveu–Schwarz string having further fermionic degrees of freedom.

The idea of Lars Brinks and mine [BN73a, BN73b] for understanding the critical dimensions was based on the by then well-known fact that the first excitation of the string had only transverse polarizations and therefore was massless. We calculated the zero-point energy contribution to the energy of a string that is not excited, i.e. is in the ground state. This contribution to the mass squared of the string state became – after a renormalization of the speed of light – the number $(d-2)$ of true, transverse, degrees of freedom of the position field X^μ, multiplied by $-1/24$ for the mass squared distance between successive bunches of string states. Here, d denotes the number of spacetime dimensions.

The zero-point energy actually shifts the mass squared for all the string states by the same quantity $-(d-2)/24 \times 1/\alpha'$ where $1/\alpha'$ is the mass squared spacing between successive levels. For the first excited level to have vanishing mass, as required for consistency, it is necessary that the shift $-(d-2)/24 * 1/\alpha'$ equals the required shift $-1/\alpha'$. This reproduces the famous value for the spacetime dimension $d = 26$.

Acknowledgements

It is a pleasure to thank Don Bennett and Bendt Gudmander for comments and improvements.

Holger B. Nielsen was born in 1941 in Copenhagen. He studied physics and mathematics at the Niels Bohr Institute (Copenhagen University) in 1961. He received his Cand Scient degree (PhD) in 1968 with Ziro Koba, in high energy physics theory. He is presently Professor of High Energy Physics at the Niels Bohr Institute. Besides the Koba–Nielsen extension of the Virasoro formula, and the Nielsen–Olesen vortex line description of strings, he has worked on Koba–Nielsen–Olesen scaling, the no-go theorem for chiral fermions on a lattice, the Froggatt–Nielsen mechanism for suppressing quark and lepton masses, and random dynamics.

References

[BR68] Bardakci, K. and Ruegg, H. (1968). Reggeized resonance model for production amplitude, *Phys. Lett.* **B28**, 342–347.

[BFN74] Brink, L., Frampton, P. H. and Nielsen, H. B. (1974). A study of the Hilbert space properties of the Veneziano model operator formalism, *J. Math. Phys.* **15**, 1599–1605.

[BN73a] Brink, L. and Nielsen, H. B. (1973). A physical interpretation of the Jacobi imaginary transformation and the critical dimension in dual models, *Phys. Lett.* **B43**, 319–322.

[BN73b] Brink, L. and Nielsen, H. B. (1973). A simple physical interpretation of the critical dimension of spacetime in dual models, *Phys. Lett.* **B45**, 332–336.

[Cha69] Chan, H. M. (1969). A generalised Veneziano model for the N-point function, *Phys. Lett.* **B28**, 425–428.

[CT69] Chan, H. M. and Tsou, S. T. (1969). Explicit construction of the N-point function in the generalized Veneziano model, *Phys. Lett.* **B28**, 485–488.

[FN70] Fairlie, D. B. and Nielsen, H. B. (1970). An analog model for KSV theory, *Nucl. Phys.* **B20**, 637–651.

[GS69] Goebel, C. J. and Sakita, B. (1969). Extension of the Veneziano form to n-particle amplitudes, *Phys. Rev. Lett.* **22**, 257–260.

[KN69a] Koba, Z. and Nielsen, H. B. (1969). Reaction amplitude for n-mesons a generalization of the Veneziano–Bardakci–Ruegg–Virasoro model, *Nucl. Phys.* **B10**, 633–655.

[KN69b] Koba, Z. and Nielsen, H. B. (1969). Manifestly crossing-invariant parameterization of n-meson amplitude, *Nucl. Phys.* **B12**, 517–536.

[KN69c] Koba, Z. and Nielsen, H. B. (1969). Generalized Veneziano model from the point of view of manifestly crossing-invariant parameterization, *Z. Phys.* **229**, 243–263.

[Nam70] Nambu, Y. (1970). Lectures at the Copenhagen Symposium, unpublished; 'Dual model of hadrons', EFI-70-07 February 1970, 14pp.

[Nie68] Nielsen, H. B. (1968). Cand Scient thesis (only existing copy is stored in the Niels Bohr Institute Library), formal advisor Joergen Kalckar, in reality more Ziro Koba.

[Nie70] Nielsen, H. B. (1970). An almost physical interpretation of the integrand of the n-point Veneziano model, preprint at Niels Bohr Institute, paper presented at the *XV Int. Conf. on High Energy Physics, Kiev, 1970* (see p. 445 in Veneziano's talk). There was an earlier version (Nordita Preprint, 1969) without the word 'almost' in the title.

[NO73] Nielsen, H. B. and Olesen, P. (1973). Vortex-line models for dual strings, *Nucl. Phys.* **B61**, 45.

[Sus70a] Susskind, L. (1970). Dual symmetric theory of hadrons I. *Nuovo Cimento* **A69S10**, 457–496.

[Sus70b] Susskind, L. (1970). Structure of hadrons implied by duality, *Phys. Rev.* **D1**, 1182–1186.

[Ven68] Veneziano, G. (1968). Construction of a crossing-symmetric, reggeon-behaved amplitude for linearly rising trajectories, *Nuovo Cimento* **A57**, 190–197.

23
From the S-matrix to string theory

YOICHIRO NAMBU

String theory traces its origin to the Veneziano model of 1968. It also happens that the Weinberg–Salam model was born about the same time. The latter has led to the successful Standard Model. The descendants of the former, on the other hand, are still struggling to be relevant to the real world in spite of their enormous theoretical appeal. Indeed there exist two pathways in the development of theoretical particle physics since its beginnings in the Thirties. I will call them the quantum field theory and the S-matrix theory respectively. In its historical lineage, the Standard Model belongs to the former, whereas the superstring theory belongs to the latter. Even though the former has turned out to be the Royal Road of particle physics, this was not entirely clear before its final triumph, and the latter has also played very important contributing roles which continue to this day. The purpose of this note is to follow this other pathway and discuss the topics that have influenced my thoughts.

In the Thirties, when nuclear physics was developing, there were uncertainties in the minds of physicists about the efficacy of quantum field theory which was still in its early stages of development. For one thing, quantum field theory had inherited the self-energy difficulty from classical theory although the degree of divergence was found to be milder. For another, the unknown nature of nuclear forces and the much higher energy range involved than in atomic phenomena made people suspect that quantum mechanics might fail in dealing with nuclear phenomena, just as the classical theory failed with atomic phenomena. It was under such circumstances that Heisenberg [Hei43, Hei44] (and others) introduced the S-matrix theory. He argued that quantum mechanics was reliable only far away from the nuclear region, and wanted to deal with general relations that hold for observable quantities where quantum mechanics was secure. His S-matrix describes the scattering amplitude for a process $a + b \to c + d$ in the form $\psi_{out} = S\psi_{in}$ with the unitarity constraint $SS^\dagger = 1$. Other properties, such as time reversal invariance and causality (i.e. analyticity) were further added to S, and the formalism was applied to such problems as nuclear reactions and propagation in wave guides (see for example, Tomonaga [Tom48]). Heisenberg argued that the S-matrix could also give information about the bound states

---

The Birth of String Theory, ed. Andrea Cappelli, Elena Castellani, Filippo Colomo and Paolo Di Vecchia.
Published by Cambridge University Press. © Cambridge University Press 2012.

through its poles as a function of energy. In the post-war explosion of new particles, the analyticity of the S-matrix as a consequence of causality was exploited extensively in the form of dispersion relations with a view to analyzing hadronic reactions.

In 1954 I joined the group of M. L. Goldberger, one of the founders of the modern dispersion approach, at the University of Chicago [GGT54]. We collaborated closely with G. F. Chew and F. Low at the University of Illinois, Urbana-Champaign. Goldberger the big-hearted boss, Chew the charismatic thinker, and Low the quick mind and able technician had considerable influence on me. The Fifties were a period of the search for internal symmetries, that is, horizontal relations among different hadrons of similar masses, as well as the search for dynamical principles that govern vertical relations as a function of mass and reaction energy. The dispersion relations were the first step for the latter purpose. Although the divergence difficulties of quantum field theory were resolved by renormalization, the faith in quantum field theory still seemed to be a bit shaky. For one thing, the perturbation theory was not adequate to pin down the nature of strong interactions. For another, the nature of the nonrenormalizable weak interaction sector was still unclear. I collaborated with Goldberger for immediate applications, but I also kept personal interests in confirming and generalizing the dispersion-like relations to many variables for many-point functions in quantum field theory. I was pleased when my work was quoted by Landau, who kindly received me in 1959 on my visit to Moscow.

A forward scattering amplitude $A(\omega)$ satisfies the generic relation

$$\operatorname{Re} A(\omega) = \frac{1}{\pi} P \int_{-\infty}^{\infty} \frac{\operatorname{Im} A(\omega)}{\omega' - \omega} dw' + C, \qquad (23.1)$$

where the integrand includes contributions from poles and cuts. The poles are due to stable single particles, and the cuts are from scattering states. The cut contributions are related by the optical theorem to the cross-section in the direct channel $(a + b \to c + d)$ for $\omega > 0$, and the 'crossed' channel $(a + \bar{c} \to d + \bar{b}$, or $a + d \to c + \bar{b})$ for $\omega < 0$, respectively. A simple even or odd property between positive and negative parts (the 'crossing relation' derived from quantum field theory), enables one to combine them and express the integral in ω^2 running from 0 to ∞. The most successful example of experimental tests was the case of pion–nucleon scattering (Anderson, Davidon and Kruse [ADK55]). The relation was then generalized to non-forward scattering amplitudes in the form of a double dispersion relation (the Mandelstam representation [Man58]) in terms of the direct and crossed channel variables

$$s = (p_a + p_b)^2, \qquad t = (p_a + p_c)^2, \qquad u = (p_a + p_d)^2,$$
$$\sum_i p_i = 0, \qquad s + t + u = \sum_i m_i^2, \qquad (23.2)$$

where the four-momenta p_i are all taken as incoming. All channels appear symmetrically in this formula, reflecting a duality between direct and crossed channels. Possible departures from conventional quantum field theory also emerged at this point.

Since the left-hand (t- or u-channel, $\omega < 0$) contributions correspond to particle exchange between a and b, i.e. the generalized Yukawa-type interaction, Chew [Che68] proposed to solve the dynamical problem of strong interactions as a self-consistent solution to the dispersion relations between the left-hand side (input) and the right-hand side (output). In the crossed channels the roles of left and right would be reversed. This led him to his bootstrap duality hypothesis that the hadrons are self-consistent composites of each other: the sum over the resonances in the s-channel should actually be equivalent to the sum over the t-channel exchanges.

In the meantime the number of mesons of higher mass and spin kept increasing as the accelerator energy went up. A new element Chew added to his attractive philosophy was the Regge pole formalism [Reg59, Reg60]. The sum of Feynman diagrams for the scattering amplitude for two particles from exchanges of a series of 'mesons' carrying the same internal quantum numbers but with increasing spin and mass may be turned into a sum over masses by regarding spin as a function of mass: $l = \alpha(m)$, which suggests a dispersion integral for the variable t. Regge applied the Sommerfeld–Watson transformation to the scattering amplitude A and obtained the asymptotic formula:

$$A(s,z) = \sum_{l=0}^{\infty}(2l+1)a_l(s)P_l(z) \sim s^{\alpha(t)-h}, \quad (23.3)$$

where

$$z = \cos\theta \sim 1 + t/s,$$

and h is the change in helicity, which describes the near-forward scattering amplitude at large energy. As it turned out, the so-called Regge trajectories $\alpha(s)$ for various quantum numbers all seemed to rise linearly in parallel with s, with a universal slope $\alpha' \sim 1$ per GeV, and intercept $\alpha(0) < 1$. An exception was the 'Pomeron' trajectory, named after Pomeranchuk, which carried no quantum numbers, had an intercept $\alpha(0) \sim 1$, the highest of all, and a slope $\alpha' \sim 1/2$. The intercept implied that the total cross-sections approached a constant ($\sim$50 mb) at large s. The Regge pole formalism had great impact and led to various testable consequences. For example, the dispersion integral for an amplitude may be viewed as consisting of the observed resonances and the background represented by Regge poles and the Pomeron. The bootstrap duality (Dolen, Horn and Schmid [DHS67]) would lead to the expression (finite energy sum rule):

$$A(s) = A_{reson} - \langle A_{reson}\rangle + A_{Regge} + A_{Pomer}, \quad (23.4)$$

which was tested successfully for the process $\pi^- + p \leftrightarrow \pi^0 + n$, with

$$\text{Im } A(\omega, t=0) \propto \sigma_{tot}(\pi^- p) - \sigma_{tot}(\pi^+ p). \quad (23.5)$$

Before coming to the Veneziano model, I would like to digress a bit and talk about a different but related subject that I was pursuing for a few years [Nam67]. As the steadily rising Regge trajectories suggested there was an infinite number of them, I thought of describing all of them synthetically in a single 'master equation' which would reproduce the observed mass spectrum, just as the Schrödinger equation did for the hydrogen atom. A particle in an infinitely rising harmonic potential would be a simple nonrelativistic example. But here one must deal with different masses and spins. Such attempts have been around since the Thirties. Unitary representations of noncompact groups like the Lorentz group are necessarily infinite dimensional. I spent a year reading up on group theory, and as an exercise wrote down wave equations using an infinite sum of finite nonunitary representations as well as single infinite-dimensional representations of a dynamical (or spectrum-generating) group containing the Lorentz group. The results were disappointing for my purposes.

A typical example is an equation originally proposed by Majorana [Maj32] as an alternative to the Dirac equation in order to avoid negative energy states. It uses a unitary representation of $SO(3,2)$, and substitutes the Dirac gamma matrices with the generators for rotation between the fifth direction and the first four. The mass spectrum is inversely proportional to spin, going down to zero asymptotically. It is actually a generic property of a linear Hamiltonian of the Dirac type,

$$H = A \cdot p + Bm, \tag{23.6}$$

which is Hermitian, Lorentz invariant, and has only positive rest mass states. This may be understood as follows. For zero momentum, diagonalize B, and it gives a spectrum of positive rest masses. Take one of them, say m_1, and compute its nonrelativistic kinetic energy K_1 for momentum p_x using perturbation theory:

$$K_1 = \sum_n \frac{|\langle m_n|A|m_1\rangle|^2}{m_1 - m_n} p_x^2. \tag{23.7}$$

One expects this to yield $p_x^2/2m_1 > 0$. For this to be true, at least one of the intermediate states, say m_2, must have its mass less than m_1. Then going to state m_2 and repeating the process, one finds that the mass spectrum must go down indefinitely to a limit ≥ 0. The situation can be different if the equation is of different structure from Eq. (23.6). I discovered that the Schrödinger equation for hydrogen, having $SO(4,2)$ as a dynamical group, could be recast in such form and reproduce its energy spectrum algebraically (the bound state spectrum goes up to 0 asymptotically). This approach became a minor industry for a while, and I competed intensely with others, but it had to be abandoned eventually. The reason was that infinite-dimensional representations as a field theory were intrinsically nonlocal, and violated the proven or desired general properties of quantum field theory, like CPT, micro-causality, absence of tachyons, crossing symmetry, etc.

I became disappointed with my efforts along infinite-dimensional representations and was about to call it quits when there popped up Veneziano's work [Ven68]. This also described an infinite number of states in a miraculous beta function formula which manifestly embodied the duality. I decided to find out what physics was behind it. My first questions were (1) whether it was possible to factorize the poles in the formula into Breit–Wigner resonances, (2) how many states there were at each pole and what their spins were, and (3) whether or not their residues were all positive so that the states were physical. At that time Paul Frampton arrived as a fresh postdoc from Oxford, so we started to investigate together. Doing this for the first few states and for asymptotically large mass states, we were reasonably convinced that the states were all physical, and also found that the degeneracy of states increased exponentially with energy. We published our results in a volume dedicated to our senior theorist Gregor Wentzel who was retiring at that time [NF70].

The next question was then what kind of physical system was behind the Veneziano formula. In the meantime there appeared the Koba–Nielsen representation of the beta function (Koba and Nielsen [KN69]),

$$B(\alpha(s), \alpha(t)) = \int_0^1 (1-x)^{-\alpha' t - \alpha_t - 1} x^{\alpha' s - \alpha_s - 1} dx, \tag{23.8}$$

for linear trajectories. Factorizing the amplitude means rewriting the t-dependence as a sum over products of a function depending on p_1 times one depending on p_2. After several trials I did the following manipulation:

$$(1-x)^{-\alpha' t - \alpha_t - 1} = \exp[-(\alpha' t + \alpha_t + 1)\ln(1-x)]$$
$$= \exp[\alpha'(p_1 \cdot p_2 - C)(x + x^2/2 + x^3/3 + \cdots)], \tag{23.9}$$

where $C = m_1^2 + m_2^2 + \alpha_t + 1$. Then I recalled the contraction formula for a field ϕ in terms of creation and annihilation operators,

$$\phi(\xi) = \sum_k \frac{1}{\sqrt{2E_k}} a_k \exp(ik \cdot \xi) + \frac{1}{\sqrt{2E_k}} a_k^\dagger \exp(-ik \cdot \xi),$$

$$[a_k, a_l^\dagger] = \delta_{kl}, \qquad \langle \exp(\lambda a) \exp(\mu a^\dagger) \rangle = \exp(\lambda \mu), \tag{23.10}$$

$$\langle \phi(x)\phi(y) \rangle = \sum_k \frac{1}{2E_k} \exp[ik \cdot (x-y)].$$

So the p-dependent part of Eq. (23.10) could be re-expressed in terms of the operators $a_n^\mu, a_n^{\mu\dagger}$ for the Fourier components of a vector field $\phi^\mu(\xi)$, with internal coordinate ξ and discrete energy spectrum $E_n = (n + \frac{1}{2})/(2\alpha'), n = 0, 1, 2, \ldots$:

$$\exp(\alpha' p_1 \cdot p_2 \ln(1-x)) = \langle \exp(ip_1 \cdot \phi(0)) \exp(-ip_2 \cdot \phi(x)) \rangle. \tag{23.11}$$

The x-integration would then bring down the energy denominator in a propagator corresponding to the squared total energy $s = \sum_n E_n$ (incidentally the use of creation and annihilation operators was also a trick I learned in handling infinite-component equations). The spectrum immediately suggested a one-dimensional harmonic oscillator system, like an oscillating string of some length moving in four dimensions, whose points $x^\mu(\sigma, \tau) = \phi^\mu(\sigma, \tau)$ could be labelled by intrinsic spacetime coordinates σ and τ. Equation (23.11) could be interpreted as a string getting excited by an external impact p_1 hitting an end of it, and subsequently de-excited by emitting an amount p_2. The number of states for a given total energy E indeed increased exponentially, reminding one of the Hagedorn model of hadrons. Since the external hadrons should also be strings, I formed a picture that the scattering is a process of two incoming strings joining ends and separating again. There was still the problem of how to deal with the term C dependent on masses and the intercept in Eq. (23.9) in a similar fashion. I imagined expressing it in terms of coordinates in extra spacetime dimensions, as a mathematical trick. At that time, however, I had not connected it with the Kaluza–Klein theory. I presented the results at a conference at Wayne University in 1969, and the proceedings came out the next year [Nam70a].

I do not quite remember in detail what I did after this initial idea of the string. The general picture was quite appealing. A meson was a string with quarks attached to its two ends and carrying the quantum numbers. They annihilate each other when two strings join, and are created when they separate, but each quark cannot be isolated, like the two poles of a magnet. For baryons there were two possibilities for assigning a chemical bond structure. Those points were developed and elaborated by various people, too. But I also encountered difficulties. The fifth term in the factorization was still a problem. Another was the form factors of hadrons. The wave function of a string being of the harmonic oscillator type, the charge distribution due to the quarks is necessarily Gaussian instead of the observed power type. I struggled for a year to fix the difficulties without success.

In the next year, 1970, I received an invitation to a summer school in Copenhagen organized by Koba (my old friend), Nielsen, and others. The traditional High Energy Physics Conference was also scheduled in Kiev, and I planned to attend both. I drafted my lecture notes for the summer school and sent them to Copenhagen in advance, to be refined later. Since this was for lectures, I was relaxed and wrote down my half-baked ideas in a casual fashion. One such idea was the world-sheet area action:

$$I = \int (d\sigma_{\mu\nu} d\sigma^{\mu\nu})^{1/2}. \tag{23.12}$$

A quadratic action for the string did not have a nice geometric meaning and I thought of imitating the world-line of a point particle. But I did not have time to investigate how to handle the square root in the area formula.

Unfortunately I was not able to attend the conferences. Before going to Europe I drove to the West to leave my family in California with a friend of ours and fly from San Francisco. But our car engine broke down when crossing Death Valley, and it took three days to fix it there. Having missed the flight, we stayed in California for a couple of weeks and returned to Chicago. I was in a depressed mood and did not make an effort to follow up on my ideas. My notes would be published in the proceedings anyway, so I thought, but this did not happen. I must thank the good will of fellow physicists who kindly attached my name to the area action even though few people may have seen the notes [Nam70b].

I will stop here, and skip the significant contributions of many which continued for some years in refining the string model. I myself tried hard as did my students and associates (L. Clavelli, J. Willemsen, L. N. Chang and F. Mansouri in particular). In the summer of 1974 many of the string theorists of the day got together in a workshop in Aspen, Colorado. A few years ago, K. Kikkawa, one of the prominent contributors to the theory and a participant in the workshop, told me that he had felt then that the string as a viable mathematical theory of hadrons had to be given up, although it could capture some prominent qualitative features of hadron physics that eventually should be the consequences of QCD.

In preparing this Chapter, I consulted *Dual Resonance Models* by P. Frampton (W. A. Benjamin, 1974) and an article by V. Fitch and J. Rosner (*Twentieth Century Physics*, Vol. II, Chapter 9, ed. Brown, L. M. *et al.*, American Institute of Physics, New York, 1995), but I apologize for following mostly my own thinking on the string model as I remember it and not making due references to contributions by other people.

Yoichiro Nambu was born in Tokyo in 1921. He graduated from the University of Tokyo in 1942. After the war he joined the group of people working on renormalization theory with S. Tomonaga. In 1951 he moved to Osaka City University to set up a theoretical group. In 1952 he moved to the Institute for Advanced Study in Princeton, and in 1954 to the University of Chicago, where he is currently Emeritus Professor. In 2008 he was awarded the Nobel Prize in physics.

References

[ADK55] Anderson, H. L., Davidon, W. C. and Kruse, U. E. (1955). Causality in the pion-proton scattering, *Phys. Rev.* **100**, 339–343.
[Che68] Chew, G. F. (1968). 'Bootstrap': a scientific idea?, *Science* **161**, 762–765.
[DHS67] Dolen, R., Horn, D. and Schmid, C. (1967). Prediction of Regge parameters of ρ poles from low-energy πN data, *Phys. Rev. Lett.* **19**, 402–407.
[GGT54] Gell-Mann, M., Goldberger, M. L. and Thirring, W. (1954). Use of causality conditions in quantum theory, *Phys. Rev.* **95**, 1612–1627.
[Hei43] Heisenberg, W. (1943). Die beobachtbaren grössen in der theorie der elenemteilchen: I, *Z. Phys.* **120**, 513–538; Die beobachtbaren grössen in der theorie der elenemteilchen: II, *Z. Phys.* **120**, 673–702.
[Hei44] Heisenberg, W. (1944). Die beobachtbaren grössen in der theorie der elenemteilchen: III, *Z. Phys.* **123**, 93–112.
[KN69] Koba, Z. and Nielsen, H. B. (1969). Reaction amplitude for n-mesons a generalization of the Veneziano–Bardakci–Ruegg–Virasoro model, *Nucl. Phys.* **B10**, 633–655.
[Maj32] Majorana, E. (1932). Teoria relativistica di particelle con momento intrinseco arbitrario, *Nuovo Cimento* **9**, 335–344.
[Man58] Mandelstam, S. (1958). Determination of the pion-nucleon scattering amplitude from dispersion relations and unitarity. General theory, *Phys. Rev.* **112**, 1344–1369. Incidentally, the representation was known to K. Symanzik who was a visitor at Chicago in 1957.
[Nam67] See, for example, Nambu, Y. (1967). *Proceedings of 1967 Int. Conf. on Particles and Fields, Rochester 1967*, ed. Hagen, C. R. *et al.* (John Wiley, New York), 347.
[Nam70a] Nambu, Y. (1970). Quark model and the factorization of the Veneziano amplitude, in *Proceedings of the International Conference on Symmetries and Quark Models, Wayne State University, June 18–20, 1969*, ed. Chand, R. (Gordon and Breach, New York), 269–277, reprinted in *Broken Symmetry, Selected Papers of Y. Nambu*, ed. Eguchi, T. and Nishijima, K. (World Scientific, Singapore, 1995), 258–277.
[Nam70b] Nambu, Y. (1970). Duality and hadrodynamics, lecture notes prepared for Copenhagen summer school, 1970, reproduced in *Broken Symmetry, Selected Papers of Y. Nambu*, ed. Eguchi, T. and Nishijima, K. (World Scientific, Singapore, 1995), 280.

[NF70] Nambu, Y. and Frampton, P. (1970). Asymptotic behavior of partial widths in the Veneziano model, in *Quanta, a Collection of Essays dedicated to Gregor Wentzel*, ed. Freund, P. G. O., Goebel, C. and Nambu, Y. (University of Chicago Press, Chicago, IL), 403–414.

[Reg59] Regge, T. (1959). Introduction to complex orbital momenta, *Nuovo Cimento* **14**, 951–976.

[Reg60] Regge, T. (1960). Bound states, shadow states and Mandelstam representation, *Nuovo Cimento* **18**, 947–956.

[Tom48] Tomonaga, S. (1947). A general theory of ultra-short wave circuits I, *J. Phys. Soc. Jpn.* **2**, 158–171. I first learned of the S-matrix through his classified paper while serving in an army radar laboratory in 1944.

[Ven68] Veneziano, G. (1968). Construction of a crossing-symmetric, Reggeon-behaved amplitude for linearly rising trajectories, *Nuovo Cimento* **A57**, 190–197.

24

The analogue model for string amplitudes

DAVID B. FAIRLIE

24.1 The beginnings

I started research in the year 1957, at the time when aspiring particle physicists were being channelled into the arid field of dispersion relations, and field theory was out of fashion. I did not find the methods of analysis employed in trying to extract information out of dispersion relations to my taste. The only tool available was the analyticity of the S-matrix, as constrained by the requirements of causality, that there should be no output before input. To give an instance of the attitude to mathematics at the time, we graduate students were advised that the only pure mathematical courses worth attending were those on functional analysis, or the theory of several complex variables! The philosophy of logical positivism reigned supreme, in which one was not allowed to talk about the unobservable features of particle interactions, but only about properties of asymptotic states. This was one of the features which inhibited the invention of the concept of quarks. I had been impressed by the tractability of electrodynamics and quantum mechanics as an undergraduate, and what Wigner has called 'The unreasonable effectiveness of Mathematics in the Natural Sciences'. In the middle of year 1968 I was feeling very pessimistic about the possibility of theorists ever being able to say anything about scattering amplitudes for hadrons, beyond the simple tree and Regge pole approximations, and was contemplating changing fields. However, to everyone's complete surprise Gabriele Veneziano came up with his famous compact form for a dual scattering amplitude, which encompassed contributions from many towers of resonances, and I felt that this was for me!

The amplitude which describes the scattering of four identical scalar particles, $A(s, t, u)$ is given by the sum of three terms,

$$A(s, t) + A(s, u) + A(t, u), \qquad (24.1)$$

where s, t, u are the energies in the three possible ways of looking at the scattering process; if the process is considered as one where the initial momenta are p_1^μ and p_2^μ and the final ones are $-p_3^\mu$ and $-p_4^\mu$ then $s = (p_1 + p_2)^2$, $t = (p_1 + p_3)^2$, and $u = (p_1 + p_4)^2$. Each

---

The Birth of String Theory, ed. Andrea Cappelli, Elena Castellani, Filippo Colomo and Paolo Di Vecchia.
Published by Cambridge University Press. © Cambridge University Press 2012.

contribution can be expressed as an integral representation,

$$A(s,t) = \int_0^1 x^{-\alpha(s)-1}(1-x)^{-\alpha(t)-1} dx = \frac{\Gamma(-\alpha(s))\Gamma(-\alpha(t))}{\Gamma(-\alpha(s)-\alpha(t))}, \quad (24.2)$$

with $\alpha(s) = \alpha_0 + \alpha' s$. The result is that the Veneziano amplitude [Ven68] with these linear trajectories implies the existence of an infinite set of poles with multiple degeneracies. The Veneziano amplitude displays the property of duality; the same amplitude may be expressed as a sum of s-channel poles with residues decomposable into positive angular functions of the scattering angle θ given by

$$t = -2p^2(1-\cos\theta), \quad (24.3)$$

where p is the centre-of-mass momentum, or a similar sum of t-channel poles.

I hit my stride in my research career after the advent of the Veneziano dual amplitude when I realized that my undergraduate education at Edinburgh University in mathematical physics would perhaps enable me to contribute to the development of these ideas where there was the hope of finding something tractable. There was the hope that the Veneziano expression would be the first term in a perturbation expansion with rapid convergence. Two avenues of research developed out of this; one was the operator approach to the subject, the foundations to which were laid by Veneziano himself, together with Fubini and Gordon [FGV69, FV69]. This arose from the realization of the set of poles of the four-point amplitude in terms of an infinite set of oscillators.

The other line of inquiry lay in the direction of generalizing the integral representation of the amplitude (24.2): Bardakci and Ruegg [BR69a, BR69b] gave an integral representation for the five-point function, which they themselves, and Chan and Tsou [CT69], generalized further to the N-point amplitude. I became interested in these activities and believe that Keith Jones and I were the first to notice the tachyon condition in a paper in which we showed that if one imposes the (unphysical) requirement that the ground state is a tachyon (i.e. possesses a particle of negative mass squared), then the four- and five-point amplitudes can be expressed as integrals of a single integrand over the whole of the real line and the plane respectively [FJ69].

I had to contend with a sceptical attitude from some of my colleagues at Durham who favoured the so-called interference model, in which s- and t-channel contributions should be added, as in Feynman tree diagrams, rather than being encoded into a single expression of an amplitude exhibiting duality. In earlier times the discovery that the ground state is a tachyon would have caused the theory to be abandoned, but as progress in physics has become more difficult, such qualms are brushed aside in the hope of a later resolution of the problem, as indeed Gliozzi, Scherk and Olive [GSO77] found a way of projecting out the tachyonic sector. An analogous situation occurred earlier, with Yang–Mills theory, which was ignored for about 20 years because of the realization that the particles described by the theory had zero mass, and thus could not describe strongly interacting particles of spin one. The problem was resolved by the discovery by Brout and Englert and Higgs of a mechanism to give mass to the particles of the theory.

The big advance came with the marvellous formula of Koba and Nielsen [KN69] giving an elegant form for the N-point tree amplitude:

$$A(s,t) = \int_{-\infty}^{\infty} \frac{d^N z}{dV_{abc}} \prod_{1}^{N-1}(z_i - z_{i+1})^{\alpha_0-1} \prod_{1}^{N-1} \theta(z_i - z_{i+1}) \prod_{i<j}(z_i - z_j)^{2\alpha' p_i \cdot p_j}, \quad (24.4)$$

with

$$dV_{abc} = \frac{dz_a dz_b dz_c}{(z_b - z_a)(z_c - z_a)(z_a - z_c)}. \quad (24.5)$$

This integration measure is introduced as a consequence of conformal invariance: the integrand is covariant under transformations of the Möbius group,

$$z' \mapsto \frac{az+b}{cz+d}, \qquad ad - bc = 1, \quad (24.6)$$

and one can fix the corresponding degeneracy in the integration by eliminating three variables (z_a, z_b, z_c) and adding a Jacobian factor. Note that the tachyon condition $\alpha_0 = 1$ makes the amplitude completely permutation invariant in all sub-energies, the generalization to amplitudes with any number of external legs of the result of [FJ69].

As an aside, while I was thumbing through 'Whittaker and Watson' [WW27] looking for inspiration and information, I came across an exercise in the chapter on hypergeometric functions, which might be paraphrased as: 'Prove that the 5 point function may be expressed in terms of the Hypergeometric Function $F_{3:2}$', referring to a paper by Dixon [Dix05] in 1905! Also the Koba–Nielsen variables first appeared in representations of cross-ratios and the permutation group by Moore and (independently) Burnside at the end of the nineteenth century. Edward Witten was fond of quoting that 'String theory is a piece of mathematics which has fallen out of the twenty first century into the twentieth'. It has seemed to me more like something dragged out of the nineteenth!

A parallel development was the study by Nambu [Nam70] and independently Goto [Got71] of the propagation of an object with a one-dimensional extension, i.e. a string instead of a particle. As a particle moves in spacetime, it traces out a curve and the action may be described by the reparameterization invariant expression

$$S = \int \sqrt{\frac{dx^\mu}{d\tau} \frac{dx_\mu}{d\tau}} d\tau. \quad (24.7)$$

A string sweeps out a world-sheet in spacetime, and the action is proportional to the area swept out by the sheet:

$$S = \int \sqrt{\left(\frac{\partial x^\mu}{\partial \tau} \frac{\partial x_\mu}{\partial \sigma}\right)^2 - \left(\frac{\partial x_\mu}{\partial \sigma}\right)^2 \left(\frac{\partial x^\nu}{\partial \tau}\right)^2} d\sigma d\tau. \quad (24.8)$$

The striking resemblance of this process to the duality diagrams of the Dual Resonance Model led to the quest for a closer connection, and an interpretation of the Koba–Nielsen amplitude. Nielsen and I developed what we called the 'analogue model'. He was motivated by path-integral representations of the string amplitudes, while I sought for an interpretation of the Koba–Nielsen amplitude in terms of an analogue in electrostatics of currents in a plate of uniform resistivity, identifying the components of momentum of each particle with an electric current entering the plate at an edge.

I hit upon the idea as a result of an advertizement for Philips in *Scientific American* in 1964 (see Figure 24.1), in the form of a research report on conformal methods in potential theory. The key property is that the shape of the plate does not matter, as long as it can be mapped conformally to a disc.

I was very familiar with the use of complex analysis in solving two-dimensional electrostatic problems as an undergraduate from Jeans' book on electromagnetism [Jea20]. The idea was a method for computing the structure of the amplitudes corresponding to the Feynman diagrams for the ground state scattering in string theory by means of an electrical analogue in which the amplitude is related to the heat generated in a plate of uniform resistivity corresponding to the world-sheet associated with each diagram, with currents related to the components of particle four-momenta.

Holger Nielsen had a rather more physical approach to the same idea, more closely related to the path-integral formalism, which he talked about at the Kiev conference in 1970. I did not speak at that conference as I assumed that one contribution on this subject would be sufficient. We got together and wrote a paper describing this idea to which I gave the deplorably recondite title, 'An analogue model for KSV theory' [FN70]! This paper demonstrated that the Veneziano amplitude describes the elementary process of string scattering, and reproduced the Koba–Nielsen multiparticle amplitude for many particles.

We also computed the one-loop contribution up to a measure and thus opened up the possibility of calculating a string perturbation theory. This one-loop amplitude had been evaluated previously, using the methods of [CT69] by Kikkawa, Sakita and Virasoro [KSV69]. This was then followed by the work of Alessandrini and Amati [AA71] and Lovelace [Lov70], who extended the calculation to the case where the world-sheet of the string has arbitrary genus, invoking the Baker Akhiezer multiply periodic functions on a Riemann surface.

Leonard Susskind described the amplitudes in the Dual Resonance Model with string scattering [Sus70b] at about the same time. In the text he acknowledges a Nordita preprint of Holger Nielsen in which Nielsen describes the fishnet approach to dual resonance theory, and takes the continuum limit. In earlier papers Susskind [Sus69, Sus70a] employs more of an operator approach to calculating amplitudes.

It is the case that as an extended model of particles Dirac [Dir62] in 1962 somewhat anticipated these ideas with his attempt to describe the electron in terms of a membrane model, with the muon as an excited state. David Olive and I were present at a seminar in Cambridge which Dirac gave about the subject, and I well remember the atmosphere of indulgent scepticism with which his talk was received!

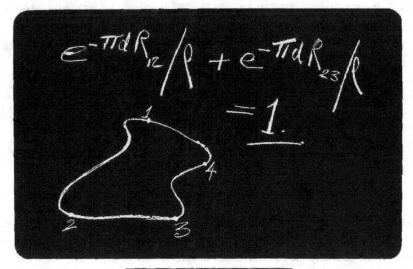

Figure 24.1 Philips advertizement.

24.2 Operator methods

As already mentioned, Veneziano himself, together with Fubini and Gordon laid the foundations of the operator approach to the subject in [FGV69, FV69]. I shall not say much about this, as I expect that it will be comprehensively treated elsewhere in this Volume. The excellent review of Paolo Di Vecchia [DiV07] contains a detailed discussion of the operator approach to the calculation of amplitudes, starting with the papers of Fubini, Gordon and Veneziano [FGV69], and Susskind [Sus69]. The idea is to introduce an operator

$$Q_\mu = Q_\mu^{(+)}(z) + Q_\mu^{(0)}(z) + Q_\mu^{(-)}(z), \tag{24.9}$$

with

$$Q_\mu^{(+)} = i\sqrt{2\alpha'} \sum_{n=1}^{\infty} \frac{a_n}{\sqrt{n} z^{-n}}, \qquad Q_\mu^{(0)} = \hat{q} + 2\alpha' \hat{p} \log(z), \tag{24.10}$$

and $Q_\mu^{(-)}(z)$ the complex conjugate of $Q_\mu^{(+)}(z)$.

Corresponding to the external leg with momentum p, a vertex operator

$$V(z; p) =: \exp(ip.Q): \tag{24.11}$$

is introduced which serves to create a string in terms of creation operators $a_n^\dagger$. Manipulation of the formalism shows that the vacuum expectation value of a product of such vertex operators gives essentially the Koba–Nielsen integrand:

$$\left\langle 0, 0 \middle| \prod_{i=1}^{N} V(z_i; p_i) \middle| 0, 0 \right\rangle = (2\pi)^d \delta\left(\sum_{i=1}^{N} p_i\right) \prod_{i<j} (z_i - z_j)^{-2\alpha' p_i.p_j}. \tag{24.12}$$

The analogue and operator methods were complementary: in the former it is easy to compute the integrand in an integral representation of the amplitude, but difficult to compute the measure over conformally equivalent diagrams; in the latter, the measure is easy, but the integrand is more difficult to calculate.

The operator method gained prominence among workers, partly, I suspect, because it is related more directly to field-theoretic Fock space quantization. It also led to the fermionic string; Pierre Ramond introduced the integrally moded d_n^μ fermionic operators to generalize the Dirac construction to fermionic string states [Ram71] and André Neveu and John Schwarz [NS71] employed half integrally moded $b_{n+1/2}^\mu$ fermionic oscillators, with the original motivation of describing pseudoscalar bosonic states. The operator approach also permitted the scattering of states in the string other than the ground state to be calculated easily.

However, Mandelstam used the analogue method to give the first calculation of the scattering amplitude for four dual fermion states [Man73] and later in the mid-Eighties used it again in his proof of finiteness of string perturbation theory. At about the same time, Corrigan, Goddard, Olive and Smith calculated the same process using the operator method [CGOS73].

24.3 Developments at CERN

In the early years of the Seventies beginning in 1970, Daniele Amati assembled most of the active European string theory workers at CERN. The high points of the next three years were the papers of Goddard and Thorn, and independently Richard Brower, proving the no-ghost theorem, the papers of Alessandrini and Amati, and of Lovelace, extending the analogue calculation to world-sheets of arbitrary genus, and the famous paper of Goddard, Goldstone, Rebbi and Thorn quantizing the bosonic string in 26 dimensions [GGRT73].

This was one of the most remarkable papers of post-war years in my opinion. By working in the light-cone gauge these authors were able to show that the Lorentz algebra constructed out of the oscillators would close only for strings moving in 26 dimensions of spacetime provided that the ground state was a tachyon (Lovelace had already shown that 26 was a significant dimension, as that in which the Pomeron singularity becomes a pole, rather than a cut, ensuring unitarity). In other words two conditions had to be satisfied.

I have always felt the Koba–Nielsen amplitude, which takes the form

$$\int d^n z \prod_{i<j} |z_i - z_j|^{-2p_i \cdot p_j}, \tag{24.13}$$

to be a very fundamental structure in string theory, and the calculation for a simply connected sheet gives this answer. The replacement of the distance function $|z_i - z_j|$ by $|z_i - z_j + \theta_i \theta_j|$, where the θ_i are Grassmannian coordinates in the Koba–Nielsen amplitude, with subsequent integration over these additional variables was shown by David Martin and myself in [FM73, FM74] to give rise to the Neveu–Schwarz tree and one-loop amplitudes. These papers incorporating superconformal invariance were early examples of the use of superfields.

The Ademollo et al. collaboration [ABDD76a, ABDD76b] had constructed a model with nontrivial internal symmetries. As Bruce, Yates and I recognized from a preprint version of their paper, and they confirmed in the published version, these all possessed ghosts except for the Abelian $U(1)$ case [FBY76].

One of the goals of the earlier period was to construct an off-shell theory, so that the string states could couple to currents. Edward Corrigan and I had a solution to this problem motivated by the analogue approach, by the introduction of Dirichlet boundary conditions; in our paper [CF75] we used both the analogue and the operator methods to construct amplitudes with the correct properties. A general bosonic state can be expressed as

$$x^\mu(\sigma, \tau) = q + 2ip\tau + \sum a_n^\mu \exp(in\tau) \cos(n\sigma), \tag{24.14}$$

with integrally moded a_n^μ bosonic oscillator creation operators; as has already been mentioned, Ramond introduced the integrally moded d_n^μ fermionic operators, and Neveu and Schwarz employed half integrally moded $b_{n+1/2}^\mu$ fermionic oscillators to describe pseudoscalars. It is obvious that what is missing to complete the table is a set of half integrally moded bosonic $c_{n+1/2}^\mu$! (see Table 24.1). These are just what is required to construct the off-shell states.

John Schwarz had described similar off-shell amplitudes earlier [Sch73], but without the half integrally moded bosonic oscillators. One of the features of our paper was that the string would stop at a finite point of spacetime and latch on to a current, or a zero-brane in present day jargon. In our idea, the stopped strings would then interact with currents.

Table 24.1 *Modes*

boson	fermion
a_n^μ	$b_{n+1/2}^\mu$
$c_{n+1/2}^\mu$	d_n^μ

This paper was written at the time of the first demise of string theory and nobody took any notice of it except for Michael Green and Joe Polchinski, who extended our theory to closed strings where the brane interpretation of the boundary is more transparent.

Almost 30 years later, Polchinski also came up with an idea which we should kick ourselves for not having thought of: namely, to impose Dirichlet boundary conditions in only a subset d of the dimensions and thus invent the concept of D-branes. In our earlier paper, we encountered a paradox; it seemed that in the bosonic case the critical dimension ought to be 16 instead of 26, till Corrigan noticed that if we included the fermionic sector then the dimension was the expected 10.

I returned to string theory with its revival in the Eighties, working with Corinne Manogue. In [FM86, FM87] we showed how the solution of the classical equations of motion of the superstring and constraints could be encoded in terms of octonions, with the aim of tackling quantization, a question which she addressed later with Anthony Sudbery.

One last paper which turned out to be of unexpected relevance, about ten years after it was written, was the one constructing a self-dual tensor in eight dimensions, which is invariant under $U(7)$ (Corrigan, Devchand, Fairlie and Nuyts [CDFN82]), and has some part to play in considerations involving manifolds of exceptional holonomy. It has always seemed to me that every revival of string theory, with its consequent reinterpretation, has taken it further away from physics than its initial conception as a theory of hadrons and their excitations. The significant feature which has been the focus of more recent research is one which was already realized in the first five years, that is the prospect of realizing a finite quantum theory of gravity. There are some who believe that exceptional structures in mathematics will have some realization in the physical world and I go along a bit of the way with this philosophy. The essential role which supersymmetry plays in the present formulations of the theory simply enhance the tractability of the theory, or may represent a feature of the physical world yet to be confirmed by experiment.

In this Chapter I have tried to describe the developments of the origins of string theory as they affected me, my students and colleagues. I have always been attracted by simple ideas. It may appear that my choice of topics to emphasize is somewhat idiosyncratic, but I believe that in a Volume dedicated to the contributions of various authors different perspectives will round out the full picture.

Reading the bibliographies of contributers to the first five years of development of string theory, or the Dual Resonance Model as it was generally called, I am astonished how prolific some have been. What is remarkable about these first few years of string theory is how many of the subsequent developments have been foreshadowed in this fruitful period.

It also led to the development of supersymmetry in the context of four-dimensional field theory, and in particular to the formulation of supergravity theory.

The change in perspective in 1984, reinterpreting the basic states of the theory as electrons, quarks, photons and gravitons etc. by pushing the excitation energy to the Planck mass has swept the excited states away into the realms of unobservability, in contrast to the original interpretation as a theory of hadrons, the excited states being identified with the Regge recurrences of the basic hadrons, the linearity of which was in approximate agreement with the experimental resonances. In its resurrected form the only function of the excited states is to guarantee finiteness of the perturbative contributions. It is my hope that some variant of the model will be discovered which allows the earlier interpretation to be reinstated as an approximate foundation for constructing states of QCD.

David B. Fairlie was born in South Queensferry, Scotland in March 1935. He received his BSc in mathematical physics in Edinburgh in 1957, and his PhD in Cambridge in 1960, under the supervision of John Polkinghorne. After postdoctoral experience at Princeton and Cambridge, he was lecturer in St Andrews (1962–1964), and at Durham University (1964), retiring as Professor in 2000.

References

[ABDD76a] Ademollo, M., Brink, L., D'Adda, A., D'Auria, R., Napolitano, E., Sciuto, S., Del Giudice, E., Di Vecchia, P., Ferrara, S., Gliozzi, F., Musto, R., Pettorino, R. and Schwarz, J. (1976). Dual string with $U(1)$ colour symmetry, *Nucl. Phys.* **B111**, 77–110.

[ABDD76b] Ademollo, M., Brink, L., D'Adda, A., D'Auria, R., Napolitano, E., Sciuto, S., Del Giudice, E., Di Vecchia, P., Ferrara, S., Gliozzi, F., Musto, R. and Pettorino, R. (1976). Dual string models with non-Abelian color and flavor symmetries, *Nucl. Phys.* **B114**, 297–316.

[AA71] Alessandrini, V. and Amati, D. (1971). Properties of dual multiloop amplitudes *Nuovo Cimento* **A4**, 793–844.

[BR69a] Bardakci, K. and Ruegg, H. (1969). Meson resonance couplings in a 5-point Veneziano model, *Phys. Lett.* **B28**, 671–675.

[BR69b] Bardakci, K. and Ruegg, H. (1969). Reggeized resonance model for arbitrary production processes, *Phys. Rev.* **181**, 1884–1889.

[FBY76] Bruce, D. J., Fairlie, D. B. and Yates, R. G. (1976). Dual models with a color symmetry, *Nucl. Phys.* **B108**, 310–316.

[CT69] Chan, H.-M. and Tsou, T.-S. (1969). Explicit construction of n-point function in generalized Veneziano model, *Phys. Lett.* **B28**, 485–488.

[CDFN82] Corrigan, E. F., Devchand, C., Fairlie, D. B. and Nuyts, J. (1982). First order equations for gauge fields in spaces of dimension greater than four, *Nucl. Phys.* **B214**, 452–464.

[CF75] Corrigan, E. F. and Fairlie, D. B. (1975). Off-shell states in dual resonance theory, *Nucl. Phys.* **B91**, 527–545.

[CGOS73] Corrigan, E. F., Goddard, P., Olive, D. and Smith, R. A. (1973). Evaluation of the scattering amplitude for four dual fermions, *Nucl. Phys.* **B67**, 477–491.

[Dir62] Dirac, P. A. M. (1962). An extensible model of electron, *Proc. R. Soc. London* **A268**, 57–67.
[DiV07] Di Vecchia, P. (2007). The birth of string theory, in *String Theory and Fundamental Interactions*, ed. Gasperini, M. and Maharana, J. (Springer, Berlin).
[Dix05] Dixon, A. C. (1905). On a certain double integral, *Proc. London Math. Soc.* **2**, 8–15.
[FJ69] Fairlie, D. B. and Jones, K. (1969). Integral representations for the complete four and five-point Veneziano amplitudes, *Nucl. Phys.* **B15**, 323–330.
[FM86] Fairlie, D. B. and Manogue, C. A. (1986). Lorentz invariance and the composite string, *Phys. Rev.* **D34**, 1832–1834.
[FM87] Fairlie, D. B. and Manogue, C. A. (1987). A parameterization of the covariant superstring, *Phys. Rev.* **D36**, 475–479.
[FM73] Fairlie, D. B. and Martin, D. (1973). New light on the Neveu–Schwarz model, *Nuovo Cimento* **18A**, 373–383.
[FM74] Fairlie, D. B. and Martin, D. (1974). Green's function techniques and dual fermion loops, *Nuovo Cimento* **A21**, 647–660.
[FN70] Fairlie, D. B. and Nielsen, H. B. (1970). An analog model for KSV theory, *Nucl. Phys.* **B20**, 637–651.
[FGV69] Fubini, S., Gordon, D. and Veneziano, G. (1969). A general treatment of factorization in dual resonance models, *Phys. Lett.* **B29**, 679–682.
[FV69] Fubini, S. and Veneziano, G. (1969). Level structure of dual resonance models, *Nuovo Cimento* **A64**, 811–840.
[GSO77] Gliozzi, F., Scherk, J. and Olive, D. (1977). Supersymmetry, supergravity theories and the dual spinor model, *Nucl. Phys.* **B122**, 253–290.
[GGRT73] Goddard, P., Goldstone, J., Rebbi, C. and Thorn, C. B. (1973). Quantum dynamics of a massless relativistic string, *Nucl. Phys.* **B56**, 109–135.
[Got71] Goto, T. (1971). Relativistic quantum mechanics of one dimensional mechanical continuum and subsidiary condition of the dual resonance model, *Prog. Theor. Phys.* **46**, 1560–1569.
[Jea20] Jeans, J. H. (1920). *The Mathematical Theory of Electricity and Magnetism* (Cambridge University Press, Cambridge).
[KSV69] Kikkawa, K., Sakita, B. and Virasoro, M. A. (1969). Feynman-like diagrams compatible with duality. I. Planar diagrams, *Phys. Rev.* **184**, 1701–1713.
[KN69] Koba, Z. and Nielsen, H. B. (1969). Reaction amplitude for n-mesons a generalization of the Veneziano–Bardakci–Ruegg–Virasoro model, *Nucl. Phys.* **B10**, 633–655.
[Lov70] Lovelace, C. (1970). M-loop generalized Veneziano formula, *Phys. Lett.* **B32**, 703–708.
[Man73] Mandelstam, S. (1973). Manifestly dual formulation of the Ramond model, *Phys. Lett.* **B46**, 447–451.
[Nam70] Nambu, Y. (1970). Quark model and the factorization of the Veneziano amplitude, in *Proceedings of the International Conference on Symmetries and Quark Models, Wayne State University, June 18–20, 1969*, ed. Chand, R. (Gordon and Breach, New York), 269–277, reprinted in *Broken Symmetry, Selected Papers of Y. Nambu*, ed. Eguchi, T. and Nishijima, K. (World Scientific, Singapore, 1995), 258–277.
[NS71] Neveu, A. and Schwarz, J. H. (1971). Factorizable dual model of pions, *Nucl. Phys.* **B31**, 86–112.
[Ram71] Ramond, P. (1971). Dual theory for free fermions, *Phys. Rev.* **D3**, 2415–2418.
[Sch73] Schwarz, J. (1973). Off-mass shell dual amplitudes without ghosts, *Nucl. Phys.* **B65**, 131–140.

[Sus69] Susskind, L. (1969). Harmonic-oscillator analogy for the Veneziano model, *Phys. Rev. Lett.* **23**, 545–547.

[Sus70a] Susskind, L. (1970). Structure of hadrons implied by duality, *Phys. Rev.* **D1**, 1182–1186.

[Sus70b] Susskind, L. (1970). Dual-symmetric theory of hadrons: 1, *Nuovo Cimento* **A69**, 457–496.

[Ven68] Veneziano, G. (1968). Construction of a crossing-symmetric, Reggeon-behaved amplitude for linearly rising trajectories, *Nuovo Cimento* **A57**, 190–197.

[WW27] Whittaker, E. T. and Watson, G. N. (1927). *A Course in Modern Analysis* (Cambridge University Press, Cambridge).

25
Factorization in dual models and functional integration in string theory

STANLEY MANDELSTAM

25.1 Introduction

This Chapter will mainly be concerned with work during the relevant period in which the author was involved, with a minimum of technicalities. Other contributions will be mentioned briefly in order to indicate the place of our work in the general development. Contributions are chosen in order to provide continuity with the contributions in which we were involved, and not for their importance. The historical development will be emphasized.

As every reader of this Volume probably knows, string theory can be traced back to the discovery by Veneziano [Ven68] of a formula for a four-point scattering amplitude (two incoming and two outgoing particles) with narrow resonances and rising Regge trajectories in both channels. Another such formula, for what are now known as closed strings, was proposed by Virasoro [Vir69a] and written as an integral by Shapiro [Sha70]. These formulae were then extended to five-point amplitudes independently by Bardakci and Ruegg [BR68] and by Virasoro [Vir69b] and then to N-point amplitudes independently by Bardakci and Ruegg [BR69], Chan and Tsou [CT69], Goebel and Sakita [GS69], and Koba and Nielsen [KN69a, KN69b]. For simplicity most of the treatment in the remainder of this Chapter will be for open strings; the closed string formulae will be very similar.

The most convenient expression for the general open string N-point amplitude is the following, which we quote here for future reference:

$$A = \int d^{N-3}z |(z_b - z_a)(z_c - z_b)(z_c - z_a)| \prod_{i>j} |z_i - z_j|^{-2 p_i p_j}. \tag{25.1}$$

The subscript i refers to the ith particle, with momentum p_i. The variables z_i are ordered cyclically along the real line. Three of them, which we have denoted by z_a, z_b, z_c, are chosen arbitrarily and held at arbitrary fixed values; we integrate over the other $N-3$ subject to the condition of cyclic ordering. By making use of the invariance of the amplitude (25.1)

The Birth of String Theory, ed. Andrea Cappelli, Elena Castellani, Filippo Colomo and Paolo Di Vecchia.
Published by Cambridge University Press. © Cambridge University Press 2012.

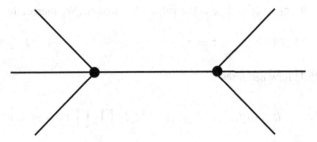

Figure 25.1 Factorization of scattering amplitudes at a pole.

under a projective transformation of the variables,

$$z' = \frac{az+b}{cz+d}, \tag{25.2}$$

we can easily see that the integral is independent of the choice and values of the constant z. We have written the formula for the case where the external particles are tachyons with μ^2, the square of their mass, equal to -1 in units of the slope of the Regge trajectories; the formula for the general case is slightly more complicated. At this stage of the development there is no reason for this choice of the mass, but later on we shall see that it is necessary in a consistent model.

The formula for the closed string amplitude is similar but not identical to (25.1); the integral is now over the whole complex plane, and the variables are not restricted by cyclic ordering.

If the variable $p_i \cdot p_j$ is sufficiently large, the above integral diverges when $(z_i - z_j)$ approaches zero. One can obtain finite results by starting from an algebraically smaller value of $p_i \cdot p_j$ and continuing analytically or, equivalently, by integrating by parts and dropping the end-point contribution. At the time, such a procedure was justifiable, since one was simply trying to obtain an S-matrix element with the correct analytic properties. We shall return to this point later.

25.2 Factorization

In order that the scattering amplitudes should together form a consistent S-matrix, they must satisfy a requirement known as factorization (the narrow-resonance equivalent of unitarity). The residue at the pole of an amplitude as a function of s, the square of the energy, should consist of a number of terms, one for each excited particle, with the square of its mass equal to s. Each term has to be the product of two factors, one depending only on the number of incoming particles and their momenta, the other depending on the corresponding variables of the outgoing particles. The factorizability of the amplitude (25.1) was shown independently by Bardakci and Mandelstam [BM69] and by Fubini and Veneziano [FV69]. In Figure 25.1, we define the variables corresponding to the particles on the left as $z_1, \ldots, z_j$, and those corresponding to the particles on the right as $z_{j+1}, \ldots, z_N$.

We fix the variables z_1, z_{j+1} and z_N at the values 0, 1, and ∞. We then define new variables

$$x_i = \frac{z_i}{z_j}, \quad 1 \le i \le j, \qquad z = z_j, \qquad y_k = \frac{z_{j+1}}{z_k}, \quad j+1 \le k \le N. \qquad (25.3)$$

The formula (25.1) now becomes:

$$A = \int dz\, dx_2 \ldots dx_{j-1} dy_{j+2} \ldots dy_{N-1} z^{-s-1} I_1 I_2 \prod_{i=1}^{j} \prod_{k=j+1}^{N} (1 - zx_i y_k)^{-2p_i \cdot p_k}, \qquad (25.4)$$

where

$$I_1 = \prod_{i=1}^{j} \prod_{k=i+1}^{j} (x_k - x_i)^{-2p_i \cdot p_k}, \qquad I_2 = \prod_{i=j+1}^{N} \prod_{k=i+1}^{N} (y_i - y_k)^{-2p_i \cdot p_k},$$

$$0 = x_1 \le x_2 \le \ldots \le x_j = 1, \qquad 1 = y_{j+1} \ge \ldots \ge y_{N-1} \ge y_N = 0,$$

and s is the square of the energy of the intermediate state in Figure 25.1.

To calculate the residue R at the pole in (25.4) when $s = n$, where n is a nonnegative integer, we first have to integrate by parts n times as described above. We then find:

$$R = -\frac{1}{n!} \int dx_2 \ldots dx_{j-1} dy_{j+2} \ldots dy_{N-1}$$

$$\times I_1 I_2 \frac{\partial^n}{\partial z^n} \prod_{i=1}^{j} \prod_{k=j+1}^{N} (1 - zx_i y_k)^{-2p_i \cdot p_k} \bigg|_{z=0}. \qquad (25.5)$$

By expanding the last factor in (25.5) in a power series and taking the coefficient of z^n, we find that R can be written as a sum of terms of the form:

$$c \int dx_2 \ldots dx_{j-1} dy_{j+2} \ldots dy_{N-1} I_1 I_2 \prod_r \sum_{i=1}^{j} \sum_{k=j+1}^{N} (2p_i \cdot p_k x_i^r y_k^r)^{\lambda_r}, \qquad (25.6)$$

where

$$\sum r \lambda_r = n. \qquad (25.7)$$

Equations (25.6) and (25.7) show that the residue is indeed a finite sum, each term corresponding to a state of an assembly of d-vector simple harmonic oscillators. The integer r labels the oscillator, the level spacing of the rth oscillator is proportional to r, and the energy of the ground state is equal to zero. In the term shown, the rth oscillator is in level λ_r. We have thus shown that the amplitude (25.1) does factorize as required, and the states of the system correspond to the assembly of simple harmonic oscillators described above. The factorization properties of the N-point amplitude were used independently by Fubini, Gordon and Veneziano [FGV69], by Nambu [Nam70] and by Susskind [Sus70a] to construct an operator formalism for the dual model, the operators being the creation and destruction operators for the above assembly of oscillators. These authors were able to construct the N-point amplitude in terms of such operators; the factorization properties were then obvious. In fact, the easiest way to describe dual models was seen to be to start from the operator formalism, thus reversing the historical order.

25.3 Further developments

Two very similar new dual models were proposed by Ramond [Ram71] (R), initially for free fermions, and by Neveu and Schwarz [NS71, NST71] (NS) for bosons; the latter model was slightly reformulated by Neveu, Schwarz and Thorn. These models both had a series of anticommuting d-vector harmonic-oscillator operators (b-operators) in addition to the a-operators mentioned at the end of the last Section. The difference was that the Ramond operators had integral mode numbers, the zero-mode operators being interpreted as gamma matrices, while the NS operators had half-integral mode numbers.

It was suggested by Kikkawa, Sakita and Virasoro [KSV69] (KSV) that the amplitude (25.1) might be the Born term of a perturbation series, and they proposed a form for the n-loop term. The simplest one-loop term was calculated independently from unitarity by Amati, Bouchiat and Gervais [ABG69], by Bardakci, Halpern and Shapiro [BHS69] and by Kikkawa, Sakita, Veneziano and Virasoro (note added in proof to [KSV69]). More general terms were calculated by Kaku and Thorn [KT70]; an improvement to their calculation was made by Gross, Neveu, Scherk and Schwarz [GNSS70]. The n-loop amplitude was calculated independently by Alessandrini [Ale71], by Kaku and Yu [KY70, KY71a, KY71b, KY71c] and by Lovelace [Lov70]; their work was further developed by Alessandrini and Amati [AA71]. Lovelace and Alessandrini, in common with all these workers, based their calculation on the operator formalism mentioned above, but they pointed out that their calculation could be understood in terms of amplitudes associated with n-hole Riemann surfaces. Their work was motivated by ideas suggested by Nielsen [Nie70] in connection with the recently proposed string interpretation of dual models, and the associated analogue model of Fairlie and Nielsen [FN70]. All these calculations, however, had ghosts,[1] which we shall treat shortly, circulating in the loops, and they were therefore incorrect. We mention them because of their relation to later work.

The authors of [GNSS70] showed that certain one-loop amplitudes possessed cuts not associated with unitarity. Such cuts were found independently by Frye and Susskind [FS70] in a calculation based on the analogue model. Lovelace [Lov71], making a reasonable conjecture about the amplitude with ghosts eliminated, showed that these cuts became poles if $d = 26$. It thus appeared that the model was inconsistent unless $d = 26$, a feature which caused considerable amusement among sceptics at the time.

Since the products in Eq. (25.6) are d-vector products, or, equivalently, the a-operators in the operator formalism are d-vectors, the states corresponding to the time component will

---

[1] By ghost we mean any particle with a negative metric. The Faddeev–Popov ghosts, of course, were not involved in dual models or string theory at this stage.

be ghosts, i.e. negative-metric states. The model will not therefore be physically acceptable unless the ghosts can be eliminated. In the case where the ground state was a tachyon with $\mu^2 = -1$, Virasoro [Vir70] found an infinite series of operators which, when applied to any particle in the operator formalism, gave a 'spurious' state, i.e. a state which would not appear as an intermediate state in Figure 25.1. (The Virasoro operators satisfy the algebra of two-dimensional conformal transformations with a central charge.) The physical Hilbert space was therefore the space orthogonal to all the spurious states, and the hope was that this Hilbert space would be ghost free. Del Giudice, Di Vecchia and Fubini [DDF72] (DDF) found a set of 'transverse' positive definite simple-harmonic oscillators, with $d - 2$ components, which commuted with the Virasoro operators and which could therefore be used to construct a ghost-free subspace of the Hilbert space, or possibly the whole Hilbert space. Finally Brower [Bro72] showed that the DDF operators, together with a new set of 'longitudinal' operators, could be used to construct the entire physical Hilbert space, and that the Hilbert space so constructed was ghost free provided $d \leq 26$.

Brower showed that the situation was particularly simple if $d = 26$, when the longitudinal operators produced null states which did not give rise to poles in Figure 25.1, so that the physical Hilbert space constructed from the DDF operators without the longitudinal operators was sufficient. The string models, as presently formulated, cannot incorporate the longitudinal modes, and, in fact, no procedure is known for allowing only the particles in the physical Hilbert physical space to circulate in loops if there are transverse and longitudinal modes. (But such a model may possibly be equivalent to the Polyakov noncritical string.) We thus again obtain $d = 26$ as a condition for consistency. We shall denote $d = 26$ as the dimension. Goddard and Thorn [GT72] have given a simpler proof of ghost elimination if $d = 26$.

The ghost-free physical Hilbert space for the NS model can be constructed in a very similar way (Brower and Friedman [BF73], Schwarz [Sch72], [GT72]). Here the square of the mass of the ground state must be equal to minus one-half of the slope of the Regge trajectory, and the critical dimension is $d = 10$.

It was observed independently by Nambu [Nam70], Nielsen [Nie70] and Susskind [Sus70b] that the oscillators of dual models could be considered as the modes of vibration of a string. Nambu and Goto [Got71] took the action for the world-sheet of the string moving in time to be the area of the sheet. Goddard, Goldstone, Rebbi and Thorn [GGRT73] (GGRT) quantized the noninteracting dual string. They first considered a classical string and used light-cone coordinates

$$X^+ = \frac{1}{\sqrt{2}}(X^0 + X^{d-1}),$$
$$X^- = \frac{1}{\sqrt{2}}(X^0 - X^{d-1}),$$
$$X^i, \quad 1 \leq i \leq d - 2.$$
(25.8)

They then made a choice of coordinates σ, τ on the string world-sheet.

Making such a choice is equivalent to fixing a gauge in a gauge theory. They showed that they could make a choice such that $\tau = X^+$, that the momentum per unit length P^+ is a constant, that X^- is determined in terms of the X^i, and that the Nambu–Goto Lagrangian is:

$$\mathcal{L} = \frac{1}{4\pi}\left[\left(\frac{\partial X^i}{\partial \tau}\right)^2 - \left(\frac{\partial X^i}{\partial \sigma}\right)^2\right]. \tag{25.9}$$

The condition of constant P^+-momentum per unit length means that the 'length' of the string is proportional to the total value of p^+. The strings satisfy the boundary condition that the slope, $\partial X^i/\partial \sigma$, is zero at the ends. The GGRT choice of coordinates is not, of course, Lorentz covariant, and the authors construct generators for the nontrivial Lorentz transformations. The transformations M_{i+}, which change X^+, are the difficult ones, since the condition $\tau = X^+$ has to be re-established. This is done by making a pseudo-conformal transformation of the $\sigma - \tau$ coordinates, since the action is pseudo-conformally invariant. The Lorentz generators for the nontrivial transformations are trilinear.

Quantization of the model with the Lagrangian (25.9) is straightforward. The Hamiltonian is normal ordered, and a constant term μ_0^2 is added. There is no problem with ghosts, since the Lagrangian only involves transverse coordinates. However, Lorentz invariance is nontrivial. The classical Lorentz generators do effect the required transformations, but anomalous terms appear in the commutator between transformations M_{i+} and M_{j+} when $i \neq j$. These terms cancel when $d - 2 = 24$ and $\mu_0^2 = -1$, which are thus the conditions for Lorentz invariance of the theory.

The Ramond and Neveu–Schwarz models were treated by Iwasaki and Kikkawa [IK73] in a similar way. As might be expected, the conditions for Lorentz invariance were $d - 2 = 8$ and $\mu_0^2 = -\frac{1}{2}$ (Neveu–Schwarz) or $\mu_0^2 = 0$ (Ramond).

25.4 Interacting strings and functional integration

We now treat the subject of interacting strings by functional integration. The problem was initially considered by Gervais and Sakita [GS73]. The amplitudes constructed by them were not manifestly factorizable and the condition $d = 26$ was not evident in their work. The approach we shall describe (Mandelstam [Man73a]) was based on the light-cone GGRT string just treated.

It was realized independently by several people that a theory of interacting strings could be constructed by allowing the free strings to split and join. The Hilbert space now consists of any number of strings. The free Lagrangian will be a sum of terms of the form (25.9), one for each string, together with an interaction which increases or decreases the number of strings by one. The interaction vertex function will be zero unless the positions of the initial and final strings coincide. In terms of operators, the vertex function V will be the infinite product:

$$V = \prod_{\sigma,i} \delta\{X^i_f(\sigma) - X^i_i(\sigma)\}. \tag{25.10}$$

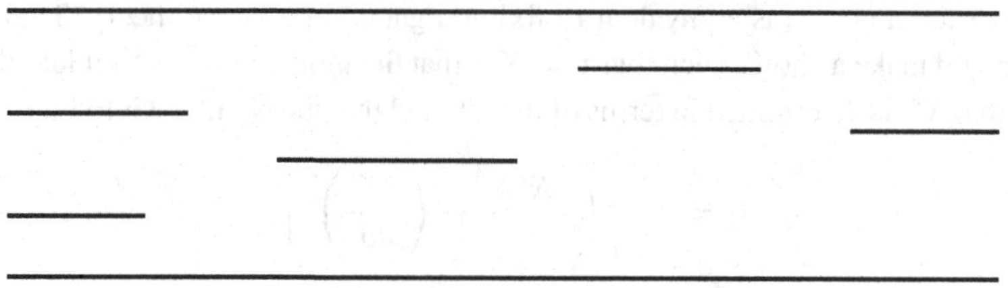

Figure 25.2 Interacting string diagram.

Two strings can also meet at a point and recombine, but we shall not encounter this term in our treatment.

The string world-sheet for a general interacting string process is shown in Figure 25.2. The horizontal axis is the light-cone time τ, the vertical axis is the length σ. The $d-2$ transverse coordinates X^i are orthogonal to the paper. Thus three strings come in from $\tau = -\infty$, after a time the lower two join, and so on until we reach $\tau = +\infty$. Note that the X^i coordinates are discontinuous across the horizontal lines, since they are coordinates of different strings. We have not separated the strings in the diagram to illustrate the fact that the total 'length' of all the strings, i.e. the total value of P^+, is constant.

For closed strings, the ends of all strings in Figure 25.2 are identified, so that the diagram consists of cylinders, each cylinder representing a closed string propagating in time. At the interaction points, two closed strings join to form one, or one string splits into two.

The scattering amplitude associated with the process shown will be found by functional integration of the transverse coordinates $X^i(\sigma, \tau)$. To begin, we integrate separately over the parts of the world-sheet associated with the different strings, with a vertex function at each interaction time. Since the vertex function requires the X^i to be continuous across the interaction times, we may simply integrate over the whole world-sheet. The X^i are discontinuous over the horizontal lines of Figure 25.2, and at the horizontal lines they satisfy Neumann boundary conditions, i.e. their derivatives orthogonal to the boundary are zero.

Before performing the functional integration, we continue analytically to Euclidean time. Thus Eq. (25.9) now becomes:

$$\mathcal{L} = -\frac{1}{4\pi}\left[\left(\frac{\partial X^i}{\partial \tau}\right)^2 + \left(\frac{\partial X^i}{\partial \sigma}\right)^2\right]. \tag{25.11}$$

After performing the functional integration, we integrate over the intermediate times in Figure 25.2, which we denote by $\tau_1, \ldots, \tau_r$. The scattering amplitude for unexcited incoming and outgoing particles is thus:

$$A = \mathcal{N} \int d\tau_1 \ldots [d\tau_i] \ldots d\tau_R \int \prod dX^i(\sigma, \tau)$$

$$\times \exp\left\{i \sum_r \frac{1}{\pi \alpha_r} p_r^i \int d\sigma X^i(\sigma, \tau_0) + \int d\sigma d\tau \mathcal{L}(\sigma, \tau) - \sum_r p_r^- \tau_0\right\}. \tag{25.12}$$

The factor $\mathcal{N}$ is a normalization factor, which is a product of an overall normalization factor and a factor for each incoming or outgoing particle. By considering the special case of a free string, it can be shown to be unity. The first sum in the exponent is over all incoming and outgoing particles; it is the wave function for the zero mode of the state. The factor $\pi \alpha_r$ is the 'length' of the rth string, and the integral is over the string. The time τ_0 is the time at the beginning or end of Figure 25.2 (large negative or large positive), depending on whether the particle is incoming or outgoing. The wave functions of the nonzero modes of the string have been omitted, since they will simply contribute to the normalization factor $\mathcal{N}$ if the particle is in its unexcited state. They must be included for general excited particles. The square brackets around τ_i indicate that one τ is omitted from the variables of integration, as the entire amplitude is invariant under time translation. The integration over this variable simply gives us conservation of energy. There are thus $N - 3$ variables of integration.

Since the functional integrand in (25.12) is a quadratic function of the X^i, the functional integral can be performed by standard methods in terms of the reciprocal of the Laplacian and its determinant. We define the Neumann function for the string world-sheet by the equation:

$$\left[\left(\frac{\partial}{\partial \sigma} \right)^2 + \left(\frac{\partial}{\partial \tau} \right)^2 \right] N(\sigma, \tau; \sigma', \tau') = 2\pi \delta(\sigma - \sigma')\delta(\tau - \tau') - \frac{1}{A}, \qquad (25.13)$$

with Neumann boundary conditions, i.e. the σ derivatives of N must be zero at the horizontal lines of Figure 25.2. A is the area of the string diagram, with large finite initial and final times. The last term is to take into account the zero mode, since the inverse of the Laplacian operator can only be defined if the zero mode is excluded.

On performing the functional integral, we obtain the result:

$$A = \int d\tau_1 \ldots [d\tau_i] \ldots d\tau_R \Delta^{(d-2)/2}$$

$$\times \exp \left\{ \sum_{r,s} \frac{1}{\pi^2 \alpha_r \alpha_s} p_r^i p_s^i \int d\sigma d\sigma' N(\sigma, \tau_0; \sigma', \tau_0') - \sum_r p_r^- \tau_0 \right\}. \qquad (25.14)$$

The factor Δ is $-1/2\pi$ times the determinant of the Laplacian on the string diagram. The first summation within the braces is now over all pairs of strings, and the σ integral is over the string at $\tau = \pm\infty$. If the two strings are different (and at $\pm\infty$), the Neumann function will be independent of the position along the string, so that the σ integrals may be written $N(r, s)$, where r and s denote the positions of the rth and sth strings. The integrals where r and s refer to the same string contribute normalization factors, independent of the shape of the string diagram. We denote the product of all these normalization factors by $\mathcal{N}$. Furthermore, it is not difficult to show that the last term in the braces of equation (25.14) simply changes the $(d-2)$-vector product $p_r^i p_s^i$ into a d-vector product. On inserting these changes in equation (25.14), we obtain our final result:

$$A = \int d\tau_1 \ldots [d\tau_i] \ldots d\tau_R \Delta^{(d-2)/2} \mathcal{N} \exp\left\{-2\sum_{r>s} p_r \cdot p_s N(r,s)\right\}. \qquad (25.15)$$

As we have been using the light-cone frame, our approach, while manifestly unitary in a positive definite Hilbert space, is not manifestly Lorentz invariant. (While the momentum-dependent factor in Eq. (25.15) is Lorentz covariant, the 'lengths' of the strings in Figure 25.2 are proportional to P^+ and therefore dependent on the Lorentz frame.) As we have already mentioned, GGRT constructed Lorentz generators for the free string and showed that they had the correct properties if $d = 26$. By applying these generators to the vertex (25.10), we have shown that the interacting-string theory is Lorentz invariant if $d = 26$ [Man73a, Man74c].

Now let us see how we can obtain the result quoted in the introduction for the Born term of the string model. In that case the only horizontal lines in Figure 25.2 will be those from the external states, i.e. the string world-sheet will have genus zero. Since the Neumann functions are conformally invariant, they can be evaluated by conformally transforming the string diagram onto the upper half-plane, with the external particles transforming onto points on the real axis. The formula for doing so is a special case of the Schwarz–Christoffel transformation, namely:

$$\rho = \sum_{r=1}^{n} \pi \alpha_r \ln(z - Z_r). \qquad (25.16)$$

The variable ρ $(= \tau + i\sigma)$ is the coordinate on the string diagram, z the coordinate on the upper half-plane. The string lengths $\pi\alpha$ are considered positive for incoming strings, negative for outgoing strings. The rth string at $\tau = \pm\infty$ transforms to the point Z_r on the real axis. It is often convenient to take $Z_N = \infty$.

The Neumann function for the upper half-plane is simply:

$$N(z, z') = \ln|z - z'| + \ln|z - z'^*|. \qquad (25.17)$$

(For closed strings, we transform the string diagram onto the whole plane. The strings at $\tau = \pm\infty$ transform to points anywhere in the plane, and the variables Z_r are complex. The second term on the right of (25.17) is absent.)

The momentum-independent factors of (25.15), i.e. the factors to the left of the exponential, may be treated in one of two ways. The simplest method is to make use of the proved Lorentz invariance of the amplitude in the critical dimension. We take the variables P^+ for all but two of the strings, one outgoing and one incoming, each at the bottom of Figure 25.2, to be equal to zero. Since the lengths of the strings are proportional to P^+, this means that the string diagram consists of one long string with several infinitely short stings entering or exiting at the top. The vertex functions, and therefore the momentum-independent factor, are then easily shown to be unity. Thus, on transforming the integral (25.15) onto the upper half-plane, inserting the formula (25.17) for the Neumann functions, and replacing the $N - 3$ variables of integration τ_r by the corresponding real Z_r, we easily obtain the formula (25.1). (For closed strings the Z_r are integrated over the entire complex plane.)

The foregoing analysis is very simple, but it cannot be extended to loops. An alternative method is to treat the string diagram in Figure 25.2 directly, without using Lorentz invariance (p. 66 in [GG85]). We then have to examine the factor Δ in (25.15), the determinant of the Laplacian. On the upper half-plane, this factor is of course a constant; however, we require it for the string diagram and, since it requires regularization, it is *not* invariant under a conformal transformation. The change of Δ under such a transformation (known to physicists as the 'conformal anomaly') has been worked out by McKean and Singer [MS67] (see also Alvarez [Alv83]). On applying their result to the conformal transformation from the string diagram to the upper half-plane, and calculating the Jacobian from the τ_r to the Z_r, which is not as simple as it was in the first method, we obtain Eq. (25.1), if and only if $d - 2 = 24$.

The calculation of the one-loop amplitude by functional integration involves new considerations, since the string diagram world-sheet is then conformally equivalent to an annulus instead of the upper half-plane. (The closed string world-sheet is conformally equivalent to a torus.) For details of the calculation, we refer the reader to the papers by Mandelstam [Man74b], and by Arfaei [Arf75], and to the book [GG85], p. 72.

Multiple loop string diagrams for processes involving open and closed strings are conformally equivalent to Riemann surfaces of higher genus, possibly with holes if some of the strings are open. Such amplitudes are most easily treated by considering their analytic properties in the moduli space of Riemann surfaces of genus g. The analytic properties have been examined by Belavin and Knizhnik [BK86], who were mainly interested in the Polyakov formalism, to be outlined in the final Section of this Chapter, but their methods can also be used in the present context. More precisely, we take a conformal metric which is the product of an analytic function on the Riemann surface and its conjugate complex, as has been suggested by Sonoda [Son87]. Such a metric necessarily has $2(g - 1)$ zeros, which are of course singular points. It can then be shown that Δ, evaluated in this metric, is an analytic function in moduli space except for known singularities where the moduli space degenerates or where two zeros of the metric coincide.

We should point out that the whole of this Section is, strictly speaking, incorrect, since we have made an illegal Wick rotation in continuing to imaginary time. A Wick rotation cannot be made if there is an intermediate state with lower energy than the initial and final states, since the integral over the exponential $\exp\{-i(E_{init} - E_{int})t\}$ diverges. Normally such an intermediate state is subtracted out and treated explicitly, but this procedure causes difficulties if applied to string models. In fact, all the amplitudes we have written here are real, and they diverge if the external energy in any channel is sufficiently large. We have already mentioned this fact in the introduction; we now see the reason for it from the point of view of string models.

In the early work on dual models, the problem was treated by analytic continuation from sufficiently low energies. For closed string amplitudes, it is not always possible to find a region where the energies in all channels are sufficiently low. Nevertheless, D'Hoker and Phong [DP93] have shown that the integral for the one-loop closed string amplitude can be divided into several parts, each of which can be continued analytically. They thereby obtain a finite amplitude with the correct singularities. It would be preferable to perform the functional integration in such a way that the amplitude is finite and unitary as it stands, without the necessity of analytic continuation. This can be done by undoing the illegal Wick rotation, as has been shown by Berera [Ber94], again for the one-loop closed string amplitude. It is not necessary to undo the entire Wick rotation; one need only consider the regions where the integral diverges. One then obtains an amplitude which is finite and unitary.

25.5 Functional integration for the RNS model

In this Section we shall summarize briefly how the functional integration of the previous Section can be applied to the Ramond–Neveu–Schwarz model [Man73b, Man74a]. The advantage of the present treatment is that the Ramond (R) and Neveu–Schwarz (NS) models now appear much more directly as two sectors of a single model.

In addition to the commuting variables of the previous model, we have two anticommuting variables, S_1^i and S_2^i with (transverse) vector indices, corresponding to the anticommuting operators in the operator treatment of the RNS model. The Lagrangian will be the sum of two terms, one being the same as the Lagrangian (25.11) of the previous model, and the other being given by the formula:

$$\mathcal{L}_2 = -\frac{1}{2\pi}\left[S_1\left(\frac{\partial}{\partial \tau} + i\frac{\partial}{\partial \sigma}\right)S_1 + S_2\left(\frac{\partial}{\partial \tau} - i\frac{\partial}{\partial \sigma}\right)S_2\right]. \quad (25.18)$$

The boundary conditions at the end of the string are either $S_1 = +S_2$ or $S_1 = -S_2$. For incoming fermions or outgoing antifermions (R) we have the plus sign at both ends; for incoming antifermions or outgoing fermions we have the minus sign at both ends, while for bosons (NS) we have different signs at the two ends. The boundary conditions are continuous along the top or bottom of the horizontal lines in Figure 25.2, but one sign changes into the other at each joining point. When expanding the S in normal modes b^i, we use the interval $-\pi\alpha \leq \sigma \leq \pi\alpha$, where $\pi\alpha$ is, as usual, the length of the string, and we take $S_2^i(-\sigma) = S_1^i(\sigma), \sigma > 0$. Thus the bosons have half-integral mode numbers, the fermions have integral mode numbers. With closed strings, S_1 and S_2 change sign after one complete rotation for bosons and do not change sign for fermions. Thus the closed string model has four sectors, NS–NS, NS–R, R–NS and R–R.

The vertex function is no longer given simply by the overlap integral, but there is an extra factor G given by the equation:

$$G = -\lim_{\sigma \to \sigma_1} i^{\pm 1/4}(\sigma - \sigma_1)^{3/4} S_1^i(\sigma)\left(\frac{\partial}{\partial \tau} + i\frac{\partial}{\partial \sigma}\right) X^i(\sigma). \quad (25.19)$$

The sign in the factor $i^{\pm 1/4}$ is plus or minus depending on whether the strings join or separate. The expression is evaluated at a point σ near the interaction point σ_1. A factor G

at each vertex will of course appear in the functional integrand. The factor G is necessary in order to prove Lorentz invariance, and in order that the functional integration gives the same result as the operator formalism.

As in the Bose case, the functional integration can be performed in terms of Neumann functions. For tree-level amplitude, the result can be conformally transformed onto the upper half-plane. Here we must bear in mind that the operators S^i, and therefore the Neumann functions involving them, have conformal weight 1/2, i.e. on transforming to the z-plane we must include an extra factor $(\partial \rho/\partial z)^{-1/2}$ for S_1 and the complex conjugate of this factor for S_2. We omit the details of the calculation. The result for scattering amplitudes of bosons is the same as that given by the operator calculation in the NS model. We can also calculate the scattering amplitude for fermion–antifermion scattering. This was the first such result as, in the operator formalism, amplitudes involving interacting Ramond fermions were much more difficult to treat than those involving bosons. The rather complicated calculations in the operator formalism were completed soon after by Schwarz and Wu [SW73], following work by Thorn [Tho71], Corrigan and Olive [CO72], Olive and Scherk [OS73], and Brink, Olive, Rebbi and Scherk [BORS73]. Their results agreed with those calculated by functional integration.

Berkovits [Ber86, Ber88] showed that the treatment of the RNS model outlined in this Section was equivalent to a functional integration over a super-world-sheet. Superfields had previously been applied to the RNS model by Fairlie and Martin [FM74] and Brink and Winnberg [BW76], and also by Polyakov [Pol81a, Pol81b] in the formalism to be mentioned in the next Section. The use of supersheets introduces a considerable simplification, since there is no longer an operator G at the interaction points.

25.6 Comparison with the Polyakov formulation for functional integration

For completeness we shall now compare briefly the formulation of the last two Sections with another formulation proposed later by Polyakov [Pol81a, Pol81b]. (For a much more detailed treatment of the Polyakov model, see [Pol98].) As we shall see, Polyakov's method can also be extended to the noncritical string, but we shall be mainly interested in its application to the critical string. Instead of starting from the Nambu–Goto string, Polyakov starts from an action defined by Brink, Di Vecchia and Howe [BDH76] and by Deser and Zumino [DZ76]. One introduces a general coordinate system on the world-sheet, with coordinates (σ_a, σ_b) and metric g_{ab}; the position of the point σ in d-dimensional space is denoted by $X^\mu(\sigma)$. The Lagrangian is then:

$$\mathcal{L} = -\frac{1}{4\pi} g^{1/2} g_{ab} \left(\frac{\partial X^\mu}{\partial \sigma_a} \frac{\partial X_\mu}{\partial \sigma_b} \right). \tag{25.20}$$

To calculate the S-matrix with N external ground-state closed strings in gth order perturbation theory, Polyakov considers a Riemann surface of genus g with N punctures (which, as we have seen, is conformally equivalent to a surface with the strings going to

$\pm\infty$). At each puncture there is a factor $\exp\{ip_r^\mu X_\mu(\sigma)\}$. The functional integrand is thus:

$$\prod_{r=1}^N \int d^2\sigma_r g^{1/2}(\sigma_r) \exp\left\{i \sum_r p_r^\mu X_\mu(\sigma_r) + \int d^2\sigma \mathcal{L}(\sigma)\right\}. \qquad (25.21)$$

Polyakov now functionally integrates over the variables $X^\mu(\sigma)$ and $g_{ab}(\sigma)$. The integration over $X^\mu(\sigma)$ proceeds as before. To integrate over g_{ab}, we first notice that we can use an Einstein coordinate transformation to convert the metric to a covariantly constant metric $g\delta_{ab}$ (in Euclidean space). Due to invariance under such transformations, this integration can be reduced to a functional Faddeev–Popov determinant, which will contain a conformal anomaly similar to the conformal anomaly in the determinant of the ordinary Laplacian mentioned in Section 25.4. Unless $d = 26$, the conformal anomaly will appear as an extra field when integrating over the remaining g, and we obtain a string theory together with an extra Liouville field. This noncritical string theory has been studied to a certain extent, but not nearly as much as that with the critical dimension $d = 26$. For this dimension the conformal anomalies in the two functional determinants cancel, and the theory possesses a Weyl conformal invariance, i.e. an invariance under a conformal change of g without a compensating change in the coordinate system. 'Most' of the integration over the remaining variable g will therefore simply be as integration over gauges and, in fact, no further Faddeev–Popov determinant is introduced. However, if the genus is not zero, not all Riemann surfaces of the same genus are conformally equivalent, and we are left with an integration over the variables characterizing the conformal class.

Not surprisingly, the Polyakov integral for genus zero reproduces the formula (25.1). Again, the amplitude is projectively invariant, so that the positions of three external particles must be fixed at arbitrary values.

The Polyakov formulation has the advantage over the light-cone formulation of Section 25.4 of being relativistically covariant. Also, the integration variables characterizing the conformal classes of Riemann surfaces correspond to the parameterization of such classes by Bers [Ber81]. Thus the Polyakov formulation can be related to the mathematical theory of Riemann surfaces. As the conformal anomaly cancels, one can avoid reference to the metric (though it may not always be advantageous to do so, see Dugan and Sonoda [DS87, Son87]). On the other hand, the Polyakov formulation gives us only the S-matrix; one cannot go off shell, whereas the light-cone formulation gives us a complete quantum mechanical theory where we are not restricted to infinite times and can go off shell, albeit noncovariantly. Also, the Polyakov formulation treats different orders in perturbation theory separately and unitarity, even perturbative unitarity, is far from obvious. In fact two inequivalent theories with the same Born term cannot both be unitary. It is therefore important that D'Hoker and Giddings [DG87] have shown that the two formulations are in fact equivalent; they are different gauges of the same theory. D'Hoker and Giddings point out that it should be possible to prove unitarity directly in the Polyakov formulation (for a possible approach, see [Pol98, Chapter 9]), but such a proof appears to be considerably more complicated than in the light-cone formulation, where unitarity is manifest.

Thus, with the two gauges of the functional integration theory, we can more easily see different aspects of the system than we could with each gauge separately.

Polyakov also considered the RNS string, and most of the general remarks we have made apply to this case as well. Aoki, D'Hoker and Phong [ADP90] have treated the unitarity of the Polyakov RNS string by relating it to Berkovits' light-cone formulation with a super-world-sheet, mentioned in the previous Section, which is manifestly unitary, but some problems remain with external fermions (Giddings [Gid92]).

The RNS formalism has spacetime supersymmetry after projecting out half of the states using the Gliozzi–Scherk–Olive (GSO) projection [GSO77] (which we have not discussed in this Chapter). However, it is not *manifestly* supersymmetric. Green and Schwarz [GS81, GS82, GS84], using the light-cone gauge, have given a formulation with manifest spacetime supersymmetry. This makes it much easier to show that the principal divergence of the theory, namely the 'dilaton' divergence, does not occur (Mandelstam [Man92]), as the absence of this divergence depends on spacetime supersymmetry. The Green–Schwarz formalism requires operators at the interaction points of Figure 25.2, and calculations, especially of multiloop amplitudes, are thereby made more difficult. At the cost of some complication, Berkovits [Ber89, Ber91, Ber93a, Ber93b] has given a manifestly Lorentz covariant formulation of the Green–Schwarz theory, which includes the 'ghosts' associated with the Faddeev–Popov determinant discussed above. His formulation thus possesses manifest Lorentz covariance and manifest spacetime supersymmetry, and avoids the necessity of operators at the interacting points. As before, unitarity is established by comparing the theory with the light-cone formulation of the previous Section.

Acknowledgements

This work is supported by the U. S. Department of Energy under contract DE-AC02-05CH11231.

Stanley Mandelstam was born in Johannesburg in 1928. After undergraduate education at the University of the Witwatersrand, Johannesburg and Cambridge University, he obtained his PhD in Rudolf Peierls' Department at the University of Birmingham in 1956. After postdocs at Birmingham, Columbia and Berkeley, he returned to Birmingham as a Professor of Mathematical Physics in 1960. In 1963 he became Professor of Physics at the University of California at Berkeley, where he is currently Emeritus Professor.

References

[Ale71] Alessandrini, V. (1971). A general approach to dual multiloop diagrams, *Nuovo Cimento* **A2**, 321–352.

[AA71] Alessandrini, V. and Amati, D. (1971). Properties of dual multiloop amplitudes, *Nuovo Cimento* **A4**, 793–844.

[Alv83] Alvarez, O. (1983). Theory of strings with boundaries: fluctuations, topology and quantum geometry, *Nucl. Phys.* **B216**, 125–184.

[ABG69] Amati, D., Bouchiat, C. and Gervais, J. L. (1969). On the building of dual diagrams from unitarity, *Lett. Nuovo Cimento* **2**, 399–406.
[ADP90] Aoki, K., D'Hoker, E. and Phong, D. H. (1990). Unitarity of closed superstring perturbation theory, *Nucl. Phys.* **B342**, 149–230.
[Arf75] Arfaei, H. (1975). Volume element for loop diagram in the string picture of dual models, *Nucl. Phys.* **B85**, 535–544.
[BHS69] Bardakci, K., Halpern, M. B. and Shapiro, J. A. (1969). Unitary closed loops in reggeized Feynman theory, *Phys. Rev.* **185**, 1910–1917.
[BM69] Bardakci, K. and Mandelstam, S. (1969). Analytic solution of the linear-trajectory bootstrap, *Phys. Rev.* **184**, 1640–1644.
[BR68] Bardakci, K. and Ruegg, H. (1968). Reggeized resonance model for production amplitude, *Phys. Lett.* **B28**, 342–347.
[BR69] Bardakci, K. and Ruegg, H. (1969). Reggeized resonance model for arbitrary production processes, *Phys. Rev.* **181**, 1884–1889.
[BK86] Belavin, A. A. and Knizhnik, V. G. (1986). Algebraic geometry and the geometry of quantum strings, *Phys. Lett.* **B168**, 201–206.
[Ber94] Berera, A. (1994). Unitary string amplitudes, *Nucl. Phys.* **B411**, 157–180.
[Ber86] Berkovits, N. (1986). Calculation of scattering amplitudes for the Neveu–Schwarz model using supersheet functional integration *Nucl. Phys.* **B276**, 650–666.
[Ber88] Berkovits, N. (1988). Supersheet functional integration and the interacting Neveu–Schwarz string, *Nucl. Phys.* **B304**, 537–556.
[Ber89] Berkovits, N. (1989). A covariant action for the heterotic superstring with manifest spacetime supersymmetry and world-sheet superconformal invariance, *Phys. Lett.* **B232**, 184–185.
[Ber91] Berkovits, N. (1991). Twistors, $n = 8$ superconformal invariance, and the Green–Schwarz superstring, *Nucl. Phys.* **B358**, 169–180.
[Ber93a] Berkovits, N. (1993). Lorentz-covariant Green–Schwarz superstring amplitudes, *Phys. Lett.* **B300**, 53–60.
[Ber93b] Berkovits, N. (1993). Calculation of Green–Schwarz superstring amplitudes using the $n = 2$ twistor-string formalism, *Nucl. Phys.* **B395**, 77–118.
[Ber81] Bers, L. (1981). Finite dimensional Teichmüller spaces and generalizations, *Bull. Am. Math. Soc.* **5**, 131–172.
[BDH76] Brink, L., Di Vecchia, P. and Howe, P. S. (1976). A locally supersymmetric and reparameterization invariant action for the spinning string, *Phys. Lett.* **B65**, 471–474.
[BORS73] Brink, L., Olive, D. I., Rebbi, C. and Scherk, J. (1973). The missing gauge conditions for the dual fermion emission vertex and their consequences, *Phys. Lett.* **B45**, 379–383.
[BW76] Brink, L. and Winnberg, J. O. (1976). The superoperator formalism of the Neveu–Schwarz–Ramond model, *Nucl. Phys.* **B103**, 445–464.
[Bro72] Brower, R. C. (1972). Spectrum-generating algebra and no-ghost theorem in the dual model, *Phys. Rev.* **D6**, 1655–1662.
[BF73] Brower, R. C. and Friedman, K. A. (1973). Spectrum-generating algebra and no-ghost theorem for the Neveu–Schwarz model, *Phys. Rev.* **D7**, 535–539.
[CT69] Chan, H. M. and Tsou, S. T. (1969). Explicit construction of the n-point function in the generalized Veneziano model, *Phys. Lett.* **B28**, 485–488.
[CO72] Corrigan, E. and Olive, D. (1972). Fermion-meson vertices in dual theories, *Nuovo Cimento* **A11**, 749–773.
[DDF72] Del Giudice, E., Di Vecchia, P. and Fubini, S. (1972). General properties of the dual resonance model, *Ann. Phys.* **70**, 378–398.

[DZ76] Deser, S. and Zumino, B. (1976). A complete action for the spinning string, *Phys. Lett.* **B65**, 369–373.

[DG87] D'Hoker, E. and Giddings, S. B. (1987). Unitarity of the closed bosonic Polyakov string, *Nucl. Phys.* **B291**, 90–112.

[DP93] D'Hoker, E. and Phong, D. H. (1993). Momentum analyticity and finiteness of the 1-loop superstring amplitude, *Phys. Rev. Lett.* **70**, 3692–3695.

[DS87] Dugan, M. J. and Sonoda, H. (1987). Functional determinants on Riemann surfaces, *Nucl. Phys.* **B289**, 227–252.

[FM74] Fairlie, D. B. and Martin, D. (1974). Green's function techniques and dual fermion loops, *Nuovo Cimento* **A21**, 647–660.

[FN70] Fairlie, D. B. and Nielsen, H. B. (1970). An analog model for KSV theory, *Nucl. Phys.* **B20**, 637–651.

[FS70] Frye, G. and Susskind, L. (1970). Removal of the divergence of a planar dual-symmetric loop, *Phys. Lett.* **B31**, 537–540.

[FGV69] Fubini, S., Gordon, D. and Veneziano, G. (1969). A general treatment of factorization in dual resonance models, *Phys. Lett.* **B29**, 679–682.

[FV69] Fubini, S. and Veneziano, G. (1969). Level structure of dual resonance models, *Nuovo Cimento* **A64**, 811–840.

[Gid92] Giddings, S. B. (1992). Punctures on super Riemann surfaces, *Commun. Math. Phys.* **143**, 355–370.

[GS73] Gervais, J. L. and Sakita, B. (1973). Ghost-free string picture of Veneziano model, *Phys. Rev. Lett.* **30**, 716–719.

[GSO77] Gliozzi, F., Scherk, J. and Olive, D. (1977). Supersymmetry, supergravity theories and the dual spinor model, *Nucl. Phys.* **B122**, 253–290.

[GGRT73] Goddard, P., Goldstone, J., Rebbi, C. and Thorn, C. B. (1973). Quantum dynamics of a massless relativistic string, *Nucl. Phys.* **B56**, 109–135.

[GT72] Goddard, P. and Thorn, C. B. (1972). Compatibility of the pomeron with unitarity and the absence of ghosts in the dual resonance model, *Phys. Lett.* **B40**, 235–238.

[GS69] Goebel, C. J. and Sakita, B. (1969). Extension of the Veneziano form to n-particle amplitudes, *Phys. Rev. Lett.* **22**, 257–260.

[Got71] Goto, T. (1971). Relativistic quantum mechanics of one-dimensional mechanical continuum and subsidiary condition of dual resonance model, *Prog. Theor. Phys.* **46**, 1560–1569.

[GG85] Green, M. B. and Gross, D. J. eds. (1985). *Unified String Theories*, Proceedings of the Workshop, Santa Barbara, CA (World Scientific, Singapore).

[GS81] Green, M. B. and Schwarz, J. H. (1981). Supersymmetrical dual string theory, *Nucl. Phys.* **B181**, 502–530.

[GS82] Green, M. B. and Schwarz, J. H. (1982). Supersymmetrical dual string theory. 2. Vertices and trees, *Nucl. Phys.* **B198**, 252–268.

[GS84] Green, M. B. and Schwarz, J. H. (1984). Properties of the covariant formulation of superstring theories, *Nucl. Phys.* **B243**, 285–306.

[GNSS70] Gross, D. J., Neveu, A., Scherk, J. and Schwarz, J. H. (1970). Renormalization and unitarity in the dual-resonance model, *Phys. Rev.* **D2**, 697–710.

[IK73] Iwasaki, Y. and Kikkawa, K. (1973). Quantization of a string of spinning material Hamiltonian and Lagrangian formulation, *Phys. Rev.* **D8**, 440–449.

[KT70] Kaku, M. and Thorn, C. B. (1970). Unitary nonplanar closed loops, *Phys. Rev.* **D1**, 2860–2869.

[KY70] Kaku, M. and Yu, L. P. (1970). The general multi-loop Veneziano amplitude, *Phys. Lett.* **B33**, 166–170.

[KY71a] Kaku, M. and Yu, L. P. (1971). Unitarization of the dual-resonance amplitude. I. Planar n-loop amplitude, *Phys. Rev.* **D3**, 2992–3007.
[KY71b] Kaku, M. and Yu, L. P. (1971). Unitarization of the dual-resonance amplitude. II. The nonplanar n-loop amplitude, *Phys. Rev.* **D3**, 3007–3019.
[KY71c] Kaku, M. and Yu, L. P. (1971). Unitarization of the dual-resonance amplitude. III. General rules for the orientable and nonorientable multiloop amplitudes, *Phys. Rev.* **D3**, 3020–3024
[KSV69] Kikkawa, K., Sakita, B. and Virasoro, M. A. (1969). Feynman-like diagrams compatible with duality. I. Planar diagrams, *Phys. Rev.* **184**, 1701–1713.
[KN69a] Koba, Z. and Nielsen, H. B. (1969). Reaction amplitude for n-mesons a generalization of the Veneziano–Bardakci–Ruegg–Virasoro model, *Nucl. Phys.* **B10**, 633–655.
[KN69b] Koba, Z. and Nielsen, H. B. (1969). Manifestly crossing-invariant parameterization of n-meson amplitude, *Nucl. Phys.* **B12**, 517–536.
[Lov70] Lovelace, C. (1970). M-loop generalized Veneziano formula, *Phys. Lett.* **B32**, 703–708.
[Lov71] Lovelace, C. (1971). Pomeron form factors and dual Regge cuts, *Phys. Lett.* **B34**, 500–506.
[Man73a] Mandelstam, S. (1973). Interacting string picture of dual resonance models, *Nucl. Phys.* **B64**, 205–235.
[Man73b] Mandelstam, S. (1973). Manifestly dual formulation of the Ramond model, *Phys. Lett.* **B46**, 447–451.
[Man74a] Mandelstam, S. (1974). Interacting-string picture of Neveu–Schwarz–Ramond model, *Nucl. Phys.* **B69**, 77–106.
[Man74b] Mandelstam, S. (1974). Dual-resonance models, *Phys. Rep.* **13**, 259–353.
[Man74c] Mandelstam, S. (1974). Dual-resonance models, *Nucl. Phys.* **B83**, 413–439.
[Man92] Mandelstam, S. (1992). The n-loop string amplitude. Explicit formulas, finiteness and absence of ambiguities, *Phys. Lett.* **B277**, 82–88.
[MS67] McKean, H. and Singer, I. M. (1967). Curvature and the eigenvalues of the Laplacian *J. Diff. Geom.* **1**, 43–69.
[Nam70] Nambu, Y. (1970). Quark model and the factorization of the Veneziano amplitude, in *Proceedings of the International Conference on Symmetries and Quark Models, Wayne State University, June 18–20, 1969*, ed. Chand, R. (Gordon and Breach, New York), 269–277, reprinted in *Broken Symmetry, Selected Papers of Y. Nambu*, ed. Eguchi, T. and Nishijima, K. (World Scientific, Singapore, 1995), 258–277.
[NS71] Neveu, A. and Schwarz, J. H. (1971). Factorizable dual model of pions, *Nucl. Phys.* **B31**, 86–112.
[NST71] Neveu, A., Schwarz, J. H. and Thorn, C. B. (1971). Reformulation of the dual pion model, *Phys. Lett.* **B35**, 529–533.
[Nie70] Nielsen, H. B. (1970). An almost physical interpretation of the integrand of the n-point Veneziano model, paper submitted to the *15th International Conference on High Energy Physics, Kiev, 26 August to 4 September, 1970* (Nordita preprint, 1969).
[OS73] Olive, D. and Scherk, J. (1973). Towards satisfactory scattering amplitudes for dual fermions, *Nucl. Phys.* **B64**, 334–348.
[Pol98] Polchinski, J. (1998). *String Theory*, Vol. 1: *An Introduction to the Bosonic String* (Cambridge University Press, Cambridge).
[Pol81a] Polyakov, A. M. (1981). Quantum geometry of bosonic strings, *Phys. Lett.* **B103**, 207–210.
[Pol81b] Polyakov, A. M. (1981). Quantum geometry of fermionic strings, *Phys. Lett.* **B103**, 211–213.

[Ram71] Ramond, P. (1971). Dual theory for free fermions, *Phys. Rev.* **D3**, 2415–2418.
[Sch72] Schwarz, J. H. (1972). Physical states and pomeron poles in the dual pion model, *Nucl. Phys.* **B46**, 61–74.
[SW73] Schwarz, J. H. and Wu, C. C. (1973). Evaluation of dual fermion amplitudes, *Phys. Lett.* **B47**, 453–456.
[Sha70] Shapiro, J. A. (1970). Electrostatic analogue for the Virasoro model, *Phys. Lett.* **B33**, 361–362.
[Son87] Sonoda, H. (1987). Functional determinants on punctured Riemann surfaces and their application to string theory, *Nucl. Phys.* **B294**, 157–192.
[Sus70a] Susskind, L. (1970). Structure of hadrons implied by duality, *Phys. Rev.* **D1**, 1182–1186.
[Sus70b] Susskind, L. (1970). Dual-symmetric theory of hadrons. 1, *Nuovo Cimento* **A69**, 457–496.
[Tho71] Thorn, C. B. (1971). Embryonic dual model for pions and fermions, *Phys. Rev.* **D4**, 1112–1116.
[Ven68] Veneziano, G. (1968). Construction of a crossing-symmetric, Reggeon behaved amplitude for linearly rising trajectories, *Nuovo Cimento* **A57**, 190–197.
[Vir69a] Virasoro, M. (1969). Alternative constructions of crossing-symmetric amplitudes with Regge behavior, *Phys. Rev.* **177**, 2309–2311.
[Vir69b] Virasoro, M. (1969). Generalization of Veneziano's formula for the five-point function, *Phys. Rev. Lett.* **22**, 37–39.
[Vir70] Virasoro, M. (1970). Subsidiary conditions and ghosts in dual-resonance models, *Phys. Rev.* **D1**, 2933–2936.

26
The hadronic origins of string theory

RICHARD C. BROWER

26.1 Not by accident

The Sixties in Berkeley were exciting times for a young graduate student, not just because of the free speech movement, but also in elementary particle physics. The strong nuclear interactions appeared to be so different from the beautiful simplicity of QED that a *systematic* programme was underway to find a new foundation to relativistic quantum interactions. A new set of axioms was sought in the so-called S-matrix programme by abstracting properties of analyticity and unitarity exhibited in Feynman diagrams without recourse to an underlying local field theory. Remarkably this ambitious programme, guided by copious experimental data for hadronic scattering, did result in an alternative solution, now referred to as 'string theory'. The fact that the nuclear or strong interaction was subsequently formulated by QCD, a local field theory, should not obscure this remarkable discovery. Only recently, with the advent of Maldacena's conjectured AdS/CFT correspondence between string and gauge field theory [Mal98], are we beginning to understand why two alternative solutions to strong interactions (gauge theory and its dual string theory) should co-exist as correct and equivalent theories. Whether string theory ultimately leads to a fundamental string/gauge duality for all forces including gravity, is not clear yet. Nevertheless the discovery of the theory, contrary to a widespread opinion, was not by accident, but was the result of the traditional interaction between theoretical ideas and experimental data.

Here let me give an intentionally anecdotal description of my recollections of early developments in string theory. I will emphasize two aspects. First, the empirical roots of string theory to fit data and satisfy the constraints of the analytic S-matrix for hadronic interactions, and second, the beginning of the transformation of the Veneziano dual model into a rigorous string theoretical framework with the proof of the 'no-ghost theorem' [Bro72, GT72]. I pick these simply because I had the opportunity to witness both at close range. For both of these subjects, the reader is referred to the Chapters by Paolo Di Vecchia, Peter Goddard, and Gabriele Veneziano, for further discussion. Here I also attempt to put the subject in the context of the still on-going effort to discover the correct QCD string theory.

As a graduate student at Berkeley with daily interactions with Geoffrey Chew and Stanley Mandelstam, the excitement was palpable. To find a theory for hadronic interactions, one needed to find a new starting point to replace the small coupling constant of weakly interacting quantum fields. Looking at the increasingly detailed spectrum of hadronic resonances with increasing intrinsic spin J, it became clear that elementary quantum fields should not be associated with all of them. Instead the spectrum suggested a new small parameter: the ratio of the resonance width to its mass, $\Gamma/M \simeq 1/10$. One talked about trying to find a 'zero-width' approximation to the S-matrix. In the context of QCD we now understand this as the leading $1/N_c$ approximation in the 't Hooft topological expansion [tHo74] at fixed $g^2 N_c$. However, at the time the phenomenon was expressed in the language of dispersion relations and Regge asymptotics, or what was referred to as Dolen–Horn–Schmid duality [DHS68]. The path from this observation to the Veneziano dual amplitudes had many interesting twists and turns. A crucial step was taken by Stanley Mandelstam in his paper on the 'Dynamics based on rising Regge trajectories' [Man68] in which he posited exactly linear trajectories, $\alpha(t) = \alpha_0 + \alpha' t$, for the zero-width approximation. Incidentally, under his guidance, my first paper as a graduate student on 'Kinematic constraints for infinitely rising Regge trajectories' (Brower and Harte [BH67]) was a modest contribution to this discussion.

With hindsight the approach appears very straightforward indeed. Amplitudes at high energies exhibit Regge behaviour: for example, it was common place to see on blackboards the Regge approximation,

$$A(s, t) \simeq g_0^2 \Gamma(1 - \alpha_\rho(t))(-\alpha' s)^{\alpha_\rho(t)}, \qquad (26.1)$$

for the ρ exchange in the t-channel. The power is dictated by Regge asymptotic at high energy, $E = \sqrt{s}$, in the centre of mass; the Γ-function prefactor provides resonance poles for integer $\alpha_\rho(t) = J \geq 1$. Only near the ρ pole, $\alpha_\rho(t) \simeq 1 + \alpha' t - \alpha'(m_\rho + i\Gamma_\rho)^2 + \cdots$, does this resemble a Breit–Wigner amplitude for the ρ resonance exchange,

$$A(s, t) \simeq g_0 \frac{s}{t - (m_\rho + i\Gamma_\rho)^2} g_0, \qquad (26.2)$$

or, in the zero-width approximation, the Feynman diagram for an elementary $J = 1$ vector particle. Next came the crucial observation from Veneziano. If we take a simple example of an elastic scattering amplitude, such as $A_{\pi^+\pi^- \to \pi^+\pi^-}(s, t)$, which is exactly crossing symmetric under s and t interchange, a natural guess for the full amplitude is:

$$A_{\pi^+\pi^- \to \pi^+\pi^-}(s, t) = g_0^2 \frac{\Gamma(1 - \alpha(t))\Gamma(1 - \alpha(s))}{\Gamma(1 - \alpha(t) - \alpha(s))}. \qquad (26.3)$$

By the application of Stirling's approximation, this reproduces the Regge form in Eq. (26.1) for a linear trajectory, $\alpha(s) \sim \alpha' s$. To be precise, Veneziano's paper [Ven68] considered a different crossing-symmetric amplitude for $\omega \to \pi^+\pi^-\pi^0$, which was extended to pion scattering by Lovelace [Lov68] and Shapiro [Sha69] shortly thereafter.

As an aside, it is worth noting the remarkable properties of this amplitude as an approximation to low energy pion scattering. There is a striking consistency with the chiral theory of pions. In the soft pion limit $p_i \to 0$, the Adler zero, $A_{\pi^+\pi^-\to\pi^+\pi^-}(s,t) \simeq g_0^2(1 - \alpha_\rho(s) - \alpha_\rho(t)) \to 0$, is imposed if we take the phenomenologically reasonable values for the ρ trajectory intercept, $\alpha_\rho(0) = 0.5$. Phenomenological six-pion Veneziano amplitudes (Brower and Chu [BC73]) constructed at this time are still roughly as successful in the infrared as modern ones based on effective five-dimensional amplitudes (Erlich, Katz, Son and Stephanov [EKSS05]) using the AdS/CFT correspondence. The generalization to the N-point pion scattering amplitude was formulated in Neveu and Schwarz's seminal paper [NS71] entitled 'Factorizable dual model of pions'. So the four-pion Veneziano amplitude turns out to be the NS superstring – ignoring the conformal constraint on the Regge intercept ($\alpha(0) = 1$) and the dimension of spacetime ($d = 10$) which were not understood at the time. The most serious phenomenological failure of this amplitude, also not understood at that time, is the lack of 'hard or valence partons' as illustrated by exponential fall-off at wide angles. As emphasized in the concluding Section, this failure can be circumvented when the strings propagate in five-dimensional anti de Sitter space as noted 30 years later by Polchinski and Strassler [PS02]!

26.2 From amplitudes to oscillators to strings

In a remarkably short period, this simple 'phenomenological' model for elastic scattering was generalized to a 'zero-width' amplitude for an N-particle S-matrix. For the simpler (but less physical) case of the bosonic open string, the N-point open string tachyon amplitude was found as a generalized beta function integral representation,

$$A_N(p_1, \ldots, p_N) = g_0^{N-2} \int \frac{dx_2 dx_3 \cdots dx_{N-2}}{(x_N - x_{N-1})(x_N - x_1)(x_{N-1} - x_1)} \times \prod_{1 \leq i < j \leq N} (x_j - x_i)^{2\alpha' p_j p_i}, \quad (26.4)$$

with the Möbius invariant integration region strictly ordered: $x_1 \leq x_2 \leq x_3 \leq \cdots \leq x_N$. Next, one can follow the pioneers (Fubini, Gordon and Veneziano [FGV69]) of the field and write down the 'old covariant quantized' string, working 'backward' from the N-point function. One needs to factorize the N-point function, i.e. introduce a complete set of states. Short circuiting the full derivation, this amounts to a free (string) field expansion,

$$X^\mu(\sigma, \tau) = \hat{q}^\mu - 2i\alpha' \hat{p}^\mu \tau + \sum_{n=1}^{\infty} \sqrt{\frac{2\alpha'}{n}} \left(a_n^\mu \exp[-n\tau] + a_n^{\mu\dagger} \exp[n\tau]\right) \cos(n\sigma), \quad (26.5)$$

into normal mode oscillators,

$$[\hat{q}^\mu, \hat{p}^\nu] = i\eta^{\mu\nu} \quad \text{and} \quad [a_n^\mu, a_m^{\nu\dagger}] = \eta^{\mu\nu} \delta_{n,m}, \quad (26.6)$$

acting on the ground state tachyon at momentum p,

$$\hat{p}^\mu |0, p\rangle = p^\mu |0, p\rangle \quad \text{and} \quad a_n^\mu |0, p\rangle = 0. \quad (26.7)$$

An algebraic exercise shows that the integrand for the N-point function, once rewritten in terms of vertex operators $V(x, p)$, does factorize as follows:

$$\langle 0, p_N | V(x_{N-1}, p_{N-1}) \cdots V(x_2, p_2) | 0, p_1 \rangle = \prod_{1 \leq i < j \leq N} (x_j - x_i)^{2\alpha' p_j p_i}, \qquad (26.8)$$

with

$$V(x, p) =: \exp[ipX(0, \tau)]:$$

$$= \exp\left[ip\hat{q} + ip\sum_n \sqrt{\frac{2\alpha'}{n}} a_n^\dagger x^n\right] \exp\left[2\alpha' p\hat{p}\log x + ip\sum_n \sqrt{\frac{2\alpha'}{n}} a_n x^{-n}\right], \qquad (26.9)$$

and $x \equiv \exp[\tau]$. To calculate the matrix element (i.e. amplitude), one should perform normal ordering of the operators, denoted by $: \cdots :$, bringing all $a_n^\dagger$ to the left, and all a_n to the right, finding factors

$$e^{-2\alpha' \sum_n \frac{p_i p_j}{n} \left(\frac{x_i}{x_j}\right)^n} = e^{2\alpha' p_i p_j \log(1 - x_i/x_j)} = \left(1 - \frac{x_i}{x_j}\right)^{2\alpha' p_i p_j}, \qquad (26.10)$$

for each pair of vertex operators. The string interpretation follows from identification of world-sheet surface coordinates (σ, τ). The above expansion for the spacetime position, $X^\mu(\sigma, \tau)$, is a solution to the two-dimensional conformal equations of motion for a free string,

$$\partial_\tau^2 X^\mu + \partial_\sigma^2 X^\mu = 0, \qquad (26.11)$$

in a Euclidean world-sheet metric. Writing down the general normal mode expansion,

$$X^\mu(z, \bar{z}) = \hat{q}^\mu - i\alpha' \hat{p}^\mu \log(z\bar{z}) + \sum_{n=-\infty, n \neq 0}^{\infty} \sqrt{\frac{2\alpha'}{|n|}} \left(a_n^\mu z^{-n} + b_n^{\dagger \mu} \bar{z}^{-n}\right), \qquad (26.12)$$

with $z = \exp[\tau + i\sigma]$, we see that the particular solution required for the open string amplitude above satisfies Neumann boundary conditions at the ends, $\sigma = 0, \pi$. The vertex functions representing tachyon emission are inserted on one side at the $\sigma = 0$ boundary. Closed strings have periodic boundary conditions in $\sigma \in [-\pi, \pi]$. Superstrings add world-sheet two-dimensional fermion fields. Superstring theory further requires summing the expectation values on the world-sheet over Riemann surfaces of all genera [Pol98].

There are many ways to motivate the string or world-sheet interpretation. One is to recognize $(x_i - x_j)^{2p_i p_j} = \exp[2p_i p_j \log(x_i - x_j)]$ as the two-dimensional electrostatic potential between charges p, and a second is to understand the oscillator spectrum as the one-dimensional harmonics of a massless relativistic string. Nambu [Nam70] and Goto [Got71] took the string interpretation of the dual model one step further by noticing that the equation of motion (26.11) is a gauge fixed form for a general coordinate invariant world-sheet action,[1]

$$S_{NG} = -\frac{1}{2\pi\alpha'} \int d^2\xi \sqrt{-\det(h)}, \qquad h_{\alpha\beta} = \partial_\alpha X^\mu \partial_\beta X_\mu, \qquad (26.13)$$

with surface tension $T_0 = 1/(2\pi\alpha')$. At the classical level this is also equivalent to the Polyakov form,

$$S_P = -\frac{1}{2\pi\alpha'} \int d^2\xi \sqrt{-\det(\gamma)} [\gamma^{\alpha\beta} \partial_\alpha X^\mu \partial_\beta X_\mu], \qquad (26.14)$$

with an auxiliary 'Lagrange multiplier' two-dimensional metric, γ_{ij}. However, the Polyakov form is easier to gauge fix and quantize using BRST technology [Pol98]. To close the circle to the original starting point, it is easy to show that the classical solution of highest angular J is linear in the squared centre-of-mass energy, $J = \alpha' E^2 + \text{constant}$, identical to the Mandelstam linear form for the trajectory function: $\alpha(s) = \alpha' s + \alpha_0$.

26.3 Eliminating ghost states

At this juncture, as I entered my postdoctoral wandering from MIT to CERN and back, there seemed to be considerable progress in finding the starting point for a 'zero-width approximation' to hadronic interactions. However, to see it as a starting point for a unitary theory, there was a major problem. The on-shell covariant oscillator basis,

$$|\{N_{i,\mu}\}, p\rangle = [(a_1^{\mu_1\dagger})^{N_{1,\mu}} (a_2^{\mu_2\dagger})^{N_{2,\mu}} \cdots]|0, p\rangle \qquad (26.15)$$

(with $\alpha' p^2 = \alpha_0 - N = \alpha_0 - \sum_{i,\mu} N_{i,\mu}$), has negative norm states due to the timelike fluctuations $[a_n^0, a_n^{0\dagger}] = -1$. A similar feature is common to Gupta–Bleuler quantization of QED but gauge invariance decouples the timelike states. A possible way out was found with the observation by Virasoro [Vir70] of an infinite number of gauge conditions, one for each timelike oscillator mode, when the intercept (of the open string) was chosen to be $\alpha(0) = 1$. All physical states satisfy the constraints,

$$(L_0 - 1)|\text{Phys}, p\rangle = L_n|\text{Phys}, p\rangle = 0 \quad \text{for} \quad n > 0, \qquad (26.16)$$

where the operators obey the conformal algebra,[2]

$$[L_n, L_m] = (n-m)L_{n+m} + \frac{d-26}{12} n(n^2 - 1)\delta_{n,-m}, \qquad (26.17)$$

generalized to d dimensions of spacetime.

The physical picture was suggestive. We may view the 'string' as composed of wee partons (Brower and Susskind [BS73]) labelled by θ, with position density $X^\mu(\tau, \theta)$ and

---
[1] Editors' note: for further details, see also the Chapter by Paolo Di Vecchia on the string action in Part VI.
[2] Actually the central charge was identified first by Joe Weis an extraordinarily talented theorist who tragically died in the Swiss Alps.

conjugate momentum density,

$$P^\nu(\tau,\theta) = \frac{i}{2\alpha'}\partial_\tau X^\nu(\tau,\theta), \qquad (26.18)$$

obeying per parton conical commutators,

$$[X^\mu(\theta,\tau), P^\nu(\theta',\tau)] = 2\pi i \eta^{\mu,\nu}\delta(\theta-\theta'). \qquad (26.19)$$

With the expansion, $:\dot{X}^2(z):= L_0 + \sum_{n\neq 0} z^{-n} L_n$, the Virasoro constraint implies that centre-of-mass proper time condition ($\dot{x}^2 = 1$) holds for each individual parton on the string:

$$\langle \text{Phys}, p| :\dot{X}(\theta)^2: |\text{Phys}, p\rangle = 1. \qquad (26.20)$$

However, the precise mathematical mechanism to eliminate all negative norm states was far from clear. To gain confidence, explicit level by level calculations were performed by Paolo Di Vecchia, Emilio Del Giudice, Charles Thorn and others to construct the physical states by applying the Virasoro constraints. For details see the Chapter by Di Vecchia in this Volume. One surprise was the discovery that when you varied the number of spacetime dimensions d above 26, ghost states in fact did appear. Fearing other surprises while at CERN, with the help of the computer division, Goddard and I ran a recursive algebraic computer program to enumerate all physical states to 30th level finding no ghosts for $d \leq 26$.

I will now discuss the proof of the 'no-ghost theorem' from my perspective, developed over a series of papers in collaboration with Goddard [BG72] and Thorn [BT71]. Indeed the Chapter by Goddard in this Volume gives an excellent account of this story, which therefore need not be repeated here. Instead I give an outline of the construction of the physical Hilbert space. The starting point is to realize that the on-shell tree amplitude 'creates' only physical states because the on-shell tachyon vertex with $\alpha' p^2 = 1$ satisfies the Virasoro constraints for a conformal spin one operator. Del Giudice, Di Vecchia and Fubini [DDF72] (DDF) defined the action on states of a vertex operator by the contour integral,

$$\langle V \rangle \equiv \frac{1}{2\pi i} \oint \frac{dz}{z} V(z,p), \qquad (26.21)$$

and went on to introduce 'gluon' vertices[3] for the $d - 2$ physical 'transverse' polarizations,

$$\vec{A}_n = \langle \vec{P}\, e^{ik_n X}\rangle. \qquad (26.22)$$

In light-cone coordinates, $k_\pm = (k^{d-1} \pm k^0)/\sqrt{2}$ and $\vec{k} = (k^1,\ldots,k^{d-2})$, and in units where $\alpha' = 1/2$ the gluon momentum is chosen to carry n units of momentum on the (+) light-cone: $k_n^\mu = n\hat{e}_+^\mu$, where $\hat{e}_\pm^\mu = (\pm 1, 0, \ldots, 1)/\sqrt{2}$ and $\hat{e}_+ \cdot \hat{e}_- = 1$. Consequently

---

[3] In modern AdS/CFT language the charges for the zero-mass gauge boson are determined by the branes to which each end is attached. For convenience I use 'colour' terminology as the example relevant to a QCD string.

$k_n X = n X_-$ and the vanishing commutators, $[X_-, X_-] = [\vec{P}, X_-] = 0$, obviate the need for normal ordering.

The operators, $A_n^\dagger = A_{-n}$, create physical states at the nth level and Goddard and I realized their algebra, $[A_n^i, A_m^j] = n\delta_{ij}\delta_{n+m,0}$, is isomorphic to the algebra for the transverse operators, $\sqrt{n}\,\tilde{a}_n$, themselves. Thus a linearly independent basis for a subspace of physical states is given by

$$|\text{Phys}, \{n_i\}, p\rangle = [(\vec{A}_1^\dagger)^{n_1} (\vec{A}_2^\dagger)^{n_2} \cdots]|0, p\rangle. \tag{26.23}$$

At this point the construction stalled because one needed not two but three physical modes at $d = 4$, for example. There was apparently a missing 'longitudinal' physical creation operator. Also with the knowledge that ghosts do appear for $d > 26$, a complete spectrum generating algebra for the physical states must somehow account for this.

26.3.1 Longitudinal operators

We now turn to the construction of the missing longitudinal generators in the physical state spectrum generating algebra. Goddard and I postulated the existence of longitudinal physical state operators of the form,

$$A_n^{(+)} = \langle : P_+ e^{inX_-} : \rangle + \cdots . \tag{26.24}$$

However, because the normal ordering destroys the conformal invariance of the first term in Eq. (26.24), a correction term was required. This construction was completed subsequently in [Bro72] by integrating a partial differential equation for the conformal Virasoro constraint. Here I give a streamlined derivation based on a trick realized soon after this publication, which is both simpler and perhaps more intuitive. The idea is to take the 'gluon' vertex,

$$A_n^\mu = \langle : P_\mu e^{ikX} : \rangle, \tag{26.25}$$

'slightly' off-shell $k^2 \neq 0$ keeping the conformal spin fixed at 1. To do this one must compensate for the conformal spin $(\alpha' k^2)$ of the off-shell 'tachyon' vertex $(: e^{ikX} :)$ by modifying the prefactor,

$$\langle (\epsilon^\mu(k) P_\mu)^{1+\alpha'(-k^2)} : e^{ik^\mu X_\mu} : \rangle. \tag{26.26}$$

This restores the operator to conformal spin one, if the 'polarization' is transverse $(k \cdot \epsilon = 0)$, so the prefactor commutes with the exponent. Without loss of generality we may choose the frame

$$k^\mu = n\hat{e}_+^\mu + (k^2/2n)\hat{e}_-^\mu, \tag{26.27}$$

in units with $2\alpha' = 1$. Then, if we choose $d - 2$ transverse polarizations $\epsilon \sim \vec{k}$, the original DDF operators [DDF72] are restored as $k^2 \to 0$. But we can also choose a longitudinal polarization orthogonal to k,

$$\epsilon^\mu = n\hat{e}_+^\mu - (k^2/2n)\hat{e}_-^\mu, \qquad (26.28)$$

to find an additional mode as $k^2 \to 0$.

The longitudinal operator is found by expanding to $O(k^2)$,

$$\langle : (\epsilon P)^{1-k^2/2} e^{ikX} : \rangle \simeq \langle P_- e^{inX_-} \rangle$$
$$+ k^2 \langle : [-P_+/2n - (n/2)P_- \ln(P_-) + nP_- iX_+/2n]e^{inX_-} : \rangle$$
$$+ O(k^4). \qquad (26.29)$$

Since $\langle P_- e^{inX_-} \rangle = 0$, the leading term at $O(k^2)$ is the physical state longitudinal operator expressed as

$$A_n^{(+)} = \langle : P_+ e^{inX_-} : \rangle + \frac{n^2}{2} \langle P_- \log(P_-) e^{inX_-} \rangle, \qquad (26.30)$$

after integrating by parts and collecting terms. This completes the spectrum generating algebra needed for the 'no-ghost theorem'.

It is also interesting that this same 'off-shell' trick has recently been used to construct an on-shell Reggeon vertex operator [BPST07], which for the leading trajectory for the flat space open string takes the form,

$$V_R^{(\pm)}(z) = [P_\pm(z)]^{1+\alpha' t} : e^{i\vec{k}\cdot\vec{X}(z)} : . \qquad (26.31)$$

This Regge string vertex allows one to simplify greatly the derivation of the high energy limits for multi-Regge amplitudes found in the earlier analysis of the Veneziano amplitudes. Just as is the case for the longitudinal operator in Eq. (26.24), the prefactor and the exponent commute so that the Regge power ($\alpha(t) = 1 + \alpha' t = 1 - \alpha' \vec{k}^2$) in the prefactor can compensate for the change in the conformal spin as one moves the zero-mass vector off-shell along the Regge trajectory.

26.3.2 Physical-state algebra

The spectrum generating algebra for these operators was understood in a series of papers leading to two independent proofs of the 'no-ghost theorem'. One important step was the awareness that at the critical dimension, $d = 26$, a new set of null states appears separating ghost-free theories for $d \le 26$ from theories with ghosts for $d > 26$. At the critical dimension Goddard and Thorn realized that the DDF operator generated [DDF72] the complete set of physical positive norm states. The classic paper by Goddard, Goldstone, Rebbi and Thorn [GGRT73] (GGRT) gave a concrete realization of this by quantizing the string in $d = 26$ in a world-sheet light-cone fixed gauge entirely in terms of 24 transverse oscillators, the need for longitudinal modes for $d < 26$ manifesting itself as a violation of Lorentz invariance. Here let us recount how the 'no-ghost theorem' appears in the covariant quantization scheme.

The algebraic approach evolved in several steps. The first step was to consider splitting the space of all on-shell string states in Eq. (26.15) into a 'spurious' subspace satisfying,

$$L_{-n}|\phi, p\rangle \quad \text{and} \quad L_0 L_{-n}|\phi, p\rangle = |\phi, p\rangle, \qquad (26.32)$$

and the orthogonal subspace of physical states in Eq. (26.35) at each level N, where $\alpha' p^2 = 1 - N$. The problem is then to establish a complete basis for each and analyze their metrics. Thorn and I were able to construct a basis of all spurious states [BT71]. By considering the metric for the spurious states basis,

$$L_{-1}^{n_1} L_{-2}^{n_2} \cdots L_{-m}^{n_m} |0, p\rangle, \qquad (26.33)$$

fixing $\alpha' p^2 = 1 - \sum_i n_i$, it was argued that eliminating all ghosts was equivalent to proving that the signature for this metric is unchanged under the replacement: $L_n \to a_n^0$. This construction is closely related to the idea of 'Verma modules' used later to classify two-dimensional conformal field theories or equivalently string vacua in nontrivial target geometries [Pol98]. In this context the physical state is a primary or highest weight state and the spurious states its descendents. A proof based on the Kac determinant for the metric of the spurious states was achieved by Charles Thorn [Tho85] many years later. The more direct approach of [BT71] considers the physical states.

The physical state generators form an elegant closed algebra:

$$[A_n^{(+)}, A_m^{(+)}] = (n - m) A_{n+m}^{(+)} + 2n^3 \delta_{n+m,0},$$
$$[A_n^i, A_m^j] = n \delta_{ij} \delta_{n+m,0}, \qquad (26.34)$$
$$[A_n^i, A_m^{(+)}] = n A_{n+m}^i.$$

Consequently, just as for the spurious states, a basis is formed by all ordered products,

$$|\text{Phys}, \{\vec{n}_i, m_i\}\rangle = \prod_i (\vec{A}_{-1})^{\vec{n}_i} (A_{-N}^{(+)})^{m_N} \cdots (A_{-1}^{(+)})^{m_1} |0, p\rangle. \qquad (26.35)$$

A complete basis including spurious states can be achieved by adding scalar operators, $\Phi_n = \langle \exp[i n X_-]\rangle$, with the extended algebra,

$$[A_n^{(+)}, \Phi_m] = -m \Phi_{n+m},$$
$$[\Phi_n, \Phi_m] = [\Phi_n, A_m^j] = 0. \qquad (26.36)$$

One can prove that these states form a complete and linearly independent basis by virtue of the map $\vec{A}_n \sim \vec{a}_n$, $A^{(+)} \sim a_n^+$, $\Phi_n \sim a_n^-$ under a continuous longitudinal boost: $X_- \to \eta X_-$, $P_+ \to P_+/\eta$. Also after dropping the Φ_n operators, it is clear by enumerations that the reduced basis must span all the physical states.

This elegant algebra was anticipated in an earlier paper with Peter Goddard [BG72] except for one small change. The correction term for the longitudinal operator doubled the central charge so that now there is an isomorphism between the Virasoro operators and the longitudinal operators at the critical dimension $d = 26$.

The 'no-ghost theorem' follows from the observation that the physical space algebra for any dimensions $j = 1, \ldots, d - 2$ is isomorphic under the substitution,

$$A_n^j \to a_n^j, \qquad A^{(+)} \to \mathcal{L}_n \doteq \sum_{j=1}^{25} a_{n+m}^j a_{-m}^j, \qquad |0, p\rangle \to |0, 0\rangle, \qquad (26.37)$$

where $\mathcal{L}_n$ are the Virasoro generators at $p = 0$ built exclusively from the 24 spatial oscillators. As a consequence, for $d \leq 26$ the metric is embedded in a (larger) positive definite norm space and all states must either have positive norm, or zero norm if there are linear dependences in this representation. The crucial details of zero norm or null states will not be repeated here. But suffice it to say that, at $d = 26$, all transverse oscillators are repeated in $\mathcal{L}_n$, which implies that the DDF operators [DDF72] are complete for positive norm states, with the extra longitudinal degrees of freedom decoupling as null states that pair up with spurious (ghost) states, as postulated in [Bro72] and realized explicitly in the classic GGRT paper [GGRT73] on light-cone string quantization. For $d < 25$, the only zero norm states begin with $A_{-1}^{(+)}$ adjacent to $|0, p\rangle$, as postulated by Charles Thorn and me in an earlier paper [BT71].

One can easily augment this algebra by conformal and anticonformal generators to repeat the 'no-ghost theorem' for the closed bosonic string, and by fermion generators for the superstring as was done in [Sch72] by Schwarz and in [BF73] by myself and Friedman. Modern approaches to the no-ghost theory use Faddeev–Popov ghosts and the BRST invariance following from the Polyakov path integral by a formalism familiar to the study of non-Abelian gauge quantization.

26.3.3 QCD and subcritical strings

Phenomenologically, the contrast between QCD and string theory hinged on the complete lack of the features that gave birth to QCD, namely hard scattering at short distances. Indeed this soft ultraviolet behaviour of string theory is central to its success as an example of a finite perturbative expansion for quantum gravity. The simplest example of very soft properties was noticed soon after the Veneziano proposal in terms of the exponential fall off in fixed angle scattering ($-t \simeq -u \simeq sf(\theta)$),

$$\lim_{s \to \infty} \frac{\Gamma[1 - \alpha(t)]\Gamma[1 - \alpha(s)]}{\Gamma[1 - \alpha(t) - \alpha(s)]} \simeq e^{-2\alpha' sf(\theta)\log(f(\theta))}, \qquad (26.38)$$

easily computed using Stirling's approximation. This fact alone discouraged some from seeking a QCD string and led to the idea that the original motivation for the discovery of string theory accidentally fell within the field of hadronic phenomenology.

However, personally one of my original motivations for exploring the strings for $d < 26$ (or better $d < 10$ superstrings) was the hope that the appearance of a unitary theory with longitudinal modes would allow for a consistent $d = 4$ string with some kind of 'Higgsing' to give the 'gluon' (or graviton/glueball for closed strings) a mass. Obviously the flat space string could not work because of the appearance of an unphysical closed string cut at one loop in the perturbative expansion. Still a few efforts in this direction were pursued by Polyakov, for example, who emphasized the appearance of the Liouville mode to 'soak up' the missing central charge when you go below the critical dimension. Nonetheless, the mathematics of subcritical strings is daunting and most rigorous developments in string theory focussed in other directions. Subsequent progress in superstring perturbation theory (in flat space) drove an increasing wedge between QCD and string theory. The superstring at the critical dimension $d = 10$ has a zero-mass graviton, six extra dimensions and extra supersymmetries not easily reconciled with QCD. The conventional wisdom turned to string theory as a theory of gravity interacting with matter. The QCD flux tube was viewed merely as an effective low energy or long distance description no more fundamental than flux tubes in a superconductor.

26.4 The counter revolution

Even at 40 years, string theory is still a young subject. With all the spectacular formal advances in the theory, the physical content and its place in nature remains very uncertain. The formal advances have been most striking in highly supersymmetric configurations with many conjectured and sometimes proven correspondences with field theory, also called 'gauge/string dualities'. This has led to two notions, pulling in opposite directions: (i) all self-consistent perturbative string theories are manifestations of single (unique) nonperturbative string field theory (sometimes called M-theory); (ii) there are an enormous number of stable or very long lived vacua that might be observed in our part of the Universe. Thus string theory is both 'unique' and hopelessly 'nonpredictive'. I think that both conclusions are not quite right. Just as with the large class of ultraviolet complete gauge theories, there may be a large class of distinct mathematically complete string theories. If QCD is dual (i.e. exactly equivalent) to a string theory, this would provide one, very special example, far from any particular supersymmetric or critical (e.g. ten dimensional) string theory.

Maldacena's AdS/CFT conjectured correspondence is in a real sense the first 'counter revolution' bringing string theory back to its roots as an alternative formulation of Yang–Mills theory, as we now view the hadronic phenomenology. Indeed in Maldacena's classic example, the $AdS_5 \times S_5$ IIB superstring is 'dual' to $\mathcal{N} = 4$ supersymmetric Yang Mills theory, and, as such, nonconfining and conformal invariant! There are no narrow flux tubes in this theory, but still a corresponding critical string theory. Strings in curved spaces can have extraordinary short-distance properties. Polchinski and Strassler noted that any string theory with similar conformal behaviour in the ultraviolet at the boundary of AdS_5 necessarily gives parton-like power fall-off at wide angles [PS02]. This observation alone changes dramatically our perspective on the historical and phenomenological origins of string theory, because now we have found a class of theories that nearly reproduce the ultraviolet behaviour of hadrons and not the long distance behaviour, exactly opposite to the original flat-space string example.

On a more personal note, Joe Weis and I spent much of our graduate school trying to incorporate local vector and axial currents into the on-shell Veneziano dual model.

This futile attempt was a dominant theme of my PhD thesis! Although we reproduced many current algebra results in the infrared, with hindsight we were fighting the extreme softness of flat space strings whose form factor (or energy distribution) has divergent root-mean-square radius, as emphasized later by Susskind. The way forward can also be seen as a result of Maldacena's AdS/CFT counter revolution. An S-matrix in d dimensions provides a map on shell ($p^2 = m^2$) to data in one less dimension at infinity. The miracle of Maldacena's string/gauge correspondence is that a five-dimensional on-shell string S-matrix amplitude can provide precisely the extra dimension needed to encode off-shell (conformal) transformation for off-shell four-dimensional currents of a gauge field theory on the boundary.

Hard scattering and local currents both require the extra radial fifth dimension of AdS_5. Where this line of reasoning leads is still uncertain. But in this era, string theory is perhaps no longer a radical departure from the rest of the Standard Model but a plausible framework for extending it. In this view, the graviton and the glueball are similar examples of low energy closed string spectra and the pion, rather than the quark, an example of a low energy state of the open string. Both Einstein–Hilbert gravity and chiral Lagrangian of pions are nice geometrical but ultraviolet incomplete effective field theories at low energies. I encourage the young string theorist to make further 'counter revolutions'. If we take Maldacena's string/gauge duality conjecture to suggest that almost all gauge theories are equivalent to string theories, perturbatively realized in the $1/N_c$ expansion, and if there are many inequivalent gauge theories, there ought to be many yet to be discovered inequivalent string theories. Finding **the** QCD string would give this substance. This is hard work of course, but constructing the QCD string does not imply solving it, anymore than finding QCD itself required finding its solutions. Numerical methods in string formulation should be explored just as they are in nonperturbative gauge theories.

Finding the QCD string should extend substantially our understanding of string theories in general. The next step should be to ask what additional 'axioms' or 'boundary conditions' define a particular string theory. In the same spirit it is also possible that we just need more patience and more experimental data to find out whether one of them is realized in nature as the correct way to combine the Standard Model gauge theories with a theory of quantum gravity. It is well to remember that string theory was discovered by input from strong interaction phenomenology before it took flight as a mathematical marvel. A lesson from the hadronic origins of string theory, I believe, is the expectation that better phenomenological insight coupled with technical advances does hold promise for finding an equivalent string theory alternative or even extension to the local gauge theories of the Standard Model.

Richard Brower received his BA in physics from Harvard College and his PhD from U.C. Berkeley. He is now Professor at Boston University. His early work included proving the no-ghost theorem for string theory. Current research includes high energy scattering in AdS/CFT and applications of lattice field theory for strong interactions beyond the Standard Model.

References

[Bro72] Brower, R. C. (1972). Spectrum-generating algebra and no-ghost theorem in the dual model, *Phys. Rev.* **D6**, 1655–1662.

[BC73] Brower, R. C. and Chu, G. (1973). Phenomenological six-pion amplitude, *Phys. Rev.* **D7**, 56–63.

[BDW74] Brower, R. C., Detar, C. E. and Weis, J. H. (1974). Regge theory for multiparticle amplitudes, *Phys. Rep.* **14**, 257–367.

[BF73] Brower, R. C. and Friedman, K. A. (1973). Spectrum-generating algebra and no-ghost theorem for the Neveu–Schwarz model, *Phys. Rev.* **D7**, 535–539.

[BG72] Brower, R. C. and Goddard, P. (1972). Collinear algebra for the dual model, *Nucl. Phys.* **B40**, 437–444.

[BH67] Brower, R. C. and Harte, J. (1967). Kinematic constraints for infinitely rising Regge trajectories, *Phys. Rev.* **164**, 1841–1844.

[BPST07] Brower, R. C., Polchinski, J., Strassler, J. M. and Tan, C. I. (2007). The pomeron and gauge/string duality, *JHEP* **12**, 005.

[BS73] Brower, R. C. and Susskind, L. (1973). Current algebra in the dual parton model, *Phys. Rev.* **D7**, 1032–1037.

[BT71] Brower, R. C. and Thorn, C. B. (1971). Eliminating spurious states from the dual resonance model, *Nucl. Phys.* **B31**, 163–182.

[DDF72] Del Giudice, E., Di Vecchia, P. and Fubini, S. (1972). General properties of the dual resonance model, *Ann. Phys.* **70**, 378–398.

[DHS68] Dolen, R., Horn, D. and Schmid, C. (1968). Finite energy sum rules and their application to $\pi-N$ charge exchange, *Phys. Rev.* **166**, 1768–1781.

[EKSS05] Erlich, J., Katz, E., Son, D. T. and Stephanov, M. A. (2005). QCD and a holographic model of hadrons, *Phys. Rev. Lett.* **95**, 261602.

[FGV69] Fubini, S., Gordon, D. and Veneziano, G. (1969). A general treatment of factorization in dual resonance models, *Phys. Lett.* **B29**, 679–682.

[GGRT73] Goddard, P., Goldstone, J., Rebbi, C. and Thorn, C. B. (1973). Quantum dynamics of a massless relativistic string, *Nucl. Phys.* **B56**, 109–135.

[GT72] Goddard, P. and Thorn, C. B. (1972). Compatibility of the pomeron with unitarity and the absence of ghosts in the dual resonance model, *Phys. Lett.* **B40**, 235–238.

[Got71] Goto, T. (1971). Relativistic quantum mechanics of one dimensional mechanical continuum and subsidiary condition of the dual resonance model, *Prog. Theor. Phys.* **46**, 1560–1569.

[Lov68] Lovelace, C. (1968). A novel application of Regge trajectories, *Phys. Lett.* **B28**, 264–268.

[Mal98] Maldacena, J. M. (1998). The large N limit of superconformal field theories and supergravity, *Adv. Theor. Math. Phys.* **2**, 231–252. [Also in Maldacena, J. M. (1999). The large N limit of superconformal field theories and supergravity *Int. J. Theor. Phys.* **38**, 1113–1133.]

[Man68] Mandelstam, S. (1968). Dynamics based on rising regge trajectories, *Phys. Rev.* **166**, 1539–1552.

[Nam70] Nambu, Y. (1970). Quark model and the factorization of the Veneziano amplitude, in *Proceedings of the International Conference on Symmetries and Quark Models, Wayne State University, June 18–20, 1969*, ed. Chand, R. (Gordon and Breach, New York), 269–277, reprinted in *Broken Symmetry, Selected Papers of Y. Nambu*, ed. Eguchi, T. and Nishijima, K. (World Scientific, Singapore, 1995), 258–277.

[NS71] Neveu, A. and Schwarz, J. H. (1971). Factorizable dual model of pions, *Nucl. Phys.* **B31**, 86–112.

[Pol98] Polchinski, J. (1998). *String Theory*. Vol. 1: *An Introduction to the Bosonic String*. Vol. 2: *Superstring Theory and Beyond* (Cambridge University Press, Cambridge).

[PS02] Polchinski, J. and Strassler, M. J. (2002). Hard scattering and gauge/string duality, *Phys. Rev. Lett.* **88**, 031601.

[Sch72] Schwarz, J. H. (1972). Physical states and pomeron poles in the dual pion model, *Nucl. Phys.* **B46**, 61–74.

[Sha69] Shapiro, J. A. (1969). Narrow resonance model with Regge behavior for $\pi\pi$ scattering, *Phys. Rev.* **179**, 1345–1353.

[tHo74] 't Hooft, G. (1974). A planar diagram theory for strong interactions, *Nucl. Phys.* **B72**, 461–473.

[Tho85] Thorn, C. B. (1985). A proof of the no-ghost theorem using the KAC determinant, in *Vertex Operators in Mathematics and Physics*, ed. Lepowsky, J., Mandelstam, S. and Singer, I. M. (Springer-Verlag, New York), 411–417.

[Ven68] Veneziano, G. (1968). Construction of a crossing-symmetric, Reggeon behaved amplitude for linearly rising trajectories, *Nuovo Cimento* **A57**, 190–197.

[Vir70] Virasoro, M. (1970). Subsidiary conditions and ghosts in dual-resonance models, *Phys. Rev.* **D1**, 2933–2936.

TOWARDS MODERN STRING THEORY

Part V

Beyond the bosonic string

27
Introduction to Part V

27.1 Introduction

Part V deals with the extensions of the Dual Resonance Model (DRM), i.e. the bosonic string, to include additional symmetries and degrees of freedom. These generalizations were originally motivated by the need to overcome the drawbacks of the DRM and obtain a more realistic model of hadrons. Such attempts were only partially successful, though, with hindsight, we can say that they added some essential elements for the construction of modern string theory.

One of the first modifications of the Koba–Nielsen amplitude aimed at incorporating the internal flavour symmetry of hadrons, and was proposed by Chan and Paton in 1969. As discussed in Section 27.2, these authors showed that an internal flavour symmetry can be introduced simply by multiplying the amplitudes by appropriate group theoretical factors. Such factors can be viewed as resulting from the presence of a quark–antiquark pair attached to the open string end-points, and carrying flavour quantum numbers.

However, the incorporation of flavour symmetry was not the only open issue. As discussed in the previous Parts, the main problems of the DRM were: (i) the presence of a tachyon; (ii) the absence of fermions, preventing the description of baryons; (iii) the presence of a critical dimension with an unrealistic value, $d = 26$. Attempts to solve these problems started very early, in fact immediately after the appearance of the Veneziano formula, and went on more or less in parallel with the understanding of the DRM and its reinterpretation as a quantum string (see Parts III and IV). From today's perspective, the cornerstone was the introduction of fermions, which led subsequently to the appearance of supersymmetry, a fundamental ingredient of modern superstring theory.

With respect to the tachyon problem (i), we have seen in Part III that in order to eliminate the nonpositive norm states from the physical subspace, one needs the value $\alpha_0 = 1$ for the intercept of the Regge trajectory. With this value, the lowest spin zero state is a tachyon and the next state is a massless vector. This is to be contrasted with the low energy spectrum of mesons: the low-lying scalar is the pion, with small mass (or zero in the so-called chiral limit of vanishing quark masses); the next state is a vector meson such as the ρ-meson, with

---

The Birth of String Theory, ed. Andrea Cappelli, Elena Castellani, Filippo Colomo and Paolo Di Vecchia.
Published by Cambridge University Press. © Cambridge University Press 2012.

a higher and certainly nonvanishing mass. Given these discrepancies, attempts were made to modify the Veneziano four-point amplitude. The most important result in this direction was the Lovelace–Shapiro amplitude for the scattering of four pions, containing a ρ-meson Regge trajectory consistent with high energy experiments, and incorporating a property of the amplitude coming from current algebra. We discuss this amplitude in some detail in Section 27.3.

Coming to problem (ii), Regge trajectories are found not only for mesons – such as the pion or the ρ-meson, which are bosons – but also, and with the same slope, for baryons – such as the proton or the neutron, which are fermions. Therefore, it was natural to search extensions of the dual amplitudes including fermionic external states. A dual model with fermions was first constructed by Ramond at the beginning of 1971. A few months later another model was proposed by Neveu and Schwarz, in the attempt to extend the Lovelace–Shapiro amplitude to any number of pions. These two models are discussed in Sections 27.4 and 27.5, respectively.

It was soon realized that the Ramond and Neveu–Schwarz models are two facets of the same theory, the Ramond–Neveu–Schwarz (RNS) model, as discussed in Section 27.6. The spectrum contains both fermions and bosons, and is much richer than that of the DRM. Unfortunately, it still contains a tachyon (the problem was eventually solved by the Gliozzi–Scherk–Olive projection, see Part VI). Nevertheless, it was soon recognized, first by Gervais and Sakita, that the RNS model had a new kind of symmetry relating bosons and fermions. This was the first occurrence of supersymmetry, as described in Section 27.7. As a matter of fact, the RNS model constitutes the backbone of modern superstring theory.

With respect to problem (iii), the quantization of the RNS model led to the critical dimension $d = 10$; this value, lower than the value $d = 26$ of critical dimension of the DRM, suggested that the construction of models with additional degrees of freedom could shift the critical dimension further down to $d = 4$. This idea was pursued in a series of papers by the 'Ademollo et al.' collaboration, in 1975–1976, but unfortunately led to values $d < 4$ of the critical dimension. Although this extension of the RNS model did not produce a more physical model, it led to the construction of the $\mathcal{N} = 2$ superconformal algebra containing one set of Virasoro generators L_n, two sets of fermionic generators G_r and $\overline{G}_r$ and one set of $U(1)$ Kac–Moody generators T_n. In the Eighties the $\mathcal{N} = 2$ minimal supersymmetric theories were constructed and they played an important role in the compactifications of the six extra dimensions after the so-called first string revolution (see the last Chapter of the Volume).

At the same time as Ramond's paper, another extension of the DRM was proposed by Bardakci and Halpern: they added fermionic oscillators that had an index transforming according to a representation of a global symmetry group G, instead of the Lorentz index of the RNS oscillators. They were called 'world-sheet fermions'. This construction, discussed briefly in Section 27.8, did not solve the problem of the tachyon either, but it yielded the first explicit realization of what are now called affine Lie algebras. This was the starting point of a series of many fruitful cross-fertilizations between mathematics and physics, which have been characterizing the developments of string theory over the years.

27.2 Chan–Paton factors

We have seen in Part II that the isotopic spin $SU(2)$ and its extension $SU(3)$ are approximate symmetries of strong interactions (the so-called flavour symmetries). Hadrons are classified according to irreducible representations of these groups: for instance, the three pions π^0 and $\pi^\pm$ form a triplet of isotopic spin $SU(2)$ and, together with the other pseudoscalar mesons, an octet of $SU(3)$. Therefore, in order to describe the hadrons correctly it was important to incorporate these symmetries in the DRM.

In 1969 Chan and Paton proposed to introduce an internal flavour symmetry by modifying the scattering amplitude. In the case of the symmetry group $U(N)$, they associated a matrix λ^a ($\lambda^a = (\lambda^a)_{ij}$, $a = 1, \ldots, N^2$, $i, j = 1, \ldots, N$), corresponding to a generator in the fundamental representation, to each of the four external particles of the scattering amplitude. Then for a given order of the external particles, e.g. 1234, the amplitude was modified as follows:

$$\frac{\Gamma(-\alpha(s))\Gamma(-\alpha(t))}{\Gamma(-\alpha(s) - \alpha(t))} \Longrightarrow \mathrm{Tr}(\lambda^{a_1}\lambda^{a_2}\lambda^{a_3}\lambda^{a_4}) \frac{\Gamma(-\alpha(s))\Gamma(-\alpha(t))}{\Gamma(-\alpha(s) - \alpha(t))}, \tag{27.1}$$

where the trace term is the so-called Chan–Paton factor.

The complete crossing-symmetric amplitude was obtained by summing over all possible permutations of the external particles that are not cyclic symmetric. In the case of the scattering of four particles there are six noncyclic permutations. The previous procedure can be easily extended to the case of M external particles, with a sum over $(M-1)!$ noncyclic permutations.

The presence of the Chan–Paton factors implies that each physical state of the DRM has N^2 internal degrees of freedom that transform among themselves according to the adjoint representation of the group $U(N)$. In fact, an equivalent way to introduce the Chan–Paton factor is to assume that an arbitrary string state is characterized not only by its momentum P and harmonic oscillator excitations α, but also by two indices i and $\bar{j}$, transforming respectively as the fundamental and the antifundamental representations of $U(N)$:

$$|P, \alpha; i, \bar{j}\rangle \Longrightarrow U_{ii'}|P, \alpha; i', \bar{j}'\rangle U^\dagger_{\bar{j}'\bar{j}}. \tag{27.2}$$

The two indices correspond to the N degrees of freedom of a quark and an antiquark located at the two end-points of the string.

Since, as a function of i and $\bar{j}$, the state is a Hermitian matrix, it can be expanded in terms of the $U(N)$ generators $\lambda^a_{i\bar{j}}$:

$$|P, \alpha; i, \bar{j}\rangle = \sum_{a=1}^{N^2} c_a \lambda^a_{i\bar{j}} |P, \alpha\rangle. \tag{27.3}$$

In conclusion, we see that, if we introduce the Chan–Paton factors, each physical state contains N^2 states, one for each $U(N)$ generator. When we compute a scattering amplitude involving M states, the scattering amplitude contains the trace of a product of M λ-matrices, each for every state, reproducing the recipe of Chan and Paton given in Eq. (27.1). Such a trace is invariant under the transformations of the group $U(N)$.

In particular, a massless gauge boson is now described by the state

$$\sum_{a=1}^{N^2} \lambda_{ij}^a A_\mu^a(P)(a_1^\dagger)^\mu |P, 0\rangle, \tag{27.4}$$

where $A_\mu^a(P)$ is the wave function corresponding to a given polarization and internal degree of freedom. This is the expression of a non-Abelian gauge boson field in Yang–Mills theory.

Although the Chan–Paton factors were originally introduced to describe a global flavour symmetry, they turned out to describe in fact a local $U(N)$ gauge symmetry. With the Chan–Paton factors, the local $U(1)$ gauge symmetry of the massless vector is upgraded to a local $U(N)$ gauge symmetry; moreover the massive string states transform according to the adjoint representation of the gauge group.

27.3 The Lovelace–Shapiro amplitude

Soon after the discovery of the dual amplitude by Veneziano, a more realistic four-pion amplitude was constructed by Lovelace and Shapiro. Since the pion has isotopic spin one, i.e. it is a triplet, consisting of π^+, π^- and π^0, the $\pi\pi$ scattering possesses three amplitudes corresponding to total isospin 0, 1, 2. In the Lovelace–Shapiro model these amplitudes are given by:

$$A^0 = \frac{3}{2}[A(s,t) + A(s,u)] - \frac{1}{2}A(t,u),$$
$$A^1 = A(s,t) - A(s,u), \tag{27.5}$$
$$A^2 = A(t,u),$$

where

$$A(s,t) = \beta \frac{\Gamma(1-\alpha_s)\Gamma(1-\alpha_t)}{\Gamma(1-\alpha_t-\alpha_s)}, \qquad \alpha_s = \alpha_0 + \alpha' s. \tag{27.6}$$

The amplitude in Eq. (27.6), discussed in the Chapters by Lovelace and Shapiro (see Part III), provides a model for $\pi\pi$ scattering with linearly rising Regge trajectories containing three parameters: the intercept of the ρ trajectory α_0, the Regge slope α' and the overall constant β. The first two parameters were determined by imposing that the Regge trajectory gives the spin of the ρ-meson $J = 1$ at $s = m_\rho^2$, and that the four-pion amplitude satisfies the so-called Adler self-consistency condition, coming from current algebra (vanishing of the amplitude for $s = t = u = m_\pi^2$, and of one pion mass).

The two previous conditions determine the Regge trajectory of the ρ-meson to be:

$$\alpha_s = \frac{1}{2}\left[1 + \frac{s - m_\pi^2}{m_\rho^2 - m_{\pi^2}}\right] = 0.48 + 0.885s. \tag{27.7}$$

These values were in reasonable agreement with experimental data. Note that for massless pions the intercept of the ρ Regge trajectory is equal to $\alpha_0 = \frac{1}{2}$ and its slope is equal to $\alpha' = 1/2m_\rho^2$. Having fixed the parameters of the Regge trajectory, the model predicts the masses and couplings of the resonances that decay in $\pi\pi$ in terms of a unique parameter β. The resulting values were in fair agreement with experiments. Moreover, the $\pi\pi$ scattering lengths could be computed and their ratio was equal to the current algebra prediction within 10%.

The construction of the Lovelace–Shapiro amplitude was an exciting result as it incorporated many experimental features of the pions. It was thus natural to try to generalize it to the scattering of N pions and from there extract the spectrum of mesons as done in the DRM. Unfortunately, all efforts in this direction were unsuccessful. As we will see in the next Sections, a consistent Dual Resonance Model for bosons and fermions was eventually constructed, but it was not suitable for the description of mesons and baryons.

27.4 The Ramond model

In a seminal paper, Ramond managed to extend the spectrum of the DRM to fermions. His argument reconsidered the procedure that leads from the Klein–Gordon to the Dirac equation for one particle and generalized it to the DRM. It is remarkable that his derivation did not use any connection with the idea of a relativistic string, but was based entirely on a 'correspondence principle' described in detail in his contribution to this Volume.

The description of the Ramond model requires some preliminary discussion of fermionic creation–annihilation operators. For a single degree of freedom, one can introduce the pair $(c, c^\dagger)$ that obeys the canonical anticommutation relations:

$$\{c, c^\dagger\} := cc^\dagger + c^\dagger c = 1, \qquad \{c, c\} = \{c^\dagger, c^\dagger\} = 0. \tag{27.8}$$

Just as for the bosonic case (Section 10.4), the physical meaning of these operators is that $c^\dagger$ adds (creates) one quantum, or excitation, to any state $|\psi\rangle$, and c removes (annihilates) it; the vacuum (or ground state) $|0\rangle$ has no excitations, i.e. $c|0\rangle = 0$. Due to the anticommutativity of operators, we have $c^2 = c^{\dagger 2} = 0$, meaning that the states can have at most one quantum: thus the anticommutation relations implement the Pauli exclusion principle. In the case of a single fermionic degree of freedom, the Hilbert space has dimension 2, being spanned by the vacuum $|0\rangle$ and the single excitation, $|1\rangle = c^\dagger |0\rangle$.

Physical observables are bilinear in fermion operators: for example, the number operator is $N = c^\dagger c$. Using (27.8), it is easily verified that the two states $|0\rangle$ and $|1\rangle$ are eigenvectors of N, with eigenvalues 0 and 1, respectively. Another observable is the energy: for the fermionic harmonic oscillator, it is given by the Hamiltonian $H = \omega c^\dagger c + E_0$. When turning to quantum field theory, just as in the bosonic case, the field operator describes the creation and annihilation of freely propagating quanta.

Having introduced the fermionic harmonic oscillator, we now briefly describe the basic idea of Ramond's model. In Part III we have seen that, in the DRM, the operator $Q_\mu(z)$ of

Fubini and Veneziano plays the role of a generalized position operator, with a corresponding generalized momentum operator $P^\mu(z)$ given by:

$$P_\mu(z) \equiv iz\frac{d}{dz}Q_\mu(z) = 2\alpha'\hat{p}_\mu + \sqrt{2\alpha'}\sum_{n\neq 0}\alpha_n z^{-n-1}, \qquad (27.9)$$

where z is the world-sheet complex coordinate. As anticipated, Ramond's observation stemmed from relativistic quantum mechanics, where the Klein–Gordon equation for relativistic spinless particles is modified into the Dirac equation for a fermionic relativistic particle by introducing, together with the momentum p^μ, the Dirac matrices γ^μ. To build a fermionic version of the DRM, Ramond applied the following 'correspondence principle': as the total momentum $\hat{p}^\mu$ corresponds to the zero mode of the generalized momentum operator $P^\mu(z)$, the Dirac matrices γ^μ should be the zero mode of some generalized fermionic operator, that was written:

$$\Gamma^\mu(z) = \sqrt{2\alpha'}\left[\psi_0^\mu + \sum_{n\neq 0}\psi_n^\mu z^{-n}\right]. \qquad (27.10)$$

The operators ψ_n^μ satisfy the algebra of the fermionic harmonic oscillator:

$$\{\psi_m^\mu, \psi_n^\nu\} = \eta^{\mu\nu}\delta_{m+n;0}, \qquad m, n \in \mathbb{Z}, \qquad \psi_{-|m|} \equiv \psi_{|m|}^\dagger; \qquad (27.11)$$

in particular, the anticommutation relations of the zero mode ψ_0^μ reproduce the Dirac algebra of gamma matrices, modulo a normalization factor.

Note that, as in the DRM case, the timelike modes of the additional fermionic operators give rise to states with negative norms (see Section 10.5) which should be removed from the physical spectrum by additional conditions. For this purpose, Ramond introduced the modes of $\Gamma \cdot P := \Gamma^\mu P_\mu$:

$$F_n = \frac{1}{2\alpha'}\oint_0 \frac{dz}{2\pi i} z^{n-1}\, \Gamma(z)\cdot P(z), \qquad n \in \mathbb{Z}, \qquad (27.12)$$

where the contour integration is carried out along a circle enclosing the origin of the complex plane. The F_n modes should be viewed as the fermionic counterpart of the Virasoro generators:

$$L_n = \frac{1}{4\alpha'}\oint_0 \frac{dz}{2\pi i} z^{n-1} P(z)\cdot P(z), \qquad n \in \mathbb{Z}, \qquad (27.13)$$

that annihilate the physical states in the DRM. In particular, in analogy to the physical condition $(L_0 - 1)|\Psi\rangle$ on states in the DRM, Ramond proposed the additional physical condition,

$$(F_0 + m)|\Psi\rangle = 0, \qquad (27.14)$$

which is now known as the 'Dirac–Ramond equation'. Indeed, given the form of the zero mode of (27.12),

$$F_0 \sim \gamma\cdot\hat{p} + \text{oscillator terms}, \qquad (27.15)$$

condition (27.14) can be interpreted as a wave-function equation for state $|\Psi\rangle$. It generalizes the Dirac equation exactly in the same way as the Virasoro operator L_0, see (27.13), generalizes the Klein–Gordon operator, $L_0 \sim p^2 +$ oscillators, and equation, $(L_0 - 1)|\Psi\rangle = 0$. In particular, as the square of the Dirac operator gives the Klein–Gordon operator, here we obtain the relation $F_0^2 = L_0$.

The F_n have the following expression in terms of the bosonic and fermionic harmonic oscillators,

$$F_n = \sum_{m=-\infty}^{\infty} \alpha_{-m} \cdot \psi_{n+m}, \qquad (27.16)$$

where the bosonic oscillators satisfy the algebra $[\alpha_n^\mu, \alpha_m^\nu] = n\, \eta^{\mu\nu} \delta_{n+m;0}$, as in the DRM. Note that the zero mode is proportional to the momentum operator: $\alpha_0^\mu \equiv \sqrt{2\alpha'}\, \hat{p}^\mu$.

With respect to the DRM, the Virasoro generators acquire a contribution from the fermions:

$$L_n = \frac{1}{2} \sum_{m \in \mathbb{Z}} :\alpha_{-m} \cdot \alpha_{n+m}: + \frac{1}{2} \sum_{m \in \mathbb{Z}} \left(\frac{n}{2} + m\right) :\psi_{-m} \cdot \psi_{m+n}: . \qquad (27.17)$$

Furthermore, we recall that the L_n are generators of the conformal transformations, the symmetry of DRM amplitudes. Together with the new generators F_n, they form an extended symmetry algebra, known as 'superconformal algebra'. Its expression can be found by using the explicit expressions for F_n and L_n in terms of the oscillators and reads:

$$\begin{aligned}
[L_m, L_n] &= (m-n)L_{m+n} + \tfrac{d}{8}m^3 \delta_{m+n,0}, \\
[L_m, F_n] &= (\tfrac{1}{2}m - n)F_{n+m}, \\
\{F_m, F_n\} &= 2L_{m+n} + \tfrac{d}{2}n^2 \delta_{m+n,0},
\end{aligned} \qquad (27.18)$$

where d is the spacetime dimension. Note the occurrence of both commutation and anticommutation relations, providing a generalization of the algebra of conformal transformations (see Part VI for further details).

This was the first occurrence of a symmetry relating bosonic and fermionic quantities, which was called supersymmetry (see the Chapter by Gervais). It was a very important result that was later extended from the two-dimensional world-sheet to four-dimensional spacetime by Wess and Zumino, leading to supersymmetric field theories (see Section 27.7).

The physical states are characterized by the following 'Virasoro-like' conditions, generalizing those obtained by Del Giudice and Di Vecchia for the DRM:

$$\begin{cases} L_n |\Psi_{\text{Phys}}\rangle = 0, & n \geq 0 \\ F_m |\Psi_{\text{Phys}}\rangle = 0, & m \geq 0. \end{cases} \qquad (27.19)$$

A remarkable feature of the Ramond model was that its spectrum of physical states did not contain a tachyon. In particular, the lowest state was a massless spin $\frac{1}{2}$ particle. However, as we shall see below, it was later recognized that the Ramond model was part of a larger model, whose spectrum still contained a tachyon.

27.5 The Neveu–Schwarz model

Soon after the Ramond paper, Neveu and Schwarz generalized the Lovelace–Shapiro amplitude to the scattering of many pions, and found that the consistency of the theory required the intercept of the leading Regge trajectory to be $\alpha_0 = 1$. Since this value was different from the phenomenological value $\alpha_0 \sim \frac{1}{2}$ found by Lovelace and Shapiro, the model lost its realistic features. For instance, a tachyon with mass $m^2 = -1/2\alpha'$ was present. The detailed construction of what Neveu and Schwarz called the 'dual pion model' is discussed in their contributions to this Volume.

The structure of the Neveu–Schwarz (NS) model is similar to that of the Ramond (R) model: together with the bosonic operator $P^\mu(z)$, a fermionic operator $H^\mu(z)$ is introduced, which is expressed in terms of fermionic oscillators, like the operator $\Gamma^\mu(z)$ of the R model. However, while $\Gamma^\mu(z)$ had an integer mode expansion, $H^\mu(z)$ has half-integer modes, as follows:

$$H^\mu(z) = \sqrt{2\alpha'} \sum_{r \in \mathbb{Z}+\frac{1}{2}} \psi_r^\mu z^{-r}. \tag{27.20}$$

The fermionic oscillators satisfy the usual anticommutation relations,

$$\{\psi_r^\mu, \psi_s^\nu\} = \eta^{\mu\nu} \delta_{r+s;0}, \qquad r, s \in \mathbb{Z}+\frac{1}{2}, \qquad \psi_{-|r|} \equiv \psi_{|r|}^\dagger, \tag{27.21}$$

differing only in the half-integer moding from those of the R model, Eq. (27.11). The half-integer mode expansion of the NS model reflects the antiperiodicity (under $z \to ze^{2\pi i}$) of the underlying fermionic operators; note, however, that all observables are bilinear in fermions, and thus still periodic (see the end of Section 27.7 for further details).

Once more, the timelike component of the fermionic oscillators led to negative norm states that needed to be discarded by imposing additional conditions for the physical states. It was found that for the value $\alpha_0 = 1$ of the intercept the physical states were annihilated by the Virasoro operators L_n and by the following additional fermionic operators,

$$G_r = \frac{1}{2\alpha'} \oint_0 \frac{dz}{2\pi i} z^{r-1} H(z) \cdot P(z)$$

$$= \sum_{n=-\infty}^{\infty} \alpha_{-n} \cdot \psi_{r+n}, \qquad r \in \mathbb{Z}+\frac{1}{2}, \tag{27.22}$$

as follows:

$$\begin{cases} L_n |\Psi_{\text{Phys}}\rangle = 0, & n > 0, \\ (L_0 - \frac{1}{2})|\Psi_{\text{Phys}}\rangle = 0, & \\ G_r |\Psi_{\text{Phys}}\rangle = 0, & r > 0, \end{cases} \tag{27.23}$$

to be compared with the conditions on physical states in the R model (27.19). In particular the term $-\frac{1}{2}$ in the second equation is a consequence of the normal ordering of bilinear operators. The corresponding term happened to be zero in the R model because of the exact

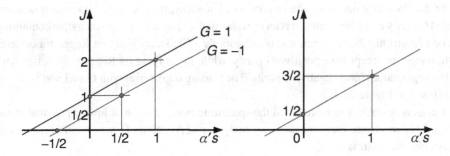

Figure 27.1 The spectrum of the RNS model: the bosonic Neveu–Schwarz sector (left), and the fermionic Ramond sector (right).

cancellation between the zero-point energy of the bosonic and fermionic oscillators, both with integer moding.

The G_r and the L_n obey a superconformal algebra,

$$[L_m, L_n] = (m-n)L_{m+n} + \tfrac{d}{8}m(m^2-1)\delta_{m+n,0},$$
$$[L_m, G_r] = (\tfrac{1}{2}m - r)G_{r+m},$$
$$\{G_r, G_s\} = 2L_{r+s} + \tfrac{d}{2}(r^2 - \tfrac{1}{4})\delta_{r+s,0},$$
(27.24)

which coincides with that of the R model, except for the central extension and the half-integer moding of G_r. As we shall see in the next Section, the difference in the central extensions can be reabsorbed into a redefinition of L_0.

27.6 The Ramond–Neveu–Schwarz model

The R and NS models, originally independent, were soon recognized to be part of a unique model, called the Ramond–Neveu–Schwarz (RNS) model. This had both spacetime bosonic and fermionic excitations, and its scattering amplitudes involved both fermions and bosons. For some time, only the amplitudes involving two fermions and an arbitrary number of bosons were known. The four-fermion amplitude was constructed later on, by means of the Corrigan–Olive fermion emission vertex (discussed here in the Chapter by Corrigan), using the operator projecting onto the subspace of physical states, discussed in the Chapters by Brink and Olive (see also the Chapter by Goddard in Part IV). Subsequently, one-loop diagrams for the RNS model were constructed. With the introduction of the N-Reggeon vertex, multiloop amplitudes were constructed as well, as discussed in the Chapter by Montonen.

With respect to the DRM, the spectrum of the RNS model is much richer, see Figure 27.1, with both bosonic and fermionic states – the would-be mesons and baryons, respectively. The fermionic (R) sector is tachyon free, with the fermions lying on linearly rising Regge trajectories spaced by integers. In the bosonic (NS) sector, the lowest state is a tachyon, with the bosons lying on Regge trajectories spaced by half-integers.

In the NS sector one can define a conserved quantity, the parity of the fermion number, called G parity in the Seventies: it is an operator with eigenvalue 1 (-1) on a state containing an odd (even) number of fermionic oscillators $\psi_r^\dagger$. The states lying on Regge trajectories with integer intercept have positive G parity, while those lying on Regge trajectories with half-integer values have negative G parity. The leading trajectories with $G = 1$ and $G = -1$ are shown in Figure 27.1.

Let us now discuss in more detail the spacetime interpretation of lowest physical states of the RNS spectrum. From the previous analysis and Appendix D, we obtain the spectrum of NS physical states:

$$\alpha'M^2 = \sum_{i=1}^{d-2}\left(\sum_{n=1}^{\infty}\alpha_{-n}^i\alpha_n^i + \sum_{r=\frac{1}{2}}^{\infty}r\psi_{-r}^i\psi_r^i\right) - \frac{1}{2} = N_{NS} - \frac{1}{2}, \qquad (27.25)$$

where r takes half-integer values. In the bosonic sector, the lowest state is a tachyon, corresponding to the oscillator vacuum of the NS sector, with $\alpha'M^2 = -1/2$. The next state is a massless spin one state, i.e. a gauge boson, that can be written

$$\epsilon(p) \cdot \psi_{\frac{1}{2}}^\dagger |0, p\rangle, \qquad (27.26)$$

where p^μ is the momentum and $\epsilon^\mu(p)$ is the vector describing the polarization of the gauge boson. The conditions in Eq. (27.23) imply $p^2 = 0$ and transverse polarization, $p \cdot \epsilon(p) = 0$, needed for the correct counting of degrees of freedom.

The fermionic spectrum (R sector) is determined by the formula

$$\alpha'M^2 = \sum_{i=1}^{d-2}\left(\sum_{n=1}^{\infty}\alpha_{-n}^i\alpha_n^i + \sum_{m=1}^{\infty}m\psi_{-m}^i\psi_m^i\right) = N_R. \qquad (27.27)$$

The lowest state in this sector is a massless spin one-half particle. This can be seen as follows: in the integer mode expansion (27.10), the zero mode ψ_0^μ obeys, modulo a factor $\sqrt{2}$, the algebra of Dirac gamma matrices in d spacetime dimensions,

$$\{\psi_0^\mu, \psi_0^\nu\} = \eta^{\mu\nu}, \qquad \psi_0^\mu = \frac{\gamma^\mu}{\sqrt{2}}. \qquad (27.28)$$

This implies that the Ramond oscillator vacuum $|A, 0\rangle_R$ has a spinor index A. Moreover, it should obey the following physical state condition:

$$F_0 u_A(p)|A, 0, p\rangle_R = \hat{p}_\mu \cdot \psi_0^\mu \, u_A(p)|A, 0, p\rangle_R = 0, \qquad (27.29)$$

where $u_A(p)$ is the spinor wave function. Since ψ_0^μ is proportional to the γ^μ matrix (with components γ_{AB}^μ), when acting on the vacuum with a spinor index A the condition (27.29) is equivalent to the massless Dirac equation for $u_A(p)$:

$$p \cdot \gamma_{BA} \, u_A(p) = 0. \qquad (27.30)$$

Therefore the Ramond oscillator vacuum is not the real vacuum of the theory, but it describes a massless Dirac particle.

Another feature of the R vacuum is its nonvanishing conformal dimension. This can be seen by comparing the two algebras in Eqs. (27.24) and (27.18). They coincide, except for the c-number central charges. It turns out that the central charge of the R sector can be brought to coincide with that of the NS sector if one redefines the operator L_0 as follows:

$$\hat{L}_0 \equiv L_0 + \frac{d}{16} = \sum_{n=1}^{\infty} (\alpha_{-n} \cdot \alpha_n + n\psi_{-n} \cdot \psi_n) + \alpha' \hat{p}^2 + \frac{d}{16}. \qquad (27.31)$$

The new operator $\hat{L}_0$ does not annihilate the R vacuum, which thus has conformal dimension equal to $d/16$.

Since the RNS model has a critical spacetime dimension $d = 10$, the dimension of the R vacuum is equal to $\frac{5}{8}$. Only in 1986 was it fully understood by Friedan, Martinec and Shenker that the NS and the R vacua are actually related by means of an operator $S^A(z)$ called the spin field, with spinor index and conformal dimension $\frac{5}{8}$. Specifically, the R vacuum is an excited state, obtained by applying the spin field to the NS oscillator vacuum. The whole spectrum of the RNS model is thus built from the (true) NS vacuum.

In conclusion, all the efforts discussed in Part V to obtain a consistent model for hadrons ended up being rather unsuccessful: in particular the model still had a tachyon. This problem was overcome in 1976, when the GSO projection was introduced. As we shall see in Part VI, this procedure consists in truncating the bosonic spectrum by projecting out the states with negative G parity (which include the tachyon, see Figure 27.1). An analogous truncation in the fermionic sector requires the lowest massless state to be a Weyl–Majorana fermion and the resulting theory is fully consistent. Anyway, the discovery of the GSO projection and of spacetime supersymmetry, although solving the tachyon problem, leads us further away from hadronic physics from which everything started.

27.7 World-sheet supersymmetry

The analysis of both the R and NS models revealed the existence of extra conditions for characterizing the physical states, in addition to the Virasoro conditions already present in the bosonic string. The question arose of the physical meaning of these additional fermionic operators F_n and G_r. In the case of the bosonic string, the Virasoro operators are the generators of the conformal symmetry of the theory; since this is the residual gauge symmetry of the Nambu–Goto action in the conformal gauge, its generators must annihilate the physical states.

Analogously, the vanishing of F_n and G_r on the physical states suggests the existence of an extended fermionic gauge symmetry present in the RNS model. This could not be proved at the time because the analogue of the Nambu–Goto action for the RNS model was not yet known. Nevertheless, proceeding in analogy with the bosonic string, where the Nambu–Goto action reduced to the quadratic scalar field action in the conformal gauge, it was assumed that, whatever the form of the generalized action, in a suitable gauge it would reduce to a quadratic expression involving both d real scalars and d real fermions. This is simply the action of d copies of Klein–Gordon and Dirac fields in two dimensions:

$$S = -\frac{T}{2} \int_M d\tau d\sigma \left(\eta^{ab} \partial_a X^\mu \partial_b X_\mu - i \bar{\psi}^\mu \rho^a \partial_a \psi_\mu \right), \qquad (27.32)$$

where $\psi^\mu(\sigma, \tau)$ is a Majorana, i.e. real, world-sheet spinor and the matrices

$$\rho^0 = \begin{pmatrix} 0 & -i \\ i & 0 \end{pmatrix}, \quad \rho^1 = \begin{pmatrix} 0 & i \\ i & 0 \end{pmatrix}, \quad \rho^3 = \rho^0 \rho^1 = \begin{pmatrix} 1 & 0 \\ 0 & -1 \end{pmatrix}, \qquad (27.33)$$

are Pauli matrices, providing a representation of the Clifford algebra in two dimensions,

$$\{\rho^a, \rho^b\} = -2\eta^{ab}. \qquad (27.34)$$

The gauge-fixed action (27.32) for the RNS model was first proposed by Gervais and Sakita in 1971; it is described fully in the Chapter by Gervais. Here we recall its main features.

The action (27.32) is invariant under the transformations

$$\delta X^\mu = \bar{\epsilon} \psi^\mu, \qquad \delta \psi^\mu = -i \rho^a (\partial_a X^\mu) \epsilon, \qquad \delta \bar{\psi} = i \bar{\epsilon} \rho^a \partial_a X^\mu, \qquad (27.35)$$

mapping the bosonic fields into the fermionic fields and vice versa (see Appendix D for a proof and further details). Historically, this was the first formulation of supersymmetry (in the Western world, see the Introduction to Part VI). The transformations (27.35) involve a fermionic parameter $\epsilon(\tau, \sigma)$, which is a two-dimensional Majorana spinor depending on the world-sheet coordinates. The action (27.32) is invariant provided that:

$$\partial_- \epsilon_+ = \partial_+ \epsilon_- = 0, \qquad \epsilon = \begin{pmatrix} \epsilon_+ \\ \epsilon_- \end{pmatrix}, \qquad (27.36)$$

where

$$\partial_\pm \equiv \frac{1}{2} (\partial_\tau \pm \partial_\sigma), \qquad (27.37)$$

are the derivatives with respect to the light-cone coordinates. This means that the upper (lower) spinor component $\epsilon_+(\epsilon_-)$ is an arbitrary function of $\xi_+ = \tau + \sigma$ ($\xi_- = \tau - \sigma$). Therefore, the gauge-fixed RNS action possesses a fermionic extension of the conformal symmetry, called superconformal, that is also expressed in terms of independent analytic/antianalytic (for Euclidean τ) reparameterizations of the light-cone variables. The generators of superconformal transformations are the F_n (R sector) and G_r (NS sector) operators introduced before, that annihilate the physical states, respectively, Eqs. (27.19) and (27.23).

Let us outline how this residual gauge symmetry is realized in the spectrum and the physical-state conditions are recovered. The Noether current corresponding to the symmetry (27.35) is the 'supercurrent'

$$J_a \sim \rho^b \rho_a \psi^\mu \partial_b X_\mu, \qquad (27.38)$$

which is clearly fermionic.

As in the case of the bosonic string, the generators of conformal transformations are constructed from the two-dimensional energy-momentum tensor. For the RNS action (27.32), this now also contains the contribution of the fermions:

$$T_{ab} = \partial_a X \cdot \partial_b X - \frac{1}{2}\eta_{ab}\partial^c X \cdot \partial_c X - \frac{i}{2}\bar\psi \left(\gamma_a \partial_b + \gamma_b \partial_a - \eta_{ab}\gamma^c \partial_c\right)\psi. \qquad (27.39)$$

Proceeding as before, we consider the Fourier transform of the energy-momentum tensor and of the supercurrent. In this way we generate a set of L_n modes from the energy-momentum tensor, and a set of fermionic oscillators F_n with integer n (R model), or G_r with half-integer r (NS model), from the supercurrent. Indeed, inserting the field expansions in terms of bosonic and fermionic oscillators, one can recover the expressions for L_n, F_n and G_r given above, see (27.17), (27.16), (27.22). In particular, being generators of a residual symmetry, they are classically vanishing. In the quantum theory, they give the physical state conditions (27.19) and (27.23).

An issue to clarify in this derivation is how to generate two models from a single Lagrangian. This can be understood by computing the equations of motion for the world-sheet fermions that read

$$\partial_+\psi_-^\mu = 0, \qquad \partial_-\psi_+^\mu = 0, \qquad \psi_\pm^\mu = \frac{1\mp\rho^3}{2}\psi^\mu, \qquad (27.40)$$

and the boundary conditions:

$$\int d\tau\, (\psi_+\delta\psi_+ - \psi_-\delta\psi_-)\,|_{\sigma=0}^{\sigma=\pi} = 0. \qquad (27.41)$$

In the case of an open string, these boundary conditions can be fulfilled in two different ways,

$$\begin{cases} \psi_-(0,\tau) = \eta_1 \psi_+(0,\tau) \\ \psi_-(\pi,\tau) = \eta_2 \psi_+(\pi,\tau) \end{cases} \qquad (27.42)$$

where η_1 and η_2 can take the values ± 1. In particular, if $\eta_1 = \eta_2$ we obtain the fermionic coordinate corresponding to the Ramond (R) model, while if $\eta_1 = -\eta_2$ we obtain those corresponding to the Neveu–Schwarz model. Indeed the previous conditions imply:

$$\begin{aligned} \psi_\pm^{NS} &= \sum_{r\in\mathbb{Z}+\frac{1}{2}} \psi_r e^{ir(\tau\pm\sigma)}, \\ \psi_\pm^R &= \sum_{n\in\mathbb{Z}} \psi_r e^{in(\tau\pm\sigma)}, \end{aligned} \qquad (27.43)$$

respectively for the NS and R models; note the integer and half-integer modings. The two boundary conditions thus originate the two sectors of the RNS theory.

Summing up, we found that the RNS model possesses an additional fermionic coordinate ψ^μ, and the superconformal symmetry, which is an extension of conformal symmetry involving fermionic generators. As the conformal symmetry of the bosonic string is derived from the reparameterization invariance of the nonlinear Nambu–Goto action, the superconformal symmetry also corresponds to the gauge symmetry of an extended fermionic action. This action was discovered in 1976: it is discussed in the Introduction to Part VI, and in the Chapters by Brink and Di Vecchia.

27.8 Affine Lie algebras

Contemporary with Ramond's paper, an another extension of the DRM was found by Bardakci and Halpern. These authors added fermionic oscillators ψ_n^I having an index I which transformed according to a representation $\mathcal{R}$ of a Lie algebra g, instead of the earlier Lorentz index μ of ψ_n^μ. The addition of these fermions did not solve the problems of the DRM (a tachyon was still present in the model), but it led to the explicit construction of affine Lie algebras, another important extension of conformal symmetries in two dimensions. The generators of this extended symmetry are the conserved currents $J^a(z)$, whose Fourier modes are built out of the fermionic oscillators ψ_n^I and Lie group generators $T^a = (T^a)^I{}_J$ as follows:

$$J_n^a = \sum_{r \in \mathbb{Z} + \frac{1}{2}} \bar{\psi}_{r;I}(T^a)^I{}_J \psi_{n-r}^J. \qquad (27.44)$$

This construction is valid for any Lie algebra g. The matrices T^a in the representation $\mathcal{R}$ have indices running over $I, J = 1, \ldots, \dim \mathcal{R}$, while a labels the generators of g. They satisfy the following commutation relations and orthonormality relations:

$$[T^a, T^b] = i f_{abc} T^c, \qquad \text{Tr}(T^a T^b) = k \delta^{ab}. \qquad (27.45)$$

It can be shown that if the ψ_n^I satisfy the usual algebra of the fermionic harmonic oscillators, then the previously constructed current modes satisfy the so-called affine Lie algebra:

$$[J_n^a, J_m^b] = f_{abc} J_{n+m}^c + nk\delta^{ab}\delta_{n+m,0},$$
$$[L_n, J_m] = n J_{n+m}, \qquad (27.46)$$

where k is an integer central charge called the 'level'. More details on this symmetry and the affine Sugawara construction relating the L_n with the J_m^a are described in the Chapter by Bardakci and Halpern. They constitute the body of what, in modern language, is called the current algebra or Kac–Moody symmetry of conformal field theory in two dimensions.

We conclude this Section by emphasizing that the above result is just one of many mathematical methods discovered and developed in string theory. Over the last three decades, this domain has been the main source of new methods in theoretical physics, which deeply influenced the discipline, and found many applications in other fields of physics and mathematics. These developments involved the latest advances in the domains of algebra, geometry and topology.

References

[DMS97] Di Francesco, P., Mathieu, P. and Sénéchal, D. (1997). *Conformal Field Theory* (Springer Verlag).
[GSW87] Green, M. B., Schwarz, J. H. and Witten, E. (1987). *Superstring Theory* (Cambridge University Press, Cambridge).
[Jac74] Jacob, M. ed. (1974). *Dual Theory* (North Holland, Amsterdam).

28
From dual fermion to superstring

DAVID I. OLIVE

I was privileged to be either a participant or an observer in many of the developments involved in string and superstring theory and my attachment to the CERN Theory Division during much of the Seventies gave me a grandstand view of much of it. In what follows I shall try to describe my own experiences, what I did, or tried to do, and what I saw and heard, because those are the things of which I am most certain. So, in particular, this does not aspire to be a comprehensive history.

I was present in the ballroom of the Hofburg when Gabriele Veneziano first presented his dual scattering amplitude to the wider world of theoretical physics during the Vienna Conference on High Energy Physics (28 August–5 September 1968). Despite the bad acoustics of the venue that experience changed my life and makes an appropriate start for my account [CERN68, Ven68], even though there is important prehistory.

The idea that there could be formulae for particle scattering amplitudes that could be, in some sense, almost exact fell on fertile ground. My scientific outlook had been formed in DAMTP, Cambridge, UK where the influence of the then charismatic figure, Geoffrey Chew, of the University of California at Berkeley still held sway. He and his school, influenced by the earlier work of Werner Heisenberg, had argued that scattering amplitudes of hadrons were the appropriate quantities to think about, rather than the quantum fields that create the particles. They had identified the important new concept of Regge trajectory but had otherwise become somewhat bogged down. The plausibility of Veneziano's formula for the amplitude involving four particles depended on the idea that all hadron Regge trajectories be (approximately) linear and have the same slope. This meant that within certain families of particles there held linear relations between the spin and the square of the mass and this implied the existence of an infinite number of particles, a situation difficult to describe in terms of fundamental fields. Very soon I set about my own research on the subject, publishing several papers that are now forgotten. However, one of my coauthors on my first paper became Provost of Boston University in 2005 (David Campbell) [COZ69]. Some time in the academic year following the Vienna Conference I was asked to give a talk on the newly developing subject to a general audience of DAMTP. I remember that during this

The Birth of String Theory, ed. Andrea Cappelli, Elena Castellani, Filippo Colomo and Paolo Di Vecchia. Published by Cambridge University Press. © Cambridge University Press 2012.

one of my colleagues, Dennis Sciama, suddenly interrupted and asked whether this was not the beginning of a new fundamental theory. I remember being taken aback by this remark but I think that it did sink in. Dennis had been the research supervisor of Stephen Hawking, Martin (now Lord) Rees and many others. Much later, after a spell in Oxford, he joined the faculty of SISSA in Trieste. A few years before he died (in 1999) I had the opportunity to ask him whether he remembered his prophetic comment and his simple answer was no!

On 5 June 1969 Richard Dalitz organized a one-day conference at the Royal Society in London, entitled 'Duality – Reggeons and resonances in elementary particle processes' and published in their proceedings [Dal70]. I attended, in particular hearing talks by Claud Lovelace and Chan Hong-Mo on the latest developments on what was now called multi-Veneziano theory.

In mid September 1969 (when I was 32) I arrived for a three month visit in the theory division at CERN and soon started a collaboration with Daniele Amati and Michel Le Bellac. I remember what an inspiring person Amati was and some impression of his charisma is preserved in thirty seconds of black and white film to be seen on the DVD published to celebrate the fiftieth anniversary of CERN. Because of the multiperipheral diagrams to be seen on the blackboard behind him, I suspect that this dates from a year or so before the Vienna Conference.

By that time important progress had been made in constructing amplitudes describing the scattering of any number of spinless particles and displaying an important property, namely the presence of simple pole singularities in intermediate channels, the lowest of whose residues factorized into products of similar amplitudes for fewer particles. This property guaranteed that the pole corresponded to the propagation of an internal particle (Olive [Oli64]). This feature interested me since I had devoted time in 1963, whilst still a research student, to deducing it from unitarity, a fundamental property that hitherto seemed to be missing from the framework. The construction of these dual multi-amplitudes had been achieved by a number of groups around the world, in CERN, Berkeley, Madison and Copenhagen.

Together Amati, Le Bellac and I wrote two papers devoted to the operator formalism for vertices and propagators in the Dual Resonance Model [ALO70a, ALO70b]. We were later joined by Victor Alessandrini from Argentina in writing a third paper [AALO70]. I do not intend to describe these papers since they are published and speak for themselves. After we had finished I remember Daniele Amati telling us of a letter he had received from yet another fellow Argentinian, his friend Miguel Virasoro, explaining a remarkable algebra of infinite dimension, now named after him. We felt suitably foolish at not having thought of this beautiful structure.

Together the four of us decided to give some lectures on the subject in CERN: 23 February (operator formalism), 2 March (factorization and the twist operation), 9 March (gauge conditions), 16 March (gauge conditions) and 23 March 1970 (construction of loops). Afterwards Maurice Jacob invited us to write them up for publication in the new startup journal *Physics Reports* (of which he was a founding editor) where they duly appeared in July 1971 [AALO71].

At the suggestion of Amati my leave from Cambridge had been extended by a second term until mid April 1970 and as a recompense I was asked to give a course of lectures in Cambridge on the Dual Resonance Models and I readily complied, giving a much extended version of the CERN lectures in May 1970, entitled 'Operator formalism in multi-Veneziano theory' (about ten lectures on Mondays and Fridays at 12). These were well attended and the audience included Ed Corrigan, Peter Goddard, Michael Green, Marc Grisaru (on leave) and possibly Jeffrey Goldstone (Jeffrey confirms that he noted the dates in his diary but no longer recalls whether he attended).

When next I briefly returned to CERN for two weeks in late June 1970, I remember Claud Lovelace giving a talk about his evaluation of the twisted loop amplitude in four dimensions. He had found what sounded like a disaster in that the result of a precise calculation revealed a branch point singularity in a certain channel. I explained to him that if only the singularity could be a simple pole, rather than a branch cut, it could be consistent with general principles of scattering matrix theory, such as unitarity, as it could be interpreted as being due to the propagation of a new sort of particle (I have to say that Claud denied all memory of this episode when I tried to check it with him in December 2007).

In any case I did not think more about this until more than a year later, when we next overlapped briefly in CERN together (probably July 1971). Then he came up to me, very excited, with the news that he could indeed make the singularity a simple pole if he assumed spacetime had 26 rather than 4 dimensions as well as making an assumption about extra gauge conditions. The singularity could be interpreted as being due to the propagation of a new particle state, then identified as the Pomeron (and later as the graviton). His paper was written up in the USA after he had left CERN and submitted in February 1971 [Lov71]. I should mention that Lovelace was a remarkable person, held in awe by all those around him. His role in CERN had been analysis of numerical data produced by the accelerators and kept as punched cards in enormous cabinets in the corridor outside his office. But his preferred recreation was the Dual Resonance Model. Of course his work was published and his strange result about 26 dimensions was very influential.

At the beginning of April in 1971 I attended a meeting at Tel Aviv University entitled 'Duality and symmetry in hadron physics'. Proceedings were published [Got71] and are very interesting to read in retrospect, particularly a round table discussion led by Murray Gell-Mann and involving Daniele Amati, Sergio Fubini and Chris Schmid who were all influential figures of that time. A highlight for me personally was my first meeting with Joël Scherk who gave a talk on what he called the zero-slope limit. The slope was that of the linear Regge trajectory and the concept showed how conventional Feynman diagrams could be related to the diagrams for the dual Veneziano amplitudes. Another highlight was meeting John Schwarz for the first time.

In June 1971 I attended part of a symposium on 'Basic questions in elementary particle physics' taking place in the Max Planck Institute in Munich and dedicated to Werner Heisenberg on the occasion of his seventieth birthday. I thought that I should take advantage of the opportunity to explain to the great man what was then called multi-Veneziano theory and its possible connection to the quark model as suggested by the Harari–Rosner diagrams. Again there are proceedings available [Oli71] but they do not record Heisenberg's reaction to my talk. Immediately after it he grabbed the microphone and protested that the quark model was not physics. (Ivan Todorov who also attended this conference remembers Heisenberg's words slightly differently as an attack on the 'parton model'.) I later learnt from Harald Fritsch and others that there was a background history of some strong disagreements between Heisenberg and physicists at Caltech, particularly, Richard Feynman.

On 1 July 1971 I arrived in CERN as a staff member at the start of a longer stay that turned out to be just over six years. I had made a momentous personal decision: I had accepted the offer by Theory Division of a staff position, for three years in the first instance. I wanted to devote all my time to working on what promised to be the theory of the future, what was still called the Dual Resonance Model, and it seemed to me that CERN Theory Division was then the centre of the universe for that.

My decision involved a gamble that not many others had made. This is because in Cambridge I had already acquired tenure in 1967 and I had to sacrifice that in view of what I saw as a wonderful opportunity. Amati had gathered together from around Europe a galaxy of young enthusiasts for this new subject as research fellows and visitors. This was possible as centres of activity had sprung up around Europe, in Copenhagen, Paris, Cambridge, Durham, Torino and elsewhere. I already knew Peter Goddard from Cambridge University who was in his second year as Fellow, Lars Brink from Chalmers in Gothenburg was just starting, as was Joël Scherk from Orsay, in Paris, all as Fellows, and destined to be collaborators and, particularly, close friends. Also present as Fellows were Paolo Di Vecchia (who arrived in January 1972), Holger Nielsen, Paul Frampton, Eugène Cremmer, Claudio Rebbi and others. Many visitors came from Italy, Stefano Sciuto, Nando Gliozzi, Luca Caneschi and so on. Visiting from the United States for the academic year were Charles Thorn and Richard Brower. Summer visitors included John Schwarz, and later Pierre Ramond, Joel Shapiro, Korkut Bardakci, Lou Clavelli and Stanley Mandelstam, all from the United States.[1]

So at that time CERN Theory Division seemed to us to be the world centre of activity on what was then called 'dual resonance theory' or 'multi-Veneziano theory' and soon to become string theory. I think Chan Hong-Mo and Claud Lovelace had just left their staff positions for the Rutherford Laboratory and Rutgers University respectively. Present for the academic year 1971–1972 was Murray Gell-Mann, and when he was not touring Europe picking up honorary degrees, he seemed to spend his time agonizing over what he came to call 'quantum chromodynamics' (QCD), though I did persuade him to participate at least once in our seminar activities.

We were a large enough group to have our own dedicated Dual Resonance Model seminar programme supplemented by discussions over lunch. Ideas on dual resonance were advancing on several fronts and it was necessary to keep abreast of different developments whilst concentrating on one chosen topic. We were driven by our shared common belief that

---

[1] I no longer remember the exact dates. As these might be of historic interest any exact information would be welcome.

we were working on the theory of the future and we were trying to work out what that was specifically. It was clearly something new and unlike conventional quantum field theory and that was attractive to us. We were well aware of the scepticism of the wider theoretical physics community. But we thought that was understandable and it was up to us to find a version of this future theory that was indeed consistent with all general principles known. Very likely the search for consistency would narrow down the range of possibilities.

One aspect of the situation was my awareness of the need to convince a wider theory audience. Having first met Arthur Wightman in 1967 at a conference in Rochester, I was aware of the axiomatic field theory community and their belief in mathematical precision. I felt that dual theory had a need for precision and that the operator formalism provided the means. It furnished a wonderful arena for proving precise relations that had a physical significance and that was very attractive. Of course some of these things could be pictured in terms of strings or rubber bands or their world-sheets and Holger Bech Nielsen kept this very much in our minds with his lively presentations. We would readily exploit all this as valuable intuition but we did not know (at that stage) how to obtain the precise, rigorous results that we sought this way. I should add what a culture shock CERN and Geneva was to someone like myself who had been brought up in a northern part of Europe (Edinburgh).

Early in the year 1971 there had been at least three exciting developments, at first sight independent. One was Lovelace's observation concerning 26 dimensions [Lov71], the second the construction of the so-called DDF states which is explained by Paolo Di Vecchia, in his article entitled 'The birth of string theory' [DiV07], and briefly below. The third was an imaginative and beguiling idea proposed by Pierre Ramond, christened the 'dual fermion' [Ram71], which indicated a possible way to remedy what was a serious flaw of the dual theory, namely the absence of fermions. These three topics provided an agenda for 1971–1972 at CERN as I shall describe. Eventually they would all fit together smoothly.

From Cambridge came my research student Ed Corrigan who had his PhD to finish off. Earlier in Cambridge (together with another student of mine, Claus Montonen, who was later to collaborate with me in a much cited paper) we had been interested in incorporating quark ideas in the theory and even wrote a paper which we decided not to submit for publication. This was based on ideas of Peter Freund and his collaborators. (I had first met Peter in 1961 before he had reached the USA.)

Ed and I decided to look at the ideas of Pierre Ramond mentioned above together with the apparently related ones of John Schwarz and André Neveu and to try to establish a precise relationship between them. Pierre's notion of the 'dual fermion' started with the Dirac equation whereas John and André had found something different that nevertheless looked very much as if it was related. I think I had heard about this in Tel Aviv. We worked on developing what was called a fermion emission vertex that in some precise sense acted as an intertwining operator between the two ideas [CO72].

After Ed returned home I started having discussions with Lars Brink from Chalmers in Gothenburg, whose physics department I had already visited in 1968. (I later learnt that William Chalmers on whose Will it was founded had a Scots father. Gothenburg and Edinburgh both used to be ports with a busy trade.)

The bosonic dual model was progressing very well thanks to our colleagues in CERN, but it lacked the crucial fermion ingredients. The dual fermion theory was very promising but had serious shortcomings which we sought to address. What existed so far were amplitudes for a single dual fermion emitting or absorbing dual mesons, or, by duality, tree diagrams with a single fermion line and otherwise made of meson lines. What was needed were tree amplitudes containing two or more dual fermion lines. The simplest goal of this kind was the construction of the four-fermion amplitude, with four external lines, all fermions. This tree diagram thus consisted of two fermion lines interacting by means of an exchange of dual meson states. This construction was not straightforward and the reason was that ghost states had to be eliminated in the exchange and that effecting this, if possible, would have an unpredictable effect in the cross channels.

It is necessary to backtrack to explain more about the operator formalism and the concomitant ghost states (that is, negative norm states) and the sense in which they might be eliminated.

The explicit formulae for the dual amplitudes were most conveniently expressed in terms of what was called an operator formalism. This involved harmonic oscillators that formed an infinite number of spacetime vectors. These oscillators were bosonic in the original dual model following from Veneziano's amplitude, and both bosonic and fermionic in the dual fermionic theory developing from the ideas of Ramond and of Neveu and Schwarz [NS71]. The bosonic and fermionic oscillators are denoted a_n^μ and b_r^μ respectively, where μ runs over the components of spacetime and n runs over the positive integers. r also runs over positive integers in the construction of Ramond and over positive half-integers in the construction of Neveu and Schwarz. These satisfy the commutation or anticommutation relations:

$$[a_m^\mu, a_n^\nu] = m\, \delta_{mn} \eta^{\mu\nu}, \quad \{b_r^\mu, b_s^\nu\} = \delta_{rs} \eta^{\mu\nu},$$

where

$$a_m^{\mu\dagger} = a_{-m}^\mu q \quad \text{and} \quad b_r^{\mu\dagger} = b_{-r}^\mu.$$

These act on a vacuum state $|0\rangle$ which is annihilated by the destruction operators a_m^μ, $m > 0$ and b_r^μ, $r > 0$, and so create a Fock space with a natural scalar product which is not positive definite because the spacetime metric $\eta^{\mu\nu}$ is not. Yet these Fock space states have something to do with a description of the different quantum states of the elementary particles of the theory. It is a fundamental principle that the scalar product on the quantum mechanical space of states is positive definite and these two facts must be reconciled.

A similar conundrum occurs with the polarization states of the photon in quantum electrodynamics and the resolution is similar. There the polarization four-vector has four components but only a two-dimensional subspace of states is realizable physically and that subspace has a positive definite metric. The unphysical components, which include negative norm states, cannot be produced in any experiment starting with electrons and positrons because of gauge conditions that have the effect of decoupling the longitudinal and timelike polarizations of the photon. This decoupling is the consequence of gauge conditions related to the gauge invariance of electromagnetic theory first clearly enunciated by Hermann Weyl.

Emilio Del Giudice and Paolo Di Vecchia realized that there can also be gauge conditions in the Dual Resonance Model and that they are related to the discovery of Virasoro already mentioned [DiV07]. In the operator formalism any state that can be prepared from scalar particles and that will, on the mass shell, satisfy the following conditions, will be known as a physical state:

$$L_n|\Psi\rangle = 0, \quad n > 0, \quad (L_0 - 1)|\Psi\rangle = 0,$$

where L_n are the generators of the Virasoro algebra. As expressed by him they are bilinear expressions in the annihilation–creation operators and satisfy the commutation relation:

$$[L_m, L_n] = (m-n)L_{m+n} + \frac{d}{24}m(m^2 - 1)\delta_{m+n,0}.$$

Of course Virasoro first wrote down this algebra and even conjectured that it had to do with the elimination of ghosts, but he overlooked something, namely the numerical term proportional to the dimension of spacetime d. That was first discovered by Joe Weis who failed to publish this crucial result (and later died in a mountaineering accident in the Alps in 1978; this death is wrongly dated as 1977 on the University of Washington Physics website).

In achieving the infinite number of gauge conditions, what had hitherto been a free parameter, namely the intercept of the leading Regge trajectory, has to be taken to equal unity. An immediate consequence of this is that the basic scalar particle of the theory has to be a tachyon and this is, of course, unsatisfactory. However, it is the only tachyon in the theory and so there might be some as yet unforeseen way of eliminating it. Another consequence is that the next particle on the leading Regge trajectory, which has unit spin, has to be massless. This very much suggests that it is a gauge particle. Likewise what had been thought of as a Pomeron looked more like a graviton. The price that the Dual Resonance Model has to pay for consistency with fundamental principles is that it looks increasingly less like a theory of strong interactions and more like a unified theory. Not only does it possess massless gauge particles but also massless gravitons. Of course the same was true of the dual fermion theory (if indeed it does really exist) and it had the innate advantage of possessing fermions. As I remember, this idea of unification of gauge and gravitational interactions was much discussed by the community in CERN Theory Division in the year 1971–1972 even though this was before the discovery of asymptotic freedom and the formulation of the Standard Model. When QCD did emerge I seem to remember that our reaction was a lack of surprise as we knew from the dual models that all fundamental interactions besides gravity had to be gauge mediated.

It was natural to try to find as many examples of physical states as possible and a remarkable construction for an infinite dimensional space of them was found by Del Giudice, Di Vecchia and Fubini, the so called DDF states [DiV07]. More remarkably it was found by Richard Brower [Bro72] and by Peter Goddard and Charles Thorn [GT72] that these comprised a complete set of physical states precisely when the dimension of spacetime, d, equalled 26, the value foreseen by Lovelace, and now called the critical dimension. Moreover, they then formed a positive definite subspace of the complete oscillator subspace. This result was called the 'no-ghost theorem' and finally put the theory on a firm footing as a consistent theoretical idea, aside from the problem of tachyons. Massless gauge particles were acceptable as they provided good candidates for fundamental interactions but the additional price of 26 dimensions to spacetime did generate considerable ridicule at the time.

Nevertheless this was an extremely nontrivial result which looked like an entirely new and unexpected piece of mathematics. Obviously there was much more to be discovered. At the Chicago conference in September 1972 I tried to give a short presentation of these results [Oli72] together with their string interpretation that had recently been achieved by Goddard, Goldstone, Rebbi and Thorn [GGRT73] but I have to say that the account that appears in volume 1 of the proceedings does not accurately represent what I said. Unfortunately I do not remember supplying a written account of what I said.

Now I can return to the four-fermion amplitude since it is after the Chicago meeting that I took up the problem in earnest with Lars Brink. As mentioned above the problem was to exchange only bosonic states that are physical between the two fermions. Their structure was also understood in the dual fermion model though the critical dimension of spacetime was 10 rather than 26 (and so closer to the reality of 4, and hence taken as encouraging).

Lars and I agreed that a suitable toy problem was the construction of what was called the planar loop amplitude, first in the nonfermionic theory and then in the fermionic version. One has to find how to ensure that only physical states with positive norm propagate around the loop. We went around in circles with this problem until a new CERN institution instigated by Amati in his new role of leader of Theory Division came to the rescue. This was the two-minute seminar sessions by means of which members of Theory Division quickly introduced themselves and their theoretical problems to each other at the beginning of the academic year. At that time Theory Division was roughly one hundred strong so even a ration of two minutes took some time.

I described our problem and someone I did not know very well, Josef Honerkamp, came up to me after with very helpful advice. He explained to me that Richard Feynman had constructed a Feynman diagram for gauge particles circulating in a loop by means of a projection operator for the physical polarizations. His account appeared only in the proceedings of a meeting in Poland and he was thinking of this as a toy problem for quantizing gravity. This was precisely the clue we needed and we were extremely grateful to Honerkamp.

The required projection operator acts in the oscillator Fock space and is known because the precise nature of the DDF states is known. However, the form that those give is of little use because one wants to express it in terms of the generators of the Virasoro algebra, in order to use the gauge conditions. Lars and I found out how to do this, obtaining what we thought was a neat and practical expression which was positive definite and manifestly collapsed to the unit operator on any physical state. Moreover we could use it to construct the desired loop amplitudes (in February 1973) [BO73a, BO73b]. Its Lorentz invariance was checked despite the projection operator not having this property. This was the first rigorous construction of a unitary, manifestly Lorentz invariant loop amplitude in string theory. I emphasize this since there were many previous constructions, all inching further towards the truth, but this did finally involve all the steps needed, at least for the single loop. It also gave a better understanding of Lovelace's construction of the Pomeron/graviton which was hitherto based on inspired guesswork. The construction was also performed for the loop with bosons propagating around a loop in the fermionic theory (in the appropriate critical dimension, now 10).

A footnote to this is that some time after, Lars and I received a letter from Richard Feynman (dated 5 June 1973) thanking us for making use of his tree theorem and explaining it 'with clarity and simplicity' (and telling us of a subsequent paper developing his idea for diagrams with more than one loop). I later learnt that we had John Schwarz to thank for spurring Feynman to do this.

Before the next step in the construction of the four-fermion amplitude there was a sideways step, the elucidation of the no-ghost theorem for the Pomeron with Joël [OS73a]. Next was an analysis of the gauge properties of the Reggeon–Pomeron vertex by Joël, Lars and myself [BOS73]. That prepared the groundwork for understanding how the gauge conditions worked relative to the fermion emission vertex. The upshot was that something unexpected happened and this was discovered by Lars Brink, Joël Scherk and myself with Claudio Rebbi (in July 1973) [BORS73]. The effect was to make the four-fermion amplitude calculation resemble the planar loop calculation just performed rather more than we had anticipated.

Two further papers on this followed, one with Joël Scherk in August [OS73b] and the other with Peter Goddard and Ed Corrigan in October [CGOS73] who had visited CERN that summer. The latter group was assisted by R. A. Smith, a pure mathematician colleague from Durham who never came to CERN that I remember. The conclusion of these papers was that the four-fermion amplitude constructed by exchanging physical states in the t-channel did, as hoped, possess poles (rather than a cut) in the s-channel which furthermore corresponded to the same mass spectrum as that exchanged. There was however one difference between the details of the two spectra namely that the input tachyon was pseudoscalar whereas the output one was scalar. Thus, in terms of the analytic properties, our result confirmed that we were constructing what looked like a consistent fermion theory but we were faced with a new mystery, the parity doubling of the scalars. Earlier, in September, Mandelstam had produced, out of the blue, a four-fermion amplitude with the same properties and also noted the parity doubling. At almost the same time there was a related paper by Schwarz and Wu. In retrospect I am puzzled as to why the problem of parity doubling and tachyons was left dangling. I think the feeling was that the really hard work had been done with the construction of the four-fermion amplitude and the result that it was indeed meromorphic.

On 13–15 September 1973 Amati, Rebbi and I held a topical meeting on Dual Resonance Models in CERN Theory Division but there were no proceedings. Participants came from Europe and America but I know of no surviving records. One significant memory is the attendance of Abdus Salam. Apart from the activity that culminated in October there was a fallow period for me personally until the next summer.

Almost a year later, 1–10 July 1974, the Particle Physics conference was held in London, at Imperial College, and I was invited to give a plenary talk on 'dual models' [Oli74]. Besides emphasizing the importance of the string concept I wanted to try to emphasize the conceptual revolution that occurred some years previously with the concept of string theory as a unified theory with both gauge and gravitational interactions. At that time I had been so immersed in the calculations I have outlined that I was not really aware of the Standard Model (nor were many other speakers at the conference as far as I remember). At any rate everything is published in the proceedings.

In his book *Rochester Roundabout* [Pol89] surveying the series of 'so-called Rochester conferences', John Polkinghorne mentioned my talk and said it would be very prophetic if confirmed. This was probably a polite way of expressing his scepticism of my contention of unification which had so inspired our work in CERN.

What was possibly another reaction to my talk was a comment Abdus Salam later made to me at Imperial College, probably in the very early Eighties. His assessment then was that the 'reason string theory failed was that it claimed too much'.

Less than a week after the London meeting I was in Aspen for a workshop on dual resonance theory organized by John Schwarz. Amongst the participants were Eugène Cremmer, Joël Scherk, Jean-Loup Gervais, Bunji Sakita, Yoichiro Nambu, David Fairlie, Ed Corrigan, Peter Goddard, Michael Green, Lars Brink, Michio Kaku, Keiji Kikkawa, Joel Shapiro besides John himself.

Other members of the Summer Institute included Richard Feynman, Murray Gell-Mann, Glennys Farrar, Heinz (and Elaine) Pagels and Itzhak Bars while other physicists around included Steven Weinberg and Lennie Susskind.

Of course there were many interesting discussions but the one that made the most impact on me personally was one that Murray Gell-Mann conducted with Itzhak Bars concerning the 't Hooft–Polyakov monopole. Of course I already knew about it, and had heard Gerard 't Hooft talk about it at the London conference just before. But I had not really understood it until I heard Gell-Mann talk about it. So for me, that was a revelation that influenced my subsequent research.

Much of the remainder of my time until I left CERN for Imperial College at the end of September in 1977 was spent thinking about magnetic monopoles, but that is another story. However, I think I should digress to explain that there were also nonscientific reasons for this change in research direction from string theory. This was that my time at CERN was drawing to an end and my family wanted to return to Britain. So I had to find a job there but Amati warned me (I no longer remember precisely when) that 'you are unemployable because you do string theory'. Such was the demise of string theory by then that I was forced to accept that this was the best advice of an enthusiast for the string theory programme. Of course many others had job problems because of their sustained interest in string theory. I remember Joël Scherk complaining later that he felt obliged to work on supergravity whereas his real conviction lay with string theory. In any case gauge theories were now more fashionable because of the undeniable triumph of asymptotic freedom and I felt that monopoles and solitons were a promising direction that could eventually link up with string theory in which I still believed. I had first met Tony Skyrme at Harwell (the Atomic Energy Research Establishment) the year before starting my research in 1960 and again in Cambridge when he presented his ideas on the quantum equivalence of Thirring and sine-Gordon theories, to some quiet ridicule from senior members of the audience. The first seminar I heard as a research student (in 1960) was Jeffrey Goldstone explaining his ideas on spontaneous breaking of symmetry.

In 1974 I started spending some time in Copenhagen at the Niels Bohr Institute. This was a very pleasant place where the staff included Holger Bech Nielsen, whom I have already mentioned, and Poul Olesen. However a more long term stay did not work with my family because my older daughter had special educational needs.

A postscript to the four-fermion amplitude story was a tidying and rendering completely concrete of all the calculations [BCO75].

In early July 1975 David Fairlie organized a followup workshop to the prior CERN and Aspen ones: 'Dual Resonance Model and its subsequent developments' in Durham. It was during this time that, prompted by worries of differing behaviours in different dimensions, I started thinking about Clifford algebras in any dimension of spacetime. In particular I realized that there was something very special about ten dimensions, which was relevant to the dual fermion theory. This was that chiral spinors were, unusually, real in spacetime of ten dimensions.

In 1976 Joël Scherk visited CERN, giving a talk on developments in supergravity, and came to tell me of his recent thoughts with Nando Gliozzi on the four-fermion amplitude. The idea of a chiral projection was not new as I remember people discussing it at Aspen in 1974 but, as Joël suggested, it did have a new advantage, that of eliminating all the parity doubled states including both tachyons. Furthermore there occurred the amazing coincidence of multiplicities of fermionic and bosonic states so suggestive of supersymmetry that they had found. For the first time there was a dual (or string) model which seemed to be totally consistent, in the sense of being totally devoid of tachyons and ghosts. I then explained to him the importance of the chiral projection maintaining reality which indeed it did in spacetime of ten dimensions which, I understood, he had not realized himself. As this was an essential ingredient, and in view of our previous collaboration in the year 1972–1973 on the subject (actually four papers), he invited me to coauthor the forthcoming paper which I was happy to do as it seemed to be a culmination of our previous work.

I saw Joël later at a 'triangle meeting' in Visegrád and afterwards in Budapest in late September 1976 and we discussed further. Later in October when I was visiting the Free University in Berlin he rang me up prior to submitting the paper. I told him that I had been thinking about the supercharges, which were implied but not explicit in the manuscript, and had realized that they could be constructed as vertex operators in terms of the operator formalism. We agreed that this was too novel to be included in the present manuscript [GSO76, GSO77] and we promised each other that we would postpone developing this idea until the future. For a variety of reasons (upheavals in each of our lives and eventually his tragic death in 1980) this never occurred until much later in other work.

In the course of our work on non-Abelian monopoles in 1975–1976 Jean Nuyts had taught me about the theory of Lie algebras and their representations which we used in a subsequent paper, written in collaboration with Peter Goddard in 1976. It occurred to Peter and me that this Lie algebra formalism was not unlike the operator formalism arising in the algebraic structure of string theory. We pursued this idea for a number of years after my return to London (on 30 September 1977) by means of discussions when we met in his office in DAMTP in Cambridge on Saturday mornings. At the same time I maintained my annual series of postgraduate lecture courses at Imperial College on string/superstring theory (ignoring discouragement from Salam).

My collaboration with Peter intensified in 1982–1983 when we were joint guests of David Brydges and Michael Fowler at the Institute for Advanced Study at the University of Virginia in Charlottesville. We met the mathematician Igor Frenkel on a visit there and were later invited to explain some of our ideas in Berkeley at the Mathematical Sciences Research Institute at the conference on 'Vertex operators in mathematics and physics' in 10–17 November 1983. We very quickly produced an account that was published in the proceedings. Amongst other things we discussed our interest in the connection between algebras and lattices (indeed that was the title of the preprint). The interesting lattices with connections with Lie algebras possessed a property of integrality and particularly interesting examples of these were even and self-dual because of the existence of associated functions with modular invariance properties. These only occurred if the metric had a signature divisible by eight. The simplest Euclidean examples were the E_8 root lattice of dimension 8 and two lattices of dimension 16, one associated with the root lattice of $E_8 \oplus E_8$ and the other a weight lattice associated with $SO(32)^*$. We suggested that these might have special significance in physics. Our preprint was much cited at the time [GO85].

Peter and I attended Aspen the following summer in 1984 when Michael Green and John Schwarz started studying anomaly cancellations in supersymmetric gauge theories coupled to gravity and circulated a preprint entitled 'Anomaly cancellations in supersymmetric $D = 10$ gauge theory require $SO(32)^*$' that was very quickly withdrawn and replaced by one entitled 'Anomaly cancellations in supersymmetric gauge theory and superstring theory', which also admitted cancellation for $E_8 \oplus E_8$. Peter and I were meanwhile developing the first version of our ideas involving the Sugawara construction and cosets.

Looking back, I can see that my time in the Theory Division at CERN was particularly fruitful and happy despite the shadow of the temporary eclipse of string/superstring theory. The need for very precise arguments concerning the structure of the fermion emission vertex suited me very well. So did the other topics concerning electromagnetic duality about which I have said much less. It was an exciting time to be there, with inspiring colleagues such as Daniele Amati and Bruno Zumino, and there was a very pleasant atmosphere, very conducive to research, and I am grateful to have had that opportunity to be there. Yet it is also obvious how fragile and transient was that environment when the participants dispersed back around Europe.

In preparing this account I have been struck by how often an important development is understated in the original, possibly because the implication of the line of thought is so surprising. This is to the extent that a casual reader from a later generation might not appreciate the significance of what has been said. Maybe the prime example is the failure of Joe Weis to write up his discovery of the c-number in the Virasoro algebra, but he did tell other people. It seems to me that other examples concern the unit intercept of the leading Regge trajectory and the critical dimension. These results forced the theory in a surprising way, away from the strongly interacting theory and towards a unified theory.

It seems to me that this early literature is of a remarkably high quality with important contributions made by a fair-sized community. Yet the literature is unlikely to be revisited now since the theories that we were working on have now been fully accepted, but we were very conscious at that time that this was not so and that therefore we had to be convincing.

Quoted dates of papers are usually the dates of acceptance of the relevant journal. The paper in question would have been typed much earlier, particularly in the case of CERN, where there could be long typing queues.

I thank Lars Brink, Peter Goddard and Claud Lovelace for help with past memories.

David Ian Olive was born in 1937. He went to school and university in Edinburgh, graduating in 1958. In Cambridge he earned his BSc and PhD, the latter with J. C. Taylor in 1963. After a period of time at the Carnegie Institute of Technology in Pittsburgh he returned to Cambridge. In 1971 he went to CERN and in 1977 to Imperial College in London. In 1992 he moved to Swansea and retired in 2002. He also spent time in Charlottesville, Berkeley, Princeton, Copenhagen, Gothenburg, Stockholm, Utrecht and San Paolo. In 1997, together with Peter Goddard, he was awarded the Dirac Medal of the International Centre for Theoretical Physics, Trieste.

References

[AALO70] Alessandrini, V., Amati, D., Le Bellac, M. and Olive, D. (1970). Duality and gauge properties of twisted propagators in multi-Veneziano theory, *Phys. Lett.* **B32**, 285–290.

[AALO71] Alessandrini, V., Amati, D., Le Bellac, M. and Olive, D. (1971). The operator approach to dual multiparticle theory, *Phys. Rep.* **C1**, 269–346, also in *Dual Theory*, ed. Jacob, M., Physics Reports Reprints Book Series, Vol. 1 (North Holland, Amsterdam).

[ALO70a] Amati, D., Le Bellac, M. and Olive, D. (1970). Twisting invariant factorisation of multi particle dual amplitudes, *Nuovo Cimento* **A66**, 815–830.

[ALO70b] Amati, D., Le Bellac, M. and Olive, D. (1970). The twisting operator in multi-Veneziano theory, *Nuovo Cimento* **A66**, 831–844.

[BO73a] Brink, L. and Olive, D. (1973). The physical state projection operator in dual resonance models for the critical dimension of spacetime, *Nucl. Phys.* **B56**, 253–265.

[BO73b] Brink, L. and Olive, D. (1973). Recalculation of the unitary single planar loop in the critical dimension of spacetime, *Nucl. Phys.* **B58**, 237–253.

[BORS73] Brink, L., Olive, D., Rebbi, C. and Scherk, J. (1973). The missing gauge conditions for the dual fermion emission vertex and their consequences, *Phys. Lett.* **B45**, 379–383.

[BOS73] Brink, L., Olive, D. and Scherk, J. (1973). The gauge properties of the dual model Pomeron–Reggeon vertex: their derivation and consequences, *Nucl. Phys.* **B61**, 173–198.
[Bro72] Brower, R. C. (1972). Spectrum-generating algebra and no-ghost theorem in the dual model, *Phys. Rev.* **D6**, 1655–1662.
[BCO75] Bruce, D., Corrigan, E. and Olive, D. (1975). Group theoretical calculation of traces and determinants occurring in dual theories, *Nucl. Phys.* **B95**, 427–433.
[COZ69] Campbell, D. K., Olive, D. and Zakrzewski, W. J. (1969). Veneziano amplitudes for Reggeons and spinning particles, *Nucl. Phys.* **B14**, 319–329.
[CERN68] 14th International Conference on High Energy Physics (CERN, 1968).
[CGOS73] Corrigan, E., Goddard, P., Olive, D. and Smith, R. A. (1973). Evaluation of the scattering amplitude for four dual fermions, *Nucl. Phys.* **B67**, 477–491.
[CO72] Corrigan, E. and Olive, D. (1972). Fermion-meson vertices in dual theories, *Nuovo Cimento* **A11**, 749–773.
[Dal70] Dalitz, R. ed. (1970). Duality – Reggeons and resonances in elementary particle processes, *Proc. R. Soc. London*, **A318**, 243–399.
[DiV07] Di Vecchia, P. (2007). The birth of string theory, in *String Theory and Fundamental Interactions*, ed. Gasperini, M. and Maharana, J. (Springer, Berlin).
[GSO76] Gliozzi, F., Scherk, J. and Olive, D. (1976). Supergravity and the spinor dual model, *Phys. Lett.* **B65**, 282–286.
[GSO77] Gliozzi, F., Scherk, J. and Olive, D. (1977). Supersymmetry, supergravity theories and the dual spinor model, *Nucl. Phys.* **B122**, 253–290.
[GGRT73] Goddard, P., Goldstone, J., Rebbi, C. and Thorn, C. B. (1973). Quantum dynamics of a massless relativistic string, *Nucl. Phys.* **B56**, 109–135.
[GO85] Goddard, P. and Olive, D. (1985). Algebras, lattices and strings, in *Vertex Operators in Mathematics and Physics*, ed. Lepowsky, J., Mandelstam, S. and Singer, I. M., Mathematical Research Institute Publications (Springer-Verlag, New York).
[GT72] Goddard, P. and Thorn, C. B. (1972). Compatibility of the pomeron with unitarity and the absence of ghosts in the dual resonance model, *Phys. Lett.* **B40**, 235–238.
[Got71] Gotsman, E. ed. (1971). *Duality and Symmetry in Hadron Physics* (Weizmann Science Press of Israel).
[Lov71] Lovelace, C. (1971). Pomeron form factors and dual Regge cuts, *Phys. Lett.* **B34**, 500–506.
[NS71] Neveu, A. and Schwarz, J. (1971). Factorizable dual model of pions, *Nucl. Phys.* **B31**, 86–112.
[Oli64] Olive, D. (1964). Exploration of S-matrix theory, *Phys. Rev.* **B135**, 745–760.
[Oli71] Olive, D. (1971). Duality approach to the strong interaction S-matrix, in *Proceedings of the Symposium on Basic Questions in Elementary Particle Physics* dedicated to W. Heisenberg (Max-Planck-Institut für Physik und Astrophysik, München), 140–147.
[Oli72] Olive, D. (1972). Clarification of the rubber string picture, in *XVI International Conference on High Energy Physics, Chicago-Batavia, September 6–13, 1972*, Vol. 1, 472–474.
[Oli74] Olive, D. (1974). Plenary report on 'dual models', in *Proceedings of the XVII International Conference on High Energy Physics, London, July 1974* (Didcot, 1974), Vol. I, 269–280.
[OS73a] Olive, D. and Scherk, J. (1973). No-ghost theorem for the pomeron sector of the dual model, *Phys. Lett.* **B44**, 296–300.

[OS73b] Olive, D. and Scherk, J. (1973). Towards satisfactory scattering amplitudes for dual fermions, *Nucl. Phys.* **B64**, 334–348.

[Pol89] Polkinghorne, J. C. (1989). *Rochester Roundabout* (Longman Scientific & Technical, Harlow).

[Ram71] Ramond, P. (1971). Dual theory for free fermions, *Phys. Rev.* **D3**, 2415–2418.

[Ven68] Veneziano, G. (1968). Construction of a crossing-symmetric, Reggeon-behaved amplitude for linearly rising trajectories, *Nuovo Cimento* **A57**, 190–197.

29

Dual model with fermions: memoirs of an early string theorist

PIERRE RAMOND

Abstract

I worked on string theory over a period of five years during the 'first string era', the most intellectually satisfying years of my scientific life. One of the early prospectors in the string theory mine, I was fortunate enough to contribute to the birth of this subject, which retains after these many years, its magical hold on our imaginations and expectations.

29.1 Graduate school

I was born in Neuilly sur Seine, a suburb of Paris, where I attended Sainte Croix de Neuilly. After the 'deuxième bachot' in 1961, I decided to spend a year with my family in New Jersey, where my civil engineer father had been working for a company that designed and manufactured concrete pipes. I enrolled at the Newark College of Engineering (NCE); there, I found spectacular teachers, especially Dr Foster, and one year turned into four; I graduated in 1965 with a Bachelor of Science in electrical engineering.

Always interested in physics, I had studied the subject on my own while at NCE. My application to Princeton graduate school in physics was rejected (and wait-listed at Yale). Fortunately, I had also applied to Syracuse University where Peter Bergmann was teaching. With the recommendations of Professors Henry Zatzkis (a student of Bergmann), Mauro Zambuto, and A. E. Foster, I was accepted, and soon afterwards, awarded a four-year fellowship.

I had wanted to study general relativity with Bergmann, but I was persuaded by Professor Alan McFarlane to switch to particle physics and join the group headed by E. C. G. Sudarshan. Graduate school was not easy. My education as an engineer had left much to be desired: the first year was a blur of exams, coffee, and feelings of inadequacy, but somehow I got through and E. C. G. Sudarshan took me on as his student. After studies of semi-leptonic kaon decays in Sudarshan's new alternative theory of the weak interactions, I decided to switch advisors, and was accepted by A. P. Balachandran. There, I spent the rest

---

The Birth of String Theory, ed. Andrea Cappelli, Elena Castellani, Filippo Colomo and Paolo Di Vecchia.
Published by Cambridge University Press. © Cambridge University Press 2012.

of my graduate school safely inside the Mandelstam triangle, trying to continue amplitudes from the s- to the t- and u-channels.

I did not think I had done very well in graduate school. My attraction to physics comes from the beauty it suggests by providing simple answers to (apparently) complicated questions. Yet, in the heyday of the S-matrix approach, I found myself wandering in the complex plane, bleeding from numerous encounters with its maze of poles and cuts. Not surprisingly, I did not get a job offer until late in the season (March 1969). Bob Wilson, the director of the National Accelerator Laboratory (NAL), the high energy collider under construction in the western suburbs of Chicago, had decided to form a small theory group. I was to be one of five junior theorists, 'The NAL Fives'. It was a gamble, as the group had no senior theorists, but it was an opportunity to pursue my dream, and I became employee number 662. NAL, known today as Fermilab, has been the site of many fundamental discoveries, most notably the top quark.

29.2 Trieste

Balachandran had kindly secured me a three-month appointment in the summer of 1969 at Professor Abdus Salam's ICTP in Trieste. It was a turning point in my scientific career. In Trieste I met Jean Nuyts with whom I had collaborated while a student. Jean and Hirotaka Sugawara were contemplating the beauties hidden behind the four-point Veneziano amplitude. It had recently been generalized to include many external legs, and intriguing regularities were emerging from the factorization of the amplitudes. I was soon hooked, and the three of us decided to extract the three-point vertex from the amplitudes. We used tensor methods, and by the end of the summer had succeeded in finding a (very ugly) expression. We were about to publish when the paper by S. Sciuto arrived [Sci69], where the same vertex was derived using the harmonic oscillator techniques of Fubini, Gordon and Veneziano [FGV69]. His expression was so much simpler than ours that we did not publish our results. In graduate school, I had already encountered the simplicity of Dirac's creation and annihilation operators, not only in quantization but also in Schwinger's treatment of $SU(2)$. They were clearly the window into the structures behind the dual models.

It was also in Trieste that I had the first glimpse of my hero, P. A. M. Dirac. A tall and lonely man, he stood aside from all, and gave the impression of being in such deep thought, that no one, well almost no one, dared to intrude.

29.3 NAL theorist

I came back to America and NAL, determined to learn more about these new techniques. In the middle of the Atlantic on the liner 'France', I was studying a Fubini–Veneziano preprint in the ship's reading room. Imagine my surprise when I saw the *same* preprint on another desk! Its owner was nowhere to be seen, and it was annotated in a different hand. I eagerly waited for the reader. This is how I met André Neveu, who was on his way to Princeton University. We decided to keep in touch about each other's progress.

NAL was more like a summer camp than a laboratory. It was a collection of houses away from the construction site. The five theorists were housed in one house, experimentalists in another.[1] The director's complex was a bunch of houses put together. One of the five theorists was Lou Clavelli, whom I had already met when he had spent a month in Trieste. David Gordon had worked with Fubini and Veneziano in developing the oscillator formalism, and I looked forward to collaborating with him. David Gordon and I wrote the *first* theory paper out of NAL (THY-1), a not very memorable attempt to include fermions in the Veneziano model.

Jim Swank, a student of Nishijima, and Don Weingarten, a student of Bob Serber, completed the quintet. David, Lou and I shared an interest in studying the Dual Resonance Model amplitudes. I did not yet know about the string connection. In fact, when I told Lou that the mass spectrum looked familiar and reminded me of something I had come across, his response was: 'Nambu says it is a string'. Lou Clavelli had been Professor Y. Nambu's graduate student at Chicago, and he soon arranged a meeting over lunch with Nambu at the Quadrangle Club. There the great man gently encouraged us in our studies, asking probing questions with no easy answers that left me totally impressed. He even treated us to the lunch! Brilliance and humility, a combination of traits not often seen in physicists.

We all visited the University of Wisconsin where we met, among others, Bunji Sakita and Miguel Virasoro. They both seemed off-scale, and so way ahead of us in their understanding of dual models. In particular, Virasoro had just constructed a set of operators which had the potential to decouple the negative-norm states found in the amplitudes (decoupling was later proved by Brower, Goddard and Thorn), but only when the lowest particle was a tachyon and the first excited state massless. He had spent a lot of time trying to get around these 'unphysical' constraints, and hesitated to publish, although all who saw his work agreed it was too beautiful not to.

Nambu was the only senior theorist with whom we had scientific discussions. Many senior theorists came to NAL to gauge the progress and visit the site, but very few showed any interest in what we theorists were doing. André and I had agreed to keep in touch and report on our mutual progress. I soon received a preprint of André Neveu and Joël Scherk, who had successfully isolated the infrared divergence of the planar one-loop dual amplitude [NS70]. It was a feat of mathematical physics that only those with French training could have achieved: they used the Jacobi imaginary transformation. I invited them to NAL. This is how I met Joël and his first wife. Joël was as quiet and introspective as André was exuberant and flashy. They came across as intellectually brilliant, and I took to both right away.

Clavelli and I found it difficult to work with David Gordon, so we continued our studies of the Dual Resonance Models on our own. It took us a long time before we published our first paper (May 1970) [CR70]. In it we put on firm group theoretical basis the vertex introduced independently by Nambu, and Fubini and Veneziano, using the $SU(1, 1)$ operators found by Gliozzi, Chiu–Matsuda–Rebbi, and Thorn. From there we understood that fixing the

---

[1] See http://bama.ua.edu/~lclavell/Weston/ for Lou Clavelli's wonderful account of these early NAL days.

(not yet known as) conformal weight of the vertex was the same as a (free) equation of motion for the emitted particle. We quickly set about trying to find the vertex for the first excited state. It had never been written down before and this was to be our first truly original work.

29.4 Aspen

Early in 1970, Bob Wilson decreed 'All theorists must go to Aspen!'. We went, although none of us knew much about the Aspen Center for Physics. This was the best advice an experimentalist ever gave me! The Center was at the time a collection of two buildings, set near the music tent at the west end of Aspen, Colorado. The oldest, Stranahan Hall, was made of bricks; the second, Hilbert Hall, was a wooden structure which had been donated to the Center by Bob Wilson after it was used for the 1968 NAL Summer Study.

It was a wonderful stay. The town was in the afterglow of the hippie era, and my days were spent playing volleyball in Wagner Park, listening to music outside the music tent in the late afternoons, and in other nonscientific activities. In my spare time, I started thinking about the particle spectrum that had been extracted from the dual amplitudes. People had already found some sort of position operator $Q_\mu(\tau)$ which appeared in the vertex, and its derivative was like a generalized momentum $P_\mu(\tau)$. Indeed, if one pursued the analogy further, the inverse propagator looked like the square of that generalized momentum. This led me to think of a 'correspondence principle' by which simple notions of point particles were related to dual models. At last, a glimpse of simplicity! Was it the altitude, was it the easy-going atmosphere at the Center that enabled me to view the problem with a different eye? I will never know, but that summer stay changed my perspective forever.

Back at NAL, Clavelli and I spent the rest of the summer refining our results on the excited vertex, and abstracting from it general rules for the construction of dual amplitudes, which we would today call conformal theory. We submitted that work to the *Physical Review* in late September [CR71]. Although interested, Clavelli did not share my enthusiasm for the correspondence principle and in October I sent the paper on the correspondence principle (applied to bosons only) to *Physics Letters*.

In the process, I realized that it could be applied to the Dirac equation: all I had to do was to generalize the Dirac matrices. To my surprise, this led to an algebra of a kind I had never seen: it contained both commutators and anticommutators, and was in essence the square-root of the Virasoro algebra. Tremendously excited, I barely ate and drank for weeks, as every derivation brought more conceptual clarity and more questions. There were some odd things. The generalization of the Dirac matrices led to fermionic harmonic oscillators $b_\mu^{(n)}$ and $b_\mu^{(n)\dagger}$ with spacetime four-vector indices, but they came with their own operators, F_n, of the right structure to decouple the negative norm states. I also realized that it was a truly novel algebraic structure, since I could now take the square root of *any* Lie algebra. I explained my results to Don Weingarten over lunch and he simply said: 'You are set for life'. Of course I did not believe him.

29.5 The correspondence principle

Nambu [Nam69] and Fubini, Gordon and Veneziano [FGV69, FV70] had taught us that the Veneziano amplitude could be written in a very suggestive way, as a linear chain of vertex operators of the form

$$V(k_\mu) =: e^{ik_\mu Q_\mu(z)} :, \tag{29.1}$$

where $z = e^{i\tau}$ is a point on the unit circle, and $Q_\mu(z)$ the two-dimensional field that plays the role of a generalized position operator,

$$Q_\mu(z) = x_\mu + 2i\alpha' \ln z\, p_\mu + \sqrt{2\alpha'} \sum_{n=0}^{\infty} \left(a_\mu^{(n)} \frac{z^n}{\sqrt{n}} + a_\mu^{(n)\dagger} \frac{z^{-n}}{\sqrt{n}} \right), \tag{29.2}$$

augmented by an infinite tower of harmonic oscillators,

$$[a_\mu^{(n)}, a_\nu^{(m)\dagger}] = -\eta_{\mu\nu} \delta^{nm}. \tag{29.3}$$

We could then define a generalized momentum:

$$P_\mu = -\frac{i}{2\alpha'} z \frac{dQ_\mu}{dz}, \tag{29.4}$$

in terms of which the propagator could be written, $\Delta = (L_0 + 1)^{-1}$, with

$$L_0 = \alpha' p_\mu p^\mu + \sum_{n=1}^{\infty} n a_\mu^{(n)\dagger} a^{(n)\mu}. \tag{29.5}$$

This operator, together with:

$$L_1 = \sqrt{\alpha'} p^\mu a_\mu^{(1)\dagger} + \cdots, \tag{29.6}$$

and its conjugate, formed the $SU(1, 1)$ algebra, found by Gliozzi [Gli69], as well as Chiu, Matsuda and Rebbi [CMR69] and also Thorn [Tho69],

$$SU(1,1): \quad [L_m, L_n] = (m-n)L_{n+m}, \quad m, n = 0, \pm 1. \tag{29.7}$$

The generalized position transformed covariantly under this algebra, as did the vertex operator (29.1); the eigenvalue of L_0 gave us $J_s = \alpha' k^2$ for its scale dimension (conformal weight).

The generalized momentum came up when Clavelli and I [CR71] constructed the vertex for the emission of a vector particle,

$$V_\rho(k_\mu) =: P_\rho(z) e^{ik_\mu Q_\mu(z)} : . \tag{29.8}$$

We checked its covariance properties and found the different scale dimension $J_v = \alpha' k^2 + 1$. By requiring that the vertex operators have the canonical dimension, $J_s = J_v = 1$, we deduced that the emitted particle is massless, and we naturally generalized the picture for all emitted particles.

The Virasoro [Vir70] operators,

$$L_n = \sqrt{\alpha'} p^\mu a_\mu^{(n)\dagger} + \cdots, \qquad (29.9)$$

obeying the algebra (I did not know of the c-number at the time),

$$[L_m, L_n] = (m-n)L_{n+m} + \frac{D}{12}m(m^2-1)\delta_{m,-n}, \qquad (29.10)$$

also reinforced the idea behind the generalized momentum, as the Virasoro operators related the unphysical operators to the rest of the theory.

This gave me some faith in taking the generalized position and momentum seriously. The question was in what sense did $P_\mu(z)$ generalize the momentum? In my first paper, I noted that the usual momentum could be obtained by varying the slope parameter, but I found that a more useful procedure would be to average out the harmonic oscillators. So I defined averaging as follows:

$$\langle \cdots \rangle_n = \frac{1}{2\pi}\int_{-\pi}^{\pi} d\tau\, e^{in\tau} \cdots, \qquad z = e^{i\tau}, \qquad (29.11)$$

such that

$$\langle P_\mu \rangle_0 = p_\mu, \qquad \langle Q_\mu \rangle_0 \approx x_\mu. \qquad (29.12)$$

The Klein–Gordon equation was just:

$$0 = p^2 + m^2 = \langle P^\mu \rangle_0 \langle P_\mu \rangle_0 + m^2, \qquad (29.13)$$

but this did not account for the richness of the Veneziano spectrum. However, by devising a *correspondence principle*,

$$\langle A \rangle \langle B \rangle \quad \rightarrow \quad \langle A\,B \rangle, \qquad (29.14)$$

we could recover the Veneziano propagator:

$$0 = p^2 + m^2 \quad \rightarrow \quad \langle P^\mu P_\mu \rangle + m^2 = L_0 + m^2. \qquad (29.15)$$

Furthermore, the conditions on the oscillators for selecting the physical states with positive norm (Gupta–Bleuler conditions) could also be written in a neat way as follows:

$$0 = p \cdot a^{(n)\dagger} \quad \rightarrow \quad \langle P^\mu P_\mu \rangle_n = L_n \approx 0 \ (n > 0). \qquad (29.16)$$

Similarly, the Lorentz generators are given by:

$$(x_\mu p_\nu - x_\nu p_\mu) \quad \rightarrow \quad \langle Q_\mu P_\nu - Q_\nu P_\mu \rangle, \qquad (29.17)$$

and so on. These results were submitted for publication in *Physics Letters B* in the fall of 1970, and rejected for publication. The paper eventually appeared in *Nuovo Cimento* [Ram71b].

In the meantime, I set upon generalizing the correspondence principle to fermions. I also knew from Dolen–Horn–Schmid duality that there had to be deep connection between bosons (ρ-channel) and fermions (s-channel). For relativistic fermions, the natural setting is the Dirac equation of motion for electrons,

$$0 = \gamma_\mu p^\mu + m, \tag{29.18}$$

but the correspondence principle did not yield anything interesting unless the Dirac matrices could be viewed as dynamical variables. Hence I considered the Dirac matrices as some average over generalized Dirac matrix fields $\Gamma_\mu(z)$ [Ram71a]:

$$\gamma_\mu = \langle \Gamma_\mu(z) \rangle_0. \tag{29.19}$$

In order to keep to the structure of the original Dirac equation, I required canonical anticommutation relations:

$$\{\Gamma_\mu(z), \Gamma_\nu(z')\} = g_{\mu\nu}\delta(z-z'). \tag{29.20}$$

The form of the generalized Dirac matrix field is then:

$$\Gamma_\mu(z) = \gamma_\mu + i\gamma_5 \sum_{n=0}^{\infty} \left(b_\mu^{(n)} z^n + b_\mu^{(n)\dagger} z^{-n}\right), \tag{29.21}$$

written in terms of an infinite number of Fermi oscillators which have spacetime vector indices,

$$\{b_\mu^{(n)}, b_\nu^{(m)\dagger}\} = -g_{\mu\nu}\delta^{n,m}. \tag{29.22}$$

The next step is to postulate the 'dual Dirac equation':

$$\langle \Gamma_\mu \rangle_0 \langle P^\mu \rangle_0 + m \quad \to \quad \langle \Gamma_\mu P^\mu \rangle_0 + m. \tag{29.23}$$

The algebraic structure of the Veneziano model was still true. We could define

$$F_n = \langle \Gamma_\mu P^\mu \rangle_n, \tag{29.24}$$

which satisfy a new kind of algebra with commutators and anticommutators,

$$\{F_n, F_m\} = 2L_{n+m},$$
$$[L_n, L_m] = (m-n) L_{m+n}, \tag{29.25}$$
$$[L_n, F_m] = (2m-n) F_{m+n},$$

that extends the Virasoro algebra of conformal transformations. I did not know at the time of the so-called central extensions (c-numbers) The F_n clearly yield the Gupta–Bleuler decoupling conditions, as follows:

$$F_n = \gamma_5 p^\mu b_\mu^{(n)\dagger} + \cdots \approx 0 \quad (n > 0). \tag{29.26}$$

The analogy continued as the spin part of the Lorentz generators had the expected form: $\gamma_{[\mu}\gamma_{\nu]} \to \langle \Gamma_{[\mu}\Gamma_{\nu]}\rangle$. So far I had found a very rich algebraic structure, but to get an interacting theory, I needed a vertex. My guess was to generalize QED, with a vertex of the form $: \Gamma_\mu \exp(ik \cdot Q(z)) :$ [Ram72], which I spent some time investigating, without success. I did not realize at the time that the theory still contained scalars, which required Yukawa couplings. Today, we know that the correct vertex is of the Yukawa type, $: \Gamma_5 \exp(ik \cdot Q(z)) :$.

29.6 Telling it around

My wife Lillian, an electrical engineer and NCE alumna, had left for a six week stay at a Western Electric training centre near Princeton, and I joined her for a few days, some time in late October, as I recall. I took advantage of the trip to visit Y. Nambu, who was spending the fall at the Institute for Advanced Study. There I also met Jamal Manassah, Satoshi Matsuda and Mike Green, and sketched my ideas on including the fermions via the correspondence principle. I was visiting Lillian, and had no time to see André at the University.

Back at NAL, a host of distractions caused me to start losing focus. First, we were informed in early November that our appointments at NAL would terminate in 1971. This came as a big surprise: when hired, we had been told to expect longer appointments because of the special circumstances, such as the lack of a senior theorist. This was devastating news for us all. My own work had not been manifestly relevant to NAL, but Lou and the others had always kept close to experiments, and worked hard in NAL-sponsored workshops. Bob Wilson was listening to senior US theorists who then had no sympathy for dual models/string theory. We all became anxious about finding another job. My last experience had been nerve wracking, and I was soon freaking out. Secondly, the correspondence principle paper was rejected by *Physics Letters*. I argued, but to no avail, and in December I withdrew the paper.

Job applications took up most of my time, preventing me from writing up the fermion paper [Ram71a], and it was not until the week after Christmas that I sent it to the *Physical Review*, as well as to André and others. At the same time, I resubmitted the correspondence principle paper, this time to *Nuovo Cimento* [Ram71b], and went back to the business of job-hunting.

In those days, postdoctoral positions were on the whole awarded at the January Meeting of the American Physical Society in New York. This January slave market was a depressing affair for Clavelli and myself: the establishment had no time for Dual Resonance Models/string theory, and showed minimal interest in what we were doing.

Then I got lucky, through a set of fortuitous circumstances orchestrated by Lou Clavelli. Before joining NAL, Lou had been a postdoc at Yale, where he had met his future wife, Estelle. As he was to visit his in-laws after the New York meeting, he suggested I join him and use the opportunity to visit the Yale high energy theory group, to which I had already applied. That Friday the Yale physicists, Sam McDowell, Charlie Sommerfield, John Harte and Dick Slansky (Feza Gürsey was out of town) asked me to give an impromptu seminar, my first seminar on the dual Dirac equation. The same night we flew back to Chicago, and three days later, Charlie Sommerfield offered me a one-year instructor position at Yale with the possibility of a second year! Finally, an early job offer, that I accepted the very same day.

Three months after my fermion paper, I received a preprint by André Neveu and John Schwarz who proposed the dual pion model, a generalization of the Veneziano model [NS71b, NS71a]. It did not have external fermions, but included fermionic ladder operators with spacetime vector indices, just like my model of fermions, with the same name but different labelling. The underlying algebraic structure they proposed was the same as mine. André told me later when I visited him in Princeton (when I first met John Schwarz), that they had indeed been motivated by my paper: they had introduced a Yukawa interaction into my model, computed the amplitude with two fermions and an arbitrary number of bosons. They had great difficulties with the zero modes in the fermion sector (who didn't?), but they factorized the amplitude in the cross channel to extract the dual pion model amplitudes. Their paper was wonderful, but imagine my surprise when I found my fermion paper mentioned only towards the end of theirs. To this day I remain baffled by their lack of acknowledgement of the seminal role of my work.

In 1969 Nambu had suggested that the Veneziano amplitudes should be derivable from a string [Nam70a]. Later, Nambu [Nam70b] and independently Goto [Got71] had proposed an action for a relativistic string. It was not linear, and seemed impossible to quantize. When Virasoro found his decoupling coefficients [Vir70], Nambu had interpreted them as evidence for his string picture, as the generators of the conformal algebra generated by the Fourier coefficients of the energy-momentum tensor of a two-dimensional theory spanned by the world-sheet of a relativistic string. His point of view, while intriguing and interesting, was not widely appreciated, as it did not offer any computational advantages over the well-developed and fruitful amplitude approach.

The spring of 1971 was spent in adding electromagnetic interactions to the dual Dirac equation, and in trying to find a meaning for the anticommutators. I understood the anti-commuting operators as generators of transformations between bosons and fermions. Such close kinship between fermions and bosons did not surprise me; it was natural in the original formulation of duality applied to the pion–nucleon amplitude, with its implied relations between its fermionic s-channel and bosonic t-channel. Unfortunately, I became confused by the parameters of these transformations (I did not know about Grassmann numbers), and did not get very far.

This symmetry between fermions and bosons was of course the first manifestation of a new type of symmetry, called (later) 'supersymmetry'. It was first found in string theory, which has proven to be an incubator for many new ideas. Unfortunately, by that time I was physically and emotionally spent, and fell ill.

I gave the second (and last) seminar on the dual fermions at the University of Chicago where I found a sympathetic reception. However, the outside world did not seem to show a similar interest. My fortunes changed when Professor Stanley Mandelstam (in whose triangle I had toiled as a graduate student) visited NAL. I excitedly told him about the fermions, expecting a reaction; there was none. Later that day, I pressed him for his opinion, and still remember his answer for its honesty: 'You claim to have done something that many have tried to do, including my colleagues at Berkeley. I need time to assess it.' True to his word, Mandelstam, together with Nambu, proved to be a most generous advocate of my work. Also, Clavelli, Neveu who had stopped by NAL on his way to Berkeley, and I discussed the F_n as decoupling conditions (never published).

In the summer, I was invited to lecture at the Boulder Summer School on the construction of dual amplitudes for vertices of arbitrary spin that Clavelli and I had developed. Many visiting senior theorists came through NAL, although the local theorists were largely ignored. There were of course exciting developments taking place elsewhere: Ben Lee, fresh from the Amsterdam conference, brought the news of a young Dutch theorist, Gerard 't Hooft, who had proved that some massive Yang–Mills theories are renormalizable. This rang a bell because, at Syracuse, Joe Schecter had suggested I read Weinberg's 1967 paper, 'A model of leptons' [Wei67], although it was believed at the time that one could not calculate anything beyond tree level. I wondered whether the tachyon of the Veneziano model was a signal of spontaneous breaking, but then of what? In the fall of the same year, at a conference in Rutgers, I tried to interest my fellow theorists in this problem, but to no avail. This was where I learned from Claud Lovelace that cuts disappear from the twisted one-loop diagram when the number of spacetime dimensions is twenty six. None of us understood the significance of this result.

29.7 Yale

I joined Yale in fall 1971 as an Instructor. I was quite taken by the friendly atmosphere of the high energy theory group headed by the Turkish physicist Feza Gürsey. He and his wife Suha provided a great intellectual and humane cocoon into which Lillian and I were readily accepted. Yale was a wonderful place, and my colleagues there have remained life-long friends.

As Yale Instructor and later Assistant Professor, I continued to work in string theory. My student Michael Kalb and I understood the fundamental role of antisymmetric tensor fields in theories of extended objects, using ideas from action-at-a-distance theories where these fields are linked to multidimensional world-sheets [KR74]. Today the two-form B-field is a mainstay of string theories. Little did I know that their study would presage branes and so many other wonderful developments! By the time of the 1974 London conference, I had started working on exceptional groups and their applications to Grand Unified Theories. My last string publication of the first string era was with another Yale student, Charles Marshall, on the covariant formulation of string field theory [MR75].

The decision to leave string theory was driven primarily by the community's lack of interest and dearth of jobs in this wonderful subject. Also, I was lured away by the intellectual promise of Feza's application of unusual algebraic structures to physics. I was, needless to say, very pleased when many years later, exceptional groups appeared in the heterotic string.

Acknowledgements

I wish to thank Sudarshan Ananth, Lars Brink and Lou Clavelli for reading the manuscript and their many helpful suggestions. This work, supported by the Department of Energy

Grant No. DE-FG02-97ER41029, was begun at the Aspen Center for Physics which I thank for its hospitality.

Pierre Ramond was born in Neuilly sur Seine (France) in 1943. In 1965, he graduated from Newark College of Engineering (NJIT) with a BS in Electrical Engineering. He obtained his PhD from Syracuse University in 1969. After a postdoctoral position with FermiLab, he was on the Yale faculty, then he was a Millikan Fellow at Caltech, and joined the University of Florida faculty in 1980. His interests lie on fundamental principles in physics. He was awarded the 2004 Oskar Klein Medal, and is a Fellow of the American Academy of Arts and Sciences.

References

[CMR69] Chiu, C. B., Matsuda, S. and Rebbi, C. (1969). Factorization properties of the dual resonance model – a general treatment of linear dependences, *Phys. Rev. Lett.* **23**, 1526–1530.

[CR70] Clavelli, L. and Ramond, P. (1970). $SU(1, 1)$ analysis of dual resonance models, *Phys. Rev.* **D2**, 973–979.

[CR71] Clavelli, L. and Ramond, P. (1971). Group theoretical construction of dual amplitudes, *Phys. Rev.* **D3**, 988–990.

[FGV69] Fubini, S., Gordon, D. and Veneziano, G. (1969). A general treatment of factorization in dual resonance models, *Phys. Lett.* **B29**, 679–682.

[FV70] Fubini, S. and Veneziano, G. (1970). Duality in operator formalism, *Nuovo Cimento* **67**, 29–47.

[Gli69] Gliozzi, F. (1969). Ward-like identities and twisting operator in dual resonance models, *Lett. Nuovo Cimento* **2**, 846–850.

[Got71] Goto, T. (1971). Relativistic quantum mechanics of one-dimensional mechanical continuum and subsidiary condition of dual resonance model, *Prog. Theor. Phys.* **46**, 1560–1569.

[KR74] Kalb, M. and Ramond, P. (1974). Classical direct interstring action, *Phys. Rev.* **D9**, 2273–2284.

[MR75] Marshall, C. and Ramond, P. (1975). Field theory of the interacting string: the closed string, *Nucl. Phys.* **B85**, 375–414.

[Nam69] Nambu, Y. (1969). University of Chicago Preprint EFI 69-64.

[Nam70a] Nambu, Y. (1970). Quark model and the factorization of the Veneziano amplitude, in *Proceedings of the International Conference on Symmetries and Quark Models, Wayne State University, June 18–20, 1969*, ed. Chand, R. (Gordon and Breach, New York), 269–277, reprinted in *Broken Symmetry, Selected Papers of Y. Nambu*, ed. Eguchi, T. and Nishijima, K. (World Scientific, Singapore, 1995), 258–277.

[Nam70b] Nambu, Y. (1970). Duality and hadrodynamics, lecture notes prepared for Copenhagen summer school, 1970, reproduced in *Broken Symmetry, Selected Papers of Y. Nambu*, ed. Eguchi, T. and Nishijima, K. (World Scientific, Singapore, 1995), 280.

[NS70] Neveu, A. and Scherk, J. (1970). Parameter-free regularization of one-loop unitary dual diagram, *Phys. Rev.* **D1**, 2355–2359.

[NS71b] Neveu, A. and Schwarz, J. H. (1971). Factorizable dual model of pions, *Nucl. Phys.* **B31**, 86–112.

[NS71a] Neveu, A. and Schwarz, J. H. (1971). Quark model of dual pions, *Phys. Rev.* **D4**, 1109–1111.

[Ram71a] Ramond, P. (1971). Dual theory of free fermions, *Phys. Rev.* **D3**, 2415–2418.
[Ram71b] Ramond, P. (1971). An interpretation of dual theories, *Nuovo Cimento* **4**, 544–548.
[Ram72] Ramond, P. (1972). Electromagnetic interaction of dual systems, *Phys. Rev.* **D5**, 2078–2084.
[Sci69] Sciuto, S. (1969). The general vertex function in dual resonance models, *Lett. Nuovo Cimento* **2**, 411–418.
[Tho69] Thorn, C. B. (1970). Linear dependences in the operator formalism of Fubini, Veneziano, and Gordon, *Phys. Rev.* **D1**, 1693–1696.
[Vir70] Virasoro, M. (1970). Subsidiary conditions and ghosts in dual-resonance models, *Phys. Rev.* **D1**, 2933–2936.
[Wei67] Weinberg, S. (1967). A model of leptons, *Phys. Rev. Lett.* **19**, 1264–1266.

30
Personal recollections

ANDRÉ NEVEU

When I look back at my involvement in the subject of string theory, what strikes me most is that the process of discoveries appears to me rather erratic in the details of who actually makes such or such discovery and when, depending on sometimes strange coincidences. On the other hand, at the level of published work the evolution of scientific knowledge is generally rather smooth and a posteriori natural. Over the decades, I have witnessed several other examples of coincidences that have been crucial in scientific progress. This Chapter is thus in some sense the opposite of Freeman Dyson's article 'Missed opportunities' [Dys72] in which he describes contributions he did not make because such coincidences which in all likelihood should have occurred actually did not occur.

In 1968–1969 I was working with Joël Scherk on our research work for our PhD in Orsay under the guidance of Claude Bouchiat and Philippe Meyer. The subject was electromagnetic and final state interaction corrections to nonleptonic kaon decays. We were classmates in our last year as students at the École Normale and good friends. We enjoyed working together a lot. While we were finishing our thesis work, we became very interested in the explosion of activity which followed the original Veneziano paper [Ven68], together with Claude Bouchiat and Daniele Amati, who was spending a sabbatical year in Orsay. We were particularly attracted by the mathematical beauty which we felt lay in this new structure, for example, the change of variables which guarantees the cyclic symmetry of the multiperipheral representation of the N-particle generalization (Bardakci and Ruegg [BR69], Goebel and Sakita [GS69], Chan and Tsou [CT69]) of the Veneziano formula. Or the proposal by Kikkawa, Sakita and Virasoro [KSV69] (on which Virasoro gave a seminar in Orsay that year) to go beyond the narrow resonance approximation. We were puzzled by the exponential divergence which was discovered (Bardakci, Halpern and Shapiro [BHS69]) in the loop diagrams when the correct level structure was taken into account, but not pessimistic like other physicists, who considered that divergence natural (and fatal) for a theory with such an exponentially growing particle spectrum and arbitrarily high spins.

The Birth of String Theory, ed. Andrea Cappelli, Elena Castellani, Filippo Colomo and Paolo Di Vecchia.
Published by Cambridge University Press. © Cambridge University Press 2012.

Now, for the year after, we were both very interested in going to the USA, and continuing working together. There was one fellowship in Princeton for a former student of the École Normale (endowed by Procter of Procter and Gamble). We knew about the existence of NATO fellowships, but General de Gaulle had just pulled France out of NATO, so we thought we were ineligible. Not true: France had only left the military part of NATO, not the cultural part. This we discovered totally by chance during a train ride back from Orsay to Paris. We happened to be seated facing two scientists discussing precisely the stay in the USA which one of them had just done with a NATO fellowship. When we asked him about that, he gave us this information together with the address where to apply. This is the first coincidence. Result: I got the Procter fellowship and Joël a NATO fellowship and we were both set for Princeton. At that time, I had already heard (in very positive terms) about Pierre Ramond from Jean Nuyts (then in Orsay), with whom he had already signed the papers (without having met, if I remember) on crossing-symmetric partial wave amplitudes which formed the basis of his PhD in Syracuse with A. P. Balachandran as adviser.

With the Procter fellowship came a Fulbright travel grant. Having the choice, I chose the ship 'France' for my first transatlantic crossing. The ship had a small and pleasant library with a few desks. I was spending many hours there, studying in detail the latest preprints on Dual Resonance Models, as they were called. Now for the second coincidence: one afternoon, I walked out of the library for a short break, leaving all my material spread on the desk. Precisely while I was absent, Pierre walked in, and looked around for a vacant desk. There appeared to be only one, mine. He walked up to it, realized that it was not really vacant, but was shocked to see on it the Fubini–Veneziano paper [FV69] on the factorization of Dual Resonance Models, the very same paper he was studying at that moment! He quickly went back to his cabin to make sure that what he had just seen was not his copy! Reassured about his sanity, he came back to the library, wondering on the way about who could be the fellow interested in such an esoteric topic. By which time, I, too, was back. You can imagine easily the next hours. This is how we became friends. Pierre had just obtained his PhD from Syracuse University and, after spending the summer in France, he was on his way to the National Accelerator Laboratory (now called Fermilab). Together with Louis Clavelli, David Gordon, Jim Swank and Don Weingarten, they formed the entire Theory Division of Fermilab.

In Princeton, Joël and I immediately realized that being an alumnus of the École Normale did not mean much, which was rather stimulating! We ended up sharing a corner of the attic of the old Palmer Lab, and it was great luck, at least for me, that the two of us were together to face this relative solitude. We quietly pursued our collaboration on Dual Resonance Models. After a few weeks, thanks to our mathematical training and to the properties of elliptic functions, we had understood how to handle the superficially catastrophic divergences of the planar one-loop diagrams of the theory. During the afternoon tea time of the physics department, we could see by what they were writing on the blackboard that David Gross and John Schwarz were also interested in these divergences, and we were amused to see them trying things which we had tried much before and knew could not work. When we showed them what we had found, our situation improved dramatically: they proposed that we should all four work together, we were treated as colleagues, and we moved to a nice office in the brand new Jadwin Hall.

I was chosen by the flip of a coin to present our results at the weekly joint informal seminar of the University, and the Institute for Advanced Study a few weeks later, when I wrote the famous formula for the Jacobi imaginary transformation applied to the partition function, in a form that would make it as impressive as possible, that is:

$$f(w) \equiv \prod (1 - w^n)^{-1}$$
$$= w^{1/24} \left(-\frac{\ln w}{2\pi}\right)^{1/6} \exp\left(-\frac{\pi^2}{6\ln w}\right) f\left(\exp\left(\frac{4\pi^2}{\ln w}\right)\right).$$

I remember that Barry Simon could not refrain from exclaiming: 'This is impossible!' Coming from him, this gives you an idea of the state of our mathematical knowledge in those days ... But it is clear that the electrostatic analogue of the Koba–Nielsen [KN69] formula meant that it was only a matter of weeks before somebody else would have discovered these elliptic functions in dual loop amplitudes and their consequences.

After that year, Joël went back via Berkeley to Orsay where he made his very important contribution about the zero-slope limit of dual amplitudes, showing that after all the model shared all the good properties of quantum field theory, and more. As for myself, I obtained the same NATO fellowship to spend another year in Princeton. In the fall, we received Claud Lovelace's preprint [Lov71] with the first appearance of the critical dimension 26, but, like everybody else, we did not take that point seriously! Simultaneously, after many other people, John Schwarz and I got interested in the problem of introducing spin one-half in the dual model. We came across the Bardakci–Halpern [BH71] paper on their attempts to build ghost-free models with fermions, which was in retrospect a very interesting paper containing what I believe are among the first examples, if not the first, of affine Lie algebras in the Sugawara construction. A pioneering paper, much too sophisticated for John and me, and we put it aside as too complicated for us.

Meanwhile, Pierre Ramond and I had kept in touch regularly. While at NAL, two important events had happened to him: (1) his three-year contract was not extended, (2) he discovered the Ramond model [Ram71]. He sent me that paper personally, which turned out to be very important. Indeed, in the Princeton University physics department, there was no preprint library. No need was felt for it: all the important people there received the important preprints themselves, and then spread around the important news. Would a short and partly speculative paper by a still relatively obscure postdoc at Fermilab have been considered important? And reached me? Perhaps, but probably after too long. So, it was most fortunate that I had met Pierre on the ship! With John we quickly discovered that from a fermion line emitting three pseudoscalar 'pions', we could factorize the first pole in the fermion–antifermion channel and obtain the Lovelace–Shapiro formula [Lov68, Sha69]. Since this meant bosonic trajectories with both integer and half-integer intercepts, it was not hard to introduce half-integer anticommuting modes and it took us only three or four weeks from there to build the bosonic sector of the Neveu–Schwarz–Ramond model. We were really naive; at first, we had not even clearly realized that a symmetry algebra larger than the Virasoro algebra was needed to get rid of the ghosts introduced by the new anticommuting modes. We were just lured by the elegance of the superconformal algebra and went ahead without further thinking. And this was most fortunate for us. How the superconformal algebra kills the ghosts was at first too clever for us and we had to discover 'experimentally' that they were absent before we could understand the ghost-killing mechanism.

Then, around Easter, stopping on the way in Fermilab to visit Pierre, I went to Berkeley, where Miguel Virasoro was. There, I met Korkut Bardakci, Marti Halpern, Stanley Mandelstam, Charles Thorn and Mike Kaku, who all made my stay most enjoyable, and so I made new life-long friends. I take this opportunity to thank them in public for their warm welcome. I told them about our model, and with Charles [NST71] we found how the ghosts were eliminated through the introduction of the 'F_2 formalism'. Marti Halpern showed me the thick pile of notes about his and Korkut's attempts at introducing spin. Indeed, the Neveu–Schwarz vertex for 'pion' emission appeared rather early in those notes, but they discarded it. They were after a Virasoro algebra enlarged with commutators as the ghost-killing mechanism, not a superalgebra. Supersymmetry did not exist then. However, by the time of my visit, they had become aware of Pierre's paper and, from conversations I had with Charles, it is clear to me that it would have been only a matter of weeks before they would have discovered that the vertex worked after all!

André Neveu was born in Paris in 1946. He attended École Normale Supérieure from 1965 to 1969, obtaining his PhD from Orsay in 1969. Then he spent two years at Princeton University, alternating during the following six years mostly between Orsay, then the École Normale and the Institute for Advanced Study. After six more years in Paris and six at CERN, he has been permanently in Montpellier since 1989. He is currently Vice President of the University of Montpellier.

References

[BH71] Bardakci, K. and Halpern, M. B. (1971). New dual quark models, *Phys. Rev.* **D3**, 2493–2506.

[BHS69] Bardakci, K., Halpern, M. B. and Shapiro, J. A. (1969). Unitary closed loops in reggeized Feynman theory, *Phys. Rev.* **185**, 1910–1917.

[BR69] Bardakci, K. and Ruegg, H. (1969). Reggeized resonance model for arbitrary production processes, *Phys. Rev.* **181**, 1884–1889.

[CT69] Chan, H.-M. and Tsou, T.-S. (1969). Explicit construction of n-point function in generalized Veneziano model, *Phys. Lett.* **B28**, 485–488.

[Dys72] Dyson, F. J. (1972). Missed opportunities, Josiah Willard Gibbs Lecture, January 17, *Bull. Am. Math. Soc.* **78**, 635–652.

[FV69] Fubini, S. and Veneziano, G. (1969). Level structure of dual resonance models, *Nuovo Cimento* **A64**, 811–840.

[GS69] Goebel, C. J. and Sakita, B. (1969). Extension of the Veneziano form to n-particle amplitudes, *Phys. Rev. Lett.* **22**, 257–260.

[KSV69] Kikkawa, K., Sakita, B. and Virasoro, M. A. (1969). Feynman-like diagrams compatible with duality. I. Planar diagrams, *Phys. Rev.* **184**, 1701–1713.

[KN69] Koba, Z. and Nielsen, H. B. (1969). Reaction amplitude for n-mesons, a generalization of the Veneziano–Bardakci–Ruegg–Virasoro model, *Nucl. Phys.* **B10**, 633–655.

[Lov68] Lovelace, C. (1968). A novel application of Regge trajectories, *Phys. Lett.* **B28**, 264–268.

[Lov71] Lovelace, C. (1971). Pomeron form factors and dual Regge cuts, *Phys. Lett.* **B34**, 500–506.

[NST71] Neveu, A., Schwarz, J. H. and Thorn, C. B. (1971). Reformulation of the dual pion model, *Phys. Lett.* **B35**, 529–533.

[Ram71] Ramond, P. (1971). Dual theory for free fermions, *Phys. Rev.* **D3**, 2415–2418.

[Sha69] Shapiro, J. A. (1969). Narrow resonance model with Regge behavior for $\pi\pi$ scattering, *Phys. Rev.* **179**, 1345–1353.

[Ven68] Veneziano, G. (1968). Construction of a crossing-symmetric, reggeon behaved amplitude for linearly rising trajectories, *Nuovo Cimento* **A57**, 190–197.

31
Aspects of fermionic dual models

EDWARD CORRIGAN

31.1 Student in Cambridge

After all this time I do not trust my memory to be accurate in every detail. Besides, I can only provide glimpses from the perspective of a beginning graduate student struggling to catch up, followed by that of a postdoc who found himself quite far from much of the action. I will not attempt to be comprehensive in either the telling or the references, rather I will restrict myself to those aspects I concentrated on at the time, some of the people I knew and who influenced me, and will try not to add any insights that have emerged over subsequent years.

In October 1969, more than forty years ago, I started my three years as a graduate student in the Department of Applied Mathematics and Theoretical Physics, Cambridge. During the previous year, as was customary then, I took Part III of the Mathematics Tripos. This was a thorough grounding in many of the tools then useful in elementary particle theory. However, by today's standards it was lacking in some respects, especially in the area of quantum field theory – the most notable omission being any mention of Yang–Mills gauge theory. This was hardly surprising, of course, because most of the Cambridge group had been actively developing S-matrix theory and 'Eden, Landshoff, Olive and Polkinghorne' was mandatory reading for prospective graduate students. Less surprising, was the omission of any reference to the Veneziano model, it being far too much of a recent innovation to make it into Part III. Instead, I heard about it in a roundabout way. On the other hand, there were compensations, such as attending Dirac's lectures on quantum mechanics, Goldstone's on statistical mechanics and Sciama's on general relativity. Though Skyrme had already written papers that would become influential, they were not mentioned; conformal field theory as we now know it was far in the future.

During the summer of 1969 I spent two months with an experimental group at the Rutherford Laboratory. The group was led by Geoff Manning and it was my entry into the world of detectors (in fact quite large spark chambers), and various other interesting pieces of electronics and machinery. I arrived during a period between data taking when the group

---

The Birth of String Theory, ed. Andrea Cappelli, Elena Castellani, Filippo Colomo and Paolo Di Vecchia.
Published by Cambridge University Press. © Cambridge University Press 2012.

was reconfiguring its detectors, and I was fascinated by the whole setup. To be frank, the intricacies of the experimental setup appeared much more interesting than the theory. Then something unexpected happened: several members of the group went off to a conference at Lund and I remember one member of the group, John Field, coming back and saying to me that what I should really start to learn about was the Veneziano model [Ven68]. He was enthusiastic about it because there was a chance of real progress in understanding the strong interactions. As a result, I found the paper and tried to work through it but I was missing too much background to really appreciate what I read.

There was a lot to learn besides the prerequisites in Regge theory, duality, current algebra and so on, and, in reality, I came to these topics after learning about the more recent developments. Earlier in 1969, Koba and Nielsen had developed their elegant representation of multiparticle amplitudes [KN69] and Fubini and Veneziano developed the vertex operator representation [FGV69, FV69], which was so useful for analyzing the spectrum. A little later, but still in 1969, around Christmas I seem to remember, these were followed by Virasoro's realization that potentially unphysical states were removed by a set of gauge conditions satisfying the beautiful and seemingly simple algebra that now bears his name [Vir70]. Each of these ideas was entirely new to me and the process of placing them in order and perspective was exceedingly daunting (as doubtless it was for almost everyone else at the time). John Polkinghorne, the then leader of the Cambridge elementary particle theory group in DAMTP, helped by giving a set of introductory lectures during my first year (followed later by David Olive, who gave a more technical set of lectures on the operator formalism backed up by the review by Alessandrini, Amati, Le Bellac and Olive [AALO71], which was circulating in preprint form). My research supervisor to begin with was Ian Drummond, replaced after a while by David Olive who had been on leave visiting CERN when I started (and he subsequently left Cambridge for CERN in July 1971). I do not now remember all the students in those years. However, I did come to know several people, some of whom became friends, colleagues or coworkers later on. In the year ahead of me there were John Cardy, David Collop and John Ellis and two years ahead, David Campbell, Peter Goddard, Alan White and Wojtek Zakrzewski. Given the labyrinthine design of DAMTP we only rarely met – at coffee time or when preprints needed assembling.

During 1969 a series of preprints from Stanley Mandelstam appeared [Man69, Man70a, Man70b] and these were suggested to me as a suitable focus for my first investigations. The thrust of the papers was an attempt to marry the quark model of elementary particles to the dual model, and the main difficulty, as I soon realized – given the dual model seemed to be geared to a description of mesons – was how to build S-matrix amplitudes involving three-quark states, or baryons. This was definitely an important question to tackle but it was not at all clear then (or perhaps even now) how best to do it. One of the sticking points was the inability at that time to describe dual amplitudes for particles with spin one-half. I also wondered about that for a while but the solution to the problem turned out to need significantly new ideas, which will be mentioned below.

Despite being attracted to Mandelstam's ideas, most of my time was spent during the years 1969–1971 trying to catch up and I did not make any identifiable progress. It was a time when articles arrived at a huge rate and it was hard to keep abreast. In fact, it was quite a depressingly slow process and I became very discouraged, coming very close to quitting altogether. The general thinking appeared to be that dual theory was going to be something new that had to be constructed from the 'bottom up', and there was tremendous activity, trying on the one hand to adapt the ideas to phenomenologically more realistic models, and on the other to render the model compatible with all the required properties of S-matrix theory, especially unitarity. There was a strong feeling that people might be groping their way towards something beyond quantum field theory. And string theory itself was still a couple of years in the future. My exact contemporaries (Peter Collins, Tony Mason and Charles Pantin) seemed to be making faster progress on other topics and I began to doubt the sense of trying to work in such a relatively new and fast-moving area. Eventually, I was rescued by David Olive who pointed me towards alternative thinking (Ellis, Frampton, Freund and Gordon [EFFG70]), and using the Koba–Nielsen formulation. This approach led to results developed with Claus Montonen (who arrived as a research student in October 1970) and David; they were never published but became the second chapter of my thesis. For a while I tried to find satisfactory vertex operator expressions for these general 'baryon–meson' amplitudes but failed to do so.

In early 1971, Pierre Ramond's article [Ram71] on dual fermions arrived, followed shortly after by an article by André Neveu and John Schwarz [NS71]. Both of these had an impact on me. The first seemed to me to be very interesting because the arguments put forward for the inclusion of fermions by cleverly extending the Dirac gamma matrices to a field $\Gamma^\mu(z)$ appeared radical and compelling. The second paper was motivated by the need to adjust the Veneziano model to include the pion and other types of mesons that appeared at first sight to have been excluded. The fact that Neveu and Schwarz introduced a set of anticommuting operators to achieve this was very interesting, but seemed less compelling; after all, at that time there were many ideas for modifying the basic amplitudes and there did not appear to be any obvious reasons why anticommuting operators should be part of a story involving mesons, unless the newly introduced operators were somehow to be associated with their constituent quarks.

Among the many ideas I gradually acquired was a realization of the importance within the dual model of the group $SU(1, 1)$ (pointed out originally by Gliozzi [Gli69]), with its Lie algebra generated by $L_{\pm 1}$, L_0, satisfying

$$[L_m, L_n] = (m - n)L_{m+n}, \quad m, n = -1, 0, +1. \tag{31.1}$$

This was soon followed by its generalization, the Virasoro algebra:

$$[L_m, L_n] = (m - n)L_{m+n} + \frac{cn(n^2 - 1)}{12}\delta_{m+n,0}, \quad m, n = 0, \pm 1, \pm 2, \ldots. \tag{31.2}$$

Note that the critically important additional 'central term' proportional to c was noticed by Joe Weis. Anyone who had studied angular momentum in quantum mechanics was aware of the finite-dimensional, unitary representations of $SU(2)$ labelled by spin. Graduate students in 1969 also knew at least a little about the unitary representations of $SU(3)$, triplets, octets, decuplets, and so on, because of the by then standard classification of hadrons and the quark model. However, it came as a surprise to me to learn about the rather more intricate, infinite-dimensional, representations of a noncompact group (or of its Lie algebra), such as $SU(1, 1)$. I spent quite some time finding out about these and rediscovering for myself some of their properties that were certainly well known to other physicists (for example, Barut and Fronsdal [BF65], Clavelli and Ramond [CR70, CR71a, CR71b]), and to mathematicians.

During that period one of the questions I asked myself was the following: which representations of (31.1) might be adapted to be representations of (31.2)? An allied question concerned the construction of N-Reggeon vertices. These were known to inherit structure from the Möbius group of conformal, linear fractional transformations (whose Lie algebra was also (31.1))

$$z \to z' = \frac{az+b}{cz+d}, \qquad ad - bc = 1, \tag{31.3}$$

and the question posed by several people, including Clavelli and Ramond, was how the representations of this group might be exploited to construct more general amplitudes than those already known at that time. Claus Montonen and I also worked together on this topic and it became the content of the paper [CM72] (for each of us our first). The main conclusion was that the only possible N-Reggeon vertices that would be permitted on the basis of satisfying all the desirable properties then required (and also ignoring spin one-half) were the standard one (Lovelace [Lov70] and Olive [Oli71]), and those associated with the Neveu–Schwarz model. Moreover, we related the then mysterious use of anticommuting annihilation and creation operators in the latter model to a specific representation of $SU(1, 1)$ that could be extended to the full Virasoro algebra only if anticommuting operators were used. Actually, it was essentially one of only two – the other being the one given in terms of the standard commuting operators used by Virasoro.

At that time, the only representations of (31.2) I was aware of were those associated with quadratic expressions in annihilation and creation operators. The unitary representations of $SU(1, 1)$ are labelled by two numbers J, k, and those that appeared to be relevant at first, in the sense that they could be extended to representations of the full Virasoro algebra (using just a single set of commuting or anticommuting annihilation and creation operators), were the representations $(0, 0)$ (Fubini–Veneziano), $(-1/2, 0)$ (Ramond) and $(-1/2, 1/2)$ (Neveu–Schwarz), and it is convenient to call the commuting or anticommuting Fock space operators a_n^μ, d_n^μ and $b_{n+1/2}^\mu$, respectively, as was customary at the time. I also noticed there was another possibility: one could also use the $(0, 1/2)$ representation of $SU(1, 1)$ and this would provide a representation of the Virasoro algebra in terms of commuting-type Fock space operators $c_{n+1/2}^\mu$, also labelled by half odd integers. Apart from mentioning its existence in my thesis, I did not explore it further because it appeared at the time to have no clear interpretation within the dual model.

If the Virasoro generators for the various representations are referred to by $L_n^{(x)}$, $x = a, b, c, d$, then the types labelled by a, b and c, d share some properties (though care needs to be taken with the Fock space operators labelled by zero). For example, if the associated ground state is represented by $|0\rangle$ then

$$L_n^{(x)}|0\rangle = 0, \quad n = 0, \pm 1, \quad x = a, b.$$

But, on the other hand,

$$L_1^{(x)}|0\rangle = 0, \quad \left(L_0^{(x)} - \frac{D}{16}\right)|0\rangle = 0, \quad L_{-1}^{(x)}|0\rangle \neq 0, \quad x = c, d.$$

Also note, to ensure the Virasoro algebra took the standard form (31.2) in each of the cases c, d the operator L_0 had to be adjusted by the addition of $D/16$ (in D spacetime dimensions). In the Ramond model of dual fermions this was understood clearly to be necessary because the operators d_0^μ were to be identified with a set of Dirac gamma matrices. However, in the extra possibility it seemed more mysterious. It was only later, and in a rather surprising way, that the commuting half odd integer labelled Fock space operators found a more or less 'natural' home (Corrigan and Fairlie [CF75a]). More will be written about that in the next Section. To conclude this paragraph, I also noticed, probably some time later, the well-known identity among partition functions

$$\prod_{n=1}^{\infty}(1 - q^{n+1/2})^{-1} = \prod_{n=1}^{\infty}(1 + q^{n+1/2}) \prod_{n=1}^{\infty}(1 + q^n),$$

which suggests an equivalence between the states created by the $c_{n+1/2}^\dagger$ and those generated by the pairs

$$(b_{n+1/2}^\dagger, d_m^\dagger).$$

Returning to the Ramond and Neveu–Schwarz story for a moment, the other remarkable discovery each made was to extend the Virasoro algebra with a set of anticommuting generators (F_N, integer labelled for Ramond, and G_r, half odd integer labelled for Neveu–Schwarz), and these were just right, as it eventually turned out, for ensuring the removal of the additional ghost states arising from the time components of the extra anticommuting creation operators. Besides noting these algebras, I did not pay much attention until the later months of 1971.

In fact, as I already mentioned, David Olive returned to CERN in the summer of 1971 and he arranged for me to accompany him for part of that year – an exciting, though daunting, prospect – and I went for about two months, from October to December. There, I renewed my acquaintances with John Cardy, who had just started as a postdoc, Peter Goddard, who was then in his second of two years at CERN, and Bruno Renner, who had recently left Cambridge for a long-term appointment at CERN, and met many other physicists for the first time. The most striking aspect of that visit was the very tangible feeling that a substantial crowd of pioneers was really making waves. I tended to be very quiet then, but I attended a great many seminars and learned hugely more than I ever could in Cambridge. It helped to put the whole subject in perspective.

The visit was particularly useful because David and I, prompted by the work of Thorn [Tho71], started to look for a convenient expression for a 'fermion emission vertex' as a preliminary to calculating amplitudes with at least four fermions. By convenient was meant an expression that respected all the gauge conditions and was as close as possible to being a standard vertex operator. The basic idea was that a suitable vertex operator would have to 'intertwine' the Ramond vertex for a fermion emitting a boson with the Neveu–Schwarz vertex for a boson emitting a boson. That this was plausible was not unreasonable because both fields were fermionic; that it would be tricky seemed likely because the two sorts of fields built out of anticommuting operators seemed to belong to inequivalent representations of $SU(1, 1)$ (and also of the Virasoro algebra – recall the remarks surrounding Eq. (31.4)). If $|A\rangle$ was a general fermion state (i.e. a spinor constructed in the d Fock space), then the fermion emission vertex $W_A(z)$ ought at least to have the $SU(1, 1)$ property

$$O_d(\gamma)W_A(z)O_b^{-1}(\gamma) = (cz + d)^{-2c_d} W_A(\gamma(z)),$$

and, setting $W_A(z) = \exp(zL_{-1}^{(d)}) \tilde{W}_A(z)$, the further properties

$$\tilde{W}_A(z)|0_b\rangle = |A\rangle,$$

$$\tilde{W}_A(z) \frac{i\sqrt{2}\gamma^{d+1} H^\mu(y)}{\sqrt{y}} = \frac{\Gamma^\mu(y - z)}{\sqrt{y - z}} \tilde{W}_A(z),$$

(31.4)

where $L_0^{(d)}|A\rangle = c_d|A\rangle$ and $m_A^2/2 = c_d - 1$. (Note, at the time it was not yet realized that the mass of the lightest Ramond fermion was zero, or that the critical dimension was ten.) Great care had to be taken with the second of the properties listed in (31.4) because of the singularities on either side. Details of the vertex operator $\tilde{W}_A(z)$ and some of its other properties are to be found in Corrigan and Olive [CO72]. While it was designed to have the transformation property (31.4), the vertex also converted (a sum of) Virasoro generators in the d Fock space to a Virasoro generator in the b Fock space, a property that was sufficient to guarantee the Virasoro gauge conditions were transmitted adequately through the vertex. At that time, the relationship between the fermion emission vertex and the F_N, G_r generators was not understood. This was clarified later.

The final part of my thesis involved an attempt to calculate the four-fermion scattering amplitude based on the ingredients known up to that point. However, the missing step was learning how to remove all the ghost states; I did not know how to do that because the fermion emission vertex did not appear to deal correctly with the gauge operators F_N and G_r (that is, to convert one set into the other). I tried out a rudimentary projection operator onto physical states but it was incomplete and unsatisfactory. For that reason I did not try to publish what I had found. In fact, the calculations were published later in [Cor74], which appeared, because of delays, after the article it was intended to precede (Corrigan, Goddard, Smith and Olive [CGOS73]). It took some time to unfold the story of how the two sets of gauges were interrelated, and that story belongs in the next Section.

To conclude this Section: during my last months as a student I had little contact with either of my supervisors but I remember John Polkinghorne being concerned and asking me from time to time what I was up to. Though I was stuck on a specific issue, in the end I wrote my thesis intensively, and quite quickly, and typed up the hand-written version in a couple of weeks.

My external examiner was Korkut Bardakci and fortunately he approved.

31.2 Research fellow in Durham

During the later part of 1971 and during 1972 I was looking for a postdoc position, though I felt it was an unlikely proposition because I had no publications at the moment I began applying. At that time the competition to go to the USA was very fierce and there were only limited possibilities in Europe or the UK. Anyway, I applied for many jobs but, in the end, it was what had seemed to be the remotest of chances that brought the best prospects. David Fairlie, who at that time I knew by name – though I had attended a seminar he gave in Cambridge – wrote to me about the A. J. Wheeler Fellowships at Durham encouraging me to apply. I applied, was asked for interview, and was very surprised (and delighted) to be offered one. It was actually a very good deal because the contract could last for up to five years, yet I had mixed feelings about returning northwards (my roots were in Birkenhead and Manchester), when CERN seemed to be the centre of activity. I was not really aware of the long-term goal to build up the Durham group following the appointments of David Fairlie, Alan Martin and Euan Squires some years previously. However, Peter Goddard, who, as I mentioned, I knew a little from Cambridge and my visit to CERN, and who I also knew had played a major role in the 'no-ghost' theorem and the development of the relativistic string (with Goldstone, Rebbi and Thorn), was also moving to Durham in October 1972. I soon appreciated that, far from being remote, Durham was very firmly on the map. Within a few months there had been visits by Lars Brink, Paolo Di Vecchia, Holger Nielsen and Charles Thorn, and probably a number of others I cannot now remember who came briefly to give talks.

For a few months, but unsuccessfully, I kept on trying to think of a way to sort out the gauge conditions for the fermion vertex and to evaluate some of the strange functions that had appeared in what I had done so far with the four-fermion scattering amplitude. Goddard was interested in the problem and we decided to look very carefully again at the intertwining property of the fermion emission vertex to see if something had been overlooked. We felt it necessary to adapt the formalism a bit (basically to make more precise the intertwining property (31.4)) and discovered that the best that could be done was to arrange for a specific Ramond anticommuting generator F_N, $N > 0$ to be converted to a linear combination $\sum_s f_{Ns} G_s$ containing generators G_s with both positive and negative labels (Corrigan and Goddard [CG73a, CG73b]). This appeared to be a negative result.

Meanwhile, Lars Brink and David Olive introduced their projection operator onto on-shell physical states [BO73a, BO73b]: this brought new insights and proved to be a very useful tool. Subsequently, Brink and Olive (with Claudio Rebbi and Joël Scherk) noted in [BORS73] that if the fermion ground state mass was zero and the critical dimension was ten, then, despite the apparently strange behaviour of the gauges noted in [CG73a], only physical states of the Neveu–Schwarz model coupled to a fermion–antifermion pair. Moreover, David and Joël [OS73] inserted the projection operator onto the Neveu–Schwarz physical states into the four-fermion calculation to derive an improved (but not evaluated) four-fermion scattering amplitude. The improvement had an interesting structure since it involved the determinant of an infinite-dimensional matrix $\Delta(x)$ defined by

$$\Delta(x) = \det\left(1 - M^2(x)\right),$$

with

$$M_{rs}(x) = -(-x)^{r+s}\frac{r}{r+s}\binom{-1/2}{r-1/2}\binom{-1/2}{s-1/2}, \qquad r, s = \frac{1}{2}, \frac{3}{2}, \ldots$$

The evaluation of the amplitude was quite complicated and was done in stages. It seemed to be an interesting calculation to do because it was not clear from the start that the result would have the same spectrum of meson states in the crossed channel (and actually it turned out not to be quite the same). The work I had done before and eventually published in [Cor74] was relevant yet incomplete. Nevertheless, it established several relationships between the ingredients, including the determinant of the antisymmetric part of M, and formed the basis of our subsequent evaluation.

Schwarz and Wu [SW73, SW74b] first guessed and checked numerically that

$$\Delta(x) = (1 - x)^{-1/4},$$

while Goddard and I (with Russell Smith – a pure mathematician at Durham whose expertise in differential equations led him to suggest some tricks we might never have thought of otherwise – and David Olive) found an analytical way of evaluating all the pieces, including this and other determinants [CGOS73]. We also analyzed carefully the interesting manner in which the Fierz transformation in ten dimensions (Case [Cas55]) meshed beautifully with the duality of the four-fermion amplitude: intriguingly, the Fierz transformation represents the transformation

$$\phi : z \to (1 - z)/(1 + z).$$

These results reinforced Mandelstam's entirely different, interacting string, approach to the same calculation [Man74], which appeared as a preprint shortly before we submitted our article. The whole story was summarized by Joël Scherk in the impressive review he wrote a few months later [Sch75].

On a side issue: Goddard and I also wrote a paper at about the same time [CG74] using the projection operator technique to extend the 'no-ghost' theorem to the Ramond model, verifying that it required the zero-mass condition and ten dimensions for the physical states to be entirely transverse (as had been suggested earlier using different arguments by Thorn, Rebbi and Schwarz).

Looking back at work that was completed thirty-five years ago is quite sobering especially as one tends to see, with the benefit of hindsight, the missed clues and mistaken paths. Nevertheless, this was a hard problem, requiring a number of complementary steps taken by different groups of people, with interdependent approaches, to reach a full understanding within the technology of the time. It would be another thirteen years before new technology enabled an alternative, and rather slicker, derivation of the same quantity (Cohn *et al.* [CFQS86]). One can also see how, in the days before electronic communication, small delays might occur in the preparation and distribution of preprints, submission dates, publication dates, and how the chronology of ideas can be affected by the pattern of conferences and personal encounters. All these factors cause small perturbations that may be magnified with time and the provenance of some ideas is lost while others are highlighted. Probably this is inevitable and it is perhaps better merely to look at the achievements as a whole and how they provided some of the steps towards where we are now. I also remember, possibly in September 1973, though I can no longer be certain precisely when, a conversation with two people in the CERN library. One of them was rather critical of the attempts to perform these complicated calculations since by that time it was already clear that the dual model, or even string theory, was going to have a hard time being phenomenologically acceptable; better in his view to be doing something else. The other, Joe Weis, was supportive, taking the view that you had best do what you can to the end because you can never be certain in advance where it might lead.

The summer of 1974 was eventful. Olive, Goddard and I were invited by John Schwarz to take part in the summer programme at Aspen and this was another novel experience. Aspen was a wonderful place to spend a few weeks. Not only did I meet many other physicists, there was always the music to accompany the seminars that were held outside. However, I did not really have much new to add to the meeting. The only idea I had had concerned the curious crossing properties of the four-fermion vertex. It was already known that the fermions had to be massless and I (and probably several others) noticed that the crossing properties made more sense if the fermions were designed to be 'left-handed' rather than Dirac particles. Since we were discussing a putative model of hadrons, that seemed odd. It occurred to me these might be regarded instead as neutrinos and I wondered for a while whether one could assemble the elements of the Standard Model. However, it did not really work because the Weinberg angle came out wrongly (to $\pi/3$) and I felt that was unsatisfactory. Apart from mentioning this in my seminar[1] and discussing it with Peter and briefly with Joël Scherk – who I knew from my stay at CERN and was also at Aspen – and realizing he had had the same idea, I did not pursue it further.

Besides the London conference in July, and the few intensive weeks at Aspen in August, I started to work on an idea of David Fairlie's that he had had some time before but had never developed. His idea was to adjust the 'analogue model' (Fairlie and Nielsen [FN70]) so that, in addition to having point sources on the rim of the disc where on-shell momentum was injected – corresponding to on-shell particles – there should also be portions of the perimeter of the disc where momentum would be 'smeared out'. These would not correspond to particles on-shell. Thus, particles would be associated with specific

---

[1] I am grateful to Peter Goddard for showing me recently his notes of that seminar!

locations z_i on the rim of the unit disc (as in the Koba–Nielsen formulation), while off-shell momentum insertions would correspond to pairs of points z'_j, z''_j.

It is perhaps easiest to express the idea by thinking about it in a conformally equivalent way, with the disc mapped to the upper half-plane and its rim lying along the real axis. From that perspective, and denoting the complex potential by $f^\mu(z)$, potentials corresponding to a set of sources along the real line should have the property that their imaginary parts jump by k_i^μ as z moves past z_i along the real line when there is a point source at z_i or, by an amount Q_j^μ as z moves from z'_j to z''_j when the momentum is smeared out. Moreover, in the latter case the real parts of $f^\mu(z)$ remain constant as z moves from z'_j to z''_j. A potential with a single pair z' and z'' and these properties has the form

$$f^\mu(z) = \sum_{i=1}^{N} k_i^\mu \ln \left(\frac{\sqrt{\frac{z_i - z'}{z_i - z''}} - \sqrt{\frac{z - z'}{z - z''}}}{\sqrt{\frac{z_i - z'}{z_i - z''}} + \sqrt{\frac{z - z'}{z - z''}}} \right). \qquad (31.5)$$

This should be compared with

$$f^\mu(z) = \sum_{i=1}^{N} k_i^\mu \ln (z - z_i),$$

which leads via the analogue procedure [FN70] to the standard Koba–Nielsen expression for a multiparticle on-shell amplitude. In (31.5) there is no requirement that $Q^\mu = -\sum_{i=1}^{N} k_i^\mu$ be on-shell. Any attempt to write a corresponding potential with two or more strips leads to elliptic functions that are difficult, if not impossible, to write down explicitly. We also noted that in the Mandelstam picture of interacting strings our strips corresponded to strings terminating at finite times. Because it looked difficult to write down multiparticle amplitudes with some particles off-shell, we decided to adopt a different strategy: to factorize the amplitude constructed using (31.5) and use the factorization to identify a vertex operator describing a process in which a string emits off-shell momentum. This was a long shot because the vertex operator had to preserve all the desirable features of the Fubini–Veneziano vertex operator (for example its transformation properties and propagating correctly the Virasoro gauge conditions), and this had been tried before without success (see, for example, Clavelli and Ramond [CR71b]). Nevertheless, we were encouraged by the close resemblance our one off-shell/many on-shell amplitude bore to the amplitudes written down previously by Schwarz [Sch73, SW74a]. He had started with a seemingly very different set of assumptions – that apparently only worked properly in sixteen dimensions. It was the latter fact, when we made the connection with Schwarz's work, that triggered something in my mind and I suddenly remembered the $(n + 1/2)$-labelled bosonic annihilation and creation operators. To my surprise and delight vertex operators constructed from these were very close to what we wanted, though by no means the whole story. Besides producing the amplitudes for us there was the advantage that there was no place for 'zero modes' because there was no momentum operator naturally linked to the vertex. In the end we developed a reasonably coherent picture and the technicalities of the story can be found in [CF75a]. Here, I will merely summarize it.

Our off-shell vertex $J(z', z'', Q)$ was a formal product of three pieces: first, an operator that switched (or intertwined) the usual modes of the string (but excluding the position operator q) with an auxiliary set labelled by $(n + 1/2)$-labelled operators, second, the operator $e^{iq \cdot Q}$ that injected an arbitrary amount of momentum into the system, and third, an operator that converted back to the usual modes. All of this was done in a manner preserving the essential features of a vertex operator yet needing no requirement on Q^2. In stringy language one may say it more picturesquely. In order to shed or gain momentum the string first switches to a state in which the momentum and the rest of the even modes are lost and effectively replaced by the $(n + 1/2)$-labelled modes. One could say, as we did at the time, that instead of Neumann conditions at both ends of an open string one should impose a Dirichlet condition at one end and a Neumann condition at the other, implying that the string no longer conserved momentum – and hence momentum could be lost or gained via the pinned down end.

The mixed conditions also required the unfamiliar modes. In other words, to go off-shell in our picture, an open string switches its boundary conditions to Neumann–Dirichlet and back again after the momentum has been inserted or extracted. The strange condition, that the dimension of spacetime ought to be sixteen for all this to work, had an explanation of sorts in the inevitable presence of the operators c_r^μ. Besides, as we also pointed out, in the Neveu–Schwarz model this peculiarity appeared to be ameliorated with the critical dimension restored to ten. It did not occur to us to apply the trickery to just sixteen of the twenty-six components of the bosonic string. With hindsight, we did not explain the idea in the best way possible – and we ought to have pursued it further (but we were distracted by generalizations of the 't Hooft–Polyakov monopole, Corrigan, Olive, Fairlie and Nuyts [COFN76]). At the time it was a persistent problem that had been looked at by many people over the years from 1969; examples, to illustrate the range of ideas but with no pretence at completeness, are given by Nambu [Nam71], Drummond [Dru71], Sato [Sat71], Neveu and Scherk [NS72] and Collins and Friedman [CF75b].

To modern eyes, doubtless, the whole question looks redundant because string/brane theory is a 'model for everything' and the questions we were trying to answer (how would a hadronic system interact with an electromagnetic or weak current whose origins were external to itself?) have now evaporated. Moreover, the language looks somewhat old-fashioned when compared with the streamlined arguments that are now commonplace. Nevertheless, for quite some time afterwards I was unreasonably elated by the picture we had uncovered, which seemed to me to be quite natural, emerging as it had in such a roundabout way without being imposed at the start. Later on, Warren Siegel [Sie76] and Mike Green [Gre77a, Gre77b] took the idea up for a while, but the idea of mixing boundary conditions had to wait more than a decade before Polchinski discovered its significance in a very different context. Some of the technology proved to be useful later in some work I did with Tim Hollowood, my then graduate student [CH88a, CH88b].

As I mentioned at the start of this part of the story, it began in the late summer of 1974, but I should also say it continued as I moved to CERN to take up a CERN Fellowship. It was completed there, but I remember vividly the hectic few days of transition when I was busy calculating during the lulls in the family move to Geneva. For what it's worth, I have noticed several times how a physical displacement, or change of job, can facilitate ideas.

There is a minor spin-off from this story: the (0, 1/2) representation of $SU(1, 1)$, which seemed to be required to enable a string to go off-shell, also provided a simple derivation of the determinant needed for the four-fermion calculation. Personally, I found this calculation and others related to it (the details were reported in Bruce, Corrigan and Olive [BCO75]) more elegant and satisfying than the earlier derivations. Unfortunately though, the ideas arrived too late to be an essential part of the stream.

Earlier, I mentioned that the fermion emission vertex resembled an intertwiner between two inequivalent representations of $SU(1, 1)$, or of the Virasoro algebra, and that this appeared to me to be strange from a representation theory perspective. Our off-shell vertex construction was also strange in the same sense because it contained a pair of intertwiners performing a similarly impossible trick. I puzzled about this for a while but was unable to resolve the apparent paradox and set it to one side. Fortunately, a proper analysis resolving these issues was provided later by Peter Goddard and Roger Horsley [GH76, Hor78]. Looked at appropriately, each type of vertex could be understood in group theoretic terms.

31.3 Epilogue

During the summer of 1975 David Fairlie organized a Dual Model workshop in Durham to which around 25 people came, including Pierre Ramond, who I had not met previously, and several others, especially from Japan, who I knew only via their articles. My memory tells me it was a successful week and perhaps the last meeting of its kind. During the meeting I was also interviewed for a lectureship at Durham, was offered the job, and instantly accepted it to start in January 1976. This meant I would have to cut short the CERN Fellowship, which was a pity but, on the other hand, I was very lucky: this was to be the only appointment in this area in Durham for a long time because the UK academic system was more or less on hold owing to severe financial problems.

Since then, I have, intermittently, worked with David Fairlie, Peter Goddard and David Olive on a series of problems in gauge theory, and later, with a sequence of my own graduate students and postdocs, on various topics in integrable quantum field theory. My time since 1976 has been divided by teaching, and increasingly by administration: I find myself looking back with some nostalgia, and wistfully, for those few years when my main preoccupation was thinking about dual models. Never afterwards has it been possible to free up such a continuous stretch of time, though I was lucky to be able to spend a year at the Ecole Normale Supérieure in Paris 1977–1978, and a year at Caltech 1978–1979. There have been other periods of intensive activity, some of which I have tried to keep up with, and there have been some personal high spots, but none of it since has been associated with similar feelings of excitement and common purpose. It could be this feeling is based on an illusion arising from my relative inexperience. That may be, but I can also say this: few of us are fortunate enough not to have need of illusions.

Acknowledgements

I am indebted to Jane for her sustained support during these years, and to the other students, especially Claus Montonen and Charles Pantin, and grateful for the guidance and help of many other people who kept me going during low moments as a graduate student – particularly Ian Drummond, David Olive and John Polkinghorne. I am especially grateful to my Durham colleagues for their kindness, enthusiasm, patient help, encouragement and collaborations, especially David Fairlie, Peter Goddard, Bob Johnson and Russell Smith, and to others I met only rarely, such as Lars Brink, Paolo Di Vecchia, Joël Scherk, Pierre Ramond and Charles Thorn. I am grateful to David Fairlie, Peter Goddard and David Olive for sharing their memories.

Edward Corrigan received his PhD in 1972 from the University of Cambridge, then held research fellowships at Durham (1972–1974) and CERN (1974–1976) before being appointed as Lecturer in mathematics at Durham, where he remained, apart from periods of leave at ENS-Paris (1977–1978), Caltech (1978–1979) and CERN (1987), until 1999. From 1999–2007 he was Professor and Head of the Department of Mathematics at the University of York, and returned to Durham in 2008 as Professor of Mathematics and Principal of Collingwood College. He has worked on aspects of string theory, gauge theory and integrable field theory. He was elected to the Royal Society in 1995.

References

[AALO71] Alessandrini, V., Amati, D., Le Bellac, M. and Olive, D. (1971). The operator approach to dual multiparticle theory, *Phys. Rep.* **C1**, 269–346.
[BF65] Barut, A. O. and Fronsdal, C. (1965). On non-compact groups. II. Representations of the 2+1 Lorentz group, *Proc. R. Soc. London* **A287**, 532–548.
[BO73a] Brink, L. and Olive, D. (1973). The physical state projection operator in dual resonance models for the critical dimension of spacetime, *Nucl. Phys.* **B56**, 253–265.
[BO73b] Brink, L. and Olive, D. (1973). Recalculation of the unitary single planar loop in the critical dimension of spacetime, *Nucl. Phys.* **B58**, 237–253.
[BORS73] Brink, L., Olive, D., Rebbi, C. and Scherk, J. (1973). The missing gauge conditions for the dual fermion emission vertex and their consequences, *Phys. Lett.* **B45**, 379–383.
[BCO75] Bruce, D., Corrigan, E. and Olive, D. (1975). Group theoretical calculation of traces and determinants occurring in dual theories, *Nucl. Phys.* **B95**, 427–433.
[Cas55] Case, K. M. (1955). Biquadratic spinor identities, *Phys. Rev.* **97**, 810–823.
[CR70] Clavelli, L. and Ramond, P. (1970). $SU(1, 1)$ analysis of dual resonance models, *Phys. Rev.* **D2**, 973–979.
[CR71a] Clavelli, L. and Ramond, P. (1971). Group theoretical construction of dual amplitudes, *Phys. Rev.* **D3**, 988–990.

[CR71b] Clavelli, L. and Ramond, P. (1971). New class of dual vertices, *Phys. Rev.* **D4**, 3098–3101.
[CFQS86] Cohn, J., Friedan, D., Qiu, Z.-A. and Shenker, S. H. (1986). Covariant quantization of supersymmetric string theories: the spinor field of the Ramond–Neveu–Schwarz model, *Nucl. Phys.* **B278**, 577–600.
[CF75b] Collins, P. V. and Friedman, K. A. (1975). Off-shell amplitudes and currents in the dual resonance model, *Nuovo Cimento* **A28**, 173–192.
[Cor74] Corrigan, E. (1974). The scattering amplitude for four dual fermions, *Nucl. Phys.* **B69**, 325–335.
[CF75a] Corrigan, E. and Fairlie, D. B. (1975). Off-shell states in dual resonance theory, *Nucl. Phys.* **B91**, 527–545.
[CG73a] Corrigan, E. and Goddard, P. (1973). Gauge conditions in the dual fermion model, *Nuovo Cimento* **A18**, 339–359.
[CG73b] Corrigan, E. and Goddard, P. (1973). The off-mass shell physical state projection operator for the dual resonance model, *Phys. Lett.* **B44**, 502–506.
[CG74] Corrigan, E. and Goddard, P. (1974). The absence of ghosts in the dual fermion model, *Nucl. Phys.* **B68**, 189–202.
[CGOS73] Corrigan, E., Goddard, P., Olive, D. and Smith, R. A. (1973). Evaluation of the scattering amplitude for four dual fermions, *Nucl. Phys.* **B67**, 477–491.
[CH88a] Corrigan, E. and Hollowood, T. J. (1988). A bosonic representation of the twisted string emission vertex, *Nucl. Phys.* **B303**, 135–148.
[CH88b] Corrigan, E. and Hollowood, T. J. (1988). Comments on the algebra of straight, twisted and intertwining vertex operators, *Nucl. Phys.* **B304**, 77–107.
[CM72] Corrigan, E. and Montonen, C. (1972). General dual operatorial vertices, *Nucl. Phys.* **B36**, 58–72.
[CO72] Corrigan, E. and Olive, D. (1972). Fermion-meson vertices in dual theories, *Nuovo Cimento* **A11**, 749–773.
[COFN76] Corrigan, E., Olive, D., Fairlie, D. B. and Nuyts, J. (1976). Magnetic monopoles in $SU(3)$ gauge theories, *Nucl. Phys.* **B106**, 475–492.
[Dru71] Drummond, I. T. (1971). Dual amplitudes for currents, *Nucl. Phys.* **B35**, 269–286.
[EFFG70] Ellis, S., Frampton, P. H., Freund, P. G. O. and Gordon, D. (1970). Hadrodynamics and quark structure, *Nucl. Phys.* **B24**, 465–489.
[FN70] Fairlie, D. B. and Nielsen, H. B. (1970). An analog model for KSV theory, *Nucl. Phys.* **B20**, 637–651.
[FGV69] Fubini, S., Gordon, D. and Veneziano, G. (1969). A general treatment of factorization in dual resonance models, *Phys. Lett.* **B29**, 679–682.
[FV69] Fubini, S. and Veneziano, G. (1969). Level structure of dual resonance models, *Nuovo Cimento* **A64**, 811–840.
[Gli69] Gliozzi, F. (1969). Ward-like identities and twisting operator in dual resonance models, *Lett. Nuovo Cimento* **2**, 846–850.
[GH76] Goddard, P. and Horsley, R. (1976). The group theoretic structure of dual vertices, *Nucl. Phys.* **B111**, 272–296.
[Gre77a] Green, M. B. (1977). Dynamical point-like structure and dual strings, *Phys. Lett.* **B69**, 89–92.
[Gre77b] Green, M. B. (1977). Point-like structure and off-shell dual strings, *Nucl. Phys.* **B124**, 461–499.
[Hor78] Horsley, R. (1978). The group theoretic structure of the dual current vertex, *Nucl. Phys.* **B138**, 474–492.

[KN69] Koba, Z. and Nielsen, H. B. (1969). Manifestly crossing-invariant parameterization of n-meson amplitude, *Nucl. Phys.* **B12**, 517–536.

[Lov70] Lovelace, C. (1970). Simple N-reggeon vertex, *Phys. Lett.* **B32**, 490–494.

[Man69] Mandelstam, S. (1969). Relativistic quark model based on the Veneziano representation. I. Meson trajectories, *Phys. Rev.* **184**, 1625–1639.

[Man70a] Mandelstam, S. (1970). Relativistic quark model based on the Veneziano representation. II. General trajectories, *Phys. Rev.* **D1**, 1734–1744.

[Man70b] Mandelstam, S. (1970). Relativistic quark model based on the Veneziano representation. III. Baryon trajectories, *Phys. Rev.* **D1**, 1745–1753.

[Man74] Mandelstam, S. (1974). Interacting string picture of the Neveu–Schwarz–Ramond model, *Nucl. Phys.* **B69**, 77–106.

[Nam71] Nambu, Y. (1971). Electromagnetic currents in dual hadrodynamics, *Phys. Rev.* **D4**, 1193–1202.

[NS72] Neveu, A. and Scherk, J. (1972). Currents and Green's functions for dual models – I, *Nucl. Phys.* **B41**, 365–387.

[NS71] Neveu, A. and Schwarz, J. H. (1971). Factorizable dual model of pions, *Nucl. Phys.* **B31**, 86–112.

[Oli71] Olive, D. (1971). Operator vertices and propagators in dual theories, *Nuovo Cimento* **A3**, 399–411.

[OS73] Olive, D. and Scherk, J. (1973). Towards satisfactory scattering amplitudes for dual fermions, *Nucl. Phys.* **B64**, 334–348.

[Ram71] Ramond, P. (1971). Dual theory for free fermions, *Phys. Rev.* **D3**, 2415–2418.

[Sat71] Sato, H. (1971). Current amplitudes in the dual resonance model, *Prog. Theor. Phys.* **45**, 1592–1606.

[Sch75] Scherk, J. (1975). An introduction to the theory of dual models and strings, *Rev. Mod. Phys.* **47**, 123–163.

[Sch73] Schwarz, J. H. (1973). Off-mass shell dual amplitudes without ghosts, *Nucl. Phys.* **B65**, 131–140.

[SW73] Schwarz, J. H. and Wu, C. C. (1973). Evaluation of dual fermion amplitudes, *Phys. Lett.* **B47**, 453–456.

[SW74a] Schwarz, J. H. and Wu, C. C. (1974). Off mass shell dual amplitudes. 2, *Nucl. Phys.* **B72**, 397–412.

[SW74b] Schwarz, J. H. and Wu, C. C. (1974). Functions occurring in dual fermion amplitudes, *Nucl. Phys.* **B73**, 77–92.

[Sie76] Siegel, W. (1976). Strings with dimension-dependent intercept, *Nucl. Phys.* **B109**, 244–254.

[Tho71] Thorn, C. B. (1971). Embryonic dual model for pions and fermions, *Phys. Rev.* **D4**, 1112–1116.

[Ven68] Veneziano, G. (1968). Construction of a crossing-symmetric, Reggeon behaved amplitude for linearly rising trajectories, *Nuovo Cimento* **A57**, 190–197.

[Vir70] Virasoro, M. (1970). Subsidiary conditions and ghosts in dual-resonance models, *Phys. Rev.* **D1**, 2933–2936.

32

The dual quark models

KORKUT BARDAKCI AND MARTIN B. HALPERN

Abstract

We briefly recall the historical environment around our 1971 and 1975 constructions of current algebraic internal symmetry on the open string. These constructions included the introduction of world-sheet fermions, the independent discovery of affine Lie algebra in physics (level one of affine $su(3)$), the first examples of the affine-Sugawara and coset constructions, and finally – from compactified spatial dimensions on the string – the first vertex-operator constructions of the fermions and level one of affine $su(n)$.

32.1 Introduction

In this Chapter we describe the environment around our 1971 paper 'New dual quark models' [BH71] and the two companion papers 'The two faces of a dual pion–quark model' [Hal71c] in 1971, and 'Quantum "solitons" which are $SU(N)$ fermions' [Hal75] in 1975, which laid the foundations of non-Abelian current-algebraic internal symmetry on the string.

The background for our contributions included many helpful discussions with other early workers, including H. M. Chan, C. Lovelace, H. Ruegg and C. Schmid (during KB's 1969 visit to CERN), as well as R. Brower, S. Klein, C. Thorn, M. Virasoro and J. Weis (with MBH at Berkeley). Both of us also acknowledge many discussions with Y. Frishman, G. Segre, J. Shapiro and especially S. Mandelstam. Later discussions with I. Bars and W. Siegel are also acknowledged by MBH. With apologies to many other authors then, we reference here only the work which was most influential in our early thinking: before Veneziano, there had been widespread interest in the quark model of Gell-Mann and Zweig, including the four-dimensional current algebra (Adler and Dashen [AD68]) of quarks.

32.2 The bosonic dual model

In 1970, the work of Fubini and Veneziano [FV69] together with the Virasoro algebra [Vir70] provided an *algebraic formulation* of open bosonic string theory which we found

---

The Birth of String Theory, ed. Andrea Cappelli, Elena Castellani, Filippo Colomo and Paolo Di Vecchia.
Published by Cambridge University Press. © Cambridge University Press 2012.

quite exciting. The first ingredients of this formulation are the bosonic oscillators

$$[J_\mu(m), J_\nu(n)] = -m\, g_{\mu\nu}\, \delta_{m+n,0},$$
$$m, n \in \mathbb{Z}, \quad g = \text{diag}(1, -1, \ldots, -1), \quad \mu, \nu = 0, 1, \ldots, 25, \tag{32.1}$$

which we have rescaled here as the modes of an Abelian current algebra on the circle. The conformal field theoretic convention for the local currents associated with these modes is

$$J_\mu(z) = \sum_{m \in \mathbb{Z}} J_\mu(m)\, z^{-m-1} = i\, \partial_z Q_\mu(z), \tag{32.2}$$

where

$$Q(z) = q - i J(0)\ln(z) + i \sum_{m \neq 0} J(m) \frac{z^{-m}}{m}, \tag{32.3}$$

are the Fubini–Veneziano fields and $\{q\}$ is canonically conjugate to the momenta $\{p = J(0)\}$. Normal ordered exponentials of the Fubini–Veneziano fields are the string vertex operators, to which we shall return below.

In our discussions we shall emphasize the mode algebras, but for various modern applications the complex variable z in the local fields can be taken in many forms, including $\exp(i\theta)$ (the circle or loop), $\exp(i(\tau + \sigma))$ (left-movers on a closed string), the Euclidean version $\exp(\tau + i\sigma)$ for conformal field theory (CFT), or the real line for open strings. These alternative forms led to some confusion in the early days of string theory, but in fact most of the early work reviewed in this Chapter was in the context of the open string, using the parameterization $z = \exp(i\theta)$ on the circle.

Finally, the Virasoro algebra of the critical open bosonic string is

$$L(m) = -\frac{1}{2} g^{\mu\nu} \sum_{n \in \mathbb{Z}} : J_\mu(n)\, J_\nu(m - n) :,$$
$$[L(m), L(n)] = (m - n)\, L(m + n) + \frac{26}{12} m\, (m^2 - 1)\, \delta_{m+n,0}, \tag{32.4}$$

where the symbol $: \ldots :$ is bosonic normal ordering. Under these generators the Abelian currents $\{J(z)\}$ transform covariantly with conformal weight equal to one.

Taken together, the algebraic statements in Eqs. (32.1)–(32.3) provided the first example of modern CFT (see Belavin, Polyakov and Zamolodchikov [BPZ84]). We have here taken an historical liberty, because the central extension of the Virasoro algebra was not observed until 1974, and even then only in a private communication from J. Weis quoted by Chodos and Thorn [CT74]. Throughout this Chapter we adopt CFT conventions, including an extra factor $z^{-\Delta}$ (see Eq. (32.1)) for the mode expansions of Virasoro primary fields with conformal weight Δ. The CFT convention guarantees that a given Virasoro primary field creates a Virasoro primary state of the same conformal weight as z goes to 0 on the vacuum (in this case the zero-momentum vacuum $|\{p = J(0) = 0\}\rangle$ of the open bosonic string). On the other hand, the extra factor $z^{-\Delta}$ must be removed for exact correspondence of the local fields with $z = \exp(i\theta)$ in early work on the open string.

32.3 World-sheet fermions

As it appeared then, the open bosonic string had two deficiencies beyond the tachyonic ground state. In the first place, Lie-algebraic internal symmetries were implemented only by the multiplicative Chan–Paton factors [PC69], which live on the boundary of the string. Second, the bosonic string had no spin, that is, no spacetime fermions. Our first attempt to solve the second problem was a generalization of Chan–Paton factors to include spin [BH69], but this multiplicative approach was unsuccessful because it included negative-norm states.

It seemed natural then to approach the problems of spin and internal symmetry in a different way, by introducing new local 'quark' fields on the string – now called 'world-sheet fermions' [BH71]:

$$\bar{\psi}^I(z) = \sum_{p \in \mathbb{Z}+\frac{1}{2}} \bar{\psi}^I(p) z^{-p-\frac{1}{2}}, \qquad \psi_I(z) = \sum_{p \in \mathbb{Z}+\frac{1}{2}} \psi_I(p) z^{-p-\frac{1}{2}}, \qquad (32.5)$$

whose modes obey anticommutation relations

$$[\psi_I(q), \psi_J(p)]_+ = [\bar{\psi}^I(q), \bar{\psi}^J(p)]_+ = 0,$$

$$[\psi_I(q), \bar{\psi}^J(p)]_+ = \delta_I^J \delta_{q+p,0}, \qquad q, p \in \mathbb{Z}+\frac{1}{2}. \qquad (32.6)$$

As in the four-dimensional quark model itself, the generic indices $\{I, J\}$ were designed so that the world-sheet fermions could be spacetime spinors $\{\psi_r\}$ or carry internal symmetry or both $\{\bar{\psi} \gamma_\mu \lambda_a \psi\}$. The Virasoro generators of our world-sheet fermions are

$$L(m) = \sum_{p \in \mathbb{Z}+\frac{1}{2}} \left(p + \frac{m}{2}\right) : \bar{\psi}^I(p) \psi_I(m-p) :, \qquad (32.7)$$

with fermionic normal ordering and central charge c equal to the number of $(\psi, \bar{\psi})$ pairs. In what follows, we often refer to our world-sheet fermions as the *dual quarks*.

Under these Virasoro generators, the dual quark fields $\{\psi, \bar{\psi}\}$ are Virasoro primary fields with conformal weight one-half and trivial monodromy. The half-integer moding of the dual quark fields guaranteed a unique vacuum (or ground state) $|0\rangle$ which is annihilated by the positive modes of both fields, and the adjoint operation is defined for internal symmetry as $\psi(p)^\dagger = \bar{\psi}(-p)$ (with a γ_0 modification for spin). In either case, our Virasoro generators satisfy the generalized Hermiticity $L(m)^\dagger = L(-m)$ and the vacuum state conditions:

$$L(m)|0\rangle = 0, \qquad m \geq -1. \qquad (32.8)$$

With internal symmetry only, our dual quark models then provided the first examples of *unitary* CFT, with a positive definite Hilbert space on a nondegenerate vacuum with conformal weight $\Delta = 0$. In modern parlance, the half-integer moded complex BH fermions are related to antiperiodic Weyl fermions on the world-sheet, but it should be born in mind that fermionic world-sheet action formulations of open string theory had not been studied at that time.

Were these the first world-sheet fermions? In fact, our work was essentially simultaneous with that of Ramond [Ram71]: our paper was submitted in November 1970, while Ramond's paper with integer-moded Majorana–Weyl world-sheet fermions was submitted in January 1971; the two papers were published in the same issue of *Physical Review*. As we now know, the Ramond (R) fermions $\{b_\mu\}$, with their somewhat counterintuitive spacetime vector label μ and superconformal symmetry, provided the correct covariant description of spacetime fermions and – in a perfect match to the R construction – half-integer moded Majorana–Weyl world-sheet fermions $\{b_\mu\}$ were later introduced for the bosonic sector by Neveu and Schwarz [NS71].

For the historical record, the lure of our seemingly more physical spacetime spinor fields $\{\psi_r\}$ persisted for some time, leading to many connections (Halpern [Hal71a, Hal71b, Hal71c], Bardakci [Bar71], Halpern and Thorn [HT71a, HT71b] and Mandelstam [Man73a, Man73b]) between our fields and those of the RNS formulation (we mention in particular the first example of half-integer moded scalar fields in [HT71b]). Our original intuition was however not realized until 1982, when a world-sheet fermion with 10-dimensional Majorana–Weyl spinor indices was successfully incorporated by Green and Schwarz in the light-cone formulation of superstring theory [GS82].

In what follows, we therefore restrict the discussion primarily to our various current algebraic and CFT constructions on the open string.

32.4 Affine Lie algebra

For continuity with the bosonic string, we begin this discussion with a footnote of our 1971 paper – which observed that the spatial components $\{I = 1, \ldots, 25\}$ of the Abelian currents (32.1) and (32.2) could be constructed out of our world-sheet fermions [BH71]:

$$J_I(m) = \sum_{p \in \mathbb{Z} + \frac{1}{2}} : \bar{\psi}^I(p) \psi_I(m-p) :,$$

$$J_I(z) =: \bar{\psi}^I(z) \psi_I(z) := \partial_z Q_I(z). \qquad (32.9)$$

This result is now recognized as a simple but important part of Bose–Fermi equivalence on the open string, a subject to which we will return below.

Our next construction was the independent discovery of affine Lie algebra in physics [BH71], still sometimes called the 'dual quark model', in which the non-Abelian currents are naturally constructed as dual quark bilinears with conformal weight 1:

$$J_a(z) = \sum_{m \in \mathbb{Z}} J_a(m) z^{-m-1} =: \bar{\psi}(z) T_a \psi(z) :, \qquad (32.10)$$

where

$$J_a(m) = \sum_{p \in \mathbb{Z} + \frac{1}{2}} : \bar{\psi}^I(p) (T_a)_I{}^J \psi_J(m-p) :,$$

$$[T_a, T_b] = i f_{ab}{}^c T_c, \qquad \text{Tr}(T_a T_b) = k \eta_{ab}. \qquad (32.11)$$

Here $\{T\}$ can be any irreducible matrix representation of any simple Lie algebra g, with structure constants $\{f\}$ and Killing metric $\{\eta\}$. Then one finds that the modes of the dual quark model currents $\{: \bar{\psi} T \psi :\}$ are the generators of the affine Lie algebra

$$[J_a(m), J_b(n)] = i\, f_{ab}{}^c\, J_c(m+n) + m\, (k\, \eta_{ab})\, \delta_{m+n,0},$$
$$a, b, c = 1, \ldots, \dim(g), \qquad m, n \in \mathbb{Z}, \qquad (32.12)$$

where the quantity k in the central term is called the *level* of the affine algebra. (We assume in this Chapter that the highest root-length squared of Lie g is two, in which case the level in the dual quark models is also equal to the Dynkin index of irreducible representation T.) Affine Lie algebra is also known as centrally extended non-Abelian current algebra on the circle. In this construction, the dual quark fields $\{\psi, \bar{\psi}\}$ transform respectively in the $\{T, \bar{T}\}$ representations of Lie g – the Lie algebra being generated by the zero modes $\{J_a(0)\}$ of the affine algebra.

In fact our first example of this construction used the 3×3 Cartesian Gell-Mann matrices $\{T_a \propto \lambda_a\}$ of $su(3)$ (and the Dynkin index of the fundamental of $su(n)$ is 1) – so the *first concrete representation of affine Lie algebra* was $su(3)_1$, that is, level one of affine $su(3)$ (see also Section 32.8). In this case, as in the four-dimensional quark model itself, the dual quark fields $\{\psi, \bar{\psi}\}$ transform respectively in the $(3, \bar{3})$ representations of $su(3)$. Dual quark model constructions of the currents of $so(3, 1)_1$ and $so(4, 1)_1$ were also included in our 1971 paper, and further studied along with those of higher Lorentz algebras in the first companion paper [Hal71c]. The dual quark models of $su(n)_1$ were given later in the second companion paper [Hal75].

Because connections between open strings and field theory were not understood at the time, our interpretation of the central terms in our string theoretic constructions of affine Lie algebras on the circle was based on the presence of such a term in the Abelian mode algebra (32.1) of the open bosonic string – to which our constructions reduced in the Abelian case. This understanding was supported by our fermionic construction (see Eq. (32.7) above) of the modes of the Abelian current algebra.

On the other hand, we often think now of the central term in affine Lie algebra as an analogue on the circle of the equal-time Schwinger term [Sch59] (indeed, see Eq. (3.10) in [BH71]), so we briefly sketch the history of $su(n)$ current algebra in the two-dimensional Lorentz-invariant quantum field theory of massless free fermions. The equal-time Schwinger term in $su(n)$ current algebra on the line was first computed in 1969 (see Coleman, Gross and Jackiw [CGJ69]), and the one-dimensional $su(n)$ current algebras of the light-cone currents $\{J_a^\pm(t \pm x)\}$ – with one-dimensional Schwinger terms – were given later by Dashen and Frishman [DF73]. In fact, however, the momenta are continuous in this noncompactified, Lorentz-invariant context, so our open string mode forms of the fermions and affine Lie algebra on the circle [BH71] do not appear in these papers.

For a modern introduction to affine Lie algebras in CFT, see the work by Gepner and Witten [GW86] and also the representation theory for all integer levels in Kac [Kac90]. In physics the affine algebras have sometimes also been called Kac–Moody algebras, but this term is more properly reserved for the general algebraic systems including affine and hyperbolic algebras (see Section 32.8).

32.5 The affine-Sugawara constructions

The 'affine-Sugawara constructions' [BH71, Hal71c] are the simplest realization of Virasoro generators as quadratic forms on the generators of affine Lie algebras,

$$L_g(m) = \frac{\eta^{ab}}{2k + Q_g} \sum_{n \in \mathbb{Z}} : J_a(n) J_b(m-n) :,$$

$$[L_g(m), L_g(n)] = (m-n) L_g(m+n) + \frac{c_g}{12} m(m^2 - 1) \delta_{m+n,0}, \quad (32.13)$$

$$c_g = \frac{2k \dim(g)}{2k + Q_g}.$$

The quantity Q_g is the quadratic Casimir of Lie g, which in our present root normalization is twice the dual Coxeter number of g. The original Sugawara model [Sug68] (Sommerfield [Som68], Bardakci, Frishman and Halpern [BFH68], Bardakci and Halpern [BH68], Gross and Halpern [GH69], Coleman, Gross and Jackiw [CGJ69]) was in four dimensions on a different current algebra, the so-called algebra of fields. The affine-Sugawara (or Sugawara-like) constructions in Eq. (32.11) hold for all integer levels of any affine Lie algebra.

The first example of the affine-Sugawara construction [BH71] used the $su(3)_1$ dual quark model currents of the previous Section (the dual Coxeter number of $su(n)$ is n). The affine-Sugawara construction on the level-one currents of affine $so(4, 1)$ was also given in our 1971 paper, and extended in the first companion paper [Hal71c] to the construction on the level-one currents of affine $so(3, 1)$ and higher Lorentz algebra. The case of $su(n)_1$ was given later in [Hal75]. We emphasize that the prefactors $(2k + Q_g)^{-1}$ in these constructions follow from careful consideration of the normal ordering $: \ldots :$ of the current modes. The affine-Sugawara constructions, being quartic in our dual quarks, were called 'current–current' or 'spin–spin' interactions in those days, and models which used the dual quarks or the affine-Sugawara constructions instead of extra dimensions on the open string were known as 'additive' models.

In physics, the general form (32.11) of the affine-Sugawara construction was first given by Knizhnik and Zamolodchikov [KZ84] thirteen years after our examples, and used by these authors to find the general KZ equations. In the same year, the important paper by Witten [Wit84] gave the action formulation and non-Abelian bosonization of these operator systems, leading to their understanding – especially at higher level – as interacting quantum field theories of the Wess–Zumino–Witten type [GW86]. The independent history of the affine-Sugawara constructions in mathematics is noted in Section 32.8.

In this connection, we also sketch the history of the $u(n)$ Sugawara-like constructions in two-dimensional Lorentz-invariant field theories of massless free fermions. Using a point-splitting regularization of the products of $u(n)$ currents, Coleman, Gross and Jackiw [CGJ69] showed the equivalence of the two-dimensional analogue $\{T_{\mu\nu}\}$ of the original Sugawara stress tensor to the free-fermionic stress tensor $\{\Theta_{\mu\nu}\}$, and (building on the Abelian case by Dell'Antonio, Frishman and Zwanziger [DFZ72]) normal ordered light-cone analogues of the $u(n)$ Sugawara construction were later given by Dashen and Frishman [DF73]

(Eqs. (5.15)–(5.16) of [DF75] also gave (and solved) light-cone analogues of the $u(n)$ KZ equations for the fermionic four-point function). Again, however, these noncompactified, Lorentz-invariant results had continuous momenta, so the corresponding affine-Sugawara constructions do not appear in these papers (see also [Hal75] and [BHN76]).

32.6 The coset constructions

The affine-Sugawara construction (32.13) on the currents of affine g is invariant under Lie g, so we moved next to the study of *symmetry breaking* in current-algebraic CFT on the string.

We begin this discussion with the results in our paper [BH71]. In particular for our example on $su(3)_1$, we first studied the addition to the affine-Sugawara construction of a term $\lambda J_8(m) + \lambda^2 \delta_{m,0}$ for all constant λ, which gave Virasoro generators with *continuous* $su(3)$ symmetry breaking. This was the first example of what is today known as inner-automorphic twists or c-fixed *conformal deformations* (Freericks and Halpern [FH88]).

Next we looked for $su(3)$ symmetry breaking solutions on $su(3)_1$ when the Virasoro generators were assumed to have the more general quadratic form $L = \sum_a C_a : J_a J_a :$ with constant coefficients $\{C_a, a = 1, \ldots, 8\}$. We stated that solutions beyond the affine-Sugawara construction for the coefficients $\{C_a\}$ existed, giving Virasoro generators with *quantized* $su(3)$ symmetry breaking. This was the first *implicit* mention of coset constructions, but we gave no details about these solutions – noting only that they all had 'infinite-degeneracy' problems. This issue (amplified in our nonadditive 'spin-orbit' constructions of the next section of this reference) was that the solutions to the Virasoro conditions come in *commuting* 'K-conjugate pairs' of Virasoro generators $[L(m), \tilde{L}(n)] = 0$, which sum to the affine-Sugawara construction $L_g(m) = L(m) + \tilde{L}(m)$. This phenomenon was called 'K-conjugation covariance' [BH71, Hal71c, Man73a, Man73b], and the associated spectral degeneracy of each commuting K-conjugate partner, due to the other, was called K-degeneracy.

The first *explicit* examples of non-Abelian coset constructions were given in the companion paper 'The two faces of a dual pion–quark model' [Hal71c], which solved the Virasoro conditions for all possible (BH quartic) conformal spin–spin interactions among the four-dimensional dual quark bilinears T, V, A, P and S ($\{T_{\mu\nu} =: \bar{\psi}\sigma_{\mu\nu}\psi :\}$ and so on). All these conformal constructions were affine-Sugawara or coset constructions, and a number of these were included explicitly in the applications of the text (see Eqs. (3.7), (3.9) and (3.14)). Each of these results was presented as a decomposition of an affine-Sugawara construction into a commuting K-conjugate pair of conformal constructions – which we would today write as

$$L_g(m) = L_{g/h}(m) + L_h(m), \quad [L_{g/h}(m), L_h(n)] = 0, \qquad (32.14)$$

where g is the original symmetry algebra, h is any subalgebra of g and g/h is the coset space.

We will focus here on the Lie-algebraic identifications in the simplest case, given as $N^{NS} + N^5 = N^H$ in Eq. (3.14) of the paper. All three sets of Virasoro generators are defined in Eqs. (3.9) and (3.14) as different linear combinations of the BH quartics T^2 and A^2, and it is not difficult to see that N^{NS} and N^H are affine-Sugawara constructions respectively on $so(3, 1)_1$ and $so(4, 1)_1$. Therefore, Eq. (3.14) can be read as the K-conjugate decomposition

$$L_{so(3,1)}(m) + L_{so(4,1)/so(3,1)}(m) = L_{so(4,1)}(m), \qquad (32.15)$$

into the commuting K-conjugate partners on $so(3, 1)$ and $so(4, 1)/so(3, 1)$. In fact, this result was checked explicitly with the identities given in the text between the BH dual-quark bilinears T, A and certain five-dimensional NS bilinears (see Eq. (2.9) of the text). Using these identities, it was noted that each quartic BH Virasoro generator collapses to an NS bilinear, the affine-Sugawara constructions N^{NS} and N^H on $so(3, 1)_1$ and $so(4, 1)_1$ being equal respectively to the four- and five-dimensional quadratic NS Virasoro generators, while the non-Abelian coset construction N^5 (with central charge $c_{so(4,1)/so(3,1)} = 5/2 - 2 = 1/2$) is equal to the quadratic Virasoro generators of the fifth NS operator alone. Similar quartic-to-bilinear equivalences had been seen earlier in [CGJ69, BH71].

The general form of the coset construction for all $h \subset g$,

$$L_{g/h}(m) = L_g(m) - L_h(m), \qquad c_{g/h} = c_g - c_h, \qquad (32.16)$$

was given by Goddard, Kent and Olive [GKO85] fourteen years after our examples. The K-degeneracy of the Virasoro construction on g/h is now understood as the local gauge invariance associated with the h-currents of the commuting K-conjugate affine-Sugawara construction $\{L_h(m)\}$ on h. It should be emphasized however that the coset constructions are by no means the end of CFT constructions on the currents of affine Lie algebras: our original method of solving for the coefficients $\{L^{ab}\}$ in quadratic forms $\{L^{ab} : J_a J_b :\}$ eventually led to the completely K-conjugation-covariant 'Virasoro master equation' (Halpern and Kiritsis [HK89]), which includes the affine-Sugawara constructions and the coset constructions as very special cases (with rational central charge) of irrational conformal field theory (Halpern, Kiritsis, Obers and Clubok [HKOC96]).

32.7 The vertex-operator constructions

There is one more important strand which belongs in this story – namely the second companion paper in 1975 'Quantum "solitons" which are $SU(N)$ fermions' [Hal75], which described – using compactified spatial dimensions on the open string – the first vertex-operator constructions of world-sheet fermions and affine Lie algebra.

By 1972, interest among physicists was shifting back to field theory. We mention in particular the Bose–Fermi equivalence of the sine-Gordon and interacting Thirring model studied by Coleman [Col75] and Mandelstam [Man75] in 1975. (See also the free-field bosonization of the single fermion bilinears by Kogut and Susskind [KS75].) One of us (MBH) realized that – with two observations – this development could be applied to our 1971 non-Abelian current-algebraic constructions on the string.

The first observation was that Mandelstam's normal-ordered line-integral construction of a single interacting fermion could be easily generalized to *many* interacting fermions on the line by including appropriate Klein transformations [Kle38]. This led to bosonic realizations of centrally extended $su(n)$ current algebras on the line, and the equivalence of various generalized sine-Gordon models with corresponding sets of interacting $u(n)$ Thirring models. In particular, the $su(n)$ currents were constructed from $(n-1)$ independent two-dimensional bosons, showing the roots of $su(n)$ in the off-diagonal currents. (See also the extension of these ideas to the bosonization of two-dimensional non-Abelian gauge theories [Hal76], and in particular Eq. (3.13) of this reference.)

The second observation in the 1975 paper was that, for massless free fermions and bosons, there was a structural parallel between Mandelstam's fermions on the line and particular string vertex operators on the circle. Indeed, a formal projective map (from the line to the circle) was given between each two-dimensional left- and right-mover free-field construction in the field theories and a *pair* of our open string counterparts on the circle, a structure which corresponded in fact to closed strings. These points were emphasized in the second appendix 'Connections with dual models' of that paper, which also gave the vertex-operator construction of many world-sheet fermions [Hal75],

$$\bar{\psi}^I(z) =: \exp(i\, Q_I(z)) : \bar{\xi}^I, \qquad \psi_I(z) = \xi_I : \exp(-i\, Q_I(z)) :, \qquad (32.17)$$

where $\{Q_I\}$ are the Fubini–Veneziano fields and $\{\bar{\xi}, \xi\}$ are the Klein transformations of the text. This relation guaranteed that the world-sheet fermions were Virasoro primary fields with conformal weight $1/2$ under the open string bosonic Virasoro generators, and we identified these constructions as our half-integer moded dual quarks [BH71]. Although it was not explicitly mentioned, this identification had in fact been checked from vacuum expectation values using the natural identification of the bosonic zero-momentum vacuum $|\{p = J(0) = 0\}\rangle$ as the unique vacuum $|0\rangle$ annihilated by the positive modes of the dual quarks.

The vertex-operator construction of $su(n)_1$ can therefore be obtained immediately in a single step by direct substitution of the fermionic vertex-operators (32.14) into our dual quark model construction $\{J_a \propto\, :\bar{\psi}\lambda_a\psi:\}$ of the currents. We regret having omitted this last simple step in the appendix, thinking it implicit in the context of the paper: the off-diagonal currents $:\bar{\psi}^I(z)\psi_J(z): \propto: \exp(i(Q_I(z) - Q_J(z))):$, $I \neq J = 1,\ldots,n$ of $su(n)_1$ follow immediately, showing the roots of $su(n)$ on $(n-1)$ independent fields, and the diagonal currents of $su(n)_1$ are derivatives of the same $(n-1)$ fields, while the extra $u(1)$ current is a derivative of the normalized orthogonal sum $Q_+(z) = (1/\sqrt{n})\sum_I Q_I(z)$. We state here the result only for $su(2)_1 + u(1)$:

$$\begin{aligned}
J_\pm(z) &= /, : \bar{\psi}(z)\, \tau_\pm\, \psi(z) : = : \exp\left(\pm i\, \sqrt{2}\, Q_-(z)\right) :, \\
J_3(z) &= \frac{1}{\sqrt{2}} : \bar{\psi}(z)\, \tau_3\, \psi(z) : = i\, \partial_z Q_-(z), \\
J_0(z) &= \frac{1}{\sqrt{2}} : \bar{\psi}(z)\, \psi(z) : = i\, \partial_z Q_+(z), \\
Q_\pm(z) &= \frac{1}{\sqrt{2}}\, (Q_1(z) \pm Q_2(z)),
\end{aligned} \qquad (32.18)$$

which is in fact the only case where (as discussed in the text) the Klein transformations can be chosen to cancel. The charged currents $\{J_\pm(z)\}$ of $su(2)_1$ show the roots of $su(2)$, and results for the neutral current $J_3(z)$ and the extra $u(1)$ current $J_0(z)$ are of course closely related to our earlier Bose–Fermi relation on the string given in Eq. (32.7) above.

We concluded that we had constructed a current algebraic $su(2)$ from a single compactified spatial dimension, and also that the paper was equivalent to constructing as much as current algebraic $u(22) = su(22) + u(1)$ from the compactified extra spatial dimensions of the critical open bosonic string. (Quantized extra momenta were well-known in early open string theory.)

In the following year, Banks, Horn and Neuberger [BHN76] again studied the bosonization of $su(n)$ Thirring models, this time with periodic boundary conditions as $L \to \infty$. Translating the fermionic zero modes of their solution at fixed L into vertex-operators on the string, one now understands that integer moded complex versions (CR) of world-sheet R fermions and $su(n)_1$ currents can also be obtained from the *same* vertex operators by choosing degenerate ground states with $\pm 1/2$ for each of the n momenta.

In later work (see for example Frenkel, Lepowsky and Meurman [FLM88] and Lepowsky [Lep07]), one learns that the true vacuum of the full CFT is the BH vacuum $|0\rangle = |\{p = J(0) = 0\}\rangle$ and products of BH currents on the BH vacuum live on the root-lattice of $su(n)$. The BH theory is itself reducible, containing many copies of the integrable representations of $su(n)$ – each copy living at different Δ on the weight lattice of $su(n)$.

32.8 The confluence with mathematics

So ended our three contributions [BH71, Hal71c, Hal75] to current algebraic internal symmetry on the string. Loosely speaking, these papers also marked the end of the first string era, as an increasing number of physicists – including ourselves – turned their interest to gauge theory for a decade.

We were therefore fundamentally surprised to see in 1985 that the seminal paper on the heterotic string (Gross, Harvey, Martinec and Rohm [GHMR85]) used the 1980 vertex-operator (root-lattice) construction of $E(8)_1$ by Frenkel and Kac [FK80]. Following this, we contacted James Lepowsky and Igor Frenkel, who patiently exchanged information with us about these parallel developments in mathematics and physics, including explanations of the first vertex-operator construction (twisted $su(2)_1$) in mathematics (Lepowsky and Wilson [LW78]), and the earlier Kac–Moody algebras [Kac67, Moo67] – which include the affine and hyperbolic algebras as special cases.

In mathematics, our 1971 dual quark model construction of $su(3)_1$ is known as the first concrete representation of affine Lie algebra, including the first explicit central term. In this connection we quote first from page 6304 of a 1980 paper by Frenkel [Fre80]: 'Ten years ago two physicists, Bardakci and Halpern, in Ref. [BH71] constructed a representation of the subalgebra $\tilde{gl}(l)$ of $\tilde{o}(2l)$ in the space $V((2\mathbb{Z}+1)^l)$ (see formulas 3.1–3.11 of Ref. [BH71]). At that time the theory of affine Lie algebras began to take its first steps.'

A less technical form of this statement is found on page 36 of the introduction to the review article by Frenkel, Lepowsky and Meurman [FLM88]: 'In their proposal of current-algebraic internal symmetry for strings, Bardakci and Halpern [BH71], independently of mathematicians, discovered affine Lie algebras in 1971, including the first case of a concrete representation – a fermionic realization of $\hat{sl}(3)$; irreducibility issues arose later in independent discoveries by mathematicians.'

The affine-Sugawara constructions [BH71, Hal71c, KZ84] in physics were also found in mathematics [Seg81], where they are known as the Segal operators.

The early vertex-operator constructions in physics are also well known in mathematics. In this connection, we find again on the same page of [FLM88]: 'The untwisted vertex-operator construction of $\hat{sl}(n)$ representations from compactified spatial dimensions was implicit in [Hal75], [BHN76].' A supplementary statement is found on page 365 of [Lep07]: 'It turned out that vertex operators and symmetry are closely related. Instances of this had already been discovered in physics, including the works [Hal75] and [BHN76].'

In conclusion, one might say that, for us, ignorance was bliss or – more optimistically – that physicists can create mathematics. In either case, it is clear that both fields have profited handsomely from this confluence.

32.9 The larger perspective

In our historical discussion above we have focussed on the discovery of current-algebraic CFT on the string world-sheet, but to complete the modern perspective we should emphasize that this is just one of many mathematical methods which have been found and developed in string theory. Over the last 25 years, this domain has been the main source of new methods in theoretical physics that deeply changed the discipline and had many other applications in different areas, such as statistical mechanics and condensed matter physics. These developments involved the latest mathematical advances in algebra, topology and geometry. Some of these results (such as the affine algebras described here) were first discovered in string theory and then studied systematically by mathematicians, while others were imported from mathematics and made clear and practical by the string applications.

This communication and cross-fertilization has also had a large impact in mathematics, where the field theory methods are now well appreciated (see for example the lecture notes [DEFJ99]). In the Eighties and Nineties, Edward Witten has played a crucial role in this interdisciplinary endeavour.

In our context we note in particular the important set of string theory mathematical methods given by general two-dimensional conformal field theory, which began with the analysis by Belavin, Polyakov and Zamolodchikov [BPZ84]. They used the mathematical results of Kac [Kac90] and Feigin and Fuchs [FF82] on representation theory of the Virasoro algebra, and realized their relevance for exact descriptions of two-dimensional critical phenomena (see for example the collection of important papers in [ISZ88]).

Acknowledgements

For help in the preparation of this Chapter, we thank O. Ganor, I. Frenkel, Y. Frishman and J. Lepowsky.

Korkut Bardakci was born in Yozgat, Turkey in 1936. He obtained his PhD in physics at the University of Rochester in 1962. He was Professor of Physics at the University of California, Berkeley from 1966 to 2003. He is currently Professor Emeritus at Berkeley. Main research interests are complex angular momentum, field theory, string theory, conformal models, large N expansion and field theory on the world-sheet.

Martin B. Halpern was born in Newark (NJ) in 1939. He obtained his PhD in physics at Harvard University in 1964, under W. Gilbert. He was Professor of Physics at the University of California, Berkeley from 1967 to 2005. He is currently Professor Emeritus at Berkeley. Main research interests are string theory, field theory, conformal field theory, large N expansion, matrix theory and orbifold theory.

References

[AD68] Adler, S. and Dashen, R. F. (1968). *Current Algebras* (W. A. Benjamin, New York).
[BHN76] Banks, T., Horn, D. and Neuberger, H. (1976). Bosonization of the $SU(N)$ Thirring models, *Nucl. Phys.* **B108**, 119–129.
[Bar71] Bardakci, K. (1971). New gauge identities in dual quark models, *Nucl. Phys.* **B33**, 464–474.
[BFH68] Bardakci, K., Frishman, Y. and Halpern, M. B. (1968). Structure and extensions of a theory of currents, *Phys. Rev.* **170**, 1353–1359.
[BH68] Bardakci, K. and Halpern, M. B. (1968). Canonical representation of Sugawara's theory, *Phys. Rev.* **172**, 1542–1550.
[BH69] Bardakci, K. and Halpern, M. B. (1969). Possible Born term for the hadronic bootstrap, *Phys. Rev.* **183**, 1456–1462.
[BH71] Bardakci, K. and Halpern, M. B. (1971). New dual quark models, *Phys. Rev.* **D3**, 2493–2506.
[BPZ84] Belavin, A. A., Polyakov, A. M. and Zamolodchikov, A. B. (1984). Infinite conformal symmetry in two-dimensional quantum field theory *Nucl. Phys.* **B241**, 333.
[CT74] Chodos, A. and Thorn, C. B. (1974). Making the massless string massive, *Nucl. Phys.* **B72**, 509–522.
[Col75] Coleman, S. (1975). Quantum sine-Gordon equation as the massive Thirring model, *Phys. Rev.* **D11**, 2088–2097.
[CGJ69] Coleman, S., Gross, D. and Jackiw, R. (1969). Fermion avatars of the Sugawara model, *Phys. Rev* **180**, 1359–1365.
[DF73] Dashen, R. F. and Frishman, Y. (1973). Thirring model with $U(n)$ symmetry – scale invariant only for fixed values of a coupling constant, *Phys. Lett.* **B46**, 439–442.
[DF75] Dashen, R. F. and Frishman, Y. (1975). Four-fermion interactions and scale invariance, *Phys. Rev.* **D11**, 2781–2802.
[DEFJ99] Deligne, P., Etingof, P., Freed, D. S., Jeffrey, L., Kazhdan, L., Morgan, J., Morrison, D. R. and Witten, E. eds. (1999). *Quantum Fields and Strings: A Course For Mathematicians* (American Mathematical Society, Providence, RI).

[DFZ72] Dell'Antonio, G. F., Frishman, Y. and Zwanziger, D. (1974). Thirring model in terms of currents: solutions and light-cone expansions, *Phys. Rev.* **D6**, 988–1007.

[FF82] Feigin, B. L. and Fuchs, D. B. (1982). Verma modules over the Virasoro algebra, *Funct. Anal. Appl.* **17**, 241.

[FH88] Freericks, J. K. and Halpern, M. B. (1988). Conformal deformation by the currents of affine g, *Ann. Phys.* **188**, 258–306.

[Fre80] Frenkel, I. B. (1980). Spinor representations of affine Lie algebras, *Proc. Natl. Acad. Sci. USA* **77**, 6303–6306.

[FK80] Frenkel, I. B. and Kac, V. G. (1980). Basic representations of affine Lie algebras and dual resonance models, *Inv. Math.* **62**, 23.

[FLM88] Frenkel, I. B., Lepowsky, J. and Meurman, A. (1988). *Vertex Operator Algebras and the Monster* (Academic Press, Boston, MA), Pure and Applied Mathematics, Vol. 134.

[FV69] Fubini, S. and Veneziano, G. (1969). Level structure of dual resonance models, *Nuovo Cimento* **A64**, 811–840.

[GW86] Gepner, D. and Witten, E. (1986) String theory on group manifolds, *Nucl. Phys.* **B278**, 493.

[GKO85] Goddard, P., Kent, A. and Olive, D. (1985). Virasoro algebras and coset models, *Phys. Lett.* **B152**, 88–92.

[GS82] Green, M. B. and Schwarz, J. H. (1982). Supersymmetrical string theories, *Phys. Lett.* **B109**, 444–448.

[GH69] Gross, D. J. and Halpern, M. B. (1969). Theory of currents and the non-strong interactions, *Phys. Rev.* **179**, 1436–1444.

[GHMR85] Gross, D. J., Harvey, J. A., Martinec, E. and Rohm, R. (1985). The heterotic string, *Phys. Rev. Lett.* **54**, 502–505.

[Hal71a] Halpern, M. B. (1971). Persistence of the photon in conformal dual models, *Phys. Rev.* **D3**, 3068–3071.

[Hal71b] Halpern, M. B. (1971). New dual models of pions with no tachyon, *Phys. Rev.* **D4**, 3082–3083.

[Hal71c] Halpern, M. B. (1971). The two faces of a dual pion-quark model, *Phys. Rev.* **D4**, 2398–2401.

[Hal75] Halpern, M. B. (1975). Quantum 'solitons' which are $SU(N)$ fermions, *Phys. Rev.* **D12**, 1684–1699.

[Hal76] Halpern, M. B. (1976). Equivalent-boson method and free currents in two-dimensional gauge theories, *Phys. Rev.* **D13**, 337–342.

[HK89] Halpern, M. B. and Kiritsis, E. (1989). General Virasoro construction on affine g, *Mod. Phys. Lett.* **A4**, 1373–1380.

[HKOC96] Halpern, M. B., Kiritsis, E., Obers, N. A. and Clubok, K. (1996). Irrational conformal field theory, *Phys. Rep.* **265**, 1–138. hep-th/9501144.

[HT71a] Halpern, M. B. and Thorn, C. B. (1971). Dual model of pions with no tachyon, *Phys. Lett.* **B35**, 441–442.

[HT71b] Halpern, M. B. and Thorn, C. B. (1971). The two faces of a dual pion-quark model II. Fermions and other things, *Phys. Rev.* **D4**, 3084–3088.

[ISZ88] Itzykson, C., Saleur, H. and Zuber, J.-B. (1988). *Conformal Invariance and Applications to Statistical Mechanics* (World Scientific, Singapore).

[Kac67] Kac, V. G. (1967). Simple graded Lie algebras of finite growth, *Funct. Anal. Appl.* **1**, 328–329.

[Kac90] Kac, V. G. (1990). *Infinite Dimensional Lie Algebras* (Cambridge University Press, Cambridge).

[Kle38] Klein, O. (1938). Approximate treatment of electrons in a crystal lattice, *J. Phys. Radium* **9**, 1–12.

[KZ84] Knizhnik, V. G. and Zamolodchikov, A. B. (1984). Current algebra and Wess–Zumino model in two dimensions, *Nucl. Phys.* **B247**, 83–103.

[KS75] Kogut, J. and Susskind, L. (1975). How quark confinement solves the $\eta 3 \to 3\pi$ problem, *Phys. Rev.* **D11**, 3594–3610.

[Lep07] Lepowsky, J. (2007). Some developments in vertex operator algebra theory, old and new, in *Lie Algebras, Vertex Operator Algebras and their Applications, International Conference in Honour of J. Lepowsky and R. Wilson*, ed. Huang, Y.-Z. and Misra, K. C. Contemporary Mathematics, Am. Math. Soc., Vol. 442, p. 355.

[LW78] Lepowsky, J. and Wilson, R. L. (1978). Construction of the affine Lie algebra $A_1^{(1)}$, *Commun. Math. Phys.* **62**, 43–53.

[Man73a] Mandelstam, S. (1973). K-degeneracy in non-additive dual resonance models, *Phys. Rev.* **D7**, 3763–3776.

[Man73b] Mandelstam, S. (1973). Simple non-additive dual resonance model, *Phys. Rev.* **D7**, 3777–3784.

[Man75] Mandelstam, S. (1975). Soliton operators for the quantized sine-Gordon equation, *Phys. Rev.* **D11**, 3026–3030.

[Moo67] Moody, R. V. (1967). Lie algebras associated with general Cartan matrices, *Bull. Am. Math. Soc.* **73**, 217–221.

[NS71] Neveu, A. and Schwarz, J. H. (1971). Factorizable dual model of pions, *Nucl. Phys.* **B31**, 86–112.

[PC69] Paton, J. E. and Chan, H. M. (1969). Generalized Veneziano model with isospin, *Nucl. Phys.* **B10**, 516–520.

[Ram71] Ramond, P. (1971). Dual theory of free fermions, *Phys. Rev.* **D3**, 2415–2418.

[Sch59] Schwinger, J. (1959). Field theory commutators, *Phys. Rev. Lett.* **3**, 296–297.

[Seg81] Segal, G. (1981). Unitary representations of some infinite dimensional groups, *Commun. Math. Phys.* **80**, 301–342.

[Som68] Sommerfield, C. (1968). Currents as dynamical variables, *Phys. Rev.* **176**, 2019–2025.

[Sug68] Sugawara, H. (1968). A field theory of currents, *Phys. Rev.* **170**, 1659–1662.

[Vir70] Virasoro, M. (1970). Subsidiary conditions and ghosts in dual-resonance models, *Phys. Rev.* **D1**, 2933–2936.

[Wit84] Witten, E. (1984) Nonabelian bosonization in two dimensions, *Commun. Math. Phys.* **92**, 455.

33
Remembering the dawn of relativistic strings

JEAN-LOUP GERVAIS

In this short account I would like to think about the remarkable history of the development of relativistic string theories which have gradually renewed the very foundations of theoretical high energy physics. For lack of time and precise knowledge, I will stick to my own personal contribution at the beginning of this wonderful story, and refer to other contributions in this Volume to complete the picture.

The year 1968–1969, marked a turning point in my career, not only because string theory came into the fore. I was just returning from two years as a postdoc at New York University, where I had met J. Wess (a visitor for one year), B. Zumino (then the head of the Theory Group located at the Courant Institute), K. Symanzik, W. Zimmermann and D. Zwanziger. My earlier interests were mostly on dispersion relations, Regge poles, and S-matrix theory, but at NYU, I had been fully converted to local field theory, and much impressed by the power of symmetries in that context, be they local or global, exact or (spontaneously) broken. Of course, this year saw the beginning of string theory, which was however initially developed using the covariant operator method within the context of S-matrix theory, giving what looked like a realization of G. Chew's programme. I refer to the contribution of Ademollo to this Volume for a review of the precursory works which led Veneziano to write his celebrated four-point function.

On the other hand, local field theory also made wonderful progress on its own. This was the time when Salam and Weinberg opened the way to the renormalizable theory of unified weak end electromagnetic interactions. The main problems then were the Adler–Bell–Jackiw anomaly, the spontaneous breaking of symmetries and the quantization of Yang–Mills theories. For the latter, the work of L. Faddeev and V. Popov was gradually becoming more and more popular. It is hardly necessary to say that these topics now belong to textbooks.

At that time the French Government was very generous with temporary positions, and during that wonderful year a handful of key visitors came for long visits at the Laboratoire de Physique Théorique et Hautes Energies (LPTHE) of Orsay (France), where I was working permanently then: D. Amati, the late Benjamin Lee, T. Veltman – who had already

---

The Birth of String Theory, ed. Andrea Cappelli, Elena Castellani, Filippo Colomo and Paolo Di Vecchia.
Published by Cambridge University Press. © Cambridge University Press 2012.

undertaken to renormalize Yang–Mills theories – and B. Zumino. I drew much inspiration from the very stimulating atmosphere they created, together with other permanent members. In particular, with D. Amati and C. Bouchiat [ABG69] we devised the now standard method to compute loops in string theories, using coherent states, and with B. Lee [GL69], I showed how to quantize the linear σ model correctly in the phase where the spontaneous symmetry breaking takes place.

This is all to say that, when I first met Bunji Sakita in fall 1970, I was fully motivated to apply field theory techniques to string theories. Moreover, the year before I had shown [Ger70] that the integrand of the Veneziano model is equal to the vacuum expectation of a product of field operators in string modes which are local functions of the Koba–Nielsen variables. This result, similar to an independent and better known work by S. Fubini and G. Veneziano [FV70], was indeed a strong hint of the world-sheet field theory aspect of string theory. This viewpoint is now commonplace, but at that time it was not at all popular among string theorists. A large majority preferred the operator method, which achieved striking technical success.

At this point, preparing the present text brings back wonderful memories of the time when my long lasting collaboration and friendship with the late Bunji Sakita began. He was then visiting both the Institut des Hautes Etudes Scientifiques, in Bures-sur-Yvette, and LPTHE, in Orsay. Sakita himself has already given his own view [Sak99]. Before we met, he and his collaborators had made important progress in developing world-sheet field theory techniques using path integrals. On the one hand, H. Hsue, B. Sakita, and M. Virasoro [HSV70] had shown how the analogue model of H. B. Nielsen could be derived from the path integral over a free scalar two-dimensional field. On the other hand, Sakita came with the draft of an article where he had started to discuss Feynman-like rules for the Veneziano model using the factorization of path integrals over sliced Riemann surfaces. There were many basic problems left, and at the beginning we spent a lot of time establishing a general scheme of Feynman-like perturbative expansion, where the vertices are given by functional integrals over two-dimensional surfaces. This complicated work was not so well received, although it contains many precursory results.

For what follows, the most important point was that we made essential use of the conformal invariance of the path integral representation over scalar free fields in two dimensions. Although we did not really consider the gauge-fixing problem at that time, we were pretty much convinced that conformal invariance of the free world-sheet action was at the origin of the elimination of negative norm states. In spring we came across the first article by A. Neveu and J. Schwarz [NS71] entitled 'Factorizable dual model of pions' (see their contributions to this Volume). This motivated us [GS71b] to discuss conformal field theories systematically in two dimensions, as a way to classify string theories, by defining what we called irreducible fields – now known as primary fields, following A. Belavin, A. Polyakov, and A. B. Zamolodchikov [BPZ84]. Considering only quadratic actions we recognized that only spin zero and spin one-half (world-sheet) fields were possible, covering all existing critical string models of today. We showed how Neveu–Schwarz and Ramond fields for open strings correspond to the two possible boundary conditions on the world-sheet. This was all very well except for one fact: the Neveu–Schwarz model has more ghosts and needs an additional killing mechanism for negative norm states, with respect to the Veneziano model. This had motivated Neveu and Schwarz [NS71], as well as Ramond in his seminal work [Ram71], to introduce in the operator formalism a set of operators whose anticommutators gave the Virasoro generators.

The visit of Sakita in France was about to terminate, and this deadline stimulated us a great deal. I quickly started to look for the possible symmetries of the action that would be the origin of the additional ghost killing. From the form of the RNS generators it was immediately clear that the boson had to transform into the fermion and vice versa. It was not difficult to envisage that the action could be invariant, except for the mixing between commuting and anticommuting fields which made everything very confusing. After many hesitations, Sakita and I solved the problem by introducing symmetry transformations with anticommuting parameters, from which the supersymmetry of our RNS world-sheet action followed very simply. The paper was completed [GS71c] and thus for us (conformal) supersymmetry was born just the day before Sakita departed France to fill his new prestigious position at City College in New York.

In those days, anticommuting c-numbers were far from being understood. They were essentially a formal tool to derive perturbation expansions over fermionic fields from path integrals. This was how we used them in our second article [GS71b], where we derived the tree scattering amplitudes of the Neveu–Schwarz model from world-sheet perturbation. Considering symmetry transformations with anticommuting parameters was, for us, quite a step which we took rather reluctantly. Quite some time passed before our idea was taken seriously, including by us.

In the early Seventies, the world-sheet dynamical approach to string theory was not well appreciated in our community, and our work did not receive much attention in general. In December 1971 Sakita delivered a talk about it at the 'Conference on Functional Methods in Field Theory and Statistics' at the Lebedev Institute in Moscow, organized by E. Fradkin. On the way back, he stopped over in Paris. We wrote a summary of our work for the proceedings, which was sent to the organizers and circulated as a preprint – the complete proceedings themselves were never published.[1] Through this, there was some early communication of our work to the Soviet scientific community where it was well appreciated, apparently [Sak99].

We separated at a very unfortunate moment of our research programme. At that time there was no email, phone was expensive and airmail slow. Moreover Sakita had to deal with his new life and responsibilities in New York. We nevertheless continued our collaboration and met in person as often as possible. I was happy to visit him and his friendly group in New York. Our meetings were always too short and therefore very busy. I did not pursue further the study of supersymmetry, however, to my regret. Other problems seemed to be more pressing. In the meantime the Nambu–Goto action and the Goddard, Goldstone, Rebbi and

---

[1] We later published our text in *Quantum Field Theory and Quantum Statistics, Essays in Honour of the Sixtieth Birthday of E. S. Fradkin*, Vol. 2, p. 435, Adam Hilger, 1987.

Thorn light-cone quantization had come out. Sakita and I [GS73] showed how the latter can be recovered using path integral with the Nambu–Goto action, using the Faddeev–Popov method to handle reparameterization invariance. This work raised much interest, and also some criticism. The main objection was that our gauge depended upon the external sources, and thus was not easily factorizable, in contrast with our previous path integral formulation [GS71a]. We tried hard to understand what was going on but we failed. The answer was given by S. Mandelstam [Man73] (see his contribution to this Volume): the light-cone gauge is not conformally invariant, so that there is only one (preferred) parameterization where factorization holds with our gauge fixing. With this parameterization one sees strings (with lengths proportional to their respective p_+ values) which split and join, and our work played a key role in Mandelstam's subsequent discussion of scattering amplitudes, which led to light-cone string field theory.

In spring 1973, Sakita visited the Niels Bohr Institute and I made a trip there to meet with him. On the way back home, he went to CERN and gave a talk where Zumino was present [Sak99]. Sakita reviewed a current work [IK73] of Y. Iwasaki and K. Kikkawa (at that time they were both at City College and had much interaction with him); they were trying to establish the Nambu–Goto type formulation of our world-sheet RNS dynamics. This seminar and a subsequent conversation with Zumino played a key role in leading Wess and Zumino to begin their seminal work on supersymmetry. After this the subject of supersymmetry in local field theory suddenly exploded.

Joël Scherk returned to France in 1971. I remember lunch at the cafeteria of Orsay with him and Roland Omnès who was working on Regge poles *per se* at that time. The conversation came to string theories and our senior colleague remarked that ordinary field theory should be recovered in the limit where, the slope of the Regge trajectory being very small, only one particle kept a finite mass. To my knowledge, this was the first time this crucial remark was made. J. Scherk put it to work quickly [Sch71], unravelling a key tool for later progress.

Later on, Neveu returned from Princeton. We [GN72] studied a zero-slope limit of the multi-Veneziano amplitudes with leading intercept different from one and in four dimensions. They were not acceptable as string theories but all the inconsistencies ran away in the limit which turned out to be spontaneously broken massive gauge theories. The crucial point was that, each Veneziano amplitude being invariant under circular perturbation, the resulting Feynman diagrams were lumped into terms sharing this property. Thus, the S-matrix at each order of perturbation was recovered by summing the zero-slope amplitudes only on noncyclic permutations. This is a crucial simplification of higher order computations where the number of Feynman diagrams quickly becomes very large. We further showed that this perturbation expansion corresponds to a special choice of gauge where the three- and four-gluon vertices themselves are cyclically symmetric. This was achieved by adding a non-Hermitian gauge-fixing term, a novel idea which turned out to be rather fruitful (for a pedagogical review, see Srednicki [Sre07]).

I did not participate in the subsequent developments of the operator approach: enough clever colleagues were in this business. During fall 1973, my concern became to establish its precise connection with the world-sheet viewpoint, which I believed was the key point. At that time it was recognized that string splitting and gluing gave the right picture for the interactions. However, what happened precisely at the point common to several strings remained a mystery. Eugène Cremmer was back from CERN and we started to work on it (see also his contribution to this Volume). I had an invitation to CERN and spent one month there that fall, struggling with Fourier transforms, while enjoying the stimulating and friendly scientific surrounding, and discussions in particular with D. Amati and D. Olive. Our idea was as follows [CG74]. Considering for instance the time when two open strings combine, one sees that the final string state will be obtained by considering the two incoming strings together as a single string. At the time of interaction, let $\vec{X}_r(\eta_r)$ ($r = 1, 2$ for the initial strings, $r = 3$ for the final one) denote the 24 components (transverses) of the string position operators. The total longitudinal momenta (p_r^+) are conserved ($p_3^+ = p_1^+ + p_2^+$) and assumed to be positive. In the light-cone formalism, p_r^+ is precisely the length of the string r. If one takes η_r to be the geometrical string parameter divided by p_r^+, the gluing will be smooth if we choose $0 \leq \eta_r \leq \pi$. We thus started from overlap conditions of the type:

$$\vec{X}_3(\eta_3) = \vec{X}_1(\eta_1), \quad \text{if } p_3^+ \eta_3 = p_1^+ \eta_1, \quad 0 \leq \eta_1 \leq \pi \qquad (33.1)$$
$$\vec{X}_3(\eta_3) = \vec{X}_2(\eta_2), \quad \text{if } p_3^+ \eta_3 = p_1^+ \pi + p_2^+ \eta_2, \quad 0 \leq \eta_2 \leq \pi. \qquad (33.2)$$

These conditions are simple in the basis of free string states where the $\vec{X}_r(\eta_r)$ are diagonal, since they are implemented by Dirac δ functions. They were already known to other authors (see e.g. [Man73]) at that time, but their connection with the three-string vertex of ADDF [ADDF74] was unknown. We explicitly changed basis in the free string Fock space by going to the coherent state basis for the oscillators. A somewhat tedious discussion showed [CG74] that the above overlap conditions re-expressed in the new basis do give the vertex of [ADDF74] under the equivalent form derived by Mandelstam [Man73].

Generalizing this discussion to more than three legs, we developed [CG75] one of the first examples of string field theories which became popular during the second half of the Eighties. My initial motivation was to try to eliminate the tachyon by expanding around the 'right' vacuum, but this did not materialize. Note that we were competing with Kaku and Kikkawa who put forward similar ideas, but did not determine precisely the integration measure. Our work attracted much interest later on, after string theory had completed its 'crossing of the desert' [Sch85].

Acknowledgements

These recollections are dedicated to the late Bunji Sakita who played a key role in this early history.

Jean-Loup Gervais was born in Paris in 1936. He carried out his undergraduate studies at Université de Paris and was appointed at CNRS in 1961. He obtained his PhD in 1965 at

Orsay. He was a research associate at New York University (1966–1968). In 1974 he moved to École Normale Supérieure to found the new Laboratoire de Physique Théorique. He was director of this group in 1979–1983 and in 1995–1998. He retired in 2002.

References

[ADDF74] Ademollo, M., Del Giudice, E., Di Vecchia, P. and Fubini, S. (1974). Couplings of 3 excited particles in dual-resonance model, *Nuovo Cimento* **A19**, 181–203.

[ABG69] Amati, D., Bouchiat, C. and Gervais, J. L. (1969). On the building of dual diagrams from unitarity, *Lett. Nuovo Cimento* **2**, 399–406.

[BPZ84] Belavin, A. A., Polyakov, A. M. and Zamolodchikov, A. B. (1984). Infinite conformal symmetry in two-dimensional quantum field theory, *Nucl. Phys.* **B241**, 333–380.

[CG74] Cremmer, E. and Gervais, J. L. (1974). Combining and splitting relativistic strings, *Nucl. Phys.* **B76**, 209–230. Reprinted in [Sch85].

[CG75] Cremmer, E. and Gervais, J. L. (1975). Infinite component field-theory of interacting relativistic strings and dual theory, *Nucl. Phys.* **B90**, 410–460. Reprinted in [Sch85].

[FV70] Fubini, S. and Veneziano, G. (1970). Duality in operator formalism, *Nuovo Cimento* **A67**, 29–47.

[Ger70] Gervais, J. L. (1970). Operator expression for the Koba–Nielsen multi-Veneziano formula and gauge identities, *Nucl. Phys.* **B21**, 192–204.

[GL69] Gervais, J. L. and Lee, B. W. (1969). Renormalization of the σ–model. II. Fermion fields and regulators, *Nucl. Phys.* **B12**, 627–646.

[GN72] Gervais, J. L. and Neveu, A. (1972). Feynman rules for massive gauge fields with dual diagram topology, *Nucl. Phys.* **B46**, 381–401.

[GS71a] Gervais, J. L. and Sakita, B. (1971). Functional-integral approach to dual-resonance theory, *Phys. Rev.* **D4**, 2291–2308. Reprinted in [KVW99].

[GS71b] Gervais, J. L. and Sakita, B. (1971). Generalization of dual models, *Nucl. Phys.* **B34**, 477–492. Reprinted in [KVW99].

[GS71c] Gervais, J. L. and Sakita, B. (1971). Field theory interpretation of super gauges in dual models, *Nucl. Phys.* **B34**, 632–639. Reprinted in [Sch85], Vol. 1 and in [KVW99].

[GS73] Gervais, J. L. and Sakita, B. (1973). Ghost-free string picture of Veneziano-model, *Phys. Rev. Lett.* **30**, 716–719. Reprinted in [KVW99].

[HSV70] Hsue, C. S., Sakita, B. and Virasoro, M. A. (1970). Formulation of dual theory in terms of functional integrations, *Phys. Rev.* **D2**, 2857–2868. Reprinted in [KVW99].

[IK73] Iwasaki, Y. and Kikkawa, K. (1973). Quantization of a string of spinning material Hamiltonian and Lagrangian formulation, *Phys. Rev.* **D8**, 440–449.

[KVW99] Kikkawa, K., Virasoro, M. and Wadia, S. eds. (1999). *A Quest For Symmetry, Selected Works of Bunji Sakita*, World Scientific Series in 20th Century Physics, Vol. 22 (World Scientific, Singapore).

[Man73] Mandelstam, S. (1973). Interacting string picture of dual resonance models, *Nucl. Phys.* **B64**, 205–235; and this Volume.

[NS71] Neveu, A. and Schwarz, J. H. (1971). Factorizable dual model of pions, *Nucl. Phys.* **B31**, 86–112.

[Ram71] Ramond, P. (1971). Dual theory of free fermions, *Phys. Rev.* **D3**, 2415–2418.

[Sak99] Sakita, B. (1999). *Reminiscences*, published in [KVW99], reprinted in hep-th/0006083.

[Sch71] Scherk, J. (1971). Zero-slope limit of the dual resonance model, *Nucl. Phys.* **B31**, 222–234.
[Sch85] Schwarz, J. ed. (1985). *Superstrings, The First Fifteen Years of Superstring Theory* (World Scientific, Singapore).
[Sre07] Srednicki, M. A. (2007). *Quantum Field Theory* (Cambridge University Press, Cambridge).

34
Early string theory in Cambridge: personal recollections

CLAUS MONTONEN

34.1 In from the periphery

At the University of Helsinki, where I took my BSc and MSc degrees, research in and teaching of modern theoretical physics had started in the early Sixties after a long period of stagnation. In particle physics, introduced to Finland by amongst others K. V. Laurikainen, the emphasis was on phenomenology.

Enthusiastic young teachers like Keijo Kajantie taught us the basics of S-matrix and Regge theory. When Veneziano's paper appeared, it was greeted with much interest. Visits to Finland by Chan Hong-Mo, David Fairlie and Holger Bech Nielsen helped to spread the gospel. In the spring of 1970, Eero Byckling organized a study group on the Koba–Nielsen formalism, and in the summer of 1970 I was able to participate in the legendary Summer School in Copenhagen, where Nambu was not present in person, but in the form of his string action; his lecture notes were distributed among the participants. Victor Alessandrini's lectures on the multiloop amplitudes of the Veneziano model [Ale71, AA71] made a deep impression on me, and I long cherished his beautiful handwritten notes on the subject.

I was fortunate enough to receive a three-year scholarship for doctoral studies at 'an established British university' from the Osk. Huttunen Foundation, the very first institution in Finland which started awarding grants for long-term postgraduate studies. Being convinced that S-matrix theory was the answer to the riddle of the strong interactions, I naturally applied to Cambridge, the European Mecca of S-matrix theory. I went to Cambridge at the beginning of Michaelmas term in 1970. My supervisor was to be David Olive, recently returned from a stay at CERN.

34.2 Vertices and loops

Ed Corrigan was also a student of David Olive, and we all three started a joint project attempting to generalize the Koba–Nielsen [KN69] amplitudes to what we called a rubber band model of (spinless) baryons. (Note that the term string theory came into common use only after the seminal paper by Goddard, Goldstone, Rebbi and Thorn [GGRT73]; before that we talked confusedly about rubber bands (Susskind [Sus70]) and the like.) The idea was that as the Koba–Nielsen formalism supposedly described the scattering of quark–antiquark mesons joined by a 'rubber band' mapped onto a circle in the complex plane, so should a three-quark baryon in a Mercedes star configuration correspond to three intersecting circles. The amplitudes we constructed had a nice plane geometry of circles intersecting at specified angles, and other attractive properties; unfortunately we did not succeed in developing an operator formalism for them, and the work remained unpublished [Cor72, Mon73b].

In Cambridge, Olive repeated some of his CERN lectures (Alessandrini, Amati, Le Bellac and Olive [AALO71]), especially the part concerning the N-Reggeon vertices of the Veneziano model. He had developed further [Oli71] Lovelace's original construction [Lov70a, Lov70b], where the oscillator operators for the external legs are coupled by infinite-dimensional matrices of certain Möbius $(SL(2, \mathbb{C})/\mathbb{Z}_2)$ transformations. This formalism allowed the construction of arbitrary loop amplitudes by inserting propagators – i.e. two-Reggeon vertices – and sewing vertices together using the over-completeness of coherent states. It should be noted that the answers were not yet quite right – spurious states propagated in the loops, the correct projection operators eliminating them were still some years away in the future.

Serendipity now intervened. I happened to pick up a preprint by Tamiaki Yoneya [Yon71], who remarked *en passant* that the representation matrices defined by Lovelace were in fact a 'spin zero' representation of the $SU(1, 1)$ group, which is the subgroup of the Möbius group leaving a given circle invariant. The representations of the $su(1, 1)$ algebra had previously been used by Clavelli and Ramond [CR70], but Yoneya was the first to spot the usefulness of the group representations.

Corrigan and I asked ourselves whether other representations could also be used: indeed we found that the recently proposed Neveu–Schwarz model [NS71] employed, in addition to the spin zero representation, a spin one-half representation coupled to anticommuting oscillator operators. Consequently we were able to construct the N-Reggeon vertices for this model [CM72]. This seemed to indicate that many dual resonance models could be constructed by combining the spin zero representation, which carried the momentum dependence, with any other representation(s). However, we found that demanding that the $su(1, 1)$ algebra could be extended to a full Virasoro algebra, a condition already then seen as necessary for the elimination of ghosts, essentially precluded other representations than the spin zero and spin one-half ones – one of the first results showing the uniqueness properties of string constructions.

David Olive now left for CERN and Ed Corrigan followed soon after. I remained in Cambridge with Ian Drummond as my new supervisor. I made a basically failed attempt to construct off-shell amplitudes in the Neveu–Schwarz model [Mon73a], but my real interest remained in using the vertices discovered by Corrigan and myself to construct multiloop amplitudes for the Neveu–Schwarz model. The technical problem was the following: the sewing together of the vertices was elegantly done for the bosonic oscillators by using functional integrals over coherent states. It was not clear how to proceed for the fermionic oscillators. I read all the literature I could find on fermionic integrals. The departmental library in DAMTP was rather small, but the University Library, being a deposit library, had everything published in English. To save space, the books in the University Library are ordered by size rather than by any other principle. My second serendipitous streak of luck came when, a few books away from some mathematical opus I was looking for, I found Berezin's fabulous book *The Method of Second Quantization* [Ber66], which contained all the answers I needed.

Thus, I was able to construct the multiloop amplitudes, by introducing anticommuting coordinates to integrate over [Mon74]. It is now quite surprising that this formalism was virtually unknown, at least in the Dual Resonance Model community. The answer was still not quite correct because of the remaining spurious states inside the loops.

After finishing this paper in March 1973, I was invited to give a talk in Durham on my results. David Fairlie in the audience was quick to pick up the idea. He and Martin [FM73] had the temerity to combine each commuting coordinate with a pair of anticommuting coordinates into what is probably the first example of a superspace coordinate, something I had not thought of. Later this idea was generalized by Brink and Winnberg [BW76] to a superoperator formalism. In retrospect, my most important contribution to the first phase of string theory seems to have been making Berezin's techniques common knowledge among the practitioners.

34.3 Epilogue

For my first postdoc, I went to Orsay. Unfortunately for me, both André Neveu and Joël Scherk were spending the academic year 1973–1974 in the USA, so I used most of my time there to catch up on my neglected studies of quantum field theory, which was emerging as the winner in the strong interaction theory stakes. Seminars by Gerard 't Hooft on his newly discovered monopoles and by André Neveu (he returned for some weeks in the summer of 1974) on solitons left seeds that were to germinate some time later, and even to find use in much later developments in string theory. But that is another story.

> Claus Montonen was born in 1946. He received his MSc at the University of Helsinki in 1968, and his PhD at the University of Cambridge in 1974. He held research fellowships at CNRS, Orsay, in 1973–1974 and at CERN in 1977–1978. From 1974 he has held various research and teaching positions at the Department of Theoretical Physics, Department of Physics, Research Institute for Theoretical Physics of the University of Helsinki and at the Helsinki Institute of Physics. He is currently Researcher in Theoretical Physics at the University of Helsinki.

References

[Ale71] Alessandrini, V. (1971). A general approach to dual multiloop diagrams, *Nuovo Cimento* **A2**, 321–352.

[AA71] Alessandrini, V. and Amati, D. (1971). Properties of dual multiloop amplitudes, *Nuovo Cimento* **A4**, 793–844.

[AALO71] Alessandrini, V., Amati, D., Le Bellac, M. and Olive, D. (1971). The operator approach to dual multiparticle theory, *Phys. Rep.* **C1**, 269–346.

[Ber66] Berezin, F. A. (1966). *The Method of Second Quantization* (Academic Press, New York).

[BW76] Brink, L. and Winnberg, J. O. (1976). The superoperator formalism of the Neveu–Schwarz–Ramond model, *Nucl. Phys.* **B103**, 445–464.

[CR70] Clavelli, L. and Ramond, P. (1970). $SU(1,1)$ analysis of dual resonance models, *Phys. Rev.* **D2**, 973–979.

[Cor72] Corrigan, E. F. (1972). *Amplitudes and vertices in dual models*. University of Cambridge, PhD thesis.

[CM72] Corrigan, E. and Montonen, C. (1972). General dual operatorial vertices, *Nucl. Phys.* **36**, 58–72.

[FM73] Fairlie, D. B. and Martin, D. (1973). New light on the Neveu–Schwarz model, *Nuovo Cimento* **A18**, 373–383.

[GGRT73] Goddard, P., Goldstone, J., Rebbi, C. and Thorn, C. B. (1973). Quantum dynamics of a massless relativistic string, *Nucl. Phys.* **B56**, 109–135.

[KN69] Koba, Z. and Nielsen, H. B. (1969). Manifestly crossing invariant parameterization of n meson amplitude, *Nucl. Phys.* **B12**, 517–536.

[Lov70a] Lovelace, C. (1970). Simple n-Reggeon vertex, *Phys. Lett.* **B32**, 490–494.

[Lov70b] Lovelace, C. (1970). M-loop generalized Veneziano formula, *Phys. Lett.* **B32**, 703–708.

[Mon73a] Montonen, C. (1973). One-current amplitudes in the model of Neveu and Schwarz, *Nuovo Cimento* **17A**, 366–374.

[Mon73b] Montonen, C. (1973). *Theory of dual resonance models*. University of Cambridge, PhD thesis.

[Mon74] Montonen, C. (1974). Multiloop amplitudes in additive dual-resonance models, *Nuovo Cimento* **A19**, 69–89.

[NS71] Neveu, A. and Schwarz, J. H. (1971). Factorizable dual model of pions, *Nucl. Phys.* **B31**, 86–112.

[Oli71] Olive, D. (1971). Operator vertices and propagators in dual theories, *Nuovo Cimento* **A3**, 399–411.

[Sus70] Susskind, L. (1970). Dual-symmetric theory of hadrons. 1, *Nuovo Cimento* **A69**, 457–496.

[Yon71] Yoneya, T. (1971). General Reggeon vertex in dual theory, *Prog. Theor. Phys.* **46**, 1192–1206.

Part VI
The superstring

35
Introduction to Part VI

35.1 Introduction

By 1973 two ghost-free dual models had been constructed, the Dual Resonance Model and the Ramond–Neveu–Schwarz model. The structure underlying the DRM was that of a relativistic string described by the Nambu–Goto action, but it was not clear yet which kind of string was underlying the RNS model. These models were not suitable for describing strong interactions because they both had massless particles with spins one and two in their spectra (together with a tachyon). An additional problem was raised by the deep inelastic experiments: probing the structure of protons at short distances, they showed the existence of pointlike particles that could be interpreted as the quarks. As already mentioned, string theory scattering amplitudes were too soft at large momentum transfer to explain these experiments. Therefore many researchers went back to field theory; in particular, quantum chromodynamics (QCD), the non-Abelian gauge theory for quarks and gluons formulated in 1972, was intensively studied in the following years obtaining convincing experimental support.

Although these events implied that dual theories could not correctly describe the hadronic processes, they did not completely put an end to their study. In fact, some of the early workers in this field were so attracted by the consistent and deep structure of these models that they continued to study them, sometimes at the price of putting their careers at risk.

The DRM was constructed by applying the principles of S-matrix theory and planar duality, and the resulting scattering theory seemed very different from that of field theory, suggesting that the two approaches were completely unrelated (see the Introduction to Part II). However, the study of the zero-slope limit, $\alpha' \to 0$, corresponding to the limit of infinite tension, showed that the DRM was an extension rather than an alternative to field theory. In fact, in this limit, the string reduces to a pointlike object, which could be described by quantum field theory; all massive states become infinitely heavy and decouple, leaving the massless states only, with spins one and two in the open and closed sectors of the string, respectively.

The Birth of String Theory, ed. Andrea Cappelli, Elena Castellani, Filippo Colomo and Paolo Di Vecchia.
Published by Cambridge University Press. © Cambridge University Press 2012.

It turns out that the dynamics of these massless particles is completely determined, namely that the result of the limit $\alpha' \to 0$ leads to the unique consistent field theories for them: gauge theory for spin one and general relativity for spin two. In particular, the scattering amplitudes computed in string theory depend on α' and, in the limit $\alpha' \to 0$, reproduce the Feynman diagrams of non-Abelian gauge theory and an extension of Einstein's theory of general relativity. This limit was originally investigated by Cremmer, Gervais, Scherk, Schwarz and Yoneya; it is described in Section 35.2 and discussed in detail in the respective Authors' Chapters.

The most surprising thing is that, in this limit, one gets exactly the two theories that are used to describe, on the one side the strong and electro-weak interactions (QCD and the Glashow–Weinberg–Salam model), and on the other side, gravity.

The understanding of the field theory limit opened the possibility of a change of perspective: instead of a theory of hadrons, string theory could be a unified theory of all interactions including gravity. Correspondingly, the Regge slope should not be related to a hadronic length, but to the Planck length, 19 orders of magnitude smaller. This new perspective was proposed by Scherk and Schwarz in 1974, and is described in Section 35.3.3 and in the Chapters by Yoneya and by Schwarz on Scherk's contributions. In the light of this new setting of string theory, it is appropriate to illustrate its developments in parallel with those of the Standard Model and Grand Unified Theories, and discuss the issue of quantization of gravity. This is done in Sections 35.3.1 and 35.3.2, respectively.

With the advent of QCD, string theory was abandoned as a fundamental theory of strong interactions. However, it still provided a description of hadrons at low energies (large distances): for example, the spectrum of resonances lying on linear Regge trajectories. The study of QCD also showed the emergence of an effective string, the so-called QCD string which confines quarks inside hadrons. This is discussed in Section 35.4.

At the time of the Scherk–Schwarz proposal, string theory was not fully consistent because it still contained a tachyon. In 1976, Gliozzi, Scherk and Olive showed that the tachyon could be eliminated from the spectrum by a suitable truncation of the Ramond–Neveu–Schwarz model. With this projection, the theory became supersymmetric also in 10-dimensional Minkowski spacetime, leading to the first superstring theory. This is presented in Section 35.7.

As discussed in the Chapter by Gervais, supersymmetry, namely the symmetry relating bosonic to fermionic particles, was discovered in string theory in the mid-Seventies. Subsequently, supersymmetric field theories in four and higher dimensions were developed independently from string theory, starting with the work by Wess and Zumino. These theories were intensively studied and, at the beginning of the Eighties, supersymmetric extensions of the Standard Model were proposed to explain in a natural way the large gap, of about 16 orders of magnitude, between the Fermi scale and the Planck scale. Nowadays, supersymmetric theories are among the most appealing candidates for describing physics beyond the Standard Model. Spacetime supersymmetry is introduced in Section 35.6 and Appendix D.

In the second half of the Seventies, supergravity theories, i.e. theories with local supersymmetry generalizing Einstein's theory of general relativity, were also constructed. In general, supersymmetric field theories are less divergent than ordinary ones, since the loop contributions of fermions and bosons tend to cancel each other. This property suggested that supergravity could provide a finite (renormalizable) theory of gravity without the need to introduce string theory. An account of these attempts is given in the Introduction to Part VII. In Section 35.10, we provide a short description of supergravity, since its construction was instrumental in building the nonlinear action for the superstring, which is discussed in Section 35.9.

Although superstring theory was a fully consistent theory, it still presented problems with experiments. First, we observe only four and not ten spacetime dimensions. To cure this problem, the Kaluza–Klein reduction was reconsidered for compactifying the six extra dimensions and making them sufficiently small to escape detection in current experiments. This was suggested in a paper by Cremmer and Scherk and is discussed in the Chapter by Cremmer. Second, supersymmetric partners of elementary particles are not observed, and thus supersymmetry is certainly not an exact symmetry: as other symmetries, it may hold at very high energies but be broken at low energies. A mechanism of supersymmetry breaking was proposed by Scherk and Schwarz immediately after the construction of the superstring. These were, in the second part of the Seventies, the first attempts to construct realistic string theories for unification. They are described briefly in Section 35.8.

In the following Sections, we first discuss in some detail the zero-slope limit, the developments that led to the construction and to the study of the Standard Model and of Grand Unified Theories, and the Scherk–Schwarz proposal of unification of all interactions including gravity. Next we discuss various kinds of fermions in different dimensions, we give an elementary introduction to spacetime supersymmetry, and describe the GSO projection. Then, in Section 35.8 we consider the possibility of compactifying the six extra spacetime dimensions by the Kaluza–Klein reduction and the associated mechanism for breaking supersymmetry. Finally, in Section 35.9 we discuss briefly the construction of the nonlinear Lagrangian for the superstring and in Section 35.10 we introduce supergravity theory.

35.2 The field theory limit

While the DRM and the SVM looked rather different from quantum field theory, their zero-slope limit $\alpha' \to 0$ showed that they were in fact an extension of it, rather than an alternative. As already said, in this limit, the two models reproduce the Feynman diagrams of non-Abelian gauge theories and general relativity, respectively. The parameter α' is the slope of the Regge trajectory, with dimension of length squared, and its inverse is the string tension $T = 1/(2\pi\alpha')$. In the zero-slope limit, the string reduces to a point-particle because the infinite force acting along the string cannot be compensated by the centrifugal force due to the rotation of the string. In the following we briefly explain how this comes about (see Appendix E for a more technical discussion).

Let us consider the mass spectrum of an open string in the light-cone gauge derived in the Introduction to Part IV and in Appendix D:

$$M^2 = \frac{1}{\alpha'}\left[\sum_{i=1}^{d-2}\sum_{n=1}^{\infty} n a_{ni}^\dagger a_{ni} - 1\right]. \tag{35.1}$$

When we take the limit $\alpha' \to 0$ all massive string states acquire a very large mass and disappear from the spectrum: the remaining massless oscillator states $a_{1i}^\dagger|0\rangle$ describe a spin one particle. Since a massless vector is a gauge boson, the mediator of gauge interactions, we expect that string theory reduces to Yang–Mills theory, the generalization of quantum electrodynamics to a non-Abelian symmetry group, typically $SU(N)$. Actually it can be shown that the open-string tree scattering amplitudes (the disc) involving vector particles reproduce, in the zero-slope limit, the Yang–Mills tree Feynman diagrams of vector bosons, called gluons; furthermore, the string loop diagrams involving vectors as external particles reproduce the higher order Feynman diagrams.

Let us first analyze the three-point amplitude of the bosonic open string. The amplitudes involving massless vector particles are given by the expectation values of their vertex operators $(\epsilon \cdot \partial X) e^{ip \cdot X}$, where ϵ_μ is the polarization vector. They can be calculated by extending the methods presented in the Introduction to Part III. The general result for the M-particle colour-ordered amplitude reads (see Appendix E for more details):

$$A^{c.o.}(p_1, \ldots, p_M) = 4\,\mathrm{Tr}(\lambda^{a_1}\cdots\lambda^{a_M})\, g_{YM}^{M-2}\, (2\alpha')^{M/2-2}$$

$$\times \int \frac{\prod_{i=1}^{M} dz_i}{dV_{abc}} \left\{\prod_{i<j}(z_i - z_j)^{2\alpha' p_i \cdot p_j}\right.$$

$$\left.\times \exp\left[\sum_{i<j}\left(\sqrt{2\alpha'}\,\frac{p_j \cdot \epsilon_i - p_i \cdot \epsilon_j}{(z_i - z_j)} + \frac{\epsilon_i \cdot \epsilon_j}{(z_i - z_j)^2}\right)\right]\right\}_{\mathrm{m.l.}}, \tag{35.2}$$

where 'm.l.' means that we have to expand the exponential and to keep only the terms that are linear in all polarizations. The prefactor in (35.2) includes the trace over the generators in the fundamental representation of the gauge group $U(N)$, corresponding to the Chan–Paton factors, discussed in the Introduction to Part V. The complete amplitude is obtained by summing over all colour orderings corresponding to noncyclic permutations of the external particles. The gauge coupling constant g_{YM} is given by

$$g_{YM} \propto g_s (2\alpha')^{(d-4)/4}, \tag{35.3}$$

in terms of the string coupling constant g_s, times a numerical factor that is not relevant here. The field theory limit is computed by sending $\alpha' \to 0$ while keeping g_{YM} fixed.

In the case of three particles, there is no integration to perform in (35.2) and we immediately obtain the amplitude corresponding to the ordering 123:

$$A^{c.o.}(p_1, p_2, p_3) = -4 g_{YM} \text{Tr}(\lambda^{a_1}\lambda^{a_2}\lambda^{a_3}) (\epsilon_1 \cdot \epsilon_2 \, p_2 \cdot \epsilon_3$$
$$+ \epsilon_2 \cdot \epsilon_3 \, p_3 \cdot \epsilon_1 + \epsilon_3 \cdot \epsilon_1 \, p_1 \cdot \epsilon_2 + O(\alpha')), \quad (35.4)$$

where $O(\alpha')$ stands for the string corrections, proportional to α', which are negligible in the field theory limit. The complete three-gluon amplitude is obtained by adding the term with anticyclic ordering 321. This has the same expression as (35.4), except for a minus sign and a Chan–Paton factor $\text{Tr}(\lambda^{a_3}\lambda^{a_2}\lambda^{a_1})$; thus the complete three-gluon amplitude is

$$A(p_1, p_2, p_3) = -2i \, g_{YM} \, f^{a_1 a_2 a_3} (\epsilon_1 \cdot \epsilon_2 \, p_2 \cdot \epsilon_3$$
$$+ \epsilon_2 \cdot \epsilon_3 \, p_3 \cdot \epsilon_1 + \epsilon_3 \cdot \epsilon_1 \, p_1 \cdot \epsilon_2 + O(\alpha')), \quad (35.5)$$

where f_{abc} are the structure constants of the Lie algebra of the $U(N)$ generators, $[\lambda^a, \lambda^b] = i f_{abc} \lambda_c$, normalized by $\text{Tr}(\lambda^a \lambda^b) = \frac{1}{2}\delta^{ab}$.

Equation (35.5) should be compared with the three-gluon amplitude that is obtained at tree level from the Lagrangian of non-Abelian gauge theory,

$$L = -\frac{1}{4} \sum_{a=1}^{N^2} F^a_{\mu\nu} F^a_{\mu\nu}, \quad (35.6)$$

when the external particles are on the mass shell, $p_i^2 = 0$, $p_i \cdot \epsilon_i = 0$ for $i = 1, 2, 3$, and the four-momentum is conserved, $p_1 + p_2 + p_3 = 0$. The non-Abelian field strength is nonlinear in the gauge field,

$$F^a_{\mu\nu} = \partial_\mu A^a_\nu - \partial_\nu A^a_\mu - g_{YM} f_{abc} A^b_\mu A^c_\nu. \quad (35.7)$$

Thus, insertion of (35.7) into (35.6) yields interaction vertices with three gluons ($AA\partial A$ term) and four gluons ($AAAA$ term). It is easy to see that the amplitude that is obtained from the three-gluon vertex ($AA\partial A$) exactly reproduces the string result (35.5).

The $\alpha' \to 0$ limit can be analyzed similarly for amplitudes of the closed bosonic string. In this case, the massless state is the two-index tensor $a^\dagger_{1i} \bar{a}^\dagger_{1j} |0\rangle$: once decomposed into irreducible parts – the traceless symmetric, the antisymmetric and the trace parts – it describes the spin two graviton, the antisymmetric field and the dilaton scalar, respectively. It turns out that the string tree and loop diagrams reproduce in the zero-slope limit those coming from the Einstein theory of general relativity coupled to these two extra fields. In this case the field theory limit is performed by sending $\alpha' \to 0$ while keeping the Newton constant fixed. Since the gauge coupling constant and the Newton constant are given, respectively, by $g_{YM}^2 \sim (\alpha')^{\frac{d-4}{2}}$ and $G_N \sim (\alpha')^{\frac{d-2}{2}}$, their ratio is $G_N / g_{YM}^2 \sim \alpha'$. This means that gauge interactions and gravity are unified within string theory, but the two coupling constants are independent in the field theory limit.

The remarkable correspondence between string theory and field theory follows from the unique consistent description of massless particles of spins one and two. Indeed, it is known that a quantum field theory for such particles should possess gauge invariance for spin one particles and spacetime reparameterization invariance for spin two particles. Since string theory is a consistent quantum mechanical system, it necessarily had to reproduce these quantum field theories.

The field theory limit can also be considered as the low energy limit of string theory: since α' is the only dimensional parameter, the limit $\alpha' \to 0$ is relative to the energy scale of the physical process being considered. We can equivalently keep α' constant and focus on the behaviour of string theory in the limit of very small energies, where the higher states in Regge trajectories are never excited. This is actually the modern view of string theory, to be discussed in the following, where α' is not the hadron scale of 1 GeV, but the much higher Planck scale of 10^{19} GeV.

Equation (35.1) and the corresponding equation for the closed string show that the type of states surviving the $\alpha' \to 0$ limit depends crucially on the values $\alpha_0 = 1$ and $\alpha_0 = 2$ of the intercepts of the Regge trajectory of the open and closed strings, respectively. If the intercept vanished, the massless states would be scalar (the naive semi-classical result) and the limiting field theory very different – that of a massless self-interacting scalar. The question arises of what determines the above values for α_0, crucially affecting the low energy behaviour of string theory.

In the Introduction to Part IV, we have seen that, in the case of the open string, the value $\alpha_0 = 1$ is a quantum effect, corresponding to the ground state energy of string oscillators (see also Appendix D). Its physical (finite) value is determined by consistency conditions: $\alpha_0 = 1$ implies that dual amplitudes have conformal invariance (the residual gauge symmetry) and that the spectrum is free from negative-norm states. In short, the gauge symmetry of the string, i.e. reparameterization invariance, implies conformal invariance of amplitudes, fixes the intercept, and finally determines the low energy limit into gauge theories and general relativity. This is the direct relation between the gauge symmetries of string theory and those of the quantum field theories.

In order to see in a more explicit way the dynamics of the zero-slope limit, it would be instructive to consider the four-gluon amplitude; however, this requires some technical details and is therefore given in Appendix E. Here, we consider a simplified case: going back to the DRM with $\alpha_0 = -\alpha' m^2$ and not equal to 1, we perform the field theory limit on the four-point amplitude of scalar particles, keeping their mass m fixed. Although the DRM with arbitrary intercept α_0 is not consistent for $\alpha' \neq 0$, the field theory resulting from the zero-slope limit is of course fully consistent: this limit is considered here as an illustration of the correct $\alpha_0 = 1$ case.

The crossing-symmetric amplitude of four identical scalar particles is given by the Veneziano formula summed over the three orderings, that we call st, su and tu, as follows (see Introduction to Part II):

$$A(s, t, u) = \beta \left[\frac{\Gamma(-\alpha(s))\Gamma(-\alpha(t))}{\Gamma(-\alpha(s) - \alpha(t))} + \frac{\Gamma(-\alpha(u))\Gamma(-\alpha(t))}{\Gamma(-\alpha(u) - \alpha(t))} + \frac{\Gamma(-\alpha(s))\Gamma(-\alpha(u))}{\Gamma(-\alpha(s) - \alpha(u))} \right], \quad (35.8)$$

where β is a constant, $\alpha(s) = \alpha_0 + \alpha' s = -\alpha'(m^2 - s)$ and we have assumed that the external states are the spin zero states of the leading Regge trajectory. Using the property $\Gamma(z) \sim 1/z$ for $z \to 0$, we can approximate

$$\frac{\Gamma(-\alpha(s))\Gamma(-\alpha(t))}{\Gamma(-\alpha(s)) - \alpha(t))} \sim \frac{m^2 - s + m^2 - t}{\alpha'(m^2 - s)(m^2 - t)}, \qquad \alpha' \to 0, \qquad (35.9)$$

and obtain from Eq. (35.8):

$$A(s, t, u) \Longrightarrow \frac{2\beta}{\alpha'} \left[\frac{1}{m^2 - s} + \frac{1}{m^2 - t} + \frac{1}{m^2 - u} \right]. \qquad (35.10)$$

Assuming that, for $\alpha' \to 0$, the ratio $2\beta/\alpha'$ is fixed and equal to g^2, the standard sum of single poles in the three channels of Φ^3 field theory is recovered. This is the way in which the tree Feynman diagrams are reproduced by the bosonic string. Further details on the zero-slope limit are given in Appendix E.

35.3 Unification of all interactions

In this Section, we describe the developments that led to the Standard Model of electro-weak and strong interactions, to Grand Unified Theories, and to the idea of unifying all interactions, including gravity, in a unique consistent quantum theory. First, we remind the reader briefly of the rise of the Standard Model in late Sixties and early Seventies; then we discuss the quantization of gravity and finally the unification of all interactions proposed by Scherk and Schwarz in the framework of string theory.

35.3.1 Standard Model and Grand Unified Theories

In the Introduction to Part II, we discussed quantum electrodynamics (QED) and its successful comparison with experiments; for weak interactions, we mentioned the four-fermion Fermi theory, which was also in agreement with experiments but not fully consistent, lacking unitarity and renormalizability.

At the end of the Sixties a unified theory of weak and electromagnetic interactions, based on the non-Abelian gauge group $SU(2) \times U(1)$, was proposed by Glashow, Salam and Weinberg, where the left-handed parts, respectively of up and down quarks, and of electrons and neutrinos, form doublets of $SU(2)$, while the right-handed parts are singlets. Knowing the experimental values of the Fermi constant G_F from weak decays and of the electric charge e, and assuming unification of these interactions, one could easily estimate the value of the mass of the W boson, the gauge boson mediating electro-weak interaction: $M_W \sim e/\sqrt{G_F} \sim 100\,\text{GeV}$.

After the deep inelastic experiments had showed the presence of pointlike quarks inside hadrons, originally called 'partons', a gauge interaction for them was proposed, based on the $SU(3)$ colour group. The colour is a new quantum number that is hidden, but whose existence allowed the solution of some phenomenological puzzles in hadron physics, such as the rate of the decay $\pi^0 \to 2\gamma$ and the size of the total cross-section $e^+e^- \to$ hadrons.

In the Seventies, these gauge theories were actively investigated and produced a steady flux of results that nicely matched experiments. Non-Abelian gauge theories were shown to be consistent by 't Hooft and Veltman, because the ultraviolet divergences in their loop diagrams could be removed consistently by the process of renormalization. Moreover, other theoretical tools found their physical application: in electro-weak theory, the Higgs mechanism of spontaneous symmetry breaking was used to give mass to the W and Z gauge boson mediating the weak force and, in fact, to all particles of the Standard Model except the photon and the neutrinos. In QCD, asymptotic freedom, i.e. the fact that quarks are almost free at short distance, was proved by Gross, Politzer and Wilczek in 1973, thus matching the results of deep inelastic experiments and the parton model. That implied that perturbative QCD was correctly describing high energy experiments.

In conclusion, all results fitted into the Standard Model as a non-Abelian gauge theory based on the group $SU(3) \times SU(2) \times U(1)$, providing a satisfactory description of the electro-weak and strong interactions up to the available energies. However the unification was not complete: the Standard Model contains three different and independent coupling constants g_1, g_2 and g_3, one for each group, instead of just one. The electromagnetic, weak and strong interactions are certainly different at low energies. Can we say something about their unification at higher energies?

In 1974 Georgi and Glashow proposed a unified gauge theory based on the larger group $SU(5)$, whose symmetry is realized above the unification scale, called M_{GUT}, and is broken into the three Standard Model subgroups at lower energies. In other words, the three interactions are unified in a single interaction at M_{GUT}, while they show different properties below it. This theory was called Grand Unified Theory (GUT), and the unification scale was estimated to be $M_{GUT} \sim 10^{16}$ GeV. Following the proposal of Georgi and Glashow, other unified gauge theories based on different groups were also constructed. An important consequence of these theories was the prediction of proton decay, which stimulated the construction of experimental facilities to detect such an effect.

35.3.2 Quantization of gravity

The unification of three fundamental forces at sufficiently high energy was a very appealing idea, but it did not consider the fourth interaction, namely gravity. This is described by Einstein's theory of general relativity, where the force is due to the curvature of spacetime and is encoded in the metric tensor $g_{\mu\nu}(x)$. The dynamics is governed by the Einstein–Hilbert action:

$$S = \frac{c^3}{16\pi G_N} \int d^4x \sqrt{-g} R, \tag{35.11}$$

where R is the scalar curvature of spacetime, g is the determinant of the metric and G_N is the Newton constant. In the limit of small velocities $v \ll c$ and small energies, called the Newtonian limit, the gravitational force reduces to the well-known attraction between two particles with mass M at a distance R:

$$F = -\frac{G_N M^2}{R^2}. \tag{35.12}$$

Note the similarity with the electrostatic force upon comparing the dimensionless quantity $G_N M^2/(\hbar c)$ with the fine-structure constant $\alpha = e^2/(\hbar c) \sim 1/137$. For masses corresponding to the current accelerator energies, $M \sim 10^3$ GeV, the ratio of the two dimensionless couplings is extremely small, 10^{-36}, meaning that gravity is totally irrelevant for particle physics at present energies.

On the contrary, the gravitational attraction between two masses can become strong when the dimensionless coupling constant $G_N M^2/\hbar c$ reaches a value of order one. This occurs at the Planck mass,

$$M_P = \sqrt{\frac{\hbar c}{G_N}} = 1.2 \cdot 10^{19} \text{ GeV}. \tag{35.13}$$

Thus, at energies of the order of the Planck mass, the gravitational interaction is strong and quantum effects due to loop corrections become important. Such high energies were present at the origin of the Universe, where quantum gravity effects were surely relevant.

The quantization of gravity is relevant also in the physics of Black Holes. One main difference between electrostatics and gravity is that the latter possesses only attractive 'charges'; this implies an instability of gravitation to stick matter together without ever reaching an equilibrium. A sufficiently large amount of matter, of the order of three solar masses, compresses itself to extremely high density, possibly infinite, forming a Black Hole. At very high densities, gravity becomes strong again and thus Black Holes require quantum treatment.

Unfortunately quantum gravity effects cannot be described consistently by Einstein's theory with the action (35.11), because this theory is nonrenormalizable and incomplete, somehow analogously with the Fermi theory of weak interactions. In gravity, quantum loops are more divergent at high energy than in gauge theories and cannot be cured by standard renormalization procedures. This result became clear in the Seventies, after renormalization had been achieved for non-Abelian gauge theories. In other words, in the case of gravity, the pointlike structure of field theory creates such divergences that the theory cannot be given a meaning at arbitrarily large energies (at least within perturbation theory). The theory is correct at the classical and semi-classical level: as such, it describes the evolution of the Universe a fraction of a second after the Big Bang; however, a fundamental quantum theory is needed to understand earlier times (as well as the stability of Black Holes).

35.3.3 String theory unification

The failure of perturbative renormalization in general relativity could be just of a technical nature or could hide a more fundamental physical issue: maybe gravity requires a completion at higher energies, where other degrees of freedom could come into play and cure it, as happened for weak interactions. Scherk, Schwarz and Yoneya opted for the second point of view and proposed to construct a quantum theory of gravity by incorporating it in string theory.

Unlike Einstein's theory, string theory is a consistent quantum theory: since the string is an extended object, the theory does not suffer from ultraviolet divergences. This is due to the presence of an intrinsic ultraviolet cutoff $l_s \sim \sqrt{\alpha'}$, corresponding to the shortest distance that can be probed at high energy in string theory. The softness of the string at small distances, which was a problem for hadron physics, now becomes a virtue, as it provides a theory free from ultraviolet divergences. Since its closed sector contains a massless spin two particle, string theory yields a finite quantum theory of gravity.

Following the early studies by Scherk in 1971 and by Neveu and Scherk in 1972 on the zero-slope limit of open strings reproducing gauge theories, in 1974 Scherk and Schwarz, and independently Yoneya, considered the same limit for closed strings and obtained Einstein's general relativity. As anticipated, the result was hardly unexpected, because this is the unique consistent theory of a massless graviton. It became clear that string theory, having an open sector with massless gauge bosons and a closed sector with massless gravitons, was an extension of both gauge field theories and general relativity.

Scherk and Schwarz then proposed to consider string theory for unifying gauge and gravitational interactions into a consistent quantum theory, where the unique interaction is among strings. However, this proposal was only taken up by the very few people who still worked in string theory; while most of the theoreticians were understanding the properties of the Standard Model. Moroever, it seemed premature to think about gravitational interactions, negligible at the accelerator energies.

Let us now discuss the virtues of the unification by string theory. Some advantages were apparent at the time of the proposal, turning old drawbacks into benefits; others needed some years of evolution in theoretical physics to be fully appreciated. They can be summarized as follows:

(i) the presence of massless spin one and two particles which are the mediators of the two physically relevant interactions;
(ii) the softness of the string at high energy, solving the problem of divergences in quantum gravity, i.e. the lack of renormalizability;
(iii) the six extra spacetime dimensions, which can be curved and have very small size, leading to additional excitations by the so-called Kaluza–Klein compactification mechanism;
(iv) the spacetime supersymmetry, following from the GSO projection, to be discussed below.

At the time of this proposal in 1974, string theory was still plagued by tachyons: however, they were to be eliminated by the GSO projection in 1976, as described in Section 35.7. Before explaining these results, we discuss the emergence of the QCD string, make a small detour to introduce spinors in four and higher dimensions, and then present an elementary introduction to supersymmetry.

35.4 The QCD string

In this Section we discuss how string theory, abandoned as a fundamental theory of hadrons in favour of quantum chromodynamics, reappears in this context as an effective description at large distances.

In the previous Sections we have seen that the Dual Resonance Model and the Ramond–Neveu–Schwarz model could not describe some experimental features of hadrons. In the meantime, QCD was formulated and intensively studied. Its property of asymptotic freedom implied that quarks inside the protons, i.e. at short distances, behave as free particles. In this regime, perturbative field theory methods allowed the computation of physical quantities, with experimentally confirmed predictions, such as the logarithmic corrections to Bjorken scaling of scattering amplitudes [HBRD97].

However, in the framework of QCD, a puzzle remained: why are quarks 'confined' inside hadrons, i.e. why could they not be taken out and observed as free particles? An answer to this question was not easy because it required knowledge of the QCD behaviour at large distances, i.e. low energies, where its coupling constant becomes large and nonperturbative methods are needed.

In order to study the nonperturbative behaviour, Yang–Mills theory was formulated on a spacetime lattice by Wilson in 1974. The inverse of the lattice distance a played the role of a high energy cutoff: in this setting, the strong coupling limit of the theory could be investigated and it led to a linearly rising potential $V(R) \sim R$ between static quarks at distance R, and thus to quark confinement. In order to compare with the continuum theory, one must send the lattice distance $a \to 0$ while keeping fixed a suitably chosen physical quantity, such as the coefficient of the linear potential, called string tension. Using Monte Carlo numerical techniques, in 1980 Creutz computed the lattice-gauge-theory path integral and showed that, indeed, the confinement property remains valid in the continuum limit $(a \to 0)$.

Furthermore, in 1975 Eichten, Gottfried, Kinoshita, Kogut, Lane and Yan constructed a nonrelativistic potential model for heavy quarks that reproduced correctly the properties of charmed mesons. In this model, the quarks interacted through a linear potential with string tension precisely equal to $1/(2\pi\alpha')$, where α' is the slope of the Regge trajectory measured in hadronic experiment.

These results could be interpreted as the quarks being attracted by a string, thus suggesting that an effective string theory was at work in QCD, at least for sufficiently large distances. On the other hand, it was also clear that this could not be the same string as the one discussed in this book because, for instance, it had no massless spin one or spin two states. The solution to this puzzle is that the quark–antiquark potential is described by an effective string theory and, thus, is linear for large quark separations, but becomes the nonconfining potential $V(R) \sim 1/R$ of perturbative QCD for small quark separations.

Another element that contributed to this picture was the formulation of quark confinement by 't Hooft and Mandelstam in analogy with the physics of superconductors. It is well known that, in vacuum, the electric field lines spread all over and lead to the $V(R) \sim 1/R$ potential between charges. Inside a superconductor, however, the magnetic field is expelled, apart from thin tubes where the field lines concentrate. Therefore, magnetic charges inside a superconductor are connected by a flux tube, which is an effective string causing a linear potential between them. The picture suggested by this analogy was that the QCD vacuum behaves as a (electric-magnetic) 'dual superconductor' for the chromoelectric field, leading to an effective string between quarks and to confinement, in agreement with the results of the lattice approach.

In 1980 Lüscher, Symanzik and Weisz proposed an effective string theory for describing the flux tube joining a quark–antiquark pair and computed the quark–antiquark potential. A simple and effective way of computing this potential was proposed by Alvarez in 1981 and by Arvis in 1983 using the bosonic string. In today's language, they computed the potential between two D0 branes at distance R, with the following result:

$$V(R) = \sqrt{\frac{R^2}{(2\pi\alpha')^2} - \frac{d-2}{24\alpha'}} \sim \frac{R}{2\pi\alpha'} - \frac{\pi}{R} \cdot \frac{d-2}{24} + \cdots . \qquad (35.14)$$

In this expression, the leading term is the linear potential while the next-to-leading one reproduces the so-called Lüscher term, actually observed in Monte Carlo simulations of lattice gauge theories in $d = 3$ and $d = 4$. At shorter distances the potential becomes imaginary and does not make sense, signalling the fact that the effective string description breaks down.

Additional evidence for the presence of strings in gauge theories came from the proposal by 't Hooft in 1974 of studying QCD for an arbitrary number N of colours (instead of just three), and performing the large-N expansion while keeping the product Ng_{YM}^2 fixed. As shown by 't Hooft, in this nonperturbative limit the Feynman diagrams of QCD are classified according to their topology precisely as those of the loop expansion of string theory.

A new insight on the connection between gauge theories and strings appeared in the late Nineties, with the Maldacena conjecture on the correspondence between a particular gauge theory in four dimensions and a closed superstring in a suitably compactified ten-dimensional space (see the final Chapter of this Volume for further details).

35.5 A detour on spinors

A Dirac fermion in four dimensions has four complex components corresponding to eight real parameters. When the Dirac equation is imposed there remain four real on-shell degrees of freedom. These are the physical degrees of freedom of a propagating Dirac fermion, corresponding to the two states of electron and positron, and to the two helicity states ($h = \pm\frac{1}{2}$). Remember that the helicity of a particle is defined as the projection of its spin along the momentum direction $h = (\vec{s} \cdot \vec{p})/|\vec{p}|$.

In Dirac theory, the charge-conjugation operator maps a particle into its antiparticle, and vice versa. Acting on a spinor ψ_D, it produces its charge conjugate ψ_D^c as follows:

$$\psi_D^c = C\bar{\psi}_D^T, \qquad C\gamma_\mu^T C^{-1} = -\gamma_\mu, \qquad (35.15)$$

where γ_μ are Dirac gamma matrices, and T denotes the transposed matrix.

A Majorana fermion coincides with its antiparticle and is, therefore, neutral. It has half of the degrees of freedom of the Dirac spinor, i.e. the two on-shell degrees of freedom corresponding to the two helicities $\pm\frac{1}{2}$. A Majorana spinor satisfies the property of self-conjugation:

$$\psi_M^c = \psi_M. \qquad (35.16)$$

A massless Dirac fermion can be decomposed in two independently propagating parts, called Weyl fermions, with definite 'chirality'. A left(right)-handed Weyl spinor describes a particle with helicity $\frac{1}{2}$ ($-\frac{1}{2}$) and an antiparticle with helicity $-\frac{1}{2}$ ($+\frac{1}{2}$).

They are obtained from a Dirac spinor by the projections:

$$\psi_L = \frac{1-\gamma_5}{2}\psi_D, \qquad \psi_R = \frac{1+\gamma_5}{2}\psi_D, \qquad \gamma_5 = i\gamma^0\gamma^1\gamma^2\gamma^3, \qquad (35.17)$$

where $(\gamma_5)^2 = 1$. All types of fermions can be expressed in terms of two-dimensional Weyl spinors:

$$\psi_D = \begin{pmatrix} \chi_\alpha \\ \bar{\psi}^{\dot{\alpha}} \end{pmatrix}, \qquad \psi_M = \begin{pmatrix} \chi_\alpha \\ \bar{\chi}^{\dot{\alpha}} \end{pmatrix}, \qquad \psi_W = \begin{pmatrix} \chi_\alpha \\ 0 \end{pmatrix}. \qquad (35.18)$$

Here undotted (α) and dotted ($\dot{\alpha}$) indices of the Weyl spinor χ_α and of its conjugate $\bar{\chi}^{\dot{\alpha}}$, respectively, refer to the two possible fundamental representations ($\frac{1}{2}$, 0) and (0, $\frac{1}{2}$) of the Lorentz group.

The previous considerations can be generalized to a d-dimensional spacetime with d even. An on-shell Dirac spinor has $2^{d/2}$ degrees of freedom. On-shell Majorana and Weyl spinors have $2^{(d/2)-1}$ physical degrees of freedom. In certain dimensions, $d = 2, 10, 18, \ldots$, the Weyl and the Majorana conditions can be imposed simultaneously. In particular, in ten dimensions we can have Weyl–Majorana spinors that have only eight on-shell degrees of freedom. We will see that these spinors occur in the superstring.

35.6 Spacetime supersymmetry

In the Introduction to Part V we have seen that a new algebra was found in the RNS model that involved both bosonic and fermionic operators obeying commutation and anticommutation relations, respectively. It turned out that these operators generated an extension of the conformal transformations involving bosonic and fermionic parameters that left invariant the free action of d real scalar bosons and d Majorana fermions in two dimensions. Since the two-dimensional space corresponded to the string world-sheet, this new symmetry was called world-sheet supersymmetry. This is the gauge symmetry of the RNS model.

In 1973 Zumino gave some lectures at a conference in Capri where he discussed the supersymmetry of the RNS model. Soon after, he started to investigate, together with Wess, whether this symmetry could be extended to four-dimensional field theory. Their study led to the construction of the four-dimensional Wess–Zumino model involving two real scalars and a Majorana fermion. Although the same kind of supersymmetry was formulated independently by Golfand and Likhtman in 1971 and by Volkov in 1972, the Wess–Zumino model was really the starting point of supersymmetric field theories in four dimensions. The fact that Wess and Zumino were inspired by the properties of the RNS model is apparent in the title of their paper: 'Supergauge transformations in four dimensions'.

In the Wess–Zumino model, supersymmetry is not a gauge symmetry as in the RNS model, but rather a classification symmetry for its spectrum of states, like the isotopic spin or the $SU(3)$ flavour symmetry; the important difference, however, is that supersymmetry does not act in the internal field space, but in spacetime, extending in a nontrivial way the invariance under the Poincaré group. The supersymmetry multiplets include both integer and half-integer spins: for instance, in the Wess–Zumino model there are two spin zero and one spin one-half fields. It is a general property of all supersymmetric theories that they have an equal number of bosonic and fermionic physical degrees of freedom (number of field polarizations) that are degenerate in mass.

Let us give a short description of the Wess–Zumino model and its spacetime supersymmetry, leaving the details to Appendix D. The action of the Wess–Zumino model (restricted to the free case for simplicity) contains two real bosonic fields, A and B, and one fermionic field ψ as follows:

$$S = \int d^4x \, \frac{1}{2}\left[-\left(\partial_\mu A\right)^2 - \left(\partial_\mu B\right)^2 + i\bar{\psi}\gamma^\mu \partial_\mu \psi\right]. \tag{35.19}$$

Note the similarity with the RNS world-sheet action discussed in the Introduction to Part V: apart from the change of spacetime dimension, there are the same kinds of kinetic terms, written with the same conventions. The supersymmetry transformations read:

$$\delta A = \bar{\epsilon}\psi, \qquad \delta B = -i\bar{\epsilon}\gamma_5\psi,$$
$$\delta\psi = -i\partial_\mu(A + i\gamma_5 B)\gamma^\mu \epsilon, \qquad (35.20)$$

where the parameter ϵ is now a constant Majorana spinor. The invariance of the action is proved following the same steps as for the RNS string action, as described in Appendix D.

In the years that followed, the supersymmetric extension of gauge theories was also formulated. In this theory, the gauge boson with spin one is accompanied by a spin one-half Majorana field called a gaugino, while electrons and quarks are both accompanied by pairs of complex scalars called selectrons and squarks.

Let us now write the spacetime supersymmetry algebra. In the two-dimensional Weyl spinor notation of the previous Section, the supersymmetry generators, denoted by Q_α and $\bar{Q}_{\dot\alpha}$, satisfy the following anticommutation relations:

$$\{Q_\alpha, Q_\beta\} = \{\bar{Q}_{\dot\alpha}, \bar{Q}_{\dot\beta}\} = 0, \qquad \{Q_\alpha, \bar{Q}_{\dot\beta}\} = 2\sigma^\mu_{\alpha\dot\beta} P_\mu, \qquad (35.21)$$

where P_μ is the generator of spacetime translations, $\sigma^\mu_{\alpha\dot\beta} = (-1, \vec{\sigma})$, and $\vec{\sigma}$ are the three Pauli matrices. The supersymmetry generators commute with the translations

$$[P_\mu, Q_\alpha] = [P_\mu, Q_{\dot\alpha}] = 0, \qquad (35.22)$$

and transform as Weyl spinors under the Lorentz transformations. It can be shown that, using the field equations of motion, the composition of two supersymmetry transformations (35.20) gives a translation, as implied by the anticommutator in Eq. (35.21). To obtain the algebra (35.21) without using the equations of motion (i.e. off-shell), one needs to introduce an additional auxiliary (i.e. nondynamical) complex field F that contributes to match the bosonic and fermionic degrees of freedom off-shell. This was not included in the action (35.19) for simplicity.

The supersymmetry transformations have a natural geometric interpretation: as the Poincaré group describes translations and rotations in spacetime, they also describe translations in an extended space, called 'superspace', involving both bosonic coordinates x^μ and fermionic spinorial coordinates $(\theta, \bar{\theta})$ (see Appendix D).

Some simple supersymmetric particle multiplets, i.e. representations, can be easily derived from the algebra (35.20). In the case of a massive multiplet, we can perform a Lorentz transformation and bring the particle at rest ('go to the rest frame') with four-momentum $P^\mu = (M, 0, 0, 0)$; the algebra (35.21) reduces to that of fermionic oscillators (discussed in the Introduction to Part V):

$$\{b_\alpha, b_\beta\} = \{b^\dagger_\alpha, b^\dagger_\beta\} = 0, \qquad \{b_\alpha, b^\dagger_\beta\} = \delta_{\alpha\beta}, \qquad (35.23)$$

where $b_\alpha \equiv Q_\alpha/\sqrt{2M}$ and $b^\dagger_\alpha \equiv \bar{Q}_{\dot\alpha}/\sqrt{2M}$. A representation of this algebra consists of the vacuum $|0\rangle$ and of the states obtained by acting on it with creation operators. In this way we get the following multiplet of states:

$$|0\rangle, \qquad |\alpha\rangle = b_\alpha^\dagger |0\rangle, \qquad |1\rangle = \epsilon^{\alpha\beta} b_\alpha^\dagger b_\beta^\dagger |0\rangle, \qquad (35.24)$$

where $\epsilon^{\alpha\beta}$ is the two-dimensional antisymmetric tensor (recall that $b_\alpha^2 = (b_\alpha^\dagger)^2 = 0$). If we choose the vacuum to be a bosonic state, then the next two states are fermionic and the last is again bosonic, i.e. the representation contains two bosonic and two fermionic states. If the vacuum is a Lorentz scalar, then these states have spin zero and one-half: this is the multiplet of the Wess–Zumino model.

In the case of massless particles, we can go instead to the frame where $P^\mu = (E, E, 0, 0)$, and the algebra (35.21) becomes:

$$\{b_\alpha, b_\beta\} = \{b_\alpha^\dagger, b_\beta^\dagger\} = 0, \qquad \{b_\alpha, b_\beta^\dagger\} = (1 + \sigma_3)\delta_{\alpha\beta}, \qquad (35.25)$$

with $b_\alpha \equiv Q_\alpha/\sqrt{2E}$ and $b_\alpha^\dagger \equiv \bar{Q}_{\dot\alpha}/\sqrt{2E}$. This algebra can be truncated consistently by putting $b_2 = b_2^\dagger = 0$, and thus leaving only one fermionic oscillator. Therefore, the smallest representation consists of two states: $|0\rangle$ and $b_1^\dagger |0\rangle$, characterized by the values s and $s + 1/2$ of the helicity, respectively. For $s = 1/2$ this is a multiplet with helicities $(1/2, 1)$ describing the gaugino and the gauge boson; instead, for $s = 3/2$, the helicity $(3/2, 2)$ multiplet contains the gravitino and graviton.

The first supergravity theory was constructed in 1976: it indeed consisted of the graviton, described by the metric tensor as in Einstein gravity, and the gravitino. Unlike the Wess–Zumino model and supersymmetric gauge theories, supergravity possesses local supersymmetry as in the RNS model, with parameters that are arbitrary functions of the spacetime coordinates.

Gauge and gravity theories with extended supersymmetry were also constructed, corresponding to $\mathcal{N} > 1$ copies of the $(Q_\alpha, \bar{Q}_{\dot\alpha})$ generators. They involve multiplets of fields with more spin values: for example the $\mathcal{N} = 2$ extended supersymmetric gauge theory possesses a single multiplet of massless fields, including the gauge boson, the gaugino, a Majorana fermion and a complex scalar, all transforming in the adjoint representation of the gauge group (see the Introduction to Part VII).

At the beginning of the Eighties, supersymmetric extensions of the Standard Model were formulated. In particular, in the so-called Minimal Supersymmetric Standard Model, a supersymmetric partner was introduced for each particle: gauginos for gauge bosons, Higgsinos for the Higgs particles, squarks and sleptons for quarks and leptons, and so on. In a supersymmetric theory these partners have the same mass as that of the Standard Model particles; since they have not been observed, supersymmetry breaking terms must be added to the Lagrangian to give them higher masses. For several reasons, an appealing value for the supersymmetry energy scale would be of the same order, or slightly higher, as the scale corresponding to the breaking of the electro-weak gauge group, $E_{SM} \sim 10^3$ GeV. If this is correct, then some supersymmetric partners could be observed at the Large Hadron Collider (LHC) now operating at CERN, Geneva.

In the second half of the Seventies and at the beginning of the Eighties, supergravity and supersymmetric gauge theories were developed and became very appealing for the programme of unification of fundamental interactions. These issues will be discussed further in the Introduction to Part VII.

35.7 The GSO projection

In Part V, we have seen that the RNS model contains a bosonic sector whose mass spectrum is given by:

$$\alpha' M^2 = \sum_{i=1}^{8} \left(\sum_{n=1}^{\infty} n a_n^{\dagger i} a_n^i + \sum_{r=1/2}^{\infty} r \psi_r^{\dagger i} \psi_r^i \right) - \frac{1}{2} \equiv N - \frac{1}{2}, \qquad (35.26)$$

where N is the sum of the number operators of the bosonic and fermionic coordinates. We are working in the light-cone gauge where only eight of the ten components of the oscillators are physical. From the previous formula we see that there are states with an even number of fermionic oscillators having a half-integer value of $\alpha' M^2$ and states with an odd number of fermionic oscillators having an integer value of $\alpha' M^2$. One can introduce a fermion number operator that counts the number of fermionic oscillators:

$$(-1)^F \quad \text{where} \quad F = \sum_{r=1/2}^{\infty} \psi_r^\dagger \cdot \psi_r - 1. \qquad (35.27)$$

It is easy to see that $(-1)^F$ is equal to 1 (-1) when acting on a state with an odd (even) number of fermionic oscillators. In particular, the tachyon represented by the oscillator vacuum with a mass equal to $\alpha' M^2 = -\frac{1}{2}$ is odd under the action of $(-1)^F$, while the gauge boson described by the state $\psi_{\frac{1}{2}}^{\dagger i}|0, k\rangle$ is even under $(-1)^F$.

The mass spectrum in the R sector is given by:

$$\alpha' M^2 = \sum_{i=1}^{8} \left(\sum_{n=1}^{\infty} n a_n^{\dagger i} a_n^i + \sum_{n=1}^{\infty} n \psi_n^{\dagger i} \psi_n^i \right) \equiv N_R. \qquad (35.28)$$

The lowest state of the spectrum is a Majorana massless spinor in ten dimensions that has 16 physical degrees of freedom. In the Ramond model we can also introduce a fermion number operator given by:

$$(-1)^F = \Gamma_{11}(-1)^{F_R}, \qquad F_R = \sum_{n=1}^{\infty} \psi_n^\dagger \cdot \psi_n,$$

$$\Gamma_{11} \equiv 2^5 \psi_0^0 \psi_0^1 \ldots \psi_0^9 = \Gamma^0 \Gamma^1 \ldots \Gamma^9, \qquad (35.29)$$

where Γ_{11} is the chirality operator in ten dimensions. In particular, the lowest massless state has both ten-dimensional chiralities.

In 1976, Gliozzi, Scherk and Olive proposed truncating the spectrum of the RNS model by keeping only the states that are even under the action of $(-1)^F$. This means that all the states of the NS model with half-integer squared mass, including the tachyon, are eliminated from the physical spectrum. In the Ramond sector, all the massive states remain, but their degeneracy is halved, while the massless state becomes a Majorana–Weyl spinor with only eight physical degrees of freedom. These developments are described in the Chapters by Gliozzi and Olive.

The most convenient way to count the number of physical states at each level is by introducing the function $Z = \sum_n c_n q^n$ that counts the degeneracy c_n at the level with mass squared $\alpha' M^2 = n$. It can be shown that Z is actually the partition function of the string on the world-sheet geometry of the annulus. In the NS sector, the partition function is:

$$\begin{aligned}
Z_{NS}^{GSO} &= \mathrm{Tr}\left(q^{N-1/2} \frac{1+(-1)^F}{2}\right) \\
&= \frac{1}{2} q^{-1/2} \left[\prod_{n=1}^{\infty} \left(\frac{1+q^{n-1/2}}{1-q^n}\right)^8 - \prod_{n=1}^{\infty} \left(\frac{1-q^{n-1/2}}{1-q^n}\right)^8\right] \\
&= \sum_{n=0}^{\infty} c_n q^n = 8 + c_1 q + c_2 q^2 + \cdots.
\end{aligned} \qquad (35.30)$$

In the R sector we have:

$$Z_R^{GSO} = \mathrm{Tr}\left(q^{N_R} \frac{1+(-)^F}{2}\right) = 8 \prod_{m=1}^{\infty} \left(\frac{1+q^m}{1-q^m}\right)^8, \qquad (35.31)$$

where the term with $(-1)^F$ gives no contribution. It turns out that the two partition functions are actually equal

$$Z_{NS}^{GSO} = Z_R^{GSO}, \qquad (35.32)$$

as a consequence of the 'aequatio identica satis abstrusa' found by Jacobi in the nineteenth century. In this way, a model without tachyons was obtained, having an equal amount of bosonic and fermionic states at each mass level. This is a necessary condition for realizing a supersymmetric theory in ten-dimensional Minkowski space and, in fact, it turned out that the model was supersymmetric. This spacetime supersymmetry is also called 'target-space' supersymmetry in order to stress the difference with the world-sheet supersymmetry.

For the first time a consistent string model without tachyons had been constructed. Moreover, it had spacetime supersymmetry: superstring theory was born. This was a remarkable achievement, but at that time only a few people appreciated it, since most of the community was busy studying QCD and the electro-weak interactions or constructing supersymmetric theories in four dimensions after the work of Wess and Zumino. As discussed in earlier Sections, superstring theory was relevant for the unification of all interactions, but had to wait eight more years before being accepted and appreciated by the vast majority of people.

A final remark on the GSO projection: it was later recognized that it is required by the consistency of the theory. In fact, the GSO projected annulus partition function, as a function of the annulus parameter, has the correct transformation properties to ensure open–closed string duality. In the analogous calculation of the partition function of closed strings one considers the doubly periodic world-sheet geometry of the torus. The corresponding amplitude must be invariant under the modular transformations of the torus, that express the residual symmetry of reparameterizations in the conformal gauge. It turns out that the modular invariance of the torus partition function is only achieved in the GSO projected theory.

If we take superstring theory seriously as a consistent quantum theory unifying the gauge theories with gravity, we are immediately confronted with two problems. The first one is that the theory is consistent in $d = 10$, while we only observe four spacetime dimensions. The second one is that spacetime supersymmetry is not apparent since we do not find the supersymmetric partners of, for example, the electron or the quarks. This implies that supersymmetry, if present, must be broken below a certain energy scale. In the next Section, we discuss the ideas that were immediately developed to cope with these two problems.

35.8 The Kaluza–Klein reduction and supersymmetry breaking

35.8.1 Kaluza–Klein theory

The presence of extra spacetime dimensions in superstring theory was first addressed within the framework of the Kaluza–Klein theory, originally formulated just after the theory of general relativity. Kaluza–Klein theory was a pure gravity theory in five spacetime dimensions: by reducing it to four dimensions, one obtains a theory with gravity in four spacetime dimensions coupled to a scalar and a vector field. It turns out that the vector field can be identified with the electromagnetic field and, in this way, electromagnetism in four dimensions follows directly from general covariance of gravity in five dimensions.

In order to explain why the fifth dimension is not observed Kaluza–Klein theory assumed that this is small and compact, thus escaping detection at large distances, i.e. at the 'low' energies available in experiments. Let us now see in a simple example how this comes out.

We consider a five-dimensional free scalar theory described by the following action:

$$S = \int d^4x \int_0^{2\pi R} dy \, \frac{1}{2} \left[-\partial_\mu \phi \partial^\mu \phi + \partial_y \phi \partial_y \phi \right], \tag{35.33}$$

where the four coordinates x^μ are those of Minkowski spacetime, while the extra coordinate y describes a compact space, a circle of length $2\pi R$, leading to the identification:

$$y + 2\pi R \sim y. \tag{35.34}$$

The equations of motion for the real scalar field involve the five-dimensional d'Alembert operator:

$$\left[\partial^\mu \partial_\mu + \partial_y^2 \right] \phi(x, y) = 0. \tag{35.35}$$

Furthermore, because of the identification in Eq. (35.34), $\phi(x, y)$ must be a periodic function in the compact dimension:

$$\phi(x, y + 2\pi R) = \phi(x, y). \tag{35.36}$$

The equation of motion of the action (35.33) can be easily solved by going to momentum space and expanding $\phi(x, y)$ in terms of plane waves $e^{ip\cdot x + ip_y y}$. The periodicity (35.36) implies

$$e^{ip_y(y+2\pi R)} = e^{ip_y y} \implies p_y = \frac{n}{R}, \tag{35.37}$$

where n is an integer. As a consequence, the momentum along the compact dimension is quantized in terms of the inverse of the radius R. The field can be expanded in Fourier components of the fifth dimension:

$$\phi(x, y) = \sum_n \phi_n(x) \, e^{i\frac{n}{R}y}. \tag{35.38}$$

Inserting this expansion back in the equation of motion (35.35), one has:

$$\left[\partial^\mu \partial_\mu - \left(\frac{n}{R}\right)^2\right] \phi_n(x) = 0. \tag{35.39}$$

This shows that a massless scalar field in a five-dimensional spacetime with a compactified dimension is equivalent to an infinite number of fields $\phi_n(x)$ from the four-dimensional point of view: $\phi_0(x)$ is still massless, while the other fields with $n \neq 0$ have mass given by

$$m_n = \frac{n}{R}. \tag{35.40}$$

This procedure is called Kaluza–Klein dimensional reduction. The massive states that are generated in four dimensions, called Kaluza–Klein excitations, are too heavy to be observed in experiments at energies much smaller than the inverse of the compactification radius of the extra dimension: $E \ll 1/R$. Only when energies $E \sim 1/R$ are reached, we start to see the effect of an extra dimension through the presence of an infinite tower of states with equally spaced masses.

Going back to superstring theory, we conclude that the presence of the six extra dimensions is not a priori in contradiction with experiments if they are compact and sufficiently small. The natural compactification scale for superstring theory is $R \sim 1/M_P \sim 10^{-33}$ cm, which is formidably small. However, recent proposals based on string configurations with D-branes have suggested the possibility of much larger extra dimensions, which could be detectable in low energy laboratory experiments and at the LHC. The existence of extra dimensions should modify, at sufficiently short distances, Newton's law of gravitational attraction between two bodies. The current upper experimental bound of about one hundredth of a millimetre on the size of the extra dimensions is obtained from the absence of observed deviations.

In conclusion, the existence of extra dimensions is a definite prediction of superstring theory; it is consistent with experiments if the compactification radius is sufficiently small.

Another effect of the Kaluza–Klein reduction combined with the spacetime supersymmetry is to generate four-dimensional supersymmetric theories with extended supersymmetry (see the Introduction to Part VII).

35.8.2 Scherk–Schwarz supersymmetry breaking mechanism

Since the observed spectrum of particles is definitely not supersymmetric, a mechanism for breaking supersymmetry was suggested by Scherk and Schwarz already in 1979 and it is described in the Chapter by Schwarz on Scherk's work in Part VI. We present here the basic idea. Let us consider an extension of the five-dimensional theory of the previous subsection, involving a complex (charged) scalar field ϕ, and a Weyl fermion χ, the simplest supersymmetric multiplet. The corresponding action is invariant under phase transformations:

$$\phi(x, y) \to e^{i\alpha} \phi(x, y), \qquad \bar\phi(x, y) \to e^{-i\alpha} \bar\phi(x, y), \qquad (35.41)$$

corresponding to charge conservation. Let us perform a Kaluza–Klein reduction but, this time, with a 'twisted' periodic boundary condition on the compact dimension:

$$\phi(x, y + 2\pi R) = e^{imy} \phi(x, y). \qquad (35.42)$$

In this case the field ϕ is not periodic, but this is still acceptable because the theory is invariant under phase transformations (this was not possible for the real, neutral field of the previous Section). The Fourier expansion of the field satisfying the boundary condition (35.42) has, instead of (35.38), the following expression:

$$\phi(x, y) = \sum_n \phi_n(x) \, e^{i(\frac{n}{R}+m)y}. \qquad (35.43)$$

When inserted in the five-dimensional equation of motion, it yields:

$$\left[\partial^\mu \partial_\mu - \left(\frac{n}{R}+m\right)^2\right]\phi_n(x) = 0, \qquad (35.44)$$

corresponding to Kaluza–Klein excitations with masses

$$m_n = \frac{n}{R} + m. \qquad (35.45)$$

In the present case, the lowest bosonic mode, corresponding to $n = 0$, acquires a nonvanishing mass m.

In the noncompactified five-dimensional supersymmetric theory the masses of the scalar and the fermion are degenerate; they are taken to be zero for simplicity. If we now impose a Kaluza–Klein reduction where the scalar field transforms as in Eq. (35.42), while the fermion field is periodic, then the spectrum of excitations will no longer be degenerate and supersymmetry is broken. The breaking of the symmetry is achieved by requiring different periodicities for the members of the same multiplet (boson and fermion) in going around the fifth dimension. In this way we have a supersymmetry breaking mechanism in field theory and superstring theory.

A drawback in this approach is that, for compactification radii corresponding to Planck energies, the natural scale of supersymmetry breaking is of the same order of magnitude. This is too high to be phenomenologically appealing; however, recent developments with D-branes allow for a supersymmetry breaking scale higher than the Standard Model scale $M_{SM} \approx 10^3$ GeV, but not as high as the Planck scale (see the last Chapter of this Volume).

35.9 The local supersymmetric action for the superstring

We have seen in previous Sections that the GSO projection of the RNS model yielded a ten-dimensional string theory, called superstring theory, which is supersymmetric in spacetime and does not contain tachyons. In the construction of the theory there was still a missing element, namely a locally supersymmetric and reparameterization invariant action, generalizing the Nambu–Goto action for the bosonic string to the superstring.

The problem was that no locally supersymmetric Lagrangian was known and, therefore, it was not clear how to proceed. A supersymmetric extension of Einstein's theory of general relativity, called supergravity, was constructed in 1976 providing the local supersymmetry transformations of the vierbein (and the metric tensor) and of a spin $\frac{3}{2}$ particle, called the gravitino, the supersymmetric partner of the graviton. Because of its importance in the construction of the local supersymmetric Lagrangian for the RNS model, the action and the supersymmetry transformations of supergravity are described briefly in the next Section.

As explained in the Chapters by Brink and Di Vecchia in this Part, the supersymmetry transformations of the vierbein and the gravitino were then specialized to the one- and two-dimensional cases and, together with the extension to curved space of the supersymmetry transformations of the particle and string bosonic and fermionic coordinates, they allowed for the construction of a locally supersymmetric action for the spin one-half pointlike particle and for the RNS string, respectively.

In 1976, Brink, Deser, Di Vecchia, Howe and Zumino first constructed the point-particle Lagrangian. Thanks to this work, it became clear that the Nambu–Goto action for the bosonic string could be written equivalently in the form:

$$S = -\frac{1}{4\pi\alpha'} \int d\tau \int d\sigma \sqrt{-g} g^{ab} \partial_a X \cdot \partial_b X. \tag{35.46}$$

This action, later used by Polyakov for the path integral covariant quantization of the bosonic string (see the Introduction to Part VII), is known as the σ-model action, or Polyakov action; it is a functional of both the string coordinate $X^\mu(\tau, \sigma)$ and the metric $g_{ab}(\tau, \sigma)$, the latter being considered now as an independent variable.

Starting from the action (35.46), the Nambu–Goto action can be recovered by eliminating the metric as follows. The equations of motion for the metric read

$$\theta_{ab} \equiv \partial_a X \cdot \partial_b X - \frac{1}{2} g_{ab} g^{cd} \partial_c X \cdot \partial_d X = 0, \tag{35.47}$$

giving the identity:

$$\sqrt{-\det(\partial_a X \cdot \partial_b X)} = \frac{1}{2}\sqrt{-g}\,g^{ab}\partial_a X \cdot \partial_b X. \tag{35.48}$$

Upon substituting it in the action (35.46), one readily recovers the Nambu–Goto action (see the Chapter by Di Vecchia in this Part). Since the two actions have the same stationary points, they are classically equivalent. However, the σ-model action (35.46) has the advantage of being considerably simpler: it is quadratic in the string coordinates, thus describing a free world-sheet bosonic field in curved two-dimensional spacetime.

In the conformal gauge, characterized by the world-sheet metric given by $g_{ab}(\sigma, \tau) = \rho(\sigma, \tau)\eta_{ab}$, the action (35.46) reduces to the free action of d bosonic fields in two dimensions:

$$S = -\frac{1}{4\pi\alpha'}\int d\tau \int d\sigma\, \eta^{ab}\partial_a X \cdot \partial_b X, \tag{35.49}$$

supplemented by Eq. (35.47), which now plays the role of a constraint.

Turning to the case of the fermionic string, no equivalent of the σ-model action was initially available. The quadratic Lagrangian, proposed by Gervais and Sakita in 1971 contained a scalar field X^μ, and a Majorana fermion ψ^μ, as follows:

$$L = -\frac{1}{4\pi\alpha'}\left[\partial_a X \cdot \partial^a X - i\bar\psi\gamma^a \cdot \partial_a\psi\right]. \tag{35.50}$$

Note that the bosonic part coincides with the expression, given in Eq. (35.49), of the σ-model action after fixing the conformal gauge.

As discussed in Part V, the previous Lagrangian is invariant under supersymmetry and conformal transformations in the two-dimensional world-sheet of the string. As a consequence of this symmetry there is a fermionic current, the supercurrent J_a, which is also conserved. By analogy with the case of the bosonic string, it was assumed that the unknown Lagrangian for the RNS model reduced to the quadratic expression (35.50) in a suitable gauge and that the currents corresponding to the symmetries gave rise to the two constraints:

$$\theta_{ab} = \partial_a X \cdot \partial_b X - \frac{1}{2}\eta_{ab}\eta^{cd}\partial_c X \cdot \partial_d X$$
$$- \frac{i}{2}\bar\psi \cdot \left[\gamma_a\partial_b + \gamma_b\partial_a - \eta_{ab}\gamma^c\partial_c\right]\psi = 0, \tag{35.51}$$
$$J_a = \gamma^b\gamma_a\partial_b X \cdot \psi = 0,$$

which generalize the bosonic string constraint (35.47).

The problem of building the supersymmetric extension of σ-model action (35.46) was that of writing a Lagrangian for free bosonic and fermionic fields in a two-dimensional curved background that is locally supersymmetric. As discussed above, after the construction of the supergravity action in 1976 it became clear how to build such an action for the

RNS model: the following expression was obtained,

$$S \propto \int d^2\xi \sqrt{-g} \left[-\frac{1}{2} \partial_a X \cdot \partial_b X \, g^{ab} - \frac{i}{2} \bar{\psi} \gamma^a \cdot \partial_a \psi \right.$$
$$\left. + \frac{i}{2} \bar{\chi}_a \gamma^b \partial_b X \cdot \gamma^a \psi + \frac{1}{8} (\bar{\chi}_a \gamma^b \gamma^a \psi) \cdot (\bar{\chi}_b \psi) \right], \quad (35.52)$$

where g_{ab} is the metric and χ_a its supersymmetric partner, the gravitino field. The construction of this action is described in the Chapter by Di Vecchia in this Part. Here, we write the local supersymmetry transformations that leave invariant the action:

$$\delta e_\alpha^a = i \bar{\epsilon} \rho^a \chi_\alpha, \qquad \delta \chi_\alpha = 2 \nabla_\alpha \epsilon,$$
$$\delta X^\mu = i \bar{\epsilon} \psi^\mu, \qquad \delta \psi^\mu = \rho^\alpha \left(\partial_\alpha X^\mu - \frac{i}{2} \bar{\chi} \psi^\mu \right) \epsilon, \quad (35.53)$$

where $\epsilon(\tau, \sigma)$ is an arbitrary Majorana fermion in two dimensions.

It is interesting to see how the RNS model and the four-dimensional supersymmetric field theories were developed in the Seventies, strongly influencing each other. In fact, the study of the RNS model led to supersymmetric theories in four dimensions; in turn, the construction of supergravity theories was essential for writing the local supersymmetric generalization of the string action that was underlying the RNS model.

35.10 Supergravity

Einstein's theory of general relativity is formulated in terms of the spacetime metric tensor $g_{\mu\nu}(x)$. In order to describe fermionic fields in curved space, it is necessary to introduce a more general and natural formalism, based on the vierbein $e_\mu{}^a$ and the spin connection $\omega_\mu{}^{ab}$. The relation between the vierbein and the metric is:

$$g_{\mu\nu} = e_\mu{}^a e_\nu{}^b \eta_{ab}, \qquad g^{\mu\nu} = e^\mu{}_a e^\nu{}_b \eta^{ab}, \quad (35.54)$$

where $e^\mu{}_a$ is the inverse of $e_\mu{}^a$. The vierbein $e_\mu{}^a$ has two kinds of indices: a curved one μ and a flat one a. The curved one transforms as a covariant vector under arbitrary changes of coordinates of the curved spacetime, while the flat one transforms as a vector under local Lorentz rotations. In fact, although the spacetime is curved, Einstein's principle of equivalence implies that at each point of the spacetime we can identify a local inertial frame with flat coordinates, where Lorentz invariance holds and acts on the index a. Therefore, in this formulation of general relativity one has curved indices $\mu, \nu \ldots$ and flat ones $a, b \ldots$. The spinors and the Dirac matrices γ^a can then be defined in the local inertial frame and transformed into curved covariant vectors by means of the vierbein: $\gamma_\mu = e_\mu{}^a \gamma^b \eta_{ab}$, where η_{ab} is the flat metric.

From the geometrical point of view, the spin connection $\omega_\mu{}^{ab}$ can be regarded as the non-Abelian gauge field of the Lorentz group. One can then compute the associated curvature tensor:

$$R_{\mu\nu}^{ab} = \partial_\mu \omega_\nu^{ab} - \partial_\nu \omega_\mu^{ab} + \omega_\mu^{ac} \omega_{\nu c}^{\ b} - \omega_\nu^{ac} \omega_{\mu c}^{\ b}. \tag{35.55}$$

It has a structure similar to that of the non-Abelian field strength $F_{\mu\nu}^a$ in (35.7). The curvature scalar is obtained by contracting the indices:

$$R(e, \omega) = e_a^\mu e_b^\nu R_{\mu\nu}^{ab}, \tag{35.56}$$

and in terms of this one can write the Einstein action,

$$S = -\frac{1}{2\kappa^2} \int d^4x \, e \, R(e, \omega), \tag{35.57}$$

where $e \equiv \det(e_\mu^a)$ is the determinant of the vierbein. The previous action is invariant under arbitrary local transformations of the coordinates x^μ and arbitrary local Lorentz rotations acting on the flat indices a, b. However, its form is different from that of the Yang–Mills action discussed in Section 35.2, thus leading to different dynamics.

When the various four-dimensional supersymmetric theories were formulated in the Seventies, an obvious question was whether it was possible to construct a supersymmetric extension of the Einstein action in Eq. (35.57) that is also invariant under local supersymmetry transformations.

The answer to this question was affirmative and such an action was constructed in 1976, in two seminal papers by Ferrara, Freedman and Van Nieuwenhuizen, and by Deser and Zumino. They introduced a spin $\frac{3}{2}$ particle, called the gravitino, described by a spinor field ψ_μ and added to the action (35.57) the kinetic term for the gravitino, as follows:

$$S = -\frac{1}{2\kappa^2} \int d^4x \, e \left[R(e, \omega) + \frac{1}{2} \bar{\psi}_\mu \Gamma^{\mu\nu\rho} D_\nu(\omega) \psi_\rho \right]. \tag{35.58}$$

In this action, the covariant derivative contains the spin connection $D_\mu(\omega) = \partial_\mu - \frac{1}{4} \omega_\mu^{ab} \Gamma_{ab}$ and $\frac{1}{2} \Gamma_{ab} = \frac{1}{4}[\Gamma_a, \Gamma_b]$. As already mentioned, the Γ_μ matrices with a curved index can be obtained from those with a flat index by means of the vierbein, $\Gamma_\mu = e_\mu^a \Gamma_a$, and $\Gamma^{\mu\nu\rho}$ is the totally antisymmetric product of three Γ-matrices.

The action in Eq. (35.58) is invariant under the following local transformations:

$$\delta e_\mu^a = \frac{1}{2} \bar{\epsilon} \Gamma^a \psi_\mu, \qquad \delta \psi_\mu = D_\mu(\omega) \epsilon, \tag{35.59}$$

where $\epsilon(x)$ is an arbitrary Majorana spinor. The gravitino is the analogue of the gauge field for local supersymmetry.

As described in the previous Section, the construction of the supergravity action in Eq. (35.58) and of the local supersymmetry transformations in Eq. (35.59) opened the way to the extension of the Nambu–Goto action to the fermionic string, leading to the construction of the action (35.52).

References

[C04] Carroll, S. M. (2004). *Space-Time and Geometry. An Introduction to General Relativity* (Benjamin Cummings, San Francisco, CA).

[GSW87] Green, M. B., Schwarz, J. H. and Witten, E. (1987). *Superstring Theory* (Cambridge University Press, Cambridge).

[HBRD97] Hoddeson, L., Brown, L. M., Riordan, M. and Dresden, M. (1997). *The Rise of the Standard Model: Particle Physics in the 1960s and 1970s* (Cambridge University Press, Cambridge).

[W96] Weinberg, S. (1996). *The Quantum Theory of Fields* (Cambridge University Press, Cambridge).

[WB92] Wess, J. and Bagger, J. (1992) *Supersymmetry and Supergravity* (Princeton University Press, Princeton, NJ).

36
Supersymmetry in string theory

FERDINANDO GLIOZZI

Abstract

I describe the early developments from the formulation of the theory of the relativistic string to the construction of the first consistent superstring theory, which I witnessed from a very short distance.

36.1 The relativistic string

The story begins in October 1972, when in a CERN preprint Goddard, Goldstone, Rebbi and Thorn (GGRT) formulated a complete theory of the quantized free relativistic string [GGRT73] describing the physical states of the Dual Resonance Model.

At that time this model of strong interactions, arisen from the Veneziano amplitude proposed in 1968, was already completely developed. Fubini and Veneziano [FGV69, FV69, FV70] and Bardakci and Mandelstam [BM69] had shown that the single-particle states of the Dual Resonance Model (DRM) could be described consistently by an infinite collection of harmonic oscillators. Nambu, Nielsen and Susskind had formulated independently the conjecture that the underlying microscopic structure of these physical states was a vibrating string. Nambu was apparently the first one to use this term in such a context, writing: 'This equation suggests that the internal energy of a meson is analogous to that of a quantized string of finite length' [Nam70a]. Susskind used a funny paraphrase: 'a continuum limit of a chain of springs' [Sus69], and in a note added in proof, where the works on the factorization of the N-point amplitude by Fubini and Veneziano and by Bardakci and Mandelstam were explicitly mentioned, he used the term 'rubber band'. Nielsen instead emphasized in [Nie70] an analogue electrostatic model defined on a disc, with sources on the boundary associated with the N-particle amplitude, and wrote explicitly that 'hadrons are conceived of as one dimensional structures'.

Subsequently, in 1970, Nambu [Nam70b] and Goto [Got71] proposed the action:

$$S = \frac{A}{2\pi\alpha'}, \tag{36.1}$$

The Birth of String Theory, ed. Andrea Cappelli, Elena Castellani, Filippo Colomo and Paolo Di Vecchia.
Published by Cambridge University Press. © Cambridge University Press 2012.

which is proportional to the area A swept by the string in the external target space as a function of the string coordinates $x_\mu(\tau;\sigma)$, $\mu = 1,\ldots,d$. As in the DRM, the spectrum of the string states was organized in linear Regge trajectories characterized by the intercept α_o and the universal slope α'

$$J = \alpha_o + \alpha' M^2, \qquad (36.2)$$

emerging when plotting masses squared of the string states versus the angular momentum. The slope α' was the only dimensional parameter of the theory. It was already observed that the Virasoro conditions on the physical states of the DRM were directly related to intrinsic geometrical properties of the string world-sheet (Chang and Mansouri [CM72] and Mansouri and Nambu [MN72]). However, only with the GGRT paper were all the consequences of this Nambu–Goto action correctly derived and it became completely clear, even at the quantum level, that the relativistic string was not simply an analogue model used to help intuition, but that it described the underlying microscopic structure of the DRM.

The crucial remark in the string formulation of the DRM is that the string action, i.e. the area of the surface swept out by the string in its motion, does not depend on its parameterization. The physical properties of the relativistic string should depend only on the intrinsic geometrical properties of the world-sheet and not on the choice of the parameters σ and τ used to describe it. The first consequence of this reparameterization invariance is the identical vanishing of the two-dimensional energy-momentum tensor $T_{\pm\pm} \equiv 0$ which at the quantum level, using the orthonormal (or conformal) gauge $\frac{\partial x}{\partial \tau} \cdot \frac{\partial x}{\partial \sigma} = (\frac{\partial x}{\partial \tau})^2 + (\frac{\partial x}{\partial \sigma})^2 = 0$, generated the full set of Virasoro conditions on the physical states, first written down by Del Giudice and Di Vecchia [DD70]:

$$L_0|\text{Phys}\rangle = |\text{Phys}\rangle, \qquad L_n|\text{Phys}\rangle = 0, \qquad n > 0. \qquad (36.3)$$

The Virasoro generators can be written as:

$$L_n = \frac{1}{2i\pi} \oint dz\, z^{n+1} T_{++}, \qquad z = e^{-i\tau}; \qquad (36.4)$$

they generate the infinite conformal algebra

$$[L_n, L_m] = (n-m) L_{n+m} + \frac{c}{24} n(n^2 - 1)\delta_{n+m,0}, \qquad (36.5)$$

where, in the bosonic string, the central charge c coincides with the dimension d of the spacetime. The Virasoro constraints had already played an essential role in the elimination of negative norm states from the spectrum of physical states in the DRM. The choice of the conformal parameterization implied the d'Alembert equation of motion for the string coordinates,

$$\frac{\partial^2 x_\mu}{\partial \tau^2} - \frac{\partial^2 x_\mu}{\partial \sigma^2} = 0, \qquad (36.6)$$

and hence its quantization in terms of free harmonic oscillators describing the normal modes of vibration of the string. In the case of open strings, i.e. strings with free end-points, an infinite family of annihilation–creation operators $a_n^\mu, a_n^{\dagger\mu}$ ($n = 1, 2, \ldots$) suffices to describe the excitation spectrum of the string, while in the case of closed strings two such sets are necessary, to describe both left and right movers. Thanks to the Virasoro constraints, the spectrum of the open string coincided exactly with the physical states of the generalized Veneziano model, once called Reggeons, while the closed string states reproduced the Shapiro–Virasoro model.

Classically, the only dynamically independent degrees of freedom are the transverse coordinates $x_i(\tau; \sigma)$, $i = 1, 2, \ldots, d - 2$. Quantizing only these spoils the Lorentz invariance of the theory unless the spacetime dimension and the intercept of the Regge trajectory coincide with the values pointed out by Claude Lovelace [Lov71] in his study of loop corrections of the DRM, namely, $d = 26$ and $\alpha_0 = 1$. This is the way the critical dimension emerged in the string formulation of the DRM spectrum [GGRT73].

A few months later Brink and Nielsen [BN73] proposed another very intuitive way to understand the critical dimension in string theory by observing that the intercept of the Regge trajectory is directly related to the zero-point (or Casimir) energy E_0 of the string. Each normal mode of frequency n (in suitable units) contributes with an $n/2$ term, Therefore for an open string one has:

$$\alpha_0 = -E_0 = -\frac{d-2}{2} \sum_{n=1}^{\infty} n. \tag{36.7}$$

This quantity is obviously infinite and needs regularization. They introduced a cutoff on the frequencies that they eliminated through a suitable renormalization of the speed of light. I proposed later, in an unpublished work [Gli76], two simple methods to regularize sums of this kind. One of them seems obvious nowadays and is known as ζ-function regularization:

$$\alpha_0 = -\lim_{s \to -1} \frac{d-2}{2} \sum_{n=1}^{\infty} n^{-s} = -\frac{d-2}{2} \zeta(-1) = \frac{d-2}{24}, \tag{36.8}$$

where $\zeta(s)$ denotes the Riemann ζ-function. Combining this expression with the fact that the first excited state of the open string spectrum has only transverse components $a_1^{\dagger i}|0\rangle$ ($i = 1, \ldots, d - 2$) and hence it is massless, yields again $\alpha_0 = 1$ and $d = 26$. The same reasoning works in the Ramond–Neveu–Schwarz (RNS) model as well, giving in that case the critical dimension $d = 10$. The closed string intercept is twice that of the open string because there are two infinite families of harmonic oscillators.

There is also an elementary argument relating the slopes of the open and closed strings (Ademollo et al. [ADDN74a]): the leading Regge trajectory can be realized by a very simple set of open string motions as rigid rotations of straight strings of length $2r$, with the end-points moving at the speed of light. Both the mass squared M^2 and the angular momentum J turn out to be proportional to r^2 so that, at the classical level, $J = \alpha' M^2$, where the slope α' is related to the string tension T_o by $\alpha' = 1/2\pi c T_o$. The analogous motion of a closed string is made with two superposed open straight strings, hence the total string tension is twice that of the open strings, so the Regge trajectory of the closed string reads

$$J = 2\alpha_0 + \frac{\alpha'}{2}M^2. \qquad (36.9)$$

These are exactly (Olive and Scherk [OS73]) the values of the slope and the intercept of the additional poles emerging from the nonplanar loop contributions at the critical dimension, once called the Pomeron sector rather than closed string sector, since it necessarily carried vacuum quantum numbers.

The formulation of a geometric action for the RNS model (the spinning string) which gave the full set of constraints on the physical states was introduced only four years later (by Brink, Di Vecchia and Howe [BDH76] and by Deser and Zumino [DZ76b]), after the discovery of supersymmetry.

36.2 The Ademollo et al. collaboration

Shortly after the appearance of the GGRT paper (30 October 1972) a group of former students and young collaborators of Sergio Fubini in Florence, Naples, Rome and Turin decided to join their efforts to understand the Dual Resonance Model in the light of this new mechanical model.

There was no recognized leader inside the group and the ideas circulated freely (by ordinary mail and/or extemporaneous meetings) without any care for questions of priority (May 1968 was not too far!)

The initial collaboration was composed of Marco Ademollo from Florence, Alessandro D'Adda, Riccardo D'Auria, Ernesto Napolitano and Stefano Sciuto from Turin, Renato Musto and Francesco Nicodemi from Naples, Paolo Di Vecchia and myself who were at that time both fellows at CERN. Roberto Pettorino and Emilio Del Giudice, both from Naples, joined us later. We also collaborated with Sergio Ferrara, Lars Brink and John Schwarz. We wrote six papers in four years [ADDN74a]–[ABDD76c]. Ideally, we continued the line of thought of the Fubini–Veneziano collaboration which was concluded that year, combining it with the new physical insight coming from the string picture.

The first paper was directly inspired by the GGRT work. The purpose was clearly indicated in the introduction: 'If the string is not only an analogue model of the DRM, one should be able to understand the interaction among the hadrons starting from the string picture.' We began to study the motion of an open string in an external electromagnetic field interacting with pointlike charges of strength g placed at the free ends $x_\mu(\tau; \sigma = 0, \pi)$. We observed that, in the case of a plane wave $A_\mu(x) = \epsilon_\mu e^{ik \cdot x}$, the minimal coupling $A_\mu j^\mu$ to the current $j_\mu = g \partial_\tau x(\tau; 0)$ associated with the end-points coincided in form with the emission vertex $V(\epsilon, k)$ of a 'strong' photon in the DRM. Reparameterization invariance of the string world-sheet required $k^2 = \epsilon \cdot k = 0$, i.e. the external field had to be the massless photon state of the open string. Under these conditions it turned out that the interacting open string had the same mass spectrum as the free case. The probability amplitude for the emission of a number of photons from an initial string state to a final one coincided exactly

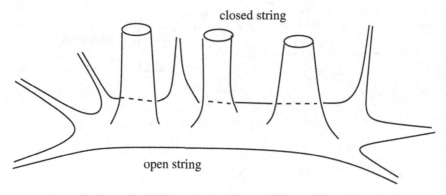

Figure 36.1 A mixed amplitude describing scattering of open and closed strings.

with the corresponding N-point DRM amplitude. Alternative approaches to the interacting string based on functional integration were proposed by Gervais and Sakita [GS73] and Mandelstam [Man73].

The argument was also extended to excited external fields. Again the reparameterization invariance implied the conformal invariance of the excited vertex. Denoting with L_f the generator of the infinitesimal conformal transformation $\delta\tau = \epsilon f(\tau)$, we obtained:

$$i[L_f, V] = \frac{d}{d\tau} \{f(\tau) V\}. \tag{36.10}$$

This established in turn a one-to-one correspondence between the excited vertices and the open string spectrum at $d = 26$ (it was explicitly verified up to the level $N = 2$). We also established a precise dictionary between the Fubini–Veneziano operator formalism and the corresponding string variables.

This game was extended also to the case of external gravitons by coupling the Nambu–Goto action to a target-space metric $g_{\mu\nu}$. Reparameterization invariance yielded the right vertex operator of the 'strong' graviton, i.e. the massless closed string state; as in the case of the photon, the external gravitational field was required to be on shell. This appears to be a precursor of the equations of motion obtained much later by Lovelace [Lov84] by requiring the vanishing of the β-function in the formulation of the string action in a curved space.

A simple but important observation made in the conclusions of our first paper is that the external gravitational field does not distinguish between open and closed strings: 'The problem is whether the gravitational interaction may apply to an open string as well as to a closed one. In principle there is no difficulty in solving this problem and it would be interesting [...] to have some information on the couplings between an open string and the strong graviton, which is a particular state of a closed string.' This anticipated the second paper [ADDN74b], where we constructed a 'Unified model for interacting open and closed strings' and gave a general recipe for the tree amplitudes containing both open and closed string states, known today as disc amplitudes (see Figure 36.1). These mixed

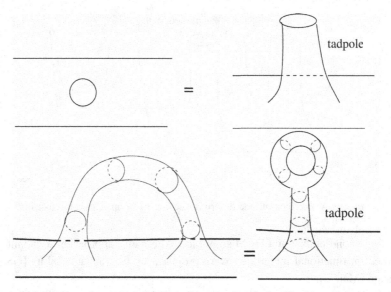

Figure 36.2 Dual transformation of a planar loop (top) and a handle (bottom) into tadpoles.

amplitudes could also be factorized [ADDN74b] in a closed string channel (using a method that nowadays is called boundary state formalism), leading to the proof of the complete equivalence, even from the point of view of the interaction, among the Pomeron sector, the closed string and the Shapiro–Virasoro model, see D'Adda et al. [DDND77].[1]

Our formulation of mixed amplitudes was the main ingredient of the third paper of the collaboration on a general way of resumming a class of divergent contributions due to unitarity corrections of string tree amplitudes [ADDD75]. These terms can be expressed as a sum over topologically inequivalent surfaces describing the string world-sheet. The topology of a (orientable) surface is determined by the number of holes and handles. The planar holes and the handles can be viewed in a dual channel as closed strings going into the vacuum (see Figure 36.2). They are called tadpole diagrams. Their contribution to string amplitudes has a divergent piece due to two different sources. One can be attributed to the tachyon and was already isolated some years before (Neveu and Scherk [NS70] and Frye and Susskind [FS70]).

The other divergent part can be thought of as the contribution of the on-shell *soft dilaton* decaying into the vacuum. Thanks to our unified open–closed string formulation we were able to rewrite it as the infrared limit of the on-shell covariant dilaton vertex (this was verified directly in the case of the planar loop, using the factorized form of the Pomeron, see Cremmer and Scherk [CS72] and Clavelli and Shapiro [CS73]).

As expected, it turned out to be proportional to the trace of the energy-momentum tensor of the string, hence to the string action. As a consequence, we were also able to sum over

---

[1] For some mysterious reason, this paper, written and accepted for publication in the spring 1974, was only published in 1977.

all the possible soft dilaton insertions. The net effect was simply a renormalization of the slope α' and of the open and closed string couplings.

In the mid-Seventies a new successful and convincing theory of strong interactions, the QCD, was constructed, while the wave of popularity of string theory was coming to an end. The principal drawback of the string as a theory of hadrons was the unrealistic spacetime dimension, $d = 26$ for the bosonic string or $d = 10$ for the RNS (or spinning) string.

The last three papers by the Ademollo et al. collaboration [ABDD76a]–[ABDD76c] can be seen as a systematic attempt to construct 'more realistic' string models, with physical spacetime dimension and a value of the Regge intercept in agreement with the hadron spectrum. These works were directly inspired by the latest developments of supersymmetry which in those years attracted a sudden surge of interest.

The birth of supersymmetry is strictly related to the history of the string: in 1971 Gervais and Sakita [GS71] pointed out that the RNS algebra (the generalization of Virasoro algebra in the RNS model) can be thought of as infinitesimal generators of new field transformations among fermions and bosons (i.e. supersymmetric transformations) in a two-dimensional field theory. Two years later, Wess and Zumino [WZ74] successfully introduced supersymmetry transformations in the context of four-dimensional field theory. A few months later Salam and Strathdee developed the elegant formalism of superfields and introduced the notion of extended supersymmetry, i.e. enlarged algebras with more than one supercharge. Most of us were fascinated by these new developments and we sought to construct and apply extended supersymmetries back in string theory by enlarging the superconformal RNS algebra. As we noted in [ABDD76b], 'the mass of the ground state, the critical dimension as well as other physical properties of the string system crucially depend on the underlying gauge algebra. As a consequence, a very natural way to provide new examples of string models with possible better properties than the previous ones, is to exhibit new examples of gauge algebras and to construct strings which are invariant under these algebras.' In other terms, it was natural to expect more interesting string theories to emerge when one introduced more structure on the world-sheet.

We constructed a whole set of extensions of the superconformal algebra, but only two of them could be realized in terms of free fields and therefore could be used to build up novel string theories. One was the extended $\mathcal{N} = 2$ superconformal algebra [ABDD76b], which also involved a Kac–Moody $U(1)$ current as a new unexpected ingredient. The other was the $\mathcal{N} = 4$ superconformal algebra [ABDD76c], which contained an $SU(2)$ Kac–Moody algebra (these extended superconformal algebras were to have a new, more glorious life after the First String Revolution of 1984). We felt bitterly disappointed when we realized that the critical dimension was $d = 2$ for the $\mathcal{N} = 2$ string and even negative for the $\mathcal{N} = 4$ string. This failure was one of the reasons, I think, for the dissolution of our collaboration.

Many years later, when string theory had again become a very active field of research as the Theory of Everything, the $\mathcal{N} = 2$ string was reconsidered and it was realized (D'Adda and Lizzi [DL87], Mathur and Mukhi [MM88], Ooguri and Vafa [OV90, OV91]) that the critical dimension is in fact two complex dimensions, i.e. $d = 4$ with two spatial and two temporal coordinates. It was subsequently argued (Siegel [Sie92]) that the $\mathcal{N} = 4$ string is the same as the $\mathcal{N} = 2$ string because of the reducibility of the constraints; more specifically, if one first solves the $\mathcal{N} = 2$ subset of the $\mathcal{N} = 4$ constraints, the remaining constraints turn out to be redundant.

36.3 Towards superstring

I spent the 1975–1976 academic year at the École Normale Supérieure in Paris, in the 'Laboratoire de Physique Théorique', at that time directed by Philippe Meyer. There I found a very stimulating environment. Still vivid in my memory are the lively, open discussions in some *café* of the *quartier latin* with the *normaliens*. Among them I remember Claude Bouchiat, Eugène Cremmer, Pierre Fayet, Jean-Loup Gervais, Jean Iliopoulos, André Neveu, Joël Scherk and Nicolas Sourlas.

I occupied one of the three offices inside the library. Next to my office Daniel Freedman, on leave from Stony Brook, was working on the properties of the Rarita–Schwinger field. At the beginning of the new year the third office was occupied by Sergio Ferrara. They started to collaborate with Peter van Nieuwenhuizen, just arrived from Stony Brook, to construct (March 1976) the $\mathcal{N} = 1$ supergravity [FNF76]. It was based on the requirement that the Rarita–Schwinger field, carrying spin $\frac{3}{2}$, now known as the gravitino, could be coupled to gravity with a local supersymmetric interaction. Soon afterwards the same theory was formulated by Deser and Zumino at CERN [DZ76a].

At that time, under the influence of these exciting results, I began to discuss with Joël Scherk the possibility of having a similar structure in the RNS string. Soon we realized that in the closed string sector, besides the bosonic states already studied by Clavelli and Shapiro [CS73], there was also a fermionic sector (with Neveu–Schwarz left movers and Ramond right movers) with a massless gravitino; moreover in the $\alpha' \to 0$ limit the massless sector did define a $d = 10$ version of supergravity. Compactifying pure supergravity in ten dimensions yields supergravity coupled to matter in four. We used this kind of string theory consideration as a sort of secret tool to extract information on the matter couplings in $d = 4$ supergravity. As a result we took part in the early developments of that theory (Ferrara, Scherk and van Nieuwenhuizen [FSN76], Ferrara, Gliozzi, Scherk and van Nieuwenhuizen [FGSN76], Ferrara *et al.* [FFNB77]).

The presence of a gravitino in the string spectrum is very constraining, as it can be consistently coupled only to supersymmetric matter. As a consequence, in the RNS model each physical state of the Neveu–Schwarz sector should have a supersymmetric partner degenerate in mass in the Ramond sector. A challenging aspect of the RNS model that puzzled us for a while was that neither the tachyon nor the whole NS sector having $\alpha' M^2 = n - \frac{1}{2}$, known as the odd G-parity sector, had a supersymmetric partner. However we discovered that this sector transformed a right-handed fermion into a left-handed fermion, therefore it decoupled altogether if the right-handed fermions were projected out using Weyl spinors. Moreover the fermion–fermion and the fermion–antifermion states had the same spectra as bosonic bound states. In order to avoid infinite degeneracy of the bosonic spectrum we were led to require that the fermions satisfy also the Majorana condition. The resulting projected model, as tachyons had been removed, was the first example of a totally consistent string theory. Only later, thanks to the contribution of David Olive, we realized that the requirement of the Majorana–Weyl condition is very constraining and is possible only if d is 2 modulo 8. The first nontrivial dimension is thus $d = 10$, which is a nice way to recover the critical dimension of the RNS model.

I still remember the morning (probably June 1976) in which Joël and I began to enumerate on the blackboard the physical states left after the above projection. The quantization in masses was given by $\alpha' M^2 = n$ with $n = 0, 1, 2, \ldots$. We verified that in the first few levels the numbers of physical states of the Bose and Fermi sectors were equal; it was therefore natural to conjecture this equality for all the levels. At lunch I discussed our results with Eugène Cremmer (Joël was momentarily busy). The discussion was very illuminating for me: going back to my office I knew I had to look for some identity involving Jacobi theta functions. I took from the shelf of mathematical books near my door a copy of 'Whittaker and Watson' [WW27]. It opened magically on page 470, where the following exercise is proposed:

'Shew that $\dfrac{1}{2}\left[\displaystyle\prod_{n=1}^{\infty}(1+q^{2n-1})^8 - \prod_{n=1}^{\infty}(1-q^{2n-1})^8\right] = 8q\prod_{n=1}^{\infty}(1+q^{2n})^8$.'

It was exactly the sought after formula! A footnote let the reader know that 'Jacobi describes this result as "aequatio identica satis abstrusa".' For us its meaning was less abstruse as it expressed the equality, level by level, of the number of bosons and fermions in the open string spectrum, an important requirement for supersymmetry of the projected RNS model. The left-hand side is the relevant part of the generating function of the NS physical states after removing the odd G-parity sector, while the right-hand side is the analogous function for the Ramond states, after projection on the Majorana–Weyl spinors; the factor 8 comes from the degeneracy of the ground state fermion.

The following day we went to the central library of ENS to consult the original book by Jacobi [Jac29]. Imagine the surprise of Joël, who was then fascinated by the idea of metempsychosis, when we discovered that in the introduction Jacobi acknowledged an assistant named Scherk[2]!

After a rapid visit of Joël at CERN, David Olive joined our collaboration, providing general theorems on Clifford algebras in any dimension of spacetime, showing the special role of ten dimensions, where it is possible to define Majorana–Weyl spinors. In this way, the so-called GSO projection acquired a sound mathematical basis.

Together the three of us wrote two papers devoted to the GSO projected RNS model [GSO76, GSO77]. The better known paper is the latter, where, besides the construction of the projected model, we wrote through dimensional reduction of the $\alpha' \to 0$ limit the $\mathcal{N} = 4$ super Yang–Mills theory in four spacetime dimensions. The first, less well known,

---

[2] 'Ut a typographorum mendis, quantum fieri potuit, mundus aderet liber, Cl. Scherk curare voluit, cui ea de re valde me strictum esse profiteor.'

paper outlined instead a general programme for constructing, starting from the RNS string theory, a consistent, possibly finite, short-distance modified theory of fundamental forces. The last sentence of that paper was 'So the whole scheme does not look too unreasonable, although many details are still to be worked out.' This could apply as well to the present string era.

> Ferdinando Gliozzi was born in Turin in 1940. He became researcher at INFN in 1968. He has been Professor of Theoretical Physics at the Turin University since 1987. Most relevant results in string theory are the operatorial formulation of the projective invariance of the DRM amplitudes, $\mathcal{N} = 2$ and $\mathcal{N} = 4$ superconformal algebras, and the GSO projection.

References

[ADDD75] Ademollo, M., D'Adda, A., D'Auria, R., Di Vecchia, P., Gliozzi, F., Napolitano, E. and Sciuto, S. (1975). Soft dilatons and scale renormalization in dual theories, *Nucl. Phys.* **B94**, 221–259.

[ADDN74a] Ademollo, M., D'Adda, A., D'Auria, R., Napolitano, E., Sciuto, S., Di Vecchia, P., Gliozzi, F., Musto, R. and Nicodemi, F. (1974). Theory of an interacting string and dual-resonance model, *Nuovo Cimento* **A21**, 77–145.

[ADDN74b] Ademollo, M., D'Adda, A., D'Auria, R., Napolitano, E., Di Vecchia, P., Gliozzi, F. and Sciuto, S. (1974). Unified dual model for interacting open and closed strings, *Nucl. Phys.* **B77**, 189–225.

[ABDD76a] Ademollo, M., Brink, L., D'Adda, A., D'Auria, R., Napolitano, E., Sciuto, S., Del Giudice, E., Di Vecchia, P., Ferrara, S., Gliozzi, F., Musto, R. and Pettorino, R. (1976). Supersymmetric strings and color confinement, *Phys. Lett.* **B62**, 105–110.

[ABDD76b] Ademollo, M., Brink, L., D'Adda, A., D'Auria, R., Napolitano, E., Sciuto, S., Del Giudice, E., Di Vecchia, P., Ferrara, S., Gliozzi, F., Musto, R., Pettorino, R. and Schwarz, J.H. (1976). Dual string with $u(1)$ color symmetry, *Nucl. Phys.* **B111**, 77–110.

[ABDD76c] Ademollo, M., Brink, L., D'Adda, A., D'Auria, R., Napolitano, E., Sciuto, S., Del Giudice, E., Di Vecchia, P., Ferrara, S., Gliozzi, F., Musto, R. and Pettorino, R. (1976). Dual string models with nonabelian color and flavor symmetries, *Nucl. Phys.* **B114**, 297–316.

[BM69] Bardakci, K. and Mandelstam, S. (1969). Analytic solution of the linear-trajectory bootstrap, *Phys. Rev.* **184**, 1640–1644.

[BDH76] Brink, L., Di Vecchia, P. and Howe, P. S. (1976). A locally supersymmetric and reparameterization invariant action for the spinning string, *Phys. Lett.* **B65**, 471–475.

[BN73] Brink, L. and Nielsen, H. B. (1973). A simple physical interpretation of the critical dimension of spacetime in dual models, *Phys. Lett.* **B45**, 332–336.

[CM72] Chang, L. N. and Mansouri, F. (1972). Dynamics underlying duality and gauge invariance in the dual resonance models, *Phys. Rev.* **D5**, 2535–2542.

[CS73] Clavelli, L. and Shapiro, J. A. (1973). Pomeron factorization in general dual models, *Nucl. Phys.* **B57**, 490–535.

[CS72] Cremmer, E. and Scherk, J. (1972). Factorization of the pomeron sector and currents in the dual resonance model, *Nucl. Phys.* **B50**, 222–252.

[DDND77] D'Adda, A., D'Auria, R., Napolitano, E., Di Vecchia, P., Gliozzi, F. and Sciuto, S. (1977). Equivalence between the unified dual model and pomeron-reggeon amplitudes, *Phys. Lett.* **B68**, 81–87.

[DL87] D'Adda, A. and Lizzi, F. (1987). Space dimensions from supersymmetry for the $n = 2$ spinning string: a four-dimensional model, *Phys. Lett.* **B191**, 85–90.

[DD70] Del Giudice, E. and Di Vecchia, P. (1970). Characterization of the physical states in dual-resonance models, *Nuovo Cimento* **A70**, 579–591.

[DZ76a] Deser, S. and Zumino, B. (1976). Consistent supergravity, *Phys. Lett.* **B62**, 335–337.

[DZ76b] Deser, S. and Zumino, B. (1976). A complete action for the spinning string, *Phys. Lett.* **B65**, 369–373.

[FFNB77] Ferrara, S., Freedman, D. Z., van Nieuwenhuizen, P., Breitenlohner, P., Gliozzi, F. and Scherk, J. (1977). Scalar multiplet coupled to supergravity, *Phys. Rev.* **D15**, 1013–1018.

[FGSN76] Ferrara, S., Gliozzi, F., Scherk, J. and van Nieuwenhuizen, P. (1976). Matter couplings in supergravity theory, *Nucl. Phys.* **B117**, 333–355.

[FSN76] Ferrara, S., Scherk, J. and van Nieuwenhuizen, P. (1976). Locally supersymmetric Maxwell–Einstein theory, *Phys. Rev. Lett.* **37**, 1035–1037.

[FNF76] Freedman, D. Z., van Nieuwenhuizen, P. and Ferrara, S. (1976). Progress toward a theory of supergravity, *Phys. Rev.* **D13**, 3214–3218.

[FS70] Frye, G. and Susskind, L. (1970). Removal of the divergence of a planar dual-symmetric loop, *Phys. Lett.* **B31**, 537–540.

[FGV69] Fubini, S., Gordon, D. and Veneziano, G. (1969). A general treatment of factorization in dual resonance models, *Phys. Lett.* **B29**, 679–682.

[FV69] Fubini, S. and Veneziano, G. (1969). Level structure of dual resonance models, *Nuovo Cimento* **A64**, 811–840.

[FV70] Fubini, S. and Veneziano, G. (1970). Duality in operator formalism, *Nuovo Cimento* **A67**, 29–47.

[GS71] Gervais, J. L. and Sakita, B. (1971). Field theory interpretation of supergauges in dual models, *Nucl. Phys.* **B34**, 632–639.

[GS73] Gervais, J. L. and Sakita, B. (1973). Ghost-free picture of Veneziano model, *Phys. Rev. Lett.* **30**, 716–719.

[Gli76] Gliozzi, F. (1976), unpublished. See also Di Vecchia, P. (1978). *Many Degrees of Freedom in Particle Physics*, ed. Satz, H. (Plenum, New York), 493.

[GSO76] Gliozzi, F., Scherk, J. and Olive, D. (1976). Supergravity and the spinor dual model, *Phys. Lett.* **B65**, 282–286.

[GSO77] Gliozzi, F., Scherk, J. and Olive, D. (1977). Supersymmetry, supergravity theories and the dual spinor model, *Nucl. Phys.* **B122**, 253–290.

[GGRT73] Goddard, P., Goldstone, J., Rebbi, C. and Thorn, C. B. (1973). Quantum dynamics of a massless relativistic string, *Nucl. Phys.* **B56**, 109–135.

[Got71] Goto, T. (1971). Relativistic quantum mechanics of one-dimensional mechanical continuum and subsidiary condition of dual resonance model, *Prog. Theor. Phys.* **46**, 1560–1569.

[Jac29] Jacobi, C. G. J. (1829). *Fundamenta Nova*, Könisberg.

[Lov71] Lovelace, C. (1971). Pomeron form factors and dual regge cuts, *Phys. Lett.* **B34**, 500–506.

[Lov84] Lovelace, C. (1984). Strings in curved space, *Phys. Lett.* **B135**, 75–77.

[Man73] Mandelstam, S. (1973). Interacting string picture of dual resonance model, *Nucl. Phys.* **B64**, 205–235.

[MN72] Mansouri, F. and Nambu, Y. (1972). Gauge conditions in dual resonance models, *Phys. Lett.* **B39**, 375–378.

[MM88] Mathur, S. D. and Mukhi, S. (1988). The $N = 2$ fermionic string: path integral, spin structures and supermoduli on the torus, *Nucl. Phys.* **B302**, 130–148.

[Nam70a] Nambu, Y. (1970). Quark model and the factorization of the Veneziano amplitude, in *Proceedings of the International Conference on Symmetries and Quark Models, Wayne State University, June 18–20, 1969*, ed. Chand, R. (Gordon and Breach, New York), 269–277, reprinted in *Broken Symmetry, Selected Papers of Y. Nambu*, ed. Eguchi, T. and Nishijima, K. (World Scientific, Singapore, 1995), 258–277.

[Nam70b] Nambu, Y. (1970). Duality and Hadrodynamics, lecture notes prepared for Copenhagen summer school, 1970, reproduced in *Broken Symmetry, Selected Papers of Y. Nambu*, ed. Eguchi, T. and Nishijima, K. (World Scientific, Singapore, 1995), 280–301.

[NS70] Neveu, A. and Scherk, J. (1970). Parameter-free regularization of one-loop unitary dual diagram, *Phys. Rev.* **D1**, 2355–2359.

[Nie70] Nielsen, H. B. (1970). An almost physical interpretation of the integrand of the n-point Veneziano model, paper submitted to the *15th International Conference on High Energy Physics, Kiev, 26 August to 4 September, 1970* (Nordita preprint, 1969).

[OS73] Olive, D. I. and Scherk, J. (1973). No ghost theorem for the pomeron sector of the dual model, *Phys. Lett.* **B44**, 296–300.

[OV90] Ooguri, H. and Vafa, C. (1990). Selfduality and $N = 2$ string magic, *Mod. Phys. Lett.* **A5**, 1389–1398.

[OV91] Ooguri, H. and Vafa, C. (1991). Geometry of $N = 2$ strings, *Nucl. Phys.* **B361**, 469.

[Sie92] Siegel, W. (1992). $N = 4$ string is the same as the $N = 2$ string, *Phys. Rev. Lett.* **69**, 1493–1495.

[Sus69] Susskind, L. (1969). Harmonic-oscillator analogy for the Veneziano model, *Phys. Rev. Lett.* **23**, 545–547.

[WZ74] Wess, J. and Zumino, B. (1974). Supergauge transformations in four dimensions, *Nucl. Phys.* **B70**, 39–50.

[WW27] Whittaker, E. T. and Watson, G. N. (1927). *A Course in Modern Analysis* (Cambridge University Press, Cambridge).

37
Gravity from strings: personal reminiscences of early developments

TAMIAKI YONEYA

Abstract

I discuss the early developments of string theory with respect to its connection with gauge theory and general relativity from my own perspective. The period covered is mainly from 1969 to 1974, during which I became involved in research on dual string models as a graduate student. My thinking towards the recognition of string theory as an extended quantum theory of gravity is described. Some retrospective remarks on my later works related to this subject are also given.

37.1 Prologue: an encounter with the dual string model

I entered graduate school at Hokkaido University, Sapporo, in April 1969. My advisor, Akira Kanazawa, who was an expert in the dispersion theoretic approach to strong interactions, proposed to have a series of seminars on Regge pole theory. However, the Regge pole theory was somewhat disappointing for me. I felt that it was too formal and phenomenological in its nature. Looking for some more favourable topics, I began studying the quantum field theory of composite particles, which, I thought, might be useful to explain the Regge behaviour from the dynamics of fundamental particles. I read many papers related to this problem, such as those on compositeness criteria, on the definition of the asymptotic field for a composite particle, the Bethe–Salpeter equation, and so on. Although I felt that these subjects themselves were not yet what I really wanted to pursue, I enjoyed learning various different facets of quantum field theory.

While still seeking subjects for my research, some senior students told me that a spectacular new development, triggered by a proposal made about a year before by Veneziano [Ven68], was springing up. After reading the paper of Veneziano and some others which extended the Veneziano amplitude to various directions, I gradually became convinced that this had to be the subject I should choose. In particular, when I was exposed to a short but remarkable preprint by Susskind [Sus69] giving a physical interpretation of the Veneziano formula in terms of vibrating strings (or 'rubber band' in Susskind's terminology), I was

---

The Birth of String Theory, ed. Andrea Cappelli, Elena Castellani, Filippo Colomo and Paolo Di Vecchia. Published by Cambridge University Press. © Cambridge University Press 2012.

struck by the simplicity of the idea. My interest was further strengthened by reading the paper by Nambu [Nam70], containing similar discussions. I was intrigued especially by the concept of 'master field' (in today's language, string field), which had been emphasized by Nambu. A little later, I also came to be fascinated by the very attractive world-sheet picture discussed by Fairlie and Nielsen [FN70].

By the beginning of the autumn of 1969, I had started to study various available preprints, which were increasing in number very rapidly. They included works on factorization, operator formalism, and the unitarization programme. One of the papers which I was most interested in was the one by Fubini and Veneziano [FV70]. They introduced an extremely elegant formulation of general N-point amplitudes in terms of the vertex operator $:\exp i k_\mu Q^\mu(z):$ for the tachyon as the ground state of the dual string. I thought that this formalism was suggestive not only from the physical viewpoint of vibrating strings, but also was quite important in making manifest the key symmetry structure, Möbius symmetry or $SL(2, \mathbb{R})$ invariance; this underlies the basic duality property of amplitudes, as exhibited in the Koba–Nielsen representation [KN69]. On the other hand, from the viewpoint of the unitarization programme, which had been initiated in a remarkable work by Kikkawa, Sakita and Virasoro (KSV) [KSV69], it was a crucial step toward constructing amplitudes or vertices with the most general external lines corresponding to general excited states of strings. Attempts towards such 'Reggeon' vertices were started by Sciuto [Sci69]. During the years from 1969 to 1971, there appeared a large number of papers discussing this problem.

In the Japanese system of graduate school, we usually have to present a thesis in order to finish the first two years, 'master course', of graduate study before entering the next stage, 'doctor course', of three years. As the subject of my master thesis, I decided to try to use the Fubini–Veneziano formalism to establish a formulation of the Reggeon vertices where the duality symmetry and factorization property were as manifest as possible. While preparing the thesis, I became aware of other papers, such as those by Lovelace [Lov70] and Olive [Oli71], where similar ideas were being pursued. I was able to achieve such a goal by adopting a general method proposed earlier by Shapiro [Sha70b] for factorizing the generalized Veneziano amplitudes. I applied this method to a multiple factorization of the N-point amplitudes, using a representation theory of the $SL(2, \mathbb{R})$ group. The resulting general Reggeon vertex satisfied the desired symmetry properties with respect to gauge, twisting, factorization and duality transformations. It also contained the previous results due to Lovelace and Olive as special cases. This small work [Yon71] became my first full paper.

37.2 Connection of dual models to field theories

From these early days of my studies on dual models, there had been a basic question which was increasingly occupying my mind. That was on the relationship between dual models and ordinary field theory. In characterizing the Veneziano amplitude, it had been emphasized that the amplitude could be expanded into a sum over an infinite set of either

s-channel poles or t-channel poles, but not of both. Adding both would amount to double counting. The unitarization programme started by the KSV paper was to be regarded as an extension of the usual Feynman-graph expansion in ordinary field theory. In the latter, however, we have to sum both s- and t-channel diagrams. One of my questions was whether it would be possible in principle to find a field-theory-like formalism from which the KSV rule would be properly derived, perhaps by introducing some appropriate self-couplings for the master field.

Another related question was on the interpretation of massless spinning excitations. The existence of such states in dual string models is an inevitable consequence of conformal symmetry associated with the Virasoro operators. In contrast to this, the existence of massless spinning fields in the framework of ordinary field theory is related to local gauge symmetries.

During my master-course years, I had also become fascinated by the beautiful geometrical structure of local gauge principle which was reminiscent of general relativity, through reading the original papers by Yang and Mills and by Utiyama. Because my main interest had been in strong interactions, I was naturally seeking works aiming to apply the local gauge principle to strong interactions. Among others, the paper by J. J. Sakurai [Sak60] impressed me with its strong persuasive power. I thought that if the local gauge symmetry could be such a fundamental principle for strong interactions, it should also play some role in dual models, and then the existence of massless spinning states in the latter had to be a clue.

37.2.1 Local gauge principle for the master field

After submitting my master thesis in the beginning of 1971, I decided to pursue this last idea as my next theme. I attempted first to interpret the interaction vertices of the massless vector states of open strings in terms of a local gauge transformation for string wave functions. I soon realized that all those vertices in the Veneziano, Neveu–Schwarz [NS71] and Ramond [Ram71] models were obtained precisely by a minimal substitution $p^\mu(\sigma) \to p^\mu(\sigma) - gA^\mu(x(0))\delta(\sigma)$ with $\Box A^\mu = \partial_\mu A^\mu = 0$ corresponding to a particular class of local gauge transformations on the master field $|\Psi\rangle$, such as

$$|\Psi\rangle \to e^{ig\lambda(x^\mu(0))}|\Psi\rangle, \tag{37.1}$$

under an on-shell condition $\Box \lambda(x) = 0$ for the gauge function $\lambda(x^\mu)$, where $x^\mu(0)$ is the string coordinate at the end-point $\sigma = 0$ of open strings. Namely, the vertices obtained from the Virasoro or super-Virasoro operators in this way coincided with those obtained by factorizing the pole residues at corresponding massless poles of dual amplitudes. I interpreted this simple fact as clear evidence that these massless vector states themselves should be regarded as gauge bosons and that the world-sheet conformal symmetry was the mechanism which made the local gauge symmetry intrinsic to dual strings. In modern language, this formulation is equivalent to the vertex operators introduced as a deformation of the world-sheet action by background gauge field at the boundary of open string world-sheets.

When I was pursuing this idea, I became aware of some prior works which were trying to apply local gauge principles to dual models in different ways, such as those by Kikkawa and Sato [KS70] and by Manassah and Matsuda [MM71]. The standpoint of these works was remote from mine because they were aiming at constructing off-shell currents which would make possible the coupling of electromagnetic and weak interactions to dual strings. Hence, in their works, the gauge bosons coupled to the Veneziano amplitudes had nothing to do with the massless spinning states of dual strings themselves. I wrote up a paper on my results, emphasizing that the conformally invariant dual models of open strings should be regarded as embodying a local gauge principle intrinsically, rather than having it as some external structure to be adjoined artificially to the dual models. I submitted it to *Nuovo Cimento*. Unfortunately this manuscript was rejected. Perhaps the paper was not sufficiently well written to persuade the referee of my viewpoint. This rejection disappointed me because to my knowledge no one had ever formulated such a viewpoint on the meaning of the local gauge principle in string theory before.

A few months later, I received a letter commenting on my preprint from M. Minami, who was at the Research Institute of Mathematical Sciences (RIMS) in Kyoto and was actively engaged in dual models. To his letter, there was attached a copy of a preprint by Neveu and Scherk [NS72] which applied Scherk's previous work [Sch71] on the zero-slope limit to the dual pion model of Neveu and Schwarz. They also made a connection with gauge theory by identifying the massless ρ-meson in the Neveu–Schwarz model to be the gauge boson. Minami also mentioned some works by Nakanishi [Nak71] on a crossing-symmetric decomposition of the integral representation of the Veneziano-type amplitudes, where each decomposed piece corresponded to a usual Feynman graph with pole singularities in a single definite set of channels only. All these works were a big surprise to me. I decided to rewrite the rejected manuscript in a more concise form, with a note emphasizing that the gauge symmetry was manifested even for finite slope parameter α'. I sent it to *Progress in Theoretical Physics* [Yon72a].

Minami's letter encouraged me to pursue my interpretation further. I immediately hit upon the idea of extending it to the case of closed strings, namely to the Virasoro–Shapiro amplitudes [Vir69, Sha70a]. At first, however, making a connection to gravity seemed to be a rather bizarre thing to do, since at that time everyone thought that the dual string models were aiming at the construction of a definitive theory for hadrons and their strong interactions. During my undergraduate years, I had been quite serious about general relativity and had studied some of the original papers. In fact, Einstein had been a basic impetus to me, since I had started out by trying to comprehend his book *The Meaning of Relativity* during my high school days. But after entering graduate school and being influenced by the atmosphere of the particle physics community, which was dominated by the 'S-matrix approach', I became quite strongly prejudiced against general relativity, regarding it as a relic of the past. In the case of the vector gauge principle, I was able to convince myself of its relevance by reading the Sakurai paper. I hoped that the connection to gauge theory would be useful in trying to give masses to the unwanted massless states, in the way the ρ-mesons were related to the gauge principle as Sakurai had advocated. To my knowledge at that time, there had been no such attempt in the case of gravity.

37.2.2 A failed attempt toward string field theory

Then, postponing my idea of extending previous work to closed strings as possible future work, I turned my attention to Nakanishi's work. His papers on the decomposition of the integral representation for the dual amplitudes prompted me to my first question concerning the relationship between dual models and field theory. From the viewpoint of my question whether the dual amplitudes could be derived from some generalized field theory, I thought that Nakanishi's work was relevant.

I soon realized that the particular method of factorization which I had formulated in my master thesis was quite suitable for reformulating Nakanishi's rule for the decomposition of integration regions as a generalized Feynman-like rule; this would make it possible to extend the decomposition to arbitrary loop diagrams. It was known that a naive application of the operator formalism, in the case of nonplanar diagrams, led to a divergence [KT70], called the 'periodicity' problem, due to integration over the same contribution infinitely many times.

I was able to confirm that the rule I had derived on the basis of my previous Reggeon vertex paper resolved this difficulty as expected, because my new Feynman-like rule automatically gave a definite prescription for the integration region for arbitrary diagrams. Also this rule only involved a generalized three-point vertex and hence naturally reduced to that of the ϕ^3 theory in the zero-slope limit discussed by Scherk.

I was certainly excited by this result, because this strongly suggested that the rule I had derived might correspond precisely to the Feynman rule of a desired string field theory. While I was finishing the paper on this result in May 1972, I was staying for about a month at the Yukawa Institute in Kyoto, having discussions on this and related problems with Nakanishi at RIMS. As soon as I completed the paper [Yon72b] on this result there, I seriously turned to an attempt at constructing such a formalism.

Seeking some hint for a new kind of field theory underlying the dual string models, I studied some old works on infinite-component field theories and especially on nonlocal field theories, including papers by Yukawa [KY68]. Unfortunately, however, I did not obtain really useful insights from them, since most of these works had been essentially restricted to free theories. More crucially, the rule I had presented was using a very cumbersome language, and hence did not allow simple interpretation in terms of an intuitive worldsheet picture which should underlie the string field interaction.[1] What I was trying to do is equivalent to obtaining a covariant triangulation of the moduli space of Riemann surfaces with boundaries and punctures such that it corresponds to a second quantized field theory

---

[1] A few years later I came to know that Goto and his student Naka had been trying to formulate such a field theory from a different direction [GN74]. But they were not successful either.

of interacting open strings. More than a decade later, Witten achieved this goal [Wit86] in his proposal of a covariant open string field theory, on the basis of the BRST operator formalism: I shall return to this work below.

A version of fully interacting string field theories first appeared in 1974, when Kaku and Kikkawa published their work [KK74]. Their formulation was based on Mandelstam's light-cone representation [Man73] of the dual amplitudes, which had appeared about a year after my failed attempt towards string field theory. The light-cone string field theory provided another possible, albeit *noncovariant*, triangulation of the moduli space of Riemann surfaces.

37.2.3 Connection of dual models to general relativity

Because of the failure of my attempt toward string field theory, I was quite exhausted. To recover, I had to spend almost half a year from the summer of 1972 wandering about various other topics. During this time many people were departing from the dual string models. But I had some feeling that the dual models contained something more to be uncovered, despite their immediate inapplicability to the physics of hadronic interactions. I was also influenced by a rapid resurgence of gauge theories, which had been initiated by 't Hooft's famous work on the renormalizability of non-Abelian field theories.

I was now thinking seriously about my previous idea on a possible connection between dual string models in the case of closed strings and general relativity. In studying the progress in gauge theories, I was gradually freeing myself from my prejudice toward general relativity. I thought that as the Yang–Mills theory had turned out to be useful in this way, general relativity might also become important for particle physics some day in some unexpected way. A possible connection of dual strings to gravity by itself would also be very suggestive of a unification of quantum theory and general relativity, independently of the interpretation within hadronic physics.

My thinking was something like this: one of the useful viewpoints on dual string models which had been made clear by Scherk's work was that the slope parameter α' should be treated as a fundamental constant as the Planck constant h had been for quantum theory. It would be very exciting if we had some kind of concrete and essential correspondence principle between dual string models and general relativity, analogous to the relation between, say, the canonical commutation relation in quantum mechanics and the Poisson bracket in classical mechanics.

In any case, I had first to study a nontrivial example of the zero-slope limit for the Virasoro–Shapiro amplitudes. I decided to compute a four-point amplitude involving two massive scalars and two gravitons, since the coupling of graviton to scalars was certainly the simplest nontrivial case, as had been treated in most of the old literature which I had come across in this subject. All such papers of course only discussed the case of four dimensions, and I was not attempting to generalize them to the critical dimension which was not essential for my purpose: for discussing only the properties of massless modes of strings without loop corrections we were allowed to restrict ourselves to lower dimensions without encountering any real difficulty. Consequently, the Newton constant was identified as proportional to $g_s^2 \alpha'$ with g_s being the three-point coupling constant of closed strings. Staying in four dimensions allowed for an arbitrary real mass, as opposed to an imaginary mass in the critical dimension, to the scalar ground state which was treated passively as a matter field, by using the conserved components of momentum along the extra spatial directions. In the early summer of 1973 I decided to publish the result [Yon73] of this computation in letter form.

In parallel with this computation, I was trying to establish the correspondence to all orders with respect to graviton interactions in the tree approximation, or in other words to the tree amplitudes with arbitrary number (N) of gravitons and with a fixed number (2) of massive scalars. Since general relativity is a nonpolynomial field theory, I needed a certain general theorem for the purpose of establishing the connection in a recursive way on the basis of my explicit computation of the $(2+2)$-point amplitude. I found a paper which was quite suitable for this aim, by W. Wyss [Wys65], who had given a very clear argument showing how the nonpolynomial Einstein Lagrangian is obtained, starting from the lowest order coupling between a massive scalar field and a massless spin two field. For this argument, I thought the presence of the scalar matter field with finite real mass was useful.

The second and more ambitious question was, as alluded to already, to find an appropriate correspondence principle which would lead from general relativity to the dual strings in a more or less unique way. I was recalling P. Ramond's argument [Ram71] in his proposal of the dual fermion model as a generalization of the Dirac equation. Unfortunately, however, I was not able to conceive such a correspondence principle in any satisfactory form. I only managed to give a modest discussion, which was essentially a generalization of the equivalence principle, about the generating functional for the S-matrix for $(n+2)$-point scatterings: the S-matrix on the string side can be obtained from that of field theory by making a simple reinterpretation for the spacetime energy-momentum tensor of a massive scalar field.

The third question was whether there could exist an extension of the Nakanishi decomposition to the general Virasoro–Shapiro amplitudes. This could have provided a clue for some concrete mechanism explaining the correspondence between general relativity and dual string theory at the level of Feynman rules. I was not able to reach a definitive conclusion on this question either, but I was convinced that the decomposition was likely to require an infinite number of higher-point vertices as we go to the higher-point graviton amplitudes. Many years later, I presented some of those considerations in a review talk [Yon86] on string field theory in a workshop held at Komaba in 1986. In the late Eighties, essentially the same problem was discussed by several groups in attempting to extend Witten's covariant open string field theory to closed strings. See for instance [Zwi92] for a systematic discussion.

Although these failures somewhat disappointed me, I decided to write a full paper on the connection between the dual closed strings and general relativity. The title was 'Connection of dual models to electrodynamics and gravidynamics', which was sent to

Progress in Theoretical Physics in the early autumn of 1973 [Yon74a]. I included the word 'electrodynamics' because the paper contained an argument for the similar case of the open string scattering of N massless vectors and two massive scalars. That was intended to emphasize the similarity and simultaneously the contrast of the arguments between electrodynamics and gravidynamics. I just concentrated on presenting concrete results with respect to the connection of string theories to general relativity and (electromagnetic) gauge theory, without emphasizing deeper questions which I had been bothering about.

37.2.4 Problem of interpretation: 'principles behind', or 'strings and gravity from gauge theory'?

I remember that, while writing this paper, my mind was in a somewhat perplexed state. One of the reasons was that I was not completely sure about my standpoint of interpreting the connection between dual models and gravity. After my first seminar on this paper in our high energy laboratory, a senior student privately raised a somewhat pressing question, 'So, are you claiming that the dual model is more fundamental than the Einstein theory, replacing the latter by the dual model?' My answer was 'Well, as I have said, I was not able to find any definite principle, relating the dual model to general relativity. If I could have provided such a deeper explanation, I might have a right to claim that. All I can say at this point is simply a fact that we can extract general relativity from the dual model in this way.'

Instead of thinking toward the application of my results to the real world, I was rather occupied by a strong desire to understand a deeper reason for why the theory of relativistic strings happened automatically to encompass general relativity, the latter being based on such deep principles as the equivalence principle and general covariance. It seemed to me at that time that possible attempts to apply my results to realistic and unified models of elementary particles should be considered after gaining appropriate understanding of these questions.

Actually, my mind was also entangled with another competing viewpoint on the relationship between the dual strings and field theory. Already in 1970, there had been suggestive discussions on planar 'fishnet' diagrams by Sakita and Virasoro [SV70] and Nielsen and Olesen [NO70]. These works were precursors to the seminal work of 't Hooft on planar diagrams in the limit of large numbers of colours $N_c \to \infty$ [tHo74]. Their elegant observation had also been quite attractive to me as another possible approach to the relation between dual models and field theories. Moreover, in 1973, Nielsen and Olesen had published two papers [NO73a, NO73b], in which the possibilities of embedding relativistic strings in field theories were discussed, either as a gauge theory with a nonlinear theory corresponding to the Lagrangian $-\sqrt{F^2}/2\alpha'$ or as a vortex solution to a gauge Higgs system. I became aware of these preprints about the time when I was working on the connection of dual models to gravity on the basis of the zero-slope limit. These works were suggesting that the dual string might well be a consequence from nonlinear gauge field theories, resulting from their nontrivial dynamics. This is a view which is apparently opposite to the derivation of field theories from the dual models in the zero-slope limit. I was asking myself, 'Does this then imply that general relativity can be contained in gauge theory?'

I was not able to resolve this puzzle for many years. Later, I wrote a brief report [Yon76a] for *Butsuri*, a publication of the Physical Society of Japan, on the recent development of dual models in connection with field theory. I wrote: 'Since the dual model has started out without any explicit connection to field theory, it has been one of the basic questions how the relationship between the dual model and field theory should be understood. There are two possible answers. One is that the dual string is something which is beyond the framework of ordinary local field theory, and the latter should be interpreted as a particular limiting case of the dual string model. Another possibility would be that although the dual model cannot be understood using the traditional perturbative methods of field theory, it may be properly derived from field theory by means of some nonperturbative treatments of its complicated nonlinear dynamics. In the present report, I will discuss the recent developments from the first viewpoint, relegating the second to other opportunities.'

In the next year, several months after the submission of my papers, I received a preprint by Scherk and Schwarz [SS74a] (see also [SS74b]) in which they had expressed their viewpoint on the dual strings as models for nonhadrons including the graviton. They also studied the zero-slope limit of closed strings. In contrast to my approach of studying the coupling of gravitons to an artificial massive scalar state mainly for the methodological reason as explained above, they treated the amplitudes with only massless states. In 1975, furthermore, they made a bold proposal [SS75] that the dual string models should be interpreted as a more realistic theory of quarks and gluons by studying appropriate compactifications from ten to four dimensions. I was quite impressed and surprised by their solid attitude with respect to the interpretation of the connection of the dual models to gravity, especially, in their last work.

At any rate, there were several reactions to my work. In the beginning of 1974 I was invited to give a talk in a workshop organized by the GRG (general relativity and gravitation) group in Japan, and I met some of the leading Japanese physicists working in this field, including R. Utiyama. In the early summer of the same year, I received a letter from Y. Hara who was then staying in Europe, telling me that the connection of dual models to gravity was mentioned by D. Olive in his review talk at the London conference. Hara encouraged me to go to the United States for my postdoctoral studies. Subsequently, I received a letter from B. Sakita asking me whether I was interested in joining his group at the City College, New York. I also met K. Kikkawa who came back to Japan from there. Very stimulating conversations with him were most encouraging. Since I was actually about to be offered a faculty position at Hokkaido University, I had to fulfill my teaching duty at Hokkaido for one and a half years. I then joined Sakita's group, from 1976 to 1978 during which time I met A. Jevicki and S. Wadia. In 1980, I moved to the Komaba campus of the University of Tokyo.

In these times my interest in dual strings was fading, though I continued to believe that the connection of strings to general relativity would turn out to be useful some day. I was sparked by the fascinating developments of gauge theories, especially by various new discoveries in their nonperturbative aspects. In fact, the idea of supergravity, which was also first proposed in 1976, certainly had some impact on me.

However, interacting with Sakita and his group in New York, I became occupied more and more with the dynamics of non-Abelian gauge theories, especially the problem of quark confinement on which I was mainly working from the late Seventies to the early Eighties. Actually, my works during this period included an attempt [Yon81] to derive a version of string field theory as an effective theory of hadronic strings directly from lattice Yang–Mills theory: I considered a generalized gauge transformation, which was not so dissimilar to (37.1), of the string field associated with Wilson loops to apply it for a reformulation of the topological structure of the algebra of Wilson and 't Hooft loops. In regard to the latter, I had previously discussed [Yon78] its realization on the lattice for the purpose of applying it to the vortex condensation mechanism of quark confinement. In retrospect, these works were in fact implicitly related to the earlier basic question on possible emergence of strings and gravity from gauge theory. The same can be said of some of the other later works; starting from the early Nineties, I have been trying to relate matrix models or gauge theories in the large-N limit to gravity and strings in collaboration with A. Jevicki and also with several students.

Then starting in the late summer of 1983 I stayed at CERN for one year. During my stay at CERN, M. Green made a visit and gave a seminar. Green's talk was quite stimulating to me, especially his enthusiasm for string theory. As far as I remember, he gave a review of his works with Schwarz, but not yet anomaly cancellation. From about this time, I became inclined again toward string theory, and the impact of their epoch-making work [GS84] on the anomaly cancellation brought me back to this field in the mid-Eighties. In fact, the works which I published during my stay at CERN were on the Liouville field theory and two-dimensional gravity, close cousins of string theory. After coming back to my home institution, I gradually turned my mind to my fundamental and unsolved questions which I had been thinking about during the mid-Seventies.

37.2.5 Further related works

Before concluding, I would like here to describe my further works which are closely related to the present subject. After submitting the first full paper on the gravity–string connection in 1973, I continued for some time to think about the possibility of establishing a Nakanishi-type decomposition for closed strings. It might, I thought, help me formulate the desired correspondence principle between string theory and general relativity. But this task was too difficult, and I soon turned to more modest problems.

First I went back to a simpler case, namely the connection of open strings to non-Abelian gauge theory, and established a detailed connection at the level of Feynman rules

[Yon74b], on the basis of my works in the early Seventies. Then, I published a mathematical work [Yon75a] on an extension of the Fubini–Veneziano operator formalism introducing superfields, perhaps for the first time in this context, which naturally arose from the Fairlie–Martin representation [FM73] of the Neveu–Schwarz amplitudes.

I also studied the coupling of graviton and other massless closed string states to fermionic open strings, by extending the formalism which I had developed earlier in the case of massless vectors. I demonstrated [Yon75b, Yon76b] that the vertex operators for these states were obtained by means of an extended local gauge principle, which was similar to the ordinary tetrad formalism of local Lorentz gauge symmetry for the Dirac field, and discussed the associated geometric theory in the zero-slope limit. In [Yon76b], I made a conjecture on how these properties would be realized in full-fledged string field theories. Ten years later, I came to a realization of this conjecture, in discussing the symmetry structure associated with massless closed string states in the light-cone string field theory [Yon85, Yon87a]: this result could have been obtained in the mid-Seventies if I had devoted myself more seriously to the problem.

A few months after sending the paper on this subject to *Physical Review Letters*, I received the astonishing preprint by Witten on a covariant open string field theory [Wit86]: he gave an impressive answer to the problem against which I had struggled more than 10 years earlier. But I was pleased to see that he had cited my latest work on the light-cone string field theory. The connection of Witten's open string field theory to the Nakanishi decomposition was noticed in a later paper by Giddings and Martinec [GM86].

The idea behind the paper [Yon85] is also related to my conjecture of a purely cubic action which I proposed [Yon87b] in the spring of 1986 at the ICOBAN '86 international workshop on Grand Unification and also at a smaller internal workshop in Kyoto at the end of 1985. My consideration was aiming for a background-independent formulation of string theory, which is still among the deepest unsolved questions in string theory. In this workshop, there were also other talks related to string theory, including those by D. Gross, A. Strominger and K. Kikkawa. To my surprise, only one or two months after the ICOBAN workshop, I received preprints on possible realizations of this conjecture by the Santa Barbara group [HLTS86] and by the Kyoto group [HIKK86], both mentioning my previous talk.

Finally, the expectation of a hidden correspondence principle, which I had been seeking without success in the Seventies for characterizing the departure of string theory from general relativity, was basically the motivation for my later proposal [Yon87c, Yon89] in 1987 of an uncertainty relation for spacetime. On the basis of this relation, I tried for some time to derive string theory as a quantization of spacetime itself with string amplitudes being interpreted as a sort of Moyal product corresponding to a spacetime commutator, but could not succeed. A decade later, this idea turned out to be of some relevance for a qualitative understanding [LY97] of the characteristic scales of D-brane dynamics, which could be smaller than the usual string scale $\ell_s = \sqrt{\alpha'}$ in the weak coupling regime $g_s \ll 1$. I wrote several papers [Yon00] related to this subject from the mid-Nineties to the early 2000s.

37.3 Epilogue

Recalling what I was pursuing during my student years from 1969 to 1974, I find somewhat surprisingly how often my later works have been rooted in this period; I am still on a quest [Yon08] for better formulation of strings and D-branes along various lines which germinated, at least partially, from these early ideas. I was working almost alone, but was supported by the environment of great progress occurring in both string and gauge theories. I was fortunate to be living in such a fruitful period of particle physics.

After more than three decades have passed, the developments in string theory of the recent 10 years are now throwing a new light, through the conception of the gravity–gauge correspondence,[2] on my old question of whether 'from strings to field theory' or 'from gauge theory to strings and general relativity'. With respect to this puzzle, personally, a work [OY99] which I did with Y. Okawa in 1998 on a derivation of the gravitational three-body force for D0-branes from the D0 gauge theory (or M(atrix) theory [BFSS97]) was a small revelation to me. This work enabled me to convince myself that general relativity may indeed be contained in the quantum dynamics of gauge field theories, when interpreted as the theory for D-branes. The connection between the two viewpoints is now conjectured to be a new kind of duality relation (or 'holography'), which may be regarded as an extension, or a reinterpretation, of the old channel duality between open and closed strings.

In a sense, as if we were watching a revolving stage, string theory came back in a new perspective to a scene with which it had started in the early Seventies, and is providing various new handles for gauge theory and hadron physics from general relativity. However, we have to admit that we are still far from grasping the fundamental principles behind string theory. Hopefully, we will see further dramatic turns in the next 30 years.

Acknowledgements

I would like to thank S. Wadia and A. Jevicki for reading through the manuscript and for useful comments.

> Tamiaki Yoneya was born in Hokkaido, Japan, in 1947. He obtained his Doctor of Science degree (physics) from Hokkaido University in 1974. His professional carreer includes Fellow of Japan Society for the Promotion of Science (1974–1975), Assistant Professor, Hokkaido University (1975–1980), Associate Professor, University of Tokyo (1980–1991), and Professor, University of Tokyo (1991–present). He has also carried out research at various institutions including City University of New York, CERN, ITP (now KITP) at Santa Barbara, Brown University, Stanford University and MIT.

References

[BFSS97] Banks, T., Fischler, W., Shenker, S. and Susskind, L. (1997). M theory as a matrix model: a conjecture, *Phys. Rev.* **D55**, 5112–5128.

---

[2] For a review of my works related to the gauge–gravity correspondence, I would like to invite the reader to look at [Yon06].

[FM73] Fairlie, D. B. and Martin, D. (1973). New light on the Neveu–Schwarz model, *Nuovo Cimento* **A18**, 373–383.

[FN70] Fairlie, D. B. and Nielsen, H. B. (1970). An analog model for KSV theory, *Nucl. Phys.* **B20**, 637–651.

[FV70] Fubini, S. and Veneziano, G. (1970). Harmonic-oscillator analogy for the Veneziano model, *Nuovo Cimento* **A67**, 29–47.

[GM86] Giddings, S. B. and Martinec, E. (1986). Conformal geometry and string field theory, *Nucl. Phys.* **B278**, 91–120.

[GN74] Goto, T. and Naka, S. (1974). On the vertex function in the string model of hadrons, *Prog. Theor. Phys.* **51**, 600–612.

[GS84] Green, M. B. and Schwarz, J. H. (1984). Anomaly cancellation in supersymmetric $d = 10$ gauge theory and superstring theory, *Phys. Lett.* **B149**, 117–122.

[HIKK86] Hata, H., Itoh, K., Kugo, T., Kunitomo, H. and Ogawa, K. (1986). Pregeometrical string field theory: creation of spacetime and motion, *Phys. Lett.* **175**, 138–144.

[HLTS86] Horowitz, G. T., Lykken, J., Rohm, R. and Strominger, A. (1986). A purely cubic action for string field theory, *Phys. Rev. Lett.* **57**, 283–286.

[KK74] Kaku, M. and Kikkawa, K. (1974). The field theory of relativistic strings, Pt. 1. trees, *Phys. Rev.* **D10**, 1110–1133.

[KT70] Kaku, M. and Thorn, C. B. (1970). Unitary nonplanar closed loops, *Phys. Rev.* **D1**, 2860–2869.

[KY68] Katayama, Y. and Yukawa, H. (1968). Field theory of elementary domains and particles. 1, *Prog. Theor. Phys. Suppl.* **41**, 1–21.

[KSV69] Kikkawa, K., Sakita, B. and Virasoro, M. A. (1969). Feynman-like diagrams compatible with duality. I: Planar diagrams, *Phys. Rev.* **184**, 1701–1713.

[KS70] Kikkawa, K. and Sato, H. (1970). Nonhadronic interactions in the dual resonance model, *Phys. Lett.* **B32**, 280–284.

[KN69] Koba, Z. and Nielsen, H. B. (1969). Manifestly crossing invariant parameterization of n meson amplitude, *Nucl. Phys.* **B12**, 517–536.

[LY97] Li, M. and Yoneya, T. (1997). D particle dynamics and the spacetime uncertainty relation, *Phys. Rev. Lett.* **78**, 1219–1222.

[Lov70] Lovelace, C. (1970). Simple n-Reggeon vertex, *Phys. Lett.* **B32**, 490–494.

[MM71] Manassah, J. T. and Matsuda, S. (1971). Hadron electrodynamics in the resonance model – deep-inelastic e-p scattering, e-e colliding-beam processes, and other applications, *Phys. Rev.* **D4**, 882–898.

[Man73] Mandelstam, S. (1973). Interacting string picture of dual resonance models, *Nucl. Phys.* **B64**, 205–235.

[Nak71] Nakanishi, N. (1971). Crossing-symmetric decomposition of the n-point Veneziano formula into tree-graph integrals. 2. Koba–Nielsen representation, *Prog. Theor. Phys.* **45**, 919–926 and references therein.

[Nam70] Nambu, Y. (1970). Quark model and the factorization of the Veneziano amplitude, in *Proceedings of the International Conference on Symmetries and Quark Models, Wayne State University, June 18–20, 1969*, ed. Chand, R. (Gordon and Breach, New York), 269–277, reprinted in *Broken Symmetry, Selected Papers of Y. Nambu*, ed. Eguchi, T. and Nishijima, K. (World Scientific, Singapore, 1995), 258–277.

[NS72] Neveu, A. and Scherk, J. (1972). Connection between Yang–Mills fields and dual models, *Nucl. Phys.* **B36**, 155–161.

[NS71] Neveu, A. and Schwarz, J. H. (1971). Quark model of dual pions, *Phys. Rev.* **D4**, 1109–1111.

[NO70] Nielsen, H. B. and Olesen, P. (1970). A parton view on dual amplitudes, *Phys. Lett.* **B32**, 203–206.
[NO73a] Nielsen, H. B. and Olesen, P. (1973). Local field theory of the dual string, *Nucl. Phys.* **B57**, 367–380.
[NO73b] Nielsen, H. B. and Olesen, P. (1973). Vortex line models for dual strings, *Nucl. Phys.* **B61**, 45–61.
[OY99] Okawa, Y. and Yoneya, T. (1999). Multibody interactions of D particles in supergravity and matrix theory, *Nucl. Phys.* **B538**, 67–99.
[Oli71] Olive, D. I. (1971). Operator vertices and propagators in dual theories, *Nuovo Cimento* **A3**, 399–411.
[Ram71] Ramond, P. (1971). Dual theory for free fermions, *Phys. Rev.* **D3**, 2415–2418.
[SV70] Sakita, B. and Virasoro, M. A. (1970). Dynamical model of dual amplitudes, *Phys. Rev. Lett.* **24**, 1146–1149.
[Sak60] Sakurai, J. J. (1960). Theory of strong interactions, *Ann. Phys.* **11**, 1–48.
[Sch71] Scherk, J. (1971). Zero-slope limit of the dual resonance model, *Nucl. Phys.* **B31**, 222–234.
[SS74a] Scherk, J. and Schwarz, J. H. (1974). Dual models for nonhadrons, *Nucl. Phys.* **B81**, 118–144.
[SS74b] Scherk, J. and Schwarz, J. H. (1974). Dual models and the geometry of spacetime, *Phys. Lett.* **B52**, 347–350.
[SS75] Scherk, J. and Schwarz, J. H. (1975). Dual field theory of quarks and gluons, *Phys. Lett.* **B57**, 463–466.
[Sci69] Sciuto, S. (1969). The general vertex function in dual resonance models, *Lett. Nuovo Cimento* **3**, 411.
[Sha70a] Shapiro, J. A. (1970). Electrostatic analogue for the Virasoro model, *Phys. Lett.* **B33**, 361–362.
[Sha70b] Shapiro, J. A. (1970). General factorization of the n-point beta function into two parts in a Moebius-invariant manner, *Phys. Rev.* **D1**, 3371–3376.
[Sus69] Susskind, L. (1969). Harmonic-oscillator analogy for the Veneziano model, *Phys. Rev. Lett.* **23**, 545–547.
[tHo74] 't Hooft, G. (1974). A planar diagram theory for strong interactions, *Nucl. Phys.* **B72**, 461–473.
[Ven68] Veneziano, G. (1968). Construction of a crossing-symmetric, Reggeon behaved amplitude for linearly rising trajectories, *Nuovo Cimento* **A57**, 190–197.
[Vir69] Virasoro, M. A. (1969). Alternative constructions of crossing-symmetric amplitudes with Regge behavior, *Phys. Rev.* **177**, 2309–2311.
[Wit86] Witten, E. (1986). Non-commutative geometry and string field theory, *Nucl. Phys.* **B268**, 253–294.
[Wys65] Wyss, W. (1965). Zur Unizität der Gravitationstheorie, *Helv. Phys. Acta* **38**, 469.
[Yon71] Yoneya, T. (1971). General Reggeon vertex in dual theory, *Prog. Theor. Phys.* **46**, 1192–1206.
[Yon72a] Yoneya, T. (1972). Note on the local gauge principle in conformal dual models, *Prog. Theor. Phys.* **48**, 616–624.
[Yon72b] Yoneya, T. (1972). Feynman-like rules for the dual-resonance model on the basis of the Nakanishi decomposition, *Prog. Theor. Phys.* **48**, 2044–2065.
[Yon73] Yoneya, T. (1973). Quantum gravity and the zero slope limit of the generalized Virasoro model, *Lett. Nuovo Cimento* **8**, 951.
[Yon74a] Yoneya, T. (1974). Connection of dual models to electrodynamics and gravidynamics, *Prog. Theor. Phys.* **51**, 1907–1920.

[Yon74b] Yoneya, T. (1974). Gauge freedom in dual Feynman like rules and its relation to the massless Yang–Mills field theory, *Prog. Theor. Phys.* **52**, 1355–1368.

[Yon75a] Yoneya, T. (1975). Grassmann algebraic approach to the Neveu–Schwarz model and representation of supermöbius algebra, *Prog. Theor. Phys.* **54**, 526–541.

[Yon75b] Yoneya T. (1975). Interacting fermionic and pomeronic strings: gravitational interaction of the Ramond fermion, *Nuovo Cimento* **A27**, 440–457.

[Yon76a] Yoneya, T. (1976). Dual model and field theory, *Butsuri* **31**, 46.

[Yon76b] Yoneya, T. (1976). Geometry, gravity and dual strings, *Prog. Theor. Phys.* **56**, 1310–1317.

[Yon78] Yoneya, T. (1978). $Z(n)$ topological excitations in Yang–Mills theories: duality and confinement, *Nucl. Phys.* **B144**, 195–218.

[Yon81] Yoneya, T. (1981). A path-functional field theory of lattice gauge models and the large-N limit, *Nucl. Phys.* **B183**, 471–496.

[Yon85] Yoneya, T. (1985). Space-time local symmetry of string field theory, *Phys. Rev. Lett.* **55**, 1828–1830.

[Yon86] Yoneya, T. (1986). String field theory, in Proceedings of the Workshop published in *Soryuushiron-Kenkyuu* **74** (1986–1987), No. 4, p. DS9.

[Yon87a] Yoneya, T. (1987). String coupling constant and dilaton vacuum expectation value in string field theory, *Phys. Lett.* **B197**, 76–80.

[Yon87b] Yoneya, T. (1987). Approaches to string field theory, in *Seventh Workshop on Grand Unification/ICOBAN'86* (World Scientific, Singapore), 508.

[Yon87c] Yoneya, T. (1987). Duality and indeterminacy principle in string theory, *Wandering in the Field, Festschrift for Professor Kazuhiko Nishijima on the Occasion of his Sixtieth Birthday* (World Scientific, Singapore), 419.

[Yon89] Yoneya, T. (1989). On the interpretation of minimal length in string theories, *Mod. Phys. Lett.* **A4**, 1587–1595.

[Yon00] Yoneya, T. (2000). String theory and spacetime uncertainty principle, *Prog. Theor. Phys.* **103**, 1081–1125.

[Yon06] Yoneya, T. (2006). Holography in the large-J limit of AdS/CFT correspondence and its applications, *Prog. Theor. Phys.* **164**, 82–101.

[Yon08] Yoneya, T. (2008). An attempt towards field theory of D0 branes – quantum M-field theory, *Int. J. Mod. Phys.* **A23**, 2343–2351.

[Zwi92] Zwiebach, B. (1992). Closed string field theory: an introduction, in *Les Houches Summer School*, p. 647, hep-th/9305026.

38
From the Nambu–Goto to the σ-model action

LARS BRINK

38.1 Introduction

My generation of string theorists was very fortunate. We were there when the first ideas leading up to string theory were proposed, and we were young and inexperienced enough not to ask too deep questions. We could accept working in 26 dimensions of spacetime, even when more experienced people laughed at it (and us). We were not more clever than they were, not at all, rather we became so attached to the ideas that we did not listen to good advice. The average age of the active people was probably well under thirty, and it was one of the rare occasions where a young generation could form its scientific future. There were a number of older heroes, most notably Yoichiro Nambu, Stanley Mandelstam, Sergio Fubini and Daniele Amati. Also, the leading theoretical physicist of those days, Murray Gell-Mann, was sympathetic. His words, always carefully phrased, were listened to by everyone in particle physics. This blend made the field so exciting that once hooked it was difficult to leave it. After some years many had to change field in order to find positions, but most of us had the secret wish to return to this subject.

38.2 The formative years

I started as a graduate student in 1967. Sweden still had the old system, which meant that there were no graduate schools. You had to study on your own, and you had to work on your own. Every year the department of theoretical physics in Göteborg accepted a few graduate students. My advisor Jan Nilsson soon told me to work on phenomenology, and I got in contact with an experimental group in Stockholm, since we had no particle experimentalists in our physics department. For some forgotten reason, I came across the paper by Dolen, Horn and Schmid [DHS68] on finite energy sum rules during my first year, and I gave a seminar on it. I tried to follow the subject and collected preprints but had no one to discuss it with in our group, who mostly worked on various forms of group theory.

I was in Stockholm in September 1968 when the professor of the experimental group, Gösta Ekspong, came in one day very excited showing everybody a paper in which some

Italian colleague had found a formula for pion–pion scattering with no free parameters other than a coupling constant. That was the paper by Gabriele Veneziano [Ven68]. Ekspong came straight from the Vienna conference. Again I tried to follow the subject and to study all the new concepts that appeared but back home no one was interested. Instead I had to concentrate on my studies of proton–proton scattering to explain the 'Deck peak', which was essentially the $\Delta(1236)$ resonance, and to use OPE, which everybody knows means one-pion exchange. I wrote a few papers on it and got my licentiate degree, which is a lower doctor's degree. After that I felt freer to study more theoretical subjects and my advisor encouraged me to do so and, mainly together with some visitors in the department, I wrote papers on current algebra and eventually on dual models. These were very simple calculations with long forgotten results, but it was a training ground and I learned a lot. I was encouraged to apply to CERN and, to my surprise and enormous happiness, I was accepted and offered a fellowship from June 1971.

38.3 The CERN years

Life is often formed by accidental events. I came to CERN in the beginning of the summer and met only people who were already established in Geneva and at CERN. One month after me David Olive came to take up a staff position leaving his job in Cambridge. We became good friends almost immediately. David was, of course, already famous having been one of the leaders of the Cambridge school in the analytic S-matrix. He was one of the old-timers (he was over 30!) who had moved into dual models, seeing it as a realization of an S-matrix theory. Also during the summer I met John Schwarz, forming a lifelong friendship; he was visiting CERN for some weeks.

When I came to CERN I was still very hesitant about what problems to work on. I spent the summer working with some short-term visitors on 'inclusive cross-sections', but I also followed all the seminars on dual models. Two more lucky events happened when the new crowd arrived at the end of the summer. One was that I got a new officemate, Joël Scherk, whom I came to share an office with for almost two years. Joël had already made a name for himself with his work in Princeton with André Neveu using the Jacobi imaginary transformation to isolate the divergence in the one-loop graphs and also with their subsequent work with John Schwarz and David Gross [GNSS70]. When he arrived he had just invented the 'zero-slope limit' [Sch71]. One of the first days after he arrived he gave a seminar about it in the small seminar room, and I still remember Bruno Zumino's excitement afterwards. I overheard him say to Mary K. Gaillard that this must have something to do with quantum field theory. (This was the starting point for Bruno's interest in dual models and led to his and Julius Wess's discovery of four-dimensional supersymmetry a few years later. Bruno, who had an office near ours, used to come to us and borrow all the important papers on dual models.) Joël looked like a genius, talked like a genius and indeed was a genius. He had long hair and some fantastic clothes. He spoke very softly and was always very nice to talk to. We forged a deep friendship that was very close all the time up to his too early death in 1980 at only 33 years of age. We always had a nice discussion when he arrived in the morning, usually about physics but often about life or Chinese history, which he was studying on the bus to and from Geneva.

The other event that happened was that Holger Bech Nielsen reluctantly came to CERN. He could stay for a year or more if he so wanted. He came with his mother. When she left, he stayed on in a hotel for nine months until he went home. Holger was regarded as *the* genius in the field. He could concentrate completely on a problem; they could have dropped an atomic bomb in the next room without disturbing him in the slightest. He had the most remarkable ideas, which nobody else had ever thought about. He spent all his time at CERN, eating all his meals there and going back to town on the last bus, if he did not miss it. Then he walked. I had met Holger before, and he became my entrance ticket to the dual model community at CERN. We started to work together – mostly on his ideas. Our main aim was to find new more realistic dual models. I did learn a lot but our progress was not great. At some stage we used duality to get sum rules for meson masses assuming a string with quarks at the ends. They worked pretty well but were very sensitive to details, since we used partition functions that involved sums of exponentials. After a year at CERN, I had learned a lot but not written any really good papers, and then Holger left.

It had, of course, been a very successful year at CERN in dual models with the no-ghost theorem proved by Peter Goddard and Charles Thorn [GT72] (Charles had the third desk in our office for the year he spent at CERN) and then their work with Claudio Rebbi and Jeffrey Goldstone on the string [GGRT73]. There were lots of seminars and lots of discussions. There were several collaborations going on, but by the end of the summer David Olive and I found ourselves a bit left out. We started to discuss and David then had the brilliant idea of trying something really hard. (David always wants to study deep and hard problems.) He suggested that we should try to compute the one-loop graphs correctly. After the marvellous paper by Lovelace [Lov71] in 1971 on one-loop graphs, where he saw that by dividing out two powers of the partition function and taking the spacetime dimension to be 26, the twisted loop contains a series of poles instead of unphysical branch cuts, it was assumed that this should be the rule for all one-loop graphs, but it was not proven. Nobody at that time had a clear idea how to prove it.

This was just a year after the gauge theory revolution and my generation and the S-matrix one, which was slightly older, knew very little about gauge theories. We had learned QED, but our knowledge of non-Abelian gauge theories was rudimentary. The wonderful talk by Feynman in Poland in 1963 [Fey63] and the subsequent work by Faddeev and Popov [FP67] on the construction of one-loop graphs in non-Abelian gauge theories were not known. The longer version of the Russian paper was in fact only written in Russian. After the gauge theory revolution it was quickly translated into English by David Gordon. David Olive and I started to study that paper in detail as well as some marvellous lecture notes by Abdus Salam, who as usual had immediately grasped the importance of the subject. At the same time we studied Gerard 't Hooft's papers.

A funny story is that the Faddeev–Popov paper discusses two different gauge choices and concludes that they give the same one-loop graph but not necessarily the same higher loop graphs. We were rather mystified by that argument and asked 't Hooft, who had just arrived as a Fellow, for a discussion. 't Hooft then said very emphatically that in his paper he had shown that it worked to all orders. No more discussion. Our problem was that we had no Lagrangian formulation of dual models – only tree diagrams. We then devised a method in non-Abelian gauge field theory of starting with a naive one-loop graph and then deriving corrections by implementing gauge invariance at the one-loop level. In this way we found the ghost contributions correctly, and we then tried this method on dual model loops. We worked heroically with enormous algebras, but we could not finish it. (Many years later we realized that we had used the wrong Virasoro generators. We had not thought of also introducing ghosts at the two-dimensional string level.) We then went back to square one and were told about Feynman's lecture in Poland by Josef Honerkamp. Feynman had been interested in quantum gravity, but Murray Gell-Mann had suggested to him that he study non-Abelian gauge theories first as a warm-up exercise. He talked about this at the conference. In the discussion session after the talk he was asked by Bryce DeWitt how to compute one-loop graphs. Feynman then described in words a method where you sew together tree diagrams using a projection operator onto the physical states. He said that one could interpret the result as if two scalar ghost fields propagated through the loop. This became the starting point for us.

Since there was no literature on this method except Feynman's words we started by redoing it in field theory. There the projection operator was known and easy to construct. However, for dual models we had to construct such an operator. We knew that the physical states in the critical dimension were given by the Del Giudice–Di Vecchia–Fubini operators $A_n^i(k)$ and their conjugates, where $n = 1, \ldots, \infty$ and $i = 1, \ldots, d-2$. The vector k is lightlike. The projection operator could then be formally written as

$$T(k) = \oint \frac{dy}{2\pi i} y^{\mathcal{L}_0 - H - 1}, \tag{38.1}$$

where

$$\mathcal{L}_0 = \sum_{n=1}^{\infty} \sum_{i=1}^{d-2} A_n^{i\,\dagger}(k) A_n^i(k), \tag{38.2}$$

and

$$H = \sum_{n=1}^{\infty} \sum_{\mu=1}^{d} \alpha_n^{\mu\,\dagger} \alpha_{n\mu}, \tag{38.3}$$

with α the ordinary harmonic oscillators of the bosonic string, which create all the excited-string states including the negative norm ones.

By the use of operator product expansions and the shifting of integration contours, we could prove the following identity for $d = 26$:

$$\mathcal{L}_0 - H = (D_0 - 1)(L_0 - 1) + \sum_{n=1}^{\infty} (D_{-n} L_n + L_{-n} D_n), \tag{38.4}$$

with L_n the ordinary Virasoro generators and D_n a new set of operators. It is then easy to see that the projection operator is equal to one on a physical state and zero otherwise, giving us our own proof of the no-ghost theorem [BO73a]. With this projection operator we could set up and prove Feynman's tree theorem in detail and then apply the same technique to the one-loop dual model loops. After a lengthy calculation we could prove that it did divide out two powers of the partition function in the measure as Lovelace had anticipated [BO73b].

We did not dare to send the paper to Feynman, but some months later we got a letter from him that I still have on my office wall. He was extremely nice to us and thanked us for writing up his theorem 'with clarity and simplicity'. John Schwarz, who had moved to Caltech at that time, had shown him the paper and he had read it carefully.

The construction of the projection operator was like opening the tap. We quickly redid the same calculations for the Neveu–Schwarz model, and Corrigan and Goddard computed the projection operator for the Ramond model. We also did the calculations for the closed string (or 'Pomeron') sector at that time and proved that the Reggeon–Pomeron vertex respected unitarity [BOS73]. We did this by commuting the projection operator from the Reggeon (open string) sector through the vertex to the Pomeron (closed string) vertex, showing that the correct projection operator appeared on that side. This work we did with Joël Scherk, who was now brought into our collaboration. We also did the same calculation for the fermion emission vertex, showing that indeed the Ramond and the Neveu–Schwarz sectors were unitarily related. For this work Claudio Rebbi also joined in [BORS73].

All through this period I still had contact with Nielsen. In the spring of 1973 he came to CERN for a week. He brought with him a mathematical way to compute $-1/12$ as the regularized sum of all positive integers. We thought hard how to connect this to strings and realized quickly that by summing up all the zero-point fluctuations of all the harmonic modes of the bosonic string, we got just the sum of all integers. We invented a physical way to regularize by renormalizing the velocity of light which is a parameter of the Nambu–Goto string. In this way we got an alternative proof of $d = 26$ (my third by then) [BN73]. It was obvious to me that the Ramond fermions must be massless since the zero-point fluctuations cancelled between the bosonic and the fermionic ones. Olive, even though he is a born gentleman, persuaded me not to publish this, since we were in the midst of our fermion calculations and the common belief at the time was that the fermion had a nonzero mass. I wrote it up in my thesis later that year.

At the beginning of the summer of 1973, my time was up at CERN. So I moved home to Sweden becoming very depressed. Only when I had left did I realize what a fantastic place CERN had been for the development of string theory. This was due to Daniele Amati who tirelessly defended us, as we understood much later, and who gave so much of his time and of himself to us Fellows. At home I tried to communicate with Olive, but this was before the Internet. Fortunately, it was also before the demise of the postal services, so we could get letters through in less than 24 hours. The final paper in this stage of our collaboration was the construction of the four-fermion amplitude that Olive and Scherk constructed during that summer [OS73].

Back home I also had to finish my thesis, which came to consist of fourteen papers and an introduction of one hundred pages. It was the longest thesis in the history of the department. After defending it I resumed the collaboration with Olive, and we worked hard to understand the fermions in dual models. In the summer of 1974 John Schwarz organized a workshop in Aspen and most people who had been involved in the developments were there except for Pierre Ramond, Charles Thorn and Holger Nielsen. Thorn was already working on the MIT-bag and Ramond was busy with the birth of his second daughter. Nielsen's interest in string theory had started to fade, and he was full of other interesting ideas. When I came to Aspen Bunji Sakita told me that Nambu wanted to talk to me. I was quite excited and thought he would comment on all the work that Olive and I had done. No, he congratulated me instead on the paper with Nielsen about zero-point fluctuations. This was very flattering, because I consider Nambu to be one of the greatest scientists of all time.

38.4 Collaborations at Nordita

I was lucky in one sense compared to my friends and collaborators. There was no pressure on me to change to a more fashionable subject. My situation in Sweden was stable but not very stimulating. I had had a research position with the research council even before going to CERN and I took it up again when I got home. It was renewed every year; on the other hand, there were no more permanent jobs to apply for.

When I came home after the 1974 summer at Aspen, Paolo Di Vecchia arrived at Nordita in Copenhagen as an assistant professor. That was to become very important for me. Nordita is a Nordic institute and its mission is to promote theoretical physics in the Nordic countries. I could travel to Copenhagen more or less whenever I wanted as long as Paolo agreed. In the beginning I still worked with Nielsen finishing up some old ideas, but Di Vecchia and I discussed more and more. Di Vecchia wanted to construct a fermionic string by starting with x^μ and a spacetime spinor θ^α. He constructed the obvious invariant and was on his way to constructing the superstring. I wanted to have a string action for the Ramond–Neveu–Schwarz model, so I was sceptical. In the early months of 1975, I visited CERN and had a long discussion with Bruno Zumino. Exactly at the same time Zumino and I said that we should have a two-dimensional spinor instead and try to have reparameterization invariance on the world-sheet. Several people, including Gervais and Sakita and Mandelstam, had worked earlier with such spinors, but they had not constructed an action from which the full constraint algebra follows. This seemed like a good problem to work on, and I convinced Di Vecchia to join me. We had no understanding of Grassmann algebras and had to start from scratch. Fortunately, there was the wonderful book by Berezin [Ber66]. Of course, we wanted to extend the Nambu–Goto string to two-dimensional superspace but found no way of doing it. We wanted to construct a square-root of something but always stumbled on the strange properties of the Grassmann variables. We also knew so little about general relativity and the difficulty of including fermions, since neither of us had studied any courses in general relativity. After a while we realized that we first ought to solve the corresponding point-particle problem, but we ran into the same problem there.

In the summer of 1975 there was a workshop in Durham that David Fairlie organized. That was the first time that I met Pierre Ramond, and we instantly became close friends. At the meeting I realized that the superfield formalism that we had developed was ideal for a super-operator formalism and later that year I worked it out with my student and formulated the superconformal formalism that later was reinvented in the second string era [BW76]. At the meeting there was a crowd of Italian colleagues, all good friends of Paolo Di Vecchia, and we started to discuss the superfield representation we had so far for the RNS string. We could write a free action for the superfield and then implement the Noether current as the constraints. It worked but was hand-waving. Soon we realized that we could extend the supersymmetry and we all met at CERN in September to work out the $SO(2)$ case. In this way we got an extended super-Virasoro algebra, the first one. The key was that the Noether current also involved a Kac–Moody $U(1)$ current. We went on and constructed an infinite sequence of extensions [ABDD76a] but only the $\mathcal{N}=2$ and $\mathcal{N}=4$ cases were interesting. By using $SO(4) = SU(2) \times SU(2)$ we could construct the $SU(2)$ algebra. This one and the $SO(2)$ algebra were the only ones with canonical operators. For the rest of that year and the beginning of the next one we were busy formulating models for these cases. After our first paper on the new super-Virasoro algebras, I got a letter from John Schwarz who had immediately understood our formalism and also constructed the $SO(2)$ model. We invited him to join us on that paper, and then the authorship consisted of eleven Italians, one American and one Swede [ABDD76b]. Shortly after John wrote me that Caltech had some money free for the coming academic year and wondered if I was interested. I accepted readily.

In the spring of 1976, we wanted to go back to the problem of finding a reparameterization invariant action. We slowly got to understand the supergravity action [FNF76, DZ76a] that had been constructed in the beginning of that year. Sergio Ferrara, who had been a member of the our huge collaboration, had moved to Paris for a year, and we should have connected more quickly to what they had done, but that spring I was very busy. Only in the summer did we manage to meet and work out the particle action. In fact, we met at CERN for a few days to finish it and were joined by Paul Howe who was a postdoc at Nordita and by Stanley Deser and Bruno Zumino [BDZD76]. The key point, which we had missed before, was the use of vierbeins (or, rather, as Murray Gell-Mann named those in general dimensions, 'vielbeins'). Once we understood it, it was rather straightforward to solve the particle problem and, as a result, we got a world-line action that leads to the Dirac equation.

We realized that we could now construct the string action, too, but it took some time before we could meet to finalize it. One problem for me was that I was planning the trip to Caltech and I had to make lots of preparations for that. Finally, we met for a week and worked like mad to construct the action and to prove all its local symmetries and then to show that it leads to the constraints of the RNS model [BDH76]. Again, we were inexperienced with the use of spinors and knew nothing about Fierz rearrangements, so we had to do it the hard way. Anyhow, I went home to get the manuscript typed by our secretary, and then I sent it out. More or less with return mail I received the paper by Deser and Zumino [DZ76b] that contains the same action (of course).

The following week I left for Caltech. The first week there John and I (very jet-lagged) constructed the $\mathcal{N} = 2$ action using the same technique [BS77]. It is interesting that it was this paper that Sasha Polyakov read when he came to Caltech the next year and learned about these actions. He had in fact been in Copenhagen for an extended period when we constructed the action, but he was so intensely engaged in his magnificent work on instantons and confinement that he had missed our work. Of course, few people took notice of it, since it was so far from the mainstream. However, our short meeting was the beginning of one of these rare lifelong friendships.

38.5 Leaving strings for a while

After our first paper together at Caltech, John and I felt that we must work on more modern stuff. We wanted to use all our insight from string theory on supersymmetric field theories. Naturally, we started with the ten-dimensional super Yang–Mills theory and by compactifying it to various dimensions we found other maximally supersymmetric gauge theories including the $\mathcal{N} = 4$ theory in $d = 4$. We got in touch with Joël in Paris, who had been doing the same thing, and we wrote it up together [BSS77]. We did feel that this was an important model but little did we know that it would be one of the cornerstones of modern theory. (Some five years later I returned to it with Bengt Nilsson and Olof Lindgren, when we finally found a way to prove perturbative finiteness [BLN83].) We wanted though to reformulate supergravity, and we teamed up with Pierre Ramond and Murray Gell-Mann and worked hard on supergravity in superspace. I was very insistent on using superspace, since I had fallen in love with it in our studies of supersymmetric strings. We worked on this for quite some time and reconstructed the first supergravities this way [BGRS78a, BGRS78b]. Eventually in 1979 Paul Howe and I managed to construct the $\mathcal{N} = 8$ supergravity in superspace [BH79]. When he and Ulf Lindström [HL81], using our formalism, the year after showed that there were possible counterterms in that theory, I went back to string theory and joined up with John Schwarz and Michael Green. My attachment to strings was a love for life.

Acknowledgements

I wish to thank Pierre Ramond, John Schwarz and Paolo Di Vecchia for reading the manuscript and for their helpful suggestions.

Lars Brink was born in 1943. He obtained his PhD at Chalmers University of Technology, under Jan S. Nilsson. He has been a research fellow at CERN (1971–1973) and Caltech (1976–1977).

He has held research positions at Chalmers since 1969. He has been Professor of Physics at Chalmers since 1986. Research interests include string theory, supersymmetric field theories and supergravity.

References

[ABDD76a] Ademollo, M., Brink, L., D'Adda, A., D'Auria, R., Napolitano, E., Sciuto, S., Del Giudice, E., Di Vecchia, P., Ferrara, S., Gliozzi, F., Musto, R. and Pettorino, R. (1976). Supersymmetric strings and color confinement, *Phys. Lett.* **B62**, 105–110.

[ABDD76b] Ademollo, M., Brink, L., D'Adda, A., D'Auria, R., Napolitano, E., Sciuto, S., Del Giudice, E., Di Vecchia, P., Ferrara, S., Gliozzi, F., Musto, R., Pettorino, R. and Schwarz, J. (1976). Dual string with $U(1)$ colour symmetry, *Nucl. Phys.* **B111**, 77–110.

[Ber66] Berezin, F. A. (1966). *The Method of Second Quantization* (Academic Press, New York).

[BDZD76] Brink, L., Deser, S., Zumino, B., Di Vecchia, P. and Howe, P. S. (1976). Local supersymmetry for spinning particles, *Phys. Lett.* **B64**, 435–438.

[BDH76] Brink, L., Di Vecchia, P. and Howe, P. S. (1976). A locally supersymmetric and reparameterization invariant action for the spinning string, *Phys. Lett.* **B65**, 471–474.

[BGRS78a] Brink, L., Gell-Mann, M., Ramond, P. and Schwarz, J. H. (1978). Supergravity as geometry of superspace, *Phys. Lett.* **B74**, 336–340.

[BGRS78b] Brink, L., Gell-Mann, M., Ramond, P. and Schwarz, J. H. (1978). Extended supergravity as geometry of superspace, *Phys. Lett.* **B76**, 417–422.

[BH79] Brink, L. and Howe, P. S. (1979). The $N = 8$ supergravity in superspace, *Phys. Lett.* **B88**, 268–272.

[BLN83] Brink, L., Lindgren, O. and Nilsson, E. W. (1983). The ultraviolet finiteness of the $N = 4$ Yang–Mills theory, *Phys. Lett.* **B123**, 323–328.

[BN73] Brink, L. and Nielsen, H. B. (1973). A simple physical interpretation of the critical dimension of spacetime in dual models, *Phys. Lett.* **B45**, 332–336.

[BO73a] Brink, L. and Olive, D. (1973). The physical state projection operator in dual resonance models for the critical dimension of spacetime, *Nucl. Phys.* **B56**, 253–265.

[BO73b] Brink, L. and Olive, D. (1973). Recalculation of the unitary single planar loop in the critical dimension of spacetime, *Nucl. Phys.* **B58**, 237–253.

[BORS73] Brink, L., Olive, D., Rebbi, C. and Scherk, J. (1973). The missing gauge conditions for the dual fermion emission vertex and their consequences, *Phys. Lett.* **B45**, 379–383.

[BOS73] Brink, L., Olive, D. and Scherk, J. (1973). The gauge properties of the dual model Pomeron–Reggeon vertex: their derivation and consequences, *Nucl. Phys.* **B61**, 173–198.

[BS77] Brink, L. and Schwarz, J. H. (1977). Local complex supersymmetry in two-dimensions, *Nucl. Phys.* **B121**, 285–295.

[BSS77] Brink, L., Schwarz, J. H. and Scherk, J. (1977). Supersymmetric Yang–Mills theories, *Nucl. Phys.* **B121**, 77–92.

[BW76] Brink, L. and Winnberg, J. O. (1976). The superoperator formalism of the Neveu–Schwarz–Ramond model, *Nucl. Phys.* **B103**, 445–464.

[DZ76a] Deser, S. and Zumino, B. (1976). Consistent supergravity, *Phys. Lett.* **B62**, 335–337.

[DZ76b] Deser, S. and Zumino, B. (1976). A complete action for the spinning string, *Phys. Lett.* **B65**, 369–373.

[DHS68] Dolen, R., Horn, D. and Schmid, C. (1968). Finite energy sum rules and their application to πN charge exchange, *Phys. Rev.* **166**, 1768–1781.

[FP67] Faddeev, L. D. and. Popov, V. N. (1967). Feynman diagrams for the Yang–Mills field, *Phys. Lett.* **B25**, 29–30.

[Fey63] Feynman, R. P. (1963). The quantum theory of gravitation, *Acta Phys. Pol.* **24**, 697–722.

[FNF76] Freedman, D. Z., van Nieuwenhuizen, P. and Ferrara, S. (1976). Progress toward a theory of supergravity, *Phys. Rev.* **D13**, 3214–3218.

[GGRT73] Goddard, P., Goldstone, J., Rebbi, C. and Thorn, C. B. (1973). Quantum dynamics of a massless relativistic string, *Nucl. Phys.* **B56**, 109–135.

[GT72] Goddard, P. and Thorn, C. B. (1972). Compatibility of the pomeron with unitarity and the absence of ghosts in the dual resonance model, *Phys. Lett.* **B40**, 235–238.

[GNSS70] Gross, D. J., Neveu, A., Scherk, J. and Schwarz, J. H. (1970). Renormalization and unitarity in the dual-resonance model, *Phys. Rev.* **D2**, 697–710.

[HL81] Howe, P. S. and Lindström, U. (1981). Higher order invariants in extended supergravity, *Nucl. Phys.* **B181**, 487–501.

[Lov71] Lovelace, C. (1971). Pomeron form factors and dual Regge cuts, *Phys. Lett.* **B34**, 500–506.

[OS73] Olive, D. and Scherk, J. (1973). Towards satisfactory scattering amplitudes for dual fermions, *Nucl. Phys.* **B64**, 334–348.

[Sch71] Scherk, J. (1971). Zero-slope limit of the dual resonance model, *Nucl. Phys.* **B31**, 222–234.

[Ven68] Veneziano, G. (1968). Construction of a crossing-symmetric, Reggeon-behaved amplitude for linearly rising trajectories, *Nuovo Cimento* **A57**, 190–197.

39

Locally supersymmetric action for the superstring

PAOLO DI VECCHIA

In this Chapter I describe the ideas that led Lars Brink, Paul Howe and myself to the formulation of the locally supersymmetric Lagrangian for the fermionic string. This Chapter goes together with that of Lars Brink, where the construction of this Lagrangian is also discussed.

By the beginning of 1973, everybody was convinced that the dynamics of a quantum relativistic string, described by the nonlinear Nambu–Goto action, was underlying the Dual Resonance Model. The Nambu–Goto action is the following functional of the string coordinate $X^\mu(\tau, \sigma)$:

$$S_{NG} = -\frac{1}{2\pi\alpha'} \int d\tau \int d\sigma \sqrt{-\det(\partial_\alpha X \cdot \partial_\beta X)}, \qquad (39.1)$$

where the two-dimensional world-sheet of the string is described by the coordinates $\xi^\alpha \equiv (\xi^0, \xi^1) \equiv (\tau, \sigma)$. Being proportional to the area spanned by the string in Minkowski spacetime, the Nambu–Goto action does not depend on the choice of coordinates chosen to parameterize the string world-sheet. In other words, it is invariant under any reparameterization of the world-sheet coordinates. This is the same kind of symmetry that one has in Einstein's theory of general relativity, where one describes the four-dimensional spacetime by its metric $g_{\mu\nu}$ and requires an invariance under any choice of coordinates. Analogously, a two-dimensional world-sheet metric $g_{\alpha\beta}$ can also be introduced by rewriting the Nambu–Goto action in the following equivalent form:

$$S = -\frac{1}{4\pi\alpha'} \int d\tau \int d\sigma \sqrt{-g}\, g^{\alpha\beta} \partial_\alpha X \cdot \partial_\beta X. \qquad (39.2)$$

This action is now a functional of both the string coordinate $X(\tau, \sigma)$ and the metric $g_{\alpha\beta}(\tau, \sigma)$. Equation (39.1) can be recovered from Eq. (39.2) as follows. We write the equation of motion for the metric, corresponding to the vanishing of the world-sheet energy-momentum tensor,

$$\theta_{\alpha\beta} \equiv \partial_\alpha X \cdot \partial_\beta X - \frac{1}{2} g_{\alpha\beta} g^{\gamma\delta} \partial_\gamma X \cdot \partial_\delta X = 0, \qquad (39.3)$$

The Birth of String Theory, ed. Andrea Cappelli, Elena Castellani, Filippo Colomo and Paolo Di Vecchia.
Published by Cambridge University Press. © Cambridge University Press 2012.

that leads to the following identity:

$$\sqrt{-\det(\partial_\alpha X \cdot \partial_\beta X)} = \frac{1}{2}\sqrt{-g}\,g^{\alpha\beta}\partial_\alpha X \cdot \partial_\beta X. \tag{39.4}$$

Using it in the action (39.2) one gets back (39.1). Therefore, the two actions are classically equivalent. That in Eq. (39.2) is the action of scalar fields in curved spacetime and has the advantage of being quadratic in the string coordinate. As later shown by Polyakov, this property allows for its complete quantization in the path-integral formalism.

In the mid-Seventies, it was clear that one could choose a particular gauge, the orthonormal gauge (later called conformal gauge) characterized by $g_{\alpha\beta} \sim \eta_{\alpha\beta}$, in which Eq. (39.2) reduces to the action of a free scalar theory:

$$S = -\frac{1}{4\pi\alpha'}\int d\tau \int d\sigma\, \partial_\alpha X \cdot \partial^\alpha X, \tag{39.5}$$

while the equation of motion for $g_{\alpha\beta}$ imposes the vanishing of the world-sheet energy-momentum tensor in Eq. (39.3). This vanishing is a consequence of the invariance under reparameterizations of the world-sheet coordinates: since these coordinates can be chosen arbitrarily, there can be neither energy nor momentum on the world-sheet. Having understood all this, one could more easily work in the conformal gauge, where the Lagrangian was that of a free theory with the constraint (39.3).

On the other hand, for the fermionic string, no equivalent of the Nambu–Goto action, nor of its extension in the form (39.2), was known. What was known, however, was a linear Lagrangian, (see Gervais and Sakita [GS71] and also Zumino [Zum73]), generalizing Eq. (39.5) and containing, together with a scalar field X^μ, also a Majorana fermion ψ^μ, namely:

$$L = -\frac{1}{4\pi\alpha'}\left[\partial_\alpha X \cdot \partial^\alpha X - i\bar{\psi}\gamma^\alpha \cdot \partial_\alpha \psi\right]. \tag{39.6}$$

This Lagrangian has a Bose–Fermi symmetry [GS71], lately called supersymmetry, in the two-dimensional world-sheet of the string, and was the starting point for the generalization by Wess and Zumino of this symmetry to four dimensions [WZ74]. As a consequence of this symmetry, there is a fermionic current, the supercurrent J_α, that is conserved.

In analogy with what was done in the case of the bosonic string, one assumed here that, whatever the fermionic string Lagrangian was, it reduced in a particular gauge to the form (39.6) with the addition of the constraints:

$$\theta_{\alpha\beta} = \partial_\alpha X \cdot \partial_\beta X - \frac{1}{2}\eta_{\alpha\beta}\eta^{\gamma\delta}\partial_\gamma X \cdot \partial_\delta X$$

$$- \frac{i}{2}\bar{\psi}\cdot\left[\gamma_\alpha\partial_\beta + \gamma_\beta\partial_\alpha - \eta_{\alpha\beta}\gamma^\gamma\partial_\gamma\right]\psi = 0, \tag{39.7}$$

$$J_\alpha = \gamma^\beta\gamma_\alpha\partial_\beta X \cdot \psi = 0,$$

that imply the vanishing of both the total energy-momentum tensor and the supercurrent. In this way, in analogy with the bosonic string, one could proceed without knowing the underlying nonlinear Lagrangian (attempts to construct such a Lagrangian were made by

Iwasaki and Kikkawa [IK73] and Chang, Macrae and Mansouri [CMM75, CMM76], but none was fully satisfactory).

On the other hand, discussing with Lars Brink, who came regularly to Copenhagen, profiting from the fact that I was working at the Nordic institution, Nordita, and with Paul Howe, who was a postdoc at the Niels Bohr Institute from September 1975, we felt that we should make an effort to construct the nonlinear Lagrangian for the fermionic string. However, it was not an easy task because at that time nobody had yet constructed a Lagrangian with local supersymmetry. Therefore, we decided to practise first with the fermionic particle, i.e. the Dirac particle which was the zero-slope limit of the fermionic string. This was the limit in which one neglects all oscillators keeping only the zero modes. For the scalar particle it was known that one had two equivalent formulations, in terms of the two following actions:

$$S_1 = -m \int d\tau \sqrt{-\dot{x}^2}, \quad S_2 = \frac{1}{2} \int d\tau \left[\frac{\dot{x}^2}{e} - m^2 e \right]. \tag{39.8}$$

The first action is proportional to the length of the trajectory spanned in Minkowski space-time and is the analogue of the Nambu–Goto action for a pointlike particle, while the second corresponds to the other string action in Eq. (39.2), where e is the einbein, related to the one-dimensional metric by $g = e^2$. Proceeding as in the case of the string, one can easily show that they are classically equivalent. The next step was to generalize the two previous actions to the case of a Dirac point particle, as described in the following.

In the case of the massless fermionic particle, the Lagrangian in the proper-time gauge is given by:

$$L = \frac{1}{2} \left(\dot{x}^2 - i\psi \cdot \dot{\psi} \right), \tag{39.9}$$

supplemented by the constraints

$$\dot{x}^2 = \dot{x} \cdot \psi = 0. \tag{39.10}$$

The task was then to write a Lagrangian for the Dirac particle that reduced to the one in (39.9) in a particular gauge and also gave the constraints (39.10). We were working on this problem when Stanley Deser visited Copenhagen in the spring of 1976, to give a seminar on the newly found action for supergravity [DZ76a] that he had worked out together with Bruno Zumino. We discussed our problem with him and rather quickly we found the locally supersymmetric and reparameterization invariant Lagrangian for the massless Dirac particle (Brink, Deser, Zumino, Di Vecchia and Howe [BDZD76]):

$$L = \frac{1}{2} \left[\frac{\dot{x}^2}{e} - \psi \dot{\psi} - \frac{i}{e} \chi \psi \cdot \dot{x} \right]. \tag{39.11}$$

If we compare this equation with (39.9) we see that it involves two additional world-line 'fields' e and χ. The first one is the einbein, discussed above, that makes the Lagrangian (39.11) invariant under an arbitrary reparameterization of the world-line coordinate, while the second one is the 'gravitino' that makes it invariant under local supersymmetry. The introduction of these two world-line fields is essential to have full reparameterization and local supersymmetry invariance.

A mass could also be given to the particle by introducing an additional Grassmann variable ψ_5 and writing the following Lagrangian:

$$L = \frac{1}{2}\left[\frac{\dot{x}^2}{e} - i(\psi\cdot\dot\psi + \psi_5\dot\psi_5) - \frac{i}{e}\chi\psi\cdot\dot{x} - m^2 e - im\chi\psi_5\right]. \qquad (39.12)$$

By varying the previous action with respect to e and χ, one gets the two equations:

$$-\frac{\dot{x}^2}{e^2} + \frac{i}{e^2}\chi\psi\cdot\dot{x} - m^2 = 0,$$
$$\frac{i}{e}\psi\cdot\dot{x} + im\psi_5 = 0. \qquad (39.13)$$

These equations, when put back into the Lagrangian in Eq. (39.12), together with the equation of motion for ψ_5, give the Lagrangian for the Dirac particle that was found independently by Barducci, Casalbuoni and Lusanna [BCL76]:

$$L = -m\sqrt{(\dot{x} + \frac{i}{m}\psi\psi_5)^2} - \frac{i}{2}\psi\cdot\dot\psi - \frac{i}{2}\psi_5\dot\psi_5. \qquad (39.14)$$

Since the Lagrangian in Eq. (39.12) is invariant under reparameterizations and local supersymmetry transformations, we can always choose a gauge where $e = $ constant and $\chi = 0$ and with such a choice the constraints in Eq. (39.14) for the massless case reduce to those in Eq. (39.10) and the Lagrangian becomes that in Eq. (39.9).

Having understood the nonlinear Lagrangian for the pointlike Dirac particle, it was quite clear how to extend it to the case of the fermionic string. We had to construct a two-dimensional Lagrangian containing the coordinates X^μ and ψ^μ together with a zweibein and a gravitino that was invariant under arbitrary reparameterizations and local supersymmetry transformations. We agreed with Lars Brink that he would visit Copenhagen for a week during the month of August 1976, with the purpose of writing the fermionic string Lagrangian. I remember that we worked for the whole week, adding terms both to the Lagrangian and to the supersymmetry transformations, and by the end of the week we were able to write down the following Lagrangian [BDH76]:[1]

$$L \sim e\left[-\frac{1}{2}\partial_\alpha X\cdot\partial_\beta X\, g^{\alpha\beta} - \frac{i}{2}\bar\psi\gamma^\alpha\cdot\partial_\alpha\psi + \frac{i}{2}\bar\chi_\alpha\gamma^\beta\partial_\beta X\cdot\gamma^\alpha\psi \right.$$
$$\left. + \frac{1}{8}(\bar\chi_\alpha\gamma^\beta\gamma^\alpha\psi)\cdot(\bar\chi_\beta\psi)\right]. \qquad (39.15)$$

The first three terms were an obvious and direct generalization of the three terms in Eq. (39.11), while the last term was not obvious at all and required a lot of spinor relations known as Fierz identities. This was the desired result: variation of the action with respect

---

[1] The same Lagrangian was also constructed independently by Deser and Zumino [DZ76b].

to $g^{\alpha\beta}$, or more precisely the zweibein, and χ_α gives the constraints (39.7); moreover, the gauge choice $g_{\alpha\beta} \sim \eta_{\alpha\beta}$ and $\chi_\alpha \sim 0$ yields the quadratic action (39.6).

After the completion of this work, I decided not to continue studying string theory because practically nobody was interested in it anymore and, if I had remained in this field, I would never have got a permanent job. I profited from the fact that Sasha Polyakov was visiting the Niels Bohr Institute for three months during the fall of 1976 to learn the new developments on instantons and a few months later I started to work in this new field. When I left Copenhagen in October 1978 I threw away all my notes on string theory, and thought that I would never work in this field again in the future. This was not because I believed that string theory was uninteresting. On the contrary, I was still very fascinated by it, but it was the reaction due to my frustration that I could not continue to work on it by myself when research in particle physics moved in other directions.

I heard again something about activity in string theory in 1981, from Lars Brink, who came back from Moscow with the famous paper by Polyakov [Pol81] where, starting from the Lorentz covariant action in Eq. (39.2), he computed the path integral and found the dimension $d = 26$.

Given my past frustrations after having worked for several years on early string theory, I remember that it took me several months before I could look at the paper and understand what Sasha had done. This was made easier by the fact that in the summer of 1981 I visited the Niels Bohr Institute and found there Bergfinnur Durhuus, Jens Lyng Petersen and Poul Olesen studying Polyakov's paper. From that period I returned to work on string theory.

Nowadays I think that it was good for me to have learnt several aspects of perturbative and nonperturbative field theory before going back to string theory in the Eighties. Or maybe even better, as correctly stressed by Lars Brink, we did not know that in fact we were not leaving string theory because all this is nowadays part of string theory.

Acknowledgements

I thank Lars Brink for reading the manuscript and for his useful comments and suggestions.

> Paolo Di Vecchia was born in Terracina (Italy) in 1942. He obtained his 'Laurea' in physics at the University of Rome in 1966, under the supervision of Bruno Touschek. He then worked in Frascati, MIT, CERN, Nordita and Berlin. He became Professor at Wuppertal University in 1980 and at Nordita in Copenhagen in 1986. Since the move of Nordita to Stockholm, he has divided his time between the Niels Bohr Institute in Copenhagen and Nordita in Stockholm. He has worked on several aspects of theoretical particle physics using both perturbative and nonperturbative methods in field and string theory.

References

[BCL76] Barducci, A., Casalbuoni, R. and Lusanna, L. (1976). Supersymmetries and the pseudoclassical relativistic electron, *Nuovo Cimento* **A35**, 377–399.

[BDZD76] Brink, L., Deser, S., Zumino, B., Di Vecchia, P. and Howe, P. S. (1976). Local supersymmetry for spinning particles, *Phys. Lett.* **B64**, 435–438.

[BDH76] Brink, L., Di Vecchia, P. and Howe, P. S. (1976). A locally supersymmetric and reparameterization invariant action for the spinning string, *Phys. Lett.* **B65**, 471–474.

[CMM75] Chang, L. N., Macrae, K. and Mansouri, F. (1975). A new supersymmetric string model and the supergauge constraints in the dual resonance model, *Phys. Lett.* **B57**, 59–62.

[CMM76] Chang, L. N., Macrae, K. and Mansouri, F. (1976). Geometrical approach to local gauge invariance: local gauge theories and supersymmetric strings, *Phys. Rev.* **D13**, 235–249.

[DZ76a] Deser, S. and Zumino, B. (1976). Consistent supergravity, *Phys. Lett.* **B62**, 335–337.

[DZ76b] Deser, S. and Zumino, B. (1976). A complete action for the spinning string, *Phys. Lett.* **B65**, 369–373.

[GS71] Gervais, J. L. and Sakita, B. (1971). Field theory interpretation of supergauges in dual models, *Nucl. Phys.* **B34**, 632–639.

[IK73] Iwasaki, Y. and Kikkawa, K. (1973). Quantization of a string of spinning material Hamiltonian and Lagrangian formulation, *Phys. Rev.* **D8**, 440–449.

[Pol81] Polyakov, A. M. (1981). Quantum geometry of bosonic strings, *Phys. Lett.* **B103**, 207–210.

[WZ74] Wess, J. and Zumino, B. (1974). Supergauge transformations in four dimensions, *Nucl. Phys.* **B70**, 39–50.

[Zum73] Zumino, B. (1973). Lectures at the 1973 Capri Summer School, published as *Renormalization and Invariance in Quantum Theory*, ed. Caianiello, E. (Plenum Press, New York, 1974).

40
Personal recollections

EUGÈNE CREMMER

In 1969, I was finishing my PhD in Orsay under the supervision of Michel Gourdin on phenomenological works related to e^+e^- annihilations. These works allowed me to deepen my knowledge of particle physics as well as to learn how to master difficult calculations. However, I was more attracted by more formal research (which we would like to be able to say was more fundamental!). The theoretical physics laboratory in Orsay was also hit by the explosion of activity which followed the paper of Veneziano, and a group of people began to work on dual models: Bouchiat, Gervais, Nuyts, Amati who was spending a sabbatical in Orsay, as well as younger researchers Neveu, Scherk and Sourlas. My first encounter with dual models was to rebel against this too fashionable growing activity. With Jean Nuyts I asked whether there could be some other example of an s–t dual amplitude with only poles in both channels: we found an example with poles lying on a logarithmic trajectory instead of a linear one [CN71]. However, this amplitude led to many unsatisfactory physical conclusions and I joined the mainstream.

40.1 Factorizing the Pomeron with Joël Scherk at CERN

In 1970, much work was devoted to the dual multiloop amplitude, in particular by Kaku and Yu [KY70]. Lovelace [Lov70] and Alessandrini [Ale71] showed the relation between these amplitudes and the Neumann function associated with a sphere with handles. I then began a rather technical work, the study of a multiloop operator and its dual properties [Cre71]. I had the opportunity to obtain a fellow's position at the Theory Division at CERN for two years (1971 and 1972). This was the first lucky event of my career. The second one was to 'integrate' the dual group at CERN which was led by Daniele Amati. In this group, there was not only a very stimulating scientific atmosphere, but also a very friendly one. At that time the Theory Division had more than a hundred people and the fellows who were not associated with a group could easily become isolated. Then my third and most important lucky event occurred, the arrival of Joël Scherk and the beginning of our collaboration. Of course I knew Joël from Orsay and had had discussions with him although we did not

The Birth of String Theory, ed. Andrea Cappelli, Elena Castellani, Filippo Colomo and Paolo Di Vecchia.
Published by Cambridge University Press. © Cambridge University Press 2012.

belong to the same 'team' (Bouchiat and Meyer for Joël, Gourdin for me), but I had had no opportunity to work with him. As early as that stage, Joël was a profound physicist. He was very quiet and efficient. His notes were very clear and are still readable many years later (contrary to mine!), and the detailed calculations were always accompanied by some commentary.

After some works on multicurrent dual amplitudes [CS72a, CS73], where we had some mathematical fun with domain variational theory on Riemann surfaces, we came back to a less technically demanding problem but more irritating one. The one-loop diagrams (and multiloop) are constructed from unitarity and should have analytic properties resulting from unitarity (namely unitarity cuts and eventually poles). However, in the nonplanar orientable case a new singularity violating unitarity appears in the channel with zero isospin (associated with the 'Pomeron'). Lovelace [Lov71] showed that when the intercept is 1, this singularity is factorizable; furthermore he conjectured that, in 26 dimensions, two sets of oscillators are cancelled by gauge conditions leading to a modification of the amplitude which has no longer cuts but poles, so that unitarity is recovered. When the intercept is 1 (implying the existence of a massless vector particle) and the dimension is less than or equal to 26 (for the pure bosonic model), the Dual Resonance Model is ghost free (the time dimension is eliminated by the gauge invariance associated with the intercept 1). In 26 dimensions, null states appear that should also be eliminated. The complete formula for the loop amplitude was proven later, when the projector on the physical states had been constructed.

Starting with the conjectured form of the amplitude [CS72b], we factorized the Reggeon and the Pomeron poles simultaneously. The Pomeron sector looks quite similar to the Shapiro–Virasoro model with a slope half that of the Reggeon trajectory and an intercept equal to 2 (implying a massless spin two particle). In string language, this will correspond to the transition from an open string to a closed string. From the two nonplanar loop diagram with three external Reggeons, it is possible to extract a three-Pomeron vertex. The appearance of this sector of new particles shows that the Veneziano model (open strings) is not consistent alone and that we must also include the Shapiro–Virasoro model (closed strings). This result was obtained independently by Clavelli and Shapiro [CS73], who extended it to the Ramond–Neveu–Schwarz model in 10 dimensions.

40.2 Combining and splitting strings with Jean-Loup Gervais in Orsay

Back in Orsay in 1973, I began a collaboration with Jean-Loup Gervais. The introduction of the string picture had improved tremendously our physical understanding of dual models. Associated with the string picture there was the functional approach to dual theories. The initial works of Hsue, Sakita and Virasoro [HSV70] and of Gervais and Sakita [GS71a, GS71b] were plagued by their inability to project out the ghost states; however, they explicitly exhibited two important properties of dual models which were not so transparent in the operator formalism, namely duality and the connection between loop amplitudes and Neumann functions. As far as the Veneziano model is concerned, crucial progress was made by Goddard *et al.* [GGRT73]: they showed that because the Lagrangian of the free relativistic string is gauge invariant, if one chooses the appropriate gauge, then one needs only to quantize the transverse components of the string variable.

Gervais and Sakita [GS73] subsequently wrote the path integral associated with transition probabilities of strings with this gauge condition, in such a way that one can perform the functional integration and obtain the original Veneziano amplitude. Later Mandelstam [Man74], starting from this amplitude, gave a complete prescription for dealing with external excited states as described in [GGRT73]. He proved that the resulting amplitude is Lorentz invariant only in 26 spacetime dimensions and that the three-Reggeon vertex coincides with the one given by Ademollo *et al.* [ADDF74], thus establishing complete correspondence between the string and the operator formalisms of the Veneziano model. Kaku and Kikkawa [KK74] introduced a multistring formalism such that the topological structure of the perturbation series is identical to that of the dual theory, each dual amplitude being obtained as a sum of several Feynman graph contributions. This formalism, based on a functional treatment of the string variable, remained ambiguous because of the lack of precise definition of the functional integration; actually, a careful determination of the functional integration measure is necessary to obtain Lorentz invariant amplitudes that coincide with dual amplitudes.

Gervais and I overcame this problem by developing an infinite component field theory of interacting relativistic strings starting from the operator approach. In particular, we introduced a three-string vertex either for two incoming strings → one outgoing string (combining strings) or for one incoming string → two outgoing strings (splitting strings). It was simply defined as the overlapping of the three strings at a given time. We showed that it was related to the ADDF three-Reggeon vertex by allowing each of the three strings to propagate for a very long time [CG74]. We then showed [CG75] that in order to get the correct two incoming strings → two outgoing strings amplitude, it was necessary to add a direct four-string interaction to the sum of two Feynman diagrams constructed from the three-string vertex and a propagator. This defined the four-string vertex introduced by Kaku and Kikkawa, who had shown that three- and four-string vertices were sufficient in the tree approximation. The formalism could be extended to all orders and was satisfactory from the point of view of precise definitions, but it was very hard to use in practice, as had already been observed in the four-string amplitude.

40.3 Compactifying strings with Joël Scherk at LPTENS

In October 1974, a group of physicists of the theoretical physics laboratory in Orsay (essentially the Bouchiat and Meyer group that I joined on my return from CERN) moved to Paris and founded the 'Laboratoire de Physique Théorique de l'École Normale Supérieure'.

A remarkable feature of dual models is the prediction that the dimension of spacetime is 26 for the Veneziano and Shapiro–Virasoro models, and 10 for the Ramond–Neveu–Schwarz model. Unfortunately these predictions are rather unphysical, moreover these models predict zero-mass particles and are therefore incompatible with hadronic physics. It is worth remembering that the same 'avatar' occurred to Yang–Mills theory. This led Scherk and Schwarz [SS74a, SS74b] as well as Yoneya [Yon73, Yon74] to study the connection between dual models and general relativity, in particular in the zero-slope limit, where the connection with field theory was known. In 1975, Scherk and Schwarz [SS75] made the really daring proposal that dual models should be interpreted as a quantum theory of gravity unified with the other forces between quarks and gluons. They suggested that considering some of the dimensions to be compact does not lead to any contradiction within the framework of dual models.

Scherk and I proved that this assertion was indeed correct [CS76]. This was the beginning of a new and very fruitful collaboration with Joël. We defined the theory of open and closed strings on a compact space (chosen to be a hypertorus). In a field theory on a compact space, the momenta in the compactified directions (hypertorus of radii R_i) are quantized ($p_i = n_i/R_i$) and to any field is associated an infinite Kaluza–Klein tower of fields living in the remaining uncompactified dimensions. The same results are obtained in the compactification of open strings. For closed strings the effect is less trivial, although simple. One must introduce another integer number (winding number) m_i corresponding to how many times a string wraps around the torus before closing. The new states corresponding to the quantum numbers n_i and m_i now have the mass M_i given by:

$$M_i^2 = n_i^2/R_i^2 + m_i^2 R_i^2/\alpha'^2. \qquad (40.1)$$

The corresponding modifications in the computation of loops (such as replacing the integration on the momentum flowing in the loop with summations on the quantized momenta) did not affect all the good results, namely, the absence of nonphysical singularities in the nonplanar orientable loop and the appearance of new particles associated with the compactified closed string. I must confess that our field theory prejudice on the difference between the two limits, $R \to \infty$, and $R \to 0$, prevented us from discovering the T-duality of closed string theory, namely the complete symmetry:

$$n_i \to m_i, \qquad m_i \to n_i, \qquad R_i \to \alpha'/R_i. \qquad (40.2)$$

Generalized to the torus associated with a group, this compactification of closed strings was to lead to the construction of the heterotic string (Gross, Harvey, Martinec and Rohm [GHMR85a, GHMR85b, GHMR86]). It is important also to note that this kind of compactification is equivalent to the introduction of quantum numbers in string theory by Bardakci and Halpern [BH71].

At that time, I turned to a new direction of work still with Joël Scherk – until his unfortunate death in 1980 – and others. This became my Supergravity Era, but that is another story.

Acknowledgements

It is a pleasure to thank Nicole Ribet for a careful reading of the manuscript as well as for useful comments.

Eugène Cremmer was born in Paris in 1942. He carried out his undergraduate and graduate studies at École Normale Supérieure in Paris, and obtained his PhD at Orsay in 1967. He has held research positions at CNRS, Orsay (1966–1970, 1973–1974), and at CERN Theory Division (1971–1972). From 1974 he was CNRS Researcher at the Laboratoire de Physique Théorique de l'École Normale Supérieure, which he directed from 2002 to 2006, and where he is currently Emeritus CNRS researcher.

References

[ADDF74] Ademollo, M., Del Giudice, E., Di Vecchia, P. and Fubini, S. (1974). Couplings of 3 excited particles in dual-resonance model, *Nuovo Cimento* **A19**, 181–203.

[Ale71] Alessandrini, V. (1971). A general approach to dual multiloop diagrams, *Nuovo Cimento* **A2**, 321–352.

[BH71] Bardakci, K. and Halpern, M. B. (1971). New dual quark models, *Phys. Rev.* **D3**, 2493–2506.

[CS73] Clavelli, L. and Shapiro, J. (1973). Pomeron factorization in general dual models, *Nucl. Phys.* **B57**, 490–535.

[Cre71] Cremmer, E. (1971). Some properties of a dual multiloop planar operator free of spurious states, *Nucl. Phys.* **B31**, 477–511.

[CG74] Cremmer, E. and Gervais, J. L. (1974). Combining and splitting relativistic strings, *Nucl. Phys.* **B76**, 209–230.

[CG75] Cremmer, E. and Gervais, J. L. (1975). Infinite component field-theory of interacting relativistic strings and dual theory, *Nucl. Phys.* **B90**, 410–460.

[CN71] Cremmer, E. and Nuyts, J. (1971). Dual resonance models with residues having a maximum fixed finite number of monomials in one Mandelstam variable: logarithmic poles, *Nucl. Phys.* **B26**, 151–166.

[CS72a] Cremmer, E. and Scherk, J. (1972). Currents and green functions for dual models (ii) off-shell dual amplitudes, *Nucl. Phys.* **B48**, 29–77.

[CS72b] Cremmer, E. and Scherk, J. (1972). Factorization of pomeron sector and currents in dual resonance model, *Nucl. Phys.* **B50**, 222–252.

[CS73] Cremmer, E. and Scherk, J. (1973). Regge limit and scaling in a dual model of currents, *Nucl. Phys.* **B58**, 557–597.

[CS76] Cremmer, E. and Scherk, J. (1976). Dual models in four dimensions with internal symmetries, *Nucl. Phys.* **B103**, 399–425.

[GS71a] Gervais, J. L. and Sakita, B. (1971). Functional-integral approach to dual-resonance theory, *Phys. Rev.* **D4**, 2291–2308.

[GS71b] Gervais, J. L. and Sakita, B. (1971). Generalization of dual models, *Nucl. Phys.* **B34**, 477–492.

[GS73] Gervais, J. L. and Sakita, B. (1973). Ghost-free string picture of Veneziano-model, *Phys. Rev. Lett.* **30**, 716–719.

[GGRT73] Goddard, P., Goldstone, J., Rebbi, C. and Thorn, C. B. (1973). Quantum dynamics of a massless relativistic string, *Nucl. Phys.* **B56**, 109–135.

[GHMR85a] Gross, D. J., Harvey, J. A., Martinec, E. J. and Rohm, R. (1985). Heterotic string, *Phys. Rev. Lett.* **54**, 502–505.

[GHMR85b] Gross, D. J., Harvey, J. A., Martinec, E. J. and Rohm, R. (1985). Heterotic string theory 1. The free heterotic string, *Nucl. Phys.* **B256**, 253–284.

[GHMR86] Gross, D. J., Harvey, J. A., Martinec, E. J. and Rohm, R. (1986). Heterotic string theory 2. The interacting heterotic string, *Nucl. Phys.* **B267**, 75–124.

[HSV70] Hsue, C. S., Sakita, B. and Virasoro, M. A. (1970). Formulation of dual theory in terms of functional integrations, *Phys. Rev.* **D2**, 2857–2868.
[KK74] Kaku, M. and Kikkawa, K. (1974). The field theory of relativistic strings, Pt. 1. trees, *Phys. Rev.* **D10**, 1110–1133.
[KY70] Kaku, M. and Yu, L. (1970). The general multi-loop Veneziano amplitude, *Phys. Lett.* **B33**, 166–170.
[Lov70] Lovelace, C. (1970). *M*-loop generalized Veneziano formula, *Phys. Lett.* **B32**, 703–708.
[Lov71] Lovelace, C. (1971). Pomeron form factors and dual Regge cuts, *Phys. Lett.* **B34**, 500–506.
[Man74] Mandelstam, S. (1974). Interacting-string picture of Neveu–Schwarz–Ramond model, *Nucl. Phys.* **B69**, 77–106.
[SS74a] Scherk, J. and Schwarz, J. H. (1974). Dual models for nonhadrons, *Nucl. Phys.* **B81**, 118–144.
[SS74b] Scherk, J. and Schwarz, J. H. (1974). Dual models and the geometry of spacetime, *Phys. Lett.* **B52**, 347–350.
[SS75] Scherk, J. and Schwarz, J. H. (1975). Dual field theory of quarks and gluons, *Phys. Lett.* **B57**, 463–466.
[Yon73] Yoneya, T. (1973). Quantum gravity and the zero slope limit of the generalized Virasoro model, *Lett. Nuovo Cimento* **8**, 951–955.
[Yon74] Yoneya, T. (1974). Connection of dual models to electrodynamics and gravidynamics, *Prog. Theor. Phys.* **51**, 1907–1920.

41
The scientific contributions of Joël Scherk

JOHN H. SCHWARZ

Abstract

Joël Scherk (1946–1980) was an important early contributor to the development of string theory. Together with various collaborators, he made numerous profound and influential contributions to the subject throughout the decade of the Seventies. On the occasion of a conference at the École Normale Supérieure in 2000 that was dedicated to the memory of Joël Scherk, I gave a talk entitled 'Reminiscences of collaborations with Joël Scherk' [Sch00]. The present Chapter, an expanded version of that presentation, also discusses work in which I was not involved.

41.1 Introduction

Joël Scherk was one of the most brilliant French theoretical physicists who emerged in the latter part of the Sixties. Together with André Neveu, he was educated at the École Normale Supérieure in Paris, and in Orsay. Together, they studied electromagnetic and final-state interaction corrections to nonleptonic kaon decays [NS70b] under the guidance of Claude Bouchiat and Philippe Meyer. They both defended their 'thèse de troisième cycle' (the French equivalent of a PhD) in 1969, and they were hired together by the CNRS that year (tenure at age 23!). In September 1969 the two of them headed off to Princeton University.

In 1969 my duties as an Assistant Professor in Princeton included advising some assigned graduate students. The first advisees, who came together to see me, were André Neveu and Joël Scherk. I had no advance warning about them, and so I presumed they were just another pair of entering students. I did not know that they had permanent jobs in France. Because their degrees were not called PhDs, Princeton University classified them as graduate students, and so (by the luck of the draw) they were assigned to me. At our first meeting, I asked the usual questions: 'Do you need to take a course on electrodynamics?', 'Do you need to take a course on quantum mechanics?', etc. They assured me that they had already learned all that, so it would not be necessary. I said okay, signed their cards, and they left.

41.2 Loop amplitudes

Veneziano discovered his famous formula for a four-particle amplitude in 1968 [Ven68]. In 1969 various groups constructed N-particle generalizations of the Veneziano amplitude (Bardakci and Ruegg [BR69], Goebel and Sakita [GS69], Koba and Nielsen [KN69a, KN69b] and Chan and Tsou [CT69]) and showed that they could be factorized consistently on a well-defined spectrum of single-particle states as required for the tree approximation of a quantum theory (Fubini and Veneziano [FGV69, FV69, FV70] and Bardakci and Mandelstam [BM69]). In those days the theory in question was called the 'Dual Resonance Model'. Today we would refer to it as the bosonic string theory. Knowing the tree approximation spectrum and couplings, it became possible to construct one-loop amplitudes. The first such attempt was made by Kikkawa, Sakita and Virasoro [KSV69]. They did not have enough information in hand to do it completely right, but they pioneered many of the key ideas and pointed the way for their successors. Around this time (the fall of 1969) I began studying these one-loop amplitudes in collaboration with David Gross, who was also an Assistant Professor at Princeton.

A couple months after our first meeting, Joël and André reappeared in my office and said that they had found some results they would like to show me. They proceeded to explain their analysis of the divergence in the planar one-loop amplitude. They had realized that by performing a Jacobi transformation of the theta functions in the integrand they could isolate the divergent piece and propose a fairly natural counterterm [NS70a]. I was very impressed by this achievement. It certainly convinced me that they did not need to take any basic courses! The modern interpretation of their result is that, viewed in a dual channel, there is a closed string going into the vacuum. The divergence can be attributed to the tachyon in that channel, and its contribution is the piece that they subtracted. This interpretation explains why such divergences do not occur in theories without closed string tachyons.

Since André and Joël were working on problems that were very closely related to those that David and I were studying, we decided to join forces. Even though we were not yet thinking in terms of string world-sheets, we were able to relate the classification of DRM amplitudes to the topological classification of Riemann surfaces with boundaries (Gross, Neveu, Scherk and Schwarz [GNSS70a]). Another discovery by the four of us was that the nonplanar loop amplitude contains unexpected singularities [GNSS70b]. (Essentially the same calculation was carried out independently and simultaneously by Susskind and Frye [FS70].) These appeared in addition to the expected two-particle threshold singularities. We assumed, of course, that the dimension of spacetime is four, since nobody had yet suggested otherwise. The Virasoro constraints, which should be taken into account for the internal states circulating in the loop, had not yet been discovered. As a result, the singularities that we found were unitarity violating branch points. We wanted to identify the leading Regge trajectory associated with these singularities with the Pomeron, since it carried vacuum quantum numbers, but clearly something was not quite right.

About a year later Lovelace observed that if one chooses the spacetime dimension to be 26, and supposes that the Virasoro conditions imply that only transverse excitations contribute, then instead of branch points the singularities would be poles [Lov71], which could be interpreted as new states in the spectrum without violating unitarity. As we now know, these are the closed string states in the nonplanar open string loop. Nowadays this is interpreted as 'open string/closed string duality' of the cylinder diagram. This calculation showed that unitarity requires choosing the dimension to be 26 and the Regge intercept value to be one, since these are requirements for the Virasoro conditions to be satisfied.

Joël left Princeton in the spring of 1970 to spend about six months in Berkeley. While he was there he collaborated with Michio Kaku, who was a student of Stanley Mandelstam at the time, studying divergences in multiloop planar amplitudes [KS71a, KS71b]. This was a very ambitious project, given the state of the art in those days.

41.3 Orsay, CERN, and NYU

Scherk's NATO Fellowship was only good for one year, so following his visits to Princeton and Berkeley, he returned to Orsay for a year followed by two years at CERN. While he was at Orsay, Joël pioneered the idea of considering string theories (or dual models) in a zero-slope limit, which is equivalent to a low energy limit [Sch71]. In particular, Joël and André studied the massless open string spin one states and showed that in a suitable low energy limit they interacted precisely in agreement with Yang–Mills theory [NS72a], and they studied the associated gauge invariance that this implied for the full string theory [NS72b]. These studies made it clear that open strings and their interactions could be viewed as short-distance modifications of Yang–Mills theory. This important observation certainly influenced the future evolution of the subject.

During the two-year period that Joël spent at CERN, beginning in August 1971, he shared an office with Lars Brink. At first, he collaborated with Eugène Cremmer on the study of currents and off-shell amplitudes in string theory as well as factorization in the closed string (or 'Pomeron') sector [CS72a, CS72b, CS73]. After that, he collaborated with Lars Brink, David Olive and Claudio Rebbi on the study of scattering amplitudes for Ramond fermions [BORS73, BOS73, OS73].

By now, Joël was very well known and in much demand. He decided to accept invitations from NYU and Caltech for the academic year 1973–1974, spending the fall term at NYU and the winter and spring terms at Caltech. During his visit to NYU, he wrote a very elegant review paper on string theory [Sch75]. This article was based on a course that he taught at NYU. He also completed a paper with Cremmer concerning gauge symmetry breaking [CS74] in the string theory context.

41.4 String theory for unification

In 1972 I left Princeton and moved to Caltech. At Caltech, Murray Gell-Mann provided funds for me to invite collaborators of my choosing. One of them was Joël Scherk, who spent the first half of 1974 visiting Caltech. The timing could not have been better.

The hadronic interpretation of string theories was plagued not only by the occurrence of massless vector particles in the open string spectrum, but by a massless tensor particle in the closed string spectrum, as well. Several years of effort were expended on trying to modify each of the two string theories (bosonic and RNS) so as to lower the leading open string Regge intercept from 1 to 1/2 and the leading closed string Regge intercept from 2 to 1, since these were the values required for the leading meson and Pomeron Regge trajectories. Some partial successes were achieved, but no fully consistent scheme was found. Efforts to modify the critical spacetime dimension from 26 or 10 to four also led to difficulties.

By 1974, almost everyone who had been working on string theory had dropped it and moved to greener pastures. The Standard Model had been developed, and it was working splendidly. Against this backdrop, Joël and I stubbornly decided to return to the nagging unresolved problems of string theory. We felt that string theory has such a compelling mathematical structure that it ought to be good for something. Before long our focus shifted to the question of whether the massless spin two particle in the spectrum interacted at low energies in accordance with the dictates of general relativity, so that it might be identified as a graviton. As was mentioned previously, Joël and André Neveu had studied the massless open string spin one states a few years earlier and showed that in a suitable low energy limit they interacted precisely in agreement with Yang–Mills theory [Sch71, NS72a]. Now we wondered about the analogous question for the massless spin two closed string state. Roughly, what we proved was that critical string theories have the gauge invariances required to decouple unphysical polarization states [SS74a]. Then it followed on general grounds, which had been elaborated previously by Weinberg [Wei65], that the interactions at low energy must be those of general relativity.

Once we had digested the fact that string theory inevitably contains gravity we were very excited. We knew that string theory does not have ultraviolet divergences, because the short-distance structure is smoothed out, but any field theoretic approach to gravitation inevitably gives nonrenormalizable ultraviolet divergences. Evidently, the way to make a consistent quantum theory of gravity is to posit that the fundamental entities are strings rather than point particles [SS74a]. (For a more detailed discussion, see my other contribution in this Volume.)

Adopting this viewpoint essentially meant that the fundamental length scale of string theory, called the string scale, should be identified with the Planck scale in order to give the correct value for Newton's constant. The Planck scale thus replaces the QCD scale, which was the natural choice for the string scale in the context of hadron physics: a change of some 19 or 20 orders of magnitude.

Our 1974 paper proposed changing the goal of string theory to the problem of constructing a consistent quantum theory of gravity. Since we already knew from the earlier work of André and Joël that string theory also contains Yang–Mills gauge interactions, it was natural for us to propose further that string theory should describe all the other forces at the same time. This means interpreting string theory as a unified quantum theory of all fundamental particles and forces – an explicit realization of Einstein's dream. Moreover, we realized that the existence of extra dimensions could now be a blessing rather than a curse. After all, in a gravity theory the geometry of spacetime is determined dynamically, and one could imagine that this would require, or at least allow, the extra dimensions to form a very small compact space. We attempted to construct a specific compactification scenario in a subsequent paper [SS75b]. From today's vantage point, that construction looks rather primitive.

Tamiaki Yoneya realized independently that the massless spin two state of string theory interacts at low energy in accordance with the dictates of general relativity [Yon74]. Indeed, his paper appeared first, though we were not aware of it at the time. However, as Yoneya graciously acknowledges, Joël and I were the only ones to take the next step and to propose that string theory should be the basis for constructing a unified quantum theory of all forces. The recognition of that possibility represented a turning point in my research career. I found the case compelling, and I became committed to exploring its implications. I believe that Joël felt the same way. I still do not understand why it took another decade until a large segment of the theoretical physics world became convinced that string theory was the right approach to unification. (There were some people who caught on earlier, of course.) One of my greatest regrets is that Joël was not able to witness the impact that this idea eventually would have.

During Joël's Caltech visit, we also explored some aspects of gravity that could be affected by the string interpretation. One paper interpreted the three-form flux $H = dB$ in terms of spacetime torsion [SS74b]. Since we knew that the light-cone gauge is convenient for exploring certain aspects of string theory, we also attempted to formulate general relativity in the light-cone gauge [SS75a].

After leaving Caltech, Joël participated in a summer 1974 workshop on string theory at the Aspen Center for Physics, which I organized. This had been planned a year earlier, when there were still quite a few people working in string theory. My memory is fading, but I do not recall the participants showing much interest in our proposal to use string theory for unification.

After Aspen, Joël returned to Paris, since his group at Orsay had moved from Orsay to the Laboratoire de Physique Théorique of the École Normale Supérieure in Paris. Aside from a few months in Cambridge in 1977 and brief visits elsewhere, this is where Joël spent the remainder of his career. At the LPTENS, Joël resumed his collaboration with Eugène Cremmer. Soon, they turned out a pair of well-known papers [CS76a, CS76b] that grappled with issues raised by the unification interpretation of string theory. The first paper [CS76a] introduced the notion of 'winding numbers' for the first time. As is now well known, closed strings can wrap on cycles in the compact dimensions. This possibility was important in the later construction of the heterotic string as well as for the discovery of T-duality almost a decade later.[1] The second paper [CS76b] emphasized the idea that the compact internal spaces cannot be chosen arbitrarily; instead, they are fixed by the

---

[1] Eugène informs me that they failed to discover T-duality because of their field theory prejudices. This shows that even dedicated string theorists had such prejudices. See the Chapter by Eugène Cremmer for further discussion of this point.

mechanism that they called 'spontaneous compactification'. This means that they must be stable or metastable solutions of the equations of motion. Note that metastability implies massless moduli, which are ruled out experimentally by tests of the tensor nature of the gravitational force.

41.5 Spacetime supersymmetry

The super-Virasoro gauge symmetry of the Ramond–Neveu–Schwarz (RNS) model [NS71, Ram71] describes the superconformal symmetry of the two-dimensional worldsheet theory. The supersymmetry of the two-dimensional world-sheet action was described later in 1971 by Gervais and Sakita [GS71]. This was the first example of a supersymmetric quantum field theory. For about five years, the supersymmetry considered by string theorists only pertained to the two-dimensional world-sheet theory. It did not occur to us that there could also be supersymmetry in ten-dimensional spacetime.

Bruno Zumino, who was also at CERN when Joël was there, became very interested in the RNS string's gauge conditions associated with the two-dimensional superconformal algebra and discussed it at length with Joël and Lars Brink. His work on this subject is described in [Zum74]. Following that, he and Julius Wess began to consider the possibility of constructing four-dimensional field theories with analogous features. This resulted in their famous work [WZ74] on globally supersymmetric field theories in four dimensions. As a consequence of their paper (note that earlier results by Golfand and Likhtman [GL71] in the USSR were not known in the West at that time), supersymmetry quickly became an active research topic. A few years later came the discovery of supergravity theories (Freedman, van Nieuwenhuizen and Ferrara [FNF76], Deser and Zumino [DZ76]).

Lars Brink, Joël, and I constructed supersymmetric Yang–Mills theories in various dimensions [BSS77]. When this work was carried out, Lars and I were at Caltech and Joël was in Paris. We discovered that the requisite gamma-matrix identity required by these theories, $\gamma^m_{(ab}\gamma^m_{c)d} = 0$, is satisfied in dimensions 3, 4, 6, and 10. Dimensional reduction was also discussed, giving (among other things) the first construction of $\mathcal{N}=4$ super Yang–Mills theory in four dimensions.

At the same time as the work described above, Joël was collaborating with Ferdinando Gliozzi and David Olive on some closely related ideas [GSO76, GSO77]. This threesome is now referred to as GSO. The GSO papers also explored super Yang–Mills theories. Moreover, they took the next major step, which concerned the RNS string theory. They proposed a projection of the RNS spectrum – the 'GSO projection' – that removes roughly half of the states (including the tachyon). Specifically, in the bosonic (NS) sector they projected away the odd G-parity states, a possibility that was discussed earlier, and in the fermionic (R) sector they projected away half the states, keeping only certain definite chiralities. Then they counted the remaining physical degrees of freedom of the free string at each mass level. After the GSO projection the masses of open string states, for both bosons and fermions, are given by $\alpha' M^2 = n$, where $n = 0, 1, 2, \ldots$ Denoting the open-string degeneracies of states in the GSO-projected theory by $d_{NS}(n)$ and $d_R(n)$, they showed that these are encoded in the generating functions

$$f_{NS}(w) = \sum_{n=0}^{\infty} d_{NS}(n) w^n$$

$$= \frac{1}{2\sqrt{w}} \left[\prod_{m=1}^{\infty} \left(\frac{1 + w^{m-1/2}}{1 - w^m} \right)^8 - \prod_{m=1}^{\infty} \left(\frac{1 - w^{m-1/2}}{1 - w^m} \right)^8 \right],$$

and

$$f_R(w) = \sum_{n=0}^{\infty} d_R(n) w^n = 8 \prod_{m=1}^{\infty} \left(\frac{1 + w^m}{1 - w^m} \right)^8.$$

In 1829, Jacobi proved the remarkable identity [Jac29]

$$f_{NS}(w) = f_R(w),$$

though he used a different notation (and acknowledged an assistant named Scherk!). Thus, there is an equal number of bosons and fermions at every mass level, as required. This was compelling evidence (though not a proof) for 'ten-dimensional spacetime supersymmetry' of the GSO-projected theory. Prior to this work, one knew that the RNS theory has worldsheet supersymmetry, but the realization that it could have spacetime supersymmetry was a major advance. I was very delighted by this result. One could now envisage a tachyon-free string theory that would make sense as a starting point for a unified theory.

At about the same time as the GSO breakthrough, Joël collaborated with Sergio Ferrara, Dan Freedman, Peter van Nieuwenhuizen and Bruno Zumino on various different topics in supergravity [FFNB77]–[FSZ77b]. In the following year (1977) he continued studying supergravity, now collaborating mostly with Cremmer [CSF77]–[CS78]. Two of these papers [CS77a, CS77b] were written while Joël spent several months in the spring of 1977 at DAMTP in Cambridge.

Cremmer, Ferrara and Scherk formulated $\mathcal{N} = 4$ supergravity in a manifestly $SU(4)$-invariant form that was motivated by $\mathcal{N} = 1$ supergravity in ten dimensions, which was itself motivated by superstring theory. It exhibited an on-shell $SU(1, 1)$ duality invariance [CSF78]. This was the first discovery of the duality symmetries of extended supergravity theories. S-duality in string theory is such a duality, as was understood some 17 years later. The discovery of such dualities was very influential in the construction of $\mathcal{N} = 8$ supergravity by Cremmer and Bernard Julia [CJ78, CJ79]. The duality group in that case is $E_{7,7}$. Cremmer informs me that health problems prevented Joël from participating in that collaboration.

One of the most beautiful results in supergravity, and Joël's most cited paper, is the March 1978 construction of the action of eleven-dimensional supergravity, with Cremmer and Julia [CJS78]. It was immediately clear that eleven-dimensional supergravity is very beautiful, and it aroused a lot of interest. However, it was puzzling for a long time how it fits into the greater scheme of things and whether it has any connection to string theory. It took more than fifteen years to find the answer, with the discovery, by Townsend [Tow95] and Witten [Wit95], that type IIA superstring theory contains an eleventh dimension of nonperturbative origin (see my other contribution in this Volume).

Later in 1978 Cremmer, Julia, and Joël teamed up with Ferrara, Girardello and van Nieuwenhuizen on a pair of papers studying the super-Higgs effect in four-dimensional supergravity theories coupled to matter [CJSN78, CJSF79]. This work has been used a great deal in subsequent studies.

41.6 Supersymmetry breaking

I spent the academic year 1978–1979 visiting the LPTENS, on leave of absence from Caltech. I was eager to work with Joël on supergravity, supersymmetrical strings, and related matters. He was struggling with rather serious health problems during that year, so he was not able to participate as fully as when we were in Caltech five years earlier, but he was able to work about half time. On that basis we were able to collaborate successfully.

After various wide-ranging discussions we decided to focus on the problem of supersymmetry breaking. We wondered how, starting from a supersymmetric string theory in ten dimensions, one could end up with a nonsupersymmetric world in four dimensions. The specific supersymmetry breaking mechanism that we discovered can be explained classically and does not really require strings, so we explored it in a field theoretic setting [SS79a, SS79b]. The idea is that in a theory with extra dimensions and global symmetries that do not commute with supersymmetry (R symmetries and $(-1)^F$ are examples), one could arrange for a twisted compactification, and that this would break supersymmetry. For example, if one extra dimension forms a circle, the fields when continued around the circle could come back transformed by an R-symmetry group element. If the gravitino, in particular, is transformed, then it acquires mass in a consistent manner.

An interesting example of our supersymmetry breaking mechanism was worked out in a paper we wrote together with Cremmer [CSS79]. We started with maximal supergravity in five dimensions. This theory contains eight gravitinos that transform in the fundamental representation of a $USp(8)$ R-symmetry group. We took one dimension to form a circle and examined the resulting four-dimensional theory keeping the lowest Kaluza–Klein modes. The supersymmetry breaking R-symmetry element is a $USp(8)$ element that is characterized by four real mass parameters, since this group has rank four. These four masses give the masses of the four complex gravitinos of the resulting four-dimensional theory. In this way we were able to find a consistent four-parameter deformation of $\mathcal{N} = 8$ supergravity.

Even though the work that Joël and I did on supersymmetry breaking was motivated by string theory, we only discussed field theory applications in our articles. The reason I never wrote about string theory applications was that in the string theory setting it did not seem possible to decouple the supersymmetry breaking mass parameters from the compactification scales. This was viewed as a serious problem, because the two scales are supposed to be hierarchically different. In recent times, people have been considering string theory brane-world scenarios in which much larger compactification scales are considered. In such a context our supersymmetry breaking mechanism might have a role to play. Indeed, quite a few authors have explored various such possibilities.

41.7 Concluding comments

When I left Paris in the summer of 1979, I visited CERN for a month. There I began a collaboration with Michael Green. In September 1979, when I spoke at a conference on supergravity that was held in Stony Brook, I reported on the work that Joël and I had done on supersymmetry breaking [Sch79]. Joël gave a talk entitled 'From supergravity to antigravity' at the Stony Brook conference [Sch79a, Sch79b]. He was intrigued by the fact that graviton exchanges in string theory are accompanied by antisymmetric tensor and scalar exchanges that can cancel the gravitational attraction. Nowadays we understand that the effect that he was discussing is quite important. For example, parallel BPS D-branes form stable supersymmetric systems precisely because the various forces cancel. The Stony Brook conference was the last time that I saw Joël.

In March 1980 Joël attended a meeting in Erice, Sicily. Lars Brink, who was also there, recalls being very worried about Joël's health. Six weeks after that meeting he passed away, which came as a great shock to his many friends and colleagues. The ideas that Joël pioneered during the decade of the Seventies have been very influential in the subsequent decades. It is a pity that he could not participate in these developments and enjoy the recognition that he would have received.

Acknowledgements

I am grateful to Lars Brink, Eugène Cremmer, André Neveu and David Olive for reading this manuscript and making numerous helpful suggestions. This work was supported in part by the U.S. Dept. of Energy under Grant No. DE-FG03-92-ER40701.

Joël Scherk (1946–1980) was born in France. He attended École Normale Supérieure from 1965 to 1969, obtaining his PhD from Orsay in 1969. He became researcher at CNRS in 1969. He was a NATO postdoctoral fellow at Princeton University and Berkeley (1969–1970), and research fellow at CERN (1971–1973). He was invited professor at NYU and Caltech in 1973–1974. He moved to École Normale Supérieure in 1974. He made major contributions to string theory and to supersymmetry.

John H. Schwarz was born in North Adams, Massachusetts in 1941. He did his undergraduate studies at Harvard, and received his PhD in physics at Berkeley in 1966. He is currently the Harold Brown Professor of Theoretical Physics at the California Institute of Technology. He has worked on superstring theory for almost his entire professional career. Schwarz has been awarded a Guggenheim Fellowship and a MacArthur Fellowship. He is a Fellow of the American Physical Society as well as a member of the National Academy of Sciences and the American Academy of Arts and Sciences. He received the Dirac Medal in 1989, and the Dannie Heineman Prize in Mathematical Physics in 2002.

References

[BM69] Bardakci, K. and Mandelstam, S. (1969). Analytic solution of the linear-trajectory bootstrap, *Phys. Rev.* **184**, 1640–1644.

[BR69] Bardakci, K. and Ruegg, H. (1969). Reggeized resonance model for arbitrary production processes, *Phys. Rev.* **181**, 1884–1889.

[BORS73] Brink, L., Olive, D. I., Rebbi, C. and Scherk, J. (1973). The missing gauge conditions for the dual fermion emission vertex and their consequences, *Phys. Lett.* **B45**, 379–383.

[BOS73] Brink, L., Olive, D. and Scherk, J. (1973). The gauge properties of the dual model pomeron–reggeon vertex: their derivation and consequences, *Nucl. Phys.* **B61**, 173–198.

[BSS77] Brink, L., Schwarz, J. H. and Scherk, J. (1977). Supersymmetric Yang–Mills theories, *Nucl. Phys.* **B121**, 77–92.

[CT69] Chan, H. M. and Tsou, S. T. (1969). Explicit construction of the N-point function in the generalized Veneziano model, *Phys. Lett.* **B28**, 485–488.

[CJ78] Cremmer, E. and Julia, B. (1978). The $N = 8$ supergravity theory. 1. The Lagrangian, *Phys. Lett.* **B80**, 48–51.

[CJ79] Cremmer, E. and Julia, B. (1979). The $SO(8)$ supergravity, *Nucl. Phys.* **B159**, 141–212.

[CJS78] Cremmer, E., Julia, B. and Scherk, J. (1978). Supergravity theory in 11 dimensions, *Phys. Lett.* **B76**, 409–412.

[CJSF79] Cremmer, E., Julia, B., Scherk, J., Ferrara, S., Girardello, L. and van Nieuwenhuizen, P. (1979). Spontaneous symmetry breaking and Higgs effect in supergravity without cosmological constant, *Nucl. Phys.* **B147**, 105–131.

[CJSN78] Cremmer, E., Julia, B., Scherk, J., van Nieuwenhuizen, P., Ferrara, S. and Girardello, L. (1978). Superhiggs effect in supergravity with general scalar interactions, *Phys. Lett.* **B79**, 231–234.

[CS72a] Cremmer, E. and Scherk, J. (1972). Currents and Green's functions for dual models (II) Off-shell dual amplitudes, *Nucl. Phys.* **B48**, 29–77.

[CS72b] Cremmer, E. and Scherk, J. (1972). Factorization of pomeron sector and currents in dual resonance model, *Nucl. Phys.* **B50**, 222–252.

[CS73] Cremmer, E. and Scherk, J. (1973). Regge limit and scaling in a dual model of currents, *Nucl. Phys.* **B58**, 557–597.

[CS74] Cremmer, E. and Scherk, J. (1974). Spontaneous dynamical breaking of gauge symmetry in dual models, *Nucl. Phys.* **B72**, 117–124.

[CS76a] Cremmer, E. and Scherk, J. (1976). Dual models in 4 dimensions with internal symmetries, *Nucl. Phys.* **B103**, 399–425.

[CS76b] Cremmer, E. and Scherk, J. (1976). Spontaneous compactification of space in an Einstein Yang–Mills Higgs model, *Nucl. Phys.* **B108**, 409–416.

[CS77a] Cremmer, E. and Scherk, J. (1977). Algebraic simplifications in supergravity theories, *Nucl. Phys.* **B127**, 259–268.

[CS77b] Cremmer, E. and Scherk, J. (1977). Modified interaction of the scalar multiplet coupled to supergravity, *Phys. Lett.* **B69**, 97–100.

[CS78] Cremmer, E. and Scherk, J. (1978). The supersymmetric nonlinear sigma model in four-dimensions and its coupling to supergravity, *Phys. Lett.* **B74**, 341–343.

[CSF77] Cremmer, E., Scherk, J. and Ferrara, S. (1977). $U(N)$ invariance in extended supergravity, *Phys. Lett.* **B68**, 234–238.

[CSF78] Cremmer, E., Scherk, J. and Ferrara, S. (1978). $SU(4)$ invariant supergravity theory, *Phys. Lett.* **B74**, 61–64.

[CSS79] Cremmer, E., Scherk, J. and Schwarz, J. H. (1979). Spontaneously broken $N = 8$ supergravity, *Phys. Lett.* **84**, 83–86.

[DZ76] Deser, S. and Zumino, B. (1976). Consistent supergravity, *Phys. Lett.* **B62**, 335–337.

[FFNB77] Ferrara, S., Freedman, D. Z., van Nieuwenhuizen, P., Breitenlohner, P., Gliozzi, F. and Scherk, J. (1977). Scalar multiplet coupled to supergravity, *Phys. Rev.* **D15**, 1013–1018.

[FCSN76] Ferrara, S., Gliozzi, F., Scherk, J. and van Nieuwenhuizen, P. (1976). Matter couplings in supergravity theory, *Nucl. Phys.* **B117**, 333–355.

[FNS76] Ferrara, S., Scherk, J. and van Nieuwenhuizen, P. (1976). Locally supersymmetric Maxwell–Einstein theory, *Phys. Rev. Lett.* **37**, 1035–1037.

[FSZ77a] Ferrara, S., Scherk, J. and Zumino, B. (1977). Supergravity and local extended supersymmetry, *Phys. Lett.* **B66**, 35–38.

[FSZ77b] Ferrara, S., Scherk, J. and Zumino, B. (1977). Algebraic properties of extended supergravity theories, *Nucl. Phys.* **B121**, 393–402.

[FNF76] Freedman, D. Z., van Nieuwenhuizen, P. and Ferrara, S. (1976). Progress toward a theory of supergravity, *Phys. Rev.* **D13**, 3214–3218.

[FS70] Frye, G. and Susskind, L. (1970). Non-planar dual symmetric loop graphs and pomeron, *Phys. Lett.* **B31**, 589–591.

[FGV69] Fubini, S., Gordon, D. and Veneziano, G. (1969). A general treatment of factorization in dual resonance models, *Phys. Lett.* **B29**, 679–682.

[FV69] Fubini, S. and Veneziano, G. (1969). Level structure of dual resonance models, *Nuovo Cimento* **A64**, 811–840.

[FV70] Fubini, S. and Veneziano, G. (1970). Duality in operator formalism, *Nuovo Cimento* **A67**, 29–47.

[GS71] Gervais, J. L. and Sakita, B. (1971). Field theory interpretation of supergauges in dual models, *Nucl. Phys.* **B34**, 632–639.

[GSO76] Gliozzi, F., Scherk, J. and Olive, D. (1976). Supergravity and the spinor dual model, *Phys. Lett.* **B65**, 282–286.

[GSO77] Gliozzi, F., Scherk, J. and Olive, D. (1977). Supersymmetry, supergravity theories and the dual spinor model, *Nucl. Phys.* **B122**, 253–290.

[GS69] Goebel, C. J. and Sakita, B. (1969). Extension of the Veneziano form to n-particle amplitudes, *Phys. Rev. Lett.* **22**, 257–260.

[GL71] Golfand, Yu. A. and Likhtman, E. P. (1971). Extension of the algebra of Poincaré group generators and violation of P invariance, *JETP Lett.* **13**, 323–326 [*Pisma Zh. Eksp. Teor. Fiz.* **13**, 452 (1971)].

[GNSS70a] Gross, D. J., Neveu, A., Scherk, J. and Schwarz, J. H. (1970). The primitive graphs of dual resonance models, *Phys. Lett.* **B31**, 592–594.

[GNSS70b] Gross, D. J., Neveu, A., Scherk, J. and Schwarz, J. H. (1970). Renormalization and unitarity in the dual-resonance model, *Phys. Rev.* **D2**, 697–710.

[HPCS77] Horvath, Z., Palla, L., Cremmer, E. and Scherk, J. (1977). Grand unified schemes and spontaneous compactification, *Nucl. Phys.* **B127**, 57–65.

[Jac29] Jacobi, C. G. J. (1829). *Fundamenta Nova*, Könisberg.

[KS71a] Kaku, M. and Scherk, J. (1971). Divergence of the two-loop planar graph in the dual-resonance model, *Phys. Rev.* **D3**, 430–439.

[KS71b] Kaku, M. and Scherk, J. (1971). Divergence of the n-loop planar graph in the dual-resonance model, *Phys. Rev.* **D3**, 2000–2004.

[KSV69] Kikkawa, K., Sakita, B. and Virasoro, M. A. (1969). Feynman-like diagrams compatible with duality. I. Planar diagrams, *Phys. Rev.* **184**, 1701–1713.
[KN69a] Koba, Z. and Nielsen, H. B. (1969). Reaction amplitude for n-mesons a generalization of the Veneziano–Bardakci–Ruegg–Virasoro model, *Nucl. Phys.* **B10**, 633–655.
[KN69b] Koba, Z. and Nielsen, H. B. (1969). Manifestly crossing-invariant parameterization of n-meson amplitude, *Nucl. Phys.* **B12**, 517–536.
[Lov71] Lovelace, C. (1971). Pomeron form factors and dual Regge cuts, *Phys. Lett.* **B34**, 500–506.
[NS70a] Neveu, A. and Scherk, J. (1970). Parameter-free regularization of one-loop unitary dual diagram, *Phys. Rev.* **D1**, 2355–2359.
[NS70b] Neveu, A. and Scherk, J. (1970). Final-state interaction and current algebra in $K \to 3\pi$ and $\eta \to 3\pi$ decays, *Ann. Phys.* **57**, 39–64.
[NS72a] Neveu, A. and Scherk, J. (1972). Connection between Yang–Mills fields and dual models, *Nucl. Phys.* **B36**, 155–161.
[NS72b] Neveu, A. and Scherk, J. (1972). Gauge invariance and uniqueness of the renormalisation of dual models with unit intercept, *Nucl. Phys.* **B36**, 317–331.
[NS71] Neveu, A. and Schwarz, J. H. (1971). Factorizable dual model of pions, *Nucl. Phys.* **B31**, 86–112.
[OS73] Olive, D. and Scherk, J. (1973). Towards satisfactory scattering amplitudes for dual fermions, *Nucl. Phys.* **B64**, 334–348.
[Ram71] Ramond, P. (1971). Dual theory for free fermions, *Phys. Rev.* **D3**, 2415–2418.
[Sch71] Scherk, J. (1971). Zero-slope limit of the dual resonance model, *Nucl. Phys.* **B31**, 222–234.
[Sch75] Scherk, J. (1975). An introduction to the theory of dual models and strings, *Rev. Mod. Phys.* **47**, 123–163.
[Sch79a] Scherk, J. (1979). Antigravity: a crazy idea?, *Phys. Lett.* **B88**, 265–267.
[Sch79b] Scherk, J. (1979). From supergravity to antigravity, in *Supergravity*, ed. van Nieuwenhuizen, P. and Freedman, D. Z. (North-Holland, Amsterdam), 43.
[SS74a] Scherk, J. and Schwarz, J. H. (1974). Dual models for nonhadrons, *Nucl. Phys.* **B81**, 118–144.
[SS74b] Scherk, J. and Schwarz, J. H. (1974). Dual models and the geometry of spacetime, *Phys. Lett.* **B52**, 347–350.
[SS75a] Scherk, J. and Schwarz, J. H. (1975). Gravitation in the light-cone gauge, *Gen. Rel. Grav.* **6**, 537–550.
[SS75b] Scherk, J. and Schwarz, J. H. (1975). Dual field theory of quarks and gluons, *Phys. Lett.* **B57**, 463–466.
[SS79a] Scherk, J. and Schwarz, J. H. (1979). Spontaneous breaking of supersymmetry through dimensional reduction, *Phys. Lett.* **B82**, 60–64.
[SS79b] Scherk, J. and Schwarz, J. H. (1979). How to get masses from extra dimensions, *Nucl. Phys.* **B153**, 61–88.
[Sch79] Schwarz, J. H. (1979). How to break supersymmetry, in *Supergravity*, ed. van Nieuwenhuizen, P. and Freedman, D. Z. (North-Holland, Amsterdam), 19.
[Sch00] Schwarz, J. H. (2000). Reminiscences of collaborations with Joël Scherk, arXiv:hep-th/0007117.
[Tow95] Townsend, P. K. (1995). The eleven-dimensional supermembrane revisited, *Phys. Lett.* **B350**, 184–188.
[Ven68] Veneziano, G. (1968). Construction of a crossing-symmetric, Reggeon behaved amplitude for linearly rising trajectories, *Nuovo Cimento* **A57**, 190–197.

[Wei65] Weinberg, S. (1965). Photons and gravitons in perturbation theory: derivation of Maxwell's and Einstein's equations, *Phys. Rev.* **138**, 988–1002.
[WZ74] Wess, J. and Zumino, B. (1974). Supergauge transformations in four dimensions, *Nucl. Phys.* **B70**, 39–50.
[Wit95] Witten, E. (1995). String theory dynamics in various dimensions, *Nucl. Phys.* **B443**, 85–126.
[Yon74] Yoneya, T. (1974). Connection of dual models to electrodynamics and gravidynamics, *Prog. Theor. Phys.* **51**, 1907–1920.
[Zum74] Zumino, B. (1974). Relativistic strings and supergauges, in *Renormalization and Invariance in Quantum Field Theory*, ed. Caianiello, E. (Plenum Press, New York), 367.

Part VII

Preparing the string renaissance

42
Introduction to Part VII

42.1 Introduction

The period between the years 1976 and 1984 shows very little activity in string theory. As we mentioned in the previous Part, a lot of work went into developing both perturbative and nonperturbative aspects of QCD, which established itself as the theory of strong interactions. Lattice gauge theory was formulated and the idea of confinement was developed. These were also the years when supersymmetry was used to construct the Minimal Supersymmetric Standard Model, and the various supergravities were obtained. The most fundamental of them, constructed by Cremmer, Julia and Scherk, was the eleven-dimensional one. The fact that the various supergravities showed better ultraviolet behaviour than the original gravity theory gave the hope that one could unify gauge interactions with gravity in the framework of supergravity without nonrenormalizable divergences. On the other hand, the machinery of string theory seemed too complicated and unnecessary for unification. We review these developments in Section 42.2.

Although research in string theory was very limited in these years, it led to three fundamental developments. The first one, discussed in the Chapter by Green, was the reformulation of the fermionic string in terms of a light-cone fermionic coordinate S^a, that is an $SO(8)$ spinor, instead of the light-cone $SO(8)$ vector ψ^i of the RNS model. This allowed the complete construction of type IIA, IIB and I superstring theories. It is described in Section 42.3.

Another important result, obtained at the beginning of the Eighties and discussed in the Chapter by Polyakov, was the modern Lorentz covariant path-integral quantization of string theory, presented in Section 42.4.

The third important achievement, discussed in the Chapter by Green (see also the Chapter by Schwarz in Part I), was the understanding of gauge and gravitational anomalies in supergravity and superstring theories. This will be presented in Section 42.5. In particular, the proof of anomaly cancellation in type I superstring theory, among other results, opened the way to the renaissance of string theory after 1984, the so-called 'first string revolution',

---

The Birth of String Theory, ed. Andrea Cappelli, Elena Castellani, Filippo Colomo and Paolo Di Vecchia.
Published by Cambridge University Press. © Cambridge University Press 2012.

as described in the last Section. Subsequent developments from 1985 to the present will be described briefly by Cappelli and Colomo in the last Chapter of this Volume.

42.2 Supergravity unification of all interactions

In the late Seventies, it became clear that supersymmetric theories were less ultraviolet divergent than their nonsupersymmetric counterparts. The presence of an equal number of fermionic and bosonic degrees of freedom implies opposite-sign contributions in loop amplitudes, leading to cancellations of some divergences. This suggested that, unlike Einstein's theory of general relativity, its supersymmetric extension called supergravity could be renormalizable and, actually, finite.

All the four-dimensional supersymmetric theories that we have considered in the Introduction to Part VI have four supercharges, denoted by Q_α and $\bar{Q}_{\dot\alpha}$, and are called $\mathcal{N} = 1$ supersymmetric. Theories with extended supersymmetry, i.e. with $\mathcal{N}$ copies of the supercharges, were also constructed and shown to be even less ultraviolet divergent. In the next Subsection, we describe these theories briefly, while in the following Subsection we discuss their application to particle phenomenology.

42.2.1 Extended supersymmetry in four dimensions

The most general supersymmetric algebra in four dimensions, unifying the Lorentz group with an internal symmetry group, was constructed by Haag, Lopuszanski and Sohnius in 1975. Its generators, Q_α^i and $\bar{Q}_{\dot\alpha}^i$, are labelled by the internal index $i = 1, \ldots, \mathcal{N}$, and satisfy the following commutation relations:

$$[Q_\alpha^i, \bar{Q}_{\dot\alpha}^j] = -2\delta^{ij}\sigma_{\alpha\dot\alpha}^\mu P_\mu,$$
$$[Q_\alpha^i, Q_\beta^j] = \epsilon_{\alpha\beta} Z^{ij}, \qquad [\bar{Q}_{\dot\alpha}^i, \bar{Q}_{\dot\beta}^j] = \epsilon_{\dot\alpha\dot\beta} \bar{Z}^{ij}, \tag{42.1}$$

where Z_{ij} are constants called central extensions of the algebra. This is called the $\mathcal{N}$-extended supersymmetry algebra. The presence of more generators allows one to construct supersymmetric multiplets with more than two consecutive spin values, as for example $(s, s + 1/2, s + 1)$ in the $\mathcal{N} = 2$ case.

The four-dimensional theories with extended supersymmetry can be most easily found by first constructing an $\mathcal{N} = 1$ supersymmetric Lagrangian in d dimensions (with $d > 4$) and then reducing it to four dimensions. The d-dimensional spinors have more components than the four-dimensional ones: under the simplest dimensional reduction, where the fields are taken to be functions of the four-dimensional coordinates and not of the remaining ones, supersymmetry is preserved and the d-dimensional spinor breaks into a number of four-dimensional spinors. Since the numbers of supercharges and spinor components are equal, several copies of the supercharges are obtained, thus realizing an extended supersymmetry.

Following this approach, Brink, Scherk and Schwarz considered super Yang–Mills theory in ten dimensions. From this theory, containing a gauge field, its Weyl–Majorana fermionic partner (the gaugino), and a Weyl–Majorana spinor supercharge with 16 degrees of freedom, they constructed the $\mathcal{N} = 4$ super Yang–Mills theory in four dimensions and the $\mathcal{N} = 2$ theory in six dimensions. These theories are both invariant under 16 supersymmetries and therefore have 16 supercharges. It turned out that the $\mathcal{N} = 4$ super Yang–Mills theory was less divergent than the $\mathcal{N} = 1$ theory, and had a vanishing renormalization-group β-function, corresponding to a coupling constant independent of the energy scale. Therefore, $\mathcal{N} = 4$ super Yang–Mills theory was a massless scale-invariant theory at the quantum level. Similarly, by dimensional reduction of $\mathcal{N} = 1$ theory in six dimensions, $\mathcal{N} = 2$ super Yang–Mills theory in four dimensions could also be constructed.

We could imagine that, starting from a supersymmetric Yang–Mills theory in more than ten dimensions, we could build even more supersymmetric four-dimensional theories. However, this is not the case; from the study of the representations of the supersymmetry algebra, Nahm showed in 1978 that one would obtain multiplets with spin higher than one, leading to a theory with local supersymmetry, like supergravity.

In conclusion, in four dimensions we have three super Yang–Mills theories containing a spin one gauge field and no higher spin field: (i) $\mathcal{N} = 1$ supersymmetric, containing also a Majorana gaugino; (ii) $\mathcal{N} = 2$ supersymmetric, containing also a Dirac gaugino and a complex scalar; (iii) $\mathcal{N} = 4$ supersymmetric, containing also four Majorana gauginos and six real scalars.

42.2.2 Kaluza–Klein reduction and the Standard Model

In this Subsection we extend the previous considerations to the construction of local supersymmetric theories, called supergravity theories, and we describe the attempts in the early Eighties to obtain a quantum theory of gravity unified with the Standard Model.

In the Introduction to Part VI, we discussed $\mathcal{N} = 1$ supergravity in four dimensions. After its proposal in 1976, a lot of work was immediately devoted to finding extensions of this theory, culminating with the construction by Cremmer, Julia and Scherk, in 1978, of eleven-dimensional supergravity. This contained the vierbein (or better the elfbein) $e^A_M(x)$, a three-index antisymmetric potential A_{MNP} with four-index antisymmetric field strength F_{MNPQ} and a gravitino Ψ_M (here capital indices run over 0, 1, ..., 10), and was invariant under 32 supersymmetries. Next, in ten dimensions there are three independent supergravities; these are actually the zero-slope limit of three superstring theories, called types IIA, IIB and I, that will be discussed in Section 42.3. The first two theories are $\mathcal{N} = 2$ supersymmetric, with 32 supercharges, while the third is $\mathcal{N} = 1$ supersymmetric, with 16 supercharges. The eleven-dimensional supergravity and the two type II ten-dimensional supergravities are the maximally extended supergravities.

The high degree of supersymmetry of the three previous supergravities, preserved by dimensional reduction to four dimensions, gave hopes of obtaining an on-shell finite four-dimensional theory of gravity. The other fundamental question was to obtain within this theory the unification of all interactions of the Standard Model, with gauge group $SU(3) \times SU(2) \times U(1)$ and matter fields representing quarks and leptons. In the following, we mention two approaches to this problem.

In the first approach, one takes the eleven-dimensional supergravity and performs a Kaluza–Klein reduction with seven compactified dimensions (a complete review of this approach can be found in the *Physics Report* by Duff, Nilsson and Pope in 1986). In this respect, Witten noted in 1981 that the seven extra dimensions are not merely the maximum number allowed by supersymmetry, but also the minimum amount needed for the Standard Model gauge symmetry. In his general study of compactifications Witten found, however, that it was difficult to obtain the correct fermion quantum numbers of the Standard Model spectrum.

In the other approach, one considers the four-dimensional $\mathcal{N} = 8$ supergravity obtained by Cremmer and Julia in 1979 by dimensional reduction of eleven-dimensional supergravity. Its $SO(8)$ gauge group was not large enough to generate the Standard Model group, but the theory also possessed an $SU(8)$ gauge group with composite and nondynamical gauge bosons. Assuming the generation of a kinetic term for the $SU(8)$ gauge bosons by loop corrections (in analogy with what happens for the two-dimensional CP^{N-1} models), the symmetry of the theory was large enough to include that of the Standard Model.

However, these approaches were not pursued for several reasons. Although much less divergent than pure gravity, it seemed that supergravity was still divergent and nonrenormalizable. Furthermore, it was not easy to obtain the correct quantum numbers of quarks and leptons in the Standard Model. Finally, the developments of 1983–1984 in string theory, to be discussed below, put forward the superstring as a much more promising unification framework and decreed the end of supergravity as an independent approach (see, however, the last Chapter in this Volume for subsequent developments in the context of M-theory).

42.3 A novel light-cone formalism

The invariance under reparameterizations of the world-sheet coordinates τ and σ discussed in Part IV involves two arbitrary functions. This implies that two of the coordinates $X^\mu(\tau, \sigma)$ of the string are redundant and can be fixed by the gauge conditions. For example, in the light-cone gauge, one light-cone coordinate is fixed to be proportional to τ:

$$X^+ = 2\alpha'\tau, \qquad X^\pm \equiv \frac{X^0 \pm X^{d-1}}{\sqrt{2}}; \qquad (42.2)$$

the independent degrees of freedom are then found to be the transverse ones, $X^i(\tau, \sigma)$ with $i = 1, \ldots, d - 2$, and $d = 26$.

Similarly, in the superstring we have the additional invariance under local supersymmetry transformations and thus the independent degrees of freedom are the transverse bosonic and fermionic coordinates, $X^i(\tau, \sigma)$ and $\psi^i(\tau, \sigma)$, where $i = 1, \ldots, 8$ is an index transforming according to the vector representation of $SO(8)$. $X^i(\tau, \sigma)$ has an expansion in terms of the bosonic oscillators with integer modes, while $\psi^i(\tau, \sigma)$ is expanded in terms of fermionic oscillators with integer modes in the R sector and half-integer modes in the NS sector. In particular, in the R sector there is a zero mode that satisfies the Clifford algebra and, as a consequence, the R sector describes spacetime fermions, while the NS sector describes spacetime bosons.

Green and Schwarz proposed an alternative light-cone description of the superstring based on the transverse bosonic degrees of freedom $X^i(\tau, \sigma)$ and the new world-sheet fermionic degrees of freedom $S^a(\tau, \sigma)$, where the index a belongs to an eight-dimensional spinor representation of $SO(8)$. We remark that this group is rather peculiar because it possesses three nonequivalent eight-dimensional representations: the vector labelled by i and denoted 8_v, and two spinors labelled by a and $\dot{a}$, and denoted 8_s and 8_c, respectively. In this formulation, both $X^i(\tau, \sigma)$ and $S^a(\tau, \sigma)$ are expanded in terms of oscillators with integer modes, respectively bosonic and fermionic. As a consequence, the zero modes satisfy the following Clifford algebra in eight dimensions:

$$\{S_0^a, S_0^b\} = \delta^{ab}. \tag{42.3}$$

The Dirac matrices in eight dimensions are sixteen dimensional and can be written in block form as follows:

$$S_0^a \sim \begin{pmatrix} 0 & \gamma_{i\dot{a}}^a \\ \gamma_{\dot{a}i}^a & 0 \end{pmatrix}, \tag{42.4}$$

where the eight-dimensional matrices $\gamma_{i\dot{a}}^a$ also satisfy the Dirac algebra:

$$\gamma_{a\dot{a}}^i \gamma_{\dot{a}b}^j + \gamma_{a\dot{a}}^j \gamma_{\dot{a}b}^i = 2\delta^{ij}\delta_{ab}, \quad i, j = 1, \ldots, 8, \tag{42.5}$$

and similarly with dotted and undotted indices interchanged. The matrix in Eq. (42.4) acts naturally on the state $|\phi_0\rangle = \begin{pmatrix} |i\rangle \\ |\dot{a}\rangle \end{pmatrix}$, corresponding to a representation space that transforms as the sum of representations $8_v + 8_c$ under $SO(8)$ transformations, because it contains both a massless vector $|i\rangle$ and a massless spinor $|\dot{a}\rangle$, each with eight physical components. This is the lowest state of the open superstring theory, also called 'type I' theory, which was obtained by the GSO projection of the RNS model.

This formalism was then used for constructing closed superstring theories, where two sets of modes are required, for right and left movers. Their massless states are thus described by the product: $|\phi_0\rangle \times |\tilde{\phi}_0\rangle$. We can distinguish two cases: one in which the two original Weyl–Majorana spinors have the same ten-dimensional chirality, and the other in which they have opposite chirality. In the first case, the corresponding product of $SO(8)$ representations has the following decomposition over irreducible representations:

$$(8_v + 8_c) \otimes (8_v + 8_c) = (1 + 28 + 35_v + 1 + 28 + 35_c)_B$$
$$+ (8_s + 8_s + 56_s + 56_s)_F. \tag{42.6}$$

Table 42.1 *Massless spectrum of type IIA and IIB superstring theories*

Type	$SO(8)$ representation	Field	Name
A/B	1	ϕ	dilaton
A/B	28	$B_{\mu\nu}$	antisymmetric potential
A/B	35_v	$h_{\mu\nu}$	graviton
A/B	$8_s, 8_s/8_c$	χ	dilatino
A/B	$56_s, 56_s/56_c$	Ψ_μ	gravitino
B	1	C_0	RR scalar
B	28	$C_{\mu\nu}$	RR antisymmetric potential
B	35_v	$C_{\mu\nu\rho\sigma}$	RR self-dual field strength
A	8_v	C_μ	RR vector
A	56_v	$C_{\mu\nu\rho}$	RR antisymmetric three form

This gives the massless spectrum of the type IIB closed string. In the second case we get:

$$(8_v + 8_c) \otimes (8_v + 8_s) = (1 + 28 + 35_v + 8_v + 56_v)_B + (8_s + 8_c + 56_s + 56_c)_F \quad (42.7)$$

corresponding to the massless spectrum of the type IIA closed superstring. The decompositions in these two equations can be understood as follows. Among the four terms obtained by the product on the left-hand side of Eq. (42.6), two of them, namely $8_v \otimes 8_v$ and $8_c \otimes 8_c$, contain spacetime bosons (B), while the other two, namely $8_v \otimes 8_c$ and $8_c \otimes 8_v$, contain spacetime fermions (F). Analogously, in the case of Eq. (42.7), the two products $8_v \otimes 8_v$ and $8_c \otimes 8_s$ contain spacetime bosons, while the other two products $8_v \otimes 8_s$ and $8_c \otimes 8_v$ contain spacetime fermions. The product $8_v \otimes 8_v$ is common to the two cases and contains the representations 1, 28 and 35_v of $SO(8)$ corresponding to the physical degrees of freedom of a dilaton (1), of a two-index antisymmetric potential B_{MN} (28) and of the graviton in ten spacetime dimensions (35_v), respectively. The complete massless spectrum of the two closed superstrings is summarized in Table 42.1.

The type I superstring also has a closed string sector, besides the open one discussed above. It is obtained by a truncation of the spectrum of the type IIB superstring, which keeps the dilaton, the graviton, the RR scalar and the RR two-index antisymmetric tensor, while it projects out the other states. This operation is called an 'orientifold projection' in technical language; it does not allow an arbitrary gauge group in the open string sector, but restricts it to either $SO(N)$ or $SP(N)$.

In the novel light-cone formalism, spacetime supersymmetry is manifest and the supersymmetric charges are expressed directly in terms of S^a as follows:

$$Q^a \sim \sqrt{p^+} S_0^a, \qquad Q^{\dot{a}} \sim \frac{1}{\sqrt{p^+}} \gamma^i_{\dot{a}a} \sum_n \alpha_n^i S_{-n}^a, \quad (42.8)$$

where p^+ is the light-cone component of the total momentum. As described in the Chapter by Green, one-loop diagrams were computed in this new formalism and the one-loop four-gluon and four-graviton amplitudes were shown to be finite. Finally, a new manifestly spacetime supersymmetric and covariant action for the superstring was constructed by Green and Schwarz in 1984.

42.3.1 Green–Schwarz superstring action

While in the RNS formalism the fermionic coordinate ψ^μ was a spacetime vector and a world-sheet spinor, in the Green–Schwarz formalism two fermionic coordinates $\theta^A(\tau, \sigma)$ ($A = 1, 2$) are introduced that are spacetime Majorana spinors and world-sheet scalars like the bosonic coordinate $X^\mu(\tau, \sigma)$. Spacetime supersymmetry acts on the coordinates as follows:

$$\delta X^\mu = i\bar{\epsilon}^A \Gamma^\mu \theta^A, \qquad \delta\theta^A = \epsilon^A, \qquad \delta\bar{\theta}^A = \bar{\epsilon}^A, \qquad (42.9)$$

where repeated indices are summed over, ϵ^A ($A = 1, 2$) are two constant spacetime Majorana spinors and world-sheet scalars, and Γ^μ are the ten-dimensional Dirac matrices. In general, the index A could run from 1 to $\mathcal{N}$, thus obtaining an $\mathcal{N}$ extended supersymmetry. Here, we limit ourselves to the $\mathcal{N} = 2$ case, which is the only one giving rise to a consistent quantum theory.

In order to obtain a spacetime supersymmetric theory, we use the combination $\partial_\alpha X^\mu - i\bar{\theta}^A \Gamma^\mu \partial_\alpha \theta^A$, which is invariant under the supersymmetry transformations (42.9), and obtain a supersymmetric generalization of the σ-model action as follows:

$$S_1 = -\frac{1}{4\pi\alpha'} \int d^2\xi \sqrt{g}\, g^{ab} \left(\partial_a X^\mu - i\bar{\theta}^A \Gamma^\mu \partial_a \theta^A\right) \eta_{\mu\nu} \left(\partial_b X^\nu - i\bar{\theta}^A \Gamma^\nu \partial_b \theta^A\right). \quad (42.10)$$

This action, originally proposed by Casalbuoni in 1976, possesses two global spacetime supersymmetries and local reparameterization invariance. However, due to the absence of a local fermionic symmetry, θ^A had twice the number of degrees of freedom of X^μ. This problem was cured by Green and Schwarz by adding the following term to the action (42.10):

$$S_2 = \frac{1}{2\pi\alpha'} \int d^2\xi \left[-i\epsilon^{ab}\partial_a X^\mu \left(\bar{\theta}^1 \Gamma_\mu \partial_b \theta^1 - \bar{\theta}^2 \Gamma_\mu \partial_b \theta^2\right) + \epsilon^{ab}\bar{\theta}^1 \Gamma^\mu \partial_a \theta^1 \bar{\theta}^2 \Gamma_\mu \partial_b \theta^2\right].$$
(42.11)

The total action $S_{GS} = S_1 + S_2$ is still $\mathcal{N} = 2$ spacetime supersymmetric, provided that θ^A is a Weyl–Majorana spinor, but also possesses a local fermionic symmetry, called κ-symmetry, that eliminates half of the degrees of freedom of θ^A and gives an equal number of fermionic and bosonic degrees of freedom.

Finally, it can be shown that the invariances under reparameterizations and κ-symmetry allow one to choose the light-cone gauge characterized by the condition (42.2) and by an analogous condition on θ^A, namely $\Gamma^+\theta^A = 0$. This gauge choice kills the nonlinear terms in the Green–Schwarz action S_{GS}; in this way, the previous analysis of type II and type I theories is recovered. In type I superstring theory, θ^1 and θ^2 are identified by the boundary conditions and only an $\mathcal{N} = 1$ supersymmetry is maintained, while in the two type II theories, there is the possibility of choosing the two spinors with the same or opposite chirality, obtaining type IIA and type IIB theories, respectively.

In conclusion, there are two equivalent formalisms for describing the various superstring theories: (i) the Green–Schwarz formalism, where the spacetime supersymmetry is manifest in the action, and a unique sector contains both spacetime fermions and bosons; (ii) the RNS formalism, described in the Introduction to Part V, where spacetime supersymmetry is not manifest in the action, and its two sectors describe spacetime fermions and bosons, respectively.

In this Section we presented the three consistent ten-dimensional superstring theories that were already known in the early Eighties, before the first string revolution. The remaining two string theories, called 'heterotic', were found in the autumn of 1984.

42.4 Modern covariant quantization

In the late Seventies σ-model actions had been found for the bosonic and fermionic string theories, which were written in terms of two independent variables, the string coordinate and the world-sheet metric tensor (see the Introduction and the Chapter by Di Vecchia in Part VI, for further details). These σ-model actions both have the nice property of disentangling the world-sheet and target-space geometries. This property turns out to be very useful for the path-integral treatment that is described below.

In 1981, Polyakov used the two σ-model actions to perform a novel Lorentz covariant quantization of both the bosonic and fermionic strings; he relied on the path-integral methods developed for non-Abelian gauge theories in the early Seventies. These results laid the foundations for the modern world-sheet methods to be developed in the Eighties and Nineties, and provided a neat treatment that also eliminated the discrepancy between the light-cone and the old covariant quantization approaches.

Let us now concentrate on the bosonic string and recall how this discrepancy comes about. The Nambu–Goto action is invariant under arbitrary reparameterizations of the world-sheet coordinates and, as a consequence, only $d - 2$ transverse coordinates are independent. This can be seen most easily in the light-cone gauge where one fixes X^+ as in Eq. (42.2) and then X^- becomes a function of the remaining transverse coordinates. As seen in the Introduction to Part IV, in this approach Lorentz invariance is recovered only if the intercept of the Regge trajectory and the dimension of the spacetime are $\alpha_0 = 1$ and $d = 26$, respectively. In the 'old' covariant quantization of the early Seventies, discussed in the Introduction to Part III, the condition $\alpha_0 = 1$ came from the requirement of positive definite physical subspace, but values of $d \leq 26$ were possible. Indeed, the condition $d = 26$

only came from requiring unitarity of nonplanar loop diagrams. This mismatch between the quantizations in the light-cone and in the covariant gauge was puzzling for a gauge invariant theory.

In the second half of the Seventies, path-integral methods offered an alternative approach to the quantization of field theories that was rather fruitful in the case of non-Abelian gauge theories, leading to many insights that were difficult to obtain otherwise. It was thus natural to apply them to string theory. Note that path-integral approaches to gauge-fixed forms of the string action had already been considered in the early days of string theory. Polyakov's analysis in 1981, besides solving the above mentioned puzzle in string quantization, showed the presence of a conformal anomaly for $d \neq 26$, and its relation with the Virasoro central charge. In the following we discuss in some detail these results in the case of the bosonic string.

The σ-model action for the bosonic string is given by:

$$S(X^\mu, g_{ab}) = -\frac{T}{2} \int d\tau \int_0^\pi d\sigma \sqrt{-g}\, g^{ab} \partial_a X^\mu \partial_b X^\nu \eta_{\mu\nu}, \qquad (42.12)$$

where $X^\mu(\tau, \sigma)$ is the string coordinate and $g_{ab}(\tau, \sigma)$ is the world-sheet metric. This action is invariant under:

(i) Poincaré transformations in spacetime acting on X^μ;
(ii) reparameterizations of the world-sheet coordinates ($\xi^a \equiv (\tau, \sigma)$) acting as follows,

$$X^\mu(\xi) = X'^\mu(\xi'), \qquad g_{ab}(\xi) = \frac{\partial \xi'^c}{\partial \xi^a} \frac{\partial \xi'^d}{\partial \xi^b} g'_{cd}(\xi'), \qquad (42.13)$$

where $\xi'^a = f^a(\xi)$ are arbitrary functions of the ξ;
(iii) local rescaling of the world-sheet metric tensor (also called Weyl transformations)

$$X^\mu(\xi) \to X^\mu(\xi), \qquad g_{ab}(\xi) \to \Lambda^2(\xi) g_{ab}(\xi), \qquad (42.14)$$

where Λ is also an arbitrary function of world-sheet coordinates.

The reparameterizations involve two arbitrary functions, while the rescaling of the metric tensor depends on one arbitrary function. These symmetries of the classical theory allow one to fix completely the world-sheet metric to be the flat, Minkowski one. However, one of the previous symmetries becomes anomalous at the quantum level and, therefore, one can only fix two of the three independent components of the metric. In the Lorentz covariant gauge, also called the conformal gauge, one can thus choose the following form for the world-sheet metric:

$$g_{ab}(\xi) = e^{\phi(\xi)} \eta_{ab}, \qquad \eta_{ab} = \text{diag}(-1, 1), \qquad (42.15)$$

where the function ϕ is the so-called conformal factor. In this gauge, the action (42.12) becomes independent of the world-sheet metric, i.e. of ϕ, and describes d free scalar particles living on the world-sheet.

Furthermore, one must impose the vanishing of the two-dimensional energy-momentum tensor as required by the equation of motion for the world-sheet metric:

$$\theta_{ab} \equiv \partial_a X \cdot \partial_b X - \frac{1}{2} g_{ab} g^{cd} \partial_c X \cdot \partial_d X = 0. \tag{42.16}$$

This was the procedure followed in the old covariant quantization. However, Polyakov showed that it was not quite correct. The quantum anomaly mentioned before manifests itself in the trace of the energy-momentum, that acquires a contribution from quantum effects and does not actually vanish. This is the trace anomaly, also called conformal anomaly because it modifies the conformal symmetry of the theory. Let us discuss this point within the path-integral approach.

The path-integral quantization of gauge theories, developed by Faddeev and Popov, is based on the following idea: the unphysical states are kept in the quantum theory in order to have manifest Lorentz invariance, but their effect is cancelled by adding further nonunitary states. The latter, called Faddeev–Popov 'ghosts' (not to be confused with the negative norm states of the DRM, also called 'ghosts' in the earlier literature), are fermionic and thus yield a minus sign in loop diagrams that cancels the contribution of unphysical degrees of freedom. The dynamics of Faddeev–Popov ghosts is completely determined by the symmetries of the problem and the gauge fixing condition.

In the case of the bosonic string, the ghosts contribute with an additional term to the energy-momentum tensor (42.16). The total tensor shows a trace anomaly for $d \neq 26$. The corresponding Virasoro algebra possesses the central charge $c = d - 26$. Therefore, the choice of critical dimension is equivalent to the condition of vanishing anomaly. This is required for enforcing the symmetries of the σ-model action, reparameterization and Weyl invariance at the quantum level.

A neat way to discuss symmetries at the quantum level is the BRST approach (after Becchi, Rouet and Stora and Tyutin). As in the case of gauge theories, the gauge fixed string action is invariant under a global symmetry called BRST invariance. Its generator, the BRST charge, is fermionic and its square vanishes. In the quantum theory such a symmetry can be maintained only if $\alpha_0 = 1$ and $d = 26$. In conclusion, the results of the covariant path-integral quantization agree with those of the light-cone quantization, as expected by the gauge invariance of the theory.

It must be said that the addition of ghosts does not affect the tree diagrams, which remain the same as those of the DRM (see the Introduction to Part III). In particular, all the results obtained for the spectrum remain correct after the inclusion of the Faddeev–Popov ghosts. Their presence is instead essential when we compute loop diagrams in a covariant gauge, because they are needed to cancel the contribution of the nonphysical states circulating in the loops.

Let us note that Polyakov also considered the bosonic string for $d < 26$, where the trace anomaly generates a dynamics for the conformal factor ϕ in (42.15). This becomes a new degree of freedom of the string, called the Liouville field. However, the corresponding 'noncritical string' involving this extra field is not fully understood at the quantum level: it has only been successfully quantized for $d \leq 1$, where it finds applications to systems of two-dimensional statistical mechanics.

42.5 Anomaly cancellation

42.5.1 Supergravity and superstrings

As mentioned in Sections 42.1 and 42.2, at the beginning of the Eighties supergravity theory was considered a promising unification framework. In addition to the fundamental eleven-dimensional supergravity, three independent supergravities were known in ten dimensions, namely types I, IIA and IIB; these could be viewed as the zero-slope limits of the three corresponding superstrings. Type I supergravity has an open string sector that is the low energy limit of the GSO projected RNS open string, discussed in the Introduction to Part VI, while types IIA and IIB supergravities are the low energy limits of closed strings with equal or opposite GSO projections of right and left modes, respectively (see Section 42.3). It is worth recalling that supergravities were originally developed irrespectively of the underlying string theories.

Theories in lower dimensions could be derived by the Kaluza–Klein compactification mechanism. The aim was to obtain a chiral theory in four dimensions including the Standard Model and gravity. Thus, it was important to check the absence of chiral anomalies, i.e. that classical gauge and gravitational symmetries are preserved at the quantum level. Anomalies of gauge symmetries would in fact make the quantum theory inconsistent.

42.5.2 Chiral fermions and chiral anomalies

Let us step back for a moment and discuss the problems that one encounters in gauge theories with chiral fermions. In the Introduction to Part II, we have seen that weak interactions are described by $V - A$ currents, where V_μ is a vector and A_μ an axial vector. While the space part of the vector current V^i changes sign under parity transformation (i.e. spatial reflections with respect to the origin), $P : x^i \to -x^i$, the axial vector A^i does not. The weak current can be expressed in terms of the left-handed component of fermions:

$$V^\mu - A^\mu \sim \overline{\psi_L} \gamma^\mu \psi_L, \qquad \psi_L = \frac{1 - \gamma^5}{2} \psi_D, \qquad (42.17)$$

where ψ_L are the left-handed 'chiral' Weyl fermions discussed in the Introduction to Part VI, with half of the degrees of freedom of Dirac fermions.

Owing to the chiral form, weak interactions are not invariant under parity transformations, as confirmed by experiments. This means that the Standard Model is a chiral gauge theory: the left- and right-handed quarks and leptons transform under different representations of the gauge group. The left-handed parts are doublets of the weak $SU(2)$ gauge

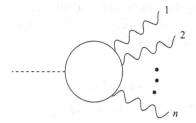

Figure 42.1 The loop diagram determining the chiral anomaly in $d = 2n$ dimensions.

group, $\psi_L = (\nu_L, e_L)$ or (u_L, d_L), in terms of the neutrino and electron, or of the up and down quarks, respectively. The right-handed parts are instead singlets of $SU(2)$. It is clear that any unifying theory should have chiral fermions in its spectrum, to reproduce the chiral form of weak interactions.

The problem with chiral gauge theories is that they can be anomalous at the quantum level. We have already seen, in the case of conformal symmetry, that a classical symmetry may be spoiled by quantum effects. Similarly, in the case of a chiral gauge theory the anomaly leads to a nonconservation of the chiral current that is coupled to the gauge field. This is not acceptable because it introduces negative-norm states: therefore, anomalies in gauge theories must be cancelled.

In odd dimensions, there is no chiral decomposition for fermions and no anomaly either; therefore, chiral anomalies are only present in even dimensions. They are typically given by the diagram shown in Figure 42.1, where one external leg corresponds to the current and the remaining legs correspond to gauge fields. The chiral current is conserved, i.e. the anomaly vanishes, if there are fermions in different representations of the gauge group that circulate in the loop, whose anomalous contributions add up to zero. For instance, in the Standard Model this cancellation occurs for an equal number of families for quarks and leptons, as observed experimentally.

Besides gauge anomalies, in supergravity there are also gravitational anomalies for chiral fermions coupled to the metric, for example the spin $3/2$ gravitino, supersymmetric partner of the graviton. In type IIB supergravity there is a four-form potential with self-dual field strength that also contributes to both gauge and gravitational anomalies.

Summing up, a sensible supergravity/string theory aiming at unifying the Standard Model with gravity should obey the double condition of having both chiral fermions coupled to gauge fields and vanishing chiral anomaly.

42.5.3 Anomaly cancellation in the type I superstring

In gauge theories, the anomaly is entirely given by the one-loop Feynman diagrams in Figure 42.1, involving $n + 1$ vertices in $d = 2n$, one for each external gauge field, and one for the current, i.e. hexagons in $d = 10$.

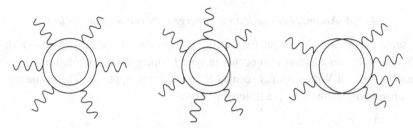

Figure 42.2 Open string diagrams contributing to the chiral anomaly in the type I superstring: the planar (left), nonorientable (centre) and nonplanar (right) hexagons.

In order to explore in a systematic way the presence of anomalies in supergravity, in 1983 Alvarez-Gaumé and Witten computed the gauge and gravitational anomalies generated by various chiral fields in an arbitrary spacetime dimension. In the case of IIB supergravity, which contains two chiral dilatinos and gravitinos and a four-index RR state with self-dual field strength, they showed that the gravitational anomaly vanished. Since type IIB supergravity does not include any gauge field, it is also trivially free from gauge anomalies. Type IIA supergravity is not a chiral theory and therefore has no anomaly. Also eleven-dimensional supergravity is anomaly free because there is no anomaly in an odd number of dimensions.

In the case of type I supergravity, however, the situation was different: using the anomaly coefficients calculated by Alvarez-Gaumé and Witten, the theory seemed to have both gauge and gravitational anomalies.

In autumn 1984, Green and Schwarz computed the gauge and gravitational anomalies in the full type I superstring and in the corresponding supergravity, and showed that these theories were, instead, both free from gauge and gravitational anomalies for a particular choice of gauge group. Let us explain this point in more detail, first in the superstring case, and then in type I supergravity in the next Section.

In the case of the type I superstring, Green and Schwarz used the RNS formalism, and computed the sum of the three one-loop string diagrams with six external gauge bosons shown in Figure 42.2. These are the planar hexagon where all gauge fields are on one boundary, the nonorientable hexagon (Möbius strip) and the nonplanar hexagon with two gauge fields on one boundary and the remaining four on the other boundary.

Green and Schwarz found that the anomalous contribution due to the chiral fermions circulating in the (planar + nonorientable) loops, equal to that of the corresponding supergravity, was cancelled by the term coming from the nonplanar diagram. This contribution is a purely stringy effect, due to the exchange of the two-index antisymmetric RR massless closed string state in the closed string channel. This is the so-called Green–Schwarz cancellation mechanism in type I superstring theory, that is only possible for the gauge group $SO(32)$.

42.5.4 Anomaly cancellation: supergravity versus superstrings

In the same paper, with hindsight from string calculations, Green and Schwarz showed how the anomaly cancellation also occurs in type I supergravity, although the use of the Alvarez-Gaumé and Witten results seemed to indicate the opposite. The bosonic part of type I supergravity is given by the following action:

$$S = \frac{1}{2\kappa_{10}^2} \int d^{10}x \sqrt{-G} e^{-2\phi} \left[R + 4\partial_\mu \phi \partial^\mu \phi - \frac{1}{12} (\tilde{F}_3)_{\mu_1\mu_2\mu_3} (\tilde{F}_3)^{\mu_1\mu_2\mu_3} \right]$$
$$- \frac{1}{2g_{10}^2} \int d^{10}x \, e^{-\phi} \text{Tr} \left(F_{2\mu_1\mu_2} F_2^{\mu_1\mu_2} \right). \qquad (42.18)$$

In this formula, F_2 is the field strength of the gauge field A_1 of $SO(32)$ and $\tilde{F}_3$ is given by:

$$\tilde{F}_3 = F_3 - \frac{\kappa_{10}^2}{g_{10}^2} \omega_3, \qquad (42.19)$$

in terms of the field strength $F_3 \equiv dC_2$ of the two-index antisymmetric field C_2 and of the Chern–Simon three-form:

$$\omega_3 \equiv \text{Tr} \left(A_1 F_2 - \frac{1}{3} (A_1)^3 \right), \qquad (42.20)$$

which is required by local supersymmetry, as shown by Chapline and Manton in 1983. To reduce the number of indices, we have introduced the differential geometry notation:

$$A_1 = A_\mu dx^\mu, \qquad F_2 = \frac{1}{2} F_{\mu\nu} dx^\mu \wedge dx^\nu, \qquad (42.21)$$

and

$$C_2 \equiv \frac{1}{2} C_{2\mu\nu} dx^\mu \wedge dx^\nu, \qquad F_3 = \frac{1}{6} \partial_\rho C_{2\mu\nu} dx^\rho \wedge dx^\mu \wedge dx^\nu. \qquad (42.22)$$

The action (42.18) contains two independent coupling constants: the gravitational, κ_{10}, and the gauge, g_{10}.

Since ω_3 transforms under a local gauge transformation, C_2 must transform as well, in order for the action in Eq. (42.18) to be gauge invariant. The fact that C_2 transforms under a gauge transformation allows one to add a local counterterm whose gauge variation cancels precisely, for the gauge group $SO(32)$, the anomaly generated by the chiral fermions and computed with the formulae provided by Alvarez-Gaumé and Witten. Diagrammatically, this local counterterm comes from the diagram in Figure 42.3, which can be viewed as the field theory limit (in the closed string channel) of the nonplanar string diagram in Figure 42.2. Thus, the anomaly cancellation is quite natural in string theory, where it is given by a stringy effect, since $\kappa_{10}^2/g_{10}^2 \sim \alpha'$; on the contrary, the cancellation requires a rather subtle parameter tuning in the limiting field theory.

Figure 42.3 The nonplanar loop diagram of Figure 42.2 in the closed string channel (left), and its field theory limit (right).

Green and Schwarz also found that, in the case of supergravity, the above mentioned cancellation is possible for another gauge group, namely $E_8 \times E_8$ where E_8 is the largest of the so-called exceptional Lie groups. In autumn 1984, however, no string theory with this gauge group was known yet: it was constructed soon after by Gross, Harvey, Martinec and Rohm and called the 'heterotic string'. This and other developments taking place after 1984 are discussed further in the final Chapter of the Volume.

42.6 A new era starts or, maybe better, continues

As discussed in Section 42.2, the discovery of supersymmetry and especially of extended supergravities with a maximal degree of supersymmetry, generated a lot of activity in the early Eighties. There was the hope of constructing a unified quantum theory of gravity and gauge interactions, that was not only internally consistent, but could also reproduce the Standard Model through Kaluza–Klein reduction. However, the difficulties in obtaining chiral fermions with the correct quantum numbers for quarks and leptons, and also the fact that one could not exclude the presence of ultraviolet divergences in high-order loop diagrams of supergravity theories, as shown in 1980 by Howe and Lindström, caused a loss of enthusiasm in pursuing this direction.

In this context, the September 1984 work of Green and Schwarz on anomaly cancellation immediately attracted a lot of attention, by suggesting that one could use superstring theory to do what seemed impossible with supergravity. In particular it stimulated two other major developments. The first one was the construction, just two months later, in November 1984, of the two heterotic strings by Gross, Harvey, Martinec and Rohm. The second one was the work of Candelas, Horowitz, Strominger and Witten, in December 1984. They considered compactifications of the $E_8 \times E_8$ heterotic string and showed that the requirement of $\mathcal{N} = 1$ supersymmetry in four dimensions selects the six-dimensional compact space to be a so-called Calabi–Yau manifold, leading to four-dimensional supersymmetric gauge theories with reasonable features for unification.

After these results, a large number of researchers moved to work on string theory. The transition was abrupt: superstring theory, which had been studied by Green, Schwarz and a few others in almost complete isolation between 1976 and 1984, became a mainstream activity by early 1985. For many people this was the beginning of an exciting new era. For those who had worked in string theory before, it was the continuation, with new important elements, of the era that had started in the mid-Seventies, with the shift of string theory from a theory of hadrons to a unified theory of all interactions.

References

[GSW87] Green, M. B., Schwarz, J. H. and Witten, E. (1987). *Superstring Theory* (Cambridge University Press, Cambridge).

[WB92] Wess, J. and Bagger, J. (1992). *Supersymmetry and Supergravity* (Princeton University Press, Princeton, NJ).

43
From strings to superstrings: a personal perspective

MICHAEL B. GREEN

Abstract

This Chapter gives an overview of my period of research in string theory up to the end of 1984. I will begin with my time as a graduate student and postdoc, which coincided with the earliest developments in dual models and string theory. However, I will not repeat the detailed history of this early period, which is covered much more completely by other authors in this Volume. The second part will concern the development of string theory with manifest spacetime supersymmetry in the late Seventies and early Eighties, a period that postdates most of the other contributions in this Volume.

43.1 String theory till 1979

The subject of string theory has its genesis in the many wonderful developments in relativity and quantum theory in the first half of the twentieth century. Two singular results of the early to mid-Sixties are particularly relevant to subsequent developments in string theory. One of these was the formulation by Dirac of a theory of the relativistic membrane [Dir62] (eight years before the formulation of the relativistic string, Nambu and Goto [Nam70, Got71]), in which he attempted to describe the μ-meson as a radial excitation of a spherical membrane whose ground state was the electron. This inspired paper was effectively ignored until the subject of supermembranes became fashionable in the late Eighties. It now plays a key role, in association with Born and Infeld's long-neglected nonlinear electrodynamics [BI34], in the Dirac–Born–Infeld description of D-branes. A second important insight of the mid-Sixties was Hagedorn's implementation of the bootstrap programme. He realized that demanding that all hadrons be 'made of each other' requires a degeneracy of single-particle states that increases exponentially with mass [Hag65]. Furthermore, such a spectrum has peculiar thermodynamics, which prevents the temperature increasing beyond an ultimate, or 'Hagedorn', temperature. In retrospect this was probably the most striking direct output of the bootstrap philosophy.

The Birth of String Theory, ed. Andrea Cappelli, Elena Castellani, Filippo Colomo and Paolo Di Vecchia.
Published by Cambridge University Press. © Cambridge University Press 2012.

The most memorable event in my period as a graduate student in Cambridge (1967–1970) was the arrival of the preprint of Veneziano's 1968 paper [Ven68]. This was a period in which Cambridge was under the spell of the bootstrap ideas promoted most notably by Chew [Che66]. This philosophy had been born out of the apparent failure of quantum field theory to describe the strong force. It was hoped that the S-matrix for the strong interactions could be determined in a self-consistent manner by incorporating a minimal set of assumptions, most notably analyticity, crossing symmetry and unitarity. Quantum field theory was used merely as a model for how various aspects of the strong force, notably Regge pole asymptotic behaviour, might emerge in such a framework. These ideas were summarized in a book (Eden, Landshoff, Olive and Polkinghorne [ELOP66]) that became the basic text for graduate students in Cambridge.

Despite the limitations of the original bootstrap programme, it promoted inventive ways of analyzing hadronic data. Together with the advent of current algebra sum rules, this eventually led to the development of finite energy sum rules (see Igi and Matsuda [IM67], Ademollo, Rubinstein, Veneziano and Virasoro [ARVV67, ARVV69], Dolen, Horn and Schmid [DHS68] and Mandelstam [Man68]), which provided evidence for a deep connection between hadronic resonances and the Regge pole description of high energy scattering. This implied that the sum of the resonance contributions to a scattering amplitude is equivalent to the sum of Regge pole contributions, which are themselves determined by the sum of resonance exchanges in the crossed channel. In other words, one should not add the resonance poles in the s-channel to those in the t-channel as one would with Feynman diagrams in conventional field theory, since that would be double counting. This was known as the 'duality' between resonances and Regge poles and subsequently became known as 'world-sheet duality'.

It was in this context that Veneziano's paper appeared. It postulated a closed-form approximation to a meson scattering amplitude, which was simply the Euler beta function as a function of the Mandelstam invariants, s and t. This can be expanded as an asymptotic series of poles in either s or t, and therefore embodies the duality principle. This ansatz was both simple and astonishing – the strong interactions had seemed so intractable and the experimental data so confusing that there had seemed to be no hope of ever obtaining a closed form expression for a strong interaction amplitude, let alone such a simple one. The Veneziano amplitude, which builds in the exponential Hagedorn density of states (Huang and Weinberg [HW70]), was eventually understood to result from the scattering of open relativistic strings, although this only followed after several years of developments. Very soon after Veneziano's paper came the generalization by Virasoro [Vir69], that was later interpreted in terms of the scattering of closed strings and generalized to the scattering of n particles by Shapiro [Sha70]. These closed strings would have been called 'glueballs' in the context of the modern quest for a QCD string, but QCD was still some years away. The impact of Veneziano's paper was profound, although it was largely limited to those, like me, who had been (mis)led to believe that quantum field theory could not possibly explain the strong interactions. It has guided my research ever since.

From 1970–1972 I was a postdoc at the Institute for Advanced Study in Princeton. It was a period of rapid advance in string theory, with a very strong group in Princeton University, including André Neveu, Joël Scherk, John Schwarz and David Gross, with whom I had remarkably little contact! The main physics event during my period at the Institute was the series of lectures by Ken Wilson on his approach to the renormalization group, which later appeared in a celebrated review article [WK74]. At the Institute I had a spirited collaboration with Bob Carlitz and Tony Zee on phenomenological aspects of hadronic duality, and it was only after my return to a postdoc in Cambridge that I began to study string theory properly.

I spent the summer of 1973 at CERN, being tutored by Pierre Ramond in the technicalities of the dual model. I decided to study properties of one-loop amplitudes in the Ramond–Neveu–Schwarz (RNS) model with external ground state tachyons. The amplitude with circulating bosons had already been evaluated (Goddard and Waltz [GW71]), so I set about constructing the amplitude with circulating fermions, which had not been studied. The loop amplitudes in the bosonic and fermionic sectors are both badly divergent but I found that the leading divergences cancelled when the amplitudes in each sector were summed, although a nonleading divergence remained [Gre73]. This seemed striking at the time but it remained a curiosity. Shortly thereafter cancellations between fermion and boson loops in supersymmetric field theory were discovered and some time later the GSO projection led to a degree of consistency that the original RNS model did not have. Later in the summer of 1973 there was a small duality conference at CERN with some very intriguing talks. I particularly remember Jeffrey Goldstone talking about three unresolved problems – in string field theory, baryonic strings and membranes. There were also talks by Lars Brink and David Olive about the construction of loop amplitudes and Ed Corrigan, Peter Goddard, Joël Scherk and others on the fermion emission vertex – a pressing problem at the time was the evaluation of the four-fermion amplitude, which involved a measure factor that was proving difficult to evaluate. Rather remarkably, John Schwarz and his student, Wu, had arrived at the correct result by a numerical study, despite the primitive state of computing in those days [SW73]. The most significant new development at that conference was the evaluation of tree amplitudes in the bosonic string and the fermionic (RNS) string by Mandelstam [Man73, Man74]. These were the first dynamical calculations making use of the light-cone gauge that had been developed in the seminal paper by Goldstone, Goddard, Rebbi and Thorn [GGRT73].

This was around the time of tremendous developments in quantum field theory: having proved the renormalizability of Yang–Mills theory, 't Hooft was developing the large-N approach to gauge theories; supersymmetry was being developed in four-dimensional theories, having originated in the fermionic string world-sheet theory; the Standard Model and asymptotic freedom were being discovered; monopoles and instantons were being discovered, etc. Connections between low energy string theory and quantum field theory were also being made, starting with scalar field theory, by Scherk [Sch71], then gauge theory, by Neveu and Scherk [NS72], and finally gravity, by Yoneya [Yon74], and Scherk and Schwarz [SS74].

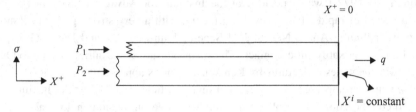

Figure 43.1 The form factor for two on-shell ground states coupling to an off-shell state at the Dirichlet boundary segment carrying momentum $q = p_1 + p_2$. The diagram is drawn in a light-cone parameterization, with $X^+ = \tau$.

Meanwhile, I pursued my interest in string theory as a potential model of mesons. For this purpose it seemed important to understand how to couple off-shell electromagnetic and weak currents to the string. Although there had been many earlier attempts to do this, the ansatz for scalar off-shell open string states made by Schwarz [Sch73] and by Corrigan and Fairlie [CF75] was particularly compelling, despite its inconsistent critical spacetime dimension, $d = 16$. I began by reinterpreting these states in the light-cone gauge string picture [Gre76], where the off-shell open string coupled at $X^+ = 0$ (X^+ is the light-cone time coordinate) to a segment of the open string boundary on which the coordinates satisfied a Dirichlet condition, $X^\mu = $ constant, which allowed momentum to flow through the boundary, as shown in Figure 43.1. Here, the consistent critical dimension was necessarily $d = 26$ but the off-shell open string states did not have definite angular momentum. The apparent inconsistency in the interpretation of these states was finally understood in terms of intersecting D-branes many years later.

However, the closed string version of this idea, which I studied in collaboration with Joel Shapiro at CERN in the summer of 1976, was free of these difficulties [GS76]. In this case we imposed the Dirichlet condition on a closed world-sheet boundary by constructing an appropriate boundary state. The consistency of this description fascinated me because it was clear that the insertion of closed Dirichlet boundaries in a world-sheet would radically change the short-distance structure of string theory. The idea was to insert pointlike boundaries in the world-sheet with Dirichlet boundary conditions so that the whole boundary was mapped to a spacetime point. In [Gre77] I interpreted the insertion of a condensate of such boundaries in a world-sheet at zero momentum as a modification of the closed string vacuum. The insertion of a single Dirichlet boundary in a closed string tree level amplitude is illustrated in Figure 43.2.

This is a spherical world-sheet with four closed strings attached, together with a boundary on which the world-sheet fields satisfy Dirichlet boundary conditions – in modern language this is the leading D-instanton correction to the four-point function. I became seduced by the fact that even a single Dirichlet boundary induces pointlike high energy fixed-angle scattering. Whereas conventional string amplitudes had exponentially suppressed behaviour at fixed angle due to their lack of parton-like substructure, the presence of Dirichlet boundaries enhances string motions in which a finite fraction of the energy

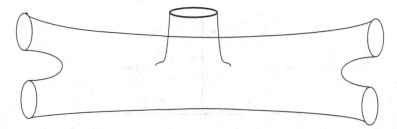

Figure 43.2 A closed Dirichlet boundary is interpreted as a closed string tadpole. A condensate of such boundaries at zero momentum defines a modified vacuum state in which pointlike fixed-angle scattering is enhanced.

in the string accumulates at points, which then scatter off each other. The amplitudes exhibit power behaviour at high energy and fixed angle. This arises from the region of integration in which the vertex operators are close to the Dirichlet boundary although far apart from each other on the world-sheet. Since the advent of D-branes in the Nineties, more general vacuum states have been studied and a more coherent picture is emerging of hadronic scattering, although it does retain a flavour of these early ideas. Hints of such pointlike conditions soon arose in the 1977 work of Charles Thorn, who was studying generalizations of known string theories by introducing spin models on a world-sheet lattice [Tho77, Tho78]. His prescient work found its natural home in the wealth of applications of two-dimensional statistical mechanics models to string theory model building of the mid-Eighties.

By the mid-Seventies most string theorists had transferred their attention to the exciting new avenues in Yang–Mills field theory and supergravity, and progress in string theory slowed dramatically. However, there were two major developments in 1976 of direct relevance to the rest of this Chapter. One of these was the formulation of the bosonic and fermionic strings in terms of the action of two-dimensional (super) gravity involving the introduction of an auxiliary world-sheet metric (and world-sheet gravitino), by Brink, Di Vecchia and Howe [BDH76] and by Deser and Zumino [DZ76]. The second development was the realization by Gliozzi, Scherk and Olive that the fermionic string, which incorporated world-sheet supersymmetry, could be truncated to a tachyon-free theory that is supersymmetric in spacetime – indeed, it was later realized that such a 'GSO' projection [GSO76, GSO77] is required for consistency of the theory. Not only were these papers important within string theory, but they also strongly emphasized the super Yang–Mills and supergravity field theory limit. One of the more surprising aspects of these papers is that in the closed string case GSO applied an inconsistent GSO projection! Indeed, they arbitrarily decreed that there should be no Ramond–Ramond sector. This is inconsistent with modular invariance and leads to a theory with 16 supercharges ($\mathcal{N} = 1$ in ten dimensions) instead of 32 supercharges. In fact, it subsequently transpired that this $\mathcal{N} = 1$ theory has gravitational anomalies unless coupled to the type I open string, while the consistent $\mathcal{N} = 2$ theories (the type IIA and IIB theories) were only developed by John Schwarz and me in the Eighties.

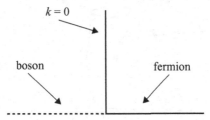

Figure 43.3 The open string fermion emission vertex at zero momentum ($k = 0$) is identified with the supercharge.

Nevertheless, the GSO paper was masterful and should have set the field alight, but it was not followed up and string theory remained a curiosity for several years.

In large part the emphasis had shifted to the development of extended and higher-dimensional supergravity. Most notably, eleven-dimensional supergravity was formulated by Cremmer, Julia and Scherk in 1978 [CJS78] and seemed to many to be a more promising avenue than string theory. With the passage of time, many important results in supergravity have been connected with the perturbative and nonperturbative structure of string theory, but such connections were by no means apparent at the time.

43.2 Superstrings 1979–1983

In January 1979 I became a lecturer at Queen Mary College in London. I spent the summer of 1979 at CERN, where I met John Schwarz, with whom I had barely interacted while I had been a postdoc at Princeton. This was a fortuitous meeting that marked the beginning of a long and exciting collaboration. We initially decided to see whether we could spot patterns in the counterterms of ten-dimensional $\mathcal{N} = 4$ super Yang–Mills theory that would illuminate its connection to the fermionic open string. Although we made very little concrete progress and published nothing that summer, we had posed a number of interesting questions concerning spacetime supersymmetry in string theory and decided to meet in Aspen the following summer to pursue our interest in string theory. In the summer of 1980 we considered how one might relate the world-sheet density of the spacetime supercharges to the world-sheet fields of the fermionic string [GS81]. Our starting observation was that the supercharge of the open fermionic string could be identified with the vertex for the emission of a zero momentum fermionic ground state from a bosonic ground state (Figure 43.3). We understood that the very complicated looking fermion emission vertex of Corrigan and Olive [CO72] decomposes in the light-cone gauge into two $SO(8)$ spinors, which led to an expression for the sixteen-component supercharge of the form

$$Q_a = \sqrt{p^+} S_a, \qquad Q_{\dot{a}} = \frac{1}{\sqrt{p^+}} \left(\gamma_i \cdot X^i S \right)_{\dot{a}}, \qquad (43.1)$$

where $i, a, \dot{a} = 1, \ldots, 8$ label the vector and two inequivalent spinor representations of $SO(8)$, p^+ is a light-cone component of the momentum and $\gamma^i_{a\dot{a}}$ are $SO(8)$

Clebsch–Gordon coefficients. The spinor S_a was expressed in a complicated manner in terms of the covariant bosonic and fermionic oscillators of the fermionic string. We were sure that this work would provide an important new insight into the structure of supersymmetric string theory which could lead to a much deeper understanding of the theory, so we decided we would meet again in Aspen the following summer and I would then spend a term at Caltech in the autumn of 1981. This was a wonderful period in which we had the luxury of not having to worry about others treading on our heels. String theory appeared to many to be a backwater, while there were enough interesting results yet to be discovered in supersymmetric field theory to occupy other researchers for many years.

Many of these non-string theory results in the end turned out to be of great importance to string theory but I have no space to mention them all here (see [BCK06] for a collection of talks on this subject). However, two papers by Witten are of particular relevance since they lie behind almost all attempts to explain real four-dimensional physics starting from a higher-dimensional supersymmetric theory. One concerned mechanisms for supersymmetry breaking [Wit81a], while the second considered conditions for obtaining realistic chiral theories by Kaluza–Klein reduction from higher dimensions [Wit81b]. This was also the year in which two seminal string theory papers by Polyakov appeared, presenting his method for summing over (super)Riemann surfaces to obtain string perturbation diagrams [Pol81a, Pol81b], starting from the actions presented in [BDH76, DZ76]. This eventually transformed the understanding of the geometry of perturbative string theory, although it received relatively little attention at the time. A key related development was the geometric understanding of the renormalization of σ-models by Friedan [Fri80a, Fri80b], who was rapidly becoming interested in string theory. Furthermore, building on Zumino's 1979 paper [Zum79], Alvarez-Gaumé and Freedman [AF80] understood the connection between supersymmetry of σ-models and Ricci flatness of the target space. This geometric understanding of two-dimensional conformal invariance is central to most subsequent developments in perturbative string theory.

In the summer of 1981, while in Aspen, John and I reformulated the fermionic string in a manner that explicitly maintained spacetime supersymmetry by use of a novel light-cone gauge formalism. We replaced the conventional light-cone gauge RNS fermionic world-sheet fields, ψ^i, which are spacetime vectors and world-sheet spinors, by fermionic world-sheet fields, S^a, which are spacetime spinors and world-sheet scalars. These are the objects that we had earlier expressed as complicated functions of the RNS fields, so the new insight was to take these fields to be the fundamental world-sheet degrees of freedom. Although the use of the light-cone gauge only manifestly preserves the transverse part of the Lorentz symmetry, this new formalism built in sufficient spacetime supersymmetry to simplify significantly the computation of the spectrum and amplitudes. We first calculated the tree amplitudes for the scattering of any four massless open string states in the sixteen-component supermultiplet [GS82a] and, upon returning to Caltech in the autumn, we constructed the corresponding one-loop amplitudes [GS82b]. The fact that the fermionic fields, S^a, are world-sheet scalars meant that there was no sum over spin structures, which significantly simplified the one-loop calculations.

Our most significant results of 1981 were contained in a short paper [GS82c] in which we used our spacetime supersymmetric formalism to construct the two ten-dimensional closed string theories, later called 'type IIA' and 'type IIB'. This rectified GSO's omission of the maximally supersymmetric cases by including the Ramond–Ramond sector (recall that they had been led to ten-dimensional $\mathcal{N} = 1$ supersymmetry instead of $\mathcal{N} = 2$ by use of an incorrect GSO projection). The spectrum of these theories is obtained as a tensor product of left-moving and right-moving supersymmetric modes, which accounts for the doubling of supersymmetry. Furthermore, we calculated the closed string loop amplitude with four external gravitons. We struggled with this since the only previous example of a closed string loop calculation was the beautiful paper by Joel Shapiro [Sha72] ten years earlier, in which he discovered the $SL(2, \mathbb{Z})$ symmetry in the one-loop tachyon amplitude of the bosonic string. However, the bosonic string has an 'infrared' divergence due to a tachyon tadpole, whereas we found that the superstring expression was finite, at least when the integral representation of the amplitude is interpreted appropriately. Supersymmetry forbids the closed string tachyon coupling to the torus. The fact that the one-loop four-graviton superstring amplitude was finite was the single most exciting discovery we had made – after all, the four-graviton amplitude of conventional ten-dimensional supergravity is quadratically divergent. This was a moment of revelation, after which there could be no doubt that the theory had the most amazing degree of consistency. For inexplicable reasons our paper [GS82c] does not highlight the finiteness result, ignoring it in the abstract and relegating it to a final short paragraph in the body of the paper.

Later that autumn we decided to use the technology we had developed in superstring theory to obtain results in its low energy approximation, namely, maximal supergravity. With the collaboration of Lars Brink, who was a frequent visitor to Caltech and had become a close colleague, we compactified our loop calculation on a d-torus in order to obtain results in low energy maximal supergravity in $10 - d$ dimensions. We evaluated the amplitude in $4 + \epsilon$ dimensions in order to regulate the four-dimensional infrared divergences [GSB82]. Although there had been work on the compactification of open string loop amplitudes (for example in Cremmer and Scherk [CS76]), compactification of the closed string amplitude presented the challenge of maintaining modular invariance. This required us to derive the modular-invariant lattice factor describing winding modes and Kaluza–Klein charges, which now seems so elementary that it is hard to remember why we found this challenging.

At the end of 1981 John Schwarz and I again arranged to meet the following year in Aspen and I would take extended leave from Queen Mary College to be at Caltech for much of the following year. This was made possible by the very supportive group at Queen Mary and by John Charap, in particular, who was a most enlightened Head of Department. The two types of orientable closed superstring theories we had described in 1981 had natural supergravity counterparts in ten dimensions, namely, the nonchiral and chiral $\mathcal{N} = 2$ supergravities. Although both of these appeared on Nahm's list of possible theories [Nah77], the chiral version had never been formulated and its existence was virtually unknown. We therefore tried to construct it, adapting our light-cone methods to the ten-dimensional field theory.

We succeeded in obtaining the kinetic terms and cubic interaction vertices [GS83a], but the clumsiness of the light-cone gauge rapidly became apparent in trying to go further. It was not until the work of Schwarz and West [SW83], of Schwarz [Sch83] and of Howe and West [HW84] that the complete description of the classical type IIB theory was formulated, using covariant methods.

We next decided to attempt a field theory description of the supersymmetric string field theory along the lines that had been developed a decade earlier by Mandelstam [Man73, Man74], Kaku and Kikkawa [KK74], Cremmer and Gervais [CG74] and others. We began to use our light-cone superspace formalism to construct interactions in the case of open strings [GS83b], and in early 1983, together with Lars Brink, we formulated the light-cone field theory of type II strings [GSB83].

This was also the time at which Alvarez-Gaumé and Witten [AW84] produced their important paper on the nature of chiral anomalies in gravitational field theory. Most strikingly, they showed that the massless chiral field content of ten-dimensional type IIB supergravity leads to a precise cancellation of gravitational anomalies. Since anomalies are infrared effects, the presence of a field theory anomaly would have implied that the corresponding string theory was also anomalous, so this cancellation was yet another string theory miracle. It meant that the type II theories were not only free of ultraviolet divergences, but were also anomaly free. However, by the summer of 1983 we were acutely aware of the possibility of inconsistencies looming when one considered chiral anomalies in theories that, at the time, seemed more likely to be of physical interest - the open string theories, which had gauge bosons and therefore the possibility of chiral gauge anomalies. Various people (most forcibly, Frank Wilczek and Edward Witten) had emphasized their concern that since anomalies are not sensitive to short-distance physics they should be as much a problem in string theory as in the low energy field theory. However, we set that embarrassing issue aside for the time being. Instead I had begun thinking of how to formulate the theory covariantly in order to get away from the cumbersome light-cone gauge.

In the autumn of 1983 John Schwarz spent a term in London where we resumed our very intense collaboration. We decided to dedicate a short period to finding a covariant description of the superstring, even though this appeared a daunting task since we were not sure that such a description should exist. Much to our delight, after pursuing various dead ends we came up with the solution. We needed the world-sheet fields to be the coordinates of ten-dimensional $\mathcal{N} = 2$ superspace. This meant that in addition to the usual bosonic fields X^μ there needed to be two world-sheet spacetime spinors, Θ^{iA} ($i = 1, 2$ labels the two supercharges of the $\mathcal{N} = 2$ theory, and $A = 1, \ldots, 16$ is a chiral spinor index). The problem was to find a formulation in which local symmetries could not only eliminate two directions in X^μ but would also cut the number of components in the spinor by a factor of two to reproduce the $SO(8)$ content of the light-cone gauge. We blindly tried all manner of ways of matching the light-cone gauge before a compelling and elegant resolution became apparent [GS84a]. The action we came upon had a structure that was very reminiscent of the action for a peculiar kind of Wess–Zumino–Witten model. This action was not only invariant under global target-space supersymmetry but embodied an unfamiliar local fermionic symmetry, κ-symmetry, which we vaguely knew about in the context of a paper by Warren Siegel on the point superparticle [Sie83]. The symmetry is associated with two Grassmann parameters, κ^1, κ^2, which are self-dual and anti-self-dual world-sheet vectors and also target-space chiral spinors. The fact that the starting action is free field theory when viewed in the light-cone gauge seemed like a miracle, although the advent of Witten's paper on non-Abelian bosonization [Wit84a] in the same year clarified this issue. Indeed, the relationship with Wess–Zumino–Witten models has been exploited for many further generalizations, some of which have a very natural home in currently exciting areas of string theory. For example, this is the key to the integrability of the theory in an $AdS_5 \times S^5$ background that is of current significance for the AdS/CFT correspondence.

A particularly obvious property of our action is that it is invariant under a discrete symmetry that interchanges the left-moving and right-moving Grassmann coordinates, Θ^1 and Θ^2. As a result there is no continuous $O(2)$ R-symmetry, as there is in the low energy supergravity. Although we noted this in our paper, if we had only pursued it further we would (perhaps!) have understood the $SL(2, \mathbb{Z})$ duality symmetry of the IIB theory long before it was discovered (Font, Ibanez, Lust and Quevedo [FILQ90], Hull and Townsend [HT95]). We struggled very hard to quantize the covariant theory using our new covariant action, but were unable to make significant progress because of the messy mixture of first and second class constraints [GS84b]. A great deal of effort has been put into covariant quantization methods since then, but the subject still holds profound mysteries, despite the powerful recent pure spinor formalism of Berkovits (first described in [Ber00]).

Later in the autumn of 1983 we decided to give a more complete formulation of light-cone gauge superstring field theory, including the explicit calculation of some scattering amplitudes. The results were contained in [GS84c, GS84d], which gave a unified description of the interactions between open and closed strings, including the open–closed superstring vertex (and containing many complicated expressions). At the end of 1983 we organized a very successful conference at Queen Mary, which brought together many of the major European groups working in supergravity and gave us an opportunity to discuss our results with a wider community.

43.3 1984

I had arranged yet another visit to Aspen in the summer of 1984, to be followed by a term's leave of absence at Caltech in the autumn. John Schwarz and I had decided we would now attack the problem of chiral anomalies in open string theory (type I theory). We were rather ignorant of the novel methods for describing non-Abelian gauge anomalies and gravitational anomalies in standard field theory, let alone in string theory, but luckily that summer a large number of experts were participating in Aspen. Among these were Bruno Zumino and Bill Bardeen, who had recently written important papers on these issues (Zumino, Wu and Zee [ZWZ84], Alvarez, Singer and Zumino [ASZ84], Bardeen and

Zumino [BZ84]), and Dan Friedan and Steve Shenker, who were also interested in many issues in string theory, including the issue of string theory anomalies. The ten-dimensional anomalies are associated with the nonconservation of the gauge and gravitational currents via their coupling to five external gauge particles or gravitons – i.e. groundstates of open or closed strings. We started with an explicit calculation of the gauge anomalies for $SU(N)$, $SO(N)$ or $Sp(N)$ gauge groups, which required the evaluation of various contributions to the open string one-loop hexagon diagram – the 'planar' cylinder world-sheet in which all gauge particles are attached to one boundary, the Möbius strip, and the 'nonplanar' cylinder with gauge particles attached to both boundaries. Our light-cone gauge formalism was inappropriate for determining such local anomalies, since in order to fix the gauge one assumes that the symmetry is nonanomalous. So we performed the calculation in the Lorentz covariant RNS formalism, with the use of an unsophisticated momentum cutoff. This gave the expected kind of anomalous terms (i.e. terms that would contribute to the anomaly of the low energy gauge theory) together with something completely unexpected – at least something that was unanticipated at the time. There was a contribution from the nonplanar cylinder diagram that would be interpreted as a tree diagram in the low energy theory. Such a higher-dimensional coupling was not present in the classical supergravity field theory, but was contained within string theory. We further found to our joy that for the gauge group $SO(32)$ the supergravity hexagon anomaly was precisely cancelled by this novel tree contribution.

Although the explicit calculation of open string gravitational and mixed gauge/gravitational anomalies seemed far too complicated for us to calculate explicitly, our general understanding of the cancellation mechanism in the low energy field theory was sufficient to determine them without the need for an explicit stringy calculation. Everything fell into place very rapidly and we were able to see how this fitted in with the general structure of gravitational anomalies and hence generalize the gauge anomaly cancellation argument. Once again, the field theory anomalies associated with hexagon diagram contributions (now including circulating gravitational particles, as well as gauge particles) could cancel with contributions from anomalous tree diagrams involving the exchange of the closed string antisymmetric tensor. Once again, this could only happen due to a higher-dimensional coupling of the antisymmetric tensor in string theory. In the gravitational case, this only arose for particular gauge groups containing 496 supermultiplets of massless particles. This is the dimension of $SO(32)$, which confirmed the consistency of the cancellation with the gauge anomaly cancellation. The compelling feature of our anomaly cancellation was that, although it could have been invented by modifying conventional supergravity field theories, it would have seemed contrived, whereas it was completely natural within string theory, in which the gauge and gravitational sectors are intrinsically related. All this was very elegant and I gave a seminar that was reported back to Princeton. A few days later we received news that Witten was writing the first paper on a phenomenological application of the type I theory compactified to four dimensions [Wit84b].

A few weeks later we wrote a paper describing the anomaly cancellation mechanism in the language of low energy ten-dimensional $\mathcal{N} = 1$ supergravity coupled to super

Yang–Mills field theory and omitting details of the open string calculation. However, the paper had to be revised since we had overlooked the other consistent gauge group, $E_8 \times E_8$. At the time there was no string theory with this symmetry and in the first version of the paper we conjectured that it would have gauge anomalies. Once various people had pointed out that E_8 has no independent fourth order and sixth order Casimirs (necessary conditions for the way our anomaly cancellation worked), we understood that $E_8 \times E_8$ is indeed consistent and the revised paper [GS84e] emphasized the two possible non-Abelian anomaly-free gauge groups – one of which was not realized in any known string theory. We very quickly realized that the $SO(32)$ open string theory was not only free of chiral anomalies, but was also free of ultraviolet divergences [GS85a]. This requires the individually divergent diagrams – the annulus, Möbius strip and Klein bottle – to be regularized in a suitable manner.

Another important line of research in Aspen that summer was the initial work by Dan Friedan and Steve Shenker, showing how to incorporate BRST ghosts that arise from gauge fixing in the Polyakov treatment of the string, which, among other things, reproduced the correct loop amplitudes in an extraordinarily simple manner. Together with Emil Martinec they later extended this to the superstring in [FSM85], laying the basis for the superconformal field theory that underlies many subsequent developments in perturbative superstring theory.

Upon returning to Caltech after the summer, John Schwarz and I wrote a longer paper containing details of the explicit calculation of the open string theory gauge anomaly and the conditions for its cancellation [GS85b]. Our anomaly cancellation argument strongly suggested that there should be an $E_8 \times E_8$ string theory, so we also set about trying to discover such a theory, with the help of Peter West, who was also visiting. Despite enthusiastic help from Murray Gell-Mann, we were floundering. We soon became aware of the work of Goddard and Olive on even self-dual lattices, where the sixteen-dimensional lattices $\Gamma_8 \oplus \Gamma_8$ (where Γ_8 is the root lattice of E_8) and Γ_{16} (the lattice associated with the roots and chiral spinor weights of $Spin(32)/\mathbb{Z}_2$) had been singled out [GO85]. However, the best we could do was to figure out how to get $(E_8 \times E_8)^2$ by compactifying the closed bosonic string to ten dimensions. We learnt about the formulation of the heterotic string by Gross, Harvey, Martinec and Rohm [GHMR85] in an informal conversation at the Santa Fe meeting later in the autumn. This took us by surprise since up to that time the only string theories we had considered were ones with equal Virasoro anomalies in the left-moving and right-moving matter sectors. The heterotic string not only produced a closed superstring theory with the $E_8 \otimes E_8$ gauge group, but also a second (this time closed) string theory with gauge group $SO(32)$.

In the autumn of 1984 we were also pursuing the idea of compactifying on manifolds of vanishing Ricci tensor, following what we had learned from Dan Friedan at Aspen. However, the only example we knew of was the four-dimensional K_3 manifold, so we could only consider compactification of the open superstring to six-dimensional theories with at least $\mathcal{N} = 1$ supersymmetry, or compactifications on $K_3 \otimes T^2$ to theories with $\mathcal{N} = 2$ supersymmetry in four dimensions [GSW85]. Once again, we were surprised – this time on hearing of the ongoing work of Candelas, Horowitz, Strominger and Witten, who compactified the heterotic string on six-dimensional Calabi–Yau spaces, about which we were ignorant, to obtain realistic four-dimensional supersymmetric models, thereby officially opening the era of superstring phenomenology.

I travelled back to the UK, in mid-December, stopping to give seminars in several places. It was while I was in Gainsville's Physics Department on the evening before my return to London that a copy of the Candelas *et al.* preprint [CHSW85] arrived via a helmeted ZAP-mail motorcycle courier in the middle of the departmental Christmas party. It was the end of 1984 and the world had changed. The explosion of subsequent developments is another story.

Acknowledgements

I am very grateful to Lars Brink for reading the manuscript and making useful suggestions for how it might be improved.

> Michael Green was born in 1946. He was educated at Cambridge University where he received his PhD in 1970. He held postdoctoral appointments at the Institute for Advanced Study, Princeton, Cambridge and Oxford before being appointed to a lectureship at Queen Mary College, London, in 1978. He became Professor of Theoretical Physics at DAMTP, Cambridge in 1993. He was elected as a Fellow of the Royal Society in 1989 when he was also awarded the Dirac Medal of the International Centre of Theoretical Physics, Trieste. He also received the Dannie Heineman Prize of the American Physical Society (2002), the Dirac Medal of the Institute of Physics (2004) and the Naylor Prize of the London Mathematical Society (2008). Since 2009 he has held the Lucasian Chair of Mathematics.

References

[ARVV67] Ademollo, M., Rubinstein, H. R., Veneziano, G. and Virasoro, M. A. (1967). Bootstraplike conditions from superconvergence, *Phys. Rev. Lett.* **19**, 1402–1405.

[ARVV69] Ademollo, M., Rubinstein, H. R., Veneziano, G. and Virasoro, M. A. (1969). Bootstrap of meson trajectories from superconvergence, *Phys. Rev.* **176**, 1904–1925.

[ASZ84] Alvarez, O., Singer, I. M. and Zumino, B. (1984). Gravitational anomalies and the family's index theorem, *Commun. Math. Phys.* **96**, 409–417.

[AF80] Alvarez-Gaumé, L. and Freedman, D. Z. (1980). Ricci-flat Kahler manifolds and supersymmetry, *Phys. Lett.* **B94**, 171–173.

[AW84] Alvarez-Gaumé, L. and Witten, E. (1984). Gravitational anomalies, *Nucl. Phys.* **B234**, 269–330.

[BZ84] Bardeen, W. A. and Zumino, B. (1984). Consistent and covariant anomalies in gauge and gravitational theories, *Nucl. Phys.* **B244**, 421–453.

[BCK06] Baulieu, L., Cremmer, E. and Kounnas, C. eds. (2006) *Conference 30 Years of Supergravity*, École Normale Supérieure, Paris, unpublished. http://www.lpthe.jussieu.fr/sugra30/

[Ber00] Berkovits, N. (2000). Super-Poincaré covariant quantization of the superstring, *JHEP* **4**, 018, arXiv:hep-th/0001035.

[BI34] Born, M. and Infeld, L. (1934). Foundations of the new field theory, *Proc. R. Soc. London* **A144**, 425–451.

[BDH76] Brink, L., Di Vecchia, P. and Howe, P. S. (1976). A locally supersymmetric and reparameterization invariant action for the spinning string, *Phys. Lett.* **B65**, 471–474.

[CHSW85] Candelas, P., Horowitz, G. T., Strominger, A. and Witten, E. (1985). Vacuum configurations for superstrings, *Nucl. Phys.* **B258**, 46–74.

[Che66] Chew, G. F. (1966). *The Analytic S Matrix* (W. A. Benjamin, New York).

[CF75] Corrigan, E. and Fairlie, D. B. (1975). Off-shell states in dual resonance theory, *Nucl. Phys.* **B91**, 527–545.

[CO72] Corrigan, E. and Olive, D. (1972). Fermion–meson vertices in dual theories, *Nuovo Cimento* **A11**, 749–773.

[CG74] Cremmer, E. and Gervais, J. L. (1974). Combining and splitting relativistic strings, *Nucl. Phys.* **B76**, 209–230.

[CJS78] Cremmer, E., Julia, B. and Scherk, J. (1978). Supergravity theory in 11 dimensions, *Phys. Lett.* **B76**, 409–412.

[CS76] Cremmer, E. and Scherk, J. (1976). Dual models in 4 dimensions with internal symmetries, *Nucl. Phys.* **B103**, 399–425.

[DZ76] Deser, S. and Zumino, B. (1976). A complete action for the spinning string, *Phys. Lett.* **B65**, 369–373.

[Dir62] Dirac, P. A. M. (1962). An extensible model of the electron, *Proc. R. Soc. London* **A268**, 57–67.

[DHS68] Dolen, R., Horn, D. and Schmid, C. (1968). Finite-energy sum rules and their application to πN charge exchange, *Phys. Rev.* **166**, 1768–1781.

[ELOP66] Eden, R. J., Landshoff, P. V., Olive, D. I. and Polkinghorne, J. C. (1966). *The Analytic S Matrix* (Cambridge University Press, Cambridge).

[FILQ90] Font, A., Ibanez, L. E., Lust, D. and Quevedo, F. (1990). Strong-weak coupling duality and nonperturbative effects in string theory, *Phys. Lett.* **B249**, 35–43.

[Fri80a] Friedan, D. H. (1980). Nonlinear models in two plus epsilon dimensions, *Phys. Rev. Lett.* **45**, 1057–1060.

[Fri80b] Friedan, D. H. (1980). Nonlinear models in two + epsilon dimensions, *Ann. Phys.* **163**, 318–419.

[FSM85] Friedan, D. H., Shenker, S. H. and Martinec, E. J. (1985). Covariant quantization of superstrings, *Phys. Lett.* **B160**, 55–61.

[GSO76] Gliozzi, F., Scherk, J. and Olive, D. (1976). Supergravity and the spinor dual model, *Phys. Lett.* **B65**, 282–286.

[GSO77] Gliozzi, F., Scherk, J. and Olive, D. (1977). Supersymmetry, supergravity theories and the dual spinor model, *Nucl. Phys.* **B122**, 253–290.

[GGRT73] Goddard, P., Goldstone, J., Rebbi, C. and Thorn, C. B. (1973). Quantum dynamics of a massless relativistic string, *Nucl. Phys.* **B56**, 109–135.

[GO85] Goddard, P. and Olive, D. (1985). Algebras, lattices and strings, in *Vertex Operators in Mathematics and Physics*, ed. Lepowsky, J., Mandelstam, S. and Singer, I. M., Mathematical Research Institute Publications (Springer-Verlag, New York).

[GW71] Goddard, P. and Waltz, R. E. (1971). One-loop amplitudes in the model of Neveu and Schwarz, *Nucl. Phys.* **B34**, 99–108.

[Got71] Goto, T. (1971). Relativistic quantum mechanics of one-dimensional mechanical continuum and subsidiary condition of dual resonance model, *Prog. Theor. Phys.* **46**, 1560–1569.

[Gre73] Green, M. B. (1973). Cancellation of the leading divergence in dual loops, *Phys. Lett.* **B46**, 392–396.

[Gre76] Green, M. B. (1976). Locality and currents for the dual string, *Nucl. Phys.* **B103**, 333–342.

[Gre77] Green, M. B. (1977). Point-like structure and off-shell dual strings, *Nucl. Phys.* **B124**, 461–499.

[GS81] Green, M. B. and Schwarz, J. H. (1981). Supersymmetrical dual string theory, *Nucl. Phys.* **B181**, 502–530.

[GS82a] Green, M. B. and Schwarz, J. H. (1982). Supersymmetrical dual string theory. 2. Vertices and trees, *Nucl. Phys.* **B198**, 252–268.

[GS82b] Green, M. B. and Schwarz, J. H. (1982). Supersymmetrical dual string theory. 3. Loops and renormalization, *Nucl. Phys.* **B198**, 441–460.

[GS82c] Green, M. B. and Schwarz, J. H. (1982). Supersymmetrical string theories, *Phys. Lett.* **B109**, 444–448.

[GS83a] Green, M. B. and Schwarz, J. H. (1983). Extended supergravity in ten-dimensions, *Phys. Lett.* **B122**, 143–147.

[GS83b] Green, M. B. and Schwarz, J. H. (1983). Superstring interactions, *Nucl. Phys.* **B218**, 43–88.

[GS84a] Green, M. B. and Schwarz, J. H. (1984). Covariant description of superstrings, *Phys. Lett.* **B136**, 367–370.

[GS84b] Green, M. B. and Schwarz, J. H. (1984). The structure of superstring field theories, *Phys. Lett.* **B140**, 33–38.

[GS84c] Green, M. B. and Schwarz, J. H. (1984). Properties of the covariant formulation of superstring theories, *Nucl. Phys.* **B243**, 285–306.

[GS84d] Green, M. B. and Schwarz, J. H. (1984). Superstring field theory, *Nucl. Phys.* **B243**, 475–536.

[GS84e] Green, M. B. and Schwarz, J. H. (1984). Anomaly cancellation in supersymmetric $d = 10$ gauge theory and superstring theory, *Phys. Lett.* **B149**, 117–122.

[GS85a] Green, M. B. and Schwarz, J. H. (1985). Infinity cancellations in $so(32)$ superstring theory, *Phys. Lett.* **B151**, 21–25.

[GS85b] Green, M. B. and Schwarz, J. H. (1985). The hexagon gauge anomaly in type I superstring theory, *Nucl. Phys.* **B255**, 93–114.

[GSB82] Green, M. B., Schwarz, J. H. and Brink, L. (1982). $N = 4$ Yang–Mills and $N = 8$ supergravity as limits of string theories, *Nucl. Phys.* **B198**, 474–492.

[GSB83] Green, M. B., Schwarz, J. H. and Brink, L. (1983). Superfield theory of type II superstrings, *Nucl. Phys.* **B219**, 437–478.

[GSW85] Green, M. B., Schwarz, J. H. and West, P. C. (1985). Anomaly free chiral theories in six-dimensions, *Nucl. Phys.* **B254**, 327–348.

[GS76] Green, M. B. and Shapiro, J. A. (1976). Off-shell states in the dual model, *Phys. Lett.* **B64**, 454–458.

[GHMR85] Gross, D. J., Harvey, J. A., Martinec, E. J. and Rohm, R. (1985). Heterotic string, *Phys. Rev. Lett.* **54**, 502–505.

[Hag65] Hagedorn, R. (1965). Statistical thermodynamics of strong interactions at high-energies, *Nuovo Cimento Suppl.* **3**, 147–186.

[HW84] Howe, P. S. and West, P. C. (1984). The complete $n = 2$, $d = 10$ supergravity, *Nucl. Phys.* **B238**, 181–220.

[HW70] Huang, K. and Weinberg, S. (1970). Ultimate temperature and the early universe, *Phys. Rev. Lett.* **25**, 895–897.

[HT95] Hull, C. M. and Townsend, P. K. (1995). Unity of superstring dualities, *Nucl. Phys.* **B438**, 109–137.

[IM67] Igi, K. and Matsuda, S. (1967). New sum rules and singularities in the complex j plane, *Phys. Rev. Lett.* **18**, 625–627.

[KK74] Kaku, M. and Kikkawa, K. (1974). The field theory of relativistic strings, Pt. 1. trees, *Phys. Rev.* **D10**, 1110–1133.

[Man68] Mandelstam, S. (1968). Dynamics based on rising regge trajectories, *Phys. Rev.* **166**, 1539–1552.

[Man73] Mandelstam, S. (1973). Interacting string picture of dual resonance models, *Nucl. Phys.* **B64**, 205–235.

[Man74] Mandelstam, S. (1974). Interacting-string picture of Neveu–Schwarz–Ramond model, *Nucl. Phys.* **B69**, 77–106.

[Nah77] Nahm, W. (1978). Supersymmetries and their representations, *Nucl. Phys.* **B135**, 149–166.

[Nam70] Nambu, Y. (1970). Duality and hadrodynamics, lecture notes prepared for Copenhagen summer school, 1970, reproduced in *Broken Symmetry, Selected Papers of Y. Nambu*, ed. Eguchi, T. and Nishijima, K. (World Scientific, Singapore, 1995), 280–301.

[NS72] Neveu, A. and Scherk, J. (1972). Connection between Yang–Mills fields and dual models, *Nucl. Phys.* **B36**, 155–161.

[Pol81a] Polyakov, A. M. (1981). Quantum geometry of bosonic strings, *Phys. Lett.* **B103**, 207–210.

[Pol81b] Polyakov, A. M. (1981). Quantum geometry of fermionic strings, *Phys. Lett.* **B103**, 211–213.

[Sch71] Scherk, J. (1971). Zero-slope limit of the dual resonance model, *Nucl. Phys.* **B31**, 222–234.

[SS74] Scherk, J. and Schwarz, J. H. (1974). Dual models for nonhadrons, *Nucl. Phys.* **B81**, 118–144.

[Sch73] Schwarz, J. (1973). Off-mass shell dual amplitudes without ghosts, *Nucl. Phys.* **B65**, 131–140.

[Sch83] Schwarz, J. H. (1983). Covariant field equations of chiral $n = 2$, $d = 10$ supergravity, *Nucl. Phys.* **B226**, 269–288.

[SW83] Schwarz, J. H. and West, P. C. (1983). Symmetries and transformations of chiral $n = 2$, $d = 10$ supergravity, *Phys. Lett.* **B126**, 301–304.

[SW73] Schwarz, J. H. and Wu, C. C. (1973). Evaluation of dual fermion amplitudes, *Phys. Lett.* **B47**, 453–456.

[Sha70] Shapiro, J. A. (1970). Electrostatic analogue for the Virasoro model, *Phys. Lett.* **B33**, 361–362.

[Sha72] Shapiro, J. A. (1972). Loop graph in the dual tube model, *Phys. Rev.* **D5**, 1945–1948.

[Sie83] Siegel, W. (1983). Hidden local supersymmetry in the supersymmetric particle action, *Phys. Lett.* **B128**, 397.

[Tho77] Thorn, C. B. (1977). On the derivation of dual models from field theory, *Phys. Lett.* **B70**, 85–87.

[Tho78] Thorn, C. B. (1978). On the derivation of dual models from field theory. 2, *Phys. Rev.* **D17**, 1073–1085.

[Ven68] Veneziano, G. (1968). Construction of a crossing-symmetric, Reggeon behaved amplitude for linearly rising trajectories, *Nuovo Cimento* **A57**, 190–197.

[Vir69] Virasoro, M. (1969). Alternative constructions of crossing-symmetric amplitudes with Regge behavior, *Phys. Rev.* **177**, 2309–2311.

[WK74] Wilson, K. G. and Kogut, J. B. (1974). The renormalization group and the epsilon expansion, *Phys. Rep.* **12**, 75–199.
[Wit81a] Witten, E. (1981). Dynamical breaking of supersymmetry, *Nucl. Phys.* **B188**, 513–554.
[Wit81b] Witten, E. (1981). Search for a realistic Kaluza–Klein theory, *Nucl. Phys.* **B186**, 412–428.
[Wit84a] Witten, E. (1984). Nonabelian bosonization in two dimensions, *Commun. Math. Phys.* **92**, 455–472.
[Wit84b] Witten, E. (1984). Some properties of $O(32)$ superstrings, *Phys. Lett.* **B149**, 351–356.
[Yon74] Yoneya, T. (1974). Connection of dual models to electrodynamics and gravidynamics, *Prog. Theor. Phys.* **51**, 1907–1920.
[Zum79] Zumino, B. (1979). Supersymmetry and Kahler manifolds, *Phys. Lett.* **B87**, 203–206.
[ZWZ84] Zumino, B., Wu, Y. S. and Zee, A. (1984). Chiral anomalies, higher dimensions, and differential geometry, *Nucl. Phys.* **B239**, 477–507.

44
Quarks, strings and beyond

ALEXANDER M. POLYAKOV

In this Chapter, I recall the sequence of ideas which led to noncritical strings and gauge/strings duality. I also comment on some promising future directions.

In the Sixties I was not much interested in string theory. The main reason for that was my conviction that the world of elementary particles should allow a field theoretic description and that this description must be closely analogous to the conformal bootstrap of critical phenomena. At the time such views were very far from the mainstream. I remember talking to one outstanding physicist. When I said that boiling water may have something to do with the deep inelastic scattering, I received a very strange look. I shall add in the parenthesis that this was the beginning of a long series of 'strange looks' which I keep receiving to this day.

Another reason for the lack of interest was actually a lack of ability. I could not follow the very complicated algebra of the early works on string theory and did not have any secret weapon to struggle with it. On the other hand, the Landau Institute, to which I belonged, was full of the leading experts in condensed matter physics. I remember that in the late Sixties to early Seventies Tolya Larkin and I discussed (many times) whether Abrikosov's vortices could be viewed as elementary particles. Nothing concrete came out of this at that time, but it helped me with my later work. With some imagination we could have related the vortex lines with strings but we missed it.

I was exploring renormalizable field theory and found the jet structure of particle production and the sum rules for deep inelastic scattering in all such theories [Pol71, Pol72]. Then, in the spring of 1973 the news about asymptotic freedom reached the Landau Institute. In a few days, after we (Sasha Migdal and I) had seen the papers on it (and checked the calculations), we had no more doubts that the field theory approach to elementary particles was the right one. My general formulae worked beautifully in this case.

It was immediately clear that the most important thing now was to study the nonperturbative phenomena. Here I was helped by the fact that I knew the dissertation of Vadim Berezinsky on the two-dimensional spin systems on the lattice. I generalized his treatment to the case of the non-Abelian gauge fields and developed lattice gauge theory. I did not

The Birth of String Theory, ed. Andrea Cappelli, Elena Castellani, Filippo Colomo and Paolo Di Vecchia.
Published by Cambridge University Press. © Cambridge University Press 2012.

expect other people to be occupied with the same subject. As Ken Wilson recently wrote [Wil05]: 'If I had not completed and published my work in a timely fashion then it seems likely that Smit, Polyakov or both would have produced publications that would have launched the subject.' To that I can only add that, while I indeed had the complete lattice gauge theory at that time, I lacked one very fundamental element – Wilson's criterion for confinement, the area law. When Wilson's paper appeared I decided not to publish my draft before finding something new. It took some more months before I added to my work the Abelian theory of quark confinement based on the idea of instantons [Pol75]. The result was quite stunning – in three dimensions the instantons (which were magnetic monopoles) led to the formation of the electric string for all couplings, while in four dimensions the instantons were the closed loops of the monopole trajectories and the confinement occurred after the coupling exceeded the critical level. A little later Gerard 't Hooft and Stanley Mandelstam arrived at the qualitative picture of dual superconductors, which is of course equivalent to the one I just described.

Things were gliding smoothly – it seemed that all that was needed was to find the non-Abelian instantons and to look at their interaction. Their disordering effect would provide the theory of quark confinement, just as in the Abelian case. And indeed the non-Abelian instantons were discovered (Belavin, Polyakov, Schwartz and Tyupkin [BPST75]). Moreover, we found a very nice self-duality equation for them and uncovered their topological origin. Many beautiful and important things were revealed in the following years: the instantons and our self-duality equation did have a big impact on physics and mathematics.

All this was great, but my efforts to build the theory of non-Abelian confinement went nowhere. The reason was that the perturbative effects were strong in the infrared and could potentially obliterate the instantons. We know today that, at best, one can build a reasonable phenomenological theory based on instantons or, if one looks for the exact theory, one has to escape to the beautiful countryside of supersymmetric gauge theories, in which the perturbative fluctuations are cancelled or controlled. Indeed, in the Nineties Seiberg and Witten [SW94] managed to guess the exact form of all instanton corrections in the case of $\mathcal{N} = 2$ supersymmetry and to discuss the (essentially Abelian) quark confinement in this case. Later Nekrasov gave a direct derivation, by summing over the instantons [Nek03]. So in this special supersymmetric case the instanton picture was fully proved.

These works had a tremendous impact in various fields, but they provided little help in nonsupersymmetric theories. By the end of 1977 it was clear to me that I needed a new strategy and I became convinced that the way to go was the gauge/string duality. This had made its appearance already in the Wilson work on the lattice gauge theory, in which the strong coupling expansion was described as a sum over random surfaces. These surfaces were the result of propagation of one-dimensional objects – electric fluxes. The major difficulty was to find the continuous limit of this picture. But already on the qualitative level I found the picture very useful. It helped me to predict the deconfining transition, leading to the quark–gluon plasma [Pol78]. This transition takes place simply because the strings are melting, as can be seen from the Peierls argument for the free energy of ordered domains.

This picture of the strings describing the flux lines is often confused with the 't Hooft picture, which suggests that the string world-sheet appears because the lines of Feynman diagrams become dense. In the normal gauge theory this certainly does not happen. These two pictures are quite different. However, 't Hooft's estimate of the strength of string interaction as $O(1/N^2)$ for the $SU(N)$ theory works in both pictures.

My concrete plan was to write the loop equation for the Wilson loop, and then to represent its solution as a sum over random surfaces. Fortunately, I grossly underestimated the depth and the difficulty of this problem. I managed to convince Migdal that the loop equation was the way to go. He joined forces with Yura Makeenko and they produced an important piece of work [MM79]. On my side, I also played with the various versions of loop equations and an idea of 'integrability in the loop space' [Pol79, Pol80]. I also thought that the string representation might help to solve the three-dimensional Ising model by reducing it to the free fermionic strings (in two dimensions, it is reduced to free fermions).

By then, the major challenge became the second part of the programme – finding, or even defining, the sum over random surfaces. In the case of paths, it has been long known (since Fock and Schwinger) that it is convenient to use a quadratic action and then to integrate over the proper time. In the case of surfaces, one can use the analogous quadratic action and introduce the independent metric on the world-sheet. Brink, Di Vecchia and Howe [BDH76] and Deser and Zumino [DZ76] ingeniously used this trick to derive the supersymmetric action for string theory. Quadratic action was also used in the Twenties in the famous work by J. Douglas on the Plateau problem.

This action is called now the Polyakov action, demonstrating the Arnold theorem which states that things are never called after their true inventors. A second application of this theorem, to which we are coming now, is the Liouville action. Namely, I found that there is a crucial difference between the vibrations of classical and quantum strings. Classically, the string is infinitely thin and has only transverse oscillations. But when I quantized it there was a surprise – an extra, longitudinal mode, which appears due to the quantum 'thickening' of the string. This new field is called the Liouville mode. It was very surprising to find that the Lagrangian for this field is proportional to $d - 26$ ($d - 10$ in the supersymmetric case); the numbers 26 and 10 were the only dimensions in which the standard string theory had been formulated. I obtained these numbers after a long struggle with ghosts and when I called Migdal and told him about this result, he was certain that I was pulling his leg.

My dream at this point was to use this noncritical string to solve both gauge theories and the three-dimensional Ising model. Even before finishing the paper, I made a one-loop estimate of the critical exponent for the Ising model. I was told by the experts that it is amazingly close to the experimental value (which I did not remember). Sadly, the last check before sending the paper for publication showed that I had made a mistake, and this 'result' was removed from the text.

Still, I was delighted to have a wonderful new playground. I hoped not only to learn more about gauge theories, but also to study two-dimensional gravity on the world-sheet as a toy model of real gravity. The fact (due to Scherk and Schwarz) that real gravity is a part of string theory added some spice to the project. This project kept me busy for the next

25 years. It started with the attempt to build a conformal bootstrap for the Liouville theory. We worked on it with my friends Sasha Belavin and Sasha Zamolodchikov. We developed a general approach to conformal field theories, something like complex analysis in the quantum domain. It worked very well in the various problems of statistical mechanics, but the Liouville theory remained unsolved. I was disappointed and inclined not to publish our results. Fortunately, my coauthors had better judgement than me, and our paper turned out to be useful in a number of fields.

The dynamics of two-dimensional gravity is very rich and even now not completely explored. One of the problems was the field-dependent cutoff which one must use in order to preserve general covariance on the world-sheet. I tried to overcome this difficulty by using a different gauge. I found, quite unexpectedly, the emergence of the $SL(2, \mathbb{R})$ current algebra and, in a subsequent joint paper by Sasha Zamolodchikov, Dima Knizhnik and myself, this symmetry allowed us to find the fractal dimensions of minimal models dressed by the gravitational field. This work had a tragic element. Dima, my fantastically talented graduate student, died of sudden heart failure before the work was done. I did not even know that he was working on this subject. But after his death Sasha and I read his notes and received a crucial insight, which allowed us to finish the work [KPZ88].

A few years before this work, Kazakov, David and others had suggested that the discrete version of two-dimensional gravity can be described by matrix models. It was hard to be certain that these models really have a continuous limit described by the Liouville theory, there were no proofs of this conjecture. To our surprise, we found that the anomalous dimensions coming from our theory coincide with those computed from the matrix model. That left no doubts that, in the case of the minimal models, the Liouville description is equivalent to the matrix description. This relation received a lot of attention. Later Witten found a third description of the same system in terms of the topological field theories.

Another aspect of our theory was a relation between gravity and $SL(2, \mathbb{R})$ gauge fields. In 1989, I wrote: 'It is possible that in this strong gravity region description in terms of the metric tensor breaks down and gauge fields should become fundamental variables. If so, we encounter one of the most exciting situations in physics' [Pol90].

I kept thinking about gauge/string dualities. Soon after the Liouville mode was discovered, it became clear to many people including myself that its natural interpretation is that random surfaces in four dimensions are described by the strings flying in five dimensions with the Liouville field playing the role of the fifth dimension. The precise meaning of this statement is that the wave function of the general string state depends on the four centre-of-mass coordinates and also on the fifth, the Liouville one. In the case of minimal models this extra dimension is related to the matrix eigenvalues and the resulting space is flat.

In 1996 I came to the conclusion that in order to describe gauge theories this five-dimensional space must be warped. The logic was as follows. In gauge/string duality the open strings describe the Wilson loop and the only allowed vertex operators in the open string sector are those corresponding to gluons (and extra fields, if present). At the same time, in the closed string sector we have an infinite number of states. So, all massive modes of the open string must go away. This can happen only if the ends of the open strings lie either at singularity or at infinity and the metric is such that this region has infinite blue shift with respect to the bulk. In this case the masses of all but massless open string states go to infinity.

Since this five-dimensional space must contain the flat four-dimensional subspace in which the gauge theory resides, the natural ansatz for the metric is just the Friedman universe with a certain warp factor. This factor must be determined from the conditions of conformal symmetry on the world-sheet. Its dependence on the Liouville mode must be related to the renormalization group flow. As a result we arrive at a fascinating picture – our four-dimensional world is a projection of a more fundamental five-dimensional string theory. As was written 25 centuries ago: 'They see only their own shadows, or the shadows of one another, which the fire throws on the opposite wall of the cave.' A small improvement on Plato – the cave has five dimensions, while the wall has four.

At this point I was certain that I had found the right language for the gauge/string duality. I attended various conferences, telling people that it is possible to describe gauge theories by solving Einstein-like equations (coming from the conformal symmetry on the world-sheet) in five dimensions. The impact of my talks was close to zero. That was not unusual and did not bother me much. What really caused me to delay the publication [Pol98, Pol99] for a couple of years was my inability to derive the asymptotic freedom from my equations. At this point I should have noticed the paper of Igor Klebanov [Kle97], in which he related D3-branes described by the supersymmetric Yang–Mills theory to the same object described by supergravity. Unfortunately I wrongly thought the paper to be related to matrix theory and I was sceptical about this subject. As a result I missed this paper which would have provided me with a nice special case of my programme. This special case was presented a little later in full generality by Juan Maldacena [Mal98] and his work opened the flood gates. The main idea was that for the supersymmetric Yang–Mills theory the geometry in five dimensions is determined by the conformal symmetry in the target space. This is the geometry of AdS_5 space, which has constant negative curvature. After that, Gubser, Klebanov and I, and Witten realized that the gauge theory should be placed at infinity in this space and gave a prescription for calculating various physical quantities.

In order to justify my picture I used intuition coming from the loop equation, while Klebanov and Maldacena appealed to the D-brane picture of the gauge fields. Both points of view are useful but neither of them led to the quantitative derivation of gauge/string duality.

In the case of D-branes the logic is as follows. We start from flat space and place there a large number of the D-branes. Their small oscillations are described by the large-N gauge theory. A nice fact about this representation of the gauge fields is that it allows us to use geometrical intuition instead of abstract field theory when considering different configurations of the D-branes. But then one has to take a major step and postulate that the collection of D-branes can be replaced by their mean gravitational field. This is a little like replacing the famous cat by its smile. While this is most probably correct, it is not clear how to justify this result. Also, there could be cases of gauge/string duality in which the flat space D-brane representation does not exist.

In the case of the loop equations, we argue that since the Wilson loop is zigzag invariant (the back and forth parts of the Wilson loop cancel), the open string must have a finite number of states (or vertex operators), corresponding to the states of the gauge theory. As we explained above, this requirement implies the warping that removes all higher states from the spectrum. The approach based on the loop equations starts from first principles. However, these equations are singular and require elaborations. With Rychkov we tried to find a nonsingular version of the equations, but we only scratched the surface of highly nontrivial technical problems. The problem of reproducing gauge perturbation theory from the string theory side remains unsolved, and extremely important.

Why should we care about the derivation from first principles? After all, in physics we value not so much the proved theorems but correct and powerful statements. However, in this case the lack of derivation really impedes progress. We do not know how far the gauge/string duality can be extended and generalized. The enormous accumulation of special cases has been useful but not sufficient for deeper understanding. This is why I think that establishing the foundations is one of the most important problems in the field.

Another important problem is integrability. In the Seventies, I was very impressed by the discovery by Belavin and Zakharov [BZ78] that the self-duality equations are completely integrable. I realized [Pol79, Pol80] that in the case of the full quantum Yang–Mills theory one can expect 'integrability in the loop space'. This means that the densities of the conserved quantities depend not on the points but on the contours. Today, due to the work of many people, we know that the dilatation operator of the super Yang–Mills theory is represented by a completely integrable spin chain. It is superficially different from the integrability which I envisaged 30 years ago. However, they must be related, since the AdS_5 string σ-model is integrable and the boundary of the world-sheet, being mapped onto the Wilson loop, must produce the integrals in the loop space. Establishing this fact is one of the still unsolved problems.

Even more important is to find the gauge theory for the de Sitter space. I conjectured that the large-N gauge theories have a fixed point at the complex gauge coupling corresponding to the radius of convergence of the planar graphs. Presumably this point is described by a nonunitary conformal field theory (CFT) corresponding to the intrinsically unstable de Sitter space. This approach will hopefully resolve the puzzle of the cosmological constant and cosmic acceleration.

Also, I confess that I still have some hopes that the three-dimensional critical phenomena can be approached by string theoretic methods. The methods of CFT and holography may also be useful in the problem of turbulence.

As for the problem of string unification, it seems to me that noncritical strings may have some future. However, it may be wise to wait for some more information about Nature (specifically about supersymmetry) which we expect to get from the LHC.

To sum up we have a large number of concrete and fascinating problems which will entertain us for many years to come. No end of physics in sight.

Acknowledgements

This work was partially supported by the National Science Foundation grant PHY-0756966.

Alexander M. Polyakov was born in the USSR in 1945. He graduated from Moscow Institute of Physics and Technology with the speciality 'experimental nuclear physics' in 1967. He received his PhD from the Landau Institute in 1968 and worked there till 1989. He has been at Princeton University since 1989.

References

[BPST75] Belavin, A. A., Polyakov, A. M., Schwartz, A. S. and Tyupkin, Y. S. (1975). Pseudoparticle solutions of the Yang–Mills equations, *Phys. Lett.* **B59**, 85–87.

[BZ78] Belavin, A. A. and Zakharov, V. E. (1978). Yang–Mills equations as inverse scattering problem, *Phys. Lett.* **B73**, 53–57.

[BDH76] Brink, L., Di Vecchia, P. and Howe, P. S. (1976). A locally supersymmetric and reparameterization invariant action for the spinning string, *Phys. Lett.* **B65**, 471–474.

[DZ76] Deser, S. and Zumino, B. (1976). A complete action for the spinning string, *Phys. Lett.* **B65**, 369–373.

[Kle97] Klebanov, I. R. (1997). World-volume approach to absorption by non-dilatonic branes, *Nucl. Phys.* **B496**, 231–242.

[KPZ88] Knizhnik, V. G., Polyakov, A. M. and Zamolodchikov, A. B. (1988). Fractal structure of $2d$-quantum gravity, *Mod. Phys. Lett.* **A3**, 819–826.

[MM79] Makeenko, Y. M. and Migdal, A. A. (1979). Exact equation for the loop average in multicolor QCD, *Phys. Lett.* **B88**, 135–137. Erratum, *Phys. Lett.* **B89**, 437.

[Mal98] Maldacena, J. M. (1998). The large N limit of superconformal field theories and supergravity, *Adv. Theor. Math. Phys.* **2**, 231–252. Also *Int. J. Theor. Phys.* **38**, 1113–1133.

[Nek03] Nekrasov, N. A. (2003). Solution of $N = 2$ theories via instanton counting, *Ann. Henri Poincaré* **4**, S129–S146.

[Pol71] Polyakov, A. M. (1971). Similarity hypothesis in strong interactions. 2. Cascade formation of hadrons and their energy distribution in e+ e− annihilation, *Zh. Eksp. Teor. Fiz.* **60**, 1572–1583 [*Sov. Phys. JETP*, **33**, 850–855, 1972].

[Pol72] Polyakov, A. M. (1972). Description of lepton hadron reactions in quantum field theory, *Zh. Eksp. Teor. Fiz.* **61**, 2193–2208 [*Sov. Phys. JETP* **34**, 1177–1183].

[Pol75] Polyakov, A. M. (1975). Compact gauge fields and the infrared catastrophe, *Phys. Lett.* **B59**, 82–84.

[Pol78] Polyakov, A. M. (1978). Thermal properties of gauge fields and quark liberation, *Phys. Lett.* **B72**, 477–480.

[Pol79] Polyakov, A. M. (1979). String representations and hidden symmetries for gauge fields, *Phys. Lett.* **B82**, 247–250.

[Pol80] Polyakov, A. M. (1980). Gauge fields as rings of glue, *Nucl. Phys.* **B164**, 171–188.

[Pol90] Polyakov, A. M. (1990). Gauge transformations and diffeomorphisms, *Int. J. Mod. Phys.* **A5**, 833–842.

[Pol98] Polyakov, A. M. (1998). String theory and quark confinement, *Nucl. Phys. B Proc. Suppl.* **68**, 1–3.
[Pol99] Polyakov, A. M. (1999). The wall of the cave, *Int. J. Mod. Phys.* **A14**, 645–657.
[SW94] Seiberg, N. and Witten, E. (1994). Electric-magnetic duality, monopole condensation, and confinement in $N = 2$ supersymmetric Yang–Mills theory, *Nucl. Phys.* **B426**, 19–52.
[Wil05] Wilson, K. G. (2005). The origins of lattice gauge theory, *Nucl. Phys. B Proc. Suppl.* **140**, 3–19.

45
The rise of superstring theory

ANDREA CAPPELLI AND FILIPPO COLOMO

45.1 Introduction

In this Chapter we review the developments of string theory from its renaissance in 1984 up to present times, as a kind of epilogue. We shall be rather brief and omit references to original work that would be unavoidably limited: extensive bibliographies can be found in the textbooks mentioned at the end.

After the mid-Eighties, research in string theory was at the forefront in theoretical physics and had a strong impact in many other domains. Even though its main goal, that of unifying fundamental interactions, has not been accomplished yet, and the chances of success are being debated, we can fairly say that string theory has had a broad and lasting influence on theoretical physics. In short, it changed the way research is done in the field. Over the last 25 years, theoreticians were more keen on using advanced mathematical tools and on resorting to abstract modelling: for example, by playing with the spacetime dimension, i.e. considering a given problem also in dimensions different than four. After being presented in the mid-Eighties as 'the theory of everything', i.e. of all fundamental interactions, string theory was later dubbed 'the theory of nothing' by its opponents; in recent times, it has also been called 'the theory for everything', for its many, often unexpected applications.

New techniques of two-dimensional conformal field theory introduced in string theory were employed to describe low-dimensional condensed matter systems and statistical models. Another main theme was the deep relation of supersymmetry with many topics of geometry and topology. Furthermore, researchers in mathematics and mathematical physics had the opportunity to learn physical applications of their methods to string theory and theoretical physics in general.

A new burst of activity in string theory, the so-called 'second string revolution', took place in the second half of the Nineties and continued in the new Millennium. The main achievement was the return of open strings and their 'fusion' with closed strings into a unique theory, called 'M-theory'. Beside the unification of fundamental forces, some results found application in the domains of Black Holes and nonperturbative gauge theories. We shall discuss these topics in turn.

The Birth of String Theory, ed. Andrea Cappelli, Elena Castellani, Filippo Colomo and Paolo Di Vecchia.
Published by Cambridge University Press. © Cambridge University Press 2012.

45.2 The quest for unification

The renewed interest in string theory at the beginning of the Eighties was a remarkable phenomenon in the theoretical physics community. Let us recapitulate from the Introduction to Part VII the developments that led to this surprising comeback. They should be framed in the programme of unification of all fundamental interactions beyond the Standard Model (SM).

45.2.1 Supersymmetric Standard Model

As described in the Introduction to Part VI, the Grand Unified Theory (GUT) was formulated in the mid-Seventies by enlarging the gauge symmetry from three independent groups $SU(3) \times SU(2) \times U(1)$ to a single group, $SU(5)$, leading to a unique coupling constant instead of three. The three coupling constants of strong, weak and electromagnetic interactions take very different values in the Standard Model, but they are not actually constant: owing to the quantum loop corrections, they acquire an energy dependence and become 'running coupling constants'. This dependence is logarithmic, i.e. very mild: by letting the couplings evolve at energies above the Standard Model limit of validity, $E_{SM} \sim 1000\,\text{GeV} = 1\,\text{TeV}$, assuming that no other dynamics takes place (the 'desert' hypothesis), they were found to converge to a common value at the GUT energy scale, $E_{GUT} \sim 10^{16}\,\text{GeV}$, although they did not meet exactly at one point.

This result was encouraging for unification, but also rather puzzling for the huge gap of 13 orders of magnitude separating the SM and GUT scales, which seemed unnatural (the so-called 'hierarchy problem'). Another problem was that the Standard Model could not be extended to energies much higher than E_{SM}, because the Higgs sector of the theory becomes strongly interacting, suggesting that another theory should come into play.

Supersymmetric extensions of the Standard Model devised at the beginning of the Eighties were found to provide an appealing solution to these problems. It was assumed that supersymmetry is broken below some energy E_{SUSY}, larger than E_{SM}, since the SM particle spectrum is not supersymmetric, but not much larger, for this symmetry to hold over a wide range, from 'low' energies up to the GUT scale. In this case, supersymmetry stabilizes the Standard Model allowing its smooth extension above 1 TeV, with logarithmic flows of all couplings. In particular, the Minimal Supersymmetric Standard Model (MSSM) leads to precise unification of the three gauge couplings at the GUT scale, as shown at the beginning of the Nineties. For these reasons, supersymmetry has been considered a rather economic principle.

Supersymmetric partners of the known particles are now searched for in LHC experiments at CERN, Geneva. In the last decade, supersymmetric partners have also been welcomed as good candidates for the nonbaryonic 'dark matter', which has been introduced to explain cosmological observations.

45.2.2 Supergravity and superstring theories

Unification of the Standard Model interactions with gravity was first attempted by using supergravity theories (SUGRA), as described earlier. In general, $d = 10$ supergravities are low energy limits of open and closed superstrings. A different case is $d = 11$ supergravity, described briefly in the Introduction to Part VII, which was not the low energy limit of a superstring; its relation with strings is rather involved and was only understood in the mid-Nineties. Theories in lower dimensions were derived by the Kaluza–Klein compactification mechanism, leading to extended supersymmetry as well as additional gauge fields.

Let us summarize the supergravity theories and the corresponding string theories, as they were known at the beginning of the Eighties:

(i) type I, the low energy limit of the GSO projected RNS open string that was discussed in the Introductions to Parts V and VI;
(ii) types IIA and IIB, corresponding to closed strings with GSO projections of left and right modes that are opposite or equal, respectively;
(iii) $d = 11$ supergravity.

These theories were investigated by computing their loop corrections and by matching their spectrum with SM gauge symmetries and matter content through suitable dimensional reductions to $d = 4$. The following features were needed for physical applications:

(i) the renormalizability of the gravitational interaction;
(ii) the presence of the SM gauge symmetries and matter multiplets, in particular the chirality of weak interactions;
(iii) the absence of chiral anomalies.

Regarding the first point, it was originally believed that supergravity could be on-shell finite, i.e. that singularities of its loops could cancel due to the opposite contributions of bosonic and fermionic intermediate states circulating in the loops. However, this cancellation was eventually found not to be complete: supergravity theories were not renormalizable either. String theories were instead well defined at the quantum level (the tachyons being cancelled by GSO projection).

As for the second point, gauge interactions were clearly present in type I string theory and supergravity in $d = 10$; gauge fields in lower dimensions were also generated in both open and closed string theories by the Kaluza–Klein reduction; however, the chirality of weak interactions was difficult to reproduce.

Turning to the third point, the analysis of anomaly cancellations showed that no anomaly was present either in $d = 11$ supergravity or in the IIA supergravity/string theory; the last theory did not possess chiral fields in $d = 10$. Regarding IIB supergravity, the cancellation of gravitational chiral anomalies was proved in a thorough study by Alvarez-Gaumé and Witten. On the other hand, the same study seemed to indicate the presence of anomalies in type I supergravity, the most promising theory for describing chiral fermions coupled to gauge fields in four dimensions.

As described in the Introduction to Part VII, Green and Schwarz addressed this problem by using both perturbative methods in supergravity and the RNS oscillator formalism in string theory. The string calculation showed anomaly cancellation for the $SO(32)$ gauge group. The field theory (supergravity) calculation also found vanishing anomaly, provided that two terms in the supergravity action were appropriately matched (the so-called 'Green–Schwarz cancellation mechanism'). This cancellation appeared to be quite natural in the superstring, but rather ad hoc in supergravity. Another relevant result was that anomaly cancellation in supergravity occurred for a second group besides $SO(32)$, namely $E_8 \times E_8$, where E_8 is the largest exceptional Lie algebra.

Taken together, these results formed the core motivation for the string theory comeback of 1984. Let us summarize them:

(i) type I superstring theory was shown to have a consistent low energy limit, allowing both SM gauge symmetries and $d = 4$ chiral fermions, and providing a viable unification framework;
(ii) the anomaly cancellation clearly showed that string theory was more fundamental and more consistent than the corresponding supergravity theory which, moreover, was not renormalizable;
(iii) the other anomaly-free group $E_8 \times E_8$ suggested the existence of another possible superstring theory yet to be found.

The missing superstring theory, found by Gross, Harvey, Martinec and Rohm, was a closed string made by left modes of the $d = 10$ fermionic string and right modes of the $d = 24$ bosonic string. It was called 'heterotic', adopting a botanic term for the vigour often observed in hybrid plants; it was spacetime supersymmetric and without tachyons. The anomaly cancellation required the symmetry groups $SO(32)$ and $E_8 \times E_8$ as expected (the corresponding string theories are often called HO and HE, respectively). The theory with $E_8 \times E_8$ symmetry, in particular, allowed for $d = 4$ chiral fermions and gauge symmetries suitable for the Standard Model, as shown by Candelas, Horowitz, Strominger and Witten soon after (as described in the next Section).

In conclusion, these results were eventually found to fit very well the expectations and hopes of a large community which had investigated supersymmetry and supergravity theories for almost a decade, aiming at post Standard Model unification of fundamental interactions.

45.3 String theory in the second half of the Eighties

With the proof of anomaly cancellation in type I superstring theory and the construction of heterotic superstrings, research on string theory boomed with a period of intense model building that continued till the end of the Eighties.

In those years, the compactifications to four dimensions were mostly analyzed in supergravity theories, where the six-dimensional compact manifold could be studied in great generality, while the corresponding string theory is too difficult to be solved on nontrivial curved backgrounds. Furthermore, at that time the techniques for open strings were less advanced than those for closed strings. Type I string theory was thus put aside, and focus switched to heterotic strings, in particular the phenomenologically promising $E_8 \times E_8$ theory. Compactifications of the extra dimensions were realized on Calabi–Yau manifolds, three-dimensional complex manifolds with vanishing Ricci tensor,

$$R_{\mu\nu} = 0, \qquad (45.1)$$

and other properties leading to $\mathcal{N} = 1$ supersymmetric gauge theories in $d = 4$. These nontrivial curved spaces provided particle contents in four dimensions analogous to that of the Standard Model. Furthermore, in the study of string theory on curved spacetime manifolds, Lovelace, Friedan and others found the remarkable result that the consistency of the theory, i.e. conformal invariance on the world-sheet, implied Einstein equations (45.1) of general relativity in the vacuum.

Another mark of string theory in the Eighties was the great technical advances following the introduction of mathematical results on representation theory of the conformal and affine algebras (current algebras). The world-sheet operator formalism was completely reformulated and generalized in the language of the two-dimensional conformal field theories introduced by Belavin, Polyakov and Zamolodchikov in 1984. This allowed a complete understanding of the covariant quantization of the RNS string and many other results on model building and the string loop expansion. One example is the quantization of string theory on compactified six-dimensional spaces that are group manifolds.

45.4 D-branes, string dualities and M-theory

At the end of the Eighties, five consistent superstring theories and a $d = 11$ supergravity theory were known, whose basic properties are summarized in Table 45.1. A series of results in the Nineties led eventually to the proof that they are manifestations of a unique theory, called 'M-theory', where M stands for 'membrane', 'mysterious', 'matrix' and so on.

In the quantization of the string discussed in the Introduction to Part IV, Neumann and periodic boundary conditions were chosen for the open and closed string, respectively; Dirichlet conditions, i.e. fixed string end-points were discarded because they would have implied a breaking of translation (and Lorentz) invariance. With hindsight, this was not necessarily unwanted. For Dirichlet conditions on $D - p$ space directions ($D = d - 1$),

$$X^\mu(\tau, 0) = X^\mu(\tau, \pi) = \Lambda^\mu, \qquad \mu = 1, \ldots, D - p, \qquad (45.2)$$

and Neumann conditions in the others, the string end-points are allowed to move on a $(p + 1)$-dimensional hyperplane in spacetime that is identified by the above conditions.

This breaking of translation invariance creates a sort of 'defect' or 'dislocation' in the nine-dimensional space. The great intuition by Polchinski in 1995 was to consider these

Table 45.1 *The superstring theories and* $d = 11$ *supergravity*

Type	Dimension	Chiral	Open/Closed	Gauge group
I	10	yes	open	$SO(32)$
IIA	10	no	closed	$U(1)$
IIB	10	yes	closed	—
HO	10	yes	closed	$SO(32)$
HE	10	yes	closed	$E_8 \times E_8$
SUGRA	11	no	—	—

defects as dynamical degrees of freedom of the theory, called 'Dp-branes', where D stands for Dirichlet and brane for membrane in the $p = 2$ case. In short, the D-brane can be viewed as physical excitation with very large mass, which acquires a (semi-classical) dynamics induced by the string motion itself.

The nature of these new degrees of freedom could be understood in the supergravity description: upon matching charges and supersymmetries, the D-brane can be associated with a nontrivial classical field configuration, solution of the nonlinear supergravity equations of motion. It is a so-called 'soliton' solution, i.e. a spatially extended collective excitation in field theory, characterized by specific topological properties and different from the basic pointlike particles. The prototype is the monopole of non-Abelian gauge theories, describing an isolated source of magnetic field (an infinitely thin and long solenoid with one end sent to infinity). Its magnetic charge m obeys the Dirac quantization condition,

$$me = hn, \quad n \text{ integer}, \qquad (45.3)$$

with e the smallest electric charge in the theory. Summarizing, the solitons of string theories were discovered and identified with the D-brane solutions; their semi-classical dynamics could be described in the supergravity theory.

Another achievement in supersymmetric gauge theories was obtained by Seiberg and Witten in 1994. Building on a series of preceding works on magnetic monopoles, they used supersymmetry nonrenormalization theorems to find exact results on the monopoles in these theories; they showed that certain gauge theories can have an alternative or 'dual' description, given by another gauge theory in which the (magnetically charged) solitons take the role of the basic (electrically charged) elementary quanta and vice versa. The respective coupling constants, e and m, are related by the Dirac condition (45.3).

From this relation, it is apparent that the dual gauge theory based on monopoles is perturbative when the original theory is strongly coupled. This nonperturbative relation, called electric–magnetic duality, or 'S-duality', involves two different field theory descriptions of the same physical system (unfortunately, the word 'duality' is ubiquitous, having several completely different meanings in closely related subjects).

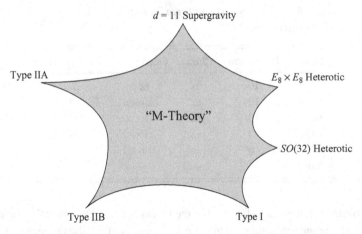

Figure 45.1 M-theory, the five $d = 10$ strings and $d = 11$ supergravity.

Research in the second half of the Nineties was able to lift S-duality from gauge theories to supergravities and string theories, leading to relations between the five superstrings. For example, it was found that HO was S-dual to type I, and that IIB was S-self-dual. Clearly, these nonperturbative relations cannot be understood within the string world-sheet formalism, which is a first-quantized language appropriate for elementary quanta but not for collective excitations. Nevertheless, these relations can be verified by matching the spectra of pairs of S-dual theories, including both elementary and solitonic states.

Another pairing relation (now, perturbative) between the five superstrings is 'T-duality', a symmetry of the spectrum of compactified strings. Finally $d = 11$ supergravity was related to type IIA and HE strings by compactifying one space direction. Altogether, the six theories in Table 45.1 were related one to another: they were all limits of a unique theory with six different incarnations, i.e. spacetime descriptions. Each of them is more convenient in a given range of parameters, for example when one coupling becomes small or a compactification radius becomes large (see Figure 45.1). The unique theory was called 'M-theory', because it included membranes besides strings. The use of M for mysterious originated from the fact that the theory is nonperturbative and thus difficult to solve. Furthermore, its formulation presents a puzzle: what are the basic degrees of freedom of a theory encompassing zero-branes (particles), one-branes (strings), two-branes (membranes) etc.? Nonetheless, a rewarding result was obtained in the programme of string unification of fundamental interactions: the 'string' theory itself is unique in $d = 10$ dimensions.

45.4.1 The landscape

These achievements came with a side effect: the possible models to which string theory reduces in four dimensions are many more than previously conceived. Allowing the D-brane solutions into the game opened a Pandora's box: the number of consistent compactifications with D-branes, defining the 'low-dimensional' theory, can be estimated to be of the order 10^{500} (at least in one specific setting). Each of these so-called 'string vacua' has its own 'low energy' physics, gauge symmetries, particles, masses and coupling constant values, giving rise to a different physical world.

This result, the so-called 'landscape' of string vacua, is rather embarrassing for a fundamental theory, such as string theory, that as such should predict the Standard Model, its spectrum and our Universe. Nowadays, some believe that a new physical principle is needed for selecting among these vacua, while others think that their abundance could even be a virtue of the theory. According to the second point of view, the landscape should be simply accepted: our Universe is only one of the many possibilities, either real (the 'Multiverse') or potential.

As an analogy, we should think that today we do not try to predict the sizes of the orbits of planets around the Sun from first principles, and from this prove the favourable conditions for life on Earth. The orbits are just what they are, a consequence of some accidental initial condition during the origin of the solar system; we were lucky that the Earth was well suited for biochemistry. Similarly, some scientists say that our Universe is as it is for no reason, beside the fact that it allowed Mankind to develop and study it (this is the 'anthropic principle').

45.5 Black Hole entropy

An important recent application of string theory has been found in the physics of Black Holes. As mentioned in the Introduction to Part VI, a sufficiently large amount of matter, over about three solar masses, compresses itself to extremely high density and forms a Black Hole. As first shown by Chandrasekhar in the Thirties, smaller masses find an equilibrium with nuclear forces, while nothing can stop the gravitational collapse of large enough masses. Nowadays, astronomical evidence of Black Holes has been reported and it is likely that one gigantic Black Hole exists at the centre of each galaxy.

At very high densities, gravity becomes strong and quantum effects are relevant. In 1975, Hawking showed that Black Holes are not really black because they emit thermal radiation at a temperature T inversely proportional to their mass M,

$$T = \frac{\hbar c^3}{8\pi G_N M k_B}, \qquad (45.4)$$

where k_B and G_N are the Boltzmann and Newton constants, respectively. This quantum effect was found through a rather simple quantum field theory argument, assuming a semi-classical theory of gravity which is correct for large masses M.

Hawking's remarkable result was the culmination of a series of studies on classical Black Holes in the Seventies showing that they have thermodynamic properties analogous to those of statistical systems, as if they were a gas of particles rather than a single macroscopic body. The fact that they also have a temperature fitted well with the thermodynamic analogy.

The physical mechanism by which Black Holes can radiate is rather simple. Around them, one identifies a spherical surface of radius $r \sim G_N M$ called the 'event horizon', because – classically – nothing can escape from its interior. On the other hand, at the quantum level any matter field has vacuum fluctuations, i.e. virtual processes of particle–hole creation and annihilation, as for example in loop corrections. In the vicinity of the horizon, the geometry of spacetime causes the virtual particle–hole pairs to become real and be emitted: one quantum of the pair falls inside the Black Hole, while the other one leads to so-called Hawking radiation. It turns out that the spectrum of this radiation is thermal, with a temperature given by (45.4).

As further observed by Bekenstein, an entropy can also be associated with the Black Hole, as with any thermal state. The expression for this entropy S is:

$$S \sim G_N M^2 \sim \frac{A}{4G_N}, \tag{45.5}$$

which is proportional to the area A of the event horizon.

Entropy in statistical mechanics is a measure of the number W of microscopic states of, say, a gas, that are possible for given values of macroscopic parameters, such as temperature, pressure etc., entering the thermodynamic description. This state counting is expressed by the Boltzmann formula:

$$S = k_B \log W. \tag{45.6}$$

Since a Black Hole is a macroscopic body in a single state at the classical level, its entropy is a puzzling result, and general relativity gives no indication of what the associated microstates could be. Since it is a quantum statistical effect, one can speculate that W counts the quantum states of the gravitational field created by the Black Hole, which should possess a large degeneracy for still unknown reasons; however, a concrete analysis requires a consistent quantum theory of gravity.

Superstring theory is a well-defined quantum theory incorporating gravity and should provide an answer. Clearly a full string theory calculation is needed, beyond the supergravity semi-classical description. The counting of string quantum states was first obtained in the mid-Nineties by Strominger and Vafa, for a particular class of rotating and charged Black Holes for which loop corrections are suppressed by supersymmetry, the so-called 'extremal Black Holes'.

The results were found to agree with the Bekenstein–Hawking formula for entropy extending (45.5) to the extremal case. Note that the necessary exponential growth of states with energy is a characteristic feature of string theory. This result suggests that the quantization of gravity within string theory could be correct.

Let us also mention two other approaches for testing quantum gravity within string theory: (i) the study of high energy string scattering, initiated by Amati, Ciafaloni and Veneziano, and by Gross and Mende; and (ii) the analysis of string corrections to the

Standard Model of cosmological evolution of the Universe, developed by Veneziano and others.

45.6 The gauge/gravity correspondence

The D-brane solutions provide another way for relating $d = 10$ string theory with four-dimensional physics, which is alternative (or complementary) to that of compactification of extra dimensions. Consider an open string configuration with D3-branes: the massless gauge degrees of freedom that are part of the low energy spectrum are constrained to live on the brane owing to the Dirichlet conditions (45.2) for the string end-points. If we identify the brane with physical space, the low energy limit of this open string configuration is a four-dimensional supersymmetric gauge theory, with gauge group $U(N)$ for N superposed branes.

On the other hand, the D-brane corresponds to a solitonic field configuration of supergravity describing a nontrivial curved spacetime, a higher-dimensional version of a Black Hole. The full string theory is defined on this background metric and is substantially different from that propagating in Minkowski flat spacetime discussed so far.

The equivalence of these two descriptions of the same theory, namely a gauge theory on the brane world-volume and a closed string theory on the $d = 10$ curved background, establishes a so-called 'gauge/gravity correspondence' or 'duality'. The prominent example is the AdS/CFT correspondence proposed by Maldacena in 1998, where the background geometry is $AdS_5 \times S^5$ (AdS_5 stands for the $d = 5$ anti de Sitter spacetime with negative constant curvature, and S^5 for the $d = 5$ sphere), and the four-dimensional gauge theory is $\mathcal{N} = 4$ supersymmetric and conformal invariant. The main features of this correspondence are the following.

(i) AdS_5 spacetime is an open manifold and its boundary at infinity, equivalent to $d = 4$ Minkowski space, can be identified with the gauge theory spacetime. This geometrical setting allows observables to be mapped between the two theories.
(ii) The gauge/gravity correspondence is said to be 'holographic' because the $d = 5$ gravitational theory in AdS_5 is equivalent to the boundary 'image', which is the $d = 4$ gauge theory.
(iii) The semi-classical solutions of supergravity field equations can be made to correspond to nonperturbative quantum effects in the dual gauge theory, owing to the strong–weak mapping of the respective coupling constants.

This last property, i.e. the possibility of studying strong-coupling quantum effects in gauge theories, has fuelled intense research activity over the last ten years. In this approach, string theory is used as a tool for understanding gauge theories beyond perturbation theory and the issues related to unification are disregarded. In the following, we shall mention some of the topics addressed using the AdS/CFT correspondence, often with rather remarkable results.

45.6.1 The hadronic string revisited

In the Introduction to Part VI, the hadronic string was seen to be quantitatively correct at large distances, matching the hadron spectrum organized in Regge trajectories. Unfortunately, full consistency of string theory as a quantum theory valid at any scale implied massless states with nonzero spin in the spectrum and other features in contradiction with experiments.

On the other hand, quantum chromodynamics (QCD) was proved correct in the opposite regime of short-distance/high-energy quark scattering where the gauge interaction is weak. Analysis of the large-distance regime in this theory was limited by the large value of the coupling constant. However, several results suggested the existence of an effective string in QCD, as for example in the formulation on a spacetime lattice, implying that quark pairs are confined by an ever-rising linear potential.

The AdS/CFT correspondence offers a new view on this problem, since it provides a precise map of QCD into a superstring theory propagating in the $AdS_5 \times S^5$ space. The correspondence is rather powerful, because results in strongly coupled QCD can be obtained by weak coupling semi-classical computations in the dual gravitational theory. In some sense, this approach amounts to a return of string theory to its origins in hadronic physics. However, as discussed in Polyakov's Chapter in this Part, the new QCD string is disguised, as it involves one extra (large) dimension and presents some features not envisaged in the past. Here are some highlights.

(i) In the simpler case of gauge/gravity correspondence, where approximations are under control, the gauge theory is $\mathcal{N} = 4$ supersymmetric and conformal invariant and does not confine quarks; therefore the dual string does not confine either. This can be explained pictorially by the fact that the quarks stay on the four-dimensional boundary of AdS_5, while the string extends in the five-dimensional interior and becomes loose by wandering around this larger space.

(ii) Nonetheless, the physical confining case can be investigated by extending the correspondence to $\mathcal{N} = 1$ gauge theories; although in this regime the results are only approximate, qualitatively correct results have been obtained for low energy hadronic physics, as for example the spectrum of glueballs.

(iii) The new string agrees qualitatively with QCD even at short distances, where the old string predicted too soft scattering amplitudes; this is due to the fact that the spectrum of states for the old string propagating in flat Minkowski space is quite different from that of new string moving in anti de Sitter curved space (see Brower's Chapter in Part IV).

(iv) Other strong coupling properties of gauge theories that are independent of confinement can be modelled by the $\mathcal{N} = 4$ theory which is under better control, as the hydrodynamic transport coefficients of the quark–gluon plasma (for example, the shear viscosity). Here, nontrivial results of the AdS/CFT correspondence have been matched favourably to experiments on heavy-ion collisions at high energy.

(v) Finally, phases of condensed matter in three dimensions characterized by strong couplings are also being actively analyzed using the gauge/gravity correspondence.

45.7 String spin-offs

The string theory renaissance in the mid-Eighties brought a wealth of new field theoretical and mathematical techniques. Some methods were first introduced in the context of string theory, like affine Lie algebras and supersymmetry, and then developed by mathematicians; others had an independent origin, such as anomalies in quantum field theory, as well as advanced topics of differential geometry and topology in mathematics, and were later extensively applied to string theory.

Thus, string theory was the cradle of many new theoretical methods, which also found physical application in other domains; today they can be considered standard tools, for example conformal field theory, supersymmetry, matrix models and topological field theory. From this point of view, we can argue that string theory was, and still is, very influential and pervasive. In this Section, we briefly introduce some of the string methods and their application to other domains.

45.7.1 Two-dimensional conformal field theory

In the Introductions to Parts III and V, we discussed the symmetry of the world-sheet theory under conformal transformations and its generalization including Lie group symmetries, leading to so-called 'current algebras'. These are more precisely called affine Lie algebras, because they involve a complex (affine) parameter or equivalently a Fourier mode index n, as for example in L_n, α_n^μ and T_n^a. The study of the representations of the Virasoro and affine Lie algebras was accomplished at the beginning of the Eighties by Kac, Peterson, Feigin and Fuchs and others (see also the Chapter by Bardakci and Halpern in Part V).

In early string theory, there occurred only integer values of the central charge c, corresponding to the number of bosonic and/or fermionic world-sheet fields; these free theories could be described directly by the operator formalism and canonical quantization. The general mathematical analysis found other representations for fractional c values, which could be used to analyze new theories that are strongly interacting.

The work by Belavin, Polyakov and Zamolodchikov in 1983 set the stage for many further developments in this domain, by introducing a new 'calculus' for interacting conformal fields. In particular, they considered the representations of the Virasoro algebra for $0 < c < 1$, in the discrete series of values,

$$c = 1 - \frac{6}{m(m+1)}, \qquad m = 3, 4, \ldots, \qquad (45.7)$$

and used them to build the so-called 'Virasoro minimal models'. These field theories could be solved exactly, since the scaling dimensions of all fields were known and the multipoint correlation functions could be obtained by solving linear differential equations.

In particular, the first value $c = 1/2$ in (45.7) corresponds to the 'Ising model', the simplest yet nontrivial model of a two-dimensional ferromagnet, at the critical point where its low energy excitations are massless and conformal invariant. Conformal field theory reproduced and extended results that had been obtained over the years by exact solutions of the statistical sum, starting from the work by Onsager in the Forties. Extension of these results to exactly solvable conformal theories with world-sheet supersymmetry were provided shortly after by Friedan, Qiu and Shenker, who obtained corresponding minimal superconformal models.

Further interacting conformal theories are solvable in the presence of extended symmetry of the affine Lie algebra G_k; these correspond to series of $c > 1$ values, according to the formula,

$$c = \frac{k \dim G}{k+h}, \qquad k = 1, 2, \ldots, \tag{45.8}$$

where k is the 'level' of the affine Lie algebra G_k, dim G is the dimension of the algebra (the number of generators) and h is the so-called dual Coxeter number (e.g. $h = N$ for $G = SU(N)$). After the pioneering work in early string theory, by Bardakci and Halpern and others, the complete understanding of their representation theory in the mathematical literature allowed for the solution of these conformal field theories by Witten, Gepner, Knizhnik and Zamolodchikov, Goddard, Kent and Olive and others. The theories were identified as those describing bosonic fields which are coordinates of the manifold of the group G, the so-called Wess–Zumino–Witten–Novikov (WZWN) models. They are generalizations of the σ-model action for the string constrained to move on the group manifold. These theories were used by Gepner to obtain nontrivial $\mathcal{N} = 2$ supersymmetric compactifications of the six extra dimensions in string theory.

Another interesting interdisciplinary development followed from the study of modular invariance. We recall that world-sheet reparameterization invariance in string theory reduces to conformal symmetry after fixing the conformal gauge. In the finite geometry of a torus, i.e. with doubly periodic boundary conditions, there is an additional residual gauge symmetry under global discrete reparameterizations respecting the boundary conditions, called modular transformations. This symmetry implies further conditions on the torus amplitude, whose implementation has far-reaching consequences. The modular invariance conditions were used extensively in heterotic string compactifications in the mid-Eighties, where they provided consistency conditions for the theory, enforcing the GSO projection among other things.

Through the work of Cardy, the conditions of modular invariance were extended to the partition function of Virasoro minimal models; here, they became a powerful tool for model building. It was found that they provide a strong condition on the spectrum of excitations and their selection rules, the so-called Verlinde fusion rules, which had several physical applications. The study of modular invariant partition functions allowed the classification of so-called 'rational conformal field theories': in particular, in 1986 Cappelli, Itzykson and Zuber found that the Virasoro minimal and $SU(2)$ WZWN models are classified in two infinite series of theories and few exceptional cases, that are in remarkable one-to-one correspondence with the Cartan–Dynkin classification of Lie algebras of A-D-E type. Following these developments, the modular conditions were applied in string theory for relating closed and open string amplitudes, by Bianchi, Pradisi and Sagnotti and others. These results were relevant for the study of D-branes in the mid-Nineties.

Altogether, conformal field theories provided a cornucopia of exact solutions for two-dimensional models that found several physical applications:

(i) critical points of statistical mechanics could be described in generalizations of the Ising model, such as the Heisenberg model, the Potts and Hubbard models, etc.;
(ii) the traditional domain of exactly solvable (integrable) systems, both in statistical mechanics and in field theory, was boosted, because the solution of critical points allowed to explore the massive phases and to span the entire phase diagram;
(iii) Some strongly coupled systems of quantum condensed matter in one and two spatial dimensions could be modelled exactly, such as the Kondo model of fermions in metals interacting with impurities, the fractional quantum Hall effect of fermions in semiconductors under a strong external magnetic field, and other systems, recently including the cold-atom gases in optical lattices.

Applications of low-dimensional field theory to domains other than particle physics developed considerably in the last twenty years, owing to the combination of exact methods of solution and of experimental advances in the fabrication of clean materials and small devices at the nanometre scale. A wealth of quantum effects could be analyzed and explained, such as the transmutation of degrees of freedom, the existence of excitations with fractional charge and fractional spin values, and the nonlinear response effects.

45.7.2 Exact S-matrices

As a matter of fact, even the original Chew–Frautschi S-matrix approach was developed successfully in two dimensions. This is more precisely a spin-off of pre-string theory methods, but it is worth mentioning it because it has been extensively applied, together with conformal field theory, to the physical systems discussed above.

Starting with the 1980 work by Zamolodchikov and Zamolodchikov, it was understood that the S-matrix bootstrap simplifies considerably in two dimensions, owing to the simplified kinematics. Furthermore, the additional requirement of integrability, i.e. the existence of further conservation laws in the theory, allowed for exact S-matrix solutions.

Integrability in two dimensions has far-reaching consequences.

(i) The scattering can only be elastic, i.e. there is no particle production.
(ii) There are only single-particle intermediate states that are stable; thus, the amplitudes possess only pole singularities with zero width – the original narrow-resonance approximation is exact in this case.

(iii) The multiparticle S-matrix is factorized into the product of elementary four-particle scattering amplitudes; as a consequence, it obeys the Yang–Baxter equation, expressing the associativity of this factorization. This equation is crucial for closing the bootstrap.

A large number of exact S-matrix solutions have been found, which describe closed, self-consistent sets of scattering particles. In many cases, these have been identified as being massive excitations of strong-interacting integrable field theories in two dimensions, although direct derivations are not possible in general.

These S-matrix solutions possess all the desired properties envisaged by Chew and others in the Sixties: complete 'democracy' between the particles, being external, intermediate or 'bound' states; strongly coupled dynamics with no quantitative field theoretic description. In other words, the S-matrix programme was indeed accomplished, but only in two dimensions and with the addition of integrability. Unfortunately, this last ingredient is specific to low dimensions.

45.7.3 Topological field theories

Another interdisciplinary application of string theory regards contemporary problems of mathematics and geometry, such as the classification of topological manifolds in three and four dimensions and the study of knots. The developments in this domain can often be associated with Edward Witten, who obtained several key results with strong inpact in both the communities of physics and mathematics. On one side, string theorists have been applying advanced results of mathematics and geometry; on the other side, mathematicians have been appreciating the methods of theoretical physics, such as path-integral quantization of field theories.

One example of these interdisciplinary developments is the study of Chern-Simons gauge theory in $d = 3$, following Witten's work of 1990. The Lagrangian of this theory is written in terms of a gauge field $A_\mu(x)$, but it does not contain a standard kinetic term for the propagation of gluons. In other words, there are no local degrees of freedom, but only global quantum effects.

As is well known, a charged particle moving in a gauge field acquires a phase factor, the so-called Dirac phase,

$$\left\langle \exp\left(i\frac{q}{\hbar c}\int_\gamma dx^\mu A_\mu\right)\right\rangle, \tag{45.9}$$

where γ is the particle trajectory and q its charge. In the Chern–Simons theory, this is a topological quantity, i.e. it is invariant under continuous deformations of the trajectory: it only depends on its topological class, namely on the number of windings around other trajectories, including self-windings, and around loops of the spacetime manifold. Since this topological property holds for all observable quantities, the Chern–Simons theory is called a 'topological field theory'; in particular, when defined on a curved spacetime, it is independent of smooth deformations of the metric.

Witten was able to solve the theory completely by establishing a correspondence with the Wess–Zumino–Witten–Novikov model of the same group G and using conformal field theory results. Since closed trajectories γ describe knots in three dimensions, the observables (45.9) of Chern–Simons theory provide a characterization for them; this led to considerable advances in the classification of knots in three-dimensional space.

Another method for obtaining topological theories is by using supersymmetry and its ability to cancel quantum effects between the contributions of bosons and fermions. In two dimensions, a topological theory can be obtained from an $\mathcal{N}=2$ supersymmetric conformal field theory by redefining the supersymmetry generators. Topological σ-models in four dimensions are similarly obtained by 'twisting' their extended supersymmetry. The latter models were applied to the classification of topological manifolds in three and four dimensions, such as the Donaldson invariants. Finally, let us mention the application of topological $\mathcal{N}=2$ supersymmetric conformal theories to build topological string theories, developed by Ooguri, Vafa and others, that provide an interesting playground for studying certain aspects of string dynamics and characterizing further topological invariants.

45.8 Conclusion

This book has covered the history of string theory from its beginnings in 1968 up to 1984. The developments had several phases: the first years of hectic activity, a period of hibernation after 1976 and the renaissance in 1984. In this last Chapter, we have outlined the later developments, that were also characterized by ups and downs: a stasis at the beginning of the Nineties and a second revolution after 1995, followed by intense activity in the new Millennium.

String theory has been at the centre of a widespread scientific debate. Some experts believe that the theory is still poorly understood, after forty years of life: in the light of the results of string dualities and M-theory, the question remains of what are the basic degrees of freedom of the theory, maybe not the strings themselves. Other researchers think that the problem of the landscape makes the theory almost useless for unification. Nonetheless, the gauge/gravity correspondence has led to remarkable results in strongly interacting gauge theories, with applications to the quark–gluon plasma, QCD and hadronic physics.

While not attempting any judgement, the editors and authors of this book have tried to raise the interest of the reader to the many facets of (early) string theory and to show its ties with several issues of contemporary theoretical physics.

Acknowledgements

In the preparation of this survey, we were helped by several colleagues with comments and suggestions. In particular, we would like to thank Denis Bernard, Francesco Bigazzi,

Elena Castellani, Leonardo Castellani, Paolo Di Vecchia, Augusto Sagnotti and Domenico Seminara.

References

[BBS06] Becker, K., Becker, M. and Schwarz, J. H. (2006). *String Theory and M-Theory: A Modern Introduction* (Cambridge University Press, Cambridge).

[DMS97] Di Francesco, P., Mathieu, P., Senechal, D. (1997) *Conformal Field Theory* (Springer Verlag, New York).

[Mus09] Mussardo, G. (2009). *Statistical Field Theory: An Introduction to Exactly Solved Models in Statistical Physics* (Oxford University Press, Oxford).

[Pol98] Polchinski, J. (1998). *String Theory*, Vol. 2 (Cambridge University Press, Cambridge).

[Tsv07] Tsvelik, A. M. (2007). *Quantum Field Theory in Condensed Matter Physics* (Cambridge University Press, Cambridge).

Appendix A
Theoretical tools of the Sixties

A.1 Quark model and current algebra

Research on strong interactions was intimately connected with that on weak interactions, and the two subjects were developed in parallel during the Sixties. In this Section we describe the developments that led to the Cabibbo theory for semi-leptonic weak processes and to the quark model for hadrons.

During the Fifties, it became clear that quantum electrodynamics (QED), constructed by applying the quantization rules to classical electrodynamics, correctly reproduced the deviations from the Dirac theory of electrons and positrons that were observed in experiments. The starting point was the QED Lagrangian:

$$\mathcal{L} = -\frac{1}{4} F_{\mu\nu} F^{\mu\nu} + \bar{\psi} \left(i\gamma^\mu D_\mu - m \right) \psi, \tag{A.1}$$

that involves the fields $A_\mu(x)$, describing the photon, and $\psi(x)$ describing the electron and positron; $F_{\mu\nu} = \partial_\mu A_\nu - \partial_\nu A_\mu$ is the electromagnetic field strength and $D_\mu = \partial_\mu - ieA_\mu$ is the covariant derivative. The interaction between photons, electrons and positrons is described by a cubic term $A_\mu j^\mu$ where $j^\mu \equiv e\bar{\psi}\gamma^\mu\psi$ is the electromagnetic current produced by the electrons and positrons.

From the previous Lagrangian, the Green functions involving the photon and electron fields can be computed perturbatively using the Feynman diagrams. The tree diagrams reproduce the classical behaviour, while the loop diagrams give the quantum corrections. It turned out that the loop corrections involved ultraviolet divergences, which were eliminated by the renormalization procedure, obtaining finite expressions that could be compared with the experimental results. The agreement was extremely good and QED was considered a major success of quantum field theory.

In the following years, a similar approach was proposed for describing weak interactions. Fermi had already proposed a current–current interaction term in 1934; Feynman and Gell-Mann suggested in the late Fifties that the weak current was the difference of a vector and an axial vector, leading to the following Lagrangian for purely leptonic weak processes:

$$\mathcal{L} = \frac{G_F}{\sqrt{2}} j_\alpha j^\alpha, \qquad j^\alpha = \bar{e}\gamma^\alpha(1-\gamma_5)\nu_e + \bar{\mu}\gamma^\alpha(1-\gamma_5)\nu_\mu, \tag{A.2}$$

where $G_F = 10^{-5}\,\text{GeV}^{-2}$ is the Fermi constant and the fields e, μ, ν_e, ν_μ represented the electron, the muon and their neutrinos, respectively. The lowest order amplitude is described by a Feynman diagram with four particles meeting in a point. The Lagrangian (A.2) successfully described processes such as for example the muon decay, $\mu^- \to e^- + \nu_\mu + \bar{\nu}_e$ (the third lepton family of the τ lepton and its neutrino ν_τ was discovered much later).

The next problem was to include the strongly interacting particles in order to describe the semi-leptonic processes, involving both leptons and hadrons, such as beta decay: $n \to p + e^- + \bar{\nu}_e$. This extension turned out to be naturally connected with the study of strong interactions. As explained in the Introduction to Part II, it was known that these interactions were, to a very good approximation, invariant under $SU(2)$ isospin transformations and that the hadrons were classified according to irreducible representations of $SU(2)$. For instance, the proton and neutron belong to a doublet of $SU(2)$, while the three pions $\pi^\pm, \pi^0$ belong to a triplet. Then, the discovery of the strange particles, the K-mesons, the Λ, the Ξ etc., led to the introduction of an additional quantum number, called strangeness, which was also conserved in the hadronic processes. The $SU(3)$ 'flavour' symmetry group, combining rotations of isotopic spin and strangeness, was introduced by Gell-Mann and Ne'eman in 1963, leading to a successful classification in octet and decuplet representations of the newly discovered mesons and baryons.

Through the Noether theorem, a symmetry implies the existence of conserved currents, whose charges correspond to the generators of the symmetry group. In the case of $SU(3)$, one has an octet of conserved vector currents V_a^μ with $a = 1, \ldots, 8$. Besides the octet of vector currents, an octet of axial vector currents A_a^μ was also introduced, which will be described later. Gell-Mann proposed that the electromagnetic current belonged to the same octet of weak vector currents V_a^μ, as follows:

electromagnetic current $\qquad V_{em}^\mu = V_3^\mu + \frac{1}{\sqrt{3}} V_8^\mu$,

weak current, $\Delta S = 0, \Delta Q = 1 \qquad V_{\Delta S=0}^\mu = V_1^\mu + i V_2^\mu$, (A.3)

weak current, $\Delta S = \Delta Q = 1 \qquad V_{\Delta S=1}^\mu = V_4^\mu + i V_5^\mu$,

where $V_{1,2,3}^\mu$ are the three vector currents of the $SU(2)$ subgroup corresponding to the isotopic spin, and $V_{4,5,8}^\mu$ are the currents corresponding to other generators of $SU(3)$.

Cabibbo introduced the complete hadronic weak current in 1963, as follows:

$$J^\mu = \cos\theta \left[(V_1^\mu + i V_2^\mu) - (A_1^\mu + i A_2^\mu) \right] \\ + \sin\theta \left[(V_4^\mu + i V_5^\mu) - (A_4^\mu + i A_5^\mu) \right]; \tag{A.4}$$

this expression has the $V - A$ form of the purely leptonic current (A.2), but it involves a rotation parameterized by the angle θ (the Cabibbo angle). Therefore, the semi-leptonic processes were described by the Lagrangian

$$\mathcal{L} = \frac{G_F}{\sqrt{2}} j_\mu J^\mu, \qquad (A.5)$$

in complete analogy with (A.2).

The two Fermi interactions (A.2) and (A.5) correctly described the weak processes in terms of Feynman diagrams to leading order in G_F. However, the current–current interaction was not renormalizable and could not be used for computing loop corrections in a meaningful way. A completely consistent theory of weak interactions, possessing renormalizable perturbative expansion, was introduced at the end of the Sixties and beginning of the Seventies with the Glashow–Weinberg–Salam model, a non-Abelian gauge theory involving the $SU(2) \times U(1)$ group. At low energy, $E \ll 100\,\text{GeV}$, this model reproduces the Fermi theory; at high energy the gauge theory predicts that the weak interaction is mediated by massive $W^\pm$ and Z^0 gauge bosons, now thoroughly confirmed experimentally [PS95].

As already discussed in the Introduction to Part II, the quarks were introduced by Gell-Mann and Zweig as 'components' of hadrons: the $SU(3)$ representations of quark–antiquark and quark triplets, $3 \times \bar{3}$ and $3 \times 3 \times 3$, respectively, were sufficient to fit all hadrons in octets and decuplets. The quark model was able to reproduce many features of the strong interactions. In the Sixties it was not at all clear which Lagrangian could describe the dynamics of quarks; since they had not been observed, they were generally considered as useful mathematical entities rather than physical particles.

The free part of the quark Lagrangian $\mathcal{L}_0$ was that of Dirac spin one-half particles:

$$\mathcal{L}_0 = \sum_{i=1}^{3} \bar{q}_i \left(i\gamma^\mu \partial_\mu - m_i \right) q_i. \qquad (A.6)$$

In this equation, q_i is the quark field that carries the index of the fundamental three-dimensional representation of $SU(3)$ and m_i, $i = 1, 2, 3$, are the masses of the three quark flavours (up, down and strange). Although quark interactions were not understood, some useful considerations could be based on the free quark Lagrangian, as described in the following.

If all quark masses are equal, this Lagrangian is symmetric under $SU(3)$ transformations. The corresponding Noether currents are the vector currents:

$$V_\mu^a = \bar{q} \gamma_\mu \frac{\lambda^a}{2} q, \qquad \partial^\mu V_\mu^a = 0, \qquad (A.7)$$

which are conserved. The λ^a are the Gell-Mann matrices satisfying the $SU(3)$ algebra:

$$[\lambda^a, \lambda^b] = 2i f^{abc} \lambda_c, \qquad a, b, c = 1, \ldots, 8. \qquad (A.8)$$

If instead the quark masses are different, the divergence of the vector currents is equal to:

$$\partial^\mu V_\mu^a = i \bar{q}_i \left(m_i - m_j \right) \frac{1}{2} \left(\lambda^a \right)_{ij} q_j. \qquad (A.9)$$

Namely, for equal masses we would have an octet of conserved vector currents, while for $m_1 = m_2 \neq m_3$, only a triplet of currents would be conserved, corresponding to the generators of the $SU(2)$ subgroup of $SU(3)$ that is left unbroken.

If all quark masses are vanishing, the free quark Lagrangian would be invariant under $SU(3)$ transformations that act independently on the right and left-handed components of the quarks: this is the chiral symmetry $SU(3)_L \times SU(3)_R$. In this case, there are two sets of conserved currents:

$$J^a_{L\mu} = \bar{q}\gamma_\mu \left(\frac{1-\gamma_5}{2}\right) \frac{\lambda^a}{2} q, \quad \partial^\mu J^a_{L\mu} = 0, \qquad (A.10)$$

$$J^a_{R\mu} = \bar{q}\gamma_\mu \left(\frac{1+\gamma_5}{2}\right) \frac{\lambda^a}{2} q, \quad \partial^\mu J^a_{R\mu} = 0, \qquad (A.11)$$

where $V^a_\mu = J^a_{R\mu} + J^a_{L\mu}$ and $A^a_\mu = J^a_{R\mu} - J^a_{L\mu}$.

In general, however, neither the vector nor the axial vector currents are exactly conserved. Assuming that the symmetry is only broken by the mass matrix, we obtain the following equations:

$$\partial^\mu V^a_\mu = i\bar{q}\left[M, \frac{\lambda^a}{2}\right] q, \quad \partial^\mu A^a_\mu = i\bar{q}\left\{M, \frac{\lambda^a}{2}\right\} q, \qquad (A.12)$$

where $M_{ij} = \delta_{ij} m_i$ is the mass matrix. These relations could be used to obtain relations of approximate chiral invariance. The vector and axial vector charges Q^α_L and Q^α_R,

$$Q^a_L = \int d^3 x J^{0a}_L, \quad Q^a_R = \int d^3 x J^{0a}_R, \qquad (A.13)$$

are bilinear in the quark fields. Even if they are not conserved, their algebra can be computed from the Lagrangian (A.6), assuming canonical anticommutation relations for the quark fields. The equal-time commutation relations for the currents (A.11) could also be computed.

In this way the so-called current algebra approach was formulated (see also the Chapter by Ademollo); it was abstracted from the quark model and assumed to be valid even if the full Lagrangian for strong interactions was not known.

Nowadays, we know that the Lagrangian (A.6) is very close to the correct form for quantum chromodynamics, which was formulated in the Seventies [PS95]; however, some aspects of the latter theory were missing in the Sixties:

(i) that the quarks have an extra $SU(3)$ colour index, and a corresponding local (gauge) symmetry, and thus interact through the exchange of an octet of spin one particles called gluons;
(ii) that the non-Abelian gauge interaction confines the quarks inside hadrons, but is weakly interacting at very high energies (asymptotic freedom).

In the Sixties, the yet unclear quark Lagrangian was assumed to be approximately $SU(3)_L \times SU(3)_R$ flavour symmetric; the algebra of the charges and the currents was then used to obtain certain relations that did not depend on the specific form of the Lagrangian.

In fact, by assuming the commutation relations between charges and currents and saturating them with one-particle states, one could obtain the sum rules, for example the Adler–Weisberger rule, which provided relations between hadronic quantities measured in scattering processes. The agreement was rather good. These developments led to the ideas that the approximate axial symmetry was 'spontaneously broken' and that the pseudoscalar mesons, the pion, kaons etc., were the corresponding Goldstone bosons; their amplitudes satisfied low energy theorems that were reproduced by phenomenological Lagrangians [Wei00].

A.2 S-matrix theory

As outlined in the Introduction to Part II, S-matrix theory provided a description of strong interactions that did not involve any field and Lagrangian. In this Section, we discuss some more technical aspects of this approach. More details can be found in [Che66, ELOP66, Col77] and in the second chapter of [DFFR73].

Let us consider the amplitude of a scattering process involving four particles (for simplicity, we take equal masses m): it is a function of the Mandelstam variables (we use the metric $(-, +, +, +)$ that is standard in string theory):

$$s = -(p_1 + p_2)^2, \qquad t = -(p_1 + p_3)^2, \qquad u = -(p_1 + p_4)^2. \tag{A.14}$$

Due to momentum conservation and mass-shell conditions $p_i^2 = -m^2$, these obey the following relations:

$$p_1 + p_2 + p_3 + p_4 = 0, \qquad s + t + u = 4m^2. \tag{A.15}$$

We have taken all momenta to be incoming, but we should remember to change the sign for the momenta of outgoing particles.

If we consider the process $1 + 2 \to 3 + 4$, then s is equal to the square of the total energy $2E$ in the centre-of-mass reference frame; t is the momentum transfer and is related to the cosine of the scattering angle θ in this reference frame:

$$s = 4E^2, \qquad t = -2|\vec{p}|^2(1 - \cos\theta), \tag{A.16}$$

where $\vec{p} = \vec{p}_1 = -\vec{p}_2$ is the three-momentum of one of the incoming particles in the centre of mass: $|\vec{p}|^2 = E^2 - m^2 = (s - 4m^2)/4$.

The scattering amplitude is a function of s and t varying inside the intervals,

$$4m^2 \le s < \infty, \qquad -(s - 4m^2) \le t \le 0, \tag{A.17}$$

called the physical region of the s-channel.

In addition to the original scattering process, we can consider two others, involving antiparticles denoted by a bar:

$$\begin{array}{lll} 1+2 \to 3+4, & s\text{-channel}, & s > 4m^2,\ t, u < 0; \\ 1+\bar{3} \to \bar{2}+4, & t\text{-channel}, & t > 4m^2,\ s, u < 0; \\ 1+\bar{4} \to \bar{2}+3, & u\text{-channel}, & u > 4m^2,\ s, t < 0. \end{array} \qquad (A.18)$$

The three processes are described by the variables s, t and u, taking values in the respective physical regions.

According to crossing symmetry, the three processes are described by the same scattering amplitude $A(s, t, u)$, which is an analytic function of the variables s, t and u extended to complex values. Paying attention to its singularities, this function can be continued analytically from one physical region to another; for example, from the physical region of the s-channel in (A.18) to the regions of the other two channels. This fundamental property of scattering amplitudes was implemented in the Veneziano solution for the scattering process $\pi\pi \to \pi\omega$.

Let us now discuss the unitarity relation of the S-matrix. Consider an initial state $|i\rangle$, a final state $|f\rangle$ and a complete set of states $|n\rangle$. The probability amplitude for the transition from the initial state to the final state defines the scattering matrix element $S_{fi} \equiv \langle f|S|i\rangle$. Since the system of states $|n\rangle$ is complete, the conservation of probability implies that:

$$\sum_n |S_{ni}|^2 = 1 = \sum_n \langle i|S^\dagger|n\rangle\langle n|S|i\rangle = \langle i|S^\dagger S|i\rangle. \qquad (A.19)$$

Since the initial state is arbitrary, then the previous relation implies the unitarity of the S-matrix:

$$SS^\dagger = 1. \qquad (A.20)$$

In addition to the S-matrix, we introduce the scattering amplitude T_{fi} as follows:

$$\begin{aligned} S_{fi} &= \delta_{fi} + i(2\pi)^4 \delta(p_f - p_i) T_{fi}, \\ S^\dagger_{fi} &= \delta_{fi} - i(2\pi)^4 \delta(p_f - p_i) T^\dagger_{fi}. \end{aligned} \qquad (A.21)$$

The unitarity relation (A.20) then becomes:

$$i\left(T_{fi} - T^\dagger_{fi}\right) = -(2\pi)^4 \sum_n \delta(p_f - p_n) T_{fn} T^\dagger_{ni}, \qquad (A.22)$$

where the sum is over a complete set of states n.

One important fact is that unitarity determines the singularity structure of the scattering amplitude T_{fi}. Let us discuss the case of four spinless particles for simplicity, where $T_{fi} = T_{if}$. We assume that, for t in the physical region, the scattering amplitude $T_{fi}(s, t)$ in the s-channel can be extended as an analytic function in the upper positive complex plane of s, $\operatorname{Im} s > 0$. Hereafter, it is written as $T(s)$ for simplicity. On the other hand, we can also consider the function $T^*(s^*)$, which is analytic in the negative lower plane. Assuming that the two functions coincide on a portion of the real axis, they can be identified:

$T^*(s^*) = T(s)$. The s-channel unitarity relation (A.22) implies that, as a function of s, $T(s)$ has only singularities along the real axis. These are simple poles lying below the two-particle threshold, $s = 4m^2$, a branch cut starting at the two-particle threshold, $s > 4m^2$, and additional branch cuts starting at the multiparticle thresholds. Because of these branch points, the scattering amplitude is defined on a multi-sheeted complex plane. The physical scattering amplitude is obtained by going on the real axis from above in the physical sheet. The discontinuity along the cut is given by:

$$\lim_{\epsilon \to 0} [T(s+i\epsilon) - T(s-i\epsilon)] = \lim_{\epsilon \to 0} [T(s+i\epsilon) - T^*(s+i\epsilon)]$$
$$= 2i \operatorname{Im} T(s). \qquad (A.23)$$

Combining this equation with the unitarity relation (A.22) we get:

$$\operatorname{Im} T_{fi}(s) = (2\pi)^4 \sum_n \delta(p_f - p_n) T_{fn}(T^\dagger)_{ni}. \qquad (A.24)$$

This determines the discontinuity of $T(s)$ along the cut in the physical sheet.

Since the scattering amplitude has only singularities on the real axis of the s plane, we can rewrite it in terms of a dispersion relation, a specific form of the Cauchy integral representation for analytic functions. We first introduce the new variable ν, which is more symmetric with respect to the s- and u-channels:

$$\nu = \frac{s-u}{4}. \qquad (A.25)$$

The amplitude $T(\nu)$ should have a branch cut along the positive real axis that starts at the two-particle threshold in the s-channel given by $\nu = \nu_0 = (4m^2 + t)/4$, and another cut along the negative ν plane starting at $\nu = -\nu_0$, corresponding to the two-particle threshold in the u-channel (see Figure A.1). Between the two branch points, there could be poles due to particles having masses below the two-particle threshold. Given this structure of singularities, we can write the Cauchy integral,

$$T(\nu) = \frac{1}{2\pi i} \oint_C d\nu' \frac{T(\nu')}{\nu - \nu'}, \qquad (A.26)$$

where the integration path C is shown in Figure A.1. If $T(\nu) \to 0$ for $\nu \to \infty$, the part of the contour at infinity can be neglected and one is left with the contributions around the two cuts and around other possible poles on the real axis. Combined with (A.23), this gives the dispersion relation:

$$T(\nu) = \frac{1}{\pi} \int_{\nu_0}^\infty d\nu' \frac{\operatorname{Im} T(\nu')}{\nu' - \nu} + \frac{1}{\pi} \int_{-\infty}^{-\nu_0} d\nu' \frac{\operatorname{Im} T(\nu')}{\nu' - \nu} + \sum_i \frac{\gamma_i}{\nu - \nu_i}, \qquad (A.27)$$

where $\nu_0 = m^2 + (t/4)$. If $T(\nu)$ is not vanishing for $|\nu| \to \infty$, then we should introduce subtraction terms which, however, will not be discussed here.

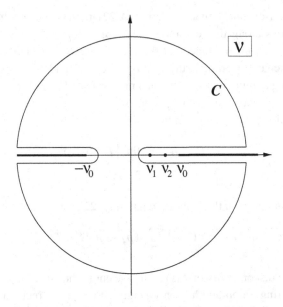

Figure A.1 The Cauchy contour for the dispersion relation.

If $T(|\nu|) \to |\nu|^{-N-1}$ for positive N then one obtains a set of superconvergence relations:

$$\int_{-\infty}^{\infty} d\nu\, \nu^n \, \mathrm{Im}\, T(\nu) = 0, \qquad n = 0, \ldots, N, \tag{A.28}$$

such as those considered for the finite energy sum rules described in Part II of the book.

Let us add some remarks on the dispersion relation (A.27). Suppose that $T(\nu)$ possesses an infinite series of simple poles on the real axis, at $\nu = \nu_i$, and no cuts, as in the case of Veneziano amplitude (the so-called narrow resonance approximation):

$$T(\nu) = \sum_i \frac{\gamma_i}{\nu - \nu_i + i\epsilon}. \tag{A.29}$$

In this equation, the poles were slightly shifted from the real axis into the lower half-plane by an infinitesimal amount ϵ. In this case, the dispersion relation (A.27), i.e. the Cauchy integral representation, is equivalent to the following substitution rule inside an ordinary integral on the real axis:

$$\frac{1}{z - z_0 + i\epsilon} = \mathrm{P}\left(\frac{1}{z - z_0}\right) + i\pi\, \delta(z - z_0), \tag{A.30}$$

where P stands for the Cauchy principal value. The relation (A.30) shows that the imaginary part of $T(\nu)$ in (A.29) is a sum of delta functions. We conclude that in the narrow resonance approximation, the superconvergence integral relations (A.28) can be written equivalently as discrete sums of powers of the pole positions ν_i multiplied by coefficients γ_i.

A.2.1 Breit–Wigner amplitude

In this Section, we discuss the unitarity relations (A.22) and (A.24) for single-particle exchanges, which are described by the Breit–Wigner amplitude.

In the case of a spin zero resonance in the s-channel with mass M_R and total width Γ, the scattering amplitude has a pole located at $s = s_R + i M_R \Gamma$ ($s_R \equiv M_R^2$), and is given by:

$$T_{fi} = \frac{2 M_R A_{fi}}{s_R - s - i\Gamma M_R}, \qquad T^\dagger_{fi} = \frac{2 M_R A_{fi}}{s_R - s + i\Gamma M_R}. \tag{A.31}$$

The left-hand side of the unitarity relation (A.22) is therefore given by:

$$i\left(T_{fi} - (T^\dagger)_{fi}\right) = -\frac{(2 M_R)^2 \Gamma A_{fi}}{(s - s_R)^2 + \Gamma^2 M_R^2}. \tag{A.32}$$

We can similarly introduce the matrix elements,

$$T_{fn} = \frac{2 M_R A_{fn}}{s_R - s - i\Gamma M_R}, \qquad (T^\dagger)_{ni} = \frac{2 M_R A_{ni}}{s_R - s + i\Gamma M_R}, \tag{A.33}$$

and use them to compute the right-hand side of (A.22), as follows:

$$(2\pi)^4 \sum_n \delta(p_f - p_n) T_{fn} (T^\dagger)_{ni} = (2\pi)^4 \sum_n \int \frac{d^4 p_n}{(2\pi)^4} \delta(p_f - p_n) \frac{(2 M_R)^2 A_{fn} A_{ni}}{(s - s_R)^2 + \Gamma^2 M_R^2}$$

$$= \sum_n \frac{(2 M_R)^2 A_{fn} A_{ni}}{(s - s_R)^2 + \Gamma^2 M_R^2}. \tag{A.34}$$

Comparison of the two expressions (A.32) and (A.34) shows that the unitarity relation implies a condition between the residue A_{fi} and the total width Γ in the Breit–Wigner formula (A.31):

$$\Gamma A_{fi} = \sum_n A_{fn} A_{ni}. \tag{A.35}$$

The solution of these conditions for all states f and i can be found by writing the residues in factorized form, as follows:

$$A_{fi} = A_f A_i, \qquad \Gamma = \sum_n A_n^2. \tag{A.36}$$

This relation has the following interesting physical interpretation. A resonance can decay in many possible channels, having index n: if we identify $\Gamma_n \equiv A_n^2$ as the decay rate (partial width) in the nth channel, the total width Γ is naturally written

$$\Gamma = \sum_n \Gamma_n, \tag{A.37}$$

as the sum of partial widths for all decay channels (being a sum of probabilities). The ratio Γ_n/Γ measures the probability that the resonance decays in the channel n.

In conclusion, unitarity implies that the residue at a pole of the scattering amplitudes must have a factorized form: $A_{fi} = A_f A_i$, where each factor square is identified as a partial decay width.

It is important to observe that the sum in (A.37) is over positive quantities, some of which are necessarily nonvanishing, otherwise the Breit–Wigner formula would vanish altogether. Therefore, unitarity implies that Γ must be nonvanishing. Within the narrow resonance approximation, the resonances were considered with $\Gamma \sim 0$, thus leading to a violation of unitarity. However, this was not a serious problem, as mentioned at the end of the Introduction to Part II. The Veneziano amplitude, derived in the narrow resonance approximation, was the tree diagram of an expansion that includes loop diagrams; the contribution of the latter restored unitarity of the theory.

References

[Che66] Chew, G. F. (1966). *The Analytic S matrix: a Basis for Nuclear Democracy* (W. A. Benjamin, New York).

[Col77] Collins, P. D. B. (1977). *An Introduction to Regge Theory and High Energy Physics* (Cambridge University Press, Cambridge).

[DFFR73] De Alfaro, V., Fubini, S., Furlan, G. and Rossetti, C. (1973). *Currents in Hadron Physics* (North Holland, Amsterdam).

[ELOP66] Eden, R. J., Landshoff, P. V., Olive, D. I. and Polkinghorne, J. C. (1966). *The Analytic S Matrix* (Cambridge University Press, Cambridge).

[HBRD97] Hoddeson, L., Brown, L., Riordan, M. and Dresden, M. (1997). *The Rise of the Standard Model: Particle Physics in the 1960s and 1970s* (Cambridge University Press, Cambridge).

[PS95] Peskin, M. E. and Schroeder, D. V. (1995). *An Introduction to Quantum Field Theory* (Addison-Wesley, New York).

[Wei00] Weinberg, S. (2000). *Quantum Field Theory* (Cambridge University Press, Cambridge).

Appendix B
The Veneziano amplitude

As discussed in Part II, the fundamental result of Veneziano in 1968 [Ven68] was to provide a closed-form analytic expression for the scattering amplitude of four mesons, realizing explicitly some of the ideas of the bootstrap programme, namely:

(i) complete crossing symmetry, i.e. invariance under permutations of Mandelstam variables s, t, u;
(ii) presence of an infinite set of simple poles in both the s- and t-channels, lying on linear Regge trajectories;
(iii) Dolen–Horn–Schmid duality;
(iv) asymptotic Regge behaviour.

In this Appendix, we provide a more detailed analysis of the Veneziano amplitude.

B.1 The Euler gamma and beta functions

The features mentioned above could be realized by employing the Euler gamma and beta functions as building blocks. We thus start by reviewing their main properties. The Euler gamma function is defined by the functional equation

$$\Gamma(z+1) = z\Gamma(z), \qquad \Gamma(1) = 1; \tag{B.1}$$

this can be seen as the generalization to complex values of the factorial, because for z equal to a positive integer n, the repeated use of (B.1) implies $\Gamma(n) = (n-1)!$. The properties of the gamma function can be most easily derived from the integral representation:

$$\Gamma(z) = \int_0^\infty t^{z-1} e^{-t} dt, \qquad \text{Re}\, z > 1. \tag{B.2}$$

For example, the functional equation (B.1) is readily verified by integrating this expression by parts.

The Birth of String Theory, ed. Andrea Cappelli, Elena Castellani, Filippo Colomo and Paolo Di Vecchia.
Published by Cambridge University Press. © Cambridge University Press 2012.

The integral representation in Eq. (B.2) is valid as long as the real part of z is positive, and shows that $\Gamma(z)$ has no singularity in this part of the complex z plane. The domain of definition of $\Gamma(z)$ can be extended to $\mathrm{Re}\, z < 0$ by using the functional equation (B.1); we can write:

$$\Gamma(z) = \frac{\Gamma(z+n)}{z(z+1)\ldots(z+n-1)}, \tag{B.3}$$

for any positive integer n. Since the right-hand side of this expression is well defined for $\mathrm{Re}\, z > -n$ by the integral representation (B.2), we can use the left-hand side of (B.3) to continue $\Gamma(z)$ analytically in this region. Since n is arbitrary, this provides the definition of $\Gamma(z)$ throughout the whole complex plane. It is apparent that its only singularities are simple poles at $z = -n = 0, -1, -2, \ldots$, with the behaviour

$$\Gamma(z) \sim \frac{1}{z+n}\frac{(-1)^n}{n!}, \qquad z \sim -n, \tag{B.4}$$

which is readily obtained from (B.3).

The gamma function satisfies many nice identities. Consider, for instance, the product $\Gamma(z)\Gamma(1-z)$; in view of the above considerations, this is a meromorphic function with simple poles at $z = n$, with n integer. Recalling the relations (B.2) and (B.4), the corresponding residue is simply $(-1)^n$. Analytic functions with the same singularities and asymptotic behaviour at infinity are equal. In the present case, a function with the same properties as $\Gamma(z)\Gamma(1-z)$ can be easily guessed and leads to the following identification:

$$\Gamma(z)\Gamma(1-z) = \frac{\pi}{\sin \pi z}. \tag{B.5}$$

This relation is known as the Euler reflection formula.

Finally, let us recall the asymptotic behaviour

$$\frac{\Gamma(z+x)}{\Gamma(z)} \sim z^x, \qquad z \to \infty, \qquad x \text{ finite}, \tag{B.6}$$

which can also be obtained from (B.2).

We now turn to the Euler beta function, defined by:

$$B(u, v) = \frac{\Gamma(u)\Gamma(v)}{\Gamma(u+v)}. \tag{B.7}$$

From the above discussion, it is apparent that $B(u, v)$ has a simple pole whenever u or v is a nonpositive integer. However, there are no double poles in (B.7), since whenever $\Gamma(u)$ and $\Gamma(v)$ have poles simultaneously the denominator develops a pole as well, leaving only a simple pole on the whole. The behaviour of $B(u, v)$ for v near $-n$, n being a nonnegative integer, is easily seen to be:

$$B(u, v) \sim \frac{1}{v+n}\frac{(-1)^n}{n!}(u-1)(u-2)\ldots(u-n), \qquad v \sim -n. \tag{B.8}$$

Note that the residue of $B(u, v)$ at the pole $v = -n$ is a polynomial in u. The Euler beta function has the following integral representation:

$$B(u, v) = \int_0^1 x^{u-1}(1-x)^{v-1}dx, \qquad \text{Re}\, u > 0,\ \text{Re}\, v > 0. \tag{B.9}$$

This integral representation can easily be extended to the whole complex plane of u and v, along the lines discussed above for the case of the gamma function; in particular, the behaviour (B.8) is reproduced.

B.2 The Veneziano amplitude and duality

We now turn to the Veneziano amplitude for the scattering amplitude of four spinless scalar particles with the same mass m. It is given by a sum of three terms:

$$A(s, t, u) = F(s, t) + F(s, u) + F(t, u). \tag{B.10}$$

It is totally symmetric in the three Mandelstam variables

$$s = -(p_1 + p_2)^2, \qquad t = -(p_1 + p_3)^2, \qquad u = -(p_1 + p_4)^2, \tag{B.11}$$

which satisfy the relation

$$s + t + u = 4m^2, \tag{B.12}$$

due to momentum conservation, $p_1 + p_2 + p_3 + p_4 = 0$. The amplitude $F(s, t)$ is given by

$$F(s, t) = B(-\alpha_s, -\alpha_t) = \frac{\Gamma(-\alpha_s)\Gamma(-\alpha_t)}{\Gamma(-\alpha_s - \alpha_t)}, \tag{B.13}$$

with linear Regge trajectories α_s and α_t,

$$\alpha_s = \alpha_0 + \alpha' s, \qquad \alpha_0 = -\alpha' m^2. \tag{B.14}$$

The second condition in (B.14) corresponds to the fact that the external particles are spinless and are the lowest state of the leading Regge trajectory. As a matter of fact, the amplitude (B.13) differs from the original formulation of the Veneziano amplitude (see e.g. (5.13), or (6.32)), which referred to the reaction $\pi\pi \to \pi\omega$, where the ω resonance has spin one. This results in a shift by one in the arguments of the beta function. Here, for simplicity, we discuss the amplitude (B.13) for four scalar particles (in today's language, four tachyons of bosonic open string theory).

$F(s, t)$ is the Euler beta function; as a consequence, it has simple poles in the variables s, t for $\alpha_{s_n} = n$ and $\alpha_{t_n} = n$, with nonnegative integer n. It can be expressed either as a sum over the s-channel poles or as a sum over the t-channel poles, but not as a sum over both. In order to see this, it is convenient to use the integral representation (B.9) of the beta function. The sum over the poles in the s-channel can be obtained by expanding the term containing $(1-x)$ around $x = 0$. We get:

$$F(s,t) = \int_0^1 dx\, x^{-\alpha_s-1} \sum_{n=0}^{\infty} (-1)^n \binom{-\alpha_t-1}{n} x^n$$

$$= \sum_{n=0}^{\infty} (-1)^n \frac{\binom{-\alpha_t-1}{n}}{n - \alpha_s}. \tag{B.15}$$

Similarly, it can be expanded around $x = 1$ obtaining the identity

$$F(s,t) = \sum_{n=0}^{\infty} (-1)^n \frac{\binom{-\alpha_t-1}{n}}{n - \alpha_s} = \sum_{n=0}^{\infty} (-1)^n \frac{\binom{-\alpha_s-1}{n}}{n - \alpha_t}, \tag{B.16}$$

showing that the amplitude $F(s,t)$ satisfies Dolen–Horn–Schmidt duality as a built-in feature. This follows directly from the properties of the beta function.

B.3 The spectrum and the critical dimension

We turn now to investigate the spectrum of the particles in the s-channel. We shall see that the analysis of the residues of the three lowest poles is sufficient to determine the value of the critical spacetime dimension, for given value of the intercept α_0 of the Regge trajectory.

From Eq. (B.8), or (B.15), we can evaluate the residue of the scattering amplitude at the pole $\alpha_{s_n} = n$,

$$R_n = (-1)^n \binom{-\alpha_t - 1}{n}$$

$$= (-1)^n \frac{(-\alpha_t - 1)(-\alpha_t - 2)\ldots(-\alpha_t - n)}{n!}, \tag{B.17}$$

which is a polynomial of order n in the momentum transfer t. Since the momentum transfer is related to the scattering angle in the centre-of-mass frame of the s-channel by the relation

$$t = -\frac{(s - 4m^2)}{2}(1 - \cos\theta), \tag{B.18}$$

we see that the residue at the pole $\alpha_{s_n} = n$ is a polynomial of order n in the variable $\cos\theta$, which can be expanded in terms of the Gegenbauer polynomials $C_\ell^{(D)}(\cos\theta)$ with $\ell = 0, 1, \ldots, n$. From angular momentum theory in quantum mechanics, we know that the polynomials $C_\ell^{(D)}$ describe the exchange of particles with spin ℓ in D space dimensions. In three space dimensions, they reduce to the more familiar Legendre polynomials $P_\ell(\cos\theta)$. Therefore, (B.17) shows that at each level $\alpha_{s_n} = n$ there is the exchange of a tower of intermediate particles with spin varying from 0 to n. It is interesting to look at the residues for $n = 0, 1, 2$. These are, respectively:

$$R_0 = 1,$$

$$R_1 = \alpha_t + 1 = \frac{1-\alpha_0}{2} + \frac{1+3\alpha_0}{2(D-2)} C_1^{(D)}(\cos\theta), \tag{B.19}$$

$$R_2 = \frac{(\alpha_t+1)(\alpha_t+2)}{2} = \left(\frac{2+3\alpha_0}{2}\right)^2 \frac{C_2^{(D)}}{D(D-2)}$$

$$+ (1-\alpha_0)\frac{2+3\alpha_0}{4(D-2)} C_1^{(D)} + \frac{1}{D}\left(\frac{2+3\alpha_0}{2}\right)^2 + \frac{\alpha_0(\alpha_0-2)}{8}.$$

We have used the following expressions of the Gegenbauer polynomials:

$$C_0^{(D)}(\cos\theta) = 1,$$

$$C_1^{(D)}(\cos\theta) = (D-2)\cos\theta, \tag{B.20}$$

$$C_2^{(D)}(\cos\theta) = \frac{D-2}{2}\left(D\cos^2\theta - 1\right).$$

From Eqs. (B.19), we see that, for the value $\alpha_0 = 1$, the contribution of the spin zero particle in R_1 and that of the spin one particle in R_2 vanish. Therefore, we get only even spins for n even and odd spins for n odd. Moreover, the contribution of the spin zero particle in R_2 is negative if $D > 25$, signalling the presence of a negative norm state (ghost), vanishes for $D = 25$ and is positive for $D < 25$. This implies that the Veneziano model is inconsistent for values of the space dimension $D > 25$. This analysis of the residues of the lowest poles in the scattering amplitude of four spinless scalar particles is the fastest way to obtain the value of the critical dimension $d = D + 1 = 26$: for this value, the physical subspace is spanned by the $D - 1$ transverse degrees of freedom only (see Appendix C).

B.4 Regge behaviour

Let us now study the asymptotic behaviour of the scattering amplitude (B.10) for large values of s and finite negative values of t. We recall that we have also assumed that the external particles are the lowest state of the Regge trajectory, namely that $\alpha_{m^2} = 0$, see Eq. (B.14). Restricting furthermore to the case $\alpha_0 = 1$, i.e. of tachyons with mass $m^2 = -1/\alpha'$, the relation (B.12) becomes

$$\alpha_s + \alpha_t + \alpha_u = -1. \tag{B.21}$$

Using the previous relation, and the Euler reflection formula (B.5) we get

$$\frac{\Gamma(-\alpha_u)\Gamma(-\alpha_t)}{\Gamma(-\alpha_u - \alpha_t)} = -\pi \frac{\Gamma(1+\alpha_s+\alpha_t)}{\Gamma(1+\alpha_s)\Gamma(1+\alpha_t)} \cdot \frac{1}{\sin\pi\alpha_t}, \tag{B.22}$$

together with

$$\frac{\Gamma(-\alpha_s)\Gamma(-\alpha_u)}{\Gamma(-\alpha_s - \alpha_u)} = -\pi \frac{\Gamma(1+\alpha_s+\alpha_t)}{\Gamma(1+\alpha_s)\Gamma(1+\alpha_t)} \cdot \frac{1}{\sin\pi\alpha_s}, \tag{B.23}$$

and

$$\frac{\Gamma(-\alpha_t)\Gamma(-\alpha_s)}{\Gamma(-\alpha_s-\alpha_t)} = -\pi \frac{\Gamma(1+\alpha_s+\alpha_t)}{\Gamma(1+\alpha_s)\Gamma(1+\alpha_t)} \left[\frac{\cos\pi\alpha_t}{\sin\pi\alpha_t} + \frac{\cos\pi\alpha_s}{\sin\pi\alpha_s}\right]. \quad (B.24)$$

Summing the three previous contributions we find:

$$A(s,t,u) = -\pi \frac{\Gamma(1+\alpha_s+\alpha_t)}{\Gamma(1+\alpha_s)\Gamma(1+\alpha_t)} \left[\frac{1+e^{-i\pi\alpha_t}}{\sin\pi\alpha_t} + \frac{1+e^{i\pi\alpha_s}}{\sin\pi\alpha_s}\right]. \quad (B.25)$$

For large values of s we can use Eq. (B.6) to obtain the following asymptotic behaviour for $A(s,t,u)$:

$$A(s,t,u) \sim \frac{(\alpha_s)^{\alpha_t}}{\Gamma(1+\alpha_t)} \left[\frac{1+e^{-i\pi\alpha_t}}{\sin\pi\alpha_t} + \frac{1+e^{i\pi\alpha_s}}{\sin\pi\alpha_s}\right]. \quad (B.26)$$

For large real s, the second term is a rapidly varying function, with many zeros and poles, reflecting the existence of resonances. The simplest way to average out this oscillating behaviour is just to give a small imaginary part to s (this choice is also physically motivated, see Appendix A); the second term in (B.26) thus vanishes. Using the reflection relation (B.5), we can rewrite the first term

$$A(s,t,u) \sim \Gamma(\alpha_t)(\alpha_s)^{\alpha_t}(1+e^{-i\pi\alpha_t}). \quad (B.27)$$

This shows that the Veneziano amplitude has Regge behaviour for large s and finite negative t [FR97].

B.5 Other models

We conclude this Appendix with a digression on the Neveu–Schwarz model [NS71], discussed in Part V. We shall analyze its spectrum along the line of Section B.3, focussing on the residue R_2 of the pole of the second excited level, and deduce the critical intercept and dimension. In the case of the Neveu–Schwarz model, the amplitude analogous to (B.13), describing the scattering of four spinless bosons, reads:

$$F^{NS}(s,t) = \frac{\Gamma(1-\alpha_s)\Gamma(1-\alpha_t)}{\Gamma(1-\alpha_s-\alpha_t)}, \quad (B.28)$$

where $\alpha_s = \alpha_0 + \alpha's$ is still the leading linearly rising Regge trajectory, but now

$$\alpha'm^2 = -\alpha_0 + \frac{1}{2}, \quad (B.29)$$

corresponding to the fact that the lowest particle does not lie on the leading Regge trajectory, but on the next to leading one shifted by $1/2$.

The residue at the pole $\alpha_{s_n} = n$ is given by

$$R_n = \frac{\alpha_t(\alpha_t+1)\ldots(\alpha_t+n-1)}{(n-1)!}. \quad (B.30)$$

The values of s and t at this pole are:

$$s_n = \frac{n - \alpha_0}{\alpha'}, \tag{B.31}$$

$$t_n = -\frac{s_n - 4m^2}{2}(1 - \cos\theta) = -\frac{n + 3\alpha_0 - 2}{2\alpha'}(1 - \cos\theta). \tag{B.32}$$

Using the previous expressions, we can compute the residue of the pole at $n = 2$ and we find:

$$R_2 = \alpha_t(\alpha_t + 1)$$
$$= \alpha_0 \left[\frac{(\alpha_0 - 2)D + 9\alpha_0}{4D} + \frac{3(1 - \alpha_0)}{2(D - 2)} C_1^{(D)} + \frac{9\alpha_0}{2D(D - 2)} C_2^{(D)} \right]. \tag{B.33}$$

If we choose the intercept $\alpha_0 = 1$ as in the Veneziano model, we see that the spin one is not contributing. Furthermore, the spin zero does not contribute either, if the space dimension is equal to $D = 9$, the critical dimension of the Neveu–Schwarz model, and corresponds to a negative norm state if $D > 9$. These results are in agreement with the general analysis of the spectrum, discussed in Appendix C.

As discussed in the Introduction to Part V, the four-point amplitude of the Lovelace–Shapiro model [Lov68, Sha69] was equal to that of the Neveu–Schwarz model, but with an intercept $\alpha_0 = \frac{1}{2}$ instead of 1. In this case the spin zero contribution to R_2 vanished for $D = 3$. Therefore, the model, with a critical space dimension $D = 3$, an intercept of the ρ trajectory $\alpha_0 = \frac{1}{2}$, close to the experimental value, and massless scalar particles interpretable as pions, was phenomenologically sound. However, the extension of the Lovelace–Shapiro dual model to more than four external pions has never been obtained, suggesting the existence of some theoretical obstruction.

References

[FR97] Forshaw, J. R. and Ross, D. A. (1997). *Quantum Chromodynamics and the Pomeron* (Cambridge University Press, Cambridge).

[Lov68] Lovelace, C. (1968). A novel application of Regge trajectories, *Phys. Lett.* **B28**, 264–268.

[NS71] Neveu, A. and Schwarz, J. H. (1971). Quark model of dual pions, *Phys. Rev.* **D4**, 1109–1111.

[Sha69] Shapiro, J. (1969). Narrow-resonance model with Regge behaviour for $\pi\pi$ scattering, *Phys. Rev.* **179**, 1345–1353.

[Ven68] Veneziano, G. (1968). Construction of a crossing-symmetric, Regge behaved amplitude for linearly rising trajectories, *Nuovo Cimento* **A57**, 190–197.

Appendix C
From the string action to the Dual Resonance Model

In this Appendix, we discuss the two actions for the bosonic string, the Nambu–Goto and σ-model actions; we show their reparameterization invariance and their equivalence at the classical level. Then, we solve the equations of motion for both the open and closed strings, quantize the theory, and obtain the spectrum of physical states of the Dual Resonance Model. We also describe the interaction of the string with external fields and show that it is consistent only for fields corresponding to states in the string spectrum. Finally, we compute the scattering amplitudes involving N strings, recovering the DRM expressions.

C.1 The string action

In the Introduction to Part IV, we showed that the relativistic actions of point particles and strings have similar features. Let us recall the line of the argument and then introduce an alternative form of these actions. The trajectory of a spinless free relativistic point particle is described by its coordinate $x^\mu(\tau)$ in Minkowski space, where τ is a parameter. The action is proportional to the length of the path and is given by:

$$S = -mc \int ds \equiv -mc \int \sqrt{-dx_\mu dx^\mu} = -mc \int d\tau \sqrt{-\dot{x}^2}. \qquad (C.1)$$

It is a geometrical object and therefore it is invariant under reparameterizations of the world-line coordinate τ. Indeed, under an arbitrary reparameterization $\tau \to \tau'$, the coordinate of the particle transforms as a scalar quantity

$$x^\mu(\tau) = (x')^\mu(\tau'), \qquad (C.2)$$

and the action remains invariant:

$$\int d\tau \sqrt{-\left(\frac{dx}{d\tau}\right)^2} = \int d\tau' \sqrt{-\left(\frac{dx'}{d\tau'}\right)^2}. \qquad (C.3)$$

The Birth of String Theory, ed. Andrea Cappelli, Elena Castellani, Filippo Colomo and Paolo Di Vecchia.
Published by Cambridge University Press. © Cambridge University Press 2012.

It is convenient to introduce the additional dynamical variable $e(\tau)$, and rewrite the action (C.1) as follows [BDZH76]:

$$S = \frac{1}{2}\int d\tau \left[\frac{\dot{x}^2}{e} - m^2 c^2 e\right]. \tag{C.4}$$

This expression does not contain τ-derivatives of e, thus e does not correspond to a propagating degree of freedom: it is just an auxiliary variable. It can be eliminated from the action (C.4) by computing its algebraic equation of motion,

$$e = \frac{\sqrt{-\dot{x}^2}}{mc}, \tag{C.5}$$

and substituting it back into the action, recovering the action (C.1). These results imply that, at the classical level, the actions (C.4) and (C.1) are completely equivalent. The new action (C.4) is reparameterization invariant if e transforms as an 'einbein':

$$e(\tau)d\tau = e'(\tau')d\tau'. \tag{C.6}$$

Therefore, the geometrical meaning of $e^2(\tau)$ is that of a one-dimensional 'metric'.

The previous considerations can be extended to the bosonic string which is described by its coordinate $X^\mu(\tau, \sigma)$, where τ and σ are the parameters describing the string world-sheet, also written as a world-sheet vector $\xi^a \equiv (\xi^0, \xi^1) = (\tau, \sigma)$. The infinitesimal area element spanned by the string in its motion is given by the wedge product of two infinitesimal tangent vectors, as follows:

$$d\sigma_{\mu\nu} = dx_\mu \wedge dx_\nu = \frac{\partial x_\mu}{\partial \xi^a}\frac{\partial x_\nu}{\partial \xi^b}d\xi^a \wedge d\xi^b = \frac{\partial x_\mu}{\partial \xi^a}\frac{\partial x_\nu}{\partial \xi^b}\epsilon^{ab}d\sigma d\tau, \tag{C.7}$$

where ϵ^{ab} is the antisymmetric tensor in two dimensions, with $\epsilon^{01} = 1$.

As seen in the Introduction to Part IV, the Nambu–Goto action for the string is proportional to the area spanned by the string in motion, and can be written:

$$S \sim \int \sqrt{-d\sigma_{\mu\nu}d\sigma^{\mu\nu}}. \tag{C.8}$$

Inserting Eq. (C.7) in (C.8) and fixing the proportionality constant, one finds [Nam70, Got71]:

$$S = -\frac{T}{c}\int_{\tau_i}^{\tau_f} d\tau \int_0^\pi d\sigma \sqrt{(\dot{X}\cdot X')^2 - \dot{X}^2 X'^2}, \tag{C.9}$$

where

$$\dot{X}^\mu \equiv \frac{\partial X^\mu}{\partial \tau}, \qquad X'^\mu \equiv \frac{\partial X^\mu}{\partial \sigma}, \tag{C.10}$$

and T is the string tension replacing the mass of the point-particle action, with dimension of an energy per unit length. The string tension can be expressed in terms of the slope of the Regge trajectory through the relation:

$$T = \frac{c}{2\pi\alpha'\hbar}. \tag{C.11}$$

In Eqs. (C.9) and (C.11) we have explicitly written the speed of light c and the Planck constant $\hbar$ which will be set to one in the following discussion.

As in the case of the point particle, we can write an equivalent action for the string, given by [BDH76, DZ76]:

$$S(X^\mu, g_{ab}) = -\frac{T}{2} \int d\tau \int_0^\pi d\sigma \sqrt{-g} g^{ab} \partial_a X^\mu \partial_b X^\nu \eta_{\mu\nu}, \quad (C.12)$$

where $\eta^{\mu\nu} = \text{diag}(-1, 1, \ldots, 1, 1)$ is the d-dimensional target space metric ($\mu, \nu = 0, 2, \ldots, d-1$). We have introduced the auxiliary field $g_{ab}(\xi)$, whose meaning is that of a two-dimensional world-sheet metric tensor, and denoted $g = \det(g_{ab})$. Viewed as a two-dimensional field theory on the world-sheet, the action (C.12) describes the interaction of a set of d massless fields with an external gravitational field. In this interpretation, the d-dimensional Lorentz index plays the role of a flavour index.

The equivalence between the two actions (C.9) and (C.12) can be shown as follows. The equations of motion for the metric tensor obtained by varying the action (C.12) with respect to g^{ab},

$$\delta S = \int d^2\xi \frac{\delta S}{\delta g^{ab}} \delta g^{ab} \sim \int d^2\xi \, T_{ab} \, \delta g^{ab} = 0, \quad (C.13)$$

imply the vanishing of the world-sheet energy-momentum tensor T_{ab}, which takes the form:

$$T_{ab} \sim \partial_a X \cdot \partial_b X - \frac{1}{2} g_{ab} g^{cd} \partial_c X \cdot \partial_d X = 0. \quad (C.14)$$

In deriving this equation, we have used the relation:

$$\frac{\delta \sqrt{-g}}{\delta g^{ab}} = -\frac{1}{2} g_{ab} \sqrt{-g}. \quad (C.15)$$

From Eq. (C.14) follows:

$$\det(\partial_a X \cdot \partial_b X) = \frac{g}{4} \left[g^{cd} \partial_c X \cdot \partial_d X \right]^2, \quad (C.16)$$

which, when inserted in Eq. (C.12), reproduces the Nambu–Goto action (C.9).

The action (C.12) is invariant under arbitrary reparameterizations of the world-sheet coordinates ξ^a. This can easily be shown using the transformations of the world-sheet 'fields':

$$X^\mu(\xi) = X'^\mu(\xi'), \qquad g_{ab}(\xi) = \frac{\partial \xi'^c}{\partial \xi^a} \frac{\partial \xi'^d}{\partial \xi^b} g'_{cd}(\xi'). \quad (C.17)$$

The last equation implies:

$$d^2\xi \sqrt{-g} = d^2\xi' \sqrt{-g'}. \quad (C.18)$$

For infinitesimal transformations $\xi' = \xi - \epsilon$, we get:

$$\delta X^\mu = \epsilon^a \partial_a X^\mu, \qquad \delta g_{ab} = \epsilon^c \partial_c g_{ab} + \partial_a \epsilon^c g_{cb} + \partial_b \epsilon^c g_{ac}. \quad (C.19)$$

The action (C.12) is also invariant under a local rescaling of the metric that leaves the coordinate of the string unchanged:

$$g_{ab}(\xi) \to \Lambda(\xi)g_{ab}(\xi), \qquad X(\xi) \to X(\xi). \tag{C.20}$$

This is called Weyl or conformal transformation in curved spacetime.

C.2 Solving the classical equations of motion

In the following, we choose a gauge condition, the so-called conformal gauge that constrains the world-sheet metric as follows:

$$g_{ab} = \rho(\xi)\eta_{ab}, \qquad \eta_{11} = -\eta_{00} = 1. \tag{C.21}$$

This gauge choice does not completely fix the gauge symmetry because we can still perform conformal transformations that leave the metric in the form (C.21), while modifying ρ. Indeed, the conformal transformations are defined as reparameterizations $\xi \to \xi' = \xi + \epsilon$, obeying:

$$\partial^a \epsilon^b + \partial^b \epsilon^a - \eta_{ab}\partial^c \epsilon_c = 0. \tag{C.22}$$

Using the second equation in (C.19), the metric is modified into

$$g_{ab} + \delta g_{ab} = (\rho + \partial_c(\epsilon^c \rho))\eta_{ab}, \tag{C.23}$$

which is still of the form (C.21).

In the conformal gauge, the constraint (C.14) of vanishing two-dimensional energy-momentum tensor reduces to the following conditions for the vectors $\dot{X}$ and X':

$$\dot{X} \cdot X' = \dot{X}^2 + X'^2 = 0. \tag{C.24}$$

Their geometrical meaning is that of orthogonality and normalization of the two vectors, that form the basis of the tangent space at each point of the world-sheet. In the old days of string theory, these conditions were called the 'orthonormal gauge'. Note that they keep the spacetime Lorentz symmetry manifest.

The conditions (C.22) are more transparent if we introduce light-cone coordinates:

$$\xi^\pm = \xi^0 \pm \xi^1, \qquad \epsilon^\pm = \epsilon^0 \pm \epsilon^1, \qquad \frac{\partial}{\partial \xi^\pm} = \frac{1}{2}\left(\frac{\partial}{\partial \xi^0} \pm \frac{\partial}{\partial \xi^1}\right). \tag{C.25}$$

In terms of these variables, the conditions (C.22) reduce to:

$$\frac{\partial}{\partial \xi^-}\epsilon^+ = \frac{\partial}{\partial \xi^+}\epsilon^- = 0. \tag{C.26}$$

In conclusion, the transformations that keep the conformal gauge are characterized by two arbitrary functions $\epsilon^+(\xi^+)$ and $\epsilon^-(\xi^-)$, that transform the variables $\xi^\pm$ as follows:

$$\delta\xi^+ = \epsilon^+(\xi^+), \qquad \delta\xi^- = \epsilon^-(\xi^-). \tag{C.27}$$

They can be viewed as holomorphic and antiholomorphic transformations with respect to the complex world-sheet coordinate z, built from the Euclidean 'time' $\xi^2 = i\xi^0$ and space ξ^1, i.e. $z = \xi^1 + i\xi^2$, as discussed in the Introductions to Parts III and IV.

In the conformal gauge, the Lagrangian (C.12) becomes

$$L = -\frac{T}{2}\partial_a X \cdot \partial^a X, \qquad (C.28)$$

which is conformal invariant. The equations of motion are:

$$\left(\frac{\partial^2}{\partial\sigma^2} - \frac{\partial^2}{\partial\tau^2}\right)X^\mu(\sigma,\tau) = 0, \qquad (C.29)$$

and will be solved with the Neumann boundary conditions,

$$\frac{\partial}{\partial\sigma}X^\mu(\tau,\sigma)|_{\sigma=0,\pi} = 0, \qquad (C.30)$$

for open strings, and periodic conditions,

$$X^\mu(\tau,\sigma) = X^\mu(\tau,\sigma+\pi), \qquad (C.31)$$

for closed strings (we do not discuss here the Dirichlet, or fixed, boundary conditions).

The most general solution for open strings can be written,

$$X^\mu(\tau,\sigma) = q^\mu + 2\alpha' p^\mu \tau + i\sqrt{2\alpha'}\sum_{n=1}^{\infty}[a_n^\mu e^{-in\tau} - a_n^{+\mu}e^{in\tau}]\frac{\cos n\sigma}{\sqrt{n}}, \qquad (C.32)$$

and for closed strings,

$$X^\mu(\tau,\sigma) = q^\mu + 2\alpha' p^\mu \tau$$
$$+ \frac{i}{2}\sqrt{2\alpha'}\sum_{n=1}^{\infty}[\tilde{a}_n^\mu e^{-2in(\tau+\sigma)} - \tilde{a}_n^{+\mu}e^{2in(\tau+\sigma)}]\frac{1}{\sqrt{n}}$$
$$+ \frac{i}{2}\sqrt{2\alpha'}\sum_{n=1}^{\infty}[a_n^\mu e^{-2in(\tau-\sigma)} - a_n^{+\mu}e^{2in(\tau-\sigma)}]\frac{1}{\sqrt{n}}. \qquad (C.33)$$

The centre-of-mass variable p^μ and the oscillators cannot be given arbitrarily, but they must satisfy the constraints (C.24). As said, they correspond to the vanishing of the following two components of the world-sheet energy-momentum tensor:

$$T^{00} \pm T^{01} \propto \frac{1}{2}(\dot{X} \pm X')^2 = 0. \qquad (C.34)$$

These are the two independent components, T_{ab} being symmetric and traceless. These equations imply that the Fourier components of (C.34) should vanish. Since these are the generators of the conformal algebra, they become the Virasoro operators L_n as follows.

In the case of an open string, the two components give the same expression and we get:

$$L_n = \frac{1}{4\pi\alpha'}\int_0^\pi d\sigma\, e^{in(\tau\pm\sigma)}(\dot{X}\pm X')^2 = \frac{1}{2}\sum_{m=-\infty}^{\infty}\alpha_{n-m}\cdot\alpha_m, \qquad (C.35)$$

where

$$\alpha_n^\mu = \begin{cases} \sqrt{n}\, a_n^\mu, & \text{if } n > 0, \\ \sqrt{2\alpha'}\, p^\mu, & \text{if } n = 0, \\ \sqrt{|n|}\, a_{|n|}^{+\mu}, & \text{if } n < 0. \end{cases} \quad (C.36)$$

For closed strings, we get two distinct sets of generators:

$$L_n = \frac{1}{4\pi\alpha'} \int_0^\pi d\sigma\, e^{in(\tau+\sigma)} \left(\frac{\dot{X}+X'}{2}\right)^2 = \frac{1}{2}\sum_{m=-\infty}^\infty \alpha_m \cdot \alpha_{n-m},$$

$$\tilde{L}_n = \frac{1}{4\pi\alpha'} \int_0^\pi d\sigma\, e^{in(\tau-\sigma)} \left(\frac{\dot{X}-X'}{2}\right)^2 = \frac{1}{2}\sum_{m=-\infty}^\infty \tilde{\alpha}_m \cdot \tilde{\alpha}_{n-m}, \quad (C.37)$$

where we have used the convention in (C.36) but for the zero modes now given by:

$$\alpha_0^\mu = \tilde{\alpha}_0^\mu = \sqrt{2\alpha'}\,\frac{p^\mu}{2}. \quad (C.38)$$

At the classical level, the coefficients in the general solutions (C.32) for the open string cannot be arbitrary, but have to satisfy the conditions $L_n = 0$, with L_n given in (C.35). Similarly for the closed string, the coefficients in the general solutions (C.33) have to satisfy the conditions $L_n = \tilde{L}_n = 0$ with L_n and $\tilde{L}_n$ given in (C.37).

The presence of these constraints is a direct consequence of the fact that the choice of the conformal gauge does not uniquely fix the gauge symmetry; we are still working with a redundant number of variables that should be constrained.

A very convenient way of fixing the gauge symmetry completely is to impose the light-cone gauge condition:

$$X^+ = 2\alpha' p^+ \tau, \quad (C.39)$$

where

$$X^\pm = \frac{X^0 \pm X^{d-1}}{\sqrt{2}}. \quad (C.40)$$

It can be shown that one can always perform a conformal transformation to bring X^+ in the form (C.39). In the notation (C.40), the scalar product of two vectors A^μ and B^μ with $\mu = +, -, i$ is given by: $A^\mu B_\mu = A^i B_i - A^+ B^- - A^- B^+$. As discussed in the Chapter by Goddard, the choice of the light-cone gauge has the advantage, with respect to the timelike gauge, of eliminating both the timelike and a longitudinal component of X^μ, keeping only the transverse physical components of the string.

The condition (C.39) implies:

$$\begin{aligned} \alpha_n^+ &= 0, & n \neq 0 & \quad \text{(open strings)}, \\ \alpha_n^+ &= \tilde{\alpha}_n^+ = 0, & n \neq 0 & \quad \text{(closed strings)}. \end{aligned} \quad (C.41)$$

On the other hand, Eqs. (C.35) and (C.37) for $n \neq 0$ imply also that the oscillators $\alpha_n^{(-)}$ and $\tilde{\alpha}_n^{(-)}$ are determined in terms of the transverse oscillators, which are those along the directions $i = 1, 2, \ldots, d-2$. One gets:

$$\sqrt{2\alpha'}\alpha_n^- = \frac{1}{2p^+} \sum_{i=1}^{d-2} \sum_{m=-\infty}^{\infty} \alpha_{n-m}^i \alpha_m^i, \quad n \neq 0, \tag{C.42}$$

for open strings, and

$$\sqrt{2\alpha'}\alpha_n^- = \frac{1}{2p^+} \sum_{i=1}^{d-2} \sum_{m=-\infty}^{\infty} \alpha_{n-m}^i \alpha_m^i, \quad n \neq 0,$$

$$\sqrt{2\alpha'}\tilde{\alpha}_n^- = \frac{1}{2p^+} \sum_{i=1}^{d-2} \sum_{m=-\infty}^{\infty} \tilde{\alpha}_{n-m}^i \tilde{\alpha}_m^i, \quad n \neq 0, \tag{C.43}$$

for closed strings. In conclusion, in the light-cone gauge the components $\alpha_n^{\pm}$ (and also $\tilde{\alpha}_n^{\pm}$ in the case of a closed string) are completely fixed in terms of the transverse components α_n^i (and $\tilde{\alpha}_n^i$). Actually, Eqs. (C.42) and (C.43) are also valid for $n = 0$. By taking into account the relation between the momentum and the zero modes α_0^μ and $\tilde{\alpha}_0^\mu$, given in (C.36) and (C.38), and the fact that $p^2 = p_i^2 - 2p^+p^-$ one gets:

$$-\alpha' p^2 = \frac{1}{2} \sum_{i=1}^{d-2} \sum_{n=1}^{\infty} n \left[a_n^{i\dagger} a_n^i + a_n^i a_n^{i\dagger} \right], \tag{C.44}$$

for open strings, and

$$-\frac{\alpha'}{4} p^2 = \frac{1}{2} \sum_{i=1}^{d-2} \sum_{n=1}^{\infty} n \left[a_n^{i\dagger} a_n^i + a_n^i a_n^{i\dagger} \right]$$

$$= \frac{1}{2} \sum_{i=1}^{d-2} \sum_{n=1}^{\infty} n \left[\tilde{a}_n^{i\dagger} \tilde{a}_n^i + \tilde{a}_n^i \tilde{a}_n^{i\dagger} \right], \tag{C.45}$$

for closed strings.

To conclude, we have shown that the equations of motion can be easily solved in the conformal gauge. Furthermore, by going into the light-cone gauge, we can fix the gauge completely and remain with independent transverse degrees of freedom only. In the next two Sections, we quantize the theory first in the light-cone gauge and then in the conformal gauge, and obtain the spectrum of physical states. In particular, the requirement of Lorentz invariance in the light-cone gauge implies that the spacetime dimension must be equal to $d = 26$ and the intercept to $\alpha_0 = 1$.

C.3 Quantization in the light-cone gauge

In the light-cone gauge, we consider the transverse oscillators only, which are quantized by requiring canonical commutation relations:

$$[a_n^i, a_m^{j\dagger}] = \delta^{ij} \delta_{nm}, \quad [a_n^i, a_m^j] = [a_n^{i\dagger}, a_m^{j\dagger}] = 0, \tag{C.46}$$

for the open string. For the closed string, one has an additional infinite set of harmonic oscillators $\tilde{a}_n^i$, $\tilde{a}_n^{i\dagger}$, that satisfy the same commutation relations and commute with the previous set. Furthermore, the zero modes, describing the position of the centre of mass and the total momentum of the string, become operators satisfying the following commutation relations:

$$[\hat{q}^i, \hat{p}^j] = i\delta^{ij}. \tag{C.47}$$

The hat reminds us that $\hat{p}$ and $\hat{q}$ are operators.

The previous canonical commutation relations can be obtained by starting from the following Lagrangian for the transverse degrees of freedom:

$$L = \frac{T}{2} \sum_{i=1}^{d-2} \partial_a X^i \partial^a X^i \tag{C.48}$$

which can be obtained from Eq. (C.28) by restricting it to the transverse oscillators and by imposing canonical commutation relations,

$$[X^i(\tau, \sigma), \Pi^j(\tau, \sigma')] = i\delta^{ij}\delta(\sigma - \sigma'), \tag{C.49}$$

between the position and the conjugate momentum $\Pi^i = T\dot{X}^i$.

In the quantum theory, owing to the commutation relations (C.46), we must be careful about the ordering of the oscillators. The only quantities for which this matters are those in Eqs. (C.44) and (C.45): in these expressions we have chosen the same ordering that is considered in the harmonic oscillator problem. Using the commutation relations, we can rewrite Eq. (C.44) as follows:

$$-\alpha' \hat{p}^2 = \sum_{i=1}^{d-2} \sum_{n=1}^{\infty} n a_n^{i\dagger} a_n^i + \frac{d-2}{2} \sum_{n=1}^{\infty} n, \tag{C.50}$$

which actually amounts to the total energy of the oscillators.

In the case of the string, the number of oscillators is infinite, and the zero-point energy is divergent and must be regularized. Brink and Nielsen [BN73] were the first to show how to obtain a finite expression from (C.50). A convenient regularization, first proposed by Gliozzi [Gli73, DiV76], is the so-called ζ-function regularization, which amounts to introducing a parameter s in the divergent sum in (C.50), evaluating the sum at values of s for which it converges, and continuing the result analytically to the original value of s. Thus we write the zero-point energy:

$$\epsilon(s) = \frac{d-2}{2} \sum_{n=1}^{\infty} n^{-s} = \frac{d-2}{2} \zeta_R(s), \tag{C.51}$$

where $\zeta_R(s)$ is the Riemann ζ-function; this is analytic for $\operatorname{Re} s > 1$, where the sum converges. Actually, $\zeta_R(s)$ is analytic in the whole complex s plane except for the simple pole at $s = 1$. In particular, at $s = -1$, we have $\zeta_R(-1) = -1/12$. Inserting this value in

Eq. (C.51), we get the following expression for the zero-point energy:

$$\epsilon \equiv \epsilon(-1) = -\frac{d-2}{24}. \tag{C.52}$$

The quantity (C.52) determines the intercept of the Regge trajectories which is a physical quantity, thus it should be independent of the regularization method. It can be shown that this is indeed the case [GSW87].

The quantization in the light-cone gauge spoils the manifest Lorentz invariance of the original action. As discussed in the Introduction to Part IV, we must check that the quantum theory is nevertheless Lorentz invariant. This can be done by constructing the Lorentz generators and checking that they indeed satisfy the Lorentz algebra.

The generators are obtained canonically from the Lagrangian (C.28) and are given by:

$$J^{ij} = \ell^{ij} - i \sum_{n=1}^{\infty} \frac{1}{n}(\alpha^i_{-n}\alpha^j_n - \alpha^j_{-n}\alpha^i_n),$$

$$J^{+-} = \ell^{+-}, \qquad J^{i+} = \ell^{i+}, \tag{C.53}$$

$$J^{i-} = \ell^{i-} - i \sum_{n=1}^{\infty} \frac{1}{n}(\alpha^i_{-n}\alpha^-_n - \alpha^-_{-n}\alpha^i_n),$$

where

$$\ell^{\mu\nu} = \hat{q}^\mu \hat{p}^\nu - \hat{q}^\nu \hat{p}^\mu, \tag{C.54}$$

and α^-_n is given in Eq. (C.42).

A rather cumbersome calculation shows that the operators in (C.53) satisfy the Lorentz algebra only if:

$$\alpha_0 = 1 \quad \text{and} \quad d = 26. \tag{C.55}$$

Therefore, only for these values of α_0 and d is Lorentz invariance maintained in the quantum theory. As mentioned in the Introduction to Part III, string theory possesses an anomaly: the Weyl symmetry (C.20) of the σ-model action (C.12) does not hold in the quantum theory, for generic values of $d \neq 26$. In the light-cone gauge quantization, this violation of symmetry manifests itself as a Lorentz violation. The anomaly in string theory became completely clear only in 1981 with Polyakov's work on the path-integral quantization of the string (see the Introduction to Part VII).

Inserting Eq. (C.55) in (C.50), (C.52) we get the final expression for the mass of the string states ($\hat{p}^2 \equiv -M^2$):

$$\alpha' M^2 = \sum_{i=1}^{d-2} \sum_{n=1}^{\infty} n a_n^{+i} a_n^i - 1 = N - 1, \tag{C.56}$$

where N is the number operator. The lowest state is given by the vacuum $|0\rangle$ that corresponds to a tachyon with $M^2 = -1/\alpha'$. The next level, corresponding to $M^2 = 0$, is given by the state $a_1^{i\dagger}|0\rangle$, describing the transverse components of a massless spin one particle ('gauge boson'). At the level $N = 2$ we find the two states:

$$a_1^{i\dagger} a_1^{j\dagger} |0\rangle, \qquad a_2^{i\dagger} |0\rangle, \tag{C.57}$$

that describe a massive spin two particle in $d = 26$ with $M^2 = 1/\alpha'$.

The degeneracy of states at an arbitrary level N can be obtained from the partition function of $(d - 2$ copies of) the one-dimensional Bose gas; we have

$$G(q) = \text{Tr}\left(q^{(\sum_{i=1}^{d-2} \sum_{n=1}^{\infty} na_n^{\dagger i} a_n^i - 1)}\right) = \frac{1}{q} \frac{1}{\prod_{n=1}^{\infty}(1 - q^n)^{d-2}}, \tag{C.58}$$

where the term -1 comes again from the zero-point energy, and $d = 26$.

From Eq. (C.58) it follows that the degeneracy $T_d(N)$ of states at level N is the coefficient of the power q^{N-1} in the expansion of Eq. (C.58) in power series around $q = 0$:

$$G(q) = \frac{1}{q} \sum_{N=0}^{\infty} T_d(N) q^N, \tag{C.59}$$

where $d - 2 = 24$. The degeneracy at the level N is just given by the number of ways in which one can express the integer N as a sum of positive integers as follows: $N = n_1 + 2n_2 + \cdots$; this is the so-called number of partitions of the integer N.

Let us recall some well-known facts about the asymptotic growth of the degeneracy of states. Equation (C.59) can be inverted as follows,

$$T_d(N) = \frac{1}{2\pi i} \oint dq\, q^{-N-1} \prod_{n=1}^{\infty}(1 - q^n)^{2-d}, \tag{C.60}$$

where the integral is performed on a small contour enclosing the origin. This integral can be evaluated approximately for large N, leading to the asymptotic behaviour:

$$T_d(N) \sim N^{-(\frac{d+1}{4})} e^{2\pi \sqrt{\frac{(d-2)N}{6}}}, \qquad N \to \infty. \tag{C.61}$$

Recalling that $N \sim \alpha' M^2$, this corresponds to a density of states per unit of mass:

$$N(M) \sim M^{-B} e^{\beta_0 M} \tag{C.62}$$

where

$$B = \frac{d-1}{2}, \qquad \beta_0 = 2\pi \sqrt{\frac{(d-2)\alpha'}{6}}. \tag{C.63}$$

This is the characteristic exponential growth of the density of states in string theory. It implies a maximal temperature for a string gas, the Hagedorn temperature $T_H = 1/\beta_0$, above which the partition function diverges because the exponential density of states dominates over the Boltzmann damping factor, i.e. entropy overwhelms energy.

We can proceed analogously for the closed string and find the spectrum of states given by the following equations:

$$\frac{\alpha'}{2} M^2 = N + \tilde{N} - 2, \tag{C.64}$$

and
$$N = \tilde{N}, \tag{C.65}$$

where

$$N = \sum_{i=1}^{24} \sum_{n=1}^{\infty} n a_n^{i\dagger} a_n^i, \qquad \tilde{N} = \sum_{i=1}^{24} \sum_{n=1}^{\infty} n \tilde{a}_n^{i\dagger} \tilde{a}_n^i. \tag{C.66}$$

The lowest state of the spectrum is a tachyon with mass $M^2 = -4/\alpha'$ described by the vacuum $|0\rangle$.

The first excited level containing massless states is described by the states:

$$a_1^{i\dagger} \tilde{a}_1^{j\dagger} |0\rangle. \tag{C.67}$$

The symmetric and traceless state corresponds to a graviton, the trace of (C.67) corresponds to a massless scalar called the dilaton and, finally, the antisymmetric state describes a two-index antisymmetric tensor. Also in the case of a closed string, the Lorentz invariance of the theory requires $d = 26$.

In conclusion, by working in the light-cone gauge, we have determined the spectrum of physical states for both open and closed strings and we have seen that the massless states correspond to a gauge boson in the case of the open string and to a graviton, a dilaton and a two-index antisymmetric tensor in the case of a closed string.

C.4 Quantization in the conformal gauge

In the previous Section, we first eliminated all redundant degrees of freedom, keeping only the physical transverse ones, and then we quantized them, obtaining the spectrum of physical states for both the open and closed strings. However, proceeding in this way, we lost the manifest Lorentz invariance of the theory. The quantum theory is nevertheless Lorentz invariant for $\alpha_0 = 1$ and $d = 26$.

In this Section, we mention the other way of quantizing the theory in the conformal gauge. In this case, one keeps manifest Lorentz invariance, but negative norm states appear in the theory. The following analysis parallels that of physical states in the DRM, already described in the Introduction to Part III.

The theory is quantized by imposing the commutation relations:

$$[a_n^\mu, a_m^{\nu\dagger}] = \eta^{\mu\nu} \delta_{nm}, \qquad [\hat{q}^\mu, \hat{p}^\nu] = i\eta^{\mu\nu}, \tag{C.68}$$

for the open string, and also

$$[\tilde{a}_n^\mu, \tilde{a}_m^{\nu\dagger}] = \eta^{\mu\nu} \delta_{nm}, \tag{C.69}$$

for the closed string, with all other commutators vanishing. The presence of the Lorentz metric $\eta^{\mu\nu}$, with the time component $\eta^{00} = -1$ being negative, implies the presence of states with negative norm.

We have seen that the invariance under reparameterization, $T_{ab} = 0$, in this gauge requires the vanishing of the Virasoro operators L_n for any n. However, this cannot be imposed at the operatorial level, because it would be in contradiction with their algebra. We can proceed as in QED and impose their vanishing on the physical states. Correspondingly, the physical Hilbert space is a subspace of the space generated by the oscillators, which is characterized by the conditions:

$$L_n|\text{Phys}\rangle = (L_0 - 1)|\text{Phys}\rangle = 0, \qquad n > 0. \tag{C.70}$$

At the quantum level, the L_n operators (C.35), are written

$$L_n = \frac{1}{2} \sum_{m=-\infty}^{\infty} :\alpha_{n-m} \cdot \alpha_m:, \tag{C.71}$$

where the double dots indicate the 'normal ordering', that amounts to putting all the creation operators to the left of annihilation operators.

As in QED, the solutions of the previous equations also contain zero-norm states which, however, can be eliminated by hand from the physical spectrum because they are decoupled from the physical states [DD70]. After the elimination of zero norm states, the most general solution of Eqs. (C.70) for $d = 26$ is provided by the DDF states [DDF72] which are in one-to-one correspondence with the transverse states discussed in the previous Section. This implies that, for $d = 26$, the two quantization procedures yield the same spectrum of physical states. As discussed in the Introduction to Part VII, a completely satisfactory covariant quantization was only obtained in the Eighties by Polyakov's path-integral approach, and the subsequent BRST approach.

C.5 Scattering of strings

In this Section, we derive the scattering amplitudes involving N strings, obtaining the N-point amplitude of the DRM. One way to proceed is to use the light-cone path integral approach, discussed in the Chapter by Mandelstam. Here we proceed instead in an alternative way that is manifestly Lorentz invariant. It consists of adding, to the free string action, a term describing the interaction of the string with an external field and from it deriving the string scattering amplitudes. This procedure is outlined in the Chapters by Gliozzi and Nicodemi.

The interaction term to be added to the free string action S_0 of the previous Sections is [ADDN74]:

$$S_{int} = \int d^d y \, \Phi_L(y) J_L(y), \tag{C.72}$$

where $\Phi_L(y)$ is the external field and J_L is the current generated by the string. The index L stands for possible Lorentz indices that are saturated in order to have a Lorentz invariant action.

In the case of a point particle, such an interaction term does not in general give any condition on self-interactions, but only describes the interaction of a particle with, for example, the electromagnetic field that is in general totally independent of the point particle. In the case of a string, S_{int} describes instead the interaction among strings because, for consistency reasons, the external fields allowed to interact with a string are only those corresponding to the string states themselves. Indeed, consistency imposes the following restrictions on S_{int}.

(i) It must be a well-defined operator in the space spanned by the string oscillators.
(ii) It must preserve the invariances of the free string theory. In particular, in the conformal gauge, it must be conformal invariant.
(iii) In the case of an open string, the interaction occurs at one of the two end-points of a string (say at $\sigma = 0$). This follows from the fact that open strings interact by attaching to each other at end-points.

Let us focus on the open string and start with some examples. The simplest scalar current generated by the motion of a string can be written as follows:

$$J(y) = \int d\tau \int d\sigma \, \delta(\sigma) \, \delta^{(d)}[y^\mu - X^\mu(\tau, \sigma)], \qquad (C.73)$$

where $\delta(\sigma)$ implies that the interaction occurs at the end-point $\sigma = 0$ of the string. For simplicity, we omit writing the coupling constant g in (C.73) and in the following interaction terms.

Inserting (C.73) in (C.72) and using for $\Phi(y) = e^{ik \cdot y}$ a plane wave, we get the following interaction:

$$S_{int} = \int d\tau : e^{ik \cdot X(\tau, 0)} : \equiv \int d\tau \, V^{(\tau)}(\tau), \qquad (C.74)$$

where the normal ordering has been introduced in order to have a well-defined operator according to requirement (i).

The invariance of (C.74) under a conformal transformation $\tau \to w(\tau)$ implies:

$$S_{int} = \int d\tau \, V^{(\tau)}(\tau) = \int dw \, V^{(w)}(w) \qquad (C.75)$$

or in other words that the integrand transforms as follows

$$V^{(\tau)}(\tau) = \frac{dw}{d\tau} V^{(w)}(w). \qquad (C.76)$$

Recalling from the Introduction to Part III that under a conformal transformation $z \to w(z)$ a primary conformal field $\Phi(z)$ transforms by

$$\Phi^{(z)}(z) = [w'(z)]^\Delta \Phi^{(w)}(w), \qquad (C.77)$$

Eq. (C.76) implies that the vertex operator $: e^{ik \cdot X(\tau, 0)} :$ must transform as a primary field with dimension $\Delta = 1$. In the following, it will be convenient to consider the vertex operator as a function of $z = e^{i\tau}$ instead of τ.

For infinitesimal transformations $z \to z + \epsilon z^{n+1}$ (ϵ being a small quantity) Eq. (C.77) becomes:

$$\delta\Phi(z) \equiv \epsilon[L_n, \Phi(z)] = \epsilon \left[z^{n+1} \frac{d}{dz} \Phi(z) + \Delta(n+1)z^n \Phi(z) \right], \qquad (C.78)$$

where we have used the fact that at the quantum level the infinitesimal conformal transformation of a field is given by its commutator with the generators of the conformal algebra L_n. In order to see whether : $e^{ik \cdot X(z)}$: is indeed a primary field with $\Delta = 1$, we compute its commutator with the Virasoro generator using the oscillator representation (C.70) and find:

$$[L_n, : e^{ik \cdot x(z)} :] = z^{n+1} \frac{d}{dz} : e^{ik \cdot X(z)} : + (n+1) z^n \alpha' k^2 : e^{ik \cdot X(z)} : . \qquad (C.79)$$

The result agrees with (C.78) for $\Delta = 1$ if:

$$\alpha' k^2 = 1. \qquad (C.80)$$

In other words, in order to satisfy the criteria (i) and (ii), the external field must correspond to the on-shell lowest state of the string, namely the tachyon. The tachyon state can be obtained in the limit:

$$\lim_{z \to 0} : e^{ik \cdot X(z,0)} : |0\rangle = |0, k\rangle \equiv e^{ik \cdot \hat{q}} |0, p = 0\rangle, \qquad (C.81)$$

by remembering that, in terms of the variable z, the normal ordering on the zero modes $\hat{p}$ and $\hat{q}$ must be such that $\hat{p}$ is put on the right of $\hat{q}$.

Another example of current generated by the string is given by:

$$J_\mu(y) = \int d\tau \int d\sigma \, \delta(\sigma) \dot{X}_\mu(\tau, \sigma) \delta^{(d)}(y - X(\tau, \sigma)). \qquad (C.82)$$

Inserting (C.82) in (C.72) we get

$$S_{int} = \int d\tau \dot{X}_\mu(\tau, 0) \epsilon^\mu e^{ik \cdot X(\tau, 0)}, \qquad (C.83)$$

if we use a plane wave for $\Phi_\mu(y) = \epsilon_\mu e^{ik \cdot y}$. Proceeding as in the case of the tachyon, it is easy to show that the vertex operator in Eq. (C.83) is a conformal field with dimension $\Delta = 1$, if the following condition is satisfied:

$$k^2 = \epsilon \cdot k = 0. \qquad (C.84)$$

Therefore, the external vector field must be the on-shell massless photon state of the string. As in the case of the tachyon, such a photon state can be obtained from the vertex operator in the following way:

$$\lim_{z \to 0} \frac{dX^\mu}{dz} \epsilon_\mu e^{ik \cdot X(z)} |0\rangle = -i(\epsilon \cdot a_1^\dagger) |0\rangle. \qquad (C.85)$$

In conclusion, the two examples have indicated that an open string can only interact with on-shell physical states of the string itself. It turns out that this property holds for arbitrary external fields. The fact that strings interact consistently only with strings (and with D-branes) makes this theory a prototype of a Theory of Everything.

The previous considerations can be extended to the closed string which, having no end-points, can interact with the external field at any point σ. This is connected to the fact that two closed strings interact with each other by touching at a particular point and there creating a closed string that contains both original strings. The point of contact is arbitrary and this is the reason why there is no privileged value of σ. In particular, the vertex operator for the graviton is obtained by introducing in the string action (C.28) an arbitrary background $G_{\mu\nu}(X)$ [ADDN74]:

$$S = -\frac{T}{2} \int d\tau \int d\sigma\, G_{\mu\nu}(X) \partial_a X^\mu \partial^a X^\nu, \quad (C.86)$$

and expanding it around the flat Minkowski metric $\eta_{\mu\nu}$,

$$G_{\mu\nu}(X) = \eta_{\mu\nu} + 2\kappa h_{\mu\nu}, \quad (C.87)$$

where κ is related to the Newton constant. Writing for $h_{\mu\nu} = \epsilon_{\mu\nu} e^{ik\cdot x}$ a plane wave as before, we get the interaction term between the string and the field of the graviton:

$$S_{int} = -T\kappa \int d\tau \int d\sigma\, \epsilon_{\mu\nu} \partial_a X^\mu \partial^a X^\nu e^{ik\cdot X}. \quad (C.88)$$

Proceeding as in the previous cases, it can be shown that the vertex preserves conformal invariance if the following equations are satisfied:

$$k^2 = k^\mu \epsilon_{\mu\nu} = k^\nu \epsilon_{\mu\nu} = 0, \quad (C.89)$$

implying that the graviton should be on the mass shell and have physical, transverse polarizations. The previous considerations are the precursor of the study of strings in an arbitrary curved background done in the Eighties [Lov84, CMPF85].

C.6 DRM amplitudes

Let us go back to the open string and compute the scattering amplitudes involving N tachyons. These can be obtained from the probability amplitude for the emission of $(N-2)$ external fields, corresponding to the tachyon state of the string, from a string that originally is in the state of the tachyon and remains in this state also after the $N-2$ emissions.

Starting from the total action $S = S_0 + S_{int}$, the probability amplitude for the transition from an initial $|i\rangle$ to a final $|f\rangle$ string state after the interaction with the external field is given in perturbation theory by the S-matrix element:

$$\langle f | S | i \rangle = \lim_{\substack{\tau_i \to -\infty \\ \tau_f \to \infty}} \langle f | T[e^{i S_{int}}] | i \rangle, \quad (C.90)$$

where in the case of a tachyonic external field, S_{int} is given by (C.74) and T stands for the time-ordered product.

The N-tachyon scattering amplitude is obtained by expanding the exponential in (C.90), keeping only the terms containing $N-2$ vertex operators and sandwiching them between an initial tachyon state with momentum k_1 and a final tachyon state with momentum k_N. The expansion of the exponential gives a sum of $(N-2)!$ terms that correspond to the different terms of the T-ordered product in (C.90). A single term is given by:

$$A(k_1, k_2, \ldots, k_N) = \int_0^\infty \prod_{i=3}^{N-1} d\tau_i \prod_{i=2}^{N-2} [\theta(\tau_{i+1} - \tau_i)]$$

$$\times \langle 0, k_1 | \prod_{i=2}^{N-1} : e^{ik_i \cdot X(\tau_i, 0)} : |0, k_N\rangle. \tag{C.91}$$

The variable τ_2 has been chosen to be equal to zero owing to translational invariance of the matrix element in (C.91). The integral over τ_i is originally defined along the positive real axis, but is made convergent by means of a Wick rotation $\tau \to i\tau$ to Euclidean time, following a standard procedure [PS95]. After the introduction of the Koba–Nielsen variables $z_i = e^{-\tau_i}$ we can rewrite (C.91) as follows:

$$A(k_1, k_2, \ldots, k_N) = \int_0^1 \prod_{i=3}^{N-2} dz_i \prod_{i=2}^{N-2} [\theta(z_i - z_{i+1})] \langle 0, k_1 | \prod_{i=2}^{N-1} : e^{ik_i \cdot X(z_i, 0)} : |0, k_N\rangle.$$
$$\tag{C.92}$$

The integral in (C.92) provides the scattering amplitude of N tachyons in the bosonic string. This derivation is, however, not completely consistent, because it gives only $(N-2)!$ terms instead of the $(N-1)!$ that are needed to have a crossing-symmetric amplitude. This is because the initial and final tachyons have been treated differently from the others. In order to get a completely crossing-symmetric amplitude, the missing terms must be added by hand.

Let us proceed to compute (C.92) explicitly. Using the expression of : $e^{ik \cdot X(z)}$: in terms of the harmonic oscillators and the Baker–Campbell–Hausdorff formula,

$$e^A e^B = e^B e^A e^{[A,B]}, \tag{C.93}$$

it is easy to prove the relation

$$: e^{ik \cdot X(z)} :: e^{ih \cdot X(\zeta)} := : e^{ik \cdot X(z)} e^{ih \cdot X(\zeta)} : (z - \zeta)^{2\alpha' k \cdot h}, \tag{C.94}$$

that allows one to compute the vacuum expectation value in (C.92) obtaining:

$$A(k_1, k_2, \ldots, k_N) = (2\pi)^d \delta^{(d)}\left(\sum_{i=1}^N k_i\right)$$

$$\times \int_0^1 \prod_{i=3}^{N-1} dz_i \prod_{i=2}^{N-1} [\theta(z_i - z_{i+1})] \prod_{i=2}^{N-1} \prod_{j=i+1}^N (z_i - z_j)^{2\alpha' k_i \cdot k_j}, \tag{C.95}$$

where $z_2 = 1$ and $z_N = 0$. The δ-function of momentum conservation is obtained from the zero modes of the vertex operators.

For $N = 4$ the previous expression becomes:

$$A(k_1, k_2, k_3, k_4) = \int_0^1 dz\, z^{2\alpha' k_3 \cdot k_4}(1 - z)^{2\alpha' k_2 \cdot k_3}. \tag{C.96}$$

In terms of the Regge trajectories in the s- and t-channels

$$\alpha_s = 1 - \alpha'(k_3 + k_4)^2, \qquad \alpha_t = 1 - \alpha'(k_3 + k_2)^2, \tag{C.97}$$

and imposing the mass-shell conditions for the external tachyons,

$$\alpha' k_i^2 = 1, \qquad i = 1, 2, 3, 4, \tag{C.98}$$

we finally recover the Veneziano amplitude:

$$A(k_1, k_2, k_3, k_4) = B(-\alpha_s, -\alpha_t) = \frac{\Gamma(-\alpha_s)\Gamma(-\alpha_t)}{\Gamma(-\alpha_s - \alpha_t)}, \tag{C.99}$$

where

$$B(x, y) = \int_0^1 dz\, z^{-x-1}(1 - z)^{-y-1}. \tag{C.100}$$

In conclusion, we have shown that the scattering of strings reproduces the DRM amplitudes.

References

[ADDN74] Ademollo, M., D'Adda, A., D' Auria, R., Napolitano, E., Sciuto, S., Di Vecchia, P., Gliozzi, F., Musto, R. and Nicodemi, F. (1974). Theory of an interacting string and dual resonance model, *Nuovo Cimento* **A21**, 77–145.

[BDZH76] Brink, L., Deser, S., Zumino, B. and Howe, P. (1976). Local supersymmetry for spinning particles, *Phys. Lett.* **B64**, 435–438.

[BDH76] Brink, L., Howe, P. and Di Vecchia, P. (1976). A locally supersymmetric and reparametrization invariant action for the spinning string, *Phys. Lett.* **B65**, 471–474.

[BN73] Brink, L. and Nielsen, H. B. (1973). A simple physical interpretation of the critical dimension of spacetime in dual models, *Phys. Lett.* **B45**, 332–336.

[CMPF85] Callan, C. G, Martinec, E. J., Perry, M. J. and Friedan, D. (1985). Strings in background fields, *Nucl. Phys.* **B262**, 593.

[DD70] Del Giudice, E. and Di Vecchia, P. (1970). Characterization of the physical states in dual-resonance models, *Nuovo Cimento* **A70**, 579–591.

[DDF72] Del Giudice, E., Di Vecchia, P. and Fubini, S. (1972). General properties of the dual resonance model. *Ann. Phys.* **70**, 378–398.

[DZ76] Deser, S. and Zumino, B. (1976). A complete action for the spinning string, *Phys. Lett.* **B65**, 369–373.

[DiV76] Di Vecchia, P. (1978). In *Many Degrees of Freedom in Particle Physics*, ed. Satz, H. (Plenum Press, New York), 493.

[Gli73] Gliozzi, F. (1973). Unpublished work.

[Got71] Goto, T. (1971). Relativistic quantum mechanics of one-dimensional mechanical continuum and subsidiary condition of dual resonance model, *Prog. Theor. Phys.* **46**, 1560–1569.

[GSW87] Green, M. B., Schwarz, J. H. and Witten, E. (1987). *Superstring Theory* (Cambridge University Press, Cambridge).

[Lov84] Lovelace, C. (1984). Strings in curved space, *Phys. Lett.* **B135**, 75.

[Nam70] Nambu, Y. (1970). Duality and hadrodynamics, lecture notes prepared for Copenhagen summer school, 1970, reproduced in *Broken Symmetry, Selected Papers of Y. Nambu*, ed. Eguchi, T. and Nishijima, K. (World Scientific, Singapore), 280.

[PS95] Peskin, M. E. and Schroeder, D. V. (1995). *An Introduction to Quantum Field Theory* (Westview Press, Boulder, CO).

Appendix D
World-sheet and target-space supersymmetry

D.1 Supersymmetry on the world-sheet and superconformal algebra

We begin by recalling the Nambu–Goto action discussed in Appendix C. The action is invariant under arbitrary reparameterizations of the world-sheet coordinates $\xi^a = (\tau, \sigma)$. In the conformal gauge, the action reduces to that of d free bosons,

$$S = -\frac{T}{2} \int d^2\xi \, \eta^{ab} \partial_a X^\mu \partial_b X^\nu \eta_{\mu\nu}, \tag{D.1}$$

where $a, b = 0, 1$, $\mu, \nu = 0, 1, \ldots, d-1$ and $\eta_{ab} = \text{diag}(-1, 1)$. Reparameterization invariance implies the vanishing of the two-dimensional world-sheet energy-momentum tensor:

$$T_{ab} = \partial_a X \cdot \partial_b X - \frac{1}{2} \eta_{ab} \eta^{cd} \partial_c X \cdot \partial_d X = 0. \tag{D.2}$$

The conformal gauge does not fix the gauge symmetry completely: we can still perform conformal transformations that are a symmetry of the action (D.1). The generators of the conformal transformations, the Virasoro generators L_n, are the Fourier components of T_{ab}; in the quantum theory, they become operators satisfying the Virasoro algebra:

$$[L_n, L_m] = (n - m)L_{n+m} + \frac{d}{12} n(n^2 - 1)\delta_{n+m,0}. \tag{D.3}$$

The vanishing of the stress tensor at the classical level becomes the following set of conditions on physical states at the quantum level:

$$\langle \text{Phys}'|L_n|\text{Phys}\rangle = 0, \quad n \neq 0,$$
$$\langle \text{Phys}'|(L_0 - 1)|\text{Phys}\rangle = 0. \tag{D.4}$$

Owing to the Hermiticity condition, $L_n^\dagger = L_{-n}$, the previous equations are satisfied if we impose the following constraints on the physical states:

$$L_n|\text{Phys}\rangle = 0, \quad n > 0, \quad (L_0 - 1)|\text{Phys}\rangle = 0. \tag{D.5}$$

The Birth of String Theory, ed. Andrea Cappelli, Elena Castellani, Filippo Colomo and Paolo Di Vecchia.
Published by Cambridge University Press. © Cambridge University Press 2012.

In Appendix C we have seen that, if $d \leq 26$, these conditions imply that the physical states span a positive definite Hilbert space. Actually, among the solutions of Eqs. (D.5), there are also zero norm states which are decoupled from the physical spectrum (see Appendix C for details).

In the Ramond–Neveu–Schwarz (RNS) model one introduces, besides the bosonic oscillators α_n, the fermionic oscillators ψ_r^μ, where r is integer and half-integer in the Ramond (R) and Neveu–Schwarz (NS) sectors, respectively. In order to eliminate the negative norm states, we need to introduce conditions generalizing (D.5), which are expressed in terms of the additional fermionic operators F_n and G_r already presented in the Introduction to Part V. The latter are generators of a local fermionic symmetry called world-sheet supersymmetry; together with conformal invariance, this leads to superconformal symmetry.

In the following, we describe these symmetries in more detail, extending the discussion in the Introductions to Parts V and VI. The starting point is the reparameterization invariant and local supersymmetric world-sheet action of the RNS model presented in the Introduction to Part VI and further discussed in the Chapters by Brink and Di Vecchia in Part VI. Similarly to the case of the bosonic string, we can choose a gauge generalizing the conformal gauge, called a superconformal gauge, in which the RNS nonlinear action reduces to the following free theory of bosons and fermions:

$$S = -\frac{T}{2} \int d^2\xi \left(\eta^{ab} \partial_a X^\mu \partial_b X^\nu - i \bar{\psi}^\mu \rho^a \partial_a \psi^\nu \right) \eta_{\mu\nu}. \tag{D.6}$$

The local symmetries of the RNS action imply the vanishing of the corresponding energy-momentum tensor and supercurrent:

$$T_{ab} = 0, \qquad J_a = 0, \tag{D.7}$$

where

$$T_{ab} \sim \partial_a X \cdot \partial_b X - \frac{1}{2} \eta_{ab} \eta^{cd} \partial_c X \cdot \partial_d X$$
$$- \frac{i}{4} \bar{\psi} \left(\rho_a \partial_b + \rho_b \partial_a - \eta_{ab} \rho^c \partial_c \right) \psi, \tag{D.8}$$
$$J_a \sim \rho^b \rho_a \bar{\psi}^\mu \partial_b X^\nu \eta_{\mu\nu}.$$

In these equations, ψ^μ is a real Majorana spinor in two dimensions and ρ^α are the Dirac gamma matrices in two dimensions, obeying $\{\rho^a, \rho^b\} = -2\eta^{ab}$; this algebra is realized by the Pauli matrices σ^2 and $i\sigma^1$, as follows:

$$\rho^0 = \begin{pmatrix} 0 & -i \\ i & 0 \end{pmatrix}, \quad \rho^1 = \begin{pmatrix} 0 & i \\ i & 0 \end{pmatrix}, \quad \rho^3 = \rho^0 \rho^1 = \begin{pmatrix} 1 & 0 \\ 0 & -1 \end{pmatrix}. \tag{D.9}$$

ρ^0 and ρ^1 are purely imaginary and satisfy the relations:

$$(\rho^\dagger)^a = \rho^0 \rho^a \rho^0, \qquad a = 0, 1, \tag{D.10}$$

showing that the matrix ρ^0 plays the role of the charge conjugation matrix, $(\rho^0)^2 = 1$; ρ^3 is instead the chirality operator, analogous to γ^5 in four dimensions. ψ is a world-sheet two-dimensional spinor whose components are anticommuting variables, called Grassmann variables: for example, the product of two spinors ψ_A and χ_B, with $A, B = 1, 2$, obeys $\psi_A \chi_B = -\chi_B \psi_A$, for any A, B. In the quantum theory, anticommuting classical fermionic fields become operators obeying canonical anticommutation relations.

Supersymmetry is, in general, a symmetry of the action under transformations that relate bosonic and fermionic fields. In our case, the action (D.6) is invariant under the following supersymmetry transformations:

$$\delta X^\mu = \bar{\epsilon} \psi^\mu, \qquad \delta \psi^\mu = -i \rho^a (\partial_a X^\mu) \epsilon, \qquad \delta \bar{\psi} = i \bar{\epsilon} \rho^a (\partial_a X^\mu), \tag{D.11}$$

where the infinitesimal parameter $\epsilon = \epsilon(\xi)$ is an anticommuting Majorana spinor satisfying the equation:

$$\rho^a \rho^b \partial_a \epsilon = 0. \tag{D.12}$$

In fact, performing the variation of the action under the previous transformations one obtains:

$$\delta S = -\frac{T}{2} \int d^2\xi \left[2\eta^{ab} \partial_a X \partial_b (\bar{\epsilon} \psi) - \bar{\psi} \rho^a \partial_a (\rho^b \partial_b X \epsilon) + \bar{\epsilon} \rho^a \partial_a X \rho^b \partial_b \psi \right]. \tag{D.13}$$

In this equation, we neglect the Lorentz indices that are irrelevant in the present discussion. Integrating the last term by parts, the previous expression becomes:

$$\delta S = -\frac{T}{2} \int d^2\xi \left[2\eta^{ab} \partial_a X \partial_b (\bar{\epsilon} \psi) + \bar{\psi} \partial^2 X \epsilon - \bar{\psi} \partial_b X \rho^a \rho^b \partial_a \epsilon \right.$$
$$\left. + \partial_b \left(\bar{\epsilon} \rho^a \partial_a X \rho^b \psi \right) - \partial_b \bar{\epsilon} \rho^a \partial_a X \rho^b \psi + \bar{\epsilon} \partial^2 X \psi \right]. \tag{D.14}$$

The second and sixth terms of the previous expression are shown to be equal as follows. Equations (D.9) and (D.10) imply the Majorana-spinors identity $\bar{\epsilon} \psi = \bar{\psi} \epsilon$, which reads in components:

$$\left(\rho^0 \right)_{AB} (\epsilon_A \psi_B - \psi_A \epsilon_B) = 0; \tag{D.15}$$

this follows from the definition $\bar{\psi} = \psi^\dagger \rho^0$, the antisymmetry of the matrix ρ^0 and the anticommutativity of spinor components. Similarly, the third and fifth terms in (D.14) are equal. All these terms can be combined together to give:

$$\delta S = \frac{T}{2} \int d^2\xi \left[\partial_b \left(\bar{\epsilon} \rho^b \rho^a \partial_a X \psi \right) + 2 \partial_b \bar{\epsilon} \rho^a \partial_a X \rho^b \psi \right]. \tag{D.16}$$

If ϵ satisfies (D.12), the second term is vanishing, while the first term, being a total derivative, leaves the action invariant. Therefore, we have shown that $\delta S = 0$ under a supersymmetry transformation with a parameter ϵ satisfying (D.12).

The Noether current, corresponding to this symmetry, is obtained by performing the variation of the action with an arbitrary ϵ. In this case δS is given by the second term in Eq. (D.16) and the Noether current is identified with the coefficient of the derivative of the arbitrary parameter ϵ obtaining, apart from an overall normalization, the spinor current in

(D.8). The spinor current J^a is of course conserved,

$$\partial_a J^a = \partial_a \left(\rho^b \rho^a \psi^\mu \partial_b X^\nu \eta_{\mu\nu} \right) = 0, \tag{D.17}$$

provided that the equations of motion for X and ψ are satisfied,

$$\eta^{ab} \partial_a \partial_b X = \rho^a \partial_a \psi = 0. \tag{D.18}$$

The condition (D.12) obeyed by the spinorial supersymmetry parameter has an interesting interpretation. Using the representation of the ρ-matrices in (D.9), it can be rewritten:

$$\left(\frac{\partial}{\partial \tau} - \rho^3 \frac{\partial}{\partial \sigma} \right) \epsilon = 0, \qquad \epsilon \equiv \begin{pmatrix} \epsilon_+ \\ \epsilon_- \end{pmatrix}. \tag{D.19}$$

In terms of light-cone variables,

$$\xi_\pm = \tau \pm \sigma, \qquad \partial_\pm = \frac{1}{2} \left(\frac{\partial}{\partial \tau} \pm \frac{\partial}{\partial \sigma} \right), \tag{D.20}$$

the conditions (D.19) read: $\partial_- \epsilon_+ = \partial_+ \epsilon_- = 0$. As in the case of conformal transformations (see the Introduction to Part III and Appendix C), there is a separation of variables in light-cone coordinates, that becomes a condition of analyticity/antianalyticity for $\epsilon_\pm$ in Euclidean world-sheets.

From the action (D.6) one can derive the equations of motion for the fermionic degrees of freedom in light-cone coordinates:

$$\partial_+ \psi^\mu_- = 0, \qquad \partial_- \psi^\mu_+ = 0, \qquad \psi^\mu_\pm = \frac{1 \mp \rho^3}{2} \psi^\mu. \tag{D.21}$$

For the sake of simplicity we limit ourselves to open strings and in this case the boundary conditions are

$$\int d\tau \, (\psi_+ \delta \psi_+ - \psi_- \delta \psi_-)|^{\sigma=\pi}_{\sigma=0} = 0. \tag{D.22}$$

They can be fulfilled in two different ways:

$$\begin{cases} \psi_-(0, \tau) = \eta_1 \psi_+(0, \tau), \\ \psi_-(\pi, \tau) = \eta_2 \psi_+(\pi, \tau), \end{cases} \tag{D.23}$$

where η_1 and η_2 can take the values ± 1. In particular, if $\eta_1 = \eta_2$ we get the Ramond (R) sector of the open string, while if $\eta_1 = -\eta_2$ we get the Neveu–Schwarz (NS) sector.

The general solution of (D.21) satisfying the boundary conditions (D.23) is given by

$$\psi^\mu_\mp \sim \sum_t \psi^\mu_t e^{-it(\tau \mp \sigma)}, \qquad \begin{cases} t \in \mathbb{Z} + \frac{1}{2}, & \text{NS sector,} \\ t \in \mathbb{Z}, & \text{R sector.} \end{cases} \tag{D.24}$$

The energy-momentum tensor has two nonzero components in light-cone coordinates,

$$\begin{aligned} T_{++} &= \partial_+ X \cdot \partial_+ X + \frac{i}{2} \psi_+ \cdot \partial_+ \psi_+, \\ T_{--} &= \partial_- X \cdot \partial_- X + \frac{i}{2} \psi_- \cdot \partial_- \psi_-, \end{aligned} \tag{D.25}$$

that obey $\partial_- T_{++} = \partial_+ T_{--} = 0$ upon using the equations of motion. The supercurrent (D.8) also reduces to

$$J_- = \psi_- \cdot \partial_- X, \qquad J_+ = \psi_+ \cdot \partial_+ X, \qquad (D.26)$$

and similarly satisfies the following equations: $\partial_+ J_- = \partial_- J_+ = 0$.

These equations for the energy-momentum tensor and the supercurrent show that their components only depend on one of the two light-cone coordinates; this result completes the proof of separation of variables for (super)conformal transformations, that is the analytic decomposition of two-dimensional Euclidean geometry. In the case of the closed string, the two components of the energy-momentum tensor and supercurrent are independent and give rise to two copies of the superconformal algebra, involving Virasoro and fermionic generators for the left and right movers, respectively (its form is given below). In the case of the open string, the two components are related by a reality condition and thus their Fourier components give one independent set of generators only.

The Virasoro generators have the following expression:

$$L_n = \frac{1}{2} \sum_{m \in \mathbb{Z}} \alpha_{-m} \cdot \alpha_{n+m} + \frac{1}{2} \sum_t \left(\frac{n}{2} + t \right) \psi_{-t} \cdot \psi_{t+n}, \qquad (D.27)$$

where the index t is half-integer (integer) in the NS (R) sector. The fermionic generators, Fourier components of the supercurrent, are:

$$G_t = \sum_{n=-\infty}^{\infty} \alpha_{-n} \cdot \psi_{t+n}. \qquad (D.28)$$

The bosonic and fermionic oscillators satisfy (anti)commutation relations:

$$[\alpha_n^\mu, \alpha_m^\nu] = n \eta^{\mu\nu} \delta_{n+m;0}, \qquad \{\psi_s^\mu, \psi_t^\nu\} = \eta^{\mu\nu} \delta_{s+t;0}, \qquad (D.29)$$

where n, m take integer values, while s, t take integer (half-integer) values in the R (NS) model. The zero mode of the bosonic oscillator is related to the momentum by $\alpha_0 \equiv \sqrt{2\alpha'} \hat{p}$, while we will see that the zero mode of the fermionic oscillators is related to the Dirac gamma matrix in target space.

As in the case of the bosonic string, the quantum Virasoro generators should be defined with the normal-ordered product of oscillators. This only affects the form of L_0, which becomes:

$$L_0 = \alpha' \hat{p}^2 + \sum_{n=1}^{\infty} \alpha_{-n} \cdot \alpha_n + \sum_{t>0} t \psi_{-t} \cdot \psi_t. \qquad (D.30)$$

The (anti)commutation relations for the operators given in (D.27), (D.28) and (D.30) give rise to the following superconformal (super Virasoro) algebra:

$$\begin{aligned}
{[L_m, L_n]} &= (m-n) L_{m+n} + \tfrac{d}{8} m(m^2 - 1) \delta_{m+n,0}, \\
{[L_m, G_r]} &= (\tfrac{1}{2} m - r) G_{r+m}, \\
\{G_r, G_s\} &= 2 L_{r+s} + \tfrac{d}{2}(r^2 - \tfrac{1}{4}) \delta_{r+s,0},
\end{aligned} \qquad (D.31)$$

for the NS sector and

$$[L_m, L_n] = (m-n)L_{m+n} + \tfrac{d}{8}m^3\delta_{m+n,o},$$
$$[L_m, G_n] = (\tfrac{1}{2}m - n)G_{n+m}, \qquad (D.32)$$
$$\{G_m, G_n\} = 2L_{m+n} + \tfrac{d}{2}n^2\delta_{m+n,o},$$

for the R sector. Note that the c-number central extensions of the algebra are different in the two sectors, but can be brought into the same standard form (of the NS sector) by redefining $L_0 \to L_0 - d/16$ in the R sector. This corresponds to the fact, already discussed in the Introduction to Part V, that the Ramond vacuum is actually an excited state obtained by applying the spin field of dimension $d/16$ to the NS vacuum.

In the RNS model, the space generated by the bosonic and fermionic oscillators (D.29) contains unphysical states with negative norm as in the case of the DRM. The conditions which select the physical states are:

$$\begin{cases} L_m|\psi_{\text{Phys}}\rangle = 0 & m > 0, \\ (L_0 - a_0)|\psi_{\text{Phys}}\rangle = 0, \\ G_t|\psi_{\text{Phys}}\rangle = 0 & t \geq 0, \end{cases} \qquad (D.33)$$

where

$$\begin{cases} a_0 = \tfrac{1}{2} & (\text{NS}), \\ a_0 = 0 & (\text{R}). \end{cases} \qquad (D.34)$$

The zero-point energy can be obtained by the ζ-function regularization introduced in Appendix C: one needs the analytic continuation of a slightly generalized expression of this function, as follows:

$$\zeta(s, \alpha) = \sum_{n=0}^{\infty}(n+\alpha)^{-s},$$
$$\zeta(-1, \alpha) = \frac{1}{2}\left(-\frac{1}{6} - \alpha^2 + \alpha\right). \qquad (D.35)$$

In the R sector the zero-point energies of the bosonic and fermionic oscillators cancel out completely leading to $a_0 = 0$. In the NS sector the following sums are found and are regularized by using (D.35):

$$\frac{d-2}{2}\sum_{n=1}^{\infty}n = -\frac{d-2}{24},$$
$$-\frac{d-2}{2}\sum_{n=0}^{\infty}(n+\tfrac{1}{2}) = \frac{d-2}{4}\sum_{n=0}^{\infty}n = -\frac{d-2}{48}, \qquad (D.36)$$

for the bosonic and fermionic coordinates, respectively. Notice that the zero-point energies involve the physical $d - 2$ transverse coordinates. For $d = 10$ the sum of the contributions in (D.36) is $-1/2$, reproducing the value of the intercept a_0 of the NS sector. Similarly to

D.2 RNS spectrum and GSO projection

In the previous Section we have seen that the RNS model is invariant under a two-dimensional local supersymmetry that ensures that the physical subspace is positive definite. In this Section, we will show that the spectrum of the theory becomes spacetime supersymmetric in ten dimensions by suitably projecting out some of the states of the RNS model (the so-called GSO projection).

In the NS sector, the spectrum of physical states is given by:

$$\alpha' M^2 = \sum_{i=1}^{d-2}\left(\sum_{n=1}^{\infty}\alpha^i_{-n}\alpha^i_n + \sum_{r=1/2}^{\infty} r\,\psi^i_{-r}\psi^i_r\right) - \frac{1}{2} \equiv N - \frac{1}{2}. \tag{D.37}$$

The degeneracy of states at a certain mass level can be obtained from the partition function:

$$Z_{NS} = \text{Tr}\,(q^{N-1/2}) = q^{-1/2}\prod_{m=1}^{\infty}\left(\frac{1+q^{m-1/2}}{1-q^{2m}}\right)^8$$

$$= \sum_{n=0}^{\infty} b_{n-\frac{1}{2}} q^{n-1/2} + \sum_{n=0}^{\infty} c_n q^n$$

$$= q^{-1/2} + 8 + c_{1/2} q^{1/2} + c_1 q + \cdots . \tag{D.38}$$

These states correspond to both half-integer and integer valued mass formulae, $\alpha' M^2 = n - \frac{1}{2}$ and $\alpha' M^2 = n$ for $n = 0, 1, 2 \ldots$, respectively; the coefficients $b_{n-1/2}$ and c_n determine the number of states present at that level. The lowest state, corresponding to the term $q^{-1/2}$ is given by the oscillator vacuum:

$$|0\rangle, \qquad \alpha' M^2 = -\frac{1}{2}, \tag{D.39}$$

and is a tachyon. The next state is a massless spacetime vector:

$$\psi^i_{-1/2}|0\rangle, \qquad i = 1, \ldots, 8, \qquad M^2 = 0. \tag{D.40}$$

In ten dimensions, a massless vector has eight independent components corresponding to the degeneracy of the second term of the last line of (D.38). This state is interpreted as the gauge boson of open string theory. We can proceed in the same way with the higher mass levels and find that the states with an odd (even) number of world-sheet fermions have integer (half-integer) values of $\alpha' M^2$. In particular, we can define a world-sheet fermion number:

$$(-1)^F, \qquad F = \sum_{i=1}^{8}\sum_{r=1/2}^{\infty} \psi^i_{-r}\psi^i_r - 1. \tag{D.41}$$

The states with an odd (even) number of world-sheet fermions are even (odd) under the action of $(-1)^F$. In particular, the tachyon is odd under $(-1)^F$, while the gauge boson is even.

Gliozzi, Olive and Scherk had the idea of performing a consistent truncation of the NS model by keeping only those states that are even under $(-1)^F$ [GSO76, GSO77]. They obtained the following partition function:

$$Z_{NS}^{GSO} = \frac{1}{2}\text{Tr}(q^{N-1/2}) + \frac{1}{2}\text{Tr}((-1)^F q^{N-1/2})$$
$$= \frac{1}{2}q^{-1/2}\left[\prod_{m=1}^{\infty}\left(\frac{1+q^{m-1/2}}{1-q^{2m}}\right)^8 - \prod_{m=1}^{\infty}\left(\frac{1-q^{m-1/2}}{1-q^{2m}}\right)^8\right]$$
$$= 8 + c_1 q + \cdots . \tag{D.42}$$

After this projection, the lowest state is a massless vector field and only the states with an integer value of $\alpha' M^2$ are left in the physical spectrum. In this way, Gliozzi, Scherk and Olive obtained for the first time a string theory without tachyons.

Let us now turn to the R sector. The fermionic oscillators have a zero mode which, in the quantum theory, satisfies the same anticommutation relations of Dirac matrices,

$$\{\psi_0^\mu, \psi_0^\nu\} = \eta^{\mu\nu}, \tag{D.43}$$

apart from an overall normalization. This means that the ground vacuum state $|0, A\rangle$ has a Dirac spinor index A and, therefore, is a spacetime fermion. The mass spectrum in the R sector is given by:

$$\alpha' M^2 = \sum_{i=1}^{d-2}\left(\sum_{n=1}^{\infty}\alpha_{-n}^i \alpha_n^i + \sum_{n=1}^{\infty} n\, \psi_{-n}^i \psi_n^i\right) \equiv N_R. \tag{D.44}$$

The lowest state of the R sector is a massless spinor given by:

$$|0, A\rangle, \qquad M^2 = 0. \tag{D.45}$$

Both this state and all excited states are ten-dimensional spinors. Let us now discuss whether they are Dirac, Majorana or Weyl spinors (see the Introduction to Part VI and the next Section for a discussion of spinor types in d dimensions). A Dirac spinor in $d = 10$ has $2^5 = 32$ physical degrees of freedom, while a Majorana or a Weyl spinor has half of them. Since the world-sheet fermionic coordinate is real, the spinor must be Majorana and have 16 physical components. This means that the partition function in the R sector is given by:

$$Z_R = \text{Tr}(q^{N_R}) = 16\prod_{m=1}^{\infty}\left(\frac{1+q^{2m}}{1-q^{2m}}\right)^8 = 16\left(1 + 16q^2 + \cdots\right). \tag{D.46}$$

On the other hand, Gliozzi, Scherk and Olive observed that in ten dimensions the spinors can simultaneously be of Weyl and Majorana type. They performed a projection also in the

R sector by introducing the following fermion number operator,

$$(-1)^F = \psi_{11}(-1)^{F_R}, \qquad \text{where} \qquad F_R = \sum_{n=1}^{\infty} \psi_{-n} \cdot \psi_n, \qquad (D.47)$$

$$\psi_{11} \equiv 2^5 \psi_0^0 \psi_0^1 \ldots \psi_0^9,$$

where ψ_{11} is the chirality operator in ten dimensions, and by keeping only the states that are even under the action of $(-1)^F$. This implies that the lowest state in (D.45) is a Weyl–Majorana spinor, while the massive states have both chiralities.

The effect of the GSO projection is to introduce the projector $(1 + (-1)^F)/2$ into the partition function (D.46). The first term of the projector gives a factor $1/2$, while the second term is found to vanish owing to cancellation of an equal number of even and odd states at each level. Therefore, one obtains:

$$Z_R^{GSO} = \frac{1}{2} \text{Tr}(q^{N_R}) = 8 \prod_{m=1}^{\infty} \left(\frac{1+q^{2m}}{1-q^{2m}} \right)^8. \qquad (D.48)$$

In conclusion, the GSO projection selects the even states under the action of $(-1)^F$, given in (D.41) and (D.47) for the NS and R sectors, respectively.

The GSO projection not only eliminates the tachyon in the bosonic NS sector but also yields the same number of bosonic and fermionic states at each mass level, $\alpha' M^2 = n$, where n is a nonnegative integer. Indeed, one can show that Eqs. (D.42) and (D.48) are identical, $Z_{NS}^{GSO} = Z_R^{GSO}$, upon using the Jacobi identity. In particular, at the massless level, we have eight degrees of freedom for both the gauge boson and the Weyl–Majorana spinor.

The presence of an equal number of degrees of freedom for bosons and fermions for each mass value is a necessary condition for the theory to be spacetime supersymmetric. It was later understood that the theory is indeed supersymmetric in ten spacetime dimensions (this is explained in the Introduction to Part VII). This supersymmetry is called target-space supersymmetry to distinguish it from the world-sheet supersymmetry discussed in the previous Section. Unlike world-sheet supersymmetry, it is not a gauge symmetry, but a classification symmetry of the spectrum: the physical states can be accommodated in supersymmetry multiplets containing the same number of boson and fermion degrees of freedom.

In the Eighties, it was also understood that the GSO projection is not a choice but is dictated by modular invariance, the residual symmetry of reparameterizations on the torus geometry after fixing the conformal gauge. This result was first found in the corresponding study of the partition function of the closed string sector of the theory, corresponding to the one-loop torus amplitude. Then, it was extended to the open string sector discussed here, corresponding to the annulus amplitudes, by requiring open/closed string duality.

D.3 Target-space supersymmetry

In this Section we describe the simplest four-dimensional supersymmetric theory, called the Wess–Zumino model, that is a generalization of the two-dimensional world-sheet theory considered in Section D.1.

We start by recalling some properties of spinors discussed in the Introduction to Part VI. A Dirac fermion in four dimensions has four complex components corresponding to eight real parameters, that reduce to four when the Dirac equation is imposed, i.e. on-shell. These correspond to the states of electron and positron with two spin components. One can reduce the degrees of freedom of a spin one-half fermion in two possible ways, leading to Majorana or Weyl fermions. A Majorana fermion is a neutral particle and coincides with its antiparticle; the corresponding spinor is unchanged under charge conjugation, $\psi_M^c = \psi_M$, and thus possesses only two degrees of freedom on-shell. Furthermore, a massless Dirac fermion has two independently propagating parts with definite chirality, i.e. given pairing of charge and helicity: these are the left- and right-handed Weyl fermions, ψ_L and ψ_R, defined by

$$\psi_L = \frac{1-\gamma_5}{2}\psi_D, \qquad \psi_R = \frac{1+\gamma_5}{2}\psi_D, \qquad \gamma_5 = i\gamma^0\gamma^1\gamma^2\gamma^3, \tag{D.49}$$

in terms of Dirac spinors. Therefore, ψ_L and ψ_R possess again only two on-shell degrees of freedom. All types of fermions can be expressed in terms of two-dimensional Weyl spinors, as follows:

$$\psi_D = \begin{pmatrix} \chi_\alpha \\ \bar{\psi}^{\dot\alpha} \end{pmatrix}, \qquad \psi_M = \begin{pmatrix} \chi_\alpha \\ \bar{\chi}^{\dot\alpha} \end{pmatrix}, \qquad \psi_L = \begin{pmatrix} \chi_\alpha \\ 0 \end{pmatrix}. \tag{D.50}$$

The undotted (α) and dotted ($\dot\alpha$) indices of the Weyl spinor χ_α and of its conjugate $\bar{\chi}^{\dot\alpha}$, respectively, refer to the two fundamental representations, $(\frac{1}{2}, 0)$ and $(0, \frac{1}{2})$ of the Lorentz group.

D.3.1 Supersymmetry in four-dimensional field theory

Let us briefly describe the spacetime four-dimensional supersymmetry transformations and their relation to the world-sheet transformations discussed before. We recall that the world-sheet string action, involving free massless scalars X^μ and Majorana fermions ψ^μ fields was given in (D.6) and its supersymmetry transformations were defined in (D.11).

This idea of transforming boson and fermion fields among themselves can be extended to any dimension, while restricting it to be space independent (constant ϵ) for the time being. The simplest supersymmetric theory in four dimensions, the Wess–Zumino model, involves two real scalar and one Majorana field that have an equal number of bosonic and fermionic degrees of freedom (on-shell field polarizations) and are degenerate in mass. The action is written in terms of the real A and B bosonic and ψ fermionic fields, as follows:

$$S = \int d^4x \left[-\frac{1}{2}(\partial_\mu A)^2 - \frac{1}{2}(\partial_\mu B)^2 + \frac{i}{2}\bar{\psi}\gamma^\mu \partial_\mu \psi + \cdots \right], \tag{D.51}$$

where γ^μ are the Dirac gamma matrices obeying $\{\gamma^\mu, \gamma^\nu\} = -2\eta^{\mu\nu}$ and the dots stand for interaction terms and terms involving auxiliary fields that will not be described here.

Note that the kinetic terms have the same form as those of the RNS action (D.6) and it is possible to devise an analogous transformation among the bosonic and fermionic fields that leaves the action invariant. This leads to the following form of the (on-shell) four-dimensional supersymmetry transformations:

$$\delta A = \bar{\epsilon}\psi, \qquad \delta B = -i\bar{\epsilon}\gamma_5\psi,$$
$$\delta \psi = -i\partial_\mu (A + i\gamma_5 B)\gamma^\mu \epsilon, \qquad \delta \bar{\psi} = i\bar{\epsilon}\gamma^\mu \partial_\mu (A + i\gamma_5 B), \tag{D.52}$$

where the parameter ϵ is now a constant Majorana spinor. The proof of the invariance of the action is done following the same steps as for the RNS string action described before. The variation of the action is:

$$\delta S = \int d^4x \bigg[-\partial_\mu A \partial^\mu (\bar{\epsilon}\psi) + i\partial_\mu B \partial^\mu (\bar{\epsilon}\gamma_5\psi)$$
$$- \frac{1}{2}\bar{\epsilon}\gamma^\mu \partial_\mu (A + i\gamma_5 B)\gamma^\nu \partial_\nu \psi + \frac{1}{2}\bar{\psi}\gamma^\mu \partial_\mu (\gamma_\nu \partial_\nu (A - i\gamma_5 B)\epsilon) \bigg]. \tag{D.53}$$

By integrating the last term by parts and using the relation

$$\bar{\chi}\gamma^\mu \gamma^\nu \partial_\nu (A - i\gamma_5 B)\epsilon = \bar{\epsilon}\partial_\nu (A - i\gamma_5 B)\gamma^\nu \gamma^\mu \chi, \tag{D.54}$$

valid for anticommuting Majorana spinors (as can easily be verified by recalling the equality $(\gamma^\mu)^\dagger = \gamma^0 \gamma^\mu \gamma^0$), we can rewrite Eq. (D.53) as follows:

$$\delta S = \int d^4x \bigg[-\partial_\mu A \partial^\mu (\bar{\epsilon}\psi) + i\partial_\mu B \partial^\mu (\bar{\epsilon}\gamma_5\psi)$$
$$- \bar{\epsilon}\gamma^\mu \partial_\mu \gamma^\nu (A - i\gamma_5 B)\partial_\nu \psi + \partial_\mu \left(\frac{1}{2}\bar{\epsilon}\gamma^\nu \gamma^\mu \partial_\nu (A - i\gamma_5 B)\psi \right) \bigg]. \tag{D.55}$$

Then, using the relation

$$\gamma^\mu \gamma^\nu = \frac{1}{2}\{\gamma^\mu, \gamma^\nu\} + \frac{1}{2}[\gamma^\mu, \gamma^\nu] = -\eta^{\mu\nu} + \frac{1}{2}[\gamma^\mu, \gamma^\nu], \tag{D.56}$$

in the third term of (D.55), this becomes:

$$\delta S = \int d^4x \bigg[\partial_\mu \left(\frac{1}{2}\bar{\epsilon}\gamma^\nu \gamma^\mu \partial_\nu (A - i\gamma_5 B)\psi \right)$$
$$- \frac{1}{2}\bar{\epsilon}[\gamma^\mu, \gamma^\nu]\partial_\mu (A - i\gamma_5 B)\partial_\nu \psi - \partial^\mu \bar{\epsilon}\, \partial_\mu (A - i\gamma_5 B)\psi \bigg]. \tag{D.57}$$

Finally, integrating by parts the term with the commutator, we obtain another total derivative that can be combined with the total derivative already appearing in Eq. (D.57), and an additional term with the derivative acting on the supersymmetry parameter $\bar\epsilon$. In conclusion, we have:

$$\delta S = \int d^4x \left[\partial_\mu \left(\frac{1}{2} \bar\epsilon \gamma^\mu \gamma^\nu \partial_\nu (A - i\gamma_5 B) \psi \right) + \partial_\mu \bar\epsilon \gamma^\nu \gamma^\mu \partial_\nu (A - i\gamma_5 B) \psi \right]. \quad (D.58)$$

Therefore, if ϵ is constant, the last term is absent, the Lagrangian transforms as a total derivative and the action is left invariant.

On the other hand, for arbitrary ϵ the term with the derivative of the supersymmetry parameter allows one to determine the supercurrent

$$J^\mu = \gamma^\nu \gamma^\mu \partial_\nu (A - i\gamma_5 B) \psi, \qquad \partial_\mu J^\mu = 0, \quad (D.59)$$

which is conserved if the equations of motion, $\gamma^\mu \partial_\mu \psi = \partial^2 A = \partial^2 B = 0$, are satisfied.

According to the Noether theorem, the generators of the supersymmetry transformations are constructed from the supercurrent:

$$Q = \int d^3x \, J^0 = \int d^3x \gamma^\nu \gamma^0 \partial_\nu (A - i\gamma_5 B) \psi. \quad (D.60)$$

They are Majorana spinors and their (anti)commutation relations with the fields generate the supersymmetry transformations. Indeed, using the canonical (anti)commutation relations of the free theories:

$$[A(x,t), \partial_0 A(y,t)] = i\delta^3(x-y),$$
$$[B(x,t), \partial_0 B(y,t)] = i\delta^3(x-y), \quad (D.61)$$
$$\{\psi_A(x,t), \psi_B(y,t)\} = i\delta_{AB}\delta^3(x-y),$$

it is easy to verify that

$$\delta A = i[\bar\epsilon Q, A] = \bar\epsilon \psi,$$
$$\delta B = i[\bar\epsilon Q, B] = -i\bar\epsilon\gamma_5 \psi, \quad (D.62)$$
$$\delta\psi = i[\bar\epsilon Q, \psi] = -i\partial_\mu (A + i\gamma_5 B) \gamma^\mu \epsilon.$$

The algebra of Q with the fields A and ψ can be found by factoring $\bar\epsilon$ from the above equations. In the third equation of (D.62) we find, by anticommuting the spinors, $\bar\epsilon Q\psi - \psi\bar\epsilon Q = \bar\epsilon(Q\psi + \psi Q)$. We conclude that the generators of the supersymmetry transformations anticommute with fermionic operators and commute with bosonic operators. In particular, Q obeys anticommutation relations with itself.

Performing two supersymmetry transformations one gets a translation, as we now check explicitly for the field A; an analogous derivation can be done for B and ψ. From the supersymmetry transformations (D.52) one finds immediately:

$$(\delta_1\delta_2 - \delta_2\delta_1)A = 2i\, \bar\epsilon_1 \gamma^\mu \epsilon_2\, \partial_\mu A. \quad (D.63)$$

On the other hand, the left-hand side can be rewritten in terms of the generators of the transformations as follows:

$$(\delta_1\delta_2 - \delta_2\delta_1)A = \delta_1[i\bar{\epsilon}_2 Q, A] - \delta_2[i\bar{\epsilon}_1 Q, A]$$
$$= [i\bar{\epsilon}_1 Q, [i\bar{\epsilon}_2 Q, A]] - [i\bar{\epsilon}_2 Q, [i\bar{\epsilon}_1 Q, A]]$$
$$= -[A, [i\bar{\epsilon}_1 Q, i\bar{Q}\epsilon_2]] = -[[\bar{\epsilon}_1 Q, \bar{Q}\epsilon_2], A], \qquad \text{(D.64)}$$

where we have used the Jacobi identity and the fact that for Majorana spinors $\bar{\epsilon}Q = \bar{Q}\epsilon$. Comparing Eqs. (D.63) and (D.64) we obtain:

$$[i\{Q_A, \bar{Q}_B\}, A] = 2\gamma^\mu_{AB}\partial_\mu A. \qquad \text{(D.65)}$$

Finally, remembering that under a translation A transforms as

$$\delta A = [iP_\mu, A] = \partial_\mu A, \qquad \text{(D.66)}$$

where P_μ is the generator of the translations, we immediately obtain the supersymmetry algebra:

$$\{Q_A, \bar{Q}_B\} = 2\gamma^\mu_{AB} P_\mu. \qquad \text{(D.67)}$$

In two-dimensional Weyl notation, Q_A and the gamma matrices are given by:

$$Q_A = \begin{pmatrix} Q_\alpha \\ \bar{Q}^{\dot{\alpha}} \end{pmatrix}, \qquad \gamma^\mu = \begin{pmatrix} 0 & \sigma^\mu_{\alpha\dot{\beta}} \\ (\bar{\sigma}^\mu)^{\dot{\alpha}\beta} & 0 \end{pmatrix}, \qquad \text{(D.68)}$$

where

$$\sigma^\mu_{\alpha\dot{\beta}} = (-1, \sigma^i), \qquad (\bar{\sigma}^\mu)^{\dot{\alpha}\beta} = \epsilon^{\dot{\alpha}\dot{\beta}}\epsilon^{\beta\alpha}\sigma^\mu_{\alpha\dot{\beta}}, \qquad \alpha, \dot{\alpha}, \beta, \dot{\beta} = 1, 2. \qquad \text{(D.69)}$$

In these relations, σ^i are the three Pauli matrices and the dotted and undotted indices are raised and lowered as follows:

$$\psi^\alpha = \epsilon^{\alpha\beta}\psi_\beta, \qquad \psi_\alpha = \epsilon_{\alpha\beta}\psi^\beta, \qquad \psi^{\dot{\alpha}} = \epsilon^{\dot{\alpha}\dot{\beta}}\psi_{\dot{\beta}}, \qquad \psi_{\dot{\alpha}} = \epsilon_{\dot{\alpha}\dot{\beta}}\psi^{\dot{\beta}}, \qquad \text{(D.70)}$$

using the antisymmetric matrices

$$\epsilon^{\alpha\beta} = \begin{pmatrix} 0 & 1 \\ -1 & 0 \end{pmatrix}, \qquad \epsilon_{\alpha\beta} = \begin{pmatrix} 0 & -1 \\ 1 & 0 \end{pmatrix}, \qquad \text{(D.71)}$$

and analogous expressions for dotted indices.

In the two-dimensional Weyl notation, the supersymmetry algebra (D.67) becomes:

$$\{Q_\alpha, Q_\beta\} = \{\bar{Q}_{\dot{\alpha}}, \bar{Q}_{\dot{\beta}}\} = 0, \qquad \{Q_\alpha, \bar{Q}_{\dot{\beta}}\} = 2\sigma^\mu_{\alpha\dot{\beta}} P_\mu. \qquad \text{(D.72)}$$

The supersymmetry generators commute with the translations,

$$[P_\mu, Q_\alpha] = [P_\mu, Q_{\dot{\alpha}}] = 0, \qquad \text{(D.73)}$$

and transform as Weyl spinors under Lorentz transformations.

The relations (D.72) and (D.73) show that target-space supersymmetry gives rise to an extension of the Poincaré algebra involving anticommutators, i.e. to a graded Lie algebra.

D.3.2 Superspace and superfields

Supersymmetry transformations have a natural geometric interpretation as being translations in 'superspace', a fermionic extension of spacetime. In the following, we give a brief presentation of this topic: a full account can be found in [WB92].

The last relation of the supersymmetry algebra (D.72) shows that the composition of two supersymmetries is equal to an ordinary translation, generated by the four-momentum operator P^μ. This suggests that supersymmetries could also correspond to translations in some kind of extended space. Another indication comes from the field transformations (D.52). We recall that under a translation $x^\mu \to x^\mu + \delta x^\mu$, a scalar function transforms with a derivative term $\phi(x) \to \phi(x) + \delta x^\mu \partial_\mu \phi(x)$: this type of increment is indeed present in the transformation of the fermion field in (D.52), with spinorial parameter ϵ, thus suggesting the translation of a 'spinorial coordinate'.

We are led to introduce the superspace with coordinates $(x^\mu, \theta^\alpha, \bar\theta^{\dot\alpha})$, with $\theta, \bar\theta$ also anticommuting Weyl spinors. A supersymmetry transformation with parameter ϵ (again a Weyl spinor) is identified with the following infinitesimal translation:

$$\begin{aligned}\theta &\longrightarrow \theta + \epsilon, \\ \bar\theta &\longrightarrow \bar\theta + \bar\epsilon, \\ x^\mu &\longrightarrow x^\mu + i\theta\sigma^\mu\bar\epsilon - i\epsilon\sigma^\mu\bar\theta.\end{aligned} \quad (D.74)$$

The variation of the bosonic coordinate is required by the supersymmetry algebra in Eq. (D.72). In fact, combining two of the previous transformations, we get:

$$\begin{aligned}(\delta_1\delta_2 - \delta_2\delta_1)\theta &= 0, \\ (\delta_1\delta_2 - \delta_2\delta_1)\bar\theta &= 0, \\ (\delta_1\delta_2 - \delta_2\delta_1)x^\mu &= 2i(\epsilon_1\sigma^\mu\bar\epsilon_2 - \epsilon_2\sigma^\mu\bar\epsilon_1);\end{aligned} \quad (D.75)$$

this is a spacetime translation, as required by the algebra in (D.72).

It is then natural to define the 'superfield' $\Phi(x^\mu, \theta, \bar\theta)$, a generalized field depending on the superspace variables. There exist several kinds of superfields, one for each supersymmetric multiplet; the Wess–Zumino model involves a so-called 'chiral' supermultiplet and a corresponding 'chiral' superfield, that depends on two coordinates only,

$$\Phi = \Phi(y^\mu, \theta), \qquad y^\mu = x^\mu + i\theta\sigma^\mu\bar\theta. \quad (D.76)$$

Thus, the $\bar\theta$ variable only enters indirectly. The chiral superfield can be expanded in series of the θ variable, as follows:

$$\Phi(y, \theta) = \phi(y) + \sqrt{2}\theta\chi(y) + \theta\theta F(y), \quad (D.77)$$

where higher polynomials in θ vanish, e.g. $\theta\theta\theta_\alpha = 0$, owing to anticommutativity of spinor components. In the following, we shall prove that the field components of the superfield ϕ

and χ are related to the fields A, B and ψ introduced before:

$$\phi = \frac{A - iB}{\sqrt{2}}, \qquad \chi = \frac{1 - \gamma_5}{2}\psi. \tag{D.78}$$

On the other hand, F is the (complex scalar) auxiliary field, that is required in the off-shell formulation of supersymmetry. This is not a propagating field, its equation of motion being algebraic, but is necessary to match the number of bosonic and fermionic degrees of freedom before imposing the equations of motion.

Under an infinitesimal translation in superspace given in (D.75), θ and y^μ transform as follows:

$$\delta_\epsilon y^\mu = 2i\theta\sigma^\mu\bar\epsilon, \qquad \delta\theta = \epsilon. \tag{D.79}$$

They induce the following transformation on the chiral superfield:

$$\begin{aligned}\delta_\epsilon \Phi(y,\theta) &\equiv \Phi(y^\mu + 2i\theta\sigma^\mu\bar\epsilon, \theta + \epsilon) - \Phi(y,\theta) \\ &= 2i(\theta\sigma^\mu\bar\epsilon)\partial_\mu\phi + \sqrt{2}(\epsilon\chi) + 2\theta\epsilon F + i\sqrt{2}\theta^2(\epsilon\bar\sigma^\mu\partial_\mu\chi),\end{aligned} \tag{D.80}$$

where in the last term we have used the relations $\theta^\alpha\theta^\beta = -\frac{1}{2}\epsilon^{\alpha\beta}\theta^2$ and $\xi\sigma^\mu\bar\psi = -\bar\psi\bar\sigma^\mu\xi$, valid for any spinor pair ξ, ψ. This explicit transformation should be matched with the variation of the field components, which, by definition (D.77), is:

$$\delta_\epsilon \Phi = \delta_\epsilon\phi(y) + \sqrt{2}\theta\,\delta_\epsilon\chi(y) + \theta\theta\,\delta_\epsilon F(y). \tag{D.81}$$

Upon comparing the two equations, term by term in the θ series, we obtain the following transformations:

$$\begin{aligned}\delta_\epsilon \phi &= \sqrt{2}\epsilon\chi, \\ \delta_\epsilon \chi &= i\sqrt{2}\sigma^\mu\bar\epsilon\,\partial_\mu\phi + \sqrt{2}\epsilon F, \\ \delta_\epsilon F &= i\sqrt{2}\bar\epsilon\bar\sigma^\mu\partial_\mu\chi.\end{aligned} \tag{D.82}$$

It can be shown with some algebra that the superfield 'translation' (D.82) are indeed equal to the supersymmetry transformations of the Wess–Zumino model described before in (D.52), once the latter are rewritten into the Weyl basis identifying the fields according to Eq. (D.78), and the infinitesimal parameters as follows:

$$\epsilon_A = -\begin{pmatrix}\epsilon_\alpha \\ \bar\epsilon^{\dot\alpha}\end{pmatrix}. \tag{D.83}$$

More precisely, the transformations (D.82) are the off-shell extension of the transformations (D.52), also involving the auxiliary field F ($F = 0$ on-shell in free theory).

In conclusion, we have presented a brief description of supersymmetry in superspace notation: this makes explicit its geometrical meaning as an extension of the Poincaré symmetry of Minkowski space into a fermionic isometry of superspace. The superfield notation is widely used because it allows one to write manifestly supersymmetric Lagrangians.

References

[GSO76] Gliozzi, F., Scherk, J. and Olive, D. I. (1976). Supergravity theories and the spinor dual model, *Phys. Lett.* **B65**, 282.

[GSO77] Gliozzi, F., Scherk, J. and Olive, D. I. (1977). Supersymmetry, supergravity theories and the dual spinor model, *Nucl. Phys.* **B122**, 253–290.

[WB92] Wess, J. and Bagger, J. (1992). *Supersymmetry and Supergravity.* (Princeton University Press. Princeton, NJ).

Appendix E
The field theory limit

E.1 Introduction

The construction, from the axioms of S-matrix theory, of the Veneziano model and of its extension to N external particles, the Dual Resonance Model, made many people believe that this theory was completely different from quantum field theory. However, it soon became clear that the DRM was actually an extension of, rather than an alternative to, the various field theories, such as ϕ^3 scalar theory [Sch71], gauge theories [NS72] and general relativity [Yon73, SS74, Yon74].

The DRM contains a parameter, the slope of the Regge trajectory α' with dimension of (length)^2, or its inverse, the string tension $T = 1/(2\pi\alpha')$. The study of DRM amplitudes in the zero-slope (or infinite string tension) limit shows that these reduce to Feynman diagrams of specific field theories. In the string picture of the DRM, an intuitive way of understanding the zero-slope limit is the following. The string tension has the tendency to make the string collapse to a point, but, if the string is moving, for example rotating, there is also the centrifugal force, which has the opposite tendency. This means that a possible string motion results from the balance between these two forces. However, in the zero-slope limit the string tension is increasingly large, the centrifugal force cannot balance it anymore, and the string collapses to a point. In this limit, string theory becomes a theory of pointlike objects that is described by ordinary quantum field theory.

In order to see how this comes about, let us consider the mass spectrum of an open string in the light-cone gauge derived in Appendix C:

$$M^2 = \frac{1}{\alpha'} \left[\sum_{i=1}^{d-2} \sum_{n=1}^{\infty} n a_{ni}^\dagger a_{ni} - 1 \right]. \tag{E.1}$$

When we take the limit $\alpha' \to 0$ all massive string states acquire a very large mass, disappearing from the spectrum, and we are left with only the massless vector boson described by the state $a_{1i}^\dagger |0\rangle$; in other words, the spectrum reduces to that of Yang–Mills theory. Furthermore, in the zero-slope limit, the string amplitudes reproduce the Feynman diagrams of gauge theory. With hindsight, the main reason for such a remarkable correspondence between string theory and field theory is consistency. Indeed, there is only one consistent field theory of massless vector bosons. String theory, being unitary, also consistently describes this particle, and thus should reproduce the field theory in the limit in which its spectrum reduces to this unique particle.

Before entering into the details of the $\alpha' \to 0$ limit for the open string, let us consider the (technically) simpler limit in which the scalar particle, i.e. the tachyon, survives instead of the vector boson. We shall momentarily disregard the consistency of string theory and allow for arbitrary values of the spacetime dimension d and intercept of the Regge trajectory. This means that we can trade -1 in Eq. (E.1) with $\alpha' m^2$ and write, instead of Eq. (E.1), the following expression:

$$M^2 = \frac{1}{\alpha'} \sum_{i=1}^{d-2} \sum_{n=1}^{\infty} n a_{ni}^\dagger a_{ni} + m^2. \tag{E.2}$$

In the zero-slope limit, we keep only the ground state $|0\rangle$ scalar particle with mass equal to m, while all other states have infinite mass and disappear from the spectrum.

In other words, if we do not care about the full consistency of string theory, we can perform the field theory limit in two different ways. In one case we get Yang–Mills theory [NS72] and in the other case ϕ^3 scalar field theory [Sch71]. Let us now discuss how the correct couplings, i.e. the Feynman diagrams of these theories, are recovered in the zero-slope limit.

E.2 Scalar theory

Let us start from the scalar field theory. The correctly normalized colour ordered scattering amplitude of N tachyons is given in the Chapter by Di Vecchia in Part III. Its expression reads:

$$B_M = \mathrm{Tr}(\lambda^{a_1} \cdots \lambda^{a_M}) \frac{1}{g_s^2 (2\alpha')^{d/2}} \left[2 g_s \left(2\alpha'\right)^{(d-2)/4} \right]^M$$

$$\times \int \frac{\prod_{i=1}^M dz_i \theta(z_i - z_{i+1})}{dV_{abc}} \prod_{i=1}^M |z_i - z_{i+1}|^{\alpha_0 - 1} \prod_{i<j} (z_i - z_j)^{2\alpha' p_i \cdot p_j}, \tag{E.3}$$

where $\alpha_0 \equiv -\alpha' m^2$, λ^a are the generators of $U(N)$ transformations in the fundamental representation, normalized as

$$\mathrm{Tr}(\lambda^a \lambda^b) = \frac{1}{2} \delta^{ab}, \qquad a, b = 1, \ldots, N^2 \tag{E.4}$$

and d is the dimension of spacetime. The complete scattering amplitude is a sum over the $(M-1)!$ noncyclic permutations of the external particles.

For $M = 3$ we get:

$$B_{M=3} = \frac{g_d}{2} \text{Tr}(\lambda^{a_1}\lambda^{a_2}\lambda^{a_3}), \qquad g_d \equiv 16g_s(2\alpha')^{\frac{d-6}{4}} \tag{E.5}$$

where g_s is the string coupling constant. This is precisely the three-point colour ordered amplitude that one gets at the tree level from the ϕ^3 theory described by the following Lagrangian:

$$L = \text{Tr}\left[-\partial_\mu \Phi \partial^\mu \Phi - m^2 \Phi^2 - \frac{g_d}{3!}\Phi^3\right], \tag{E.6}$$

which is invariant under global $U(N)$ transformations, where $\Phi = \sum_{a=1}^{N^2} \Phi^a \lambda^a$, is an $N \times N$ matrix scalar field.

For $M = 4$ Eq. (E.3) gives:

$$B_{M=4} = \text{Tr}(\lambda^{a_1}\lambda^{a_2}\lambda^{a_3}\lambda^{a_4}) \frac{g_d^2}{8} \alpha' \frac{\Gamma(\alpha'(m^2 - s))\Gamma(\alpha'(m^2 - t))}{\Gamma(\alpha'(m^2 - s) + \alpha'(m^2 - t))}. \tag{E.7}$$

Using the Γ-function relation, $x\Gamma(x) = \Gamma(x+1)$, Eq. (E.7) reduces for $\alpha' \to 0$ to [Sch71]:

$$B_{M=4} = \text{Tr}(\lambda^{a_1}\lambda^{a_2}\lambda^{a_3}\lambda^{a_4}) \frac{g_d^2}{8} \left[\frac{1}{m^2 - s} + \frac{1}{m^2 - t}\right], \tag{E.8}$$

where s, t, u are the Mandelstam variables (see Appendix A). Using the formula

$$\text{Tr}(AB) = 2\sum_a \text{Tr}(A\lambda^a)\text{Tr}(\lambda^a B), \tag{E.9}$$

we can rewrite the previous equation as follows:

$$B_{M=4} = \sum_a \text{Tr}(\lambda^{a_1}\lambda^{a_2}\lambda^a)\text{Tr}(\lambda^a\lambda^{a_3}\lambda^{a_4}) \frac{g_d^2}{4} \frac{1}{m^2 - s}$$
$$+ \sum_a \text{Tr}(\lambda^{a_2}\lambda^{a_3}\lambda^a)\text{Tr}(\lambda^a\lambda^{a_4}\lambda^{a_1}) \frac{g_d^2}{4} \frac{1}{m^2 - t}, \tag{E.10}$$

which shows that the amplitude factorizes correctly in both the s- and t-channels. Equation (E.10) matches the colour ordered amplitude that one obtains for the tree level Feynman diagrams from the Lagrangian (E.6). We have shown in some examples that the bosonic string theory reduces to ϕ^3 theory in the zero-slope limit; the proof can be generalized to higher point functions, to loop diagrams [DMLMR96, FMR00] and to ϕ^4 theory [FMR00, MP00].

As discussed in the Introduction to Parts II and VI, we recall that in string theory, one term of the scattering amplitude contains both the poles in the s- and t-channels (see (E.7)), while in field theory one must sum over two contributions, one with s-channel poles and the other with t-channel poles. This was one of the reasons for believing that string theory had nothing to do with field theory, but the previous examples show that string theory is rather an extension of field theory and reduces to it for $\alpha' \to 0$.

E.3 Yang–Mills theory

We consider now the zero-slope limit that gives Yang–Mills theory, the gauge theory of non-Abelian groups, describing vector bosons, for example the gluons of quantum chromodynamics. We start by writing the colour ordered M-vector amplitude for the bosonic string (we set $\alpha_0 = 1$):

$$A(p_1, \ldots, p_M) = 4 \operatorname{Tr}(\lambda^{a_1} \cdots \lambda^{a_M}) g_d^{M-2} (2\alpha')^{M/2-2}$$

$$\times \int \frac{\prod_{i=1}^{M} dz_i}{dV_{abc}} \left\{ \prod_{i<j} (z_i - z_j)^{2\alpha' p_i \cdot p_j} \right.$$

$$\left. \times \exp \left[\sum_{i<j} \left(\sqrt{2\alpha'} \frac{p_j \cdot \epsilon_i - p_i \cdot \epsilon_j}{(z_i - z_j)} + \frac{\epsilon_i \cdot \epsilon_j}{(z_i - z_j)^2} \right) \right] \right\}_{m.l.} \quad (E.11)$$

where m.l. means that we have to expand the exponential and keep only the term linear in all polarizations. The gauge coupling constant g_d is given by

$$g_d = 2g (2\alpha')^{\frac{d-4}{4}} \quad (E.12)$$

in terms of the string coupling constant.

From Eq. (E.11) we can immediately compute the colour ordered three-gluon amplitude:

$$A^{(0)}(p_1, p_2, p_3) = -4 g_d \operatorname{Tr}(\lambda^{a_1} \lambda^{a_2} \lambda^{a_3}) \Big(\epsilon_1 \cdot \epsilon_2 \, p_2 \cdot \epsilon_3$$

$$+ \epsilon_2 \cdot \epsilon_3 \, p_3 \cdot \epsilon_1 + \epsilon_3 \cdot \epsilon_1 \, p_1 \cdot \epsilon_2 + O(\alpha') \Big) \quad (E.13)$$

where $O(\alpha')$ stands for string corrections, proportional to α', such as $\alpha'(\epsilon_1 \cdot p_2)(\epsilon_2 \cdot p_3)(\epsilon_3 \cdot p_1)$, that are not relevant in the present discussion. As discussed in the Introduction to Part VI, Eq. (E.13) is exactly the three-point function that one obtains at tree level in non-Abelian gauge theory with $U(N)$ gauge group [NS72], whose Lagrangian is:

$$L = -\frac{1}{4} \sum_{a=1}^{N^2} F_{\mu\nu}^a F_a^{\mu\nu}, \quad (E.14)$$

where

$$F_{\mu\nu}^a = \partial_\mu A_\nu^a - \partial_\nu A_\mu^a - g_d f_{bc}^a A_\mu^b A_\nu^c, \quad (E.15)$$

with f_{bc}^a the structure constants of the Lie algebra of $U(N)$ generators. The gauge theory has interaction vertices with three gluons ($AA\partial A$ term) and four gluons ($AAAA$ term). The three-gluon vertex is reproduced by string theory in (E.13).

From Eq. (E.11), we can also compute the colour ordered four-vector string amplitude, that is given by:

$$A(p_1, p_2, p_3, p_4) = 4g_d^4 \, \text{Tr}(\lambda^{a_1}\lambda^{a_2}\lambda^{a_3}\lambda^{a_4}) \frac{\Gamma(1-\alpha's)\Gamma(1-\alpha't)}{\Gamma(1+\alpha'u)}$$

$$\left[\frac{u}{s(1+\alpha's)} \epsilon_1 \cdot \epsilon_2 \, \epsilon_3 \cdot \epsilon_4 + \frac{1}{1+\alpha'u} \epsilon_1 \cdot \epsilon_3 \, \epsilon_2 \cdot \epsilon_4 \right.$$

$$+ \frac{u}{t(1+\alpha't)} \epsilon_1 \cdot \epsilon_4 \, \epsilon_2 \cdot \epsilon_3$$

$$- \frac{2}{t} \Big(\epsilon_1 \cdot \epsilon_3 \, \epsilon_4 \cdot p_1 \, \epsilon_2 \cdot p_3 + \epsilon_1 \cdot \epsilon_4 \, \epsilon_3 \cdot p_1 \, \epsilon_2 \cdot p_4$$

$$+ \epsilon_2 \cdot \epsilon_3 \, \epsilon_1 \cdot p_3 \, \epsilon_4 \cdot p_2 + \epsilon_2 \cdot \epsilon_4 \, \epsilon_3 \cdot p_2 \, \epsilon_1 \cdot p_4 \Big)$$

$$- \frac{2}{s} \Big(\epsilon_1 \cdot \epsilon_2 \, \epsilon_4 \cdot p_2 \, \epsilon_3 \cdot p_1 + \epsilon_1 \cdot \epsilon_3 \, \epsilon_4 \cdot p_3 \, \epsilon_2 \cdot p_1$$

$$+ \epsilon_2 \cdot \epsilon_4 \, \epsilon_3 \cdot p_4 \, \epsilon_1 \cdot p_2 + \epsilon_3 \cdot \epsilon_4 \, \epsilon_1 \cdot p_3 \, \epsilon_2 \cdot p_4 \Big)$$

$$- \frac{2u}{st} \Big(\epsilon_1 \cdot \epsilon_2 \, \epsilon_4 \cdot p_1 \, \epsilon_3 \cdot p_2 + \epsilon_1 \cdot \epsilon_4 \, \epsilon_3 \cdot p_4 \, \epsilon_2 \cdot p_1$$

$$+ \epsilon_2 \cdot \epsilon_3 \, \epsilon_1 \cdot p_2 \, \epsilon_4 \cdot p_3 + \epsilon_3 \cdot \epsilon_4 \, \epsilon_2 \cdot p_3 \, \epsilon_1 \cdot p_4 \Big)$$

$$\left. + O(\alpha') \right]. \tag{E.16}$$

It can be shown that the zero-slope limit of this expression also reproduces the colour ordered four-gluon amplitude obtained from the gauge theory Lagrangian (E.14). The correspondence can be extended to the loop diagrams in this case as well [DLMMR96, FMR01].

As anticipated, the three and four-gluon couplings are the unique ones at low energy which are compatible with gauge symmetry, and they of course appear in the Yang–Mills Lagrangian (E.14). The next $O(\alpha')$ terms in (E.16) would correspond in the Yang–Mills theory to higher-order polynomials of A_μ in the Lagrangian, that are irrelevant at low energies.

The zero-slope limit of the closed bosonic string can be shown to reproduce Einstein's theory of general relativity coupled to a two-index antisymmetric tensor and a dilaton [Yon73, SS74, Yon74]. Again the reason for such correspondence is the uniqueness of the field theory description; the Einstein Lagrangian,

$$S = \frac{c^3}{16\pi G_N} \int d^4x \sqrt{-g}\, \mathcal{R}, \tag{E.17}$$

yields the unique consistent theory for gravitons (apart from higher-order terms in the curvature tensor and its derivatives), and is invariant under coordinate reparameterizations.

References

[DLMMR96] Di Vecchia, P., Lerda, A., Magnea, L., Marotta, R. and Russo, R. (1996). String techniques for the calculation of renormalization constants in field theory, *Nucl. Phys.* **B469**, 235–286.

[DMLMR96] Di Vecchia, P., Magnea, L., Lerda, A., Marotta, R. and Russo, R. (1996). Two loop scalar diagrams from string theory, *Phys. Lett.* **B388**, 65–76.

[FMR00] Frizzo, A., Magnea, L. and Russo, R. (2000). Scalar field theory limits of bosonic string amplitudes, *Nucl. Phys.* **B579**, 379–410.

[FMR01] Frizzo, A., Magnea, L. and Russo, R. (2001). Systematics of one-loop Yang–Mills diagrams from bosonic string, *Nucl. Phys.* **B604**, 92–120.

[MP00] Marotta, R. and Pezzella, F. (2000). String generated quartic scalar interactions, *JHEP PRHEP-tmr99* **033**, 01–08.

[NS72] Neveu, A. and Scherk, J. (1972). Connection between Yang–Mills fields and dual models, *Nucl. Phys.* **B36**, 155–161.

[Sch71] Scherk, J. (1971). Zero slope limit in the dual resonance model, *Nucl. Phys.* **B31**, 222–234.

[SS74] Scherk, J. and Schwarz, J. H. (1974). Dual models for non-hadrons, *Nucl. Phys.* **B81**, 118–144.

[Yon73] Yoneya, T. (1973). Quantum gravity and the zero slope limit of the generalized Virasoro model, *Lett. Nuovo Cimento* **8**, 951.

[Yon74] Yoneya, T. (1974). Connection of dual models to electrodynamics and gravidynamics, *Prog. Theor. Phys.* **51**, 1907–1920.

Index

Abrikosov–Nielsen–Olesen vortex, 267, 544
action
  bosonic string, 586–589
    Nambu–Goto, 30, 31, 156, 171–172, 230–231, 238, 250, 251, 266, 280, 285, 298, 305, 369, 409, 410, 414, 442, 443, 445, 447, 448, 451, 460, 478, 479, 484, 486, 527, 587, 604
    quadratic, 171, 238, 443, 590, 604
    σ-model, 50, 316, 321, 442–443, 485, 518, 519, 538, 546
  point particle, 228–230, 486–487, 586–587
  superstring
    Green–Schwarz, 517–518, 535–536
    quadratic, 44, 342, 409, 434, 443, 485, 605, 613
    σ-model, 442–444, 474–481, 484–488, 518, 546, 588
  supersymmetric point particle, 442
Ademollo et al. collaboration, 200, 205, 208–212, 216, 289, 332, 450–454
Ademollo, Marco, 100–115, 174, 216, 450
Adler, Stephen L., 573
Adler–Bell–Jackiw anomaly, see chiral anomaly
ADONE, at Frascati INFN laboratories, 173
AdS spacetime, 561, 562
AdS/CFT correspondence, 48, 55, 98, 119, 198, 200, 222, 312, 322, 323, 432, 536, 548, 549, 561–563
affine Lie algebra, 344–345, 375, 396–399, 556, 563, 564
Aharonov, Yakir, 262
Alessandrini, Victor, 191, 239, 253, 289, 347, 414
Alvarez, Orlando, 432
Alvarez-Gaumé, Luis, 54, 523, 524, 554
Amati, Daniele, 4, 65, 191–192, 198, 199, 239, 240, 241, 245, 253, 256, 288, 347, 349, 353, 355, 357, 373, 407, 408, 411, 474, 478, 490, 560
amplitude, see scattering amplitude
anomaly, see chiral anomaly, conformal anomaly
anti de Sitter spacetime, see AdS spacetime
anti de Sitter/conformal field theory correspondence, see AdS/CFT correspondence

anticommuting variables, 195, 335, 338, 351, 364–370, 376, 380, 381–384, 395, 409, 416, 435, 606, 611, 614, 617
Arvis, Jean-François, 432
Aspen Center for Physics, 52, 54, 184, 185, 186, 255, 256, 280, 355, 356, 357, 364, 386, 479, 500, 532, 533, 534, 536, 538
asymptotic freedom, 38, 199, 239, 256, 352, 356, 428, 431, 529, 544, 548, 572
Atiyah, Michael, 257

Baker–Campbell–Hausdorff formula, 162, 601
Balachandran, Aiyalam P., 212, 361, 374
Bardakci, Korkut, 25, 182, 196, 214, 332, 344, 349, 375, 376, 384, 393–404, 564
Bardeen, William A., 54, 536
Bars, Itzhak, 355
Becchi, Carlo M., 520
Becchi–Rouet–Stora–Tyutin quantization, see quantization, BRST
Bekenstein, Jacob D., 560
Belavin, Alexander A., 394, 403, 547, 549, 556, 563
Berezinsky, Vadim, 544
Bergmann, Peter G., 361
Berkeley, University of California, 4, 25, 38, 39, 43, 173, 181, 185, 195, 198, 246, 256, 312, 313, 346, 347, 357, 376
beta function, 24, 40, 114, 137, 268, 278, 314, 579–581
Bethe–Salpeter equation, 122, 459
Bianchi, Massimo, 565
Bjorken, James D., 31, 118, 431
Black Hole entropy, 559–561
Block–Nordsieck approximation, 172
bootstrap, see Chew–Frautschi bootstrap approach
bosonic sector, see Neveu–Schwarz sector
bosonization, 398, 400–402, 536
Bouchiat, Claude, 191, 373, 408, 454, 490, 491, 492, 496
boundary conditions, 224, 226

Index

Dirichlet, 225, 226, 289, 290, 388, 530, 531
Neumann, 91, 226, 228, 300, 301, 315, 388, 556, 590
periodic, 226, 556, 564
twisted, *see also* supersymmetry breaking, 348, 441, 476
brane, *see* Dirichlet brane
Breit–Wigner amplitude, 91, 92–93, 95, 98, 278, 313, 577–578
Brink, Lars, 9, 11, 45, 51, 53, 216, 253, 273, 349, 350, 355, 384, 442, 450, 474–482, 484, 486, 487, 498, 513, 529, 534, 593
Brink–Di Vecchia–Howe–Deser–Zumino action, *see* action, σ-model
Brookhaven National Laboratories, 193
Brout–Englert–Higgs mechanism, *see* Higgs mechanism
Brower, Richard C., 45, 76, 169, 241–249, 288, 298, 312–323, 349, 393
BRST quantization, *see* quantization, BRST

Cabibbo angle, 101, 570
Cabibbo, Nicola, 569, 570
Calabi–Yau compactification, 55, 525, 539, 556
California Institute of Technology, Pasadena, *see* Caltech
Callan, Curtis, 245
Callan–Treiman relation, 104
Caltech, 4, 45, 47, 51, 52, 129, 173, 389, 480, 481, 498, 500, 501, 503, 533, 534, 536, 538
Campbell, David, 379
Candelas, Philip, 55, 525, 539, 555
Caneschi, Luca, 27, 215, 349
Cappelli, Andrea, 564
Capri, NATO Summer School, 434
Cardy, John L., 379, 382, 564
Carlitz, Robert, 529
Casimir
  energy, 449
  invariant, 398, 538
central charge, *see* central extension
central extension, *see also* affine Lie algebra, 27, 43, 102, 147, 150, 166, 298, 316, 320, 322, 339, 341, 344, 358, 367, 380, 394, 395, 397, 400, 401, 402, 448, 512, 519, 520, 563, 609
CERN, 4, 45, 52, 65, 88, 191, 193, 198–199, 216, 239–256, 273, 288, 316, 346–357, 382, 389, 393, 410, 411, 436, 450, 468, 470, 475–479, 490–491, 498, 501, 504, 529, 532, 553
CFT, *see* conformal field theory
Chan, Hong-Mo, 214, 284, 331, 333, 347, 349, 393, 414
Chan–Paton factors, 40, 47, 160, 186, 331, 333–334, 395, 424, 425

Chandrasekhar, Subrahmanyan, 559
Chang, Lay N., 280
Chapline, George F., 524
Chern–Simons theory, 566, 567
Chew, Geoffrey F., 20–22, 38, 67, 91, 104, 117, 118, 276, 277, 313, 346
Chew–Frautschi bootstrap approach, 39, 68, 68, 91, 96, 117, 129–130, 194, 203, 565, 579
  exact solution, 565–566
Chicago, University of, 4, 276, 353, 363, 369
chiral anomaly, *see also* fermion, chiral, 52–54, 407, 521–522, 554
  cancellation, 521–525, 554–555
  Green–Schwarz cancellation mechanism, 31, 54–55, 119, 357, 468, 522–525, 534–538, 555
chiral symmetry, 572
Ciafaloni, Marcello, 560
City College, *see* NYU
Clavelli, Louis, 187, 193–197, 200, 280, 349, 363, 370, 374, 381
Coleman, Sydney R., 245
Collins, Peter V., 380
Collop, David, 379
compactification, *see* Kaluza–Klein compactification
confinement, 431
conformal
  algebra, *see* Virasoro algebra
  anomaly, 166, 252, 303, 519, 520, 594
  dimension, 149
  factor, 519, 520
  field theory, 149, 394–403, 549, 556, 563, 564, 565
  gauge, *see* gauge choice, conformal
  invariance, 46, 145–147, 166–169, 171, 182, 185, 192, 215, 223, 237, 264, 271, 285–288, 298, 302–306, 317–321, 322, 364, 369, 381, 394–403, 408, 461, 462, 533, 544, 547–549, 589, 590, 598, 600, 604
  minimal models, 547, 563, 564
  primary field, *see* vertex operator
  representation, 166, 381, 389, 397, 403, 415, 556, 563
Corrigan, Edward, 12, 253–255, 289, 290, 339, 348, 350, 355, 378–390, 414, 415, 478, 529
coset construction, *see also* affine Lie algebra, 357, 399–400
counterterms, *see also* renormalization, 481, 497, 532
Courant Institute, 407
covariance, *see also* Lorentz invariance, 53–54, 146, 149, 164, 209, 228, 237, 250–253, 285, 299, 302, 306, 307, 365, 394, 396, 439, 452, 463, 465, 488, 533, 535–536, 537, 547
CP factors, *see* Chan–Paton factors
Cremmer, Eugène, 223, 349, 355, 411, 423, 454, 455, 490–494, 498, 500, 502, 511, 513, 514
Creutz, Michael, 431

critical dimension
  bosonic string, 30, 42, 45, 48, 142, 143, 144, 145, 147, 171, 172, 196, 199, 234, 246, 247, 249, 251, 254, 273, 290, 298, 302, 320, 332, 375, 449, 464, 488, 520, 530, 582, 583
  superstring, 45, 127, 210, 247, 253, 255, 298, 332, 341, 388, 449, 453, 455, 610
crossing symmetry, *see* scattering amplitude
current algebra, *see also* affine Lie algebra, 46, 90, 102–109, 174, 208, 323, 332, 334, 335, 379, 393–403, 475, 528, 547, 569–573
curvature tensor
  Ricci, 538, 556
  Riemann, 624
Cushing, James T., 67
Cutkosky rules, 239

D'Adda, Alessandro, 174, 205, 216, 450
D'Auria, Riccardo, 174, 205, 216, 450
D-brane, *see* Dirichlet brane
Dahmen, Hans, 23
Dalitz, Richard, 119, 347
DAMTP, 239, 250, 346, 357, 379, 416, 502, 539
Dancoff, Sidney M., 116
Dashen, Roger, 245, 256
David, François, 547
DDF states, 27, 30, 152, 169–170, 234, 244–247, 249, 256, 298, 317–321, 350, 353, 597
De Alfaro, Vittorio, 107, 214
De Sitter spacetime, 549
Del Giudice, Emilio, 9, 76, 152, 173, 205, 216, 234, 317, 337, 352, 450
Del Giudice–Di Vecchia–Fubini operators, *see* DDF states
Della Selva, Angelo, 215
Department of Applied Mathematics and Theoretical Physics, Cambridge UK, *see* DAMTP
Deser, Stanley, 50, 442, 445, 486
DeWitt, Bryce, 477
DHS duality, *see* Dolen–Horn–Schmid duality
Di Vecchia, Paolo, 9, 76, 152, 156–175, 205, 216, 234, 238, 253, 337, 349, 352, 384, 442, 450, 479–481, 484–488
diffeomorphisms, *see* reparameterization invariance
dilatino, 516, 523
dilaton, 46, 307, 425, 452, 516, 596, 624
Dirac
  brackets, 256
  equation, 42, 116, 243, 278, 335, 336, 337, 340, 350, 364, 367, 465, 480, 613
  fermion, *see* fermion
  matrices, 42, 44, 85, 101, 278, 336, 340, 364, 367, 380, 382, 444, 445, 605, 608, 611, 614
  medal, 56, 258, 358, 505, 539
Dirac, Paul A. M., 100, 212, 232, 257, 286, 288, 362, 378, 527, 569
Dirac–Born–Infeld action, 527
Dirac–Ramond equation, 43, 336, 367, 368, 369
Dirichlet brane, 49, 52, 200, 290, 370, 432, 440, 442, 469, 504, 527, 529, 530, 531, 548–549, 556–557, 559, 561, 565
dispersion relation, 106, 117, 276, 277, 283, 407, 459, 575, 576
divergences, *see also* renormalization, 241, 275, 303, 307, 323, 363, 373, 374, 429, 430, 452, 475, 497, 498, 499, 512, 514, 525, 534, 535, 593
Dolen, R., 8, 95
Dolen–Horn–Schmid duality, 22, 24, 29, 68, 77, 91, 95, 96, 97, 98, 100, 108, 118, 123, 130, 142, 157, 180, 181, 204, 208, 277, 313, 366, 528, 581, 582
Dothan, Joe, 23
DRM, *see* model, dual resonance
DRM amplitude, *see also* Koba–Nielsen amplitude, Veneziano amplitude, 43, 137–143, 157, 158–163, 210, 238, 249, 294, 296, 451, 600–602
Drummond, Ian, 240, 379, 390, 415
duality, 27, 28, 39, 40, 42, 104, 110–112, 122, 158, 181, 236, 264, 278, 284, 347, 379, 460, 491, 498, 528, 529
  among string theories, *see* superstring dualities
  diagram, 31, 97, 98, 137, 138, 139, 140, 141, 142, 144, 159, 191, 214, 276, 286, 333, 351
  planar, *see also* Dolen–Horn–Schmid duality, 139, 182, 460
  self-dual fields, equations, 545, 549
  T-type, 53, 493, 500
Duff, Michael J., 514
Durham University, 240, 246, 253, 254, 256, 284, 349, 356, 384, 385, 389, 416, 480
Durhuus, Bergfin, 488
Dynkin index, 397
Dyson, Freeman, 116

École Normale Supérieure, Paris, *see* ENS and LPTENS
Eichten, Estia J., 431
einbein, *see also* vierbein, 486, 587
Einstein
  action, 428, 445
  medal, 33
  theory of gravity, 17, 48, 49, 131, 179, 180, 199, 204, 205, 211, 262, 323, 361, 378, 425, 428–429, 430, 439, 442, 444, 461, 462, 464–470, 480, 484, 493, 499, 500, 512, 527, 548, 556, 624
Einstein, Albert, 462, 499, 624
Ellis, John R., 379
energy-momentum tensor, *see also* conformal invariance, 343, 369, 443, 448, 465, 484, 485, 520, 588, 604, 605, 607
  trace anomaly, *see* conformal anomaly
ENS, 51, 373, 374, 389, 454, 455, 492, 496, 500

Erice Summer School, 22, 504
European Centre for Nuclear Research, Geneva, *see* CERN

F-theory, 49, 200
factorization, 25, 26, 27, 28, 41, 143, 147, 157, 158–163, 173, 183, 186, 191, 194, 196, 198, 200, 210, 214, 278, 280, 295–297, 314, 315, 362, 369, 374, 375, 387, 410, 447, 452, 460, 463, 491, 497, 498
Faddeev, Ludwig D., 520
Faddeev–Popov
 ghosts, 297, 321, 520
 quantization, *see* quantization, path-integral
Fairlie, David B., 11, 71, 238, 240, 246, 250, 267, 270, 271, 283–291, 355, 356, 384, 386, 389, 414, 416, 480
Farrar, Glennys, 355
Fayet, Pierre, 454
Feigin, Boris L., 563
Fermi National Acceleration Laboratory, *see* Fermilab
Fermi theory, 83, 427, 429, 569, 571
Fermi, Enrico, 569
Fermilab, 193, 194, 195, 362–368, 369, 374, 375, 376
fermion, 433–434
 chiral, 46, 50, 54, 255, 314, 356, 433–434, 521, 533
 Dirac, 85, 340, 386, 433–434, 469, 571, 611, 613
 in d dimensions, 433–434
 Majorana, 44, 433–434, 455, 485, 611
 Majorana–Weyl, 51, 200, 341, 396, 433–434, 438, 455, 513, 515, 517, 611
 Ramond, 47, 255, 383, 478, 498
 Weyl, 395, 433–434, 435, 441, 454, 521, 611
fermionic sector, *see* Ramond sector
Ferrara, Sergio, 50, 216, 445, 450, 454, 480, 502
FESR, *see* sum rule, finite energy
Feynman
 diagram, 21, 22, 40, 48, 84–88, 122–124, 182, 237, 239, 264, 266, 269–272, 277, 284, 286, 312, 313, 348, 353, 408, 410, 461, 462, 463, 468, 477, 478, 492, 528, 546, 570
 parton model, 187, 196, 205, 314, 316, 317, 322, 349, 428, 530
 path-integral, *see* quantization
Feynman, Richard P., 349, 354, 355, 476, 569
Feynman–Gell-Mann $V - A$ theory, 101, 569
Fierz identities, 385, 481, 487
fishnet diagram, 31, 122, 185, 221, 271–272, 286, 466
FNAL, *see* Fermilab
Foster, A. E., 361
Fradkin, Efim S., 409
Frampton, Paul, 253, 273, 278, 349, 380
Freedman, Daniel Z., 50, 445, 454, 502
Frenkel, Igor, 357, 402
Freund, Peter G. O., 96, 122, 127, 350, 380
Friedan, Daniel, 54, 341, 538, 556, 564

Frishman, Yitzhak, 393
Fritsch, Harald, 349
Froissart bound, 118
Fubini, Sergio, 4, 9, 22, 23, 25, 26, 27, 74, 76, 107, 118, 135, 148, 149, 173, 205, 208, 212, 214–216, 234, 245, 256, 287, 335, 348, 363, 450, 474
Fubini–Veneziano operators, 26–27, 135, 148, 149, 172, 184, 191, 210, 222, 232, 287, 335, 362, 374, 379, 387, 394, 408, 447, 451, 460, 469
Fuchs, Dmitry B., 563
functional integral, *see* quantization, path-integral
Furlan, Giuseppe, 103, 214

gamma function, 106, 137, 181, 579–581
gauge choice
 conformal, 171, 443, 448, 451, 485, 519, 589, 590, 591, 592, 596, 598, 604
 covariant, 44, 171, 223, 232, 233, 307, 352, 488, 519
 light-cone, 172, 200, 256, 298, 307, 410, 437, 529, 591, 592, 610
 orthonormal, *see also* gauge choice, conformal, 232, 233, 234, 251, 589
gauge fixing, *see* gauge choice
gauge particle, 50, 53, 169, 352, 353, 424, 425, 572, 594
gauge symmetry, 40, 48, 158, 164, 228, 230, 233, 237, 334, 341, 344, 426, 434, 461–463, 498, 499, 514, 521, 553, 554, 555, 559, 564, 589, 604, 624
 residual, 233, 341, 342, 589, 604, 612
gauge theory
 Abelian, *see also* QED, 476
 anomaly, *see* chiral anomaly
 coupling constant, 38, 104, 202, 424, 425, 428, 513, 545, 553, 621, 623, 624
 non-Abelian, 47, 130, 156, 179, 187, 239, 257, 264, 321, 322, 355, 370, 378, 401, 402, 407, 408, 410, 423, 425, 427, 431, 464, 467, 468, 476, 477, 493, 499, 529, 544, 546, 547, 549, 571, 620, 621, 623, 624
 nonperturbative, 323, 431, 432, 557, 561
 on a lattice, 431, 432, 468, 544, 545, 562
 supersymmetric, 55, 200, 322, 357, 435, 436, 455, 481, 501, 525, 545, 548, 556, 557, 561, 562
gauge/gravity correspondence, *see* AdS/CFT correspondence
Gell-Mann matrices, 101, 397, 571
Gell-Mann, Murray, 4, 22, 45, 88, 89, 101, 102, 123, 129–131, 204, 209, 253, 348, 349, 355, 393, 474, 477, 480, 481, 538, 569, 570, 571
general relativity, *see* Einstein theory of gravity
Georgi, Howard M., 428
Gepner, Doron R., 564
Gervais, Jean-Loup, 6, 44, 66, 191, 332, 355, 407–411, 443, 453, 454, 479, 490, 491–492

ghost, 150
  Faddeev–Popov, *see* Faddeev–Popov ghosts
  no-ghost theorem, 27, 44, 76, 170, 241–249, 252, 298, 312, 316–321, 354, 385, 476, 478
  state, 25, 26, 27, 158, 163–170, 181, 184, 185, 209, 210, 237, 289, 351, 382, 409, 415, 491, 520
Girardello, Luciano, 503
Glashow, Sheldon L., 38, 427, 428, 571
Glashow–Weinberg–Salam model, *see* Standard Model
Gliozzi, Ferdinando, 6, 14, 66, 73, 174, 216, 243, 349, 356, 363, 422, 437, 447–456, 501, 611
Gliozzi–Scherk–Olive projection, 30, 50, 51, 53, 212, 307, 341, 430, 437–439, 454–456, 501, 502, 515, 521, 529, 531, 554, 564, 610–612
glueball, 32, 321, 323, 528, 562
gluon, 31, 32, 255, 317, 321, 467, 493, 547, 566, 572, 623, 624
Goddard, Peter, 5, 73, 76, 150, 152, 222, 233, 236–257, 288, 317, 318, 348, 349, 355, 357, 379, 382, 384, 385, 386, 478, 529, 538, 564
Goddard–Goldstone–Rebbi–Thorn quantization, *see* quantization
Goebel, Charles J., 214
Goldberger, Marvin L., 130, 245, 276
Goldberger–Treiman relation, 208
Goldstone, Jeffrey, 5, 73, 150, 152, 222, 233, 250, 253, 257, 348, 356, 378, 476, 529
Goldwasser, Edwin L., 194
Golfand, Yuri A., 434
Gordon, David, 9, 193, 287, 363, 374, 476
Goto, Tetsuo, 5, 11, 73, 136, 221, 238
Gottfried, Kurt, 431
Gourdin, Michel, 490, 491
Grand Unified Theories, 120, 352, 427–428, 469, 553
Grassmann variables, *see also* anticommuting variables, 289, 369, 479, 487, 536, 606
gravitino, 50, 52, 436, 442, 444, 445, 454, 487, 503, 513, 516, 522, 523, 531
graviton, 47, 50, 53, 127, 131, 136, 199, 210, 212, 323, 348, 352, 425, 430, 436, 442, 451, 464, 465, 467, 499, 504, 516, 522, 534, 537, 596, 600, 624
Green, Michael B., 7, 15, 45, 52, 53, 54, 67, 119, 126, 129, 192, 197, 290, 348, 355, 357, 368, 468, 504, 515, 517, 523, 525, 527–539, 555
Green–Schwarz light-cone formalism, *see also* action, Green–Schwarz, 53, 54, 307, 514–518
Grisaru, Marc, 348
Gross, David J., 38, 41, 55, 199, 374, 428, 469, 497, 525, 529, 538, 555, 560
GSO projection, *see* Gliozzi–Scherk–Olive projection
Gubser, Steven S., 548
Gupta–Bleuler quantization, *see* quantization, 152
Gürsey, Feza, 368, 370
GUT, *see* Grand Unified Theories

Haag, Rudolf, 512
hadron spectrum, 19, 21, 46, 88–91, 453
Hagedorn transition, 26, 32, 279, 527, 528, 595
Hagedorn, Rolf, 595
Halpern, Martin B., 182, 196, 332, 344, 376, 393–404, 564
Hansen, Klud, 271
Hara, Osamu, 221
Hara, Yasuo, 467
Harari, Haim, 96
Harari–Rosner duality diagram, *see also* duality diagram, 22, 29, 96, 97, 124, 157, 186, 349
Harte, John, 368
Harvey, Jeffrey A., 55, 525, 538, 555
Hawking, Stephen W., 347, 559
Heineman prize, 33, 56, 504, 539
Heisenberg, Werner K., 4, 67, 129, 173, 203, 204, 208, 272, 275, 346, 348
helicity, 137, 158, 277, 433, 436, 613
Higgs mechanism, 267, 284, 321, 428, 466, 503
holography, 200, 470, 549, 561
holonomy, 290
Honerkamp, Josef, 353, 477
Horn, David, 8, 95
Horowitz, Gary T., 55, 525, 539, 555
Horsley, Roger, 256
Howe, Paul S., 442, 525
Huttunen Foundation, 414
hypercharge, 88, 89

IAS, 188, 199, 256, 258, 281, 357, 368, 375, 376, 529, 539
ICTP, 362
IHES, 408
Iliopoulos, Jean, 454
IMF, 262–263
infinite momentum frame, *see* IMF
instanton, 481, 488, 529, 530, 545
Institut des Hautes Études Scientifiques, Bures-sur-Yvette, *see* IHES
Institute of Advanced Study, Princeton, *see* IAS
International Center for Theoretical Physics, Trieste, *see* ICTP
Intersecting Storage Ring, CERN, *see* ISR
Ising model, 546, 564, 565
isospin, 88, 89, 90, 101, 107, 109, 109, 117, 160, 267, 268, 333, 334, 434, 491, 570
ISR, 31
Itzykson, Claude, 564
Iwasaki, Yoichi, 410

Jacobi identity, 51, 438, 455, 502, 612, 616
Jevicki, Antal, 467, 468
Jones, Keith, 284
Julia, Bernard, 511, 513, 514

Kac determinant, 242, 320
Kac, Victor G., 563
Kac–Moody algebra, *see also* affine Lie algebra, 256, 397, 402, 453, 480
Kajantie, Keijo, 414
Kaku, Michio, 355, 376, 411, 464, 492, 498
Kalb–Ramond field, 370
Kaluza–Klein compactification, 49, 54, 56, 199, 280, 430, 439–441, 493 503, 513, 514, 533, 534, 554
Kanazawa, Akira, 459
Kavli Institute of Theoretical Physics, Santa Barbara, *see* KITP
Kazakov, Vladimir A., 547
Kent, Adrian P. A., 564
Kikkawa, Keiji, 280, 297, 355, 410, 411, 467, 469, 492
Kinoshita, Toichiro, 431
KITP, 470
KK compactification, *see* Kaluza–Klein compactification
Klebanov, Igor R., 548
KN amplitude, *see* Koba–Nielsen amplitude
Knizhnik, Vadim G., 547, 564
Koba, Ziro, 8, 214, 268, 269–271, 280
Koba–Nielsen amplitude, 139–141, 158, 185, 237, 243, 269, 271–272, 285, 286, 288, 289, 331, 379, 380, 387, 408, 414, 415, 601
Kogut, John B., 431
Komaba campus, University of Tokyo, 465, 467
Kosterlitz, John M., 215
Kuhn, Thomas, 70

Laboratoire de Physique Théorique et Hautes Energies, Orsay, *see* LPTHE
Laboratoire de Physique Théorique, École Normale Supérieure, Paris, *see* LPTENS
landscape of string vacua, 558–559
Lane, Kenneth D., 431
Langlands, Robert, 257
Large Electron–Positron collider, CERN, *see* LEP
Large Hadron Collider, CERN, *see* LHC
Le Bellac, Michel, 239, 347
Lebedev Institute, Moscow, 409
Lederman, Leon M., 239
Lee, Benjamin W., 370, 407
LHC, 188, 436, 549, 553
light-cone, 153, 154
    coordinates, 172, 233, 262, 298, 300, 302, 317, 342, 411, 530, 532, 535, 589, 607
    gauge, *see* gauge choice, light-cone
Likhtman, Eugeny P., 434
Linström, Ulf, 525
Liouville theory, 306, 322, 468, 520, 546–548
loop diagram, 41, 53, 55, 87, 88, 125, 143–145, 157, 175, 180, 182–187, 198, 414–416, 478, 529, 533

nonplanar, 28, 29, 42, 97, 99, 142, 143, 144, 170, 185–187, 200, 247, 491, 523, 524, 525, 537
planar, 28, 143, 184, 270–290, 491, 523, 537
Lopuszanski, Jan T., 512
Lorentz invariance, 84, 163, 172, 173, 226, 228, 233, 234, 252, 278, 289, 299, 302, 305, 319, 354, 366, 367, 397, 398, 444, 449, 469, 492, 519, 520, 533, 556, 592, 594, 596
Lovelace, Claud, 5, 9, 42, 43, 75, 127, 144, 187, 191, 195, 198–200, 215, 241, 243, 247, 251, 252, 254, 264, 289, 297, 334, 347, 348, 349, 350, 353, 354, 370, 393, 415, 460, 497, 556
Lovelace–Shapiro amplitude, 195, 332, 334–335, 338, 375, 585
LPTENS, *see also* ENS, 492, 500, 503
LPTHE, *see also* Orsay, 407, 408
Lüscher, Martin, 432

M-theory, 49, 51, 322, 556–557
Maiani, Luciano, 245
Majorana fermion, *see* fermion
Makeenko, Yuri, 546
Maldacena conjecture, *see* AdS/CFT correspondence
Maldacena, Juan M., 98, 561
Manassah, Jamal, 368
Mandelstam variables, *see also* scattering amplitude, 24, 39, 86, 88, 98, 104, 105, 109, 137, 267, 276, 362, 528, 573, 579, 581, 622
Mandelstam, Stanley, 4, 25, 38, 39, 43, 86, 198, 223, 255, 256, 294–307, 313, 316, 349, 354, 369, 376, 379, 387, 393, 401, 410, 432, 474, 479, 498, 545
Mansouri, Freydoon, 280
Manton, Nicholas S., 256, 524
Martinec, Emil J., 55, 341, 525, 538, 555
Mason, Antony, 380
mass-shell condition, 38, 86, 130, 425, 573, 602
Massachusetts Institute of Technology, Boston, *see* MIT
Matsuda, Satoshi, 368
McDowell, Samuel W., 368
McFarlane, Alan, 361
Mende, Paul F., 560
metric tensor
    spacetime, 164, 236, 351, 428, 436, 442, 444, 451, 484
    world-sheet, 305, 306, 315, 316, 442, 443, 444, 518, 519, 520, 531, 546, 588
Migdal, Alexander A., 544, 546
Minami, Masatsugu, 462
Minimal Supersymmetric Standard Model, 436, 553
MIT, 4, 25, 27, 173, 205, 214, 316
Möbius
    invariance, 139–141, 145, 158–160, 182, 183, 186, 237, 285, 381, 415, 460
    strip, 54, 240, 523, 537, 538

model
  analogue, 141, 222, 237, 238, 243, 250, 286–290, 297, 375, 386, 408, 447, 448
  dual pion, *see* model, Neveu–Schwarz, 338
  dual quark, 395, 403
  dual resonance, 18, 25–27, 69, 98, 99, 113, 137–155, 157–163, 174, 180, 205–209, 210, 214, 221, 222, 223, 226, 231, 232, 233, 234, 254, 331, 333, 335, 337, 347, 356, 363, 368, 423, 426, 447, 585, 600–602, 609, 620
    interactions, 211, 450, 597–600
    physical states, *see also* DDF states, 73, 76, 150–153, 167–170, 174, 228, 231–234, 336
  Lovelace–Shapiro, *see* Lovelace–Shapiro amplitude
  Neveu–Schwarz, 289, 297, 299, 332, 338–339, 351, 380, 462, 584
  Ramond, 289, 297, 299, 332, 335–337, 351, 382, 396
  Ramond–Neveu–Schwarz, 42–45, 130, 157, 175, 187, 191, 196, 210, 212, 244, 255, 273, 304, 332, 339–341, 351, 375, 382, 408, 434, 436, 437, 442, 443, 444, 449, 454, 478, 479, 491, 492, 501, 511, 515, 529, 533, 537, 605, 609
    spectrum of, 339, 437, 610
  Shapiro–Virasoro, 141, 142, 143, 150, 199, 205, 210, 423, 449, 491, 492
modular invariance, 187, 199, 357, 439, 531, 534, 564, 565, 612
monodromy, 395
monopole, 355, 357, 529, 545, 557
't Hooft–Polyakov, 257, 355, 388, 416
Montonen, Claus, 350, 380, 381, 390, 414–416
Moyal product, 469
MSSM, *see* Minimal Supersymmetric Standard Model
multiloop diagram, *see* loop diagram
multiperipheral model, 347, 373
Musto, Renato, 174, 202–206, 216, 450

Nahm, Werner, 513
Nakanishi, Noboru, 463, 465, 468
NAL, *see* Fermilab
Nambu, Yoichiro, 4, 5, 11, 72, 73, 136, 238, 272, 275–281, 355, 363, 368, 369, 460, 474, 479
Nambu–Goto action, *see* action, Nambu–Goto
Napolitano, Ernesto, 174, 205, 216, 450
narrow-resonance approximation, *see* resonance width
Naylor prize, 539
Ne'eman, Yuval, 88, 570
Neveu, André, 5, 6, 12, 41, 43, 45, 70, 74, 129, 191, 256, 332, 338, 350, 362, 370, 373–376, 410, 416, 430, 454, 475, 490, 496, 499, 529, 605
Neveu–Schwarz model, *see* model
Neveu–Schwarz sector, 44, 50, 304, 339, 340, 341, 342, 396, 437, 438, 454, 478, 501, 529, 605, 607, 608, 609, 610
New York University, *see* NYU

Nicodemi, Francesco, 174, 208–212, 216, 450
Niels Bohr Institute, Copenhagen, 175, 266, 272, 273, 356, 410, 486, 488
Nielsen, Holger B., 5, 8, 11, 71, 72, 136, 198, 214, 237, 238, 250, 264, 266–273, 280, 286, 349, 350, 356, 384, 408, 414, 447, 476, 478, 479, 593
Nilsson, Bengt E.W., 514
no-ghost theorem, *see* ghost
Noether current, 342, 480, 570, 571, 606, 615
nonrenormalizability, *see* renormalization
Nordita, 175, 286, 479–481, 486, 488
normal ordering, 149, 161, 299, 315, 338, 398, 597, 598, 599, 608
NS sector, *see* Neveu–Schwarz sector
Nussinov, Shmuel, 185
Nuyts, Jean, 257, 357, 362, 374, 490
NYU, 407, 409, 410, 498

Olesen, Paul, 267, 272, 356, 466, 488
Olive, David I., 6, 9, 12, 14, 55, 66, 73, 239, 240, 253, 254, 256, 257, 286, 339, 346–358, 379, 380, 382, 385, 386, 389, 390, 411, 414, 415, 422, 437, 455, 467, 475, 476, 479, 529, 564, 611
Omnès, Roland, 410
Ooguri, Hirosi, 567
operator
  anticommuting, *see* anticommuting variables
  formalism, *see also* Fubini-Veneziano operators, 71, 147–150
  physical-state, *see* DDF states
O'Raifeartaigh, Lochlainn, 204
orientifold, 516
Orsay, *see also* LPTHE, 191, 349, 373, 374, 375, 407, 408, 410, 416, 490, 491, 492, 496, 498, 500
Oskar Klein medal, 33, 371

Pagels, Heinz R., 355
Pantin, Charles, 380, 390
partially conserved axial current, *see* PCAC
partition function, 375, 382, 438, 476, 478, 564, 595, 610, 611, 612
partitions of integers, 148, 595
Pasternack, Simon, 182
Paton, Jack E., 331, 333
Pauli matrices, 342, 605
Pauli, Wolfgang E., 116, 335
PCAC, 103, 181, 573
Peierls, Rudolf E., 307, 545
Petersen, Jens L., 488
Peterson, Dale H., 563
Pettorino, Roberto, 174, 216, 450
Picasso, Emilio, 239
planar diagram, *see* duality diagram
Poincaré invariance, *see* Lorentz invariance

Polchinski, Joseph, 290, 322, 388, 556
Politzer, David H., 428
Polkinghorne, John C., 239, 291, 355, 379, 383, 390
Polyakov action, *see* action, σ-model
Polyakov, Alexander M., 15, 50, 66, 76, 145, 147, 192, 264, 298, 303, 305, 306, 307, 322, 442, 481, 488, 518, 520, 533, 538, 544–550, 556, 563, 594
Pomeranchuk
  prize, 33
  theorem, 118
Pomeranchuk, Isaak Y., 95, 124
Pomeron, 28, 32, 46, 76, 93, 95, 96, 97, 124, 126, 144, 184, 187, 191, 194, 195, 196, 199, 241, 248, 264, 277, 289, 348, 352, 354, 450, 452, 478, 491, 497, 499
Ponzano, Giorgio, 256
Pope, Christopher N., 514
Popov, Victor N., 520
Pradisi, Gianfranco, 565
Princeton University, 40, 191, 194, 199, 256, 374, 410, 496, 497, 498, 537

QCD, 18, 19, 31, 38, 47, 120, 130, 352, 428, 499, 562, 572, 623
  string, 55–56, 119, 312, 317, 321–323, 431–432, 528, 562
QED, 17, 19, 21, 38, 83, 84–88, 100, 122, 158, 171, 173, 180, 312, 367, 476, 569
QFT, *see* quantum field theory
Qiu, Zongan, 564
quantization
  BRST, 150, 156, 215, 321, 464, 520, 538
  covariant (old), 165–170, 234, 314–319, 518, 520, 596–597
  Goddard–Goldstone–Rebbi–Thorn, 30, 42, 73, 76, 150, 152, 170–172, 210, 215, 222, 233–234, 249–254, 289, 298, 299, 302, 319, 409, 415, 447, 448, 450, 592–596
  Gupta–Bleuler, 172, 173, 252, 316, 366, 367
  in QED, *see also* quantization, Gupta–Bleuler, 163–165
  light-cone, 42, 76, 150, 152, 172, 233–234, 518, 520, 592–596
  of gravity, 46–49, 428–429, 464, 466, 467, 477, 493, 499, 500, 535, 547
  path-integral, 150, 156, 306, 307, 407, 410, 476, 520
    covariant, 305–307, 485, 518–521, 597
    light-cone, 223, 304–305, 597
  Polyakov, *see* quantization, path-integral, covariant
  with constraints, 43, 44, 53, 168, 171, 252, 450, 479, 485, 486, 487, 536
quantum chromodynamics, *see* QCD
quantum electrodynamics, *see* QED
quantum field theory, 84–88, 148, 335, 423, 569, 620

quark, 22, 31, 32, 38, 47, 89, 101, 102, 117, 118, 126, 186, 209, 283, 380, 427, 428, 467, 493, 521, 571–573
  confinement, 19, 32, 38, 118, 119, 431, 468, 481, 545, 572

R sector, *see* Ramond sector
Ramond model, *see* model
Ramond sector, 44, 50, 255, 304, 339, 340, 341, 342, 396, 438, 454, 478, 501, 529, 531, 534, 605, 607, 608, 609, 611
Ramond, Pierre, 5, 12, 43, 70, 129, 193, 195, 197, 210, 212, 243, 288, 332, 335, 349, 350, 371, 374, 375, 381, 389, 479, 480, 481, 529, 605
Ramond–Neveu–Schwarz model, *see* model
Rarita–Schwinger field, 454
Rebbi, Claudio, 5, 73, 150, 152, 215, 222, 233, 249, 250, 253, 349, 354
Rees, Martin J., 347
Regge
  behaviour, 21, 24, 39, 93–94, 95, 97, 98, 106, 107, 113, 180, 181, 194, 195, 204, 209, 239, 263, 269, 313, 459, 528
  pole, *see* Reggeon
  trajectories, 20–22, 29, 39, 40, 43, 46, 90, 91, 93–97, 106, 108, 110, 117, 137, 141, 148, 165, 180, 184, 204, 209, 210, 253, 268, 277, 294, 298, 334, 335, 338, 339, 346, 352, 358, 423, 426, 427, 431, 448, 449, 497, 499
Regge, Tullio, 84, 93, 245, 256
Reggeon, 39, 84, 92, 93–97, 106, 124–126, 130, 173, 181, 184, 194, 195, 198, 200, 214–216, 239, 241, 267, 277, 283, 347, 381, 410, 415, 449, 459, 460, 478, 491, 492, 528
regularization, *see also* renormalization
  ζ-function, 187, 449, 593
Renner, Bruno, 382
renormalization, 48, 49, 54, 87, 116, 131, 180, 187, 202, 240, 241, 276, 303, 370, 398, 407, 428, 429, 464, 478, 499, 529, 533, 538, 544, 569
renormalization group, 256, 513, 529, 548
reparameterization invariance, 50, 136, 156, 171, 172, 205, 211, 222, 230, 231, 232, 234, 285, 306, 410, 426, 439, 442, 448, 451, 479, 480, 484, 485, 486, 487, 519, 564, 586, 587, 588, 604
Research Institute for Mathematical Sciences, Kyoto, *see* RIMS
resonance, 90, 92–97, 335, 577
  width, 99, 143, 576, 578
Riemann surface, 170, 192, 198, 286, 297, 303, 306, 315, 408, 463, 491, 497
RIMS, 462, 463
RNS model, *see* model
Rohm, Ryan, 55, 525, 538, 555
Rosner, Jonathan L., 96
Rouet, Alain, 520

Rubinstein, Hector R., 23, 100, 108, 116–120, 262, 266, 269, 272
Ruegg, Henri, 214, 393
Rutgers University, 187, 193, 195, 199, 241, 349, 370
Rychkov, Vyacheslav S., 549

S-matrix theory, 19–25, 38–40, 67, 84, 91, 98, 100, 104, 117, 120, 129–130, 141, 157–169, 174, 184, 202–205, 208, 209, 210, 238–240, 262, 275–278, 283, 295, 306, 313, 362, 380, 407, 414, 465, 475, 476, 528, 573–576, 620
Sagnotti, Augusto, 565
Saito, Satoru, 215
Sakata model, 102, 118
Sakita, Bunji, 6, 44, 66, 195, 214, 272, 332, 355, 363, 408–411, 443, 467, 468, 479
Salam, Abdus, 100, 199, 245, 355, 357, 362, 407, 427, 453, 476, 571
scattering
　channel, 86, 87, 88, 92, 93, 94, 95, 96, 137, 138, 143, 574, 575, 581, 602
　deep inelastic, 38, 45, 64, 90, 92, 118, 187, 192, 205, 209, 421, 427, 428, 544
　diffractive, 124
　elastic, 85, 94, 96
　matrix, see S-matrix theory
　unitarity, 70, 84, 91, 92, 98, 143, 144, 574, 575, 577, 578
scattering amplitude, 105–106
　analyticity, 84, 87, 92, 93, 574, 575
　asymptotic behaviour, 93, 95, 97, 98, 583, 584
　Breit–Wigner, see Breit–Wigner amplitude
　crossing symmetry, 84, 86, 91, 98, 574, 579, 601
　Koba–Nielsen, see Koba–Nielsen amplitude
　loop, see loop diagram
　N-point dual, 137–139
　planar, 31, 56, 98, 122, 124, 161, 240, 264, 270, 271, 285, 460
　tree-level, 578
Schecter, Joe, 370
Scherk, Joël, 6, 14, 45, 47, 49, 51, 52, 65, 66, 73, 74, 75, 129, 191, 192, 253, 348, 349, 354, 355, 356, 373, 386, 410, 422, 423, 430, 437, 441, 454, 463, 464, 475, 478, 490–491, 492, 493, 496–504, 511, 513, 529, 611
Scherk–Schwarz supersymmetry breaking mechanism, see supersymmetry breaking
Schmid, Christoph, 8, 95, 393
Schwarz, John H., 5, 6, 7, 12, 15, 37–56, 65, 67, 70, 74, 75, 129, 192, 196, 197, 199, 200, 255, 256, 289, 332, 338, 348, 349, 350, 354, 355, 357, 369, 374, 375, 386, 387, 422, 423, 430, 441, 450, 475, 478, 480, 496–504, 513, 515, 517, 523, 525, 529, 531, 532, 534, 536, 538, 555, 605
Schwimmer, Adam, 23, 27, 185, 215, 238
Schwinger term, see central extension

Schwinger, Julian S., 116, 362
Sciama, Dennis W., 347, 378
Sciuto, Stefano, 174, 214–216, 349, 450
Scuola Internazionale Superiore di Studi Avanzati, Trieste, see SISSA
Segre, Gino C., 393
Seiberg, Nathan, 545, 557
Serber, Robert, 363
Serpukhov Laboratories, 193
Shapiro, Joel A., 9, 41, 43, 70, 141, 179–188, 195, 196, 199, 334, 349, 355, 393, 530
Shapiro–Virasoro amplitude, 210, 243, 464, 465
Shapiro–Virasoro model, see model, Shapiro–Virasoro
Shenker, Stephen H., 54, 341, 537, 538, 564
SISSA, 347
Skyrme, Tony H. R., 356, 378
SLAC, 38, 45, 64, 92, 118, 185
Slansky, Richard, 368
SM, see Standard Model
Sohnius, Martin, 512
soliton, 356, 400, 416, 557, 558, 561
Sommerfield, Charles M., 368
Sourlas, Nicolas, 454, 490
spinor, see fermion
SSC, 193
SSR, see sum rule, superconvergence
Standard Model, 18, 100, 275, 323, 386, 407, 427–428, 436, 442, 499, 513, 514, 521, 522, 529, 553, 571
Stanford Linear Accelerator Center, see SLAC
Stirling approximation, 313, 321
Stony Brook University, 454
Stora, Raymond, 520
strangeness, 88, 89, 90, 570
Strassler, Matthew J., 322
Strathdee, John A., 453
string
　closed, see also model, Shapiro–Virasoro, 97, 141, 142, 143, 144, 145, 590–597, 624
　coupling constant, 424, 453, 462, 467, 469, 623
　field theory, 322, 370, 411, 463–464, 468, 469, 529, 535, 536
　length scale, 251
　noncritical, see also Liouville theory, 544–550
　open, see also model, dual resonance, 97, 139, 141, 142, 143, 144, 590–597, 598, 600, 620
　scattering, 597–600
　supersymmetric, see superstring
　tension, 91, 98, 224, 226, 423, 431, 587, 620
　world-sheet, 97, 135, 136, 139, 141, 142, 149, 150, 221, 222, 229, 230, 434, 438, 443, 518, 519, 556, 563, 587, 588
string action, see action
string revolution, 37, 145, 357, 518, 523, 525–526, 536–538, 552, 554–555, 567
Strominger, Andrew, 55, 469, 525, 539, 555, 560
Sudarshan, Ennackal C. G., 361

Sudbery, Antony, 290
Sugawara construction, 344–345, 357, 375, 398–400
Sugawara, Hirotaka, 344, 362
SUGRA, see supergravity
sum rule
  Adler–Weisberger, 102, 573
  finite energy, 22, 39, 95–97, 100, 102–104, 106–108, 110, 119, 157, 173, 194, 204, 277, 474, 528, 576
  superconvergence, 107, 108, 576
super-Virasoro algebra, see superconformal algebra
Superconducting Super Collider, see SSC
superconformal
  algebra, 43, 337, 339, 376, 382, 383, 453, 480, 605–610
  symmetry, 44, 46, 342, 396, 443, 501
supercurrent, 342, 443, 485, 605, 608, 615
superfield, 289, 469, 480, 617–618
supergravity, 49–56, 423, 436, 442, 443, 444–445, 454, 468, 480, 486, 501, 504, 512–514, 521, 522–525, 531, 532, 535, 537, 554–555, 556, 557, 558
supergroup collaboration, see Ademollo et al. collaboration
superspace, 305, 416, 435, 453, 479, 481, 535, 617–618
superstring, 30, 52–55, 129, 130, 314, 315, 321, 322, 332, 355, 396, 438, 439, 440, 442, 514, 515–518, 521–526, 532–536, 554–555, 556–553
  dualities, 48, 55, 470, 502, 536, 547, 549, 556–557
  heterotic, 49, 55, 196, 402, 493, 500, 518, 525, 538, 539, 555, 556, 557, 558, 564
  type I, 53, 54–55, 513, 515–516, 518, 521, 523, 532, 537, 554, 557
  type IIA, 51, 53, 503, 513, 515, 516, 518, 521, 523, 532, 534, 554, 557, 558
  type IIB, 53, 513, 515–516, 518, 521, 522, 523, 532, 534, 535, 554, 557
superstring action, see action, superstring
supersymmetric partner, see also supersymmetry, multiplet, 50, 436, 454
supersymmetry, 30, 44, 45, 49–52, 130, 191, 195, 216, 256, 290, 322, 356, 369, 376, 435, 453, 549, 552, 553, 560, 564, 567
  breaking, 52, 441–442, 503–504, 553
  extended, 51, 55, 322, 453, 481, 501, 512–513, 533, 554
  multiplet, 537, 612, 617
  spacetime, 307, 410, 434–437, 438, 454–456, 476, 501–503, 529, 533, 545, 548, 613–618
  world-sheet, 44, 332, 341–344, 396, 409–410, 434, 453, 485–488, 604–610
Susskind, Leonard, 5, 11, 72, 136, 238, 262–264, 272, 286, 323, 355, 447, 459
SUSY, see supersymmetry

SVM, see model, Shapiro–Virasoro
Symanzik, Kurt, 407, 432

tachyon, 30, 42, 43, 44, 46, 50, 130, 140, 142, 153–155, 163, 170, 175, 199, 209, 237, 253, 255, 264, 278, 284, 285, 289, 298, 314, 315, 317, 331, 332, 337, 338, 339, 341, 352, 353, 354, 356, 363, 370, 411, 437, 438, 452, 454, 460, 497, 501, 502, 529, 531, 534, 581, 594, 596, 599, 600, 601, 610, 611, 612, 621
tadpole, 55, 452, *531*, 534
Tamm, Igor Y., 116
tetrad, see vierbein
Theory of Everything, 18, 120, 184, 192, 388, 453, 552, 600
Thirring model, 356, 400, 402
't Hooft, Gerard, 257, 313, 370, 428, 432, 464, 466, 468, 477, 529, 545, 546
Thorn, Charles B., 5, 43, 73, 76, 150, 152, 199, 222, 233, 241, 242, 244, 246–249, 250, 253, 288, 297, 317, 319, 320, 349, 363, 376, 384, 393, 447, 479, 531
Todorov, Ivan T., 349
TOE, see Theory of Everything
Toller, Marco, 239
Tomonaga, Sin-Itiro, 281
topological field theories, 547, 566, 567
torus geometry, 53, 136, 256, 303, 439, 493, 534, 612
Touschek, Bruno, 172, 173, 174
trace anomaly, see conformal anomaly
Treiman, Sam B., 194
Tsou, Sheung T., 214
Tyutin, Igor V., 520

unification, 427–431, 554, 555
  in field theory, see Grand Unified Theories
  in string theory, 46–49, 429–431, 464, 498–501
  in supergravity, 54, 502, 531, 532, 536
unitarity, see scattering
Utiyama, Ryoyu, 461, 467

Vafa, Cumrun, 560, 567
van Nieuwenhuizen, Peter, 445, 454, 502, 503
Varenna Summer School, 245
Veltman, Martinus J. G., 407, 428
Veneziano amplitude, 24, 68, 84, 97–99, 113–114, 119, 136, 137, 140, 149, 160, 194, 198, 240, 262–264, 266, 267, 270, 279, 313, 322, 334, 362, 408, 449, 576, 578, 579, 581, 584, 602
  DHS duality in, 40, 130, 209, 581–582
  Regge behaviour of, 583–584
  spectrum of, 113, 284, 582–583
Veneziano model, see model, dual resonance

Veneziano, Gabriele, 4, 8, 9, 17, *23*, 33, 40, 64, 97, 100, 108, 135, 148, 149, 173, 191, 198, 204, 214, 215, 221, 236, 237, 262, 267, 283, 284, 287, 313, 334, 336, 346, 560, 561

Verma module, *see also* conformal representation, 242, 320

vertex
  fermionic, 254, 339, 354, 357, 368, 382–389, 478, 529
  N-Reggeon, 145, 170, 198, 215, 319, 339, 362, 460, 463, 492
  operator, 149, 158, 161, 168, 169, 256, 299, 315, 318, 357, 365, 395, 400–403, 460, 531, 532, 549, 598, 599, 600, 601

vierbein, 442, 444–445, 480, 513

Virasoro
  algebra, 27, 147, 149, 166, 242, 247, 249, 352, 358, 380, 381, 394, 398, 400, 409, 415, 520, 563, 590, 604
  amplitude, 40, 141, 142, 491, 528
  conditions, 26, 42, 44, 47, 165–168, 185, 232, 237, 241, 251, 317, 320, 383, 384, 387, 448, 497, 597

Virasoro, Miguel A., 9, *23*, 70, 100, 108, 141, 214, 347, 363, 373, 376, 393

Volkov, Dmitry V., 434

Wadia, Spenta, 467
Waltz, Ron, 243
Weinberg, Steven, 38, 100, 255, 355, 386, 407, 427, 571
Weingarten, Donald H., 363, 364, 374
Weis, Joseph H., 9, 27, 166, 242, 322, 352, 358, 393, 394
Weisberger, William I., 573
Weisz, Peter H., 432
Wentzel, Gregor, 116
Wess, Julius, 6, 44, 66, 195, 407, 410, 422, 434, 438, 475, 501
Wess–Zumino model, 45, 49, 195, 337, 434–436, 613–616, 617, 618
Wess–Zumino–Witten model, *see* Wess–Zumino–Witten–Novikov model
Wess–Zumino–Witten–Novikov model, 398, 535, 564, 567
Weyl
  anomaly, *see* conformal anomaly
  fermion, *see* fermion
  invariance, 306, 519, 589
Weyl, Hermann K. H., 352
White, Alan R., 239, 379
Wilczek, Frank, 428, 535
Willemsen, Jorge F., 280
Wilson loop, 468, 546, 547, 549
Wilson, Kenneth G., 431, 529, 545
Wilson, Robert R., 194, 362, 364, 368
Witten, Edward, 54, 55, 200, 285, 403, 465, 469, 514, 523, 524, 525, 533, 535, 537, 539, 547, 548, 554, 555, 557, 564, 566, 567
world-sheet, *see* string world-sheet
Wray, Dennis, 215
WZ model, *see* Wess–Zumino model
WZWN model, *see* Wess–Zumino–Witten–Novikov model

Yan, Tung-Mow, 431
Yang–Mills theory, *see* gauge theory, non-Abelian
Yellin, Joel, 181
Yoneya, Tamiaki, 6, 65, 73, 75, 415, 430, 459–470, 500

Zakrzewski, Wojtek, 379
Zambuto, Mauro, 361
Zamolodchikov, Alexander B., 547, 556, 563, 564, 565
Zamolodchikov, Alexei B., 565
Zatzkis, Henry, 361
zero mode, 42, 297, 301, 336, 337, 340, 369, 387, 397, 402, 486, 515, 591, 592, 593, 599, 601, 608, 611
zero-point energy, *see also* Casimir, energy, 263, 267, 273, 339, 449, 478, 479, 593, 594, 595, 609
zero-slope limit, 47, 348, 375, 410, 423–427, 430, 462, 463, 464, 466, 467, 469, 475, 486, 493, 498, 513, 521, 620–624
zero-width approximation, *see also* resonance width, 130, 204, 209, 313, 314, 316, 576
Zimmermann, William, 407
Zuber, Jean-Bernard, 564
Zumino, Bruno, 6, 44, 54, 66, 191, 195, 357, 407, 408, 410, 422, 434, 438, 442, 445, 475, 479, 486, 501, 502, 536
Zwanziger, Daniel, 407
zweibein, *see also* vierbein, 487
Zweig, George, 89, 571